Global Change – The IGBP Series

Springer

Berlin
Heidelberg
New York
Hong Kong
London
Milan
Paris
Tokyo

Michael J. R. Fasham (Ed.)

Ocean Biogeochemistry

The Role of the Ocean Carbon Cycle in Global Change

With 130 Figures

Springer

Editor

Michael J. R. Fasham

Southampton Oceanography Centre
Waterfront Campus, Southampton, SO14 3ZH, UK
mjf@soc.soton.ac.uk

ISSN 1619-2435

ISBN 978-3-642-62691-3   ISBN 978-3-642-55844-3 (eBook)
DOI 10.1007/978-3-642-55844-3

Library of Congress Cataloging-in-Publication Data

Ocean biogeochemistry : a synthesis of the Joint Global Ocean Flux Study (JGOFS) /
Michael J.R. Fasham (ed.).
p.cm. -- (Global change--the IGBP series, ISSN 1619-2435)
Includes bibliographical references.

1. Carbon cycle (Biogeochemistry) 2. Chemical oceanography.
I. Fasham, M. J. R. II. Joint Global Ocean Flux Study. III. Series.

QH344.025 2003
551.46'01--dc21   2002044503

Bibliographic information published by Die Deutsche Bibliothek
Die Deutsche Bibliothek lists this publication in the Deutsche Nationalbibliografie;
detailed bibliographic data is available in the Internet at http://dnb.ddb.de

Photos courtesy of U.S. JGOFS Planning Office.
RVIB Nathaniel B. Palmer docks at McMurdo Station, Antarctica, at start of final U.S. JGOFS cruise. Photo by Mardi Bowles.
Launching moored instruments from RV Atlantis II. Photo by Craig Dickson.
El Niño-Southern Oscillation effects on equatorial Pacific chlorophyll *a* concentrations during January (El Niño, left) and July (La Niña, right) of 1998. (Image courtesy of the SeaWiFS Project, NASA Goddard Space Flight Center and ORBIMAGE.)
The copepod photo is courtesy of Southampton Oceanography Centre.

Cover Design: Erich Kirchner, Heidelberg
Dataconversion: Büro Stasch, Bayreuth

Printed on acid-free paper - 3111 - 5 4 3 2   SPIN 11867302

Foreword

Peter G. Brewer, Monterey Bay Aquarium Research Institute, Moss Landing, USA

Pre-History

The development of the field of 'Ocean Biogeochemistry' is a remarkable story, and one in which the JGOFS program, and the researchers whose work is presented here, have played critical leadership roles. The term 'biogeochemical cycles' became familiar in the mid 1980s when scientists first tried to describe to the policy world the complex set of interlocking processes involved in global change. Before then almost wholly physical descriptions were given, of radiative balances, heat fluxes, transport processes etc., and a few simple ocean or land CO_2 terms were added. However when attempts were then made to add the real effects and feedback terms of land and oceans a view of amazing complexity appeared. The first attempts to communicate this are fondly remembered in the 'horrendograms' produced by Francis Bretherton – wiring diagrams of computer chip complexity showing simultaneous links between warming and respiration, photosynthesis and CO_2, ocean circulation and productivity, energy balances and chlorophyll, ocean gas releases and clouds. The problem was that no one knew how to handle all this, and, since real knowledge was lacking, all kinds of claims were made for rates, reservoirs and mechanisms, with no idea as to which one was dominant or even important. It was messy, clamorous, essential, and wide open. Today it is a powerful discipline, with measured rates, innovative experiments, complex models, and vigorous testing of ideas.

The use of biogeochemical cycles as a term to describe diagenetic reactions in sediments had arisen earlier, but it was a total shock to hear in about 1986 that NASA had formally reorganized its earth science programs to highlight the new discipline. No one really knew how to react, since well-entrenched physics, chemistry, and biology programs were suddenly cast adrift. I well recall a corridor conversation with a distinguished physical oceanographer the day the news broke. It seemed incomprehensible. Today the AGU journal 'Global Biogeochemical Cycles' ranks third out of 117 titles in the Geosciences in the 2000 Journal Citation Reports.

The JGOFS program, or more accurately the scientists whose energy, dedication, and creativity are represented by that acronym, arose almost simultaneously with this transition, and its success is unassailable. How that happened is a remarkable story, and one worth telling.

A U.S. Initiative

In the US in the early 1980s the set of ocean observing programs deriving from the large scale International Decade of Ocean Exploration programs were winding down. Only the geochemical tracer efforts were truly active on a global scale, and here the first powerful measurements of the chlorofluorocarbon tracers were made. In ocean biology a program examining the processes associated with warm core rings, spun off from the Gulf Stream, occupied center stage. The first glimpses of ocean color data from space appeared from the Coastal Zone Color Scanner (CZCS) on Nimbus 7, but this was a struggle compounded not only by the technical challenge, but also by problems within the NASA team responsible for the data. The sensor was exceeding its design life, and beginning to fail. Frankly, it wasn't an impressive picture. And discussions of plans for more of the same for the future were received by NASA with little enthusiasm.

Yet there were opportunities. The unease over global change was being translated, by Presidential Science Advisor D. Allan Bromley, into a 'U.S. Global Change Program' that offered the promise of political support. There were advances being made in trace metal clean techniques that yielded new insights into ocean biogeochemical processes. Lively and assertive individuals outside NASA were pressing forward with innovative CZCS results, and extending observations into blue water far beyond the coastal zones. New sediment trap techniques were capturing rhythmic fluctuations in the rain of particles to the sea floor. And the first ice core records of large scale CO_2 fluctuations associated with glaciations, and attributed to linked changes in ocean circulation and productivity, were appearing.

The leadership to capture these opportunities came first undeniably from John Steele. John was frustrated by seeing plans emerge within the ocean physics community for a major observing program, allied to an altimetric satellite, without equivalent planning for biological and geochemical programs. His own background in marine ecosystem modeling had not previously exposed him to serious ocean geochemistry, but he sensed that alone the ocean biology community would not be able to seize the opportunity. A set of planning meetings of an 'Ad Hoc Group on Ocean Flux Experiments' (John Steele, Jim Baker, Wally Broecker, Jim McCarthy, and Carl Wunsch) took place in 1983–1984 under the auspices of the US National Academy Ocean Sciences Board on Ocean Science and Policy, and this led to a major workshop at the NAS Woods Hole Center in September 1984. The 'Global Ocean Flux Study' (GOFS) report from that workshop provided the impetus for what is now the JGOFS program.

The preface for that report emphasized the need for study of "the physical, chemical, and biological processes governing the production and fate of biogenic materials in the sea ... well enough to predict their influences on, and responses to, global scale perturbations, whether natural or anthropogenic ..." It went on to draw analogies between the Pleistocene fluctuations in climate, and "the beginning of a fossil-fuel CO_2-induced super-interglacial period."

I attended that meeting, and was soon perplexed. Few attendees knew anything of the assumed linkage to satellite ocean color, and widespread skepticism prevailed (the language alone of this community was foreign to most, and to me). The failing CZCS was apparently to be replaced by an Ocean Color Imager in 1985 or so – in practice it took 15 years of hard work for SeaWIFS to be launched! The busy sediment trap community were convinced that their technique was central; but trapping particles close to the euphotic zone was fraught with problems of technique, and mixed layer modeling was scarcely understood. The reference to paleo-climates ensured that the sediment record had to be included, but the scale mismatch of those studies with upper water column chemistry and biology was obvious. And persistent large discrepancies between productivity estimates from the oxygen balance and ^{14}C uptake, were aggressively debated. It was an interesting mess. Ocean physics was tacitly assumed to be taken care of somewhere else, most likely through WOCE. And in spite of the strong reference to the anthropogenic CO_2 signals, no one had thought to schedule any CO_2 papers into the meeting.

Building a U.S. Program

The lack of CO_2 papers in the GOFS meeting report was fixed very simply by a committee charged with editing the proceedings of the meeting. Some manuscripts were simply added. I was asked by Neil Andersen to join that editorial committee (with Ken Bruland, Peter Jumars, Jim McCarthy) and we met about a week later in Washington, D.C. It at once became clear that we were to be charged with not simply editing, but with program creation. We were briefed by Jim Baker that the most urgent item was a pitch to NASA headquarters for support of an ocean color satellite. None of us knew how to proceed with turning such a wide-ranging report into a viable and coherent research program. To break the impasse I suggested we go home, each with an editing assignment, and meet a week later to finish the report. We also had homework to do in the nature of drafting some outline or scheme for pulling the material together into a comprehensible program. This did not work as I had hoped, for when we reconvened I was the only one of the group to have prepared any semi-formal material. These were some results from a very crude North Atlantic mixed layer model comparing physical and biological forcing of CO_2, and comments on extending a similar calculation to the basin scale with some sense of what could be measured and tested. This was enough, and I was asked to Chair the group.

A very difficult period then followed. Since the constituency was broad a large committee was formed, few of whom had any prior collegial contact. But we worked at it, and I formed a 'Planning Office' of one. We had meetings and produced some reports. After some wrangling a broad plan to implement a strategy of carrying out a global survey of CO_2 and related properties, implementing time series stations, and executing a set of sophisticated process studies was set in place. It took, and it allowed for multiple roles for remote sensing on many space and time scales.

In the fall of 1986 a set of key events took place in rapid succession. Planning for WOCE had reached the point where a large-scale hydrographic program was forming. A meeting at the U.S. National Academy framed the debate; I had independently discussed with Carl Wunsch the issue of CO_2 measurements, and he felt that these were not part of the WOCE observing package. But he asked if I would present the case at the NAS meeting. This went very well, and a compromise was proposed whereby 'GOFS' would provide funding, oversight, people and tools, and WOCE would provide bunks on their cruises and access to samples and supporting data. A handshake sealed the deal, which was proposed and accepted in about 30 seconds. We could go global.

Secondly Gene Feldman, new at NASA had created the first basin scale chlorophyll image from CZCS data. It had flaws: dubbed by some as 'the ocean on fire' from the garish orange for high pseudo-chlorophyll levels, or the 'Pac-Man' image from an oddly shaped data gap in Hudson Bay, it nonetheless broke new ground. This image appeared on the cover of the November 4, 1986 issue of EOS, with the AGU Fall Meeting abstracts and brief papers by the GOFS committee, and the NASA team. It was a coveted slot, and it had great impact.

And since the WOCE connection had been made, and WOCE was formally international, we had the impetus to move beyond the U.S. This had always been intended, but without a formal opening or partially defined plan to propose, real progress had not been made. Within an hour or two of the WOCE handshake deal a letter to SCOR was drafted, and hand carried to Tasmania the next day. The letter requested that SCOR take up the challenge of sponsoring a major new initiative on an Ocean Flux Program, and cited the progress made. It was well received, and work began at once on an enabling meeting.

Creation of JGOFS

The first SCOR-sponsored meeting was held in Paris, at ICSU House, February 17–20, 1987. Gerold Siedler, as President of SCOR, kept a careful eye on proceedings, Jim Baker acted as Chair, and Elizabeth Gross facilitated. Roger Chesselet helped secure the superb location. We had all learned some hard lessons about preparation: Gene Feldman had now created the first global chlorophyll image, and this was first shown to me in dim dawn light at the luggage carousel in De Gaulle airport by Jim Baker the day before the meeting. It was superb. I had written a discussion paper especially for the meeting on the comparative North Atlantic heat, CO_2, and nutrient budgets, with David Dyrssen. It illustrated what we might gain from a survey, and an abbreviated version was later backed up by some measurements and published in Science.

There was no real understanding of how international logistics might work, but as we went round the room it was clear that the desire to create a novel and important program was strong. Jim Baker proposed a 'J' for the program; everyone said yes. Gerold Siedler nominated Bernt Zeitschel as the Chair, and this was agreed upon. It all happened quickly, and the real work began. Hugh Livingston had joined the U.S. Planning Office, and was superb in science and diplomacy in this role. Liz Gross guided the multi-national effort with grace and skill. Neil Andersen kept the thread of funding and agency sponsorship alive and well. The U.S. JGOFS Newsletter was born, and thrived. And a succession of excellent reports cataloged the evolution of scientific planning and understanding.

Fast Forward

History is fun, and important. But what did all this start up effort achieve?

There are now volumes of papers to testify to this, but perhaps I can pick on a few highlights. The 1989 multi-national North Atlantic Bloom Experiment was put together with extraordinary speed, and it combined ships and aircraft observations in new ways. U.S., German, U.K., Canadian and Dutch ships and scientists co-operated in the field. This set the tone for a whole series of successful process studies. The 1990 Fasham-Ducklow-McKelvie paper on modeling upper ocean production and the microbial loop laid the ground for a decade long renaissance in ocean biogeochemistry. The 1988 establishment of the Hawaii and Bermuda Time series Stations was essential. Critics at the time pointed to two sub-tropical locations as a deficiency: the separate signals evolving so beautifully there answer the challenge. The global CO_2 survey was a heroic effort. From that we now see the penetration of fossil fuel CO_2 to well below 1 km throughout the ocean, and the detection of sea floor carbonate dissolution from the 20th century chemical invasion. There were problems. An ocean color satellite did not fly until 1997: a full decade late, and only after endless effort. The linkage of the oceanic CO_2 problem with biogeochemical measurements and models had amazing birth pains. Most ocean chemists had no real knowledge of microbial processes; and all (so far as I could tell) ocean biologists were in disbelief that the fossil fuel CO_2 invasion was a purely inorganic/physical phenomenon. Very few people looked far to the future, say to an ocean at the end of the 21st century where the CO_2 maximum may be at the surface, and essential biogeochemical cycles may be profoundly changed. The first US JGOFS response to John Martin's proposal of an iron fertilization field experiment (at a meeting I did not attend) was to vote it down! However, funding was obtained from the DOE and the resulting iron fertilization experiments were a brilliant success.

The papers in this volume show how far we have come. Satellite ocean color images pervade the literature. Sophisticated models of biogeochemical cycles are routinely used. All participants are fluent in the CO_2 connection. Ready access to more than a decade of time series data is taken for granted. Synthesis and modeling efforts are supported, and are productive. And iron fertilization science has a strong international community.

The goals of JGOFS were carefully negotiated, and they included the need "To determine on a global scale the processes controlling the time varying fluxes of carbon and associated biogenic elements in the ocean ...", and "To develop the capability to predict on a global scale the response of oceanic biochemical processes to anthropogenic perturbations, in particular those related to climate change." This knowledge is urgently required, for mankind's influence on the carbon cycle is proceeding far faster than we usually acknowledge. In 1984, at the time of the first U.S. 'GOFS' meeting, atmospheric CO_2 levels were 344 ppm, or 64 ppm above the pre-industrial baseline. Today they are 372 ppm, or 92 ppm above the pre-industrial levels – a 43% increase while we have been planning and carrying out our research. Over these 18 years the ocean has taken up some 131 billion tons of CO_2 gas.

International ocean science of a new kind evolved with JGOFS. It is created by the efforts of individuals who do not see boundaries, only opportunities. A thriving community of students and Post Docs. emerges each year, and happily spreads across international borders seeking excellence. And they often find it in the laboratories of scientists whose work is represented here.

Contents

Contributors

Anderson, Robert

Lamont-Doherty Earth Observatory
Columbia University
P.O. Box 1000, Palisades, NY 10964, USA

Barber, Richard T.

Nicholas School of the Environment and Earth Sciences
Duke University
135 Duke Marine Lab Road, Beaufort, NC 28516-9721, USA

Bates, Nicholas R.

Bermuda Biological Station for Research, Inc.
Ferry Reach, St. George's GE01 Bermuda

Boyd, Philip W.

National Institute of Water and Atmosphere
Centre for Chemical and Physical Oceanography
University of Otago
Dunedin, New Zealand
pboyd@alkali.otago.ac.nz

Brewer, Peter G.

Monterey Bay Aquarium Research Institute
Moss Landing CA 93923, USA

Chen, Chen-Tung Arthur

Institute of Marine Geology and Chemistry
National Sun Yat-sen University
Kaohsiung 804, Taiwan, China (Taipei)
ctchen@mail.nsysu.edu.tw

Dittert, Nicolas

Université de Bretagne Occidentale
Institut Universitaire Européen de la Mer
UMR CNRS 6539, Brest, France

Doney, Scott C.

Climate and Global Dynamics
National Center for Atmospheric Research
P.O. Box 3000, Boulder, CO 80307, USA

Ducklow, Hugh W.

School of Marine Science
The College of William & Mary
Box 1346, Gloucester Point, VA 23062-1346, USA
duck@vims.edu

Emerson, Steven

Department of Oceanography
University of Washington
Seattle, WA 98195, USA

Falkowski, Paul G.

Institute of Marine and Coastal Science and Department of Geology
Rutgers, The State University of New Jersey
New Brunswick, New Jersey 08901, USA

Fasham, Michael J. R.

Southampton Oceanography Centre
Waterfront Campus, Southampton, SO14 3ZH, UK
mjf@soc.soton.ac.uk

Follows, Michael J.

Program in Atmospheres, Oceans and Climate
Department of Earth, Atmospheric and Planetary Sciences
Massachusetts Institute of Technology
Cambridge MA 02139, USA
ric@liv.ac.uk or *mick@plume.mit.edu*

Francois, Roger

Woods Hole Oceanographic Institution
Woods Hole, MA 02543, USA

Harrison, Paul J.

Department of Earth and Ocean Sciences (Oceanography)
University of British Columbia
Vancouver, BC V6T 1Z4, Canada

Jahnke, Richard A.

Skidaway Institute of Oceanography
10 Ocean Science Circle, Savannah, Georgia 31411, USA

Jeandel, Catherine

Observatoire Midi-Pyrenées
14 Ave E. Belin, 31400 Toulouse, France

Karl, David M.

Department of Oceanography, SOEST
University of Hawaii, Honolulu, HI 96822, USA
dkarl@soest.hawaii.edu

Laws, Edward A.

Department of Oceanography
University of Hawaii at Manoa
Honolulu, Hawaii 96822

Legendre, Louis

Laboratoire d'Océanographie
CNRS, Villefranche-sur-mer, France

Lindsay, Keith

National Center for Atmospheric Research
Boulder, CO 80307, USA

Liu, Kon-Kee

Institute of Oceanography
National Taiwan University
Taipei 100, Taiwan 106, ROC

Llinás, Octavio

Instituto Canario de Ciencias Marinas
Telde, Gran Canaria, Spain

Lochte, Karin

Institut für Meereskunde
Düsternbrooker Weg 20
24105 Kiel, Germany
klochte@ifm.uni-kiel.de

Macdonald, Robie

Institute of Ocean Sciences
Sidney, B.C. V8L 4B2, Canada

Marty, Jean-Claude

Laboratoire d'Océanographie de Villefranche
BP 08, F-06 238 Villefranche sur mer Cedex, France

Michaels, Anthony F.

Wrigley Institute for Environmental Studies
University of Southern California
Los Angeles, CA 90089-0371, USA

Miquel, Jean C.

Marine Environment Laboratory
International Atomic Energy Agency
4 Quai Antoine ler, BP 800, MC98000 Monaco

Moore, J. Keith

National Center for Atmospheric Research
Boulder, CO 80307, USA

Murray, James W.

School of Oceanography
University of Washington
Box 355351, Seattle, Washington, 98195-5351, USA

Neuer, Susanne

Department of Biology
Arizona State University
Tempe, AZ 85287, USA

Nojiri, Y.

National Institute for Environmental Studies
Tsukuba, Ibaraki 305-0053, Japan

Orr, James C.

Laboratoire des Sciences du Climat et de l'Environnement
Unite Mixte de Recherche CEA-CNRS, CEA Saclay
F-91191 Gif-sur-Yvette Cedex, France

Ragueneau, Olivier

Université de Bretagne Occidentale
Institut Universitaire Européen de la Mer
UMR CNRS 6539, Brest, France

Rivkin, Richard T.

Ocean Sciences Centre
Memorial University of Newfoundland
St. John's, Canada

Shimmield, Graham

Scottish Association for Marine Science
Dunstaffnage Marine Laboratory
Oban, Argyll, PA37 1QA, Scotland, UK

Tréguer, Paul

Université de Bretagne Occidentale
Institut Universitaire Européen de la Mer
UMR CNRS 6539, Brest, France
Paul.Treguer@univ-brest.fr

Vetrov, Alexander

P. P. Shirshov Institute of Oceanology
Russian Academy of Sciences
Krasikova 23, 117218 Moscow, Russia

Watson, Andrew J.

School of Environmental Sciences
University of East Anglia
Norwich NR4 7TJ, UK
a.j.watson@uea.ac.uk

Williams, Richard G.

Oceanography Laboratories
Department of Earth Sciences, University of Liverpool
Liverpool L69 7ZL, UK

Wong, Chi Shing

Climate Chemistry Laboratory, OSAP
Institute of Ocean Sciences
P.O. Box 6000, Sidney, B.C., Canada V8L 4B2

Introduction

Michael J. R. Fasham · Hugh W. Ducklow

The Joint Global Ocean Flux Study (JGOFS) had its genesis in the US where a need for a programme to study the role of the ocean in the global carbon cycle was perceived in the late 1980s. Peter Brewer was in the forefront of these developments and in the Foreword he gives his own personal view of events and the excitement generated by this new global approach to ocean biogeochemistry.

Following the lead given by US scientists the Scientific Committee on Oceanic Research (SCOR) sponsored a meeting Paris in February 1987 at which JGOFS was born and its main goal was defined, namely:

> To determine and understand on a global scale the processes controlling the time varying fluxes of carbon and associated biogenic elements in the ocean, and to evaluate the related exchanges with the atmosphere, sea floor and continental boundaries.

Later a second objective was added:

> To develop a capability to predict on a global scale the response of oceanic biogeochemical processes to anthropogenic perturbations, in particular those related to climate change.

The JGOFS Science Plan was developed through 1989–1990 (SCOR 1990) and, together with the implementation plan (IGBP 1992); this formed the basis of the JGOFS strategy. The main features were intensive process studies in areas thought to make significant contributions to the ocean-atmosphere CO_2 flux, a global survey of DIC in collaboration with WOCE, and long-term time-series measurement programmes in key ocean basins. How these plans were put into practice by the international community is described in Chap. 11.

During the period of JGOFS' genesis, Alan Longhurst was informing traditional biogeographic approaches to characterising large-scale ocean ecosystems, using CZCS data (Longhurst et al. 1995). Longhurst's province concept rests largely on the hypothesis that physical forcing is the primary factor governing ocean ecosystem structure and variability. Some aspects of JGOFS studies in Longhurst's 'biogeochemical provinces' are discussed by Ducklow in Chap. 1, while in Chap. 2 Williams and Follows address the key physical processes influencing ocean biogeochemical dynamics.

Global-scale observations of the partial pressure of CO_2 in the surface ocean and the global observations of total dissolved inorganic carbon obtained during JGOFS have provided an invaluable tool for understanding both natural and anthropogenic CO_2 exchanges in the ocean and the results are discussed in Chap. 5 by Watson and Orr. In Chap. 4, Falkowski et al. summarise what has been learnt about primary production during JGOFS and, perhaps more importantly for the carbon cycle, how estimates of export production might be derived from primary production observations.

During most of the JGOFS programme much of the scientific effort has been in the upper water column. However, the remineralisation processes in the 'twilight' zone, the midwater column between ca. 200–1 000 m, is just as important for understanding the carbon cycle and this zone is discussed by Tréguer et al. in Chap. 6. The ocean floor is the ultimate sediment trap, as well as the site of preservation and burial of the palaeo-oceanographic records of past ecosystems and climate signals. Deep-ocean sediment traps have been a feature of most JGOFS process studies, although palaeo-oceanographic observations have been carried out less frequently. Lochte et al. (Chap. 8) address these important aspects of ocean biogeochemistry.

The role that the ocean margins play in the ocean carbon and nutrient cycles has still to be fully quantified and there has been much debate about whether the margins are sources or sinks of CO_2. In Chap. 3, Chen, Liu and MacDonald provide an excellent stimulus for this work by reviewing the presently available data to derive nutrient and carbon budgets for the main shelf areas.

The time-series stations have been providing invaluable monthly data to the JGOFS community, some since 1988. A summary of the results from all the JGOFS time-series stations and a description of some of the exciting new concepts arising from this work are given in Chap. 10 by Karl et al.

Now that the observational programme is mainly complete, the emphasis of JGOFS has switched to analysing and synthesising the vast datasets that have been obtained. Reviews of what has been achieved during the last 10 years have now been published (Fasham et al. 2001; Buesseler 2001). Two important elements of the ongoing synthesis include modelling and projecting the effects of climate change on ocean ecosystems and biogeochemistry. Doney et al. assess the state of the art in ocean biogeochemical modelling (Chap. 9), while Boyd et al. discuss impacts of climate change on the ocean in Chap. 7.

Hundreds of oceanographers, students, post-docs, technicians, ships' crews and officers contributed to JGOFS. The fruits of their labours are summarized here. We hope our book conveys some of the intellectual, as well as the sometimes physical and emotional adventure that was JGOFS.

Acknowledgements

On the behalf of all the authors we would like to acknowledge the generous financial support provided by the International Council of Science (ICSU), the U.S. National Science Foundation (NSF), the Scientific Committee on Oceanic Research (SCOR), and the International Geosphere-Biosphere Programme (IGBP) for this book. We would also like to acknowledge the aforementioned organisations, the Bundesminister für Bildung, Forschung und Technologie (BMBF), Deutsche Forschungsgemeinschaft (DFG), the Research Council of Norway, and the University of Bergen for the funding of the JGOFS International Project Office over past twelve years. Without this funding the international Scientific Steering Committee could not have functioned over the years and the unique international cooperation that was such an essential feature of this book would not have happened. Finally we would like to thank Angela Bayfield for her careful copy editing of the manuscripts.

References

Buesseler K (2001) Ocean biogeochemistry and the global carbon cycle. An introduction to the US Joint Global Ocean Flux Study. Oceanography Special Issue 14(4):1–120

Fasham MJR, Baliño BM, Bowles MC (2001) A new vision of ocean biogeochemistry after a decade of the Joint Global Ocean Flux Study (JGOFS). Ambio Special Report 10, 31 pp

IGBP (1992) Joint Global Ocean Flux Study: implementation plan. IGBP Report No 23, IGBP Secretariat, Stockholm

Longhurst A, Sathyendranath S, Platt T, Caverhill C (1995) An estimate of global primary production in the ocean from satellite radiometer data. J plankton res 17:1245–1271

SCOR (1990) The Joint Global Ocean Flux Study (JGOFS) science plan. JGOFS Report No 5, Halifax Canada. Scientific Committee on Oceanic Research 61 pp

Chapter 1

Biogeochemical Provinces: Towards a JGOFS Synthesis

Hugh W. Ducklow

'The ocean is a desert with the life underground,
and the perfect disguise up above.'

America, A Horse With No Name, 1972

Most people are intuitively familiar with the existence of recognizable, bounded units of landscape with characteristic climatic regimes, land cover and animal populations – the basis of the ecosystem concept in ecology. Theophrastus (ca. 320 B.C.) documented this recognition in his 'Inquiry into Plants' and it is implicit much later in the writings of Thoreau, G. P. Marsh and others who by the mid-19th century already lamented the loss of the North American primeval forests (Cronon 1983). Thus we recognize particular terrestrial ecosystems: grasslands, savannas, deserts, temperate and tropical forests, polar tundra and so on. What about the ocean? To the uneducated eye of the non-sailor, the surface of nearly three quarters of the planet is largely homogeneous, with minor differences in surface roughness and color. The featureless nature of the ocean's upper surface is especially conspicuous offshore, away from the gradients in color resulting from terrestrial sources of organic matter and resuspended sediments found in shallow waters. Do distinct marine provinces or ecosystems analogous to the familiar terrestrial biomes exist? Many (but not all) oceanographers agree that they do, and there have been many schemes to distinguish and classify them, but there is little agreement on how many should be identified and their spatial scale. Yet most of us would agree that there are distinctive, large scale ocean regimes which also support characteristic flora and fauna, and exist in the familiar climatic regions of the planet.

In this chapter I address the question of biogeochemical provinces in the ocean. JGOFS embodied the ocean biogeochemistry paradigm, that is, the idea that ocean is an organized system of physically-driven, biologically-controlled chemical cycles which regulate the planetary climate over large spatial and temporal scales. Much of the JGOFS program over the past decade has been structured around intensive studies in particular geographic locations chosen because they exemplify different aspects of the ocean biogeochemical system (SCOR 1990 and see the many special volumes of Deep-Sea Research, Part II). Thus at this point it is important to look critically at the question of whether such locations are distinctive, and whether a province-based approach was a good way to study ocean carbon fluxes and the controls acting on them. An alternative to the province approach is the continuum model in which the ocean is viewed (and modeled) as a continuously varying biogeochemical system, structured and differentiated by the responses of organisms to regional changes in stratification, vertical mixing and advection (e.g., Sarmiento et al. 1993). Here I review how we can characterize different ocean regions using biogeochemical criteria. I am less concerned with whether biogeochemical provinces have some ultimate reality (the strong province model) or arise as emergent properties from a continuum of physical drivers and biological responses. I start by reviewing some previous attempts and schemes to partition the ocean into distinctive provinces or regions. Then I review some recent observations on primary production, bacterial activity, and the net production of dissolved organic carbon (DOC) in various ocean provinces studied by JGOFS and allied programs. These observations will serve as an introduction to JGOFS for non-JGOFS readers (and some JGOFS readers too): where we worked and what we found. Other authors in this volume examine some of these same questions in much greater detail. Finally I conclude with some thoughts on where this approach might lead in the coming decade of ocean science.

1.1 Plankton Community Structure and Distribution

Whether or not one believes in partitioning the ocean into discrete provinces or domains, it seems beyond debate that the ocean climate determines the composition and size of the regional-scale phytoplankton community, which in turn influences the structure of the plankton system in a given area. Cullen et al. (2002) review the modern theory of plankton community dynamics, based on Margalef's (1978) scheme of how turbulence and nutrient availability select phytoplankton life forms. They conclude:

> We infer from such observations that it is the characteristic physical oceanography of each region that primarily determines the functional composition and seasonal biomass of the pelagic ecosystem from plants to predators. This 'bottom-up' control of

the ecosystem is, of course, mediated by the time-dependent supply of inorganic nutrients to the euphotic zone. Consequently, 'top-down' modulation of ecosystem structure by herbivory and predation must be considered a subsidiary process. In this, the pelagic biomes resemble terrestrial biomes. A simple combination of geology, latitude, altitude, exposure and rainfall determines the characteristic vegetation of any site ashore. Though terrestrial herbivores undoubtedly modulate the final expression of forest or tundra, neither elephants nor reindeer determine which vegetation type shall develop. It can be argued that a similarly parsimonious set of factors determines the distribution of pelagic biomes, each with its characteristic type of plant growth. For the open ocean, these factors are simply those required by Sverdrup (1953) to control illumination and the vertical stratification of the water column; they may be reduced to latitude, regional winds, cloud cover and the flux (if any) of low-salinity surface water. From these factors may be deduced sufficient information to predict the seasonality and kind of phytoplankton production. Over continental shelves we must also know water depth and tidal range. Copepods and whales do not determine which groups of plants shall flourish: like the phytoplankton, they are themselves expressions of the regional physical oceanographic regime (Cullen et al. 2002, pp. 8–9).

This theory provides the mechanistic foundation for the existence of ocean provinces. Cullen et al. (2002) also make the important point that not all limiting nutrients are supplied from below by turbulent mixing processes. In particular iron is supplied to wide areas of the surface ocean principally by aeolian deposition, and elemental nitrogen (N_2) is fixed by diazotrophs like *Trichodesmium* and other cyanobacteria. Thus not all departures from the equilibrium, background state dominated by picoplankton are initiated by turbulence. It is also important to understand that multiple nutrient colimitation seems to regulate interbasin differences in plankton community structure and N vs. P limitation (Wu et al. 2000). Before proceeding, I review briefly our current understanding of plankton community structure in the context of geographic variability in ocean climate and physical forcing.

In general open ocean photosynthesis is dominated by picoplankton (diameter 0.2–2 μm; Sieburth et al. 1978) and nanoplankton (diameter 2–20 μm), with as much as 90% of the active primary producers small enough to pass through 2 μm pore-sized filters (Li et al. 1982). These microbial phytoplankton exhibit little variability in time and space (Malone 1980; Banse 1992) because their iron requirements are relatively low and they are preyed on by small grazers, principally heterotrophic nanoflagellates (HNAN) which have growth rates as fast or faster than their prey (Landry et al. 1997; Strom et al. 2000). Population outbreaks of the smallest primary producers are held in check by the rapid functional responses of their predators. Coexisting with, and supporting, oceanic primary producers through its role in nutrient regeneration is a complex assemblage of viruses, bacterioplankton and protozoans, all in the 0.02–20 μm size range (Azam et al. 1983; Ducklow and Carlson 1992; Sherr and Sherr 2000; Fuhrman 2000). Larger phytoplankton (e.g., diatoms and dinoflagellates) contribute the major source of variability in plankton biomass and production (Malone 1980) during population outbreaks occurring over a range of scales from small, sporadic local miniblooms stimulated by event-scale processes (Walsh 1976) to the basin scale spring bloom covering the North Atlantic (Ducklow and Harris 1993), seen conspicuously in ocean color imagery. It is against, or underlying, this pattern of bloom and decline of larger-celled organisms at various scales that the background, small-celled plankton system persists, sustained by nutrient recycling and held in check by intense grazing pressure.

Until 1977, the very existence of the dominant oceanic cyanobacterial primary producers was unknown (Johnson and Sieburth 1979; Waterbury et al. 1979; Chisholm et al. 1988), and even today the taxonomic affinities of many major groups are just now being identified by new molecular genomic tools (Giovannoni and Rappe 2000). The large operational grouping generically known as 'bacteria' typifies the problem and presents a good case in point. 'Bacteria' include the oxygenic, photosynthetic cyanobacteria (both *Synecchococcus* spp. and prochlorophytes), the heterotrophic 'true' *Bacteria* and the *Archaea*, newly recognized as a separate major domain of life (Giovannoni and Rappe 2000). The cyanobacteria also include the major oceanic nitrogen-fixing organism, *Trichodesmium* which is becoming dominant in a new regime or successional stage developing in the North Central Pacific Gyre (Letelier and Karl 1996, 1998; Cullen et al. 2002; but see also Wu et al. 2000). Within the heterotrophic bacteria, most of the major groups still cannot be cultivated and studied in the laboratory, so the identities and occurrence of the unculturable majority are known only from molecular probes (Giovannoni and Rappe 2000). The specific roles of these organisms and of the *Archaea* are almost completely unknown. This situation is especially pointed for the mesopelagic depths below the euphotic zone (ca 200–1000 m) where *Archaea* might predominate numerically (Karner et al. 2001). Even among the grazers, identity and role identification are not well understood, because a possibly large portion of the HNAN are mixotrophic, combining both photosynthetic and grazing trophic functions (Caron 2000).

In the open sea approximately 90% of the total net primary production (NPP) is supported by regenerated nutrients, of which the great majority is produced by the small grazers and heterotrophic bacteria (Harrison 1980). Bacterial productivity may average 20% of the NPP (Cole et al. 1988; Ducklow and Carlson 1992; Ducklow 1999) and is sustained by a flux of dissolved organic matter (DOM) arising from phytoplankton exudation, grazer feeding and metabolism, viral attack and particle decomposition (Nagata 2000; Williams 2000). Since the bacterial conversion efficiency is low (10–30%; del Giorgio and Cole 1998, 2000), the DOM flux fueling bacterial metabolism approaches the magnitude of the NPP

(Pomeroy 1974; Williams 1981, 1984). Our ignorance of microbial identity is mirrored in a similar lack of knowledge about the composition of the DOM pool, a complicated mixture of monomers, polymers and condensed heterocyclic compounds of which less than 10% is chemically identified (Benner et al. 1992). Besides serving as the sole quantitatively important agents of DOM oxidation and as important nutrient remineralizers, bacteria are an important alternative food source for HNAN, and thus a stabilizing factor in the nanoplankton foodweb (Strom et al. 2000). A large portion of oceanic respiration is bacterial, or derived from bacterial processes, and a large portion of ocean metabolism is driven by fluxes of dissolved matter. All these issues are not trivial and are intimately related to our understanding of ocean ecology and biogeography. We cannot move beyond the current simple PZND models of plankton dynamics toward more detailed, adaptive model ecosystems without a better appreciation of the identity and functional roles of these major plankton groups.

1.2 Partitioning the Oceans

Most schemes to partition the ocean into a system of bounded regions have been based on physical climate and circulation or have been biogeographic, based on the occurrence of distinctive species assemblages (e.g., van der Spoel and Heyman 1983). Longhurst (1998) pioneered a more encompassing ecological scheme, to which I return below. In his book, 'Ecological Geography of the Sea,' Longhurst (1998) reviews previous partitioning schemes in some detail. The following summary is taken largely from Longhurst's review.

Most later efforts, including Longhurst's, can be traced to Dietrich (1963), who distinguished seven major regions on the basis of global winds and the underlying current systems. Thus Bailey (1998), who erected a detailed 'ecosystem geography' of the continents based on climate, geomorphology, vegetation cover and local meteorology, also mapped out a series of oceanic ecoregions in the sea. Bailey's marine ecoregions, however, bear little similarity to the richness of ecological differentiation of his land classification. Banse (1987) and Barber (1988) took more comprehensive approaches by integrating physical and ecological processes to distinguish both larger and smaller scale partitioning. Banse (1987) showed that three previously defined hydrographic areas in the NW Arabian Sea also possessed distinctive seasonality in surface chlorophyll *a*, providing a foreshadowing of several later syntheses of the Coastal Zone Color Scanner (CZCS) imagery of ocean color (e.g., Platt and Sathyendranath 1988; Banse and English 1993; Longhurst et al. 1995). Barber (1988) considered the reality of ocean basin ecosystems, eventually distinguishing six (Table 1.1). Barber (1988, p. 171) recalled Odum's (1969) definition of an ecosystem "... as the unit of biological organization interacting with the physical environment such that the flow of energy and mass leads to a *characteristic trophic structure and material cycles*," (my emphasis) which elegantly ties together ecological and biogeochemical dynamics. Barber also argued that during ENSO events, much of the tropical and subtropical Pacific becomes a unified large scale ecosystem, blurring the distinctions evident during non-ENSO conditions. This is important: ecosystem or province boundaries in the sea are literally and figuratively fluid in time and space. The lines shown on maps are inescapable if we want to map regions on a solid medium, but those lines belie the fluidity of the actual boundaries.

The concept of a new biogeographical segmentation of the sea was proposed by Platt and Sathyendranath

Table 1.1. Typology and description of ocean basin ecosystems (after Barber 1988)

Ecosystem	Stratification		New nutrients[a]		Primary productivity[c] per unit area	Process regulating productivity[d]
	Strength	Duration	Level[b]	Source		
Coastal upwelling	Patchy	Continuous	High	Advection	Medium to high	Space
Low latitude gyre	Strong	Permanent	Low	Mixing	Low to medium	Nutrient supply
Equatorial upwelling	Strong	Permanent	High	Advection	Medium	Physical processes and grazing
Subarctic gyre	Strong	Seasonal	High	Mixing	Medium (low in winter)	Grazing (summer); mixing/light (winter)
Southern Ocean	Weak	Seasonal	Very high	Mixing	High (low in winter)	Mixing (summer); light (winter)
Eastern boundary current	Medium	Permanent	Medium	Advection	Medium	Grazing and nutrient supply

[a] New nutrients *sensu* Dugdale and Goering (1967).
[b] Relative to phytoplankton uptake characteristics (*high:* concentration always saturating uptake rate; *low:* concentrating always rate limiting; *medium:* varies between high–low).
[c] *Low:* <0.1 g C m^{-2} d^{-1}; *medium:* ≈ 0.5 g C m^{-2} d^{-1}; *high:* >1 g C m^{-2} d^{-1}.
[d] Regulation may refer to rate or yield limitation (*sensu* Caperon 1975).

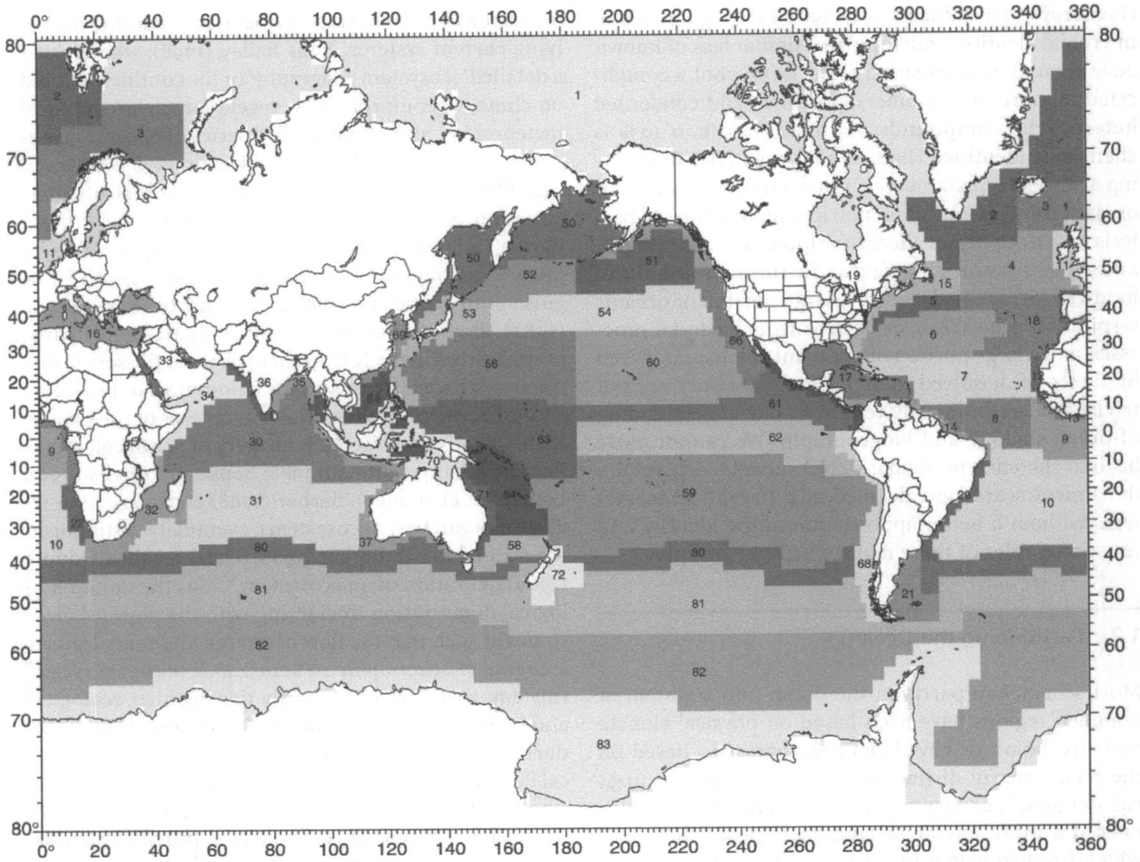

Fig. 1.1. Longhurst's 'Ecological Geography of the Sea'. This map is available from: *http://www.mar.dfo-mpo.gc.ca/science/ocean/BedfordBasin/Papers/Longhurst1998/Provinces/*, and described in detail in Longhurst (1998) although in this slightly newer version, some of the provinces have been subdivided and a few new ones have been added. A table identifying the provinces is also available at the website, and those studied in JGOFS are listed in Table 1.3

(1988). Later, possibly reflecting the influence of JGOFS on their thinking, they were apparently the first to use the term 'biogeochemical province' (Platt et al. 1991). Longhurst and colleagues (Longhurst 1995, 1998; Longhurst et al. 1995; Platt and Sathyendranath 1999) exploited the global, near-synoptic CZCS data sets on regional and basin-global scale, seasonally-resolved distributions of surface ocean pigments, along with extensive data on the vertical structure of chlorophyll *a* and photosynthesis-irradiance (P-I) relationships to produce a new ecological geography of the sea. Longhurst's scheme (Fig. 1.1) borrows from Dietrich (1963) and verifies with his own analysis of global hydrography a global ocean system with four principal *domains* (Longhurst 1995) or *biomes* (Longhurst 1998) which are the major climate regimes in which the provinces are based (Table 1.2). In the rest of this chapter I use the more generic term domain when referring to the four major divisions of the ocean shown in Table 1.2, to avoid the more strictly ecological connotation of biome. I use the term province

Table 1.2. Four primary domains or biomes in the ocean (after Longhurst 1998)

Domain	Definition
Polar	Where the mixed layer is constrained by a surface brackish layer formed each spring in the marginal ice zone (>60° latitude)
Westerlies	Where the mixed layer depth is forced largely by local winds and irradiance (ca. 30–60° latitude).
Trades	Where the mixed layer depth is forced by geostrophic adjustment on a basin scale to often-distant wind forcing (ca. 30° N to 30° S latitude).
Coastal	Where diverse coastal processes (e.g., tidal mixing, estuarine runoff) force mixed layer depth (all latitudes).

in the same sense as Longhurst, to denote the regional-scale divisions of the domains within each ocean basin (Table 1.3). Platt and Sathyendranath (1988, 1999) systematically analyzed P-I data and argue that P-I parameters are distributed discontinuously, and assume that prov-

Table 1.3. JGOFS studies in ocean biogeochemical provinces. Province names and locations from Longhurst (1998) and *http://www.mar.dfo-mpo.gc.ca/science/ocean/BedfordBasin/Papers/Longhurst1998/Provinces/*

Year	Province/domain[a] and map number[b]	Study	Nations	Reference
Atlantic Ocean				
1986–1997	Mediterranean Sea (MEDI)/W (16)	DYFAMED	France	Copin-Montegut (2000)
1988–1992	NE. Atlantic Shelves (NECS)/C (11) Mediterranean (MEDI)/W (16)	FRONTAL, ECOMARGE	France	Monaco et al. (1999)
1988–1989	N. Atlantic Subtropical Gyral – E (NASE)/W (18)	MEDATLANTE	France	Savenkoff et al. (1993)
1988–present	N. Atlantic Subtropical Gyral – W (NASW)/W (6)	BATS	USA	Karl and Michaels (1996)
1989–present	Atlantic Arctic (ARCT)/P (2) Atlantic Subarctic (SARC)/P (3) N. Atlantic Drift (NADR)/W (4) N. Atlantic Subtropical Gyral – E (NASE)/W (18)	NABE/BOFS, German JGOFS	USA, UK, Netherlands, Germany, Canada	Ducklow and Harris (1993) Harris et al. (1997) Harrison et al. (1993)
1990–1996	Atlantic Arctic (ARCT)/P (2) Atlantic Subarctic (SARC)/P (3)	Nordic Seas	Norway	Chierici et al. (1997)
1991–1992	N. Atlantic Subtropical Gyral – E (NASE) (18)	EUMELI	France	Morel (1996)
1991–1992	Canary Coastal (CNRY) (12)	EUMELI	France	Morel (1996)
1992–1994	NW Atlantic Shelves (NWCS)/C (15)	Gulf of St. Lawrence	Canada	Roy and Sundby (2000)
1992–1996	Benguela Current Coastal (BENG)/C (22)	Benguela Ecology Program (BEP)	S. Africa	Jarre-Teichmann et al. (1998)
Pacific Ocean				
1988–present	N. Pacific Tropical Gyre (NPTG)/T (60)	HOT	USA	Karl and Michaels (1996)
1989–present	Kuroshio Current (KURO)/W (53)	KEEP	China-Taipei	Wong et al. (2000)
1992	N. Pacific Tropical Gyre (NPTG)/T (60) N. Pacific Equatorial Countercurrent (PNEC)/T (61) Pacific Equatorial Divergence (PEQD)/T (62) S. Pacific Subtropical Gyre (SPSG)/W (59)	EQPAC	USA	Murray et al. (1994)
1992–1998	Pacific Subarctic Gyres – East (PSAG-E)/W (51) Alaskan Downwelling Coastal (ALSK)/C (65) California Downwelling Coastal (CCAL)/C (66)	Canadian NE Pacific JGOFS NPPS	Canada	Boyd et al. (1999)
1993–present	China Sea Coastal (CHIN)/C (69)	China JGOFS	China-Beijing	Hu and Tsunogai (1999)
1994	W. Pacific Warm Pool (WARM)/T (63) Pacific Equatorial Divergence (PEQD)/T (62)	EPOPE/FLUPAC	France	
1990–present	N. Pacific Epicontinental (BERS)/P (50) Pacific Subarctic Gyres – West (PSAG-W)/W (52) Kuroshio Current (KURO)/W (53)	Japan-JGOFS	Japan	Tsunogai (1997)
Indian Ocean				
1994–1997	Indian Monsoon Gyres (MONS)/T (30) NW Arabian Upwelling (ARAB)/C (34) W. India Coastal (INDW)/C (36) Red Sea / Persian Gulf (REDS)/C (33)	Arabian Sea	Germany, India, Netherlands, Pakistan, UK, USA	Lal (1994) Smith (1998) Burkill (1999) Pfannkuche and Lochte (2001)
Southern Ocean				
1993–1995	S. Subtropical Convergence (SSTC)/W (80) Subantarctic (SANT)/W (81) Antarctic (ANTA)/P (82)	ANTARES	France	Gaillard and Tréguer(1997) LeFevre and Treguer
1990–1995	Subantarctic (SANT)/W (81)	KERFIX	France	Pondaven et al. (2000)
1992	Subantarctic (SANT)/W (81) Antarctic (ANTA)/P (82) Austral Polar (APLR)/P (83)	BOFS/STERNA	UK	Turner et al. (1995)
1992	Subantarctic (SANT)/W (81) Antarctic (ANTA)/P (82)	POLARSTERN ANT X/6	Germany, Belgium, France, Netherlands, UK	Smetacek et al. (1997)
1996–1998	Subantarctic (SANT)/W (81) Antarctic (ANTA)/P (82) Austral Polar (APLR)/P (83)	AESOPS	USA	Smith et al. (2000)
1999	Antarctic (ANTA)/P (82)	SOIREE	New Zealand	Boyd and Law (2000)

[a] Principal domains: *P:* Polar; *W:* Westerlies; *T:* Trade Winds; *C:* coastal (after Longhurst 1998).
[b] Numbers refer to provinces shown in Fig. 1.1.

ince boundaries delineate regions within which the parameters are predictable. Provinces then provide a systematic means of using remotely sensed data to recover global estimates of primary and new production (Longhurst et al. 1995) or to parameterize large-scale models. Below, as a way to flesh out the province concept and introduce the JGOFS field program, I compare Longhurst's (1998) regional (province-based) estimates of primary production with in situ observations based on new ^{14}C measurements made during NABE and other recent research cruises.

1.3 Primary Production in Ocean Domains and Provinces

Primary production (PP) of organic matter by phytoplankton forms the foundation of life in the sea (Falkowski and Raven 1997; Falkowski et al. (2003, this volume) and also formed the basis of Longhurst's partitioning scheme. Since the ^{14}C method was one of the most widely performed core measurements of a rate process in JGOFS, the PP data set is useful for looking at differences among the domains and provinces. The following summary is preliminary but provides one of the first such syntheses of the recent observations. All these data were obtained from bottle incubations (on-deck or in situ) using ^{14}C-labelled bicarbonate and following trace-metal-free clean techniques as specified in the JGOFS Core Measurement Protocols (Knap et al. 1996). I obtained the data starting from the International JGOFS Data Management Homepage (*http://ads.smr.uib.no/jgofs/inventory/index.htm*) and following links where available. In some cases data were provided by individual investigators. Each graph shows the PP observations integrated to the base of the euphotic zone (depth of 0.1–1% of surface irradiance, I_0) and plotted against the day of the year. Observations are composited from different years in some cases. I also plotted the domain-averaged PP derived by Longhurst (1998) from the global CZCS data, which I obtained on Excel spreadsheets from *www.mar.dfo-mpo.gc.ca/science/ocean/BedfordBasin/Papers/Longhurst1998/Provinces/*, for comparison. The comparison is discussed further below.

Sathyendranath et al. (1995) concluded from their analysis of chlorophyll *a* profiles and photosynthetic parameters in 19 provinces of the North Atlantic that the most fundamental distinction among provinces was between the coastal and ocean domains. That distinction is exemplified by comparison of the adjacent coastal (ARAB) and trade winds (MONS) provinces in the Arabian Sea studied in great detail during the international Arabian Sea expeditions in JGOFS (Fig. 1.2). The new JGOFS observations support Longhurst's (1998) estimates of primary productivity approaching 4 g C m^{-2} d^{-1} during the Southwest Monsoon, driven by intense coastal upwelling and abundant inputs of iron-containing dust from the Arabian Peninsula (Tindale and Pease 1999). With the exception of a few individual measurements in the Antarctic, these are the highest PP observations recorded in JGOFS. Observations from three separate expeditions (Germany, The Netherlands and UK) all suggest that primary production responds to the onset of the monsoon more rapidly than the CZCS-based estimates from Longhurst's (1998) synthesis, with values ranging from 1–3 g C m^{-2} d^{-1} by mid-May. In contrast, PP reached about 2 g C m^{-2} d^{-1} during June–August in the offshore Indian Monsoons (MONS) Province.

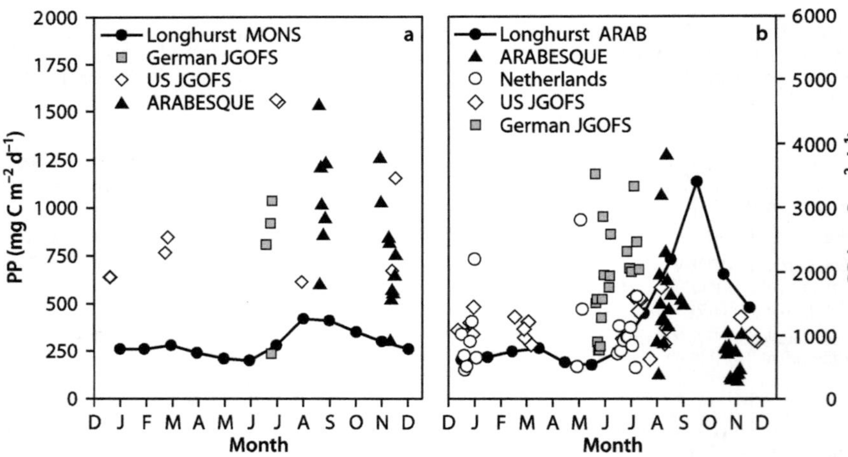

Fig. 1.2. Primary production in the Arabian Sea. **a** Indian Monsoon Gyres Province (Trade Winds Domain) with observations from the UK ARABESQUE (*triangles*), US (*diamonds*) and German JGOFS (*squares*) JGOFS cruises. **b** NW Arabian Upwelling Gyre Province (Coastal Domain; *note change in Y-axis scale*) with data from the UK ARABESQUE (*triangles*), Netherlands (*circles*), US (*diamonds*) and German JGOFS (*squares*) cruises. The domain-averaged annual cycles of PP derived by Longhurst (1998) are shown for the MONS (**a**) and ARAB (**b**) provinces for comparison (*lines with symbols*). Observational data were obtained from published reports and JGOFS databases (see text)

However there was little seasonality apparent in the oceanic province, with no clear distinction between the NE and SW Monsoons (Fig. 1.2b). Also striking is the contrast between Longhurst's (1998) estimate of PP in the MONS region and the new observations. The underestimates might be due in part to interference of dust and aerosols; and could also be an artifact of concentrated sampling in the NW part of the MONS Province which is more geographically extensive, extending throughout the northern Indian Ocean (Fig. 1.1).

Clearly the most intensively studied region by JGOFS has been the Trade Winds Domain, especially the equatorial region and tropical gyre of the N Pacific and subtropical gyre of the western Atlantic. Data from the Hawaii Ocean Time-series (HOT) and Bermuda Atlantic Time-series Study (BATS) stations are shown in Fig. 1.3. Over 100 trace-metal-free, in situ ^{14}C determinations of PP have been accomplished at each station. The physical regimes at the two sites are not comparable: HOT is a true low-latitude tropical gyre regime which is permanently stratified with sporadic mixing events (Karl and Lukas 1996; Karl 1999) whereas BATS is on the western edge of the N Atlantic subtropical gyre, and is influenced by the high eddy kinetic energy regime of the adjacent Gulf Stream. BATS experiences large interannual variability in vernal deep mixing from <100 to >400 m (Michaels and Knap 1996), and periodic nutrient enrichment during the passage of mesoscale eddies (McGillicuddy et al. 1998). Nonetheless, PP in the two areas is comparable, except during January–April, the period of the spring bloom at BATS. The two areas have similar annual mean PP (BATS, 459 ±216 mg C m^{-2} d^{-1}; HOT, 478 ±147 mg C m^{-2} d^{-1}). The seasonal cycle in PP at HOT (Fig. 1.3a) is a consequence of enhanced growth under peak summer irradiance (Winn et al. 1995; Karl et al. 1996). The new, high-quality in situ data from HOT surpass the estimates derived from CZCS by Longhurst (1998) by a factor of 2–4, suggesting global estimates of PP in the tropical gyres may require substantial upward revision (see below). The estimates are more consistent at BATS, especially if one takes into account the proximity of the BATS station to the Gulf Stream Province (GFST shown for comparison in Fig. 1.3b). PP observations in the French EUMELI Program (Fig. 1.3b) do not suggest much difference between the eastern and western gyres in the Atlantic, a conclusion also reached for PP by Harrison et al. (2001).

The Pacific Ocean has been intensively studied in JGOFS and by several other programs (Fig. 1.4). Extensive Canadian observations by C. S. Wong (Institute of Ocean Sciences, Sidney, BC; data not shown here) make Station P in the subarctic North Pacific among the most heavily sampled oceanic sites. Station P was among the first oceanic regimes where PP was measured with trace-metal-clean technique (Welschmeyer et al. 1993), and these early 'modern' estimates have now been corroborated by the Canadian JGOFS Program (Boyd and Harrison 1999; both data sets depicted in Fig. 1.4b). Although there is considerable (3-fold, presumably interannual) variability in the observations, the data are consistent with a broad summertime peak in PP, as modeled by Frost (1987), somewhat in contrast to the late-spring peak derived by Longhurst (1998) for the east and west gyres (*dashed lines* in Fig. 1.4b). The few early data now available for the eastern Pacific subarctic gyre from the Japanese time series station KNOT suggest lower PP, which is surprising since the region is closer to sources of iron-containing dust from Asia.

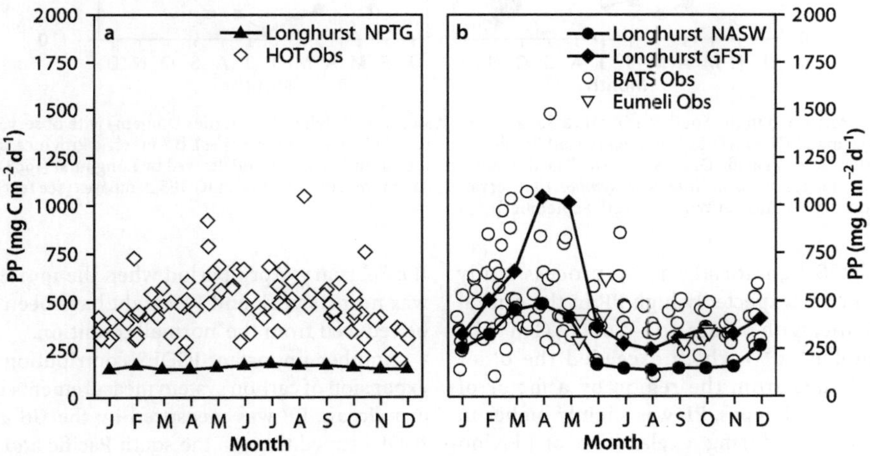

Fig. 1.3. Primary production in the Trade Winds Domain at the JGOFS Hawaii Ocean Time Series (**a** HOT, 1990–1998), the Bermuda Atlantic Time Series (**b** BATS, 1990–1998) and EUMELI Oligotrophic stations (1991–1992; *inverse triangles*, **b**). The domain-averaged annual cycles of PP derived by Longhurst (1998) for the North Pacific Tropical Gyres Province (**a**), North Atlantic Subtropical Gyre – West (**b**) and Gulf Stream (**b**) are shown for comparison (*lines with symbols*). Observational data were obtained from the JGOFS database (see text). The Hawaii and Bermuda data sets represent the most intensive oceanic primary production observations available ($n = 91$ and 105 for HOT and BATS respectively)

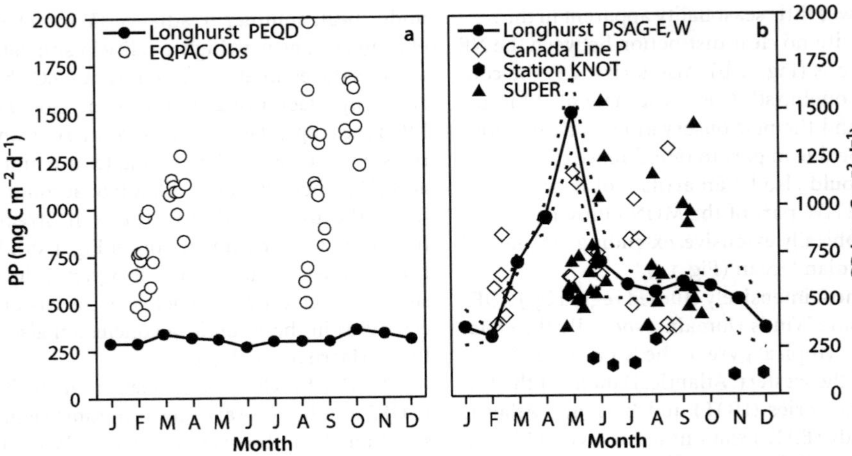

Fig. 1.4. Primary production in the Pacific Ocean. **a** Pacific Equatorial Divergence Province (Trade Winds Domain) with observations within 1° of the equator at 140° W Longitude (US EqPac, *open circles*). **b** Pacific Subarctic Gyre Province (Westerlies Domain) with data from the US SUPER (*triangles*) and Canadian JGOFS (*open diamonds*) programs and the Japanese KNOT Time Series (*closed octagons*). The domain-averaged annual cycles of PP derived by Longhurst (1998) are shown for the east and west PSAG Provinces for comparison (*lines with symbols* and *dashed lines*). Observational data were obtained from published reports and from Y. Nojiri (see text for details on data acquisition and processing)

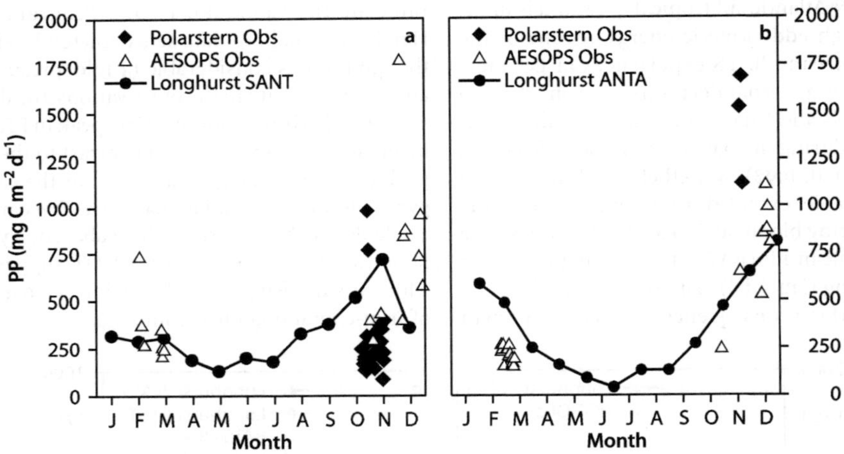

Fig. 1.5. Primary production in the Southern Ocean. **a** Subantarctic Water Ring Province (Westerlies Domain) with observations from the South Atlantic (German POLARSTERN, *diamonds*) and South Pacific (US AESOPS, *open triangles*). **b** Antarctic Province (Polar Domain) with data from Atlantic and Pacific Oceans as in (**a**). The domain-averaged annual cycles of PP derived by Longhurst (1998) are shown for these two provinces for comparison (*lines with symbols*). Observational data were obtained from JGOFS databases (see text). Note the lack of observations during the Austral Winter (April–September)

The US JGOFS Equatorial Pacific process study (EQPAC) revealed unexpectedly high PP in the central equatorial Pacific, with values 500–2 000 g C m^{-2} d^{-1} (mean 95 mmol m^{-2} d^{-1}) which exceeded the older 'climatological' data from the region by a factor of about 1.3 (Barber et al. 1996). PP was slightly higher in August–October 1992 during a relaxation of El Niño conditions (Murray et al. 1994), perhaps triggered by the passage of a tropical instability wave (Archer et al. 1997). The EQPAC PP observations are also startlingly higher than Longhurst's (1998) estimates, for reasons not entirely understood. The contrast is especially striking for the February–April period when the 1991–1992 El Niño was near its peak and PP might have been expected to be reduced from the 'normal' condition.

Another important JGOFS contribution is the great expansion of carbon system measurements in the Southern Ocean. PP was measured by the US and German JGOFS expeditions to the south Pacific and Atlantic, respectively (Fig. 1.5a,b) and by the UK STERNA cruise to the Bellingshausen Sea. The available observations suggest austral spring blooms equal in magnitude to PP observed elsewhere (≥1 000 g C m^{-2} d^{-1}), in spite of low temperature, deep mixing and severe iron depletion. This area has also been studied in detail by the New Zealand Southern Ocean Iron Enrichment Experiment (SOIREE). Figure 1.5 points out the need for wintertime measurements, a key shortcoming of many studies in the Polar and Westerlies Domains (cf. also Fig. 1.4b). Extensive PP observations have also been made by US AESOPS at the high-latitude study area in the Ross Sea (Smith et al. 2000).

1.3.1 Adding up Global PP Observations

Longhurst et al. (1995) showed the utility of a province-based partitioning of marine primary production for making estimates of the total global PP. Here I utilize the new JGOFS data to provide a preliminary updating of Longhurst et al's (1995) estimate of oceanic PP. The Coastal Domain is excluded because so few data were available for my analysis, especially considering the heterogeneity of the Coastal Domain. PP values shown in Table 1.4 were taken from Longhurst et al. (1995) and from the data compilations shown in Fig. 1.2–1.5 and a few other source regions (see Table 1.3). The observations were simply averaged without any time weighting to yield regional, annual means for the better-studied provinces. The areal estimates (g C m^{-2} d^{-1}) were multiplied by Longhurst's province areas to give new annual totals (Gt C yr^{-1}) for each province. Primarily because of the greatly increased PP estimates for the Trade Winds provinces (shaded values in Table 1.4) these few JGOFS estimates alone yield a new global total (excluding the Coastal Domain) of ca 45 Gt C yr^{-1}. This value is about equal to the global totals derived by Field et al. (1998) and Laws et al. (2000), but should be viewed with reservation since the areal coverage is patchy and the productive coastal zones are excluded. The greatly increased PP estimates in the three Trade Winds provinces, which if true, might require upward revision in other, similar but still unstudied area (e.g., South Pacific), seem very worthy of careful scrutiny. It is also important to remember that the Longhurst et al. (1995) estimates were based on CZCS imagery and presuppose that there exist data on photosynthetic parameters for each region. When information is lacking, or sparse, educated guesses were made to fill in the global picture. Divergences between observations and estimations are more a reflection of paucity of data rather than a weakness of the idea of a partition (T. Platt, pers. comm.).

1.4 Bacterial Production and DOC Flux

JGOFS caused a tremendous expansion in understanding of microheterotrophic processes fueled by dissolved organic matter flux. This is surprising since the Program was initially conceived as emphasizing CO_2 exchange, vertical flux of particles and remote sensing. But in order to interpret and model integrating, system-level fluxes like export and CO_2 exchange, or large-scale pigment distributions, clearer insight into trophodynamic processes was required (e.g., Eppley and Ducklow 1986; SCOR 1990). Among the great successes of JGOFS was a new high precision assay for the concentration of dissolved organic carbon (DOC) in seawater (Sharp et al. 1993; Hedges and Farrington 1993). This assay was initially developed by Sugimura and Suzuki (1988), and subsequently corrected and improved by Benner and Strom (1993), following a series of JGOFS-sponsored workshops and 'bake-offs' during which different DOC analytical instruments and methods were compared and intercalibrated. The perfection of a reliable assay for DOC led to an unprecedented view of the distribution and dynamics of DOC, which we can now begin to place into a global scale, geographical context. DOC is produced seasonally in the upper ocean with greater buildup in the tropics and subtropics, and lower accumulations at higher latitudes (Kumar et al. 1990; Carlson and Ducklow 1995, 1996; Kähler et al. 1997; Carlson et al. 1998). It appears that net DOC production is minimal below the Antarctic Polar Frontal Zone (Hansell and Carlson 1998b). The small seasonal build-up of DOC in the Ross Sea is entirely consumed by bacteria (or perhaps also oxidized by UV radiation – see Moran and Zepp 2000 for a review) by the end of the growing season in April (Carlson et al. 2000). It is not known if the same is true in the North Polar Domain (Arctic Ocean), where terrestrial DOC from the high freshwater input (Tomczak and Godfrey 1994) potentially obscures the marine signal and fuels exceptionally high bacterial activity (Rich et al. 1998). At this time, we can generalize a global pattern at the level of the four great biogeochemical domains, but cannot yet distinguish differences at the basin scale or province level. A global pattern in DOC distribution is also observed in the deep ocean, where DOC concentrations reflect the thermohaline conveyor belt circulation. The highest concentrations of deep ocean DOC are in the North Atlantic, whereas the lowest concentrations are in the deep North Pacific, at the opposite end of the conveyor belt (Hansell and Carlson 1998a). This large scale pattern is the result of interaction between geographically-focused inputs of DOC at sites of mode water and deepwater formation and slow bacterial decomposition. Thus fresh DOC is supplied in North Atlantic Deep Water, and slowly decays during its transit through the deep sea. The origin of the DOC in NADW is not yet clear: it may be produced locally, or it might be transported from lower latitudes in the surface circulation. There is abundant net production of semilabile DOC in the tropics and subtropics which survives bacterial decomposition over seasonal time scales so it can be exported horizontally off the equator (Archer et al. 1997; Hansell et al. 1997; Peltzer and Hayward 1996), or vertically during late win-

Table 1.4. Global partitioning of oceanic primary production (after Longhurst et al. 1995). Note that coastal provinces are not included. New in situ estimates from JGOFS and other studies are included for several well-studied provinces, and a new province-averaged annual PP has been extrapolated (see text). The three *shaded boxes* indicated regions for which a large change in the original estimate impacted the new global total

Domain	Province	Original estimation		JGOFS	
		(g C m^{-2} d^{-1})	(Gt C yr^{-1})	(g C m^{-2} d^{-1})	(Gt C yr^{-1})
Polar	BPLR	1770	1.07		1.07
Polar	ARCT	1330	1.02		1.02
Polar	SARC	830	0.7		0.70
Polar	BERS	990	1.41		1.41
Polar	ANTA	450	1.47		1.47
Polar	APLR	1090	0.77	1204	0.85
Westerlies	NADR	660	0.84		0.84
Westerlies	GFST	490	0.2		0.20
Westerlies	NASW	260	0.55	459	0.97
Westerlies	MEDI	590	0.67		0.67
Westerlies	NASE	330	0.54		0.54
Westerlies	PSAE	550	0.64	704	0.82
Westerlies	PSAW	720	0.77		0.77
Westerlies	KURO	530	0.72	489	0.66
Westerlies	NPPF	470	0.52		0.52
Westerlies	NPSE	300	0.76		0.76
Westerlies	NPSW	300	0.43		0.43
Westerlies	OCAL	320	0.28		0.28
Westerlies	TASM	450	0.27		0.27
Westerlies	SPSG	240	3.23		3.23
Westerlies	SSTC	370	2.29		2.29
Westerlies	SANT	330	3.63	393	4.32
Trades	NATR	290	0.88		0.88
Trades	WTRA	360	0.7		0.70
Trades	ETRA	430	0.84		0.84
Trades	SATL	210	1.33		1.33
Trades	CARB	520	0.85		0.85
Trades	MONS	290	1.49	876	4.50
Trades	ISSG	190	1.37		1.37
Trades	NPTG	160	1.24	478	3.70
Trades	PNEC	290	0.87		0.87
Trades	PEQD	310	1.17	1083	4.09
2Trades	WARM	220	1.38		1.38
Trades	ARCH	270	0.88		0.88
Total			35.78		45.48

ter overturning (Copin-Montegut and Avril 1993; Carlson et al. 1994). Export of DOC appears to account for about 20% of the total export production globally (Hansell and Carlson 1998b).

The principal sink for DOC is bacterial metabolism, assisted by photochemical breakdown (Anderson and Williams 1999; Ducklow 2000). PP estimates may have increased during JGOFS but estimates of bacterial production (BP), the rate at which bacteria convert DOC and inorganic nutrients into biomass, have declined. Earlier estimates of BP (which included very few oceanic measurements) indicated the BP was 20–30% of PP measured approximately simultaneously (Cole et al. 1988; Ducklow and Carlson 1992). Williams (2000) notes that many of the earlier estimates now seem unrealistic in light of a more comprehensive understanding of DOC

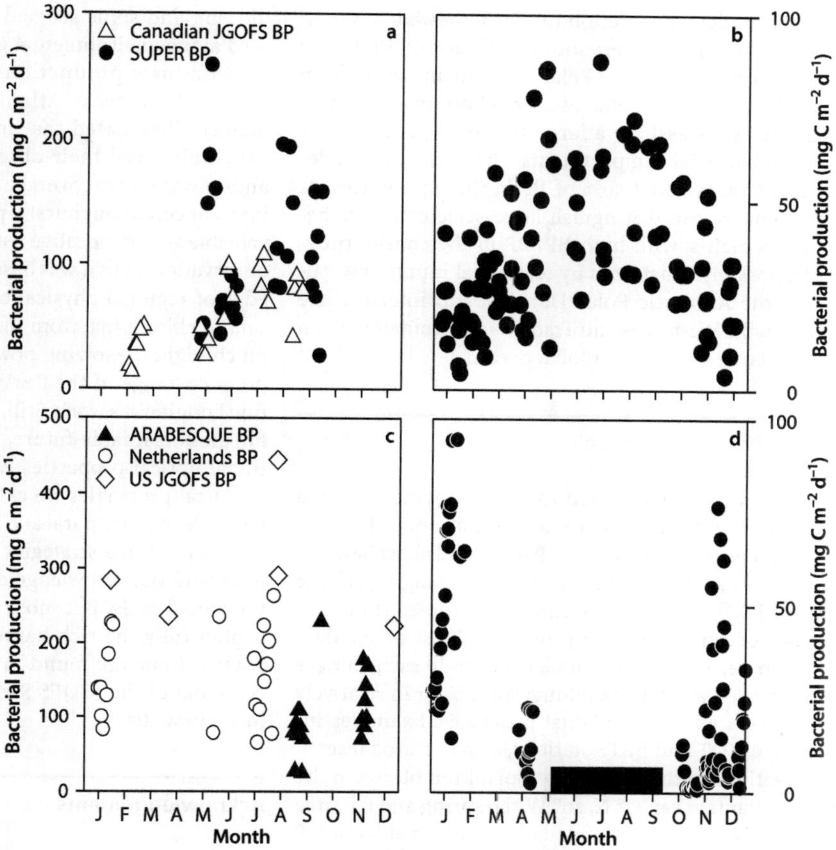

Fig. 1.6. Bacterial production in well-studied ocean domains. **a** Pacific Subarctic Gyres Province, with data from Project SUPER (*circles*) and Canadian JGOFS (*triangles*). **b** North Atlantic Subtropical Gyres – West Province (data from BATS); **c** Arabian Upwelling Province, with data from UK ARABESQUE (*triangles*), US (*diamonds*) and Netherlands (*circles*) JGOFS programs; and **d** the Austral Polar Province, with data from the US AESOPS program in the Ross Sea. Note that *Y*-axis scales differ from plot to plot

flux and bacterial conversion efficiencies. Once thought to be 50% or higher, the bacterial growth efficiency (BGE) on DOC is now estimated to be about 15–30% (del Giorgio and Cole 1998). A simple example shows the impossibility of BP = 0.3 times PP and BGE = 0.15: since BGE is BP divided by BP plus respiration, the respiration in this example is 1.7 times PP! This situation might occur when bacteria utilize accumulated products of a bloom (Ducklow et al. 1993; Azam et al. 1994) but cannot occur in the quasi-steady state ecosystems of the oceanic gyres and other Trade Winds provinces.

In fact recently synthesized JGOFS data do suggest BP is closer to 10% of PP. Figure 1.6 shows BP observations in four well-studied provinces. Strong seasonal cycles are apparent in the subarctic North Pacific (Fig. 1.6a; Kirchman et al. 1993; Sherry et al. 1999), at the BATS station (Fig. 1.6b; Carlson et al. 1996) and in the Ross Sea, Antarctica (Fig. 1.6d; Ducklow et al. 2001). In the Ross Sea, BP has an amplitude of almost 100-fold, with values increasing from about 1 mg C m^{-2} d^{-1} in late October to almost 100 in mid-January. However the mean PP is very high and the annual average BP is just 24 mg C m^{-2} d^{-1}, lowest among the provinces studied, so BP/PP in the Antarctic Polar Province is just 2% (Table 1.5). As suggested above, the low BP is possibly due

Table 1.5. Bacterial and primary production rates in several ocean provinces

Province	PP[a]	BP[a]	BP/PP	Eff[b]	Carbon utilized/PP[c]
ARAB	1479	163	0.11	0.17	0.65
MONS	998	166	0.17	0.17	0.98
PSAG	647	86	0.13	0.27	0.49
PEQD	1185	111	0.09	0.15	0.60
NASW	459	37	0.08	0.15	0.54
APLR	1234	24	0.02	0.15	0.13

[a] Primary and bacterial production in the euphotic zone (mg C m^{-2} d^{-1}).
[b] Bacterial growth (conversion) efficiency measured in the same studies.
[c] Bacterial carbon utilization is BP/Eff.

to a lack of net DOC production in Polar Domains. Interestingly BP/PP is very high (17%) in the Indian Ocean Monsoon Province (MONS, Table 1.5; Wiebinga et al. 1997; Pomroy and Joint 1999; Ducklow et al. 2001) where offshore transport of DOC produced during coastal upwelling in the southwest monsoon might subsidize offshore BP. BP in the coastal Arabian Upwelling Province (ARAB; Fig. 1.4c) is equally high (Table 1.5) but the high PP renders the fraction BP/PP somewhat lower. In general these new observations indicate BP/PP in the range 8–13%, outside the Polar Domain and away from coastal influence. Nonetheless, the relatively low growth efficiencies necessitate a large flux of organic matter through bacterial compartments – bacterial carbon demand averages about 50% of PP in these provinces. At this point we can distinguish large-scale contrasts between domains, with high BP/PP in the coastal provinces, possibly subsidized by terrestrial inputs, low values in the Antarctic Polar Domain, and intermediate values in the Westerlies and Trade Winds Domains which cover a large area of the global ocean.

1.5 A Provincial Outlook

In this chapter I have tried to use a province – based partitioning of the oceans to look at some biogeochemical processes studied in JGOFS, and show how our views of ocean carbon cycling might be changing. A large part of JGOFS emphasized intensive process studies in various ocean regimes or provinces. Most of the data shown here, for example, came from studies which were concentrated at a few individual stations or in relatively small areas within individual provinces. Examples include the HOT and BATS stations, a multinational set of observations of the spring phytoplankton bloom in the North Atlantic near 47° N, 20° W, the spring and fall time series in EQPAC and extended observations at Station P in the North Pacific. In other locations, traditional transect studies were employed, for example the NABE cruises along 20° W, the Arabian Sea expeditions and several national studies in the Southern Ocean. It remains difficult and expensive to carry out process-related studies over geographically-extended areas, although lines or arrays of sediment trap moorings were deployed in the N Atlantic, central Pacific, Arabian Sea and Southern Ocean. Intensive, but geographically concentrated process studies, moorings and time series provide the means to characterize a manifold of processes within a province, but we still need to know if the observations made at local scales are characteristic of larger areas. This problem has been attacked in the BATS program with a series of regional validation cruises (Michaels and Knap 1996). Establishing ecological continuity within provinces is in fact the acid test of the province concept – are relevant ecological and biogeochemical properties and processes consistently distributed within provinces? In most cases we still don't know. Harrison et al. (2001) examined a range of hydrographic properties, and biological rate processes during two cruises across three provinces in the North Atlantic. They found that some properties and rates differed significantly among provinces and seasons (e.g., regenerated production and bacterial production), whereas others did not, and some seemed to be continuously distributed along environmental gradients (e.g., primary production, new production and chlorophyll standing stocks along cross-Atlantic gradients of nitracline depth). They noted that meridional variability could have influenced their observations. Further transects and/or wider area coverage are required to test the province concept. Longhurst's pioneering work provided a valuable and provocative template for synthesis of JGOFS observations, but it was based almost entirely on knowledge of regional physical oceanography and remotely-sensed chlorophyll from the CZCS. Re-analysis with the much higher resolving power and more complete temporal coverage of the SeaWiFS sensors will help to refine Longhurst's work. Still, it is critical to recognize that for the foreseeable future, chlorophyll and some additional optical properties will remain the only biogeochemical properties we can observe at the global scale with relevant temporal and spatial resolution. New kinds of observational strategies and models are still needed to extend our knowledge of ocean biogeochemistry to the global scale. But most of all, we need new ideas to exploit fully the rich harvest of observations made in JGOFS. Some are found in this volume and others will come out of the JGOFS Synthesis, of which this book is just a first step.

Acknowledgements

Preparation of this chapter was supported by NSF Grant OCE 9819581. I am grateful to the following individuals who contributed data and answered questions: Nelson Sherry, Joachim Herrmann, Wolfgang Koeve, Glen Harrison, Bill Li, Beatriz Balino, Craig Carlson, Dennis Hansell, Dave Kirchman and Dave Karl.

References

Anderson TR, Williams PJ leB (1999) A one-dimensional model of DOC cycling in the water column incorporating combined biological-photochemical decomposition. Global Biogeochem Cy 13:337–349

Archer D, Peltzer ET, Kirchman DL (1997a) A timescale for dissolved organic carbon production in equatorial Pacific surface waters. Global Biogeochem Cy 11:435–452

Archer D, Aiken J, Balch W, Barber R, Dunne J, Flament P, Gardner W, Garside C, Goyet C, Johnson E, Kirchman D, McPhaden M, Newton J, Peltzer E, Welling L, White J, Yoder J (1997b) A meeting place of great ocean currents: Shipboard observations of a convergent front at 2° N in the Pacific. Deep-Sea Res Pt II 44:1827–1849

Azam F, Fenchel T, Field JG, Gray JS, Meyer-Reil LA, Thingstad F (1983) The ecological role of water-column microbes in the sea. Mar Ecol Prog Ser 10:257–263

Azam F, Steward GF, Smith DC, Ducklow HW (1994) Significance of bacteria in the carbon fluxes of the Arabian Sea. P Indian As-Earth 103:341–351

Bailey RG (1998) Ecoregions: the ecosystem geography of the oceans and continents. Springer-Verlag, New York, 176 pp

Banse K (1987) Seasonality of phytoplankton chlorophyll *a* in the central and northern Arabian Sea. Deep-Sea Res Pt I 34:713–723
Banse K (1992) Grazing, temporal changes of phytoplankton concentrations and the microbial loop in the open sea. In: Falkowski P (ed) Primary productivity and biogeochemical cycles in the sea. Plenum, New York, pp 409–440
Banse K, English DC (1993) Revision of satellite-based phytoplankton pigment data from the Arabian Sea during the Northeast Monsoon. Marine Research 2:83–103
Barber RT (1988) Ocean basin ecosystems. In: Alberts J, Pomeroy LR (eds) Concepts of ecosystem ecology: a comparative view. Springer-Verlag, New York, pp 171–193
Barber RT, Sanderson MP, Lindley ST, Chai F, Newton J, Trees CC, Foley DG, Chavez FP (1996) Primary productivity and its regulation in the equatorial Pacific during and following the 1991–1992 El Niño. Deep-Sea Res Pt II 43:933–969
Benner R, Strom S (1993) A critical evaluation of the analytical blank associated with DOC measurements by high-temperature catalytic oxidation. Mar Chem 41:153–60
Benner R, Pakulski JD, McCarthy M, Hedges JI, Hatcher PG (1992) Bulk chemical characteristics of dissolved organic matter in the ocean. Science 255:1561–1564
Boyd PW, Harrison PJ (1999) Phytoplankton dynamics in the NE subarctic Pacific. Deep-Sea Res Pt II 46:2405–2432
Boyd PW, Law CS (2000) The Southern Ocean Iron RElease Experiment (SOIREE) – introduction and summary. Deep-Sea Res Pt II 48:2425–2438
Boyd PW, Harrison PJ, Johnson BD (1999) The Joint Global Ocean Flux Study (Canada) in the NE subarctic Pacific. Deep-Sea Res Pt II 46:2345–2350
Burkill PH (1999) ARABESQUE: An overview. Deep-Sea Res Pt II 46:529–547
Caperon J (1975) A trophic level ecosystem model analysis of the plankton community in a shallow-water subtropical estuarine embayment. In: Cronin LE (ed) Estuarine research, vol. 1. chemistry, biology and the estuarine system. Academic Press, New York, pp 691–709
Carlson CA, Ducklow HW (1995) Dissolved organic carbon in the upper ocean of the central equatorial Pacific Ocean, 1992: Daily and finescale vertical variations. Deep-Sea Res Pt II 42:639–56
Carlson CA, Ducklow HW (1996) Growth of bacterioplankton and consumption of dissolved organic carbon in the Sargasso Sea. Aquat Microb Ecol 10:69–85
Carlson CA, Michaels AM, Ducklow HW (1994) Annual flux of dissolved organic carbon from the euphotic zone in the northwestern Sargasso Sea. Nature 371:405–408
Carlson CA, Ducklow HW, Sleeter TD (1996) Stocks and dynamics of bacterioplankton in the northwestern Sargasso Sea. Deep-Sea Res Pt II 43:491–516
Carlson CA, Ducklow HW, Smith WO, Hansell DA (1998) Carbon dynamics during spring blooms in the Ross Sea polynya and the Sargasso Sea: Contrasts in dissolved and particulate organic carbon partitioning. Limnol Oceanogr 43:375–386
Carlson CA, Hansell DA, Peltzer ET, Smith WO Jr. (2000) Stocks and dynamics of dissolved and particulate organic matter in the southern Ross Sea, Antarctica. Deep-Sea Res Pt II 47:3201–3226
Caron DA (2000) Symbiosis and mixotrophy among pelagic microorganisms. In: Kirchman D (ed) Microbial ecology of the oceans. John Wiley & Sons, New York, pp 495–524
Chierici M, Drange H, Anderson LG, Johannessen T (1997) Inorganic carbon fluxes through the boundaries of the Greenland Sea Basin based on in situ observations and water. J Marine Syst 22:295–309
Chisholm SW, Olson RJ, Zettler ER, Goericke R, Waterbury JB, Welschmeyer NA (1988) A novel free-living prochlorophyte abundant in the oceanic euphotic zone. Nature 334:340–343
Cole JJ, Pace ML, Findlay S (1988) Bacterial production in fresh and saltwater ecosystems: a cross-system overview. Mar Ecol Prog Ser 43:1–10
Copin-Montegut C (2000) Consumption and production on scales of a few days of inorganic carbon, nitrate and oxygen by the planktonic community: results of continuous measurements at the Dyfamed Station in the northwestern Mediterranean Sea (May 1995). Deep-Sea Res Pt I 47:447–477
Copin-Montegut G, Avril B (1993) Vertical distribution and temporal variation of dissolved organic carbon in the north-western Mediterranean Sea. Deep-Sea Res Pt I 40:1963–1972
Cronon W (1983) Changes in the land. Indians, colonists and the ecology of New England. Hill and Wang, New York, 241 pp
Cullen JJ, Franks PJS, Karl DM, Longhurst A (2002) Physical influences on marine ecosystem dynamics. In: Robinson AR, McCarthy JJ, Rothschild BJ (eds) The sea, vol. 12. John Wiley & Sons, New York, pp 297–336
Giorgio PA del, Cole JJ (1998) Bacterial growth efficiency in natural aquatic systems. Annu Rev Ecol Syst 29:503–541
Giorgio PA del, Cole JJ (2000) Bacterial bioenergetics and growth efficiency. In: Kirchman DL (ed) Microbial ecology of the oceans. John Wiley & Sons, New York, pp 289–326
Dietrich G (1963) General oceanography, an introduction. Interscience Publishers, New York, 588 pp
Ducklow HW (1999) The bacterial content of the oceanic euphotic zone. Fems Microbiol Ecol 30:1–10
Ducklow HW (2000) Bacterioplankton production and biomass in the oceans. Chap. 4, In: Kirchman DL (ed) Microbial ecology of the oceans. John Wiley & Sons, New York, pp 85–120
Ducklow HW, Carlson CA (1992) Oceanic bacterial productivity. Adv Microb Ecol 12:113–181
Ducklow HW, Harris R (1993) Introduction to the JGOFS North Atlantic Bloom Study. Deep-Sea Res 40:1–8
Ducklow HW, Kirchman DL, Quinby HL, Carlson CA, Dam HG (1993) Stocks and dynamics of bacterioplankton carbon during the spring phytoplankton bloom in the eastern North Atlantic Ocean. Deep-Sea Res Pt II 40:245–263
Ducklow HW, Smith DC, Campbell L, Landry MR, Quinby HL, Steward GF, Azam F (2001a) Heterotrophic bacterioplankton distributions in the Arabian Sea: basinwide response to high primary productivity. Deep-Sea Res Pt II 48:1303–1323
Ducklow HW, Carlson CA, Church M, Kirchman DL, Smith DC, Steward G (2001b) The seasonal development of bacterioplankton in the Ross Sea, Antarctica, 1994–97. Deep-Sea Res Pt II 47:3227–3247
Dugdale RC, Goering JJ (1967) Uptake of new and regenerated forms of nitrogen in primary production. Limnol Oceanogr 12:196–206
Eppley RW, Ducklow HW (1986) Workshop on Upper Ocean Processes. US GOFS Report 3, Woods Hole, MA. US JGOFS Planning Office, WHOI, pp 1–141
Falkowski PG, Raven JA (1997) Aquatic photosynthesis. Blackwell Scientific, Malden, MA, 375 pp
Falkowski PG, Laws EA, Barber RT, Murray JW (2003) Phytoplankton and their role in primary, new, and export production. (this volume)
Field CB, Behrenfeld MJ, Randerson JT, Falkowski PG (1998) Primary production of the biosphere: integrating terrestrial and oceanic components. Science 281:237–240
Frost BW (1987) Grazing control of phytoplankton stock in the open subarctic Pacific Ocean: a model assessing the role of mesozooplankton, particularly the large calanoid copepods *Neocalanus* spp. Mar Ecol Prog Ser 39:49–68
Fuhrman J (2000) Impact of viruses on microbial processes. In: Kirchman D (ed) Microbial ecology of the oceans. John Wiley & Sons, New York, pp 327–351
Gaillard JF, Tréguer P (eds) (1997) Antares I: France-JGOFS in the Indian sector of the Southern Ocean; benthic and water column processes. Deep-Sea Res Pt II 44:951–1176
Giovannoni SJ, Rappe M (2000) Evolution, diversity, and molecular ecology of marine prokaryotes. In: Kirchman DL (ed) Microbial ecology of the oceans. John Wiley & Sons, New York, pp 47–84
Hansell DA, Carlson CA (1998a) Deep-ocean gradients in the concentration of dissolved organic carbon. Nature 395:263–266
Hansell DA, Carlson CA (1998b) Net community production of dissolved organic carbon. Global Biogeochem Cy 12:443–453
Hansell DA, Carlson CA, Bates NR, Poisson A (1997) Horizontal and vertical removal of organic carbon in the equatorial Pacific Ocean: a mass balance assessment. Deep-Sea Res Pt II 44:2115–2130
Harris RP, Boyd P, Harbour DS, Head RN, Pingree RD, Pomroy AJ (1997) Physical, chemical and biological features of a cyclonic eddy in the region of 61° 10' N 19° 50' W in the North Atlantic. Deep-Sea Res Pt I 44:1815–1839

Harrison WG (1980) Nutrient regeneration and primary production in the sea. In: Falkowski PG (ed) Primary productivity in the sea. Plenum Publishing Co, pp 433–60
Harrison WG, Head EJH, Horne EPW, Irwin B, Li WKW, Longhurst AR, Paranjape MA, Platt T (1993) The Western North Atlantic Bloom Experiment. Deep-Sea Res Pt II 40:279–306
Harrison WG, Arístegui J, Head EJH, Li WKW, Longhurst AR, Sameoto DD (2001) Basin-scale variability in plankton biomass and community metabolism in the sub-tropical North Atlantic Ocean. Deep-Sea Res Pt II 48:2241–2270
Hedges J, Farrington J (1993) Measurement of dissolved organic carbon and nitrogen in natural waters: workshop report. Mar Chem 41:5–10
Hu D, Tsunogai S (1999) Margin fluxes in the East China Sea. China Ocean Press, Beijing, 247 pp
Jarre-Teichmann A, Shannon LJ, Moloney CL, Wickens PA (1998) Comparing trophic flows in the southern Benguela to those in other upwelling ecosystems. S Afr J Marine Sci 19:391–414
Johnson PW, Sieburth JMcN (1979) Chroococcoid cyanobacteria in the sea: a ubiquitous and diverse phototrophic biomass. Limnol Oceanogr 24:928–35
Kähler P, Bjørnsen PK, Lochte K, Antia A (1997) Dissolved organic matter and its utilization by bacteria during spring in the Southern Ocean. Deep-Sea Res Pt II 44:341–353
Karl DM (1999) A sea of change: biogeochemical variability in the North Pacific subtropical gyre. Ecosystems 2:181–214
Karl DM, Lukas R (1996) The Hawaii Ocean Time-series (HOT) program: background, rationale and field implementation. Deep-Sea Res Pt II 43:129–156
Karl DM, Michaels AF (1996) Preface: The Hawaii Ocean Time Series (HOT) and the Bermuda Atlantic Time Series (BATS). Deep-Sea Res Pt II 43:127–129
Karl DM, Christian JR, Dore JE, Hebel DV, Letelier RM, Tupas LM, Winn CD (1996) Seasonal and interannual variability in primary production and particle flux at Station ALOHA. Deep-Sea Res Pt II 43:539–56
Karner MB, DeLong EF, Karl DM (2001) Archaeal dominance in the mesopelagic zone of the Pacific Ocean. Nature 409: 507–510
Kirchman DL, Keil RG, Simon M, Welschmeyer NA (1993) Biomass and production of heterotrophic bacterioplankton in the oceanic subarctic Pacific. Deep-Sea Res 40:967–988
Knap A, Michaels A, Close A, Ducklow HW, Dickson A (eds) (1996) Protocols for the Joint Global Ocean Flux Study (JGOFS) core measurements. JGOFS Report No. (19, vi+170 pp Reprint of the IOC Manuals and Guides No. 29, UNESCO 1994
Kumar MD, Rajendran A, Somasundar K, Haake B, Jenisch A, Shuo Z, Ittekkot V, Desai BN (1990) Dynamics of dissolved organic carbon in the northwestern Indian Ocean. Mar Chem 31:299–316
Lal D (1994) Biogeochemistry of the Arabian Sea. Reprinted from P Indian As-Earth 103:99–352
Landry MR, Barber RT, Bidigare RR, Chai F, Coale KH, Dam HG, Lewis MR, Lindley ST, McCarthy JJ, Roman MR, Stoecker DK, Verity PG, White JR (1997) Iron and grazing constraints on primary production in the central equatorial Pacific: An EqPac synthesis. Limnol Oceanogr 42:405–418
Laws EA, Falkowski PG, Smith WO, Ducklow HW Jr., McCarthy JJ (2000) Temperature effects on export production in the open ocean. Global Biogeochem Cy 14:1231–1246
Le Fèvre J, Tréguer P (1998) Special issue: carbon fluxes, dynamic processes in the Southern Ocean: present, past. J Marine Syst 17:1–4
Letelier RM, Karl DM (1996) The role of *Trichodesmium* spp. in the productivity of the subtropical North Pacific Ocean. Mar Ecol Prog Ser 133:263–273
Letelier RM, Karl DM (1998) *Trichodesmium* spp. physiology and nutrient fluxes in the North Pacific subtropical gyre. Aquat Microb Ecol 15:265–276
Li WKW, Subba Rao DV, Harrison WG, Smith JC, Cullen JJ, Irwin B, Platt T (1982) Autotrophic picoplankton in the tropical ocean. Science 219:292–95
Longhurst AR (1995) Seasonal cycles of pelagic production and consumption. Prog Oceanogr 36:77–167
Longhurst AR (1998) Ecological geography of the sea. Academic, San Diego, 398 pp
Longhurst AR, Sathyendranath S, Platt T, Caverhill C (1995) An estimate of global primary production in the ocean from satellite radiometer data. J Plankton Res 17:1245–1271
Malone TC (1980) Size-fractionated primary productivity of marine phytoplankton. In: Falkowski PG (ed) Primary productivity in the sea. Plenum Publishing Co., pp 301–319
Margalef R (1978) Life forms of phytoplankton as survival alternatives in an unstable environment. Oceanol Acta 1:493–509
McGillicuddy DJ Jr., Robinson AR, Siegel DA, Jannasch HW, Johnson R, Dickey TD, McNeil J, Michaels AF, Knap AH (1998) Influence of mesoscale eddies on new production in the Sargasso Sea. Nature 394:263–265
Michaels AF, Knap AH (1996) Overview of the U.S. JGOFS Bermuda Atlantic Time-series Study and the Hydrostation S Program. Deep-Sea Res Pt II 43:157–198
Monaco A, Biscaye PE, Laborde P (1999) France-JGOFS/ECOMARGE: The ECOFER (ECOsystem du Canyon du Cap FERret) Experiment on the Northeast Atlantic Continental Margin. Deep-Sea Res Pt II 46:1944–2379
Moran MA, Zepp RG (2000) UV radiation effects on microbes and microbial processes. In: Kirchman DL (ed) Microbial ecology of the oceans. John Wiley & Sons, New York, pp 201–228
Morel A (1996) An ocean flux study in eutrophic, mesotrophic, and oligotrophic situations: the EUMELI program. Deep-Sea Res Pt I 43:1185–1190
Murray JW, Barber RT, Roman M, Bacon MP, Feely RA (1994) Physical and biological controls on carbon cycling in the equatorial pacific. Science 266:58–65
Nagata T (2000) Production mechanisms of dissolved organic matter. In: Kirchman DL (ed) Microbial ecology of the oceans. Wiley-Liss, New York, pp 121–152
Odum EP (1969) The strategy of ecosystem development. Science 164:262–270
Peltzer E, Hayward N (1996) Spatial distribution and temporal variability of total organic carbon along 140° W in the Equatorial Pacific Ocean in 1992. Deep-Sea Res Pt II 43:1155–1180
Pfannkuche O, Lochte K (eds) (2001) Biogeochemistry of the deep Arabian Sea: German research programmes in the Arabian Sea. Deep-Sea Res Pt II 47:2615–3072
Platt T, Sathyendranath S (1988) Oceanic primary production: estimation by remote sensing at local and regional scales. Science 241:1613–1622
Platt T, Sathyendranath S (1999) Spatial structure of pelagic ecosystem processes in the global ocean. Ecosystems 2:384–394
Platt T, Caverhill C, Sathyendranath S (1991) Basin-scale estimates of oceanic primary production by remote sensing: the North Atlantic. J Geophys Res 96:147–159
Pomeroy LR (1974) The ocean's food web, a changing paradigm. Bioscience 24:499–504
Pomroy A, Joint I (1999) Bacterioplankton activity in the surface waters of the Arabian Sea during and after the 1994 SW Monsoon. Deep-Sea Res Pt II 46:767–794
Pondaven P, Ruiz-Pino D, Fravalo C, Tréguer P, Jeandel C (2000) Interannual variability of Si and N cycles at the time-series station KERFIX between 1990 and 1995 – a 1-D modeling study. Deep-Sea Res Pt I 47:223–257
Rich J, Gosselin M, Sherr E, Sherr B, Kirchman DL (1998) High bacterial production, uptake and concentrations of dissolved organic matter in the central Arctic Ocean. Deep-Sea Res Pt II 44:1645–1663
Roy S, Sundby B (2000) A Canadian JGOFS Process Study in the Gulf of St. Lawrence (Canada): carbon transformations from production to burial. Deep-Sea Res Pt II 47:385–760
Sarmiento JL, Slater RD, Fasham MJR, Ducklow HW, Toggweiler JR, Evans GT (1993) A seasonal three-dimensional ecosystem model of nitrogen cycling in the North Atlantic euphotic zone. Global Biogeochem Cy 7:417–450
Sathyendranath S, Longhurst AR, Caverhill CM, Platt T (1995) Regionally and seasonally differentiated primary production in the North Atlantic. Deep-Sea Res Pt I 42:1773–1802
Savenkoff C, Lefevre D, Denis M, Lambert CE (1993) How do microbial communities keep living in the Mediterranean outflow within N.E. Atlantic intermediate waters? Deep-Sea Res 40:627–641

SCOR (1990) JGOFS science plan. JGOFS Report No. 5. Halifax NS: SCOR and JGOFS

Sharp JH, Suzuki Y, Munday WL (1993) A comparison of dissolved organic carbon in North Atlantic Ocean nearshore waters by high temperature combustion and wet chemical oxidation. Mar Chem 41:253–259

Sherr E, Sherr B (2000) Marine microbes: an overview. In: Kirchman DL (ed) Microbial ecology of the oceans. John Wiley & Sons, New York, pp 13–46

Sherry ND, Boyd PW, Sugimoto K, Harrison PJ (1999) Seasonal and spatial patterns of heterotrophic bacterial production, respiration, and biomass in the subarctic NE Pacific. Deep-Sea Res Pt II 46:2557–2578

Sieburth JMcN, Smetacek V, Lenz J (1978) Pelagic ecosystem structure: heterotrophic compartments of plankton and their relationship to plankton size fractions. Limnol Oceanogr 23:1256–1263

Smetacek V, De Baar HJW, Bathmann UV, Lochte K, Van Der Loeff MM Rutgers (1997) Ecology and biogeochemistry of the Antarctic Circumpolar Current during austral spring: a summary of Southern Ocean JGOFS cruise ANT X/6 of R.V. Polarstern. Deep-Sea Res Pt II 44:1–21

Smith SL (1998) The 1994–1996 Arabian Sea Expedition: oceanic response to monsoonal forcing, Part I. Deep-Sea Res Pt II 45:1917–2501

Smith WO Jr., Anderson RF, Moore JK, Codispoti LA, Morrison JM (2000a) The U.S. Southern Ocean Joint Global Ocean Flux Study: an introduction to AESOPS. Deep-Sea Res Pt II 47:3073–3093

Smith WO Jr., Marra J, Hiscock MR, Barber RT (2000b) The seasonal cycle of phytoplankton biomass and primary productivity in the Ross Sea, Antarctica. Deep-Sea Res Pt II 47:3119–3140

Strom SL, Miller CB, Frost BW (2000) What sets lower limits to phytoplankton stocks in high-nitrate, low-chlorophyll regions of the open ocean? Mar Ecol Prog Ser 193:19–31

Sugimura Y, Suzuki Y (1988) A high-temperature catalytic oxidation method of non-volatile dissolved organic carbon in seawater by direct injection of liquid samples. Mar Chem 14:105–131

Sverdrup HU (1953) On the conditions for the vernal blooming of phytoplankton. J Cons Perm Int Explor Mer 18:287–295

Tindale NW, Pease PP (1999) Aerosols over the Arabian Sea: Atmospheric transport pathways and concentrations of dust and sea salt. Deep-Sea Res Pt II 46:1577–1595

Tomczak M, Godfrey JS (1994) Regional oceanography: an introduction. Pergamon, Oxford, 422 p

Tsunogai S (1997) Biogeochemical processes in the North Pacific. Proceedings of the International Marine Science Symposium held on 12–14 November 1996 at Mutsu, Aomori, Japan. Tokyo: Japan Marine Science Foundation

Turner D, Owens N, Priddle J (1995) Southern Ocean JGOFS: The U.K. 'STERNA' study in the Bellingshausen Sea. Deep-Sea Res Pt II 42:905–906

Van der Spoel J, Heymann RP (1983) A comparative atlas of zooplankton. Springer-Verlag, Berlin

Walsh JJ (1976) Herbivory as a factor in patterns of nutrient utilization in the sea. Limnol Oceanogr 21:1–13

Waterbury JB, Watson SW, Guillard RR, Brand LE (1979) Widespread occurrence of a unicellular, marine, planktonic cyanobacterium. Nature 277:392–394

Welschmeyer N, Strom SL, Goericke R, diTullio G, Belvin M, Peterson W (1993) Primary production in the subarctic Pacific Ocean: project SUPER. Prog Oceanogr 32:101–135

Wiebinga CJ, Veldhuis MJW, De Baar HJW (1997) Abundance and productivity of bacterioplankton in relation to seasonal upwelling in the northwest Indian Ocean. Deep-Sea Res Pt I 44:451–476

Williams PJ leB (1981) Incorporation of microheterotrophic processes into the classical paradigm of the planktonic food web. Kieler Meeresforschung 5:1–28

Williams PJ leB (1984) Bacterial production in the marine food chain: the emperor's new suit of clothes? In: Fasham M (ed) Flows of energy and materials in marine ecosystems: theory and practice. Plenum Press, pp 271–299

Williams PJ leB (2000) Heterotrophic bacterial and the dynamics of dissolved organic material. In: Kirchman DL (ed) Microbial ecology of the oceans. John Wiley & Sons, New York, pp 153–201

Winn CD, Campbell L, Letelier R, Hebel D, Fujieki L, Karl DM (1995) Seasonal variability in chlorophyll concentrations in the North Pacific subtropical gyre. Global Biogeochem Cy 9:605–620

Wong GTF, Chao S-Y, Li Y-H, Shiah F-K (2000) The Kuroshio edge exchange processes (KEEP) study – an introduction to hypotheses and highlights. Cont Shelf Res 20:335–347

Wu J, Sunda W, Boyle EA, Karl DM (2000) Phosphate depletion in the Western North Atlantic Ocean. Science 289:759–762

Chapter 2

Physical Transport of Nutrients and the Maintenance of Biological Production

Richard G. Williams · Michael J. Follows

2.1 Introduction

The oceanic distributions of nutrients and patterns of biological production are controlled by the interplay of biogeochemical and physical processes, and external sources. Biological and chemical processes lead to the transformation of nutrients between inorganic and organic forms, and also between dissolved and particulate forms. Physical processes redistribute nutrients within the water column through transport and mixing. The combined role of biogeochemical and physical processes is reflected in the observed distributions of nitrate, phosphate and silicate (macro-nutrients). These distributions broadly reflect those of classical water masses, as defined by temperature and salinity, highlighting the important role of physical transport. However, there are also significant differences between the nutrient and water-mass distributions, notably with nutrients showing stronger vertical and basin-to-basin contrasts. Biological production leads to these greater nutrient contrasts with inorganic nutrients consumed and converted to organic matter in the surface, sunlit ocean. A small fraction of the organic matter in this euphotic zone is exported to depth, driven by the gravitational sinking of particles and subduction of dissolved organic matter. This organic fallout is eventually remineralised leading to an accumulation of inorganic nutrients in deeper and older water masses.

Biological production would eventually cease without other processes acting to supply nutrients to the surface ocean. This supply is largely achieved by ocean transport and mixing processes, redistributing nutrients within the water column, with some contribution from external inputs such as atmospheric deposition, river runoff in the coastal zone and nitrogen fixation. On long timescales, the external inputs are balanced by burial of organic nutrients in the sediments or denitrification. Traditionally, the physical transfer of nutrients from the deep to the surface ocean has been predominately viewed in terms of a vertical supply arising from large-scale vertical advection, diffusion and convection (Fig. 2.1a). Whether this picture holds outside upwelling zones depends on the strength of diapycnic mixing within the upper ocean. In order to explain the vertical profile of tracers, Munk (1966) diagnosed that the diabatic diffusivity needed to reach the order of $10^{-4}\ m^2\ s^{-1}$. However, direct observations of the mixing suggest that the diapycnic diffusivity is an order of magnitude smaller over much of the ocean interior, particularly in the subtropical thermocline[(1)]. Hence, such diffusive transfer is unlikely to dominate within the upper ocean or explain the nutrient supply to the euphotic zone.

Instead the maintenance of biological production is affected by the three-dimensional, time-varying circulation (Fig. 2.1b). The global-scale overturning plays a key role in transporting nutrients and modulating biological production. However, smaller-scale processes involving horizontal variations in convection, gyre circulations, boundary currents, eddies and fronts are also significant. For example, the spatial patterns of surface chlorophyll partly reflect those of physical processes, as shown later in Fig. 2.15 for a remotely-sensed view of the Gulf Stream.

Despite the important role of physical processes, there are significant differences in the individual nutrient and trace-metal distributions. For example, silicate differs from nitrate and phosphate distributions due to the relatively slow remineralisation of silica from sinking particles (Gnanadesikan 1998). Total nitrogen and phosphorus distributions become decoupled from each other through fixation of nitrogen gas at the surface and, perhaps, differential cycling of their dissolved organic forms (Wu et al. 2000). The processes controlling the distribution of iron, an important micronutrient, are not fully understood but appear to involve a combination of atmospheric deposition, geothermal inputs, and scavenging and complexation within the water column (e.g. Martin and Fitzwater 1988; Archer and Johnson 2000). Consequently, individual nutrients and trace metals have different distributions and each can be the limiting factor over different regions and at different times in the surface ocean (see a model illustration by Moore et al. 2002).

In this review, we address the following questions concerning the role of physical processes and their impact on biogeochemical cycling over the open ocean:

- How are observed nutrient distributions over global to kilometre scales controlled by physical transports and physical-biogeochemical interactions?
- What is the role of physical processes in maintaining and modulating biological productivity?
- How are the temporal variations in nutrient distributions and biological productivity, on interannual and longer timescales, connected to changes in atmospheric physical forcing and ocean circulation?

We adopt a mechanistic approach discussing the physical processes and their effect on biogeochemical cycles in the open ocean on horizontal scales ranging from global to frontal. The role of the overturning circulation is discussed in terms of the transport between ocean basins and the Southern Ocean. The role of convection is considered in terms of the seasonal cycle and its limited role in maintaining levels of export production. The role of gyres and boundary currents is outlined in terms of vertical and horizontal transports within ocean basins. The role of smaller-scale eddies and fronts is addressed in terms of both their local and far field transport effects. Finally, interannual and longer-term variability is discussed in terms of regional examples: the El Niño-Southern Oscillation, the North Atlantic Oscillation, the eastern Mediterranean and the glacial North Atlantic.

This review complements the reviews of Denman and Gargett (1995), who focussed on vertical and small-scale transport processes, and Barber (1992), who discussed large-scale processes and the geological record. For a more complete description of ocean circulation, modeling and data analysis, see contributions in the WOCE book edited by Siedler et al. (2001).

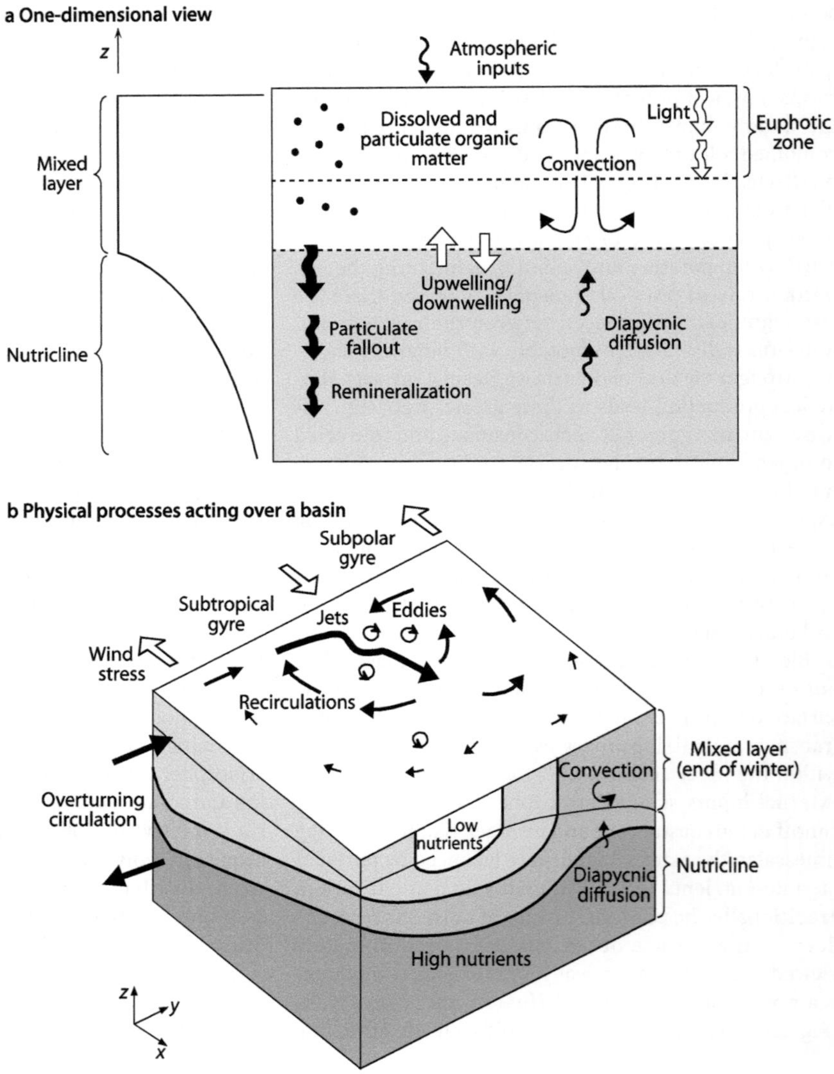

Fig. 2.1.
Schematic figure displaying one- and three-dimensional views of the physical processes affecting biological production: **a** In the vertical, there is a phytoplankton growth within the ecosystem, export of organic matter and remineralisation at depth, which is partly maintained through the physical transfer of nutrients within the ocean involving vertical advection, diapycnic diffusion and convection; **b** a more complete view includes the physical transfer of nutrients by the three-dimensional circulation involving contributions from the overturning, gyre, eddy and frontal circulations, as well as involving interactions with spatial variations in convection

2.2 Global Overturning Circulation and Nutrient Transport

2.2.1 Overturning Circulation and Water-Mass Distributions

The overturning circulation transports water masses and nutrients through individual basins and, via the Southern Ocean, around the globe; see reviews by Schmitz (1995) and Zenk (2001). The transport is primarily achieved by narrow boundary currents, jets and recirculating gyres (together with contribution from eddies, particularly over the Southern Ocean). An estimate of the global ocean circulation from Macdonald (1995) is shown in Fig. 2.2, determined from observed sections of temperature, salinity and tracers using an inverse model; also see Macdonald and Wunsch (1996) and Macdonald (1998). This inverse solution emphasises the separate overturning circulations contained within each basin and their connection with the Southern Ocean, as well as basin-scale recirculations. There is a mean northward transport of relatively warm surface waters in the Atlantic Basin (*red arrow* in Fig. 2.2), a transformation to colder, deep waters at high latitudes, and a return southwards transport of cool waters at depth (*blue arrow* in Fig. 2.2). There is a strong eastwards volume flux associated with the Antarctic Circumpolar Current. There are influxes of cold deep waters into the Indian and Pacific Oceans and return flow of warmer water, including a warm-water pathway from the Pacific to Indian Ocean.

This hydrographic solution is more complex than the widely held view of the global circulation as a large overturning cell, often described as the 'conveyor-belt' circulation (Broecker 1991). The distributions of longer-lived oceanic tracers, such as radiocarbon, and the accumulation of nutrients are consistent with the latter view. However, the global hydrographic analyses (e.g. Schmitz 1995) and inverse models (Macdonald and Wunsch 1996; Fig. 2.2) suggest a more complex global circulation, emphasizing separate overturning cells in the Atlantic and Pacific Basins each connecting independently with the Southern Ocean. This more complex circulation is also found to be consistent with the nutrient distributions in inverse models (Ganachaud and Wunsch 2002).

Meridional nutrient sections broadly reflect the classic water-mass distributions over the globe (Fig. 2.3), but have elevated vertical and basin contrasts compared with other characteristics, such as temperature. Water masses are formed within the mixed layer and then transferred into the ocean interior involving a combination of subduction, deep convection and mixing at overflows. In the North Atlantic, surface waters are nutrient depleted and are transformed to intermediate and deep waters at high latitudes. North Atlantic Deep Water (NADW) is formed through a combination of overflow waters, spreads southwards (*green* in Fig. 2.3a) and gradually increases in nutrient concentration by the accumulation of remineralised nutrients from organic fallout. In the Southern Ocean, the deep waters circulate around the globe acquiring high nutrient concentrations through exchanges with Pacific waters and further accumulation of remineralised sinking particulate matter. These nutrient-rich waters of the Southern Ocean partly return into the Atlantic through the northwards flux of Antarctic Intermediate Water (AAIW) and Antarctic Bottom waters (AABW) (*orange* in Fig. 2.3a).

In contrast, the deep and bottom waters of the Pacific are not ventilated locally. Instead the water masses are advected into the Pacific from the Southern Ocean (Fig. 2.3b) where they have acquired characteristically high concentrations in silica. The deep waters spread northwards and continue to accumulate nutrients from sinking particu-

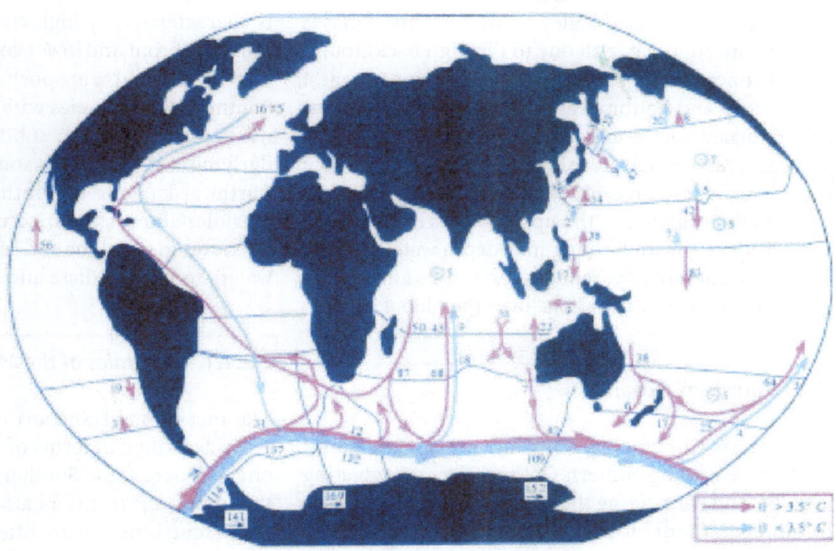

Fig. 2.2. Global inverse solution for the ocean circulation from Macdonald (1995): *red* and *blue lines* depict the volume transport (Sverdrup; 1 Sv = 10^6 m^3 s^{-1}) associated with fluid warmer or cooler than 3.5 °C respectively. The overturning cell over the Atlantic is revealed with 16 Sv of cold water exported from the North Atlantic, as well as strong zonal transports over the Southern Ocean

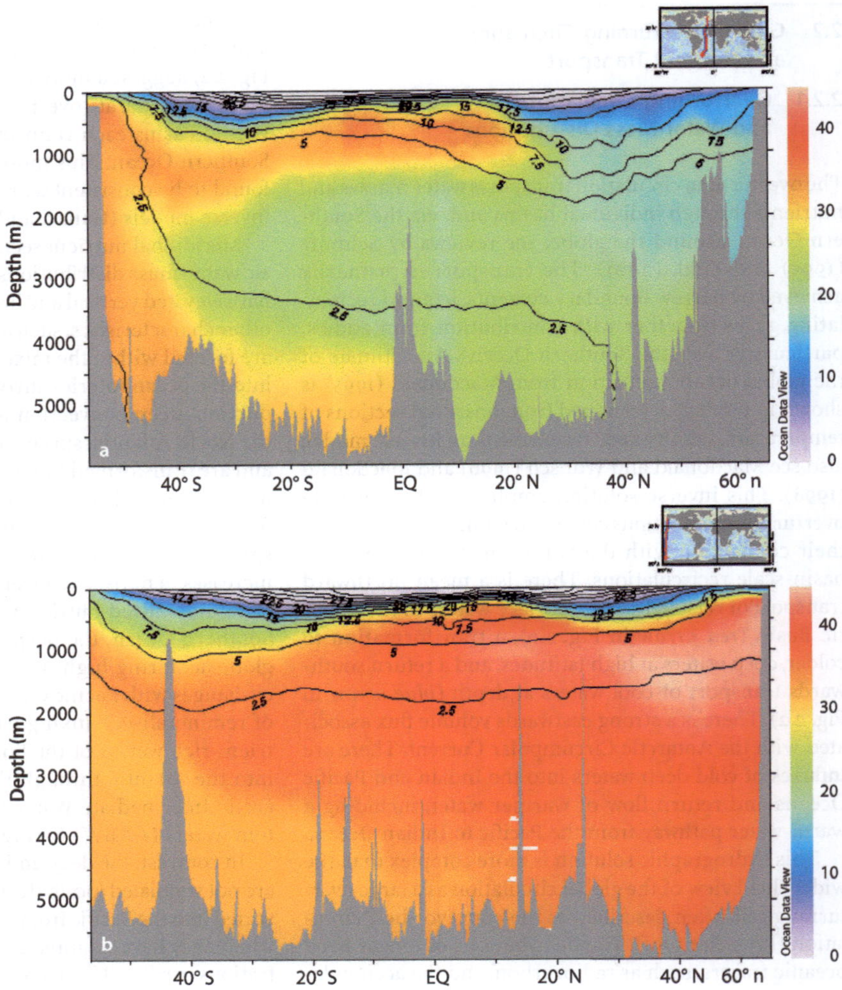

Fig. 2.3. Meridional WOCE sections of nitrate (*colour shading* in µmol kg⁻¹) and potential temperature (*contours* in °C) for **a** Atlantic (A16) and **b** Pacific (P15). In the Atlantic, there are signals of a southwards spreading of North Atlantic Deep Water (*green*), as well as northwards spreading of Antarctic Intermediate Water and Antarctic Bottom Water (*upper and lower orange plumes*). In the Pacific, there is a northwards influx of bottom and deep water from the Southern Ocean, which is probably returned southwards at mid-depth

lates. There is also a northwards spreading of AAIW in the Pacific (Hanawa and Talley 2001); but this signal is not pronounced in Fig. 2.3b due to the high background nutrient concentrations. The silica distributions suggest that the deep and bottom waters in the Pacific are eventually returned southwards at mid-depths to the Southern Ocean, rather than transformed locally into surface waters (Gnanadesikan 1998; Wunsch et al. 1983). The restriction of ventilation to the upper waters of the North Pacific leads to the underlying mid-depth waters having a local maximum in apparent oxygen utilisation and the highest nutrient concentrations over the global ocean.

2.2.2 Southern Ocean

The Antarctic Circumpolar Current (ACC) is the dominant feature of the Southern Oceans, circumnavigating the globe, and connecting the separate ocean basins; see Rintoul et al. (2001) for a comprehensive review. The surface nitrate distribution across the Southern Ocean is characterised by high concentrations to the south of the Polar Front and lower concentrations to the north[(2)]. The meridional transport of water masses across the Southern Ocean varies with depth: upper waters spread northwards including subducted AAIW, while the underlying NADW spreads southwards and AABW spreads northwards along the bottom (Fig. 2.3 and 2.4). These meridional transfers are crucial in redistributing nutrients around the globe and returning nutrients from deep waters to intermediate and surface waters.

2.2.2.1 *Dynamics of the Meridional Transport*

The meridional transport of nutrients and other tracers is described in terms of an idealised, double meridional cell across the Southern Ocean (Fig. 2.4), although the important transfers also involve zonal and meridional excursions that are often linked to topography. The

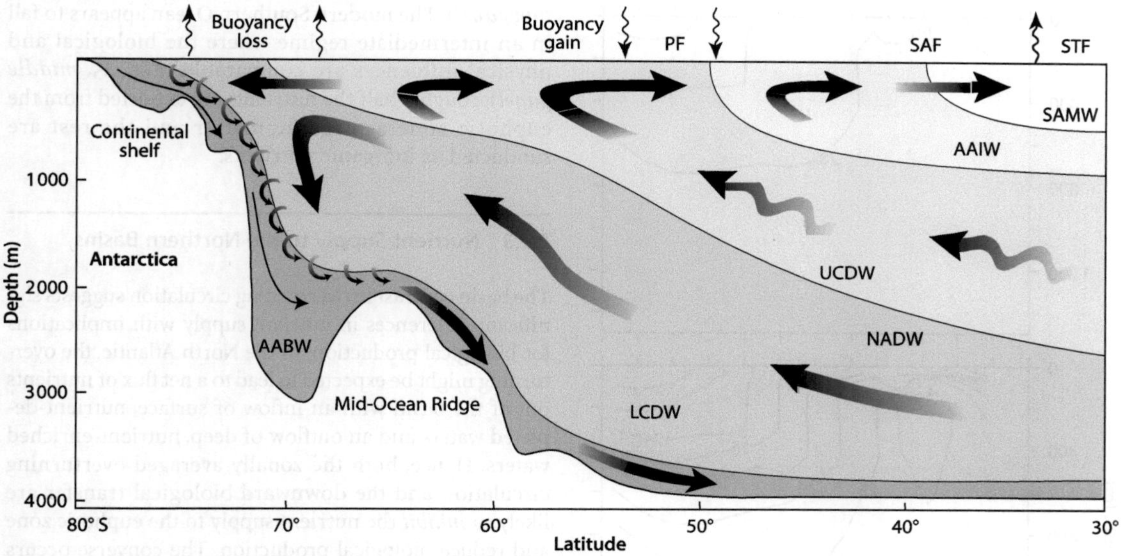

Fig. 2.4. Schematic figure of the two-cell meridional overturning in the Southern Ocean. The upper cell is primarily formed by northwards Ekman transport and gyre transport of surface and Antarctic Intermediate Water (*AAIW*). The lower cell is primarily driven by dense water formation near the Antarctic continent with a northwards transport of Antarctic Bottom Water (*AABW*). Both cells are fed by the southwards transport of Circumpolar and North Atlantic Deep Waters (*NADW*), which is partly achieved by an eddy transport. The Polar Front is denoted by *PF* (reproduced from Speer et al. (2000))

meridional transfer involves different dynamical balances throughout the water column:

i In the upper cell, the westerly wind drives a northwards Ekman transport across the ACC (Toggweiler and Samuels 1995). Nutrient-rich and fresh AAIW is formed north of the Polar Front and probably involves a freshening and cooling of Subantarctic Mode Water in the SW Pacific (McCartney 1977). AAIW is subducted and spreads northwards probably via the gyre circulation between continental barriers.

ii In the lower cell, AABW spreads northwards along the bottom through a geostrophic flow supported by pressure contrasts across topographic ridges.

iii The return flow of the upper and lower cells is associated with the southwards spreading of NADW. This NADW spreading is fed from the northern hemisphere, rather than by a local 'Deacon' cell with diabatic transfer across the thermocline (as misleadingly implied by averaging velocities at fixed points and depths). The physical balance controlling this southwards return flow is still an open question. General circulation models suggest that the southwards spreading is achieved through an adiabatic eddy transport involving contributions from standing and transient eddies (Döös and Webb 1994; Danabasoglu et al. 1994; McIntosh and McDougall 1996)(3).

iv Finally, NADW upwells south of the Polar Front and is converted to lighter surface waters through surface buoyancy gain, as well as converted to denser AABW through surface buoyancy loss (Speer et al. 2000).

2.2.2.2 *Interplay of Transport and Biology*

Nutrient distributions across the Southern Ocean, as elsewhere, are controlled by the competition between biological production and transport. Biological production is always acting to increase the nutrient concentrations in deep waters, whereas transport processes act to decrease the vertical contrast. This competition is illustrated here through idealised integrations of a global biogeochemical model. The model transport includes Ekman and parameterised eddy-induced transport contributions (as depicted later in Fig. 2.21). In the surface ocean, inorganic nutrients are converted to organic nutrients based on a prescribed lifetime. The organic nutrients are exported as sinking particles, falling rapidly relative to the vertical circulation, and are remineralised at depth based on an e-folding depth scale.

Export and remineralisation act to reduce nutrient concentrations in the surface and increase them in the thermocline, enhancing the vertical gradient. In upwelling regions, the nutrient transport acts to further enhance this gradient by supplying nutrients from below, whereas it is weakened in downwelling regions. When the efficiency of biological export is weak relative to the advective transport, the gradients in nutrients become eroded and surface nutrient concentrations become high (Fig. 2.5, *top panel*). When biological export is strong relative to transport, the gradients become strong – both vertically and horizontally – and surface nutrient concentrations become very low (Fig. 2.5, *bot-*

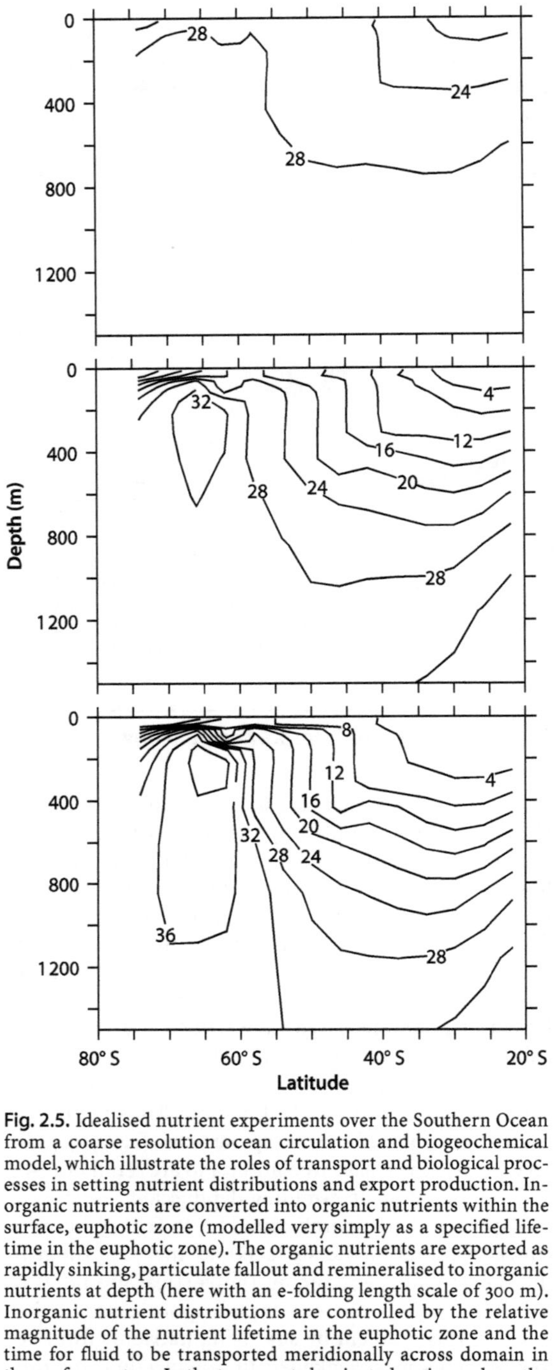

Fig. 2.5. Idealised nutrient experiments over the Southern Ocean from a coarse resolution ocean circulation and biogeochemical model, which illustrate the roles of transport and biological processes in setting nutrient distributions and export production. Inorganic nutrients are converted into organic nutrients within the surface, euphotic zone (modelled very simply as a specified lifetime in the euphotic zone). The organic nutrients are exported as rapidly sinking, particulate fallout and remineralised to inorganic nutrients at depth (here with an e-folding length scale of 300 m). Inorganic nutrient distributions are controlled by the relative magnitude of the nutrient lifetime in the euphotic zone and the time for fluid to be transported meridionally across domain in the surface waters. In the transport-dominated regime, where the biological export timescale is long relative to the transit time (10 years, *upper panel*), nutrient gradients in the horizontal and vertical are relatively weak and export production is low. For a biologically dominated regime where the export timescale is relatively short (3 months, *lower panel*), there are strong horizontal gradients of nutrients below the euphotic zone and high rates of export production. The Southern Ocean for the present day probably lies in an intermediate regime (*middle panel*) where physical transport and biological influences are comparable

tom panel). The modern Southern Ocean appears to fall in an intermediate regime where the biological and physical influences are comparable (Fig. 2.5, *middle panel*); roughly half the nutrients are exported from the euphotic zone as organic matter and the rest are subducted as inorganic nutrients.

2.2.3 Nutrient Supply to the Northern Basins

The basin contrast in overturning circulation suggests significant differences in nutrient supply with implications for biological production. In the North Atlantic, the overturning might be expected to lead to a net flux of nutrients out of the basin with an inflow of surface, nutrient-depleted waters and an outflow of deep, nutrient-enriched waters. Hence, both the zonally-averaged overturning circulation and the downward biological transfer are likely to *inhibit* the nutrient supply to the euphotic zone and reduce biological production. The converse occurs over the Pacific and Indian Oceans, where the overturning circulation is expected to enhance biological production through the inflow of nutrient-rich waters at depth.

2.2.3.1 *Nutrient Transport over the North Atlantic*

This overturning view of the global transport of nutrients overlooks the regional-scale contributions of many physical processes. For example, Fig. 2.6 shows the zonally-averaged fluxes of nitrate at 24° N and 36° N in the Atlantic estimated from hydrographic data by Rintoul and Wunsch (1991). Analysis of these sections revealed a northwards flux of nitrate in the upper part of the water column associated with the Gulf Stream, also highlighted by Pelegri and Csanady (1991), countered by a southward flux in the deeper waters associated with the NADW. In this particular inversion, the nitrate flux is southward across the basin at 24° N and northwards at 36° N, reaching -8 ± 39 kmol N s^{-1} and 119 ± 35 kmol N s^{-1} respectively. This northwards nitrate flux at 36° N is contrary to the southward flux expected from a simple overturning in the Atlantic. The northwards nitrate flux is instead due to the contribution from the horizontal gyre transport (the product of zonal deviations in nitrate concentration and volume flux) dominating over an opposing contribution from the overturning cell.

However, while the contribution of the horizontal circulation is very important, it is not yet clear whether it necessarily dominates. A more recent global inversion by Ganachaud and Wunsch (2002) shows a net southwards nitrate flux over the North Atlantic, although they were unable to include reliable data along 36° N and directly compare with the diagnostics of Rintoul and Wunsch (1991). The relative abundance of nitrate to phosphate, however, suggests that the subtropical North At-

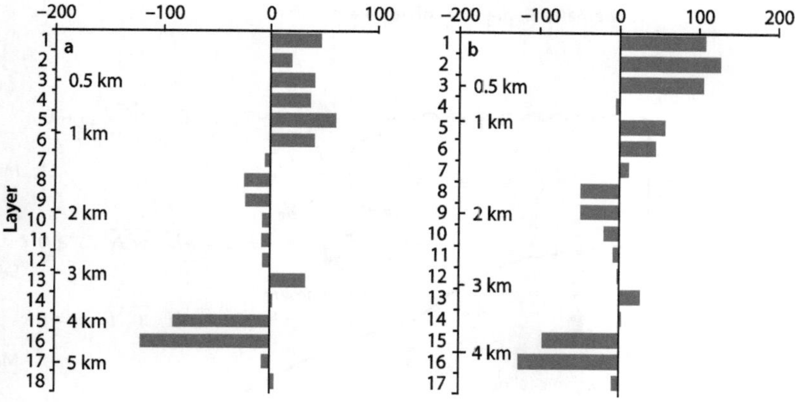

Fig. 2.6.
Zonally-averaged fluxes of nitrate over model layers (10^3 mol N s^{-1}) for **a** 24° N and **b** 36° N in the N. Atlantic from an inverse solution of Rintoul and Wunsch (1991). The net transport of nitrate across 24° N is −8 ±39 kmol N s^{-1}, which is indistin-guishable from zero, while there is net northwards transport of ni-trate across 36° N of 119 ±35 kmol N s^{-1}. Each layer corresponds approximately to a neutral surface with layer 1 at the surface and layer 18 at the seafloor; the typical depths are included (redrawn from Deep-Sea Res. 1, 38, Rintoul SR and Wunsch C, Mass, heat, and oxygen budgets in the North Atlantic Ocean, S355–S377, Copyright (1991), with permission from Elsevier Science)

lantic may be a source of new nutrients through nitrogen fixation (Michaels et al. 1996; Gruber and Sarmiento 1997), which would be qualitatively consistent with a net southwards nitrate flux out of the North Atlantic.

In summary, the nutrient transports by the vertical and horizontal large-scale circulation are opposing and both significant over the North Atlantic. The role of transport as a net source or sink of nitrogen to the basin has not yet been decisively identified. An uncertainty arising from inverse models is whether the inversion is chosen to apply globally or locally, which either leads to global constraints being satisfied or particular hydrographic sections being simulated. In addition, there is the potentially important omission of the transport of organic nitrogen, as raised by Rintoul and Wunsch (1991).

2.2.4 Summary

The overturning circulation is important in determining the global distribution of nutrients in the ocean and in supplying nutrients to the upper ocean where biological productivity occurs. On seasonal to interannual timescales, however, biological productivity is more sensitive to the basin-scale gyre circulations, which control regional upwelling and downwelling patterns close to the surface, and to convective activity, which controls the transfer of nutrients through the seasonal-boundary layer to the euphotic zone. We now focus in more detail on the influence of these physical processes including convection, the gyre circulation, and finer-scale eddy and frontal circulations.

2.3 Convection

Surface convection occurs throughout the ocean in response to a surface buoyancy loss or wind stirring; see the review by Marshall and Schott (1999). The combination of solar irradiance and atmospheric forcing induces characteristic diurnal and seasonal cycles in the mixed-layer thickness. Convective changes can alter biological production through the nutrient and trace metal supply to the euphotic zone, change the light experienced by phytoplankton, as well as impact on grazers and community structure. Here, we focus on the seasonal changes in nutrient supply and light received by phytoplankton.

2.3.1 Vertical Transfer of Nutrients

Convection increases the surface nutrient concentrations whenever the mixed layer thickens and nutrients are entrained from the underlying thermocline (Fig. 2.7a). The maximum thickness of the mixed layer usually occurs at the end of winter (defined by when the surface buoyancy loss to the atmosphere ceases) and denotes the extent of the seasonal boundary layer.

2.3.1.1 *North Atlantic Example*

Over the North Atlantic, the surface buoyancy loss at high latitudes leads to a pronounced thickening of the mixed layer at the end of winter (Fig. 2.7b). Convection redistributes nutrients vertically within this layer every year; a climatological estimate of the annual convective supply of nitrate to the euphotic zone is shown in Fig. 2.7c (Williams et al. 2000). The convective supply broadly increases polewards as the winter mixed-layer thickens and ranges from 0.05 to 1.4 mol N m^{-2} yr^{-1} over the basin. There is a similar magnitude of the convective flux diagnosed from in situ observations at Bermuda of 0.1 mol N m^{-2} yr^{-1} (Michaels et al. 1994). In addition, there are interannual variations in the mixed-layer thickness, surface nitrate and chlorophyll cycles, as discussed later in relation to Fig. 2.24 for the Bermuda Time-Series site.

2.3.2 Biophysical Interactions and Convection

The supply of macro-nutrients to the euphotic zone is only one factor in constraining the levels of biological

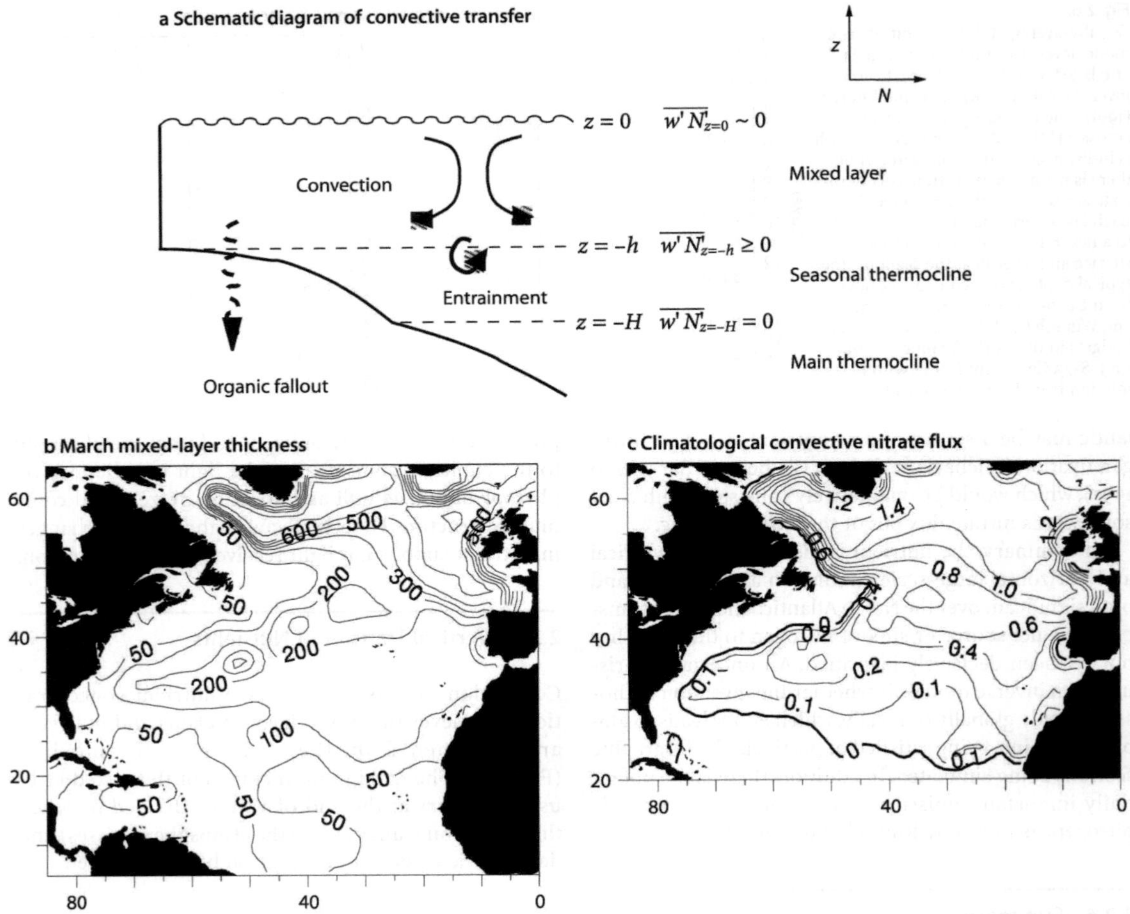

Fig. 2.7. **a** Schematic figure for the convective boundary layer denoting the variations in the vertical turbulent nutrient flux, $\overline{w'N'}$, where w is a vertical velocity and N is a nutrient concentration (a *prime* denotes a turbulent deviation and an *overbar* represents a time-average over many turbulent events). Climatological estimates of **b** end of winter mixed-layer thickness (m) and **c** convective flux of nitrate (mol N m^{-2} yr^{-1}) into the upper 100 m over the North Atlantic. The thickness is diagnosed from climatological density profiles assuming a density increase of 0.125 kg m^{-3} from the surface. The convective flux is diagnosed by combining the seasonal change in mixed-layer thickness with climatological nitrate profiles. For further details, see Williams et al. (2000)

activity, for example, as measured by the amplitude and timing of the annual maximum phytoplankton abundance. A strong spring bloom occurs over the North Atlantic, but not over the North Pacific or Southern Ocean. The lack of a spring bloom and full macro-nutrient drawdown over much of the global ocean has been explained in terms of different hypotheses involving iron and trace metal limitation, light limitation or zooplankton grazing; these competing processes are reviewed by Fasham (1995).

2.3.2.1 *Response of the North Atlantic Bloom*

There are contrasting responses in the phytoplankton growth over the subpolar and subtropical gyres of the North Atlantic; also see Strass and Woods (1991) for a description of how the passage of the bloom leads to a subsurface, deep chlorophyll maximum extending over the North Atlantic.

In the subpolar region, the winter mixed layer may be hundreds of metres thick and nutrients may be abundant. Photosynthesis is light limited during winter, so the bloom occurs in spring when the mixed layer shoals sufficiently to allow phytoplankton to remain within the sunlit region and enable net growth (Sverdrup 1953) (Fig. 2.8a, *left panel*). Here, enhanced convective mixing during the bloom period, perhaps due to the passage of storms, leads to a *weakened* bloom through phytoplankton spending less time in the euphotic zone.

In the subtropical gyres, insolation and stratification are both strong, and winter mixed layers may be only 100 m or so thick. The system is nutrient limited and the bloom occurs in winter when the mixed layer deep-

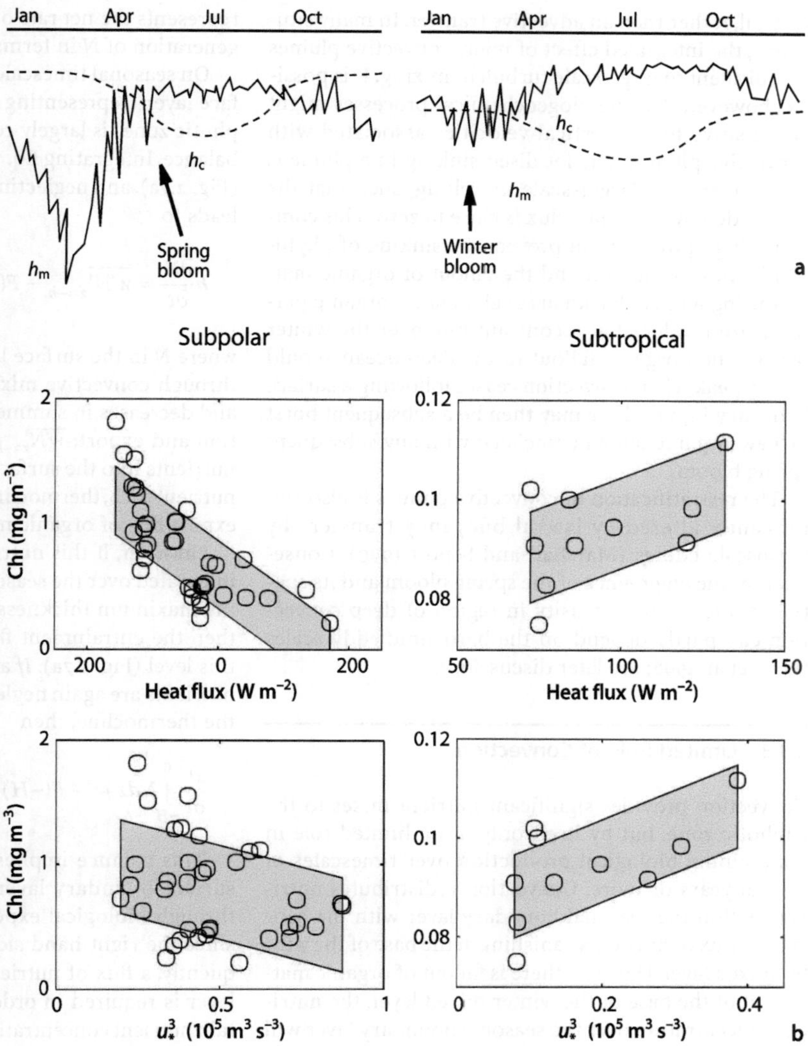

Fig. 2.8. **a** Schematic depiction of the annual cycle of mixed-layer thickness (h_m) and critical layer depth (h_c) for subpolar (*left panel*) and subtropical (*right panel*) regimes. The critical depth is the region where photosynthetic production outstrips respiration or grazing (e.g. Townsend et al. 1994) – this depth is typically up to a few 10s of metres in spring. The mixed-layer thickens in winter and shoals in spring. In the light-limited subpolar region (*left panel*) the bloom occurs in spring when phytoplankton are confined within the critical layer. In the subtropics (*right panel*) it occurs when winter convection supplies nutrients. **b** Illustration of the relationship of bloom intensity and physical forcing for 1997 bloom in the N. Atlantic. The figures show remotely observed chlorophyll (SeaWiFS, level 3) vs. heat flux out of the ocean (W m^{-2}) and wind forcing (m^{3}s^{-3}) (Follows and Dutkiewicz 2002); the latter defined by a friction velocity cubed derived from NCEP reanalysed meteorological data. The bloom-period is defined here as the two months bracketing the annual maximum surface chlorophyll at each latitude, and data points represent area averages from 5° square regions. In the subpolar region (*left panel*), the bloom is more intense where greatest heat input favours restratification. In the subtropical region (*right panel*), the bloom is intensified where there is greater surface heat loss and wind mixing, consistent with nutrient limitation. The *shading* denotes one standard deviation either side of a linear least squares fit (redrawn from Deep-Sea Res. II 49, Follows MJ and Dutkiewicz S, Meteorological modulation of the North Atlantic spring bloom, 321–344, Copyright (2002), with permission from Elsevier Science)

ening provides new nutrients (Menzel and Ryther 1961) (Fig. 2.8a, *right panel*). Here, enhanced convection during the bloom period leads to a *strengthened* bloom through the increase in nutrient supply.

These regimes and responses can be identified in the North Atlantic ocean from remote ocean colour observations and meteorological data as illustrated in Fig. 2.8b (Follows and Dutkiewicz 2002; Dutkiewicz et al. 2001). The subtropical and subpolar regimes exhibit opposing relationships to regional changes in surface heat flux out of the ocean and wind mixing. In the subpolar region, the bloom occurs during periods of restratification (Fig. 2.8b, *left panel*) when the heat flux is into the ocean (marked as negative values). The strength of the bloom is inversely proportional to the strength of wind mixing. The opposite is true in the subtropical region where there is a positive correlation between chlorophyll concentration and the heat flux out of the ocean, as well as with the strength of wind mixing (Fig. 2.8b, *right panel*). These responses are consistent with the phytoplankton growth being initially limited by light over the subpolar gyre and instead by nutrients over the subtropical gyre. The broad classification of subtropical and subpolar regimes, along with other significant processes, such as zooplankton grazing, ecological variability and mesoscale motions leads to a significant spread in the data (as marked by the *shaded regions* in Fig. 2.8b indicating one standard deviation either side of a linear least squares fit).

2.3.2.2 *Role of Convective Plumes*

Convection involves an overturning of dense water in narrow plumes with horizontal scales of a few kilometres (Marshall and Schott 1999). Convection is traditionally represented in models by turbulent mixing in the vertical, rather than an advective transfer. In many situations, the integrated effect of many convective plumes is equivalent to large-scale turbulent mixing. It is possible, however, that the biogeochemical processes might be sensitive to the vertical velocities associated with convective plumes. The localised sinking in a plume is accompanied by larger-scale upwelling, such that the overall downwards mass flux is close to zero. This compensating upwelling can prevent the sinking of phytoplankton, zooplankton and the fallout of organic matter during winter (Backhaus et al. 1999). *If* organic particles are produced at a constant rate over the winter period, then organic fallout to the deep ocean should locally peak when convection ceases following a surface buoyancy input. There may then be a subsequent burst in 'new' organic fallout associated with any subsequent spring bloom.

The restratification of convective plumes is also significantly altered by lateral buoyancy transfers by mesoscale eddies (Marshall and Schott 1999). Consequently, the emergence of the spring bloom and its spatial character and intensity in region of deep convection can partly depend on the baroclinic eddy scales (Lévy et al. 1998; see later discussion).

2.3.3 Limited Role of Convection

Convection provides significant nutrient fluxes to the euphotic zone, but by itself only has a limited role in maintaining biological production over timescales of several years or more. Convection redistributes nutrients within the seasonal boundary layer with the convective flux of nutrients vanishing at the base of the winter mixed layer. Hence, if there is fallout of organic matter out of the base of the winter mixed layer, the nutrient concentration over the seasonal boundary layer will gradually decrease in time unless there is another source of nutrients. This result is illustrated in the following derivation.

Consider the evolution of a nutrient, N, representing the sum of inorganic and dissolved organic forms, which may be written as

$$\frac{\partial N}{\partial t} + \nabla.(uN) + \frac{\partial}{\partial z}\overline{w'N'} = \frac{\partial F}{\partial z} \qquad (2.1)$$

The supply of N is controlled by the divergence of the advective fluxes, uN, and the vertical divergence of the turbulent fluxes, $\overline{w'N'}$, within the mixed layer (representing convective mixing). Here u is the three-dimensional velocity vector and w' is the turbulent vertical velocity where a prime represents a turbulent event in the mixed layer and an overline represents a time-average of these turbulent events. F represents the vertical sinking flux of particulate organic matter, and $\partial F/\partial z$ represents the net rate of biological consumption or regeneration of N in terms of a vertical flux divergence.

On seasonal timescales, the evolution of N over a surface layer (representing either a mixed layer or the euphotic zone) is largely controlled by a one-dimensional balance. Integrating Eq. 2.1 over this layer of thickness h (Fig. 2.7a), and neglecting advective and surface sources leads to

$$h\frac{\partial N}{\partial t} \approx \overline{w'N'}_{z=-h} - F(-h) \qquad (2.2)$$

where N in the surface layer increases in concentration through convective mixing and entrainment in winter and decreases in summer through biological consumption and export; $\overline{w'N'}_{z=-h}$ represents a turbulent flux of nutrients into the surface layer from the entrainment of nutrient-rich, thermocline fluid and $F(-h)$ represents the export flux of organic matter out of the surface layer.

However, if this nutrient balance (Eq. 2.1) is instead integrated over the seasonal boundary layer (defined by the maximum thickness, H, of the winter mixed layer), then the entrainment flux vanishes, $\overline{w'N'}_{z=-h} \approx 0$, below this level (Fig. 2.7a). *If* advective and surface sources of nutrients are again neglected, as well as diffusion within the thermocline, then

$$\frac{\partial}{\partial t}\int_{-H}^{0} N dz \approx -F(-H) \qquad (2.3)$$

This balance implies that the nutrients within the surface boundary layer will eventually be exhausted through biological export into the main thermocline, since the right-hand side of Eq. 2.3 is negative. Consequently, a flux of nutrients into the seasonal boundary layer is required in order to offset this loss and maintain nutrient concentrations. Wind-driven gyre and eddy transfer mechanisms may account, at least in part, for this source, and are discussed later.

2.3.4 Summary

Convection provides a nutrient flux into the euphotic zone, as well as altering the light phytoplankton receive by modifying how much time they spend in the euphotic zone. Seasonal or interannual changes in convection have contrasting effects on biological production according to whether phytoplankton growth is limited by the availability of nutrients or trace-metals, or by the light received. Convection only redistributes nutrients within the seasonal boundary layer. Consequently, other processes, such as advective transport, must provide a nutrient source to the seasonal boundary layer offsetting the loss from export of organic matter organic fallout into the underlying thermocline.

2.4 Wind-Driven Circulations: Gyres and Boundary Currents

2.4.1 Wind-Induced Upwelling and Gyre Circulations

The atmospheric winds induces a horizontal (Ekman) volume flux over the surface (Ekman) layer of the upper ocean, directed to the right of the wind stress in the northern hemisphere and to the left in the southern hemisphere[(4)]. A horizontal divergence of this volume flux in turn drives upwelling into the surface Ekman layer and, conversely, a horizontal convergence drives downwelling.

2.4.1.1 *Tropical and Coastal Upwelling*

Over the tropics, the surface Trade winds are generally directed westwards and equatorwards (as part of the atmospheric Hadley circulation). Accordingly, this wind pattern drives a polewards Ekman volume flux on either side of the equator (Fig. 2.9a) and an off-shore Ekman flux along the eastern boundary of an ocean basin. Consequently, the divergence of this horizontal Ekman volume flux drives a band of upwelling along the equator, which extends along the eastern boundary.

This equatorial and coastal upwelling supplies macronutrients to the euphotic zone elevating local productivity; see later Fig. 2.23 (*lower panel*) for an example of elevated chlorophyll in the equatorial Pacific due to equatorial and coastal upwelling. Since coastal upwelling is driven by the local winds, the nutrient supply is modulated on atmospheric synoptic timescales. A succession of phytoplankton species follow upwelling events and the frequency of such events can modulate the dominant community structure; see reviews of the physical and biogeochemical aspects of coastal upwelling systems by Smith (1995) and Hutchings et al. (1995) respectively.

2.4.2 Gyre-Scale Circulations

The surface winds drive double-gyre systems over ocean basins (Fig. 2.9a,b). Subpolar gyres are characterised by a cyclonic circulation, upwelling and a raised thermocline. Conversely, subtropical gyres are associated with an anticyclonic circulation, downwelling and a depressed, thicker thermocline. These thermocline changes are reflected in the nutrient distribution with uplifted or depressed nutriclines occurring over the subpolar and subtropical gyres respectively (see Fig. 2.3). In turn, there are generally higher concentrations in remotely-sensed estimates of surface chlorophyll and primary production over the subpolar gyre (Fig. 2.10) (Sathyendranath et al. 1995; Behrenfeld and Falkowski 1997). Higher levels of primary production generally appear to coincide with upwelling regions over the subpolar gyre and along coastal boundaries (Fig. 2.10) (Williams and Follows 1998a; McClain and Firestone 1993).

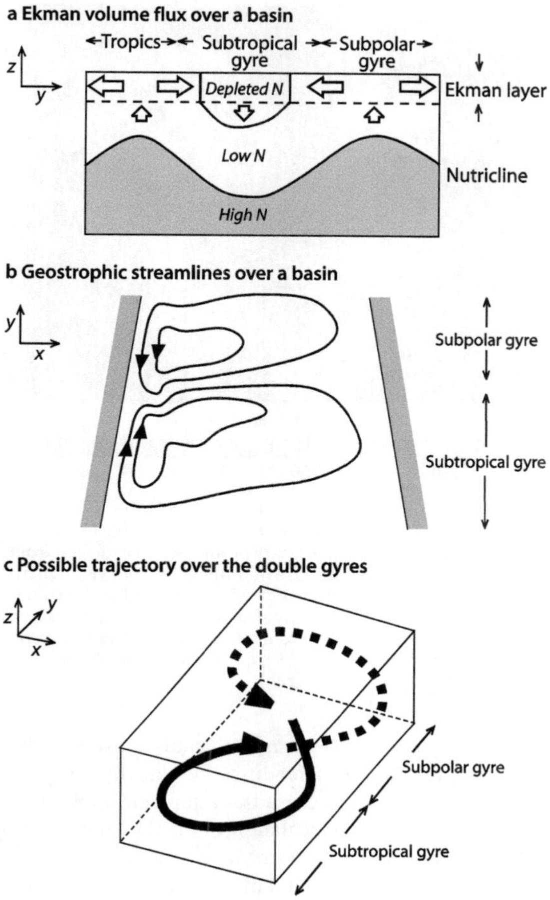

Fig. 2.9. Schematic figure of the gyre circulation within a basin: **a** Ekman volume fluxes along a vertical meridional section, **b** surface view of geostrophic streamlines, and **c** possible trajectory circuiting the double gyre system (*dashed* and *full lines* are within the subpolar or subtropical gyres respectively). Divergence of the horizontal Ekman flux induces upwelling and a raised nutricline in the tropics and subpolar gyre, as well as downwelling and a depressed nutricline in the subpolar gyre. The surface geostrophic flow follows a cyclonic pattern over the subpolar gyre and an anticyclonic pattern over the subtropical gyre. Fluid is exchanged between the gyres through a boundary current transport along the western boundary, as well as through an Ekman and time-varying eddy transport over the entire domain. This idealised gyre picture is also modified through the contribution of the overturning circulation, which can further enhance or oppose the gyre transport of upper waters along the western boundary

The subtropical and subpolar gyre circulations are directly connected to each other. For an idealised wind-driven double-gyre, fluid particles follow trajectories making up an idealised 'figure of eight' (Fig. 2.9c):

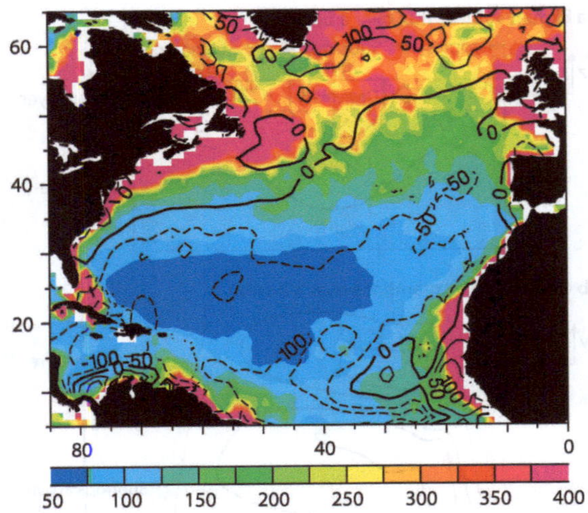

Fig. 2.10. Annual primary productivity (*colour shaded* in mol C $m^{-2} yr^{-1}$) and wind-induced (Ekman) upwelling (*solid contours* in $m\,yr^{-1}$). The annual primary productivity is inferred from satellite observations of surface chlorophyll by Sathyendranath et al. (1995) and the upwelling inferred from a wind-stress climatology. The primary productivity shows maximum values in the subpolar gyre and reduced values over the subtropical gyre, broadly following the patterns of gyre-scale upwelling (reproduced from Williams and Follows (1998b))

(*i*) fluid particles upwell and recirculate around the subpolar gyre, (*ii*) transfer into the subtropical gyre through a lateral flux across the inter-gyre boundary, (*iii*) downwell and recirculate around the subtropical gyre, and (*iv*) eventually return to the subpolar gyre via the western boundary current. This interpretation is consistent with simplified models of a subtropical gyre emphasising the balance between a volume influx from the surface Ekman transfer and a volume outflux in the western boundary (Veronis 1973). In a similar manner, nutrients should be transferred between the subtropical and subpolar gyres through the lateral fluxes across the inter-gyre boundaries. This idealised picture is also modified through the overturning circulation acting over the basin.

2.4.3 Subduction and Fluid Transfer into the Seasonal Boundary Layer

The gyre-scale circulation transfers fluid between the mixed layer/seasonal boundary layer and the underlying thermocline (Fig. 2.11a); see Marshall et al. (1993) and a review by Williams (2001). This subduction process helps to determine the properties of the interior ocean. Fluid is preferentially subducted into the main thermocline at the end of winter (due to the seasonal migration of density outcrops) leading to the interior water-mass properties matching those of the mixed layer at the end of winter (Stommel 1979; Williams et al. 1995). Local maxima in the subduction process lead to formation of 'mode' waters, weakly stratified fluid with nearly homogeneous properties, which can spread over relatively large geographic regions (e.g. Hanawa and Talley 2001).

The reverse of the subduction process is important in determining nutrient distributions over the upper ocean, where fluid is transferred from the ocean interior into the seasonal boundary layer. The annual volume flux, or induction flux, into the seasonal boundary layer from the time-mean circulation consists of vertical and horizontal contributions:

$$w_b + u_b.\nabla H \qquad (2.4)$$

where w_b and u_b are the vertical velocity and horizontal velocity vector at the base of the seasonal boundary layer with a thickness H and a horizontal gradient ∇H. This contribution of the time-mean circulation can be augmented by a rectified contribution from the time-varying circulation (Marshall 1997). This volume exchange from the time-mean and time-varying circulations determines whether nutrient-rich thermocline waters are advected into the seasonal boundary layer or nutrient-depleted surface waters are subducted into the thermocline. In turn, this advective transfer of nutrients helps to determine whether the surface waters are nutrient rich or poor.

2.4.3.1 *North Atlantic Example*

The induction flux over the North Atlantic is controlled by both the vertical and horizontal transfers between the mixed layer and thermocline (Eq. 2.4). Climatological estimates of the vertical Ekman volume flux and induction flux into the seasonal boundary layer evaluated from climatology are shown in Fig. 2.11b,c; details of the calculation are described in Marshall et al. (1993). The induction flux is evaluated assuming the thickness H is defined by the base of the end of winter mixed layer (Fig. 2.7b).

Over the subpolar gyre, fluid is transferred from the thermocline into the seasonal boundary layer. The induction flux reaches 300 m yr^{-1}, compared to a vertical Ekman flux of only about 50 m yr^{-1} (where the volume fluxes are expressed per unit horizontal area). Therefore, the flux is dominated by the horizontal transfer due to the thickening of the seasonal boundary layer, caused by the surface cooling, along the cyclonic circuit of the gyre. Hence, the nutrient-rich surface waters of the subpolar gyre are sustained through both the horizontal and vertical advective influx from the nutrient-rich thermocline.

Over the subtropical gyre, fluid is subducted from the mixed layer into the thermocline. The induction flux is

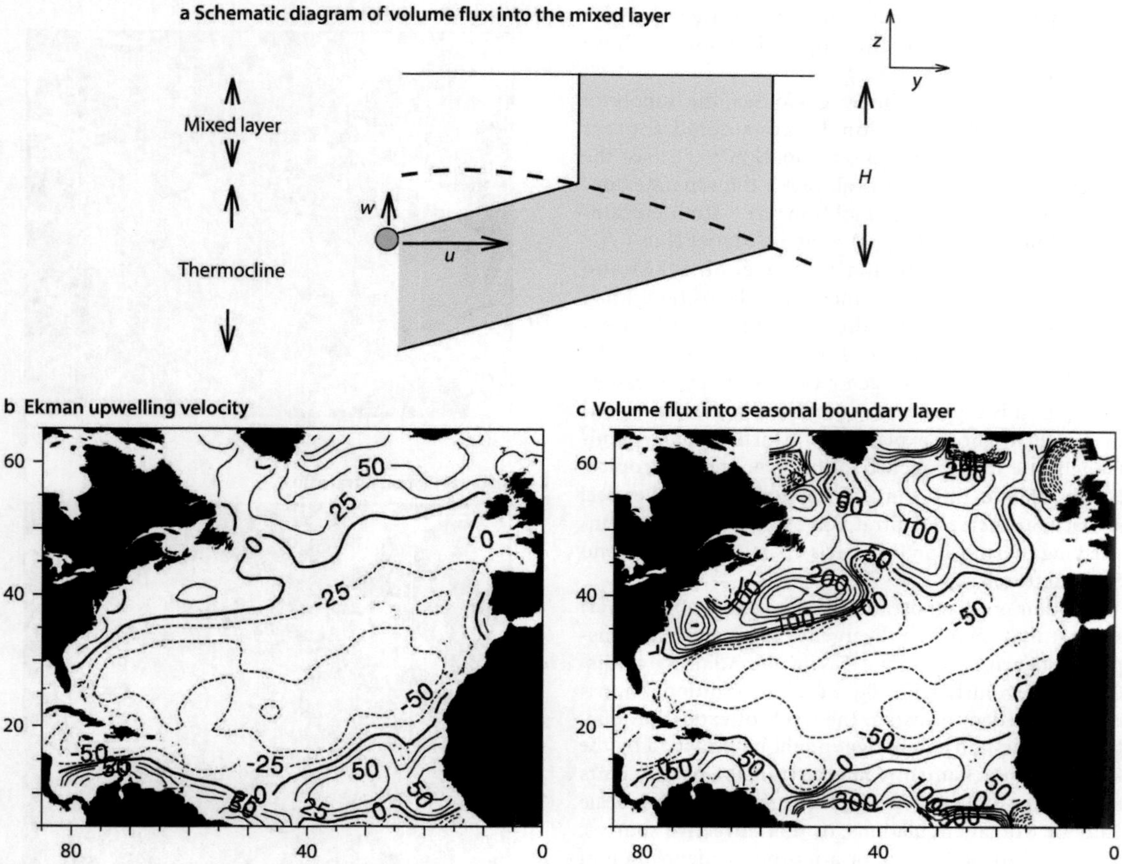

Fig. 2.11. a Schematic figure of the subduction process involving fluid being transferred between the mixed layer and stratified thermocline. A fluid particle is advected by the vertical velocity, w, and horizontal velocity, u. The volume flux into the mixed layer is given by the sum of the vertical and horizontal transfer into the mixed layer: $w + u\nabla H$ where H is the thickness of the mixed layer at the end of winter (or equivalently the extent of the seasonal boundary layer). The lateral transfer is important when the mixed layer thickens polewards, as depicted here. Climatological estimates of **b** Ekman upwelling velocity (m yr^{-1}) and **c** volume flux into the seasonal boundary layer per unit horizontal area (m yr^{-1}) for the North Atlantic; see Marshall et al. (1993) for details of the calculation. The volume flux into the seasonal boundary layer in **c** exceeds the contribution from Ekman upwelling **b** due to the lateral transfer

typically −50 m yr^{-1} with similar contributions from the vertical and horizontal transfer. Any surface nutrients are transported into the thermocline by the gyre-scale circulation. The nutrient-rich waters in the thermocline are only likely to reach the euphotic zone over the interior of the subtropical gyre through diapycnic mixing or fine-scale upwelling (discussed later). Instead, the thermocline nutrients are more likely to be transported into the western boundary current and either to recirculate around the subtropical gyre or enter the subpolar gyre.

2.4.4 Oligotrophic Subtropical Gyres

The gyre-scale subduction process leads to surface waters being relatively nutrient rich over the subpolar gyre, and nutrient poor over the subtropical gyre. This different nutrient supply is reflected in patterns of primary production as suggested by remotely-sensed estimates of surface chlorophyll (Fig. 2.10). Much of the primary production is associated with recycling of organic matter, but a fraction is associated with sinking organic particles or dissolved organic matter transported out of the euphotic zone. This fraction is referred to as export production. Estimates of export production over subtropical gyres reach 0.48 ±0.14 mol N m^{-2} yr^{-1} in the Sargasso Sea from transient-tracer and oxygen diagnostics (Jenkins 1982, 1988; Jenkins and Goldman 1985), as well as 0.19 mol N m^{-2} yr^{-1} near Hawaii from sediment-trap estimates (Emerson et al. 1997).

If these modest levels of export production are representative, then the extensive area of the subtropical gyres means that they might account for half of the global export of organic carbon to the ocean interior (Emerson et al. 1997). Consequently, this level of export production raises the question of how sufficient nutrients are supplied to the euphotic zone in these oligotrophic waters?

For example, over the Sargasso Sea, the supply of nitrate from the traditionally considered sources only amounts typically to 0.21 mol N $m^{-2} yr^{-1}$ (see the review by McGillicuddy et al. 1998); the separate contributions are 0.13 ±0.05 mol N $m^{-2} yr^{-1}$ from entrainment (Michaels et al. 1994), 0.05 ±0.01 mol N $m^{-2} yr^{-1}$ from diapycnic diffusion (Lewis et al. 1986) and 0.03 mol N $m^{-2} yr^{-1}$ from atmospheric deposition (Knap et al. 1986). Accordingly, the shortfall in the nutrient supply over the Sargasso Sea needed to explain the transient-tracer and oxygen based estimates of export production is typically 0.27 mol N $m^{-2} yr^{-1}$. Part of this mismatch might be explained by a further source of nitrogen due to nitrogen fixation over the subtropical North Atlantic; this source is implied by a geochemical signal of an increased nitrate/phosphate ratio in the underlying thermocline (Michaels et al. 1996; Gruber and Sarmiento 1997).

In addition, diagnostics of Rintoul and Wunsch (1991) suggest that there is a divergence in the nitrate transport between 24° N and 36° N in the Atlantic (as discussed previously with Fig. 2.6). The additional nutrient flux needed to sustain the levels of export production and the nutrient budget might be provided by the following mechanisms: lateral transfer of nutrients across inter-gyre boundaries, the action of the finer-scale eddy or frontal circulations, or non-advective sources, such as nitrogen fixation or atmospheric deposition of organic nitrogen. The role of the lateral transfer of nutrients is discussed next and later the role of the time-varying circulation.

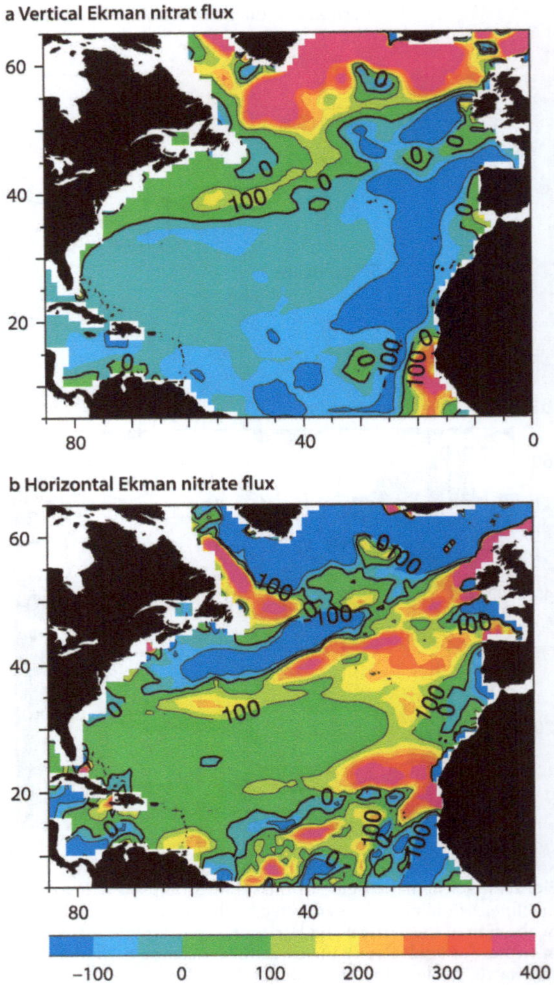

Fig. 2.12. Convergence of Ekman nitrate flux (10^{-3} mol N $m^{-2} yr^{-1}$) over the North Atlantic for April for **a** vertical and **b** horizontal components; the diagnostics are based on a climatological analysis (Williams and Follows 1998a). The vertical Ekman transfer provides a nitrate supply to the euphotic zone over the subpolar gyre, while the horizontal Ekman transfer provides a supply over the flanks of the subtropical gyre (reprinted from Deep-Sea Res. I 45, Williams RG and Follows MJ, The Ekman transfer of nutrients and maintenance of new production over the North Atlantic, 461–489, Copyright (1998), with permission from Elsevier Science)

2.4.4.1 *Lateral Transfer of Nitrate*

Downwelling over the subtropical gyre is induced by a convergence of the horizontal Ekman volume flux. These lateral Ekman fluxes likewise transfer nutrients horizontally into the subtropical gyre from the neighbouring nutrient-rich tropics and subpolar gyre (Fig. 2.9a). The convergence of the horizontal Ekman flux of nitrate is significant over the flanks of the subtropical gyre and is comparable to that from the vertical flux over upwelling regions (Fig. 2.12) (as diagnosed from climatology by Williams and Follows 1998a).

The annual Ekman flux of nitrate is dominated by the winter and spring contributions when the surface nitrate concentration is enhanced through entrainment. Over the North Atlantic, the annual convergence of the Ekman horizontal nitrate flux reaches a maximum of 0.1 mol N $m^{-2} yr^{-1}$ along the flanks of the subtropical gyre, but decreases rapidly over the gyre interior due to the limited lifetime of nitrate in surface waters.

2.4.4.2 *Lateral Transfer of Organic Nutrients*

Organic and inorganic nutrients can be transported in different directions due to their different spatial distributions and lifetimes in the euphotic zone (Jackson and Williams 1985). For example, total organic nitrogen (TON) is generally surface intensified, whereas nitrate has a higher concentration at depth, and semi-labile TON has a longer lifetime than nitrate in the euphotic zone[5]. Rintoul and Wunsch (1991) suggested that organic nitrogen might be transferred into the subtropi-

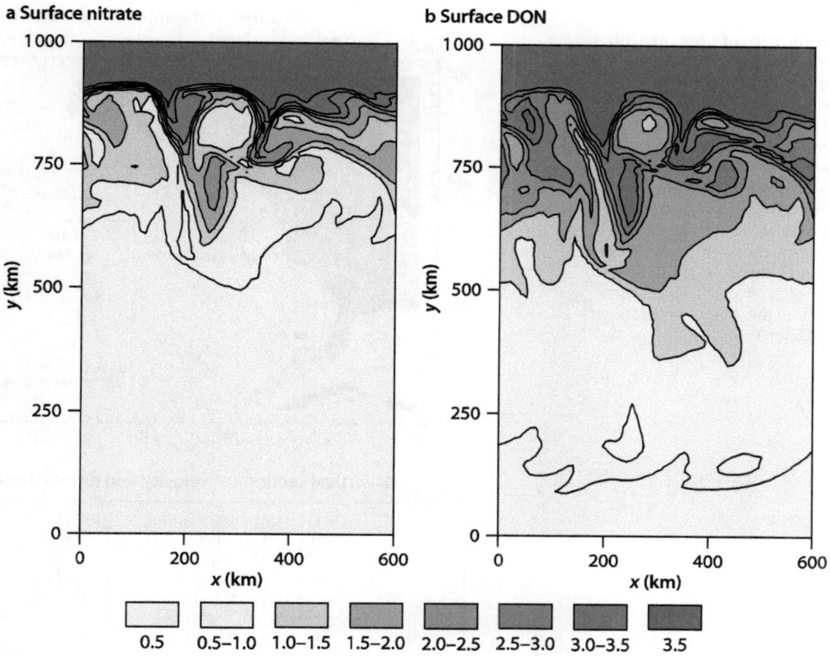

Fig. 2.13. Modelled distributions of **a** nitrate and **b** dissolved organic nitrogen (DON) (μmol kg^{-1}) in the euphotic zone from an eddy-resolving model for a wind and buoyancy-driven zonal channel. The nutrients are transferred from a northern source equatorwards through a combination of the surface Ekman and eddy circulations. The nitrate and DON are chosen to have lifetimes of 3 and 6 months respectively. Consequently, DON penetrates further equatorwards than nitrate and helps to enhance export production within the interior of the domain (reproduced from Lee and Williams (2000))

cal gyre, offsetting the loss of nitrate and maintaining the total nitrogen budget. Indirect support is provided by observations suggesting that more organic carbon is consumed by respiration in the centre of subtropical gyres, than produced by photosynthesis (Duarte and Agusti 1998). The necessary organic carbon might be laterally transferred into the subtropical gyre along with organic nutrients.

Abell et al. (2000) observe that there is a meridional gradient in surface total organic phosphorus (TOP) over the North Pacific subtropical gyre while there is no significant gradient in TON. These signals are consistent with the hypothesis that any production in the subtropical gyre is sustained by a combination of nitrogen fixation providing the required nitrogen and perhaps a lateral influx of TOP providing the required phosphorus. The organic phosphorus might be brought into the subtropical gyre through a combination of the Ekman and eddy transfer across the intergyre boundaries. This response is illustrated in an eddy-resolving model experiment shown in Fig. 2.13 (Lee and Williams 2000): dissolved inorganic and organic nutrients are released along a northern boundary and are transferred southwards through the Ekman and eddy circulations. The dissolved organic nutrient penetrates further from the source than for inorganic nutrient due to its longer lifetime in the euphotic zone.

Closing the nitrogen and phosphorus budgets requires a systematic set of hydrographic sections including the measurement of organic nutrients. In our view, the lateral transfer of organic phosphorus is likely to be important in closing the budget, but it is unclear whether a lateral transfer of organic nitrogen is needed given the potential role of nitrogen fixation at least over the North Atlantic. In explaining how export production is maintained over subtropical gyres, there is an important additional contribution from time-varying currents (discussed later in relation to Fig. 2.17 and 2.22).

2.4.5 Western Boundary Transport of Nutrients

Western boundary currents are crucial in transporting water masses over ocean basins and providing 'nutrient streams' passing from the subtropical gyre to the subpolar gyre in the upper ocean (as depicted in Fig. 2.9c).

2.4.5.1 *Gulf Stream Example*

Hydrographic sections reveal subsurface, enhanced nutrient fluxes coinciding with the core of the Gulf Stream (Pelegri and Csanady 1991; Pelegri et al. 1996); centred along $\sigma_t = 26.8$ surface at typically 500 m depth, see Fig. 2.14. The nitrate flux associated with the Gulf Stream (defined by the product of the nitrate concentration and the geostrophic velocity normal to the section) increases by a factor of 3 from Florida Strait (24° N) to the mid-Atlantic Bight (36° N) where it reaches 860 kmol N s^{-1}, and then decreases downstream. The increase in nitrate transport along the Gulf Stream can be attributed to the recirculating, nutrient-enriched thermocline waters of the subtropical gyre joining the boundary current. The nitrate flux at the mid-Atlantic Bight partly passes into

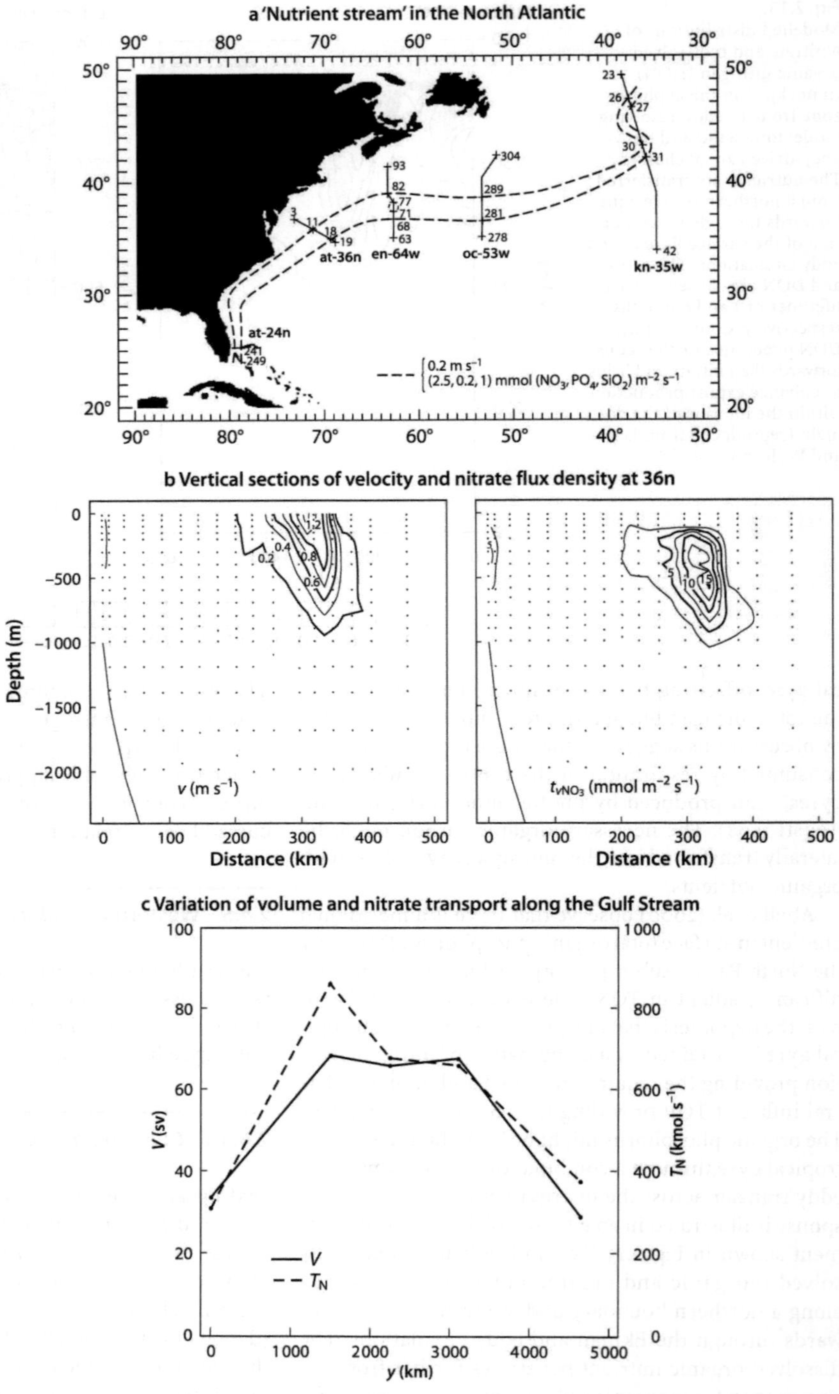

Fig. 2.14.
Diagnostics of how the Gulf Stream acts as a 'Nutrient Stream': **a** *dashed lines* depict the boundaries of the 'nutrient stream' and *full lines* depict the hydrographic sections used in the analysis; **b** velocity and nitrate flux density along 36° N; **c** volume flux (Sv), and nitrate flux (kmol s^{-1}) along the Gulf Stream (redrawn from Pelegri, JL and Csanady GT, J. Geophys. Res. 96, 2577–2583, 1991, Copyright by the American geophysical Union)

the subpolar gyre and is partly recirculated within the subtropical gyre. Hence, this western boundary flux of nitrate is important in maintaining elevated nitrate concentrations over the subpolar gyre.

The western boundary flux of the upper waters also includes an overturning contribution, which balances the export or import of dense water into the basin. The gyre and overturning contributions to the volume and heat fluxes within the western boundary reinforce each other over the North Atlantic, but oppose each other over the North Pacific. Whether the associated boundary current contributions to the nutrient flux reinforce or oppose each other depends on the nutrient distributions. For the North Atlantic, the nutrient fluxes associated with the gyre and overturning circulations oppose each other (Fig. 2.6; Rintoul and Wunsch 1991), but elsewhere either response might occur.

2.4.6 Summary

The combination of gyre and overturning circulations, and patterns of convection determine the large-scale contrasts in nutrient concentration and biological production within ocean basins. The gyre-scale circulation transfers fluid both horizontally and vertically between the thermocline and mixed layer. This transfer leads to the surface waters being nutrient rich over the subpolar gyre and nutrient poor over the subtropical gyre. Nutrients are transferred between each gyre through western boundary currents, Ekman and time-varying eddy circulations.

2.5 Smaller-Scale Circulations: Mesoscale Eddies, Waves and Sub-Mesoscale Fronts

Until recently the nutrient supply to the euphotic zone has been considered primarily in terms of the large-scale, time-mean circulation and diapycnic mixing. However, there is an energetic, time-varying circulation, particularly associated with mesoscale eddies and sub-mesoscale fronts. For example, Fig. 2.15 shows a snapshot of the chlorophyll distribution (derived from Coastal Zone Colour Scanner data) in the Northwest Atlantic Ocean. This snapshot reveals signatures of a range of physical processes: the meandering of the Gulf Stream, the formation of mesoscale eddies and finer-scale frontal filaments drawn out between the larger-scale circulations.

2.5.1 Formation of Mesoscale Eddies and Sub-Mesoscale Fronts

2.5.1.1 *Mesoscale Eddies*

Ocean eddies are formed predominately through baroclinic instability of boundary currents and density fronts. Baroclinic instability occurs preferentially on the scale of the internal Rossby radius of deformation, $L_d = NH/f$, which is typically 30–50 km in the open ocean; here N is the buoyancy frequency, f is the Coriolis parameter and H is a typical thickness scale for the motion. In the ocean, these eddies are usually referred to as mesoscale features, although they are dynamically analogous to atmospheric synoptic-scale, weather sys-

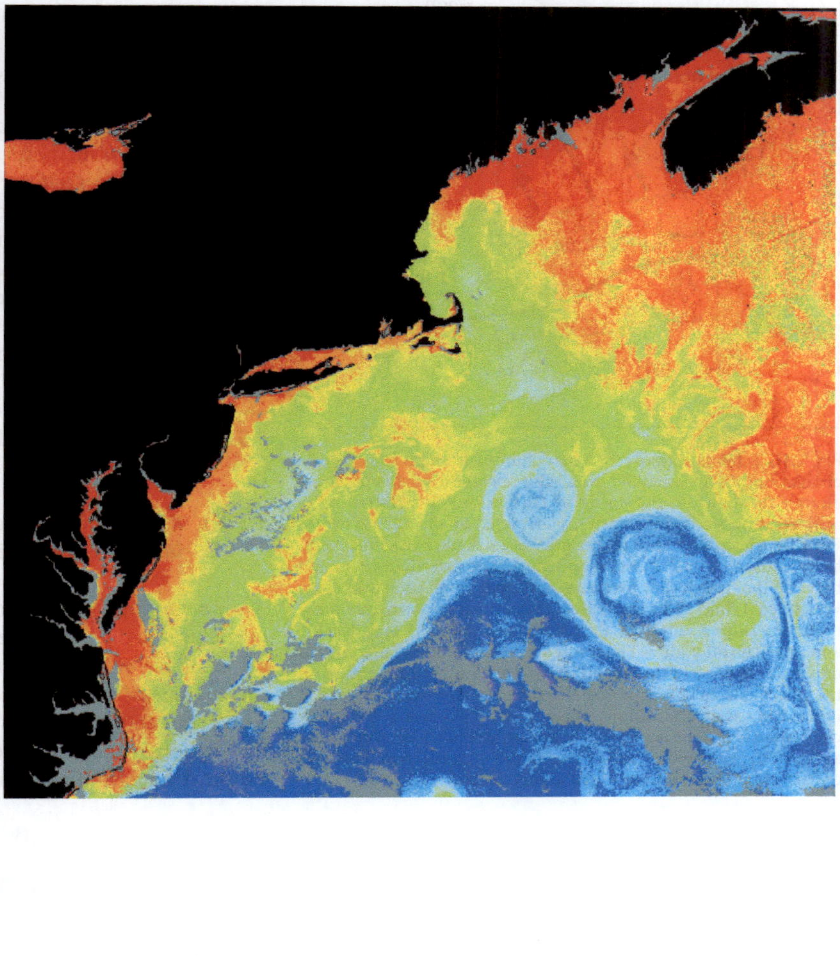

Fig. 2.15. Chlorophyll picture derived from CZCS over the Northwestern Atlantic. Higher concentrations of chlorophyll (*red*) are evident along the coastal boundary and at higher latitudes. Lower concentrations (*blue*) correlate with the Gulf Stream boundary, the subtropical gyre and anticyclonic eddies. Note the range of physical processes revealed here including boundary currents, mesoscale eddies and finer-scale fronts and filaments (figure courtesy of NASA)

tems. See Rhines (2001) for a brief review of mesoscale eddies, Green (1981) and Gill et al. (1974) for physical descriptions of the instability process and energetics for the atmosphere and ocean respectively.

Baroclinic instability involves a slantwise exchange of fluid across a jet or frontal zone. Available potential energy is released through isotherms being flattened: warm, light fluid rises and is replaced by colder, dense fluid (Fig. 2.16a). For a poleward decrease in temperature, this slantwise exchange leads to the warm fluid rising and moving poleward, and cool fluid sinking and moving equatorward, such that there is a poleward eddy heat flux. In the growth phase of a baroclinic eddy, meanders develop with the rising warm fluid occurring downstream of the trough (region of low pressure) and sinking cold fluid occurring downstream of the ridge (region of high pressure) (Fig. 2.16a). The meanders of the jet amplify and may break off as warm-core, anticyclonic and cold-core, cyclonic eddies, which preferentially occur on the colder and warmer sides of the jet respectively.

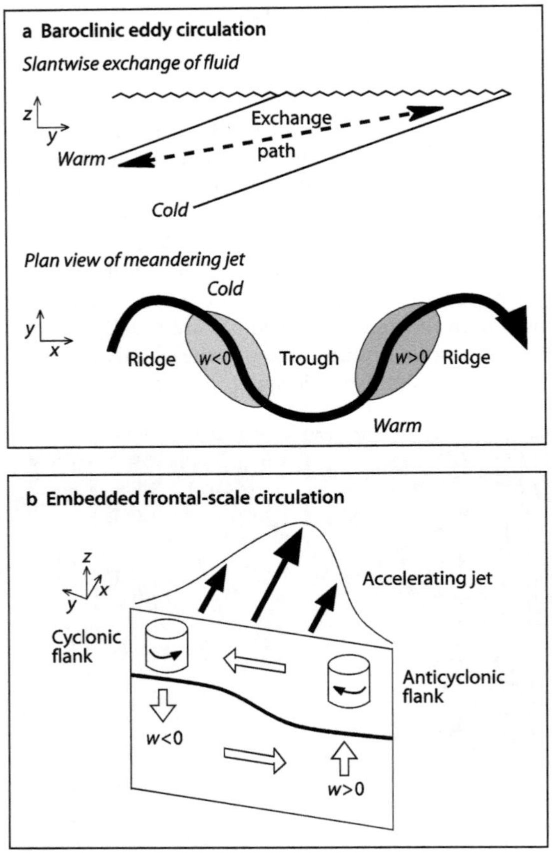

Fig. 2.16. Schematic figure for **a** baroclinic eddy and **b** frontal-scale circulations. In **a**, baroclinic instability releases available potential energy through the flattening of isotherms. This flattening is achieved by a slantwise transfer, as depicted by the *dashed arrow*. For a jet with cooler fluid on the polewards flank, baroclinic instability leads to warm fluid rising and moving polewards downstream between the trough (low pressure) and ridge (high pressure). Conversely, cold fluid sinks and moves equatorwards downstream between the ridge and trough. Eventually, the meanders may develop and form cut-off eddies. The cold-core, cyclonic eddies are formed on the warm flank of the jet and warm-core, anticyclonic eddies on the cold flank. In **b**, frontal-scale circulations are often embedded in the eddy-scale circulations. When a jet accelerates, a secondary circulation is excited across the jet with upwelling occurring on the anticyclonic side and downwelling on the cyclonic side. This secondary circulation reverses if the jet decelerates

2.5.1.2 *Sub-Mesoscale Fronts*

Sub-mesoscale features include fronts and drawn out filaments between mesoscale eddies, as seen in the satellite image in Fig. 2.15. These sub-mesoscale features have a horizontal scale across the flow which is much smaller than the internal Rossby radius of deformation, L_d. When the flow along a jet accelerates, a secondary circulation develops across the jet leading to upwelling on the anticyclonic side and downwelling on the cyclonic side[(6)]. In turn, when the flow along the jet decelerates, a secondary circulation of the reverse sign is formed.

The frontal response may also be understood in terms of conservation of a dynamical tracer, potential vorticity defined by $(\zeta + f)/h$ where h is the thickness of an isopycnic layer, $\zeta = \partial v/\partial x - \partial u/\partial y$ and $f = 2\Omega \sin\theta$ are the vertical components of relative and planetary vorticity respectively; u and v are the eastwards and northwards velocities, Ω is the Earth's angular velocity and θ is the latitude. When a jet accelerates, the absolute vorticity, $\zeta + f$, increases on the cyclonic side of the front and decreases on the anticyclonic side. Consequently, the layer thickness h increases (or decreases) wherever the absolute vorticity, $\zeta + f$, increases (or decreases) in order to conserve potential vorticity. Hence, for the upper ocean, there is downwelling on the cyclonic side and upwelling on the anticyclonic side of a front (Fig. 2.16b). This frontal circulation is discussed further by Woods (1988), as well as for observed ocean case studies by Voorhis and Bruce (1982), and Pollard and Regier (1992).

2.5.2 Local Response to Planetary Waves, Eddies and Fronts

2.5.2.1 *Rectified Transfer of Nutrients into the Euphotic Zone*

A time-varying flow can provide a rectified transfer of nutrients into the euphotic zone through the asymmetrical response of the ecosystem. Over the upper ocean, there is a usually an increase in nutrient concentrations with depth and, in regions such as the subtropical gyres, the euphotic zone is depleted in nutrients by biological export. Undulating motions associated with the time-

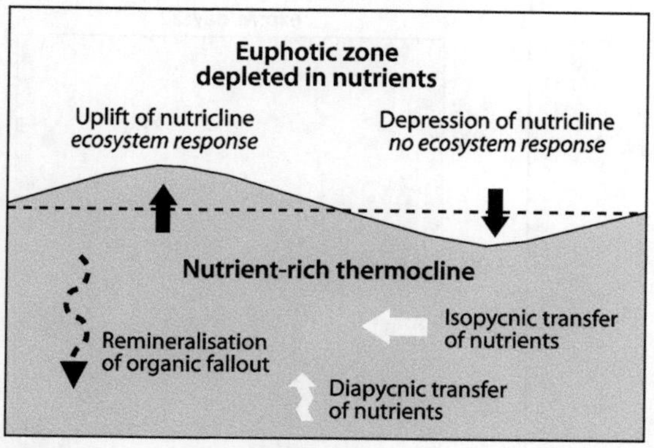

Fig. 2.17. Schematic figure depicting the ecosystem response to an uplift and depression of the nutricline. When nutrient-rich isopycnals are raised into the euphotic zone, there is biological production. Conversely, when the nutrient-rich isopycnals are pushed into the dark interior, there is no biological response. In order for the transient upwelling to persist, there needs to a process maintaining the nutrient concentrations in the thermocline, which might be achieved by remineralisation of organic fallout, diapycnal transfer or a lateral influx of nutrients from the time-mean or time-varying circulations. The schematic figure is generalised from that of McGillicuddy and Robinson (1997)

varying flow can lift nutrient-rich isopycnal surfaces into and out of the euphotic zone. When the isopycnal surface is lifted into the euphotic zone, their illumination may result in photosynthesis and the production of organic matter. In contrast, there is no biological response when the isopycnal surface with nutrient-rich waters is pushed out of the euphotic zone. Consequently, a net biological production can occur as aresult of the time-varying flow[7]. This rectification occurs as long as there is sufficient time for the phytoplankton, which have a doubling timescale of typically one day, to respond to an increased nutrient supply, and provided that all the necessary trace elements are available. This process is depicted in Fig. 2.17; the figure is a generalised version of the schematic by McGillicuddy and Robinson (1997) to include horizontal, as well as vertical transfer. Examples of elevated nutriclines and enhanced biological production in cyclonic eddies are provided by Falkowski et al. (1991) and McGillicuddy and Robinson (1997).

2.5.2.2 *Planetary Wave Signals*

The passage of planetary waves and tropical waves might induce rectified upwelling of nutrients. Uz et al. (2001) and Cipollini et al. (2001) identify signatures of planetary waves in surface chlorophyll observations at latitudes less than 40°. There is a westward propagation of surface chlorophyll anomalies and, more importantly, a more rapid propagation towards the equator. These features are broadly consistent with wave theory; e.g. see a review by Killworth (2001). However, it is presently unclear how these chlorophyll signals should be interpreted: the planetary waves might induce enhanced biological production through the rectified upwelling of nutrients into the euphotic zone or the signals might simply be due to a vertical or horizontal advection of existing chlorophyll anomalies.

2.5.2.3 *Mesoscale-Eddy Signals from Baroclinic Instability*

Baroclinic instability leads to the formation of cyclonic eddies with a raised thermocline and anticyclonic eddies with a depressed thermocline. *If* the nutricline and thermocline are coincident, then enhanced production is expected in cyclonic eddies (in accord with the schematic in Fig. 2.17).

The raised thermocline and associated nutricline in a cold-core, cyclone formed by baroclinic instability is *not* due to a simple vertical transfer, since cold fluid sinks and warm fluid rises in slantwise exchange (Fig. 2.16a). Instead in the background environment, the thermocline and associated nutricline have to be raised on the cold side of a frontal zone and depressed on the warm side. Consequently, cyclones acquire their signature of a raised thermocline and nutricline through their *horizontal* movement to a new warmer environment. Likewise, the depressed thermocline and nutricline in an anticyclone is due to its movement into a new colder environment.

Lévy et al. (2001) conduct a careful model study of the instability of a jet, examining where fluid upwells and identifying the biogeochemical response for a range of model resolutions. In their study, they initialise the model with a flat nutricline and integrate for 24 days. For a mesoscale resolution of 6 km, they find enhanced new production occurring preferentially in an anticyclonic eddy and anticylonic filaments where there is upwelling (Fig. 2.18, upper panel). Here, they only find a weaker response for a cyclonic eddy.

Consequently, in our view, baroclinic instability leads to an enhancement of biological production in cyclones due to the *lateral* transfer of cold, nutrient-rich waters, rather than a vertical transfer. However, there is a range of other processes which might provide a direct vertical transfer of nutrients. If instead of baroclinic instability, the cyclones are generated by the interaction of the large-scale flow and topographic features (as speculated by Falkowski et al. 1991), then a vertical uplift of the nutricline should provide enhanced production (Fig. 2.17). There might also be eddy-eddy interactions (as speculated by McGillicuddy and Robinson 1997), which in some cases might intensify features and lead to a local uplift of the nutricline. In addition, there might be enhanced production associated with the vertical contributions from smaller-scale, frontal upwelling (discussed subsequently).

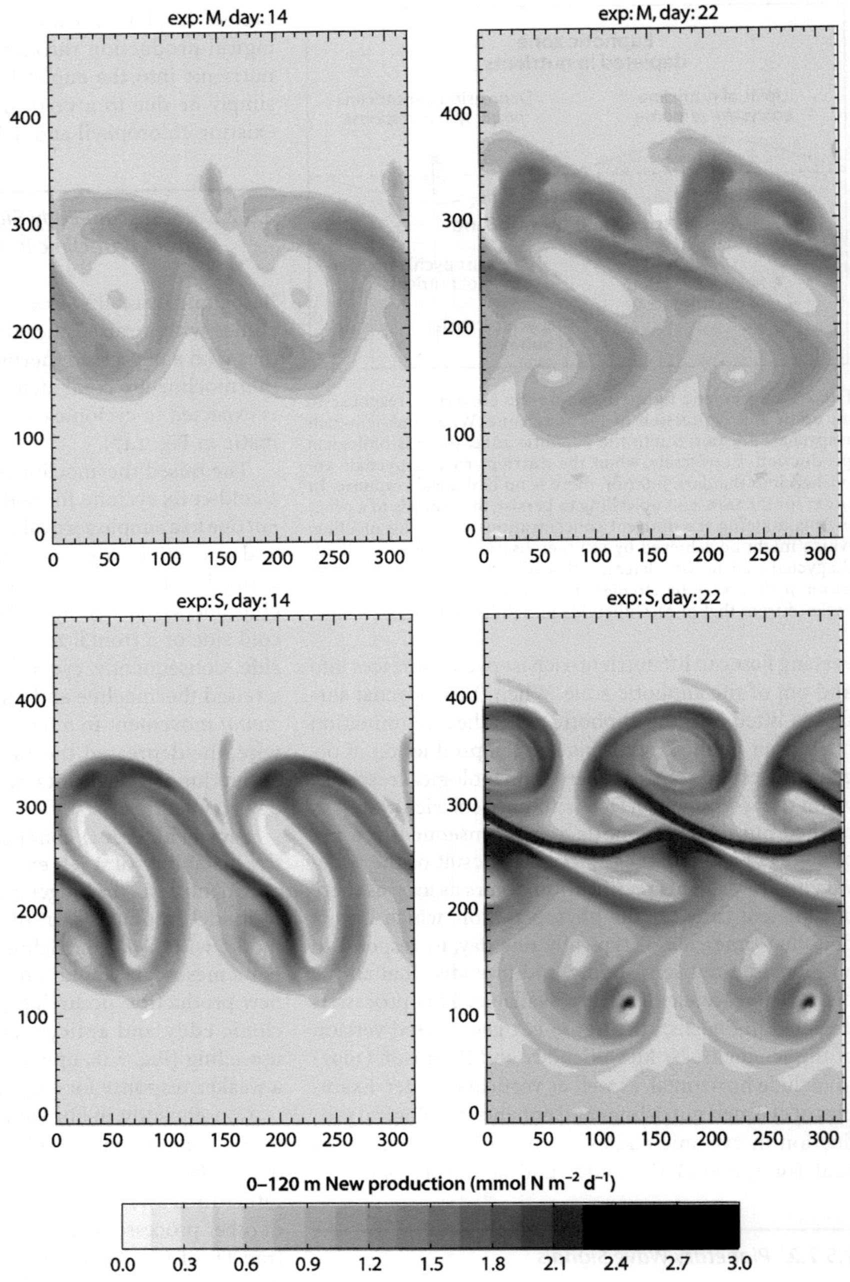

Fig. 2.18. Modelled new production (10^{-3} mol N m^{-2}) within the euphotic layer for a zonal jet undergoing baroclinic instability (Lévy et al. 2001). Snapshots of the new production are shown at days 14 and 22 for integrations at mesoscale (*upper panel*) and sub-mesoscale (*lower panel*) resolutions of 6 km and 2 km respectively. Over the integration, meanders develop leading to anticyclones to the north and cyclones to the south of the jet. The new production increases in intensity as the resolution increases and becomes concentrated along anticyclonic filaments. The area-averaged new production increases from 6.5 to 10.7 × 10^{-3} mol N m^{-2} as the resolution is increased from 6 km to 2 km; in comparison, the area-averaged new production only reaches 3.7 × 10^{-3} mol N m^{-2} if the resolution is reduced to 10 km

Eddies can also alter the ecosystem by changing the mixed-layer structure and hence altering the light limitation for phytoplankton. Baroclinic slumping of fronts leads to an increase in the stratification and hence a reduction in the thickness of the mixed layer; see modeling studies by Visbeck et al. (1996), Lévy et al. (1998) and Nurser and Zhang (2000). This eddy-induced shallowing can then lead to an onset of a spring bloom prior to the buoyancy input from the atmosphere (Lévy et al. 1998).

2.5.2.4 *Frontal-Scale Signals*

Frontal-scale circulations can also lead to a rectified supply of nutrients to the euphotic zone, since there is upwelling on the anticyclonic side and downwelling on the cyclonic side of an accelerating jet. The observational support for this process acting in the ocean is rather indirect, although maxima in surface chlorophyll have been observed on the anticyclonic side of fronts (Strass 1992).

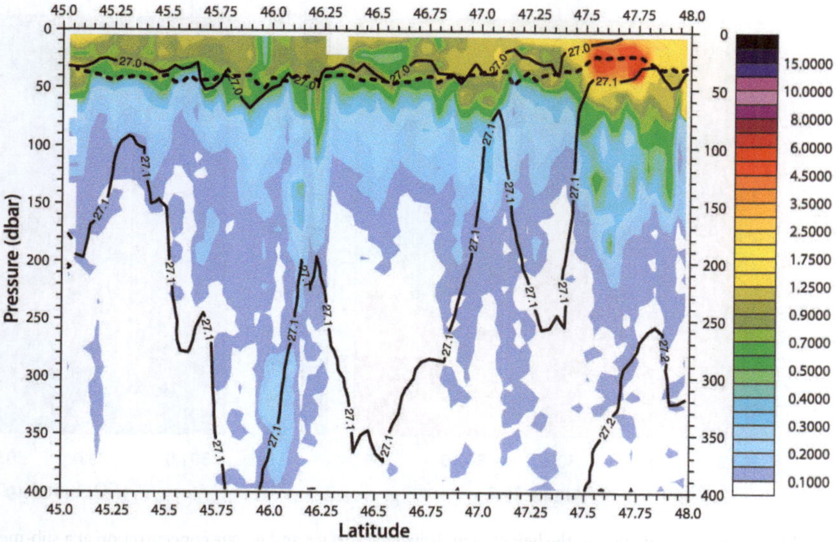

Fig. 2.19. A vertical section of chlorophyll concentration from the Vivaldi cruise through the eastern North Atlantic in spring 1991. The filaments of chlorophyll extend below the euphotic zone and mixed layer to a depth of up to 400 m and so cannot be formed by convection. Instead the filaments might be formed through frontal subduction occurring on a horizontal scale of 10 km (reprinted from Nurser AJG and Zhang JW, J. Geophys. Res. 105, 21851–21868, 2000, Copyright by the American Geophysical Union)

In addition, plumes of short-lived chlorophyll have been observed penetrating up to 400 m in the stratified thermocline on a horizontal scale of several tens of kilometres (Fig. 2.19). These chlorophyll plumes cannot be formed by convection as they penetrate below the mixed layer. Instead they are consistent with frontal-scale subduction occurring on the cyclonic side of an accelerating front (Fig. 2.16b). This response is analogous to frontal subduction in the atmosphere (Follows and Marshall 1994): ozone-rich stratospheric air is transferred into the weakly stratified, troposphere and assimilated into the troposphere through diabatic forcing (Danielson 1968).

Modeling studies suggest that there is a marked increase in biological activity when frontal circulations are resolved (e.g. Flierl and Davis 1993; Spall and Richards 2000; Mahadevan and Archer 2000; Lévy et al. 2001). The enhancement in biological production is more pronounced at the frontal scale than the ocean eddy scale. For example, the idealised experiments by Lévy et al. (2001) reveal increased levels of new production (and phytoplankton concentration) when the model resolution is increased from 6–2 km (compare upper and lower panels respectively in Fig. 2.18). In the higher-resolution integration, the maxima in new production becomes concentrated at the edges of the anticyclonic eddy, in the anticyclonic area surrounding the cyclonic eddy and within narrow anticyclonic filaments. This response is consistent with the frontal dynamics discussed in relation to Fig. 2.16b. In addition, in a more realistic environment in the vicinity of BATS, Mahadevan and Archer (2000) also demonstrate that there is enhanced nutrient supply to the euphotic zone as the sub-mesoscale is resolved for integrations lasting 120 days (Fig. 2.20).

Specific questions remain though as to how the frontal response alters the productivity over a larger basin scale and the extent that the initial enhancement is sustained over many eddy lifetimes?

2.5.2.5 *Maintenance of Rectified Supply of Nutrients*

The rectified supply of nutrients to the euphotic zone by the time-varying circulations is an appealing mechanism to sustain the export productivity of oligotrophic gyres. Estimates of the eddy supply of nitrate in the Sargasso Sea, based on a combination of hydrographic observations and satellite altimetry, typically reach 0.19 ±0.10 mol N $m^{-2} yr^{-1}$ (McGillicuddy et al. 1998) and 0.24 ±0.10 mol N $m^{-2} yr^{-1}$ (Siegel et al. 1999). These estimates are significant and might help bring estimates of nitrate supply into consistency with estimates of export production inferred from oxygen utilisation and 'age' tracers (Jenkins 1982).

If the rectified supply of nutrients is indeed a dominant process and involves a vertical transfer (irrespective of whether it is due to planetary waves, mesoscale eddies or sub-mesoscale fronts), then this raises the question of how the underlying nutrients in the thermocline are maintained? If the nutrients in the thermocline are *not* maintained, then any rectified supply will only provide an *initial* enhancement in productivity, which will gradually decreases in time until an equilibrium state is reached with lower productivity. This issue is directly analogous to the limited role convection plays in sustaining biological production over several years; see discussion related to Eq. 2.1 to 2.3.

McGillicuddy and Robinson (1997) speculate that an eddy-induced, upwards transport of inorganic nutrients is balanced locally by remineralisation of the downwards flux of sinking organic matter over eddy lifetimes lasting typically a year. However, this local vertical balance cannot hold on longer timescales if any subsequent remineralisation of organic matters occurs at depths greater than the vertical scale for the eddy displacement

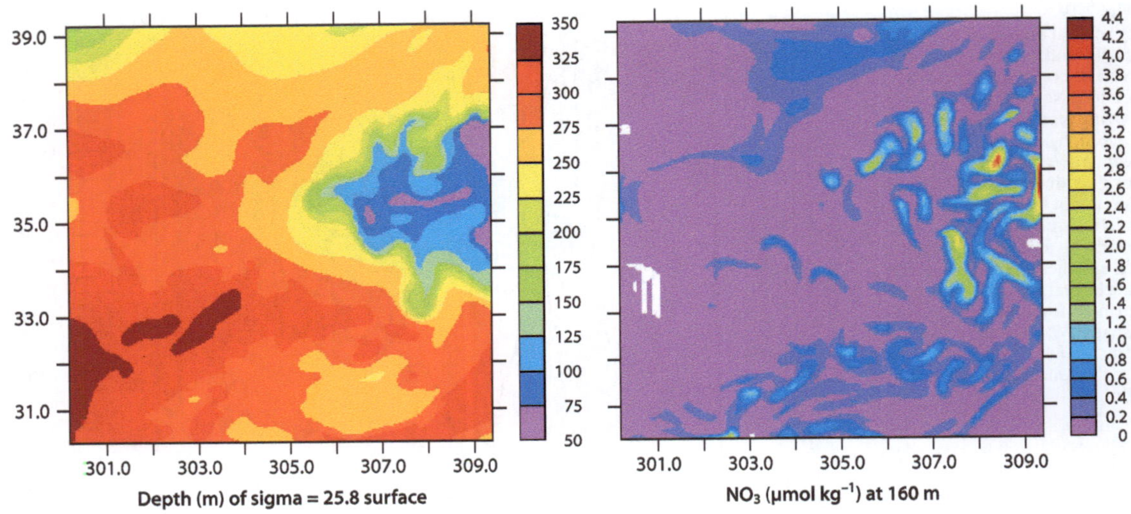

Fig. 2.20. A model simulation of the height of an isopycnal surface and nitrate concentration at a sub-mesoscale resolution of 0.1° integrated at 30–40° N, 50–60° W in the vicinity of the BATS site (Mahadevan and Archer 2000). The model is initialised and forced at the boundaries using output from a global circulation model, with nitrate initialised with data from the BATS, and the regional model integrated for 3 months. The simulation includes folds and frontal-scale undulations in the isopycnal surface. The modelled nitrate concentration reaches a maximum value of 4.3 μmol kg^{-1} at a resolution of 0.1°, which decreases to 1.4 μmol kg^{-1} when the resolution is reduced to 0.4° (not shown here). Hence, this model study suggests that the vertical transport of nitrate takes place primarily at frontal scales, rather than through larger mesoscale features (reprinted from Mahadevan A and Archer D, J. Geophys. Res 105, 1209–1225, 2000, Copyright by the American Geophysical Union)

of isopycnals. In our view, the concentration of nutrients in the upper thermocline are maintained through the *lateral* transport and diffusion of nutrients along isopycnals from surrounding nutrient-rich regions, such as the tropics or subpolar gyre, together with contributions from upwards diapycnal transfer and remineralisation of organic fallout (Fig. 2.17).

2.5.3 Far Field Effects: Eddy Transport and Diffusion

While eddies may provide important *local* impacts on nutrient supply and the ecosystem, they also provide a far field effect. Eddies systematically transfer heat, tracers and nutrients laterally along isopycnals. This larger-scale transfer of tracers consists of a diffusion and a rectified advection along isopycnals (e.g. see Andrews et al. 1987; Gent et al. 1995). The diffusion always acts to transfer tracers down gradient, whereas the rectified advection can lead to an up or down-gradient transfer of tracers.

2.5.3.1 *Eddy Transport of Tracers*

Eddies induce a transport through a correlation in the velocity, v, and vertical spacing between isopycnals, h; the eddy transport velocity or 'bolus' velocity is given by $v^* = \overline{v'h'} / \overline{h}$ where the overbar represents a temporal average and a prime represents a deviation from the temporal average. In the idealised example depicted in Fig. 2.21a, there is a greater volume flux directed to the

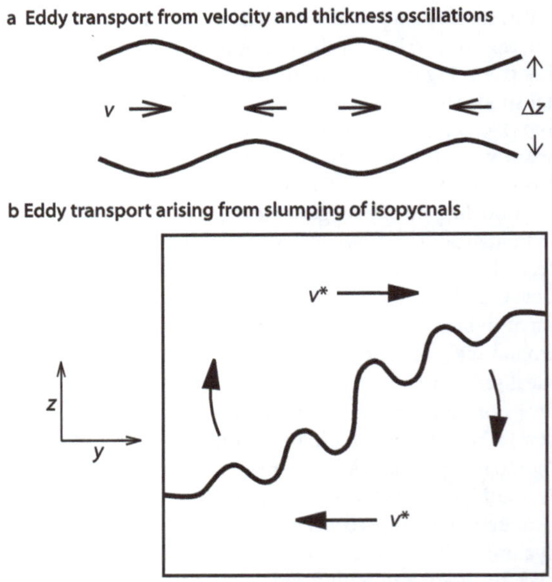

Fig. 2.21. Schematic figure of the eddy-induced transport or 'bolus' velocity, $v^* = \overline{v'h'} / \overline{h}$, for **a** a single layer and **b** a meridional section. The transport arises through a temporal correlation in the velocity, v, and vertical spacing between isopycnals, h, where a prime denotes an eddy deviation and an overbar represents a time average over many eddy events. In **a**, consider an isopycnic layer with no time-mean flow and a layer thickness and velocity oscillating in time. There is a rectified transport whenever the oscillating velocity is correlated with the layer thickness, which is directed to the right as drawn here. In **b**, the eddy-induced advection occurs when there is slumping of the interface, which is usually associated with baroclinic instability (reproduced from Lee et al. (1997))

right than to the left due to the temporal correlation between v and h. The slumping of isopycnals in baroclinic instability generates this rectified transport. For a polewards shoaling of isopycnals, as sketched in Fig. 2.21b, the eddy transport is directed polewards in the surface layer and equatorwards at depth. Conversely, for an equatorwards shoaling of isopycnals, such as on the equatorial side of a subtropical gyre, the eddy transport is directed equatorwards in the surface layer and polewards at depth.

The spreading of a tracer is initially controlled by down-gradient diffusion, but eventually on larger space scales can be controlled by advection[(8)]; this result is illustrated for idealised tracers and nutrients in eddy-resolving experiments by Lee et al. (1997) and Lee and Williams (2000). The time-varying circulation can lead to different tracer distributions according to whether eddy-induced transport and diffusion either reinforce or oppose each other. The eddy-induced transport is particularly important in controlling the spreading of water masses over the large scale of the Southern Ocean (Danabasoglu et al. 1994; Marshall 1997), as well as across intense and unstable currents, such as the Gulf Stream, and inter-gyre boundaries.

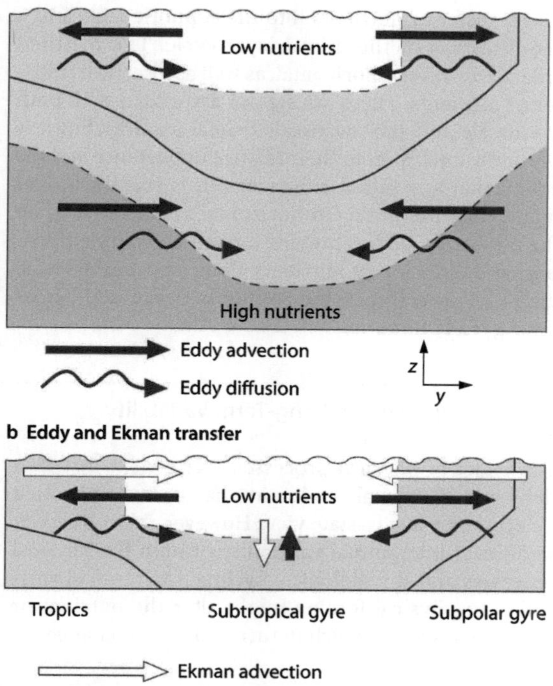

Fig. 2.22. Schematic figure of the eddy-induced nutrient transfer for a subtropical gyre. In **a**, the eddy-induced advection (*black straight arrows*) and diffusion (*curly arrows*) oppose each other at the surface, but reinforces each other at depth. In **b**, the Ekman advection (*white arrow*) is included and dominates over the opposing, eddy-induced advection. Hence, the combination of the eddy and *Ekman transfer* leads to an influx of nutrients into the interior of the subtropical gyre. The same balance should occur over the Southern Ocean where at the surface, there should be a northwards Ekman flux and eddy diffusive transfer of nutrients, which is partially opposed by a southwards eddy-induced advective transfer of nutrients (consistent with the left side of this schematic)

2.5.3.2 *Impact of Eddies on Export Production over the Basin Scale*

On the basin scale, eddies lead to a lateral transport and down-gradient diffusion of nutrients and tracers along isopycnals, which may be important in determining regional nutrient budgets. Eddy transport acts to flatten isopycnals and stratify the water column (Fig. 2.21). Consequently, the eddy transport will only increase biological production *if* the flattening of isopycnals brings nutrients to the surface. This possibility is unlikely to occur over the basin scale, since inorganic nutrient concentrations generally increase with density over the upper ocean (as revealed by gyre-scale undulations of the nutricline reflecting those of the pycnocline).

For a subtropical gyre, eddy diffusion and Ekman transfer should supply nutrients to the euphotic zone (Fig. 2.22, black curly and white arrows respectively). Conversely, the eddy transport should *reduce* the nutrient supply to the euphotic zone and inhibit biological production (Fig. 2.22, black straight arrows), while enhancing the lateral influx of nutrients at depth, and help to maintain nutrient concentrations within the thermocline.

For the North Atlantic, the eddy enhancement of export production over the basin scale has been examined over the basin scale using a simplified ecosystem model coupled with eddy-permitting circulation models at 1/3° (Oschlies and Garçon 1998; Garçon et al. 2001) and 1/9° horizontal resolution (Oschlies 2002). Resolving (at least partially) the eddy mesoscale leads only to a modest enhancement in export production of typically 1/3 over the subtropical gyre. This enhancement is principally achieved through an eddy vertical transfer around the Gulf Stream and an eddy horizontal transfer along the flanks of the subtropical gyre (Oschlies 2002). The inclusion of finer eddy scales also leads to a shallowing of the mixed layer, which reduces the convective supply of nutrients. Further modeling studies are needed to identify how the response over the basin scale alters as submesoscale fronts are fully resolved.

2.5.4 Summary

Fine-resolution observations and modeling studies suggest that nutrient supply by the time-varying circulation supports a significant fraction of biological production, although basin-scale integral estimates of their contribution are difficult to obtain. The time-varying circulation modifies the nutrient distributions and biological production on a local scale through a rectified transfer of nutrients into the euphotic zone and a modification of the mixed-layer cycle. The rectified transfer involves a horizontal, as well as a vertical transfer of nutrients. There are signals associated with both mesoscale and sub-mesoscale frontal features, but it is presently unclear as to their relative importance and the extent that any initial enhancement is sustained over many eddy lifetimes. On horizontal scales larger than the mesoscale, eddies provide a rectified volume transport and diffusion of nutrients along isopycnals, which might be crucial over the Southern Ocean and across inter-gyre boundaries.

2.6 Interannual and Long-Term Variability

The effect of physical processes on nutrient distributions and biological production has been emphasised in terms of a steady-state view. However, there is clearly significant interannual variability of both the physical processes and the biological cycling. Interannual variability in physical forcing might alter the nutrient or trace metal supply, which in turn can lead to changes in biological production or in the ecosystem through shifts in community structure. However, other ecosystem changes might occur independently of the physical forcing, for example, from result of complex, non-linear interactions in the ecosystem.

2.6.1 Coupled Atmosphere-Ocean Changes: ENSO

The most striking pattern of interannual climate change is El Niño-Southern Oscillation (ENSO) of the tropical Pacific which has far reaching climatic effects around the globe; see reviews by Philander (1990) and Godfrey et al. (2001). This coupled ocean-atmosphere phenomenon, ENSO, is characterised in the oceans by a reduction of the normal Eastern Pacific tropical upwelling related to changes in wind patterns, together with the formation of an anomalously warm surface layer and deep thermocline in the region. These physical changes have a significant impact on the biogeochemical system of the Pacific Ocean and beyond. For example, Fig. 2.23 illustrates the strong contrast in the standing stock of Equatorial Pacific chlorophyll between the January 1998 (El Niño) and January 1999 as observed from the SeaWiFS remote platform.

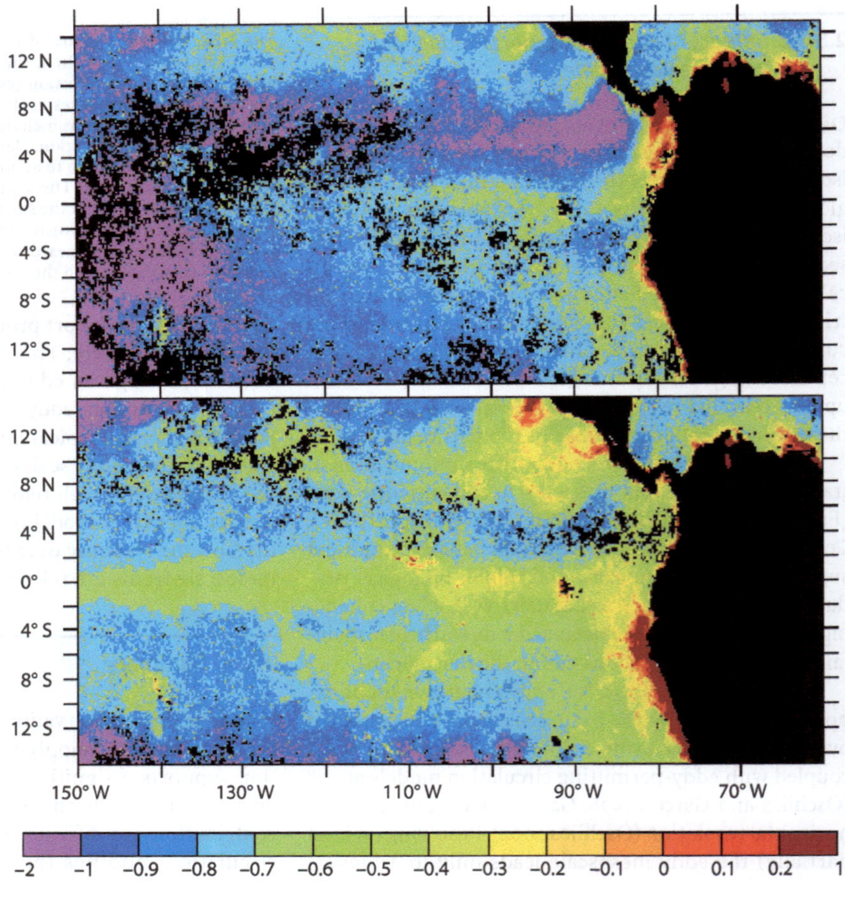

Fig. 2.23. December–January composites of Eastern Pacific chlorophyll derived from SeaWiFS ocean colour observations (level 3 processing). The *upper panel* is for 1997–1998 (during a strong El Niño event), and the *lower panel* is for 1998–1999. Note the suppressed chorophyll concentrations along the equator during the El Niño, corresponding to reduced equatorial and coastal upwelling, a warm sea-surface temperature anomaly, and depressed thermocline and nutricline (data courtesy of NASA)

In 'normal' periods the surface waters of the eastern Equatorial Pacific are relatively cool and rich in macro-nutrients, sustained by a narrow band of equatorial and coastal upwelling of nutrient bearing waters, and relatively inefficient biological export. Consequently, there is typically a band of enhanced chlorophyll associated with the upwelling region (Fig. 2.23, *lower panel*). Modeling studies have emphasised the important role of the Equatorial Undercurrent (EUC) which advects cool, oxygenated waters eastwards at depths of a few tens to hundreds of metres as part of the closure of the wind-driven upper ocean circulation. The EUC is actually depleted in nutrients relative to the adjacent waters on the isopycnic surfaces, but enriched relative to the equatorial surface. Vertical mixing brings nutrients from the EUC to the surface, sustaining local productivity (Toggweiler and Carson 1995; Chai et al. 1996). The relative inefficiency of the biological drawdown of macro-nutrients in the region, and the presence of an HNLC (High Nutrient Low Chlorophyll) regime, may be due to iron limitation. The EUC is also a source of iron to the surface Equatorial Pacific (Coale et al. 1996), although it remains depleted in iron relative to the macro-nutrients. Due to its complex biogeochemical cycling and significant atmospheric and geothermal sources (Christian et al. 2002), the role of iron is, as yet, not fully understood.

During an El Niño event, the easterly Trade Winds are reduced in strength with the equatorial and coastal upwelling weakened or suppressed; the thermocline, nutricline and EUC become anomalously deep. Consequently, the supply of macro-nutrients and iron to the surface from the upwelling and EUC is reduced. In the 1997/1998 El Niño event, even the macro-nutrients eventually became depleted in the region and the chlorophyll concentrations became very low (Fig. 2.23, *upper panel*) (Chavez et al. 1998; Chavez et al. 1999; Murtugudde et al. 1999).

El Niño events occur at irregular intervals of about three to seven years presently, but their frequency and intensity varies on decadal and longer timescales. This longer-term variability may have consequences for nutrient supply and community structure in the Pacific Basin. In this context, there may have been an abrupt shift in the ecosystem of the North Pacific subtropical gyre, during the past 30 years, which has shifted in favour of nitrogen-fixing micro-organisms (Karl et al. 1995; Karl 1999). It is speculated that such an ecosystem shift might be a consequence of the changing physical environment, though the underlying mechanisms are not yet fully revealed.

2.6.2 North Atlantic Oscillation

There is strong interannual variability over the upper ocean of the North Atlantic, which is associated with atmospheric anomalies and changes in air-sea fluxes (Bjerknes 1964). The interannual changes in sea surface temperature (SST) are probably excited by atmospheric anomalies, since the tendency in SST anomalies correlates with latent and sensible heat flux anomalies (Cayan 1992). The dominant mode of atmospheric variability over the North Atlantic is associated with the North Atlantic Oscillation (NAO) (Hurrell 1995), where the NAO index is defined in terms of a sea-level pressure difference between Iceland and Portugal. A high NAO index correlates with enhanced winter surface heat loss and deep convective mixing over the Labrador Sea (Dickson et al. 1996). The opposite is true during periods of low NAO index with anomalous surface heat loss and enhanced convection over the Greenland Sea and Sargasso Sea.

Interannual changes in the ecosystem have been statistically related to regional climate indicators over the North Atlantic. For example, in situ observations of plankton species in the Northeast Atlantic have been correlated with the NAO index (Aebischer et al. 1990) and the position of the northern wall of the Gulf Stream (Taylor et al. 1992). However, in attempting to correlate changes in the ecosystem and physical forcing, there is an inherent problem in the shortness of the temporal records, which could lead to erroneous conclusions[(9)].

Notwithstanding this reservation, atmospheric forced changes in convection and circulation should modulate the nutrient supply to the euphotic zone, as well as the irradiance phytoplankton receive in spring. Consequently, there are likely to be interannual changes in new and primary production. Variability in local winter mixing has been observed to alter nutrient supply and primary production in the Sargasso Sea (Menzel and Ryther 1961). Interannual variability is apparent in the mixed-layer thickness, and surface nutrient and chlorophyll concentrations at BATS (Bermuda Atlantic Time-Series Station; Fig. 2.24).

Bates (2001) examines the decade long record from the BATS and finds a correlation of −0.33 between the annual primary productivity anomaly there and the NAO index. To the extent that it is significant, the negative correlation indicates the enhancement of convection and nutrient supply there associated with negative anomalies of the NAO. In addition, Williams et al. (2000) predict that much of the variability in convective nitrate supply is correlated with the NAO index over the western and central Atlantic (negative and positive correlations respectively), but not over the eastern Atlantic. Their prediction is consistent with the data analysis of Bates (2001). Their model study is based on an array of one-dimensional mixed-layer models forced by a time-series of atmospheric heat fluxes for 25 years over the North Atlantic. The modelled variability is entirely due to the local atmospheric heat fluxes. Therefore, their predicted lack of correlation with the NAO over the eastern Atlantic is due to the basin-scale mode of the NAO only representing 1/3 of the atmospheric variability. Hence, it is important when interpreting biogeochemical records that the projection of the basin-scale NAO mode is known

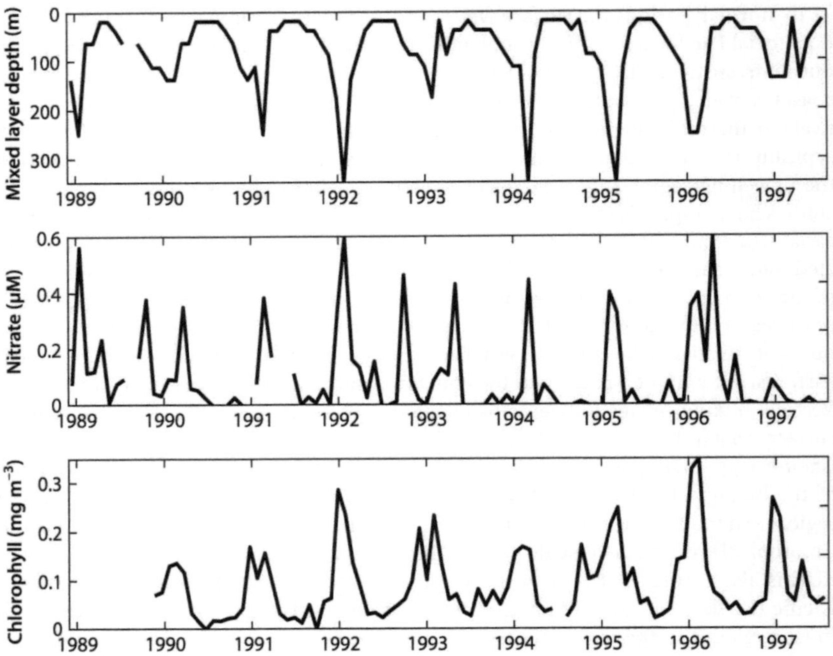

Fig. 2.24.
Time series of **a** mixed-layer thickness (m), **b** nitrate concentration (µmol kg^{-1}) and **c** chlorophyll concentration over the upper 100 m of the water column at the Bermuda Atlantic Time-Series Site (32° N, 65° W) in the North Atlantic subtropical gyre. The mixed-layer thickness is diagnosed by the depth at which the density increases by 0.125 kg m^{-3} from the surface value. The time-series illustrates the interannual variability in winter-time convection and the corresponding influence on the supply of nitrate to the euphotic zone and the response in primary production. For example, the winter-mixed layer is thin in 1990 and thick in 1992, which correlates with reduced and enhanced concentrations in nitrate and chlorophyll respectively (data from BATS)

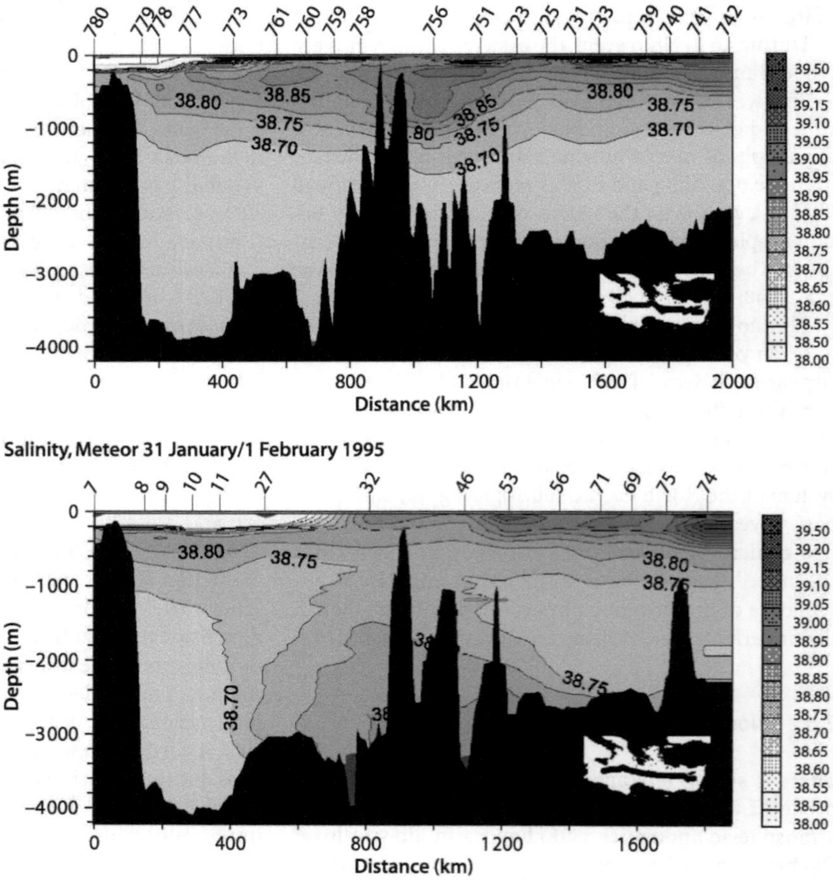

Fig. 2.25.
Deep salinity shift for 1987 (*upper panel*) and 1995 (*lower panel*) for a zonal section through the Eastern Mediterranean (see *inset*). The deep salinity is relatively uniform in 1987, but then dramatically changes by 1995 with an influx of new saline bottom water originating from the Aegean (modified from Roether et al., Recent changes in Eastern Mediterranean deep waters, Science 271, 333–335, Copyright (1996) American Association for the Advancement of Science)

for a particular region, since other more regional atmospheric modes might sometimes become important.

Interannual variations in the spring bloom of the North Atlantic are subject to the same meteorological influences that lead to regional variations illustrated and discussed in Fig. 2.8. In the subtropical gyre, much of the variability in local interannual variations of the spring bloom may be attributed to changes in meteorological forcing (Follows and Dutkiewicz 2002), although other processes, such as mesoscale motions and ecosystem changes probably become more significant at other times of year. In the subpolar gyre, the interannual variability of the spring bloom does *not* appear to exhibit any clear relationship with meteorological forcing due to the dominance of ecosystem changes or the control by mesoscale eddies of the restratification process.

2.6.3 Changes in Overturning Circulation

Major changes in the global overturning circulation might result in dramatic changes in biological production. Here, we discuss two specific examples related to changes in nutrient distributions: a present day, abrupt change over the eastern Mediterranean and a larger-scale, inferred glacial change over the North Atlantic.

2.6.3.1 *Abrupt Change in the Eastern Mediterranean*

A unique present day example of such an overturning shift has occurred in the eastern Mediterranean. There is an internal overturning cell within the eastern Mediterranean with dense waters usually formed in the Adriatic and spreading over the eastern Mediterranean. This mode of deep circulation appears to have persisted over the last century (as implied by hydrographic observations), which has led to the deep water masses being particularly homogeneous. However, between 1987 and 1995, there has been an abrupt change with the Aegean forming the most dense water mass and, temporarily, replacing the Adriatic as the source of bottom water (Roether et al. 1996). This signal is revealed by an influx of warm, salty water at depth across the eastern Mediterranean in 1995 (Fig. 2.25). The resulting uplift of the previous bottom water has moved the nutricline much closer to the euphotic zone (Fig. 2.26)

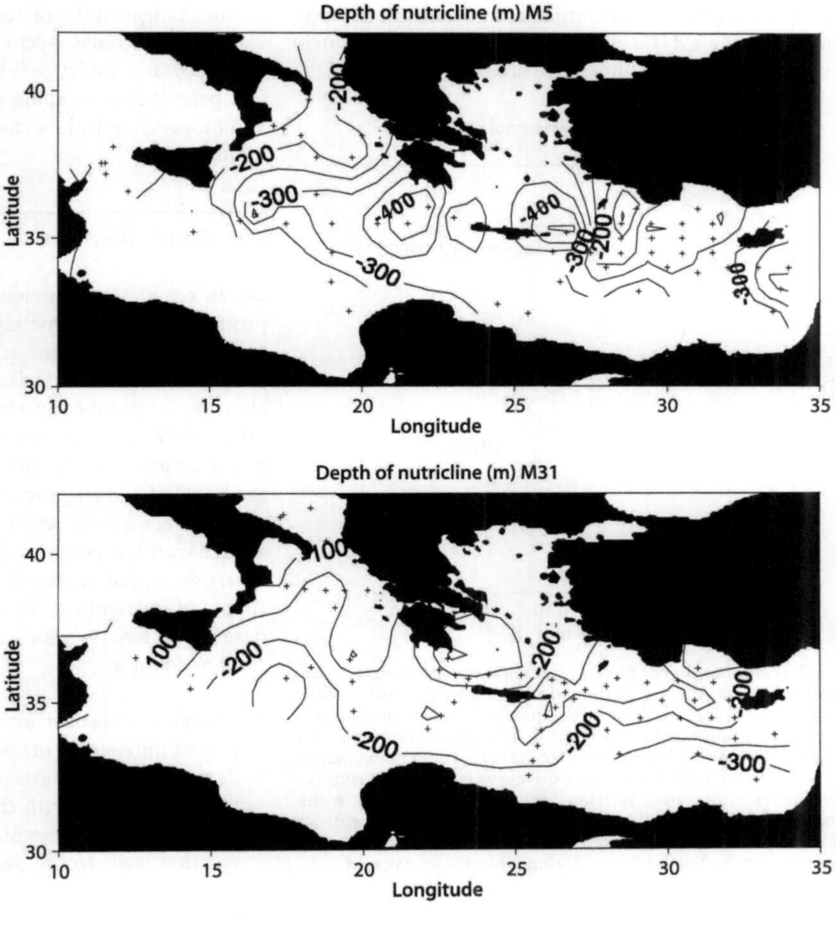

Fig. 2.26. Nutricline shift (m) for 1987 (*upper panel*) and 1995 (*lower panel*) over the Eastern Mediterranean, which corresponds to the salinity sections shown in Fig. 2.25. The nutricline is defined by the depth below the surface where the nitrate concentration reaches 3 µmol N kg^{-1}. The nutricline is typically 300 m to 400 m deep in 1987, but is uplifted to a depth of 200 m by 1995 following the influx of dense water (reprinted from Deep-Sea Res. 1, 46, Klein B et al., 371–414, Copyright (1999) with permission from Elsevier Science)

(Klein et al. 1999), which provides a 'natural fertilization experiment' for the oligotrophic surface waters. Subsequently, enhanced biological production is likely to occur in the spring following severe winters when there is increased entrainment of nutrient-rich, thermocline waters; a signal of winter-induced, interannual variability in the phytoplankton bloom is observed over the Adriatic (Gacic et al. 2002). Such a localised change might, in principle, occur elsewhere over the global ocean.

2.6.3.2 *Change in the Phosphorus Distribution over the Glacial North Atlantic*

Larger-scale shifts in the nutrient distributions, probably achieved by overturning changes, have also been suggested by the paleo-record in ocean sediments. For example, studies of the cadmium content of benthic foraminifera suggest that the ocean distribution of phosphorus has changed over glacial-interglacial timescales. Variations of cadmium around the ocean are closely correlated with the variations of phosphorus, since the two elements are assimilated into organic matter and regenerated in close accord (Boyle et al. 1976; Boyle 1988). The cadmium is assimilated into calcium carbonate shells, the Cd/Ca ratio reflecting the waters of origin and, in turn, the phosphate concentration.

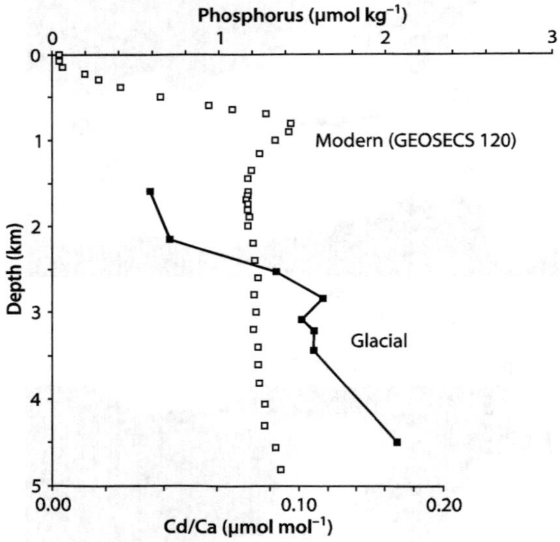

Fig. 2.27. A phosphorus profile representative of the present day, western North Atlantic compared with a glacial profile reconstructed from Cd/Ca ratios of benthic foraminifera in sediment cores (from Boyle and Keigwin 1987). The glacial reconstruction assumes that the Cd-P relationship of the glacial oceans was similar to the present day. The observed Cd-P curve is well represented as two intersecting linear relationships which is reflected in the change of scale on the Cd/Ca axis (see Boyle 1988). The reconstruction suggests a significant increase in the vertical gradient in phosphorus over the North Atlantic during the last glacial period

Boyle and Keigwin (1987) compared a phosphorus profile from the modern western North Atlantic with a reconstruction from Cd/Ca ratios in sediment cores for the last glacial period, as shown in Fig. 2.27. There appears to have been a significant shift in the phosphorus distribution over the North Atlantic with increased concentrations in the bottom waters and an enhanced vertical gradient for the last glacial period. They speculate that these changes are due to variations in the overturning circulation with an increased influx of AABW and a decreased formation of NADW occurring over the last glacial period for the North Atlantic.

2.6.4 Summary

There is marked interannual and longer-term variability in the physical forcing and biogeochemical response over the ocean. This variability can involve coupled atmosphere-ocean interactions, such as ENSO, or atmospheric-triggered changes (with probably little immediate atmospheric feedback) such as NAO. This variability is not only restricted to the upper ocean, but can involve abrupt changes in the deep ocean, as evident in recent changes in the eastern Mediterranean. All of this variability can also occur in the paleo-environment. An important challenge is to interpret both present day and paleo-oceanographic records in terms of physical and biogeochemical mechanisms in a well-constrained manner.

2.7 Conclusions

Ocean circulation, nutrient transports, and biological productivity are intimately linked over a broad spectrum of space and time scales. The physical supply of nutrients and trace metals can sometimes limit the levels of export production. However, even in regions where other processes become limiting, the physical supply of nutrients and trace metals needs to be consistent with the levels of export production for there to be a steady state. Here, we have reviewed the role of physical phenomena, on horizontal scales from global to frontal, in determining nutrient distributions and modulating the supply of nutrients to the euphotic zone over the open ocean. The key points of our review may be summarised as follows:

- The large-scale overturning circulation and its transport of nutrients is the principal physical mechanism determining the nutrient contrasts between ocean basins. Together with the biological fallout and remineralisation of organic matter, the overturning circulation leads to the deep waters of the Pacific and

Indian Oceans having much higher concentrations in inorganic nutrients than the Atlantic Ocean. Separate and independent overturning cells connect each basin with the Southern Ocean.

- Convective mixing acts to transfer nutrients from the seasonal boundary layer into the euphotic zone and is significant in modulating productivity on seasonal and interannual timescales.
- The gyre-scale circulation transfers nutrients both vertically and horizontally within ocean basins. In association with biological consumption and fallout, this advective transfer leads to the surface waters being nutrient rich over subpolar gyres and nutrient poor over subtropical gyres. The western boundary currents and lateral Ekman transfers across the inter-gyre boundaries help to maintain the nutrient concentrations within each gyre.
- Time-varying circulations may locally enhance biological production through rectified transfer of nutrients across the base of the euphotic zone. This fine-scale enhancement can occur at the ocean eddy and frontal scale, and involves both vertical and horizontal advection of nutrients. Eddies also provide a large-scale, rectified volume transport and diffusion of nutrients along isopycnals. This isopycnic transfer is particularly important in determining nutrient distributions over the Southern Ocean and in transferring nutrients across inter-gyre boundaries.
- In addition, these physical phenomena affect the inter-annual and longer-term variability of the biogeochemical system, as evident in changes associated with circulation modes: ENSO and the North Atlantic Oscillation. For example, atmospheric modulation of convection or the circulation might alter the levels of biological production or the state of the ecosystem through changes in nutrient or trace-metal supply.

While we have attempted to provide a mechanistic view of the competing physical processes for the open ocean, we are aware of several difficulties. Firstly, for clarity, we have described each process separately and independently, rather than adopt a purely regional view. However, each process is naturally connected to each other. For example, boundary currents are part of both the gyre and overturning circulations, energetic circulations contain a mixture of eddies and fronts, and atmospheric variability induces changes in convection, circulation and inputs of trace metals. Secondly, the biogeochemical response to the physical processes has only been discussed in a qualitative manner. The lack of quantitative constraints is one of the principal challenges that need to be addressed for both present day and paleo analyses. In assessing the role of different processes, a distinction needs to be made between how important a process is in perturbing the system (such as providing a short-term burst in nutrient supply to the euphotic zone) and in maintaining a long term equilibrium.

Understanding the impact of the present-day variability requires maintenance and further development of long term time-series stations in order to provide a reliable observational context. In addition, regional and global surveys of the organic forms of nutrients are required in order to quantify the fluxes and budgets of the total nutrient pools. Addressing these outstanding issues will be challenging due to the importance of rectified signals in biogeochemistry, such as those involving temporal variations in convection, the mesoscale eddy and frontal circulations. One approach is to adopt an increased use of targeted experiments, combined with models, to identify the observational signals and assess the competing processes.

Acknowledgements

We are grateful to all authors kindly providing copies of their figures, and thank Ed Boyle, Jim Christian, Harry Leach and George Nurser for their feedback. RGW is grateful for support from the UK Natural Environment Research Council (NER/O/S/2001/01245) and MJF is grateful for support from NASA grant NCC5-244.

Notes

1. The diapycnic diffusivity is typically 10^{-5} m^2 s^{-1} in the main thermocline (e.g. Ledwell et al. 1993), but reaches much higher values above rough topography (Polzin et al. 1997) and near coastal boundaries. Munk and Wunsch (1998) argue that the combination of the weak background mixing and localised enhanced mixing might lead to an effective diffusivity of 10^{-4} m^2 s^{-1} acting over the deep ocean. They suggest that this mixing is driven by the mechanical inputs of energy from the winds and tides. However, even with a higher mixing rate over the deep ocean, there is still a difficulty in explaining the nutrient transfer over the upper ocean in terms of diapycnic diffusion and vertical advection by the large-scale circulation.
2. Traditionally, frontal definitions are used to define the regions over which water masses form and spread over the Southern Ocean (Orsi et al. 1995). The surface signatures of fronts are particularly evident in maps of the gradient in sea surface temperature inferred from remotely-sensed observations (Hughes and Ash 2001). These remotely-sensed observations and general circulation model experiments suggest

though that these fronts are *not* always continuous features around the Southern Ocean, but instead merge or separate according to the topography (Pollard et al. 2002).

3. The eddy transport was first discussed in the atmosphere to explain how tracers can be transported in the opposite direction to the time-mean Eulerian circulation; e.g. Andrews et al. (1987). This eddy transport explains the polewards heat transport across the troposphere and cancellation of the 'Ferrel' cell, as well as the transport of ozone-rich, tropical air towards the high latitude, winter hemisphere in the stratosphere. For the Southern Ocean, the dominant eddy contribution across *latitude circles* occurs from standing eddies involving spatial deviations in a zonal average (Döös and Webb 1994; Karoly et al. 1997), although transient eddies control the transfer across time-mean *streamlines* (Hallberg and Gnanadesikan 2001).
4. The eastwards and northwards Ekman volume fluxes are defined by $U = \tau_y / (\varrho f)$ and $V = -\tau_x / (\varrho f)$ respectively; here τ_x and τ_y are the eastwards and northwards components of the wind stress, ϱ is the density and f is the Coriolis parameter. The upwelling at the base of the surface Ekman layer is given by the convergence of the horizontal Ekman flux or equivalently the curl of the wind stress divided by the Coriolis parameter.
5. Model estimates of the lifetime of semi-labile, dissolved organic nitrogen (DON) range from several months in the surface to years in the main thermocline. For example, Anderson and Williams (1999) model the cycling of organic matter and obtain DON lifetimes of 0.4 years at the surface and increasing to 6 years at 1 000 m as the concentration of bacteria decreases with depth.
6. Along a front, there is a balance between the acceleration of the along front velocity, Du/Dt and the cross-frontal Coriolis acceleration, $-fv$, where

$$\frac{Du}{Dt} - fv = 0$$

where, u and v are the velocities aligned along and across the front, f is the Coriolis parameter and D/Dt is the rate of change following a fluid parcel (Eliassen 1962). The secondary circulation associated with the cross-frontal flow satisfies

$$\frac{\partial v}{\partial y} + \frac{\partial w}{\partial z} = 0$$

where w is the vertical velocity. Thus, for an accelerating eastwards jet, $Du/Dt > 0$, then the secondary circulation is with a polewards surface flow ($v > 0$), inducing downwelling ($w < 0$) on the polewards, cyclonic side of the front and upwelling ($w > 0$) on the equatorwards, anticyclonic side of the front; see Newton (1978) for a discussion of both ocean and atmospheric cases.

7. The asymmetrical response of the ecosystem to transient upwelling is directly analogous to the interaction of a time-varying circulation and photochemistry in the stratosphere. Ozone-rich air in the stratosphere survives when transferred towards the darker, winter pole and is destroyed when transferred to the sunlit equator. This asymmetry leads to a rectified transport of ozone towards the winter pole and maximum concentration in total ozone occurring away from its photochemical source.
8. Consider a patch of dye spreading from a localised source through advection and diffusion. Scale analysis suggests that the dye spreads advectively over a spatial scale

$$L_{adv} \sim Vt$$

where V is the transport velocity including the rectification of the time-varying flow from the correlation of the velocity and layer thickness, and t is time. The dye will also spread diffusively over a spatial scale,

$$L_{dif} \sim (Kt)^{1/2}$$

where K is the eddy-induced diffusivity of tracer. The ratio of these length scales is

$$\frac{L_{adv}}{L_{dif}} \approx \frac{V}{K^{1/2}} t^{1/2}$$

Thus, for non-zero V and K, the initial spreading will be diffusive, since the ratio of L_{adv}/L_{dif} is small. Over longer timescales, however, advection becomes important, as the horizontal scales inflate, and may eventually dominate over diffusion. For example, assuming $K \sim 1\,000\ m^2\,s^{-1}$ and $V \sim 1\ cm\,s^{-1}$ suggests that advection and diffusion becomes comparable after a timescale of order several years.

9. The issue of misleading correlations arising from the shortness of the time record compared with the period of any oscillations is discussed by Wunsch (1999) in terms of the North Atlantic and Southern Oscillations. In addition, Wunsch (2001) reports on a climate example where the water level of Central African lakes were thought to be correlated with monthly sunspot numbers based on a 20 year record. However, this correlation turned out to be erroneous when the lake record was extended to 70 years and included more cycles in the sunspot numbers.

References

Abell J, Emerson S, Renaud P (2000) Distribution of TOP, TON, and TOC in the North Pacific subtropical gyre: implications for nutrient supply in the surface ocean and remineralisation in the upper thermocline. J Mar Res 58:203–222

Aebischer NJ, Coulson JC, Colebrook JM (1990) Parallel long-term trends across four marine trophic levels and weather. Nature 347:753–755

Anderson TR, Williams PJL (1999) A one-dimensional model of dissolved organic carbon cycling in the water column incorporating combined biological-photochemical dependence. Global Biogeochem Cy 13:337–349

Andrews DJ, Holton JR, Leovy CB (1987) Middle atmosphere dynamics. Academic Press, 489 pp

Archer D, Johnson EK (2000) A model of the iron cycle in the ocean. Global Biogeochem Cy 14:269–279

Backhaus JO, Wehde H, Hegseth EN, Kampf J (1999) 'Phyto-convection': the role of oceanic convection on primary production. Mar Ecol Prog Ser 189:77–92

Barber RT (1992) Geologic and climatic time scales of nutrient variability. In: Falkowski PG, Woodhead AD (eds) Primary productivity and biogeochemical cycles in the sea. Plenum Press, New York, pp 89–106

Bates N (2001) Interannual variability of oceanic CO_2 and biogeochemical properties in the Western North Atlantic Subtropical Gyre. Deep-Sea Res Pt II 48:1507–1528

Behrenfeld MJ, Falkowski PG (1997) Photosynthetic rates derived from satellite-based chlorophyll concentration. Limnol Oceanogr 42:1–20

Bjerknes J (1964) Atlantic air-sea interaction. Adv Geophys, 20, 1–82

Boyle EA (1988) Cadmium: chemical tracer of deepwater paleoceanography. Paleoceanography 3:471–489

Boyle EA, Keigwin L (1987) North Atlantic thermohaline circulation during the past 20 000 years linked to high-latitude surface temperature. Nature 330:35–40

Boyle EA, Sclater FR, Edmond JM (1976) On the marine geochemistry of cadmium. Nature 263:42–44

Broecker WS (1991) The great ocean conveyor. Oceanography 4: 79–89

Cayan DR (1992) Latent and sensible heat flux anomalies over the Northern Oceans: driving the sea surface temperature. J Phys Oceanogr 22:859–881

Chai F, Lindley ST, Barber RT (1996) Origin and maintenance of a high nitrate condition in the Equatorial Pacific. Deep-Sea Res Pt II 43:1031–1064

Chavez FP, Strutton PG, McPhaden MJ (1998) Biological-physical coupling in the central equatorial Pacific during the onset of the 1997–98 El Niño. Geophys Res Lett 25:3543–3546

Chavez FP, Strutton PG, Friedrich GE, Feely RA, Feldman GC, Foley DG, McFaden MJ (1999) Biological and chemical response of the Equatorial Pacific Ocean to the 1997–98 El Niño. Science 286:2126–2131

Christian JR, Verschell MA, Murtugudde R, Busalacchi AJ, McClain CR (2002) Biogeochemical modelling of the tropical Pacific Ocean. II. Iron biogoechemistry. Deep-Sea Res Pt II 49:545–565

Cipollini P, Cromwell D, Challenor PG, Raffaglio S (2001) Rossby waves detected in global ocean colour data. Geophys Res Lett 28:323–326

Coale KH, Fitzwater SE, Gordon RM, Johnson KS, RT Barber (1996) Control of community growth and export production by upwelled iron in the equatorial Pacific Ocean. Nature 379:621–624

Danabasoglu G, McWilliams JC, Gent PR (1994) The role of mesoscale tracer transports in the global ocean circulation. Science 264:1123–1126

Danielsen EF (1968) Stratospheric-tropospheric exchange based on radioactivity, ozone and potential vorticity. J Atmos Sci, 25, 502–518

Denman KL, Gargett AE (1995) Biological-physical interactions in the upper ocean: the role of vertical and small scale transport processes. Annu Rev Fluid Mech 27:225–255

Dickson R, Lazier J, Meinke J, Rhines P, Swift J (1996) Long-term coordinated changes in convective activity of the North Atlantic. Prog Oceanogr 38:241–295

Döös K, Webb DJ (1994) The Deacon cell and the other meridional cells of the Southern Ocean. J Phys Oceanogr 24:429–442

Duarte CM, Agusti S (1998) The CO_2 balance of unproductive aquatic ecosystems. Science 281:234–236

Dutkiewicz S, Follows M, Marshall J, Gregg WW (2001) Interannual variability of phytoplankton abundances in the North Atlantic. Deep-Sea Res Pt II 48:2323–2344

Eliassen A (1962) On the vertical circulation in frontal zones. Geofysiske Publikasjoner, Geophysica Norvegica XXIV:147–160

Emerson S, Quay P, Karl D, Winn C, Tupas L, Landry M (1997) Experimental determination of the organic carbon flux from open-ocean surface waters. Nature 389:951–954

Falkowski PG, Ziemann D, Kolber Z, Bienfang PK (1991) Role of eddy pumping in enhancing primary production in the ocean. Nature 352:55–58

Fasham MJR (1995) Variations in the seasonal cycle of biological production in subarctic oceans; a model sensitivity analysis. Deep-Sea Res Pt I 42:1111–1149

Flierl GR, Davis CS (1993) Biological effects of Gulf Stream meandering. J Mar Res 51:529–560

Follows MJ, Dutkiewicz S (2002) Meteorological modulation of the North Atlantic spring bloom. Deep-Sea Res Pt II 49: 321–344

Follows MJ, Marshall JC (1994) Eddy-driven exchange at ocean fronts. Ocean modelling 102:5–9

Gacic M, Civitarese G, Miserocchi S, Cardin V, Crise A, Mauri E (2002) The open-ocean convection in the Southern Adriatic: a controlling mechanism of the spring phytoplankton bloom. Cont Shelf Res 22:1897–1908

Ganachaud A, Wunsch C (2002) Oceanic nutrient and oxygen transports and bounds on export production during the World Ocean Circulation Experiment. Global Biogeochem Cy, vol. 16, 4:1057

Garçon VC, Oschlies A, Doney SC, McGillicuddy D, Waniek J (2001) The role of mesoscale variability on plankton dynamics in the North Atlantic. Deep-Sea Res Pt II 48:2199–2226

Gent PR, Willebrand J, McDougall TJ, McWilliams JC (1995) Parameterising eddy-induced tracer transports in ocean circulation models. J Phys Oceanogr 25:463–474

Gill AE, Green JSA, Simmons AJ (1974) Energy partition in the large-scale ocean circulation and the production of mid-ocean eddies. Deep-Sea Res 21:499–528

Gnanadesikan A (1998) A global model of silicon cycling: sensitivity to eddy parameterization and dissolution. Global Biogeochem Cy 13:199–220

Godfrey JS, Johnson GC, McPhaden MJ, Reverdin G, Wijffels SE (2001) The tropical ocean circulation. In: Siedler G, Church J, Gould J (eds) Ocean circulation and climate. Academic Press, pp 215–246

Green JSA (1981) Trough-ridge systems as slant-wise convection. In: Atkinson BW (ed) Dynamical meteorology: an introductory selection. Methuen & Co. Ltd, pp 176–194

Gruber N, Sarmiento JL (1997) Global patterns of marine nitrogen fixation and denitrification. Global Biogeochem Cy 11:235–266

Hallberg R, Gnanadesikan A (2001) An exploration of the role of transient eddies in determining the transport of a zonally reentrant current. J Phys Oceanogr 31:3312–3330

Hanawa K, Talley L (2001) Mode waters. In: Siedler G, Church J, Gould J (eds) Ocean circulation and climate. Academic Press, pp 373–386

Hughes CW, Ash E (2001) Eddy forcing of the mean flow in the Southern Ocean. J Geophys Res 106:2713–2722

Hurrell JW (1995) Decadal trends in the North Atlantic Oscillation: regional temperatures and precipitation. Science 269: 676–679

Hutchings L, Pitcher GG, Probyn TA, Bailey GW (1995) The chemical and biological consequences of coastal upwelling. In: Summerhayes CP, Emeis K-C, Angel MV, Smith RL, Zeitzchel B (eds) Upwelling in the ocean: modern processes and ancient records. John Wiley & Sons, New York, pp 65–81

Jackson GA, Williams PM (1985) Importance of dissolved organic nitrogen and phosphorus to biological nutrient cycling. Deep-Sea Res 32:223–235

Jenkins WJ (1982) Oxygen utilization rates in North Atlantic subtropical gyre and primary production in oligotrophic systems. Nature 300:246–248

Jenkins WJ (1988) Nitrate flux into the photic zone near Bermuda. Nature 331:521–523

Jenkins WJ, Goldman JC (1985) Seasonal oxygen cycling and primary production in the Sargasso Sea. J Mar Res 43:465–491

Karl DM (1999) A sea of change: biogeochemical variability in the North Pacific subtropical gyre. Ecosystems 2:181–214

Karl D, Letelier R, Hebel D, Tupas L, Dore J, Christian J, Winn C (1995) Ecosystem changes in the north Pacific subtropical gyre attributed to the 1991–92 El Niño. Nature 373:230–234

Karoly DJ, McIntosh PC, Berrisford P, McDougall TJ, Hirst AC (1997) Similarities of the Deacon cell in the Southern Ocean and Ferrel cells in the atmosphere. Q J Roy Meteor Soc 123:519–526

Killworth PD (2001) Rossby waves. In: Steele JH, Turekian KK, Thorpe SA (eds) Encyclopedia of ocean sciences. Academic Press, pp 2434–2443

Klein B, Roether WR, Manca BB, Bregant D, Beitzel V, Kovacevic V, Luchetta A (1999) The large deep water transient in the Eastern Mediterranean. Deep-Sea Res Pt I, 46, 371–414

Knap A, Jickells T, Pszenny A, Galloway J (1986) Significance of atmospheric-derived fixed nitrogen on productivity of the Sargasso Sea. Nature 320:158–160

Ledwell JR, Watson AJ, Law CS (1993) Evidence for slow mixing across the pycnocline from an open-ocean tracer-release experiment. Nature 364:701–703

Lee M-M, Williams RG (2000) The role of eddies in the isopycnic transfer of nutrients and their impact on biological production. J Mar Res 58:895–917

Lee M-M, Marshall DP, Williams RG (1997) On the eddy transfer of tracers: advective or diffusive? J Mar Res 55:483–505

Lévy M, Mémery L, Madec G (1998) The onset of a bloom after deep winter convection in the northwest Mediterranean sea: mesoscale process study with a primitive equation model. J Marine Syst 16:7–21

Lévy M, Klein P, Treguier A-M (2001) Impact of sub-mesoscale physics on production and subduction of phytoplankton in an oligotrophic regime. J Mar Res 59:535–565

Lewis MR, Harrison WG, Oakley NS, Hebert D, Platt T (1986) Vertical nitrate fluxes in the oligotrophic ocean. Science 234:870–873

Macdonald AM (1995) Oceanic fluxes of mass, heat and freshwater: a global estimate and perspective. PhD. thesis, Massachusetts Institute of Technology/Woods Hole Oceanographic Institute Joint program, Cambridge MA, 326 pp

Macdonald AM (1998) The global ocean circulation: a hydrographic estimate and regional analysis. Prog Oceanogr 41: 281–382

Macdonald AM, Wunsch C (1996) An estimate of global ocean circulation and heat fluxes. Nature 382:436–439

Mahadevan A, Archer D (2000) Modelling the impact of fronts and mesoscale circulation on the nutrient supply and biogeochemistry of the upper ocean. J Geophys Res 105:1209–1225

Marshall D (1997) Subduction of water masses in an eddying ocean. J Mar Res 55:201–222

Marshall J, Schott F (1999) Open-ocean convection: observations, theory and models. Rev Geophys 37:1, 1–64

Marshall JC, Nurser AJG, Williams RG (1993) Inferring the subduction rate and period over the North Atlantic. J Phys Oceanogr 23:1315–1329

Martin JH, Fitzwater SE (1988) Iron deficiency limits phytoplankton growth in the north-east Pacific subarctic. Nature 331: 341–343

McCartney MS (1977) Subantarctic mode water. In: Angel MV (ed) A voyage of discovery: George Deacon 70th anniversary volume. Deep-Sea Res (suppl.) 103–119

McClain CR, Firestone J (1993) An investigation of Ekman upwelling in the North Atlantic. J Geophys Res 98:12327–12339

McGillicuddy DJ, Robinson AR (1997) Eddy-induced nutrient supply and new production in the Sargasso Sea. Deep-Sea Res Pt I 44:1427–1449

McGillicuddy DJ, Robinson AR, Siegel DA, Jannasch HW, Johnson R, Dickeys T, McNeil J, Michaels AF, Knap AH (1998) New evidence for the impact of mesoscale eddies on biogeochemical cycling in the Sargasso Sea. Nature 394:263–266

McIntosh PC, McDougall TJ (1996) Isopycnal averaging and the residual mean circulation. J Phys Oceanogr 26:1655–1660

Menzel DW, Ryther JH (1961) Annual variations in primary production in the Sargasso Sea off Bermuda. Deep-Sea Res 7:282–288

Michaels AF, Knap AH, Dow RL, Gundersen K, Johnson RJ, Sorensen J, Close A, Knauer GA, Lohrenz SE, Asper VA, Tuel M, Bidigare R (1994) Seasonal patterns of ocean biogeochemistry at the U.S. JGOFS Bermuda Atlantic Time-series Study site. Deep-Sea Res Pt I 41:1013–1038

Michaels AF, Olson D, Sarmiento JL, Ammerman JW, Fanning K, Jahnke R, Knap AH, Lipschultz F, Prospero JM (1996) Inputs, losses and transformations of nitrogen and phosphorus in the pelagic North Atlantic Ocean. Biogeochemistry 35:181–226

Moore JK, Doney SC, Kleypas JA, Glover DM, Fung I (2002) An intermediate complexity marine ecosystem model for the global domain. Deep-Sea Res Pt II 49:403–462

Munk WH (1966) Abyssal recipes. Deep-Sea Res 13:707–730

Munk WH, Wunsch C (1998) Abyssal recipes II: energetics of tidal and wind mixing. Deep-Sea Res Pt I 45:1977–2010

Murtugudde RG, Signorini SR, Christian JR, Busalacchi AJ, McClain CR, Picaut J (1999) Ocean color variability of the tropical Indo-Pacific basin observed by SeaWiFS during 1997–1998. J Geophys Res 104:18351–18366

Newton CW (1978) Fronts and wave distributions in Gulf Stream and atmospheric jet stream. J Geophys Res 83, 9:4697–4706

Nurser AJG, Zhang JW (2000) Eddy-induced mixed-layer shallowing and mixed-layer/thermocline exchange. J Geophys Res 105:21851–21868

Orsi AH, Whitworth T, Nowlin WD (1995) On the meridional extent and fronts of the Antarctic Circumpolar Current. Deep-Sea Res Pt I 42:641–673

Oschlies A (2002) Can eddies make ocean deserts bloom? Global Biogeochem Cy (in press)

Oschlies A, Garçon V (1998) Eddy-induced enhancement of primary production in a model of the North Atlantic Ocean. Nature 394:266–269

Pelegri JL, Csanady GT (1991) Nutrient transport and mixing in the Gulf Stream. J Geophys Res 96:2577–2583

Pelegri JL, Csanady GT, Martins A (1996) The North Atlantic nutrient stream. J Oceanogr 52:275–299

Philander SGH (1990) El Niño, La Nina and the Southern Oscillation. Academic Press, 293 pp

Pollard RT, Regier LA (1992) Vorticity and vertical circulation at an ocean front. J Phys Oceanogr 22:609–624

Pollard RT, Lucas MI, Read JF (2002) Physical controls on biogeochemical zonation in the Southern Ocean. Deep-Sea Res Pt II 49:3289–3305

Polzin KJ, Toole JM, Ledwell JR, Schmitt RW (1997) Spatial variability of turbulent mixing in the abyssal ocean. Science 276: 93–96

Rhines PB (2001) Mesoscale Eddies. In: Steele JH, Turekian KK, Thorpe SA (eds) Encyclopedia of ocean sciences. Academic Press, pp 1717–1729

Rintoul SR, Wunsch C (1991) Mass, heat, oxygen and nutrient fluxes and budgets in the North Atlantic Ocean. Deep-Sea Res Pt I 38:S355–S377

Rintoul SR, Hughes C, Olbers D (2001) The Antarctic circumpolar current system. In: Siedler G, Church J, Gould J (eds) Ocean circulation and climate. Academic Press, pp 271–300

Roether WR, Manca BB, Klein B, Bregant D, Georgopoulos D, Beitzel V, Kovacevic V, Luchetta A (1996) Recent changes in Eastern Mediterranean deep waters. Science 271:333–335

Sathyendranathan S, Longhurst RSA, Caverhill CM, Platt T (1995) Regionally and seasonally differentiated primary production in the North Atlantic. Deep-Sea Res Pt I 42:1773–1802

Schmitz WJ (1995) On the interbasin-scale thermohaline circulation. Rev Geophys 33:151–173

Siedler G, Church J, Gould J (eds) (2001) Ocean Circulation and Climate: Observing and Modelling the Global Ocean. Academic Press, 712 pp

Siegel DA, McGillicuddy DJ, Fields EA (1999) Mesoscale eddies, satellite altimetry and new production in the Sargasso Sea. J Geophys Res 104:13359–13379
Smith RL (1995) The physical processes of coastal ocean upwelling systems. In: Summerhayes CP, Emeis K-C, Angel MV, Smith RL, Zeitzchel B (eds) Upwelling in the ocean: modern processes and ancient records. John Wiley & Sons, New York, pp 39–64
Spall SA, Richards KJ (2000) A numerical model of mesoscale frontal instabilities; plankton dynamics – I, model formulation and initial experiments. Deep-Sea Res Pt I 47:1261–1301
Speer K, Rintoul SR, Sloyan B (2000) The diabatic Deacon cell. J Phys Oceanogr 30:3212–3222
Stommel H (1979) Determination of watermass properties of water pumped down from the Ekman layer to the geostrophic flow below. P Natl Acad Sci Usa 76:3051–3055
Strass VH (1992) Chlorophyll patchiness caused by mesoscale upwelling at fronts. Deep-Sea Res 39:75–96
Strass VH, Woods JD (1991) New production in the summer revealed by the meridional slope of the deep chlorophyll maximum. Deep-Sea Res 38:35–56
Sverdrup HU (1953) On conditions of the vernal blooming of phytoplankton. Journal du conseil international pour l'exploration de la mer 18:287–295
Taylor AH, Colebrook JM, Stephens JA, Baker NG (1992) Latitudinal displacements of the Gulf Stream and the abundance of plankton in the North-east Atlantic. J Mar Biol Assoc Uk 72: 919–921
Toggweiler JR, Carson S (1995) What are upwelling systems contributing to the ocean's carbon and nutrient budgets? In: Summerhayes CP, Emeis K-C, Angel MV, Smith RL, Zeitzschel B (eds) Upwelling in the ocean: modern processes and ancient records. John Wiley & Sons, New York, pp 337–360
Toggweiler JR, Samuels B (1995) Effect of Drake Passage on the global thermohaline circulation. Deep-Sea Res Pt I 42:477–500
Townsend DW, Cammen LM, Holligan PM, Campbell DE, Pettigrew NR (1994) Causes and consequences of variability in the timing of the spring phytoplankton blooms. Deep-Sea Res Pt I 41:747–765
Uz BM, Yoder JA, Osychny V (2001) Pumping of nutrients to ocean surface waters by the action of propagating planetary waves. Nature 409:597–600
Veronis G (1973) Model of the world ocean circulation: 1. Wind-driven, two-layer. J Mar Res 31:228–288
Visbeck M, Jones H, Marshall J (1996) Dynamics of isolated convective regions in the ocean. J Phys Oceanogr 26:1721–1734
Voorhis AD, Bruce JG (1982) Small scale surface stirring and frontogenesis in the subtropical convergence of the western North Atlantic. J Mar Res 40 (suppl.) 801–821
Williams RG (2001) Ocean subduction. In: Steele JH, Turekian KK, Thorpe SA (eds) Encyclopedia of ocean sciences. Academic Press, pp 1982–1992
Williams RG, Follows MJ (1998a) The Ekman transfer of nutrients and maintenance of new production over the North Atlantic. Deep-Sea Res Pt I 45:461–489
Williams RG, Follows MJ (1998b) Eddies make ocean deserts bloom. Nature, 'News and Views' 394:228–229
Williams RG, Spall MA, Marshall JC (1995) Does Stommel's mixed-layer 'Demon' work? J Phys Oceanogr 25:3089–3102
Williams RG, McLaren A, Follows MJ (2000) Estimating the convective supply of nitrate and implied variability in export production. Global Biogeochem Cy 14:1299–1313
Woods JD (1988) Mesoscale upwelling and primary production. In: Rothschild BJ (ed) Toward a theory of biological physical interactions in the World Ocean. Kluwer, pp 7–38
Wu J, Sunda W, Boyle EA, Karl D (2000) Phosphate depletion in the western North Atlantic Ocean. Science 289:759–762
Wunsch C (1999) The interpretation of short climate records, with comments on the North Atlantic and Southern Oscillations. B Am Meteorol Soc 80:245–255
Wunsch C (2001) Global problems and global observations. In: Siedler G, Church J, Gould J (eds) Ocean circulation and climate. Academic Press, pp 47–56
Wunsch C, Hu D, Grant B (1983) Mass, salt, heat and nutrient fluxes in the South Pacific Ocean. J Phys Oceanogr 13:725–753
Zenk W (2001) Abyssal currents. In: Steele JH, Turekian KK, Thorpe SA (eds) Encyclopedia of ocean sciences. Academic Press, pp 12–27

Chapter 3

Continental Margin Exchanges

Chen-Tung Arthur Chen · Kon-Kee Liu · Robie Macdonald

3.1 Introduction

Biogeochemical processes principally occur in the upper 200 metres of the sea and are often associated with continental margins. Although the continental margins, with waters shallower than 200 m, occupy a mere 7% of the ocean surface and even less than 0.5% of the ocean volume, they still play a major role in oceanic biogeochemical cycling. Significantly higher rates of organic productivity occur, in fact, in the coastal oceans than in the open oceans because of rapid turnover and the higher supply of nutrients from upwelling and riverine inputs. Also, 8 to 30 times more organic carbon and 4 to 15 times more calcium carbonate per unit area accumulate in the coastal oceans than in the open oceans. Similarly, gas exchange fluxes of carbon and nitrogen are considerably higher in coastal waters than in the open oceans per unit area. As a result, it has been reported that around 14% of total global ocean production, along with 80–90% of new production and as much as up to 50% of denitrification takes place in the coastal oceans. The burial sites of 80% of the organic carbon derived from both oceanic processes and terrestrial sources, in excess of 50% of present day global carbonate deposition, are also located in the coastal oceans. The unburied portion of organic carbon may be respired on the shelf, thus forming a potential natural source of atmospheric carbon dioxide. However, how much is actually respired is unknown since much of this carbon is highly inert and only mixes conservatively with seawater (Smith and Mackenzie 1987; Mantoura et al. 1991; Wollast 1998).

Humans strongly interfere with the global biogeochemical cycle of carbon, nitrogen and phosphorus and this has led to substantially increased loadings of chemicals from their activities on land and in the atmosphere. The horizontal fluxes of these elements to the coastal oceans via rivers, groundwaters and the atmosphere have a strong impact on biogeochemical dynamics, cycling, and the metabolism of coastal waters. Potential feedback may result from the anthropogenic eutrophication of continental shelf areas due to increases in nutrient availability, which may then lead to a reduction of oxygen concentration in subsurface waters as a result of increased microbial respiration. In this way, denitrification and methanogenesis are promoted, resulting in the release of N_2, N_2O and CH_4. Eutrophication accelerates the cycle of organic synthesis and regeneration of nutrients other than silicate; however, calcareous shells and skeletons are preserved. On the other hand, the damming of major rivers may reduce freshwater output and the buoyancy effect on the shelves, which in turn reduces upwelling and nutrient input. Subsequently, productivity and eutrophication are diminished. These processes, however, cannot yet be accurately quantified (Walsh et al. 1985; Christensen 1994; Galloway et al. 1995; Kempe 1995).

The flux of carbon from the terrestrial biosphere to the oceans takes place via river transport. The global river discharge of carbon in both organic and inorganic form may approximate 1–1.4 Gt C yr^{-1} (Schlesinger and Melack 1981; Degens et al. 1991; Meybeck 1993). A substantial fraction of this transport (up to 0.8 Gt C yr^{-1}), however, reflects the natural geochemical cycling of carbon and does not in any way affect the global budget of anthropogenic CO_2 perturbation. Furthermore, to a large extent anthropogenically-induced river carbon fluxes are indicative of increased soil erosion rather than a removal of excess atmospheric CO_2. The above mentioned discharges of organic matter and nutrients from coastal communities do not have a significant impact on the open world oceans but can have important effects on the coastal oceans (Chen and Tsunogai 1998). Smith and Hollibaugh (1993) suggested that the observed invasion of fluxes of anthropogenic CO_2 in coastal oceans should be increased by 0.08 Gt C yr^{-1} to accurately represent the perturbed carbon fluxes because, under natural conditions, they are net sources of CO_2.

The discharge of excess nutrients by rivers may have significantly stimulated carbon fixation (up to 0.5–1 Gt C yr^{-1}). Bolin (1977) and Walsh et al. (1981) claimed that the acceleration in the amount of organic carbon storage in the coastal sediments can partially be attributed to the increased primary productivity due to large increases in anthropogenic nitrate inputs from rivers. They proposed that this process could account for the "missing billion metric tons of carbon" in global CO_2 budgets. Their hypotheses cannot be confirmed because the productivity increase over the continental marginal seas is still small when compared with total primary productivity. The excess organic carbon production due to nutrient input is estimated at only 3% of gross primary productivity. Furthermore, much of the unused shelf primary productivity is not available for export or deposition but is remineralized on the shelf instead (Berner 1992; Kempe and Pegler 1991; Biscaye et al. 1994). The results of the one Shelf Edge Exchange Processes Study (SEEP; Rowe et al. 1996) indeed raised the question: "Do continental shelves export organic matter?" Perhaps, in response, Bauer and Druffel (1998) suggested that dissolved organic carbon and particulate organic carbon inputs from ocean margins to the open ocean interior may be more than one order of magnitude greater than inputs of recently produced organic matter derived from the open surface ocean.

At present, it is not clear how much of this excess organic carbon is simply reoxidized and how much is permanently sequestered by export to the deep oceans or in sediments on the shelves and shallow seas. Because of the limited surface area, a burial rate significantly exceeding 0.5 Gt C yr^{-1} is not very likely, as this would require that all coastal seas be on average more than 50 μatm undersaturated in pCO_2 annually, in order to supply carbon from the atmosphere. Even though such undersaturations have been documented, as in the North Sea and the East China Sea (ECS) for example, (Kempe and Pegler 1991; Chen and Wang 1999), it is not certain to what extent these measurements are representative of all coastal oceans. For instance, Holligan and Reiners (1992) estimated a net flux of approximately 0.4 Gt C yr^{-1} to the atmosphere from the coastal oceans because of supersaturation. Ver et al. (1999) also gave a sea-to-air flux of 0.1 Gt C yr^{-1} as of 2000.

To sum up, based on the above considerations, the role of the coastal oceans has not been thoroughly, let alone accurately, assessed. In order to determine the contribution of continental margins and seas to CO_2 sequestration and the horizontal flux of carbon, nitrogen and phosphorus across the ocean-continental margin boundary, the JGOFS/LOICZ Continental Margins Task Team (CMTT) was established (Chen et al. 1994). The specific objectives are to:

i identify relevant and appropriate data sets from continental margin studies and investigate their applicability to IGBP projects;
ii develop a conceptual framework so as to integrate continental margin carbon, nitrogen and phosphorus fluxes and to assess the influence of anthropogenic factors on the fluxes;
iii quantify, as best possible, the vertical and horizontal carbon, nitrogen and phosphorus fluxes in different types of continental margins, such as: (*a*) eastern boundary currents, (*b*) western boundary currents, (*c*) marginal seas, (*d*) polar margins, and (*e*) tropical coasts;
iv establish an overall synthesis and assessment of carbon, nitrogen and phosphorus fluxes on and across continental margins to feed into the IGBP program; and
v determine major gaps and uncertainties in the current understanding of continental margin carbon, nitrogen and phosphorus fluxes and recommend a priority of needs for further observational and modeling endeavors.

Selected national and international projects are identified in Fig. 3.1. General descriptions of the continental margins and marginal seas are given in Appendix 3.1.

The first CMTT effort to study cross-shelf exchanges and regionalization was aimed at differentiating between systems where budgets for C, N, P and water already exist, or at least are close to completion. It should be noted that some of these systems fall at least partially outside the definition of 'continental margins'. The CMTT recognized that the goal of obtaining a budget for each and every system in the world is unrealistic. To compensate for this, however, key shelf characteristics have been identified and the systems have been categorized into two broad groups, namely the recycling and export systems based on specific defining characteristics (Table 3.1). It has also been recognized that in a number of regions there is fairly distinct continuum between the two extremes. For example, some important systems (notably the North Sea and the ECS) may function like recycling systems towards their landward side and like export systems towards their seaward side.

Water exchange time is perhaps the single defining difference between 'recycling' and 'export' systems with respect to CNP budgets. The recycling system is frequently influenced and often dominated by rivers, whereas the export system is frequently governed by the oceans. JGOFS (1997) has considered some of the features that control exchange time, and these include:

- shelf width, depth and shelf-edge depth: where a narrow width and deep shelf break suggest short exchange time;
- long-shore relative to cross-shelf scales: when long-shore transport is larger than the cross-shelf transport, the retention of material on the shelf is favored;
- river input: large rivers discharge beyond the shelf-break, while small rivers discharge onto the shelf, affecting fresh-water buoyancy control. If the river flow is larger than 10% of the shelf volume, the buoyancy effect is significant;
- winds and orientation to the coast: this induces entrainment, circulation, upwelling or downwelling cycles;
- insolation and heat balance: this affects stratification, making the system one- or two-layered; and
- retention time: whether it is less than or greater than one month seems to be significant in the balance between 'recycling' or 'export' system processes.

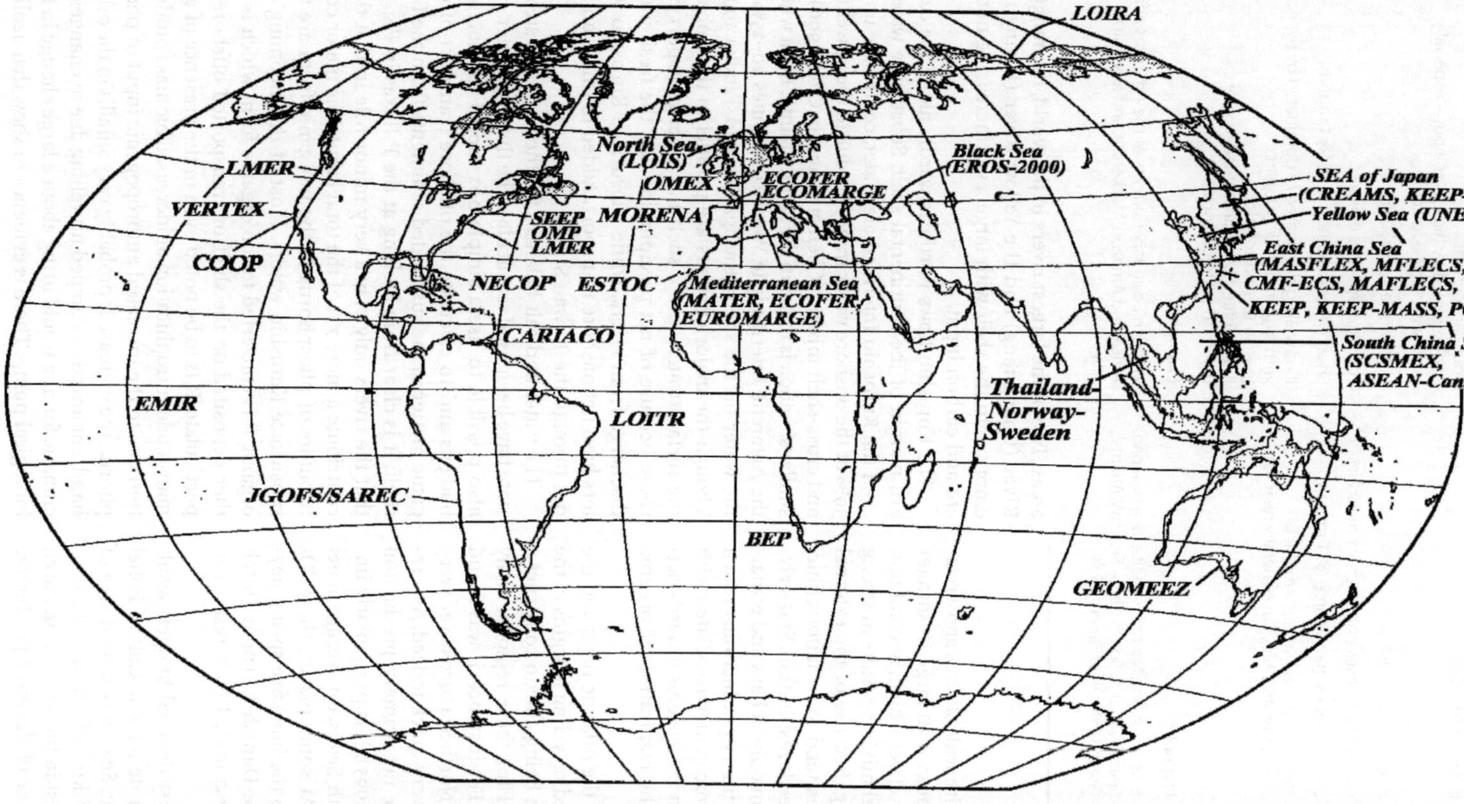

Fig. 3.1. Map of continental margins. Stippled areas represent the continental shelves with depths less than 200 m. Selected programs for continental margin studies, both completed and on-going, are shown. The acronyms are as follows: *BEP* (Benguela Ecology Programme); *CARIACO* (Carbon Retention In a Colored Ocean); *CMF-ECS* (Continental Margin Flux-East China Sea); *COPEX* (The Coastal Ocean Processes Experiment of the East China Sea); *COOP* (Coastal Ocean Processes); *CREAMS* (The Circulation Research Experiment in Asian Marginal Sea); *ECOFER* (Ecosysteme du canyon du cap Ferret program); *ECOMARGE* (Ecosystems de Marge); *EMIR* (Exportation de Carbon sur une Marge Insulaire Recifale); *EROS-2000* (European River-Ocean System); *ESTOC* (European Station for Time-Series on the Ocean, Canary Islands); *GEOMEEZ* (Marine Geolological and Oceanographic computer model for Management of Australia's EEZ); *JGOFS/SAREC* (Eastern Boundary Current Programme); *KEEP* (Kuroshio Edge Exchange Processes); *KEEP-MASS* (Kuroshio Edge Exchange Processes-Marginal Seas Studies); *LMER* (Land Margin Ecosystem Research); *LOIRA* (Land/Ocean Interactions in the Russian Arctic); *LOISE* (Land Ocean Interaction Study); *LOITRO* (Land-Ocean Interaction in Tropical Regions); *MAFLECS* (Marginal Flux in the East China Sea); *MASFLEX* (The Marginal Sea Flux Experiment); *MATER* (Mass Transfer and Ecosystem Response); *MFLECS* (The Margin Flux in the East China Sea Program); *MORENA* (Multidisciplinary Oceanographic Research in the Eastern Boundary of the North Atlantic); *NECOP* (Nutrient Enhanced Coastal Ocean Productivity); *OMEX* (The Ocean Margin Exchange); *OMP* (US Ocean Margins Programme); *POFLECS* (The Key Processes of Ocean Fluxes in the East China Sea); *SEEP* (Shelf Edge Exchange Processes); *UNEP-NOWPAP* (The United Nations Environment Program – Northern Western Pacific Area Protection); *VERTEX* (Vertical Transport and Exchange)

Table 3.1
Comparisons between systems dominated by recycling systems and systems dominated by material export (modified from JGOFS 1997)

River dominated, recycling systems	Ocean dominated, export systems
Biologically mediated; Tidal, geostrophic circulation; Benthic bioturbation	Physically forced; Ekman shelf-edge baroclinic upwelling; Boundary currents; Can be forced by high river input – especially rivers with shelf-edge deltas
Broad shelves (>50 km)	Narrow shelves (<50 km)
Forcing, responses on seasonal time scales	Forcing, responses episodic
Long water exchange time (≫1 month)	Short water exchange time (≪1 month)
May have $(p-r) = +$ (autotrophic) or – (heterotrophic), but generally near 1.0	If upwelling dominated or dominated by rivers with high inorganic nutrients, $(p-r) > 0$ (autotrophic); if dominated by sediment-laden rivers, $(p-r) < 0$ (heterotrophic)
Examples	**Examples**
East China Sea, North Sea, Baltic, Sea of Japan, Sea of Okhotsk, South China Sea, NW Atlantic, Great Barrier Reef, Barents Sea	Western Boundary Currents of the Americas and Africa, and Amazon, Mackenzie, and Mississippi Rivers

3.2 Recycling Systems

Recycling systems are those with wide shelves and a relatively long residence time. Those with large freshwater input have a strong buoyancy effect which, in combination with wind stress and tidal mixing, results in strong upwelling and vertical mixing. Because of the external input of nutrients, the shelves tend to be autotrophic, i.e., biological production exceeds respiration. Small rivers, however, trap sediments mainly in bays and estuaries where turbidity hinders primary production. In such a case, the respiration of organic matter associated with these sediments tends to push the systems towards heterotrophy; in other words, the respiration of organic matter exceeds production.

The East China Sea (ECS), the Atlantic Bight and the North Sea are the most studied recycling margins in the world. Although still far from being able to construct a fully balanced carbon model for these regions, many important fluxes are known. For instance, it was found that the organic carbon in the North Sea turns over every 1.3 days on average, which is very quick indeed. As a result, only less than 1% of the net primary production reaches the bottom (Kempe 1995). It is particularly important to learn that the North Sea is, on average, a net sink for atmospheric CO_2. At some places, the pCO_2 drops below 100 ppmv due to the intensive spring phytoplankton bloom along the Danish-German-Dutch coast. The ECS is also undersaturated with respect to CO_2 in all seasons.

Since the ECS has the most detailed budgets available, it is discussed in greater detail. The Gulf of Bohai and the Yellow and East China Seas have a total area of 1.15×10^6 km^2, about 0.9×10^6 km^2 of which is the continental shelf, one of the largest in the world. Compared with other oceans, it is also one of the most productive areas. Two of the largest rivers in the world, the Yangtze River (Changjiang) and the Yellow River (Huanghe), empty onto the shelf with large, ever increasing nutrient and carbon input.

The Kuroshio flows northeastwardly along the eastern margin of the continental shelf. Subsurface waters, even the Kuroshio Intermediate Water, contribute up to 30% of the onshore water transport due to the upwelling and cross-shelf mixing (Chen et al. 1995b; Chen 1996). A similar situation is found in the South Atlantic Bight where the Antarctic Intermediate Water contributes 20–25% of the water on the shelf (Kashgarian and Tanaka 1991). Though the major currents are parallel to the isobath, the surface water on the shelf has a net transport offshore because of net precipitation and the fresh water discharge from rivers, while subsurface Kuroshio waters have net onshore transport. Added to this is an input through the Taiwan Strait.

The water and salt balances for the shelf at a steady state have been used to calculate the fluxes of water. It is also possible to use a simple box model for nutrient budgets and to calculate the offshore transport of organic matter in the suspended sediments from the ECS shelf. It is clear from looking at the P fluxes in Fig. 3.2 that the rivers only play a very minor role in that they contribute a mere 7% of the total input. The major contributors of phosphorus in the inorganic form are the subsurface Kuroshio waters. Most of the incoming inorganic P is converted to the organic form which is either deposited on the shelf or transported offshore as particulate. It is to be noted even in the presence of any man-made eutrophication or increased biomass production due to the increased anthropogenic input of phosphorus, the increase is probably very small. On the other hand, enhanced or damped upwelling due to changes in climatic forcing would bring about a large change in the biological pump. There is evidence to show that aeolian

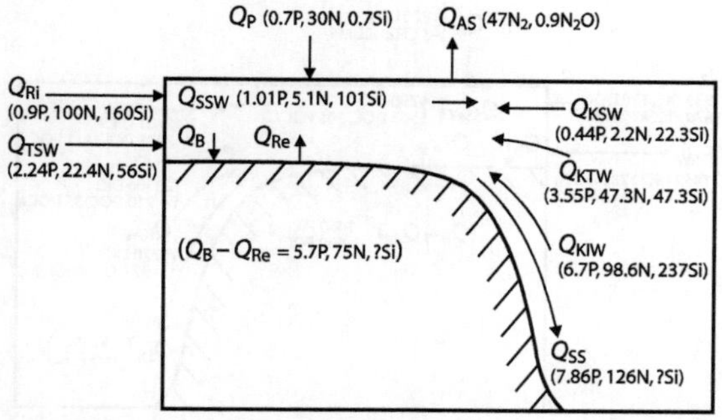

Fig. 3.2. Schematic diagram for the annual nutrient budgets (numbers in 10^9 mol yr^{-1}) in the East China Sea where Q is the flux, subscripts *Ri*, *P*, *TSW*, *KSW*, *KTW*, *KIW* and *SSW* denote river input, precipitation, Taiwan Strait Water, Kuroshio Surface Water, Kuroshio Tropical Water, Kuroshio Intermediate Water and shelf surface water, respectively; *P* denotes phosphorus, *N* denotes nitrogen, *Si* denotes silicate, *Re* denotes the release from sediments, *AS* denotes the air-sea exchange, *B* denotes the nutrients buried, and *SS* denotes suspended sediments transported offshore

fluxes may have increased during the past few hundred years (Chen et al. 2001a). Anthropogenic iron input or enhanced aeolion input of loess, being rich in iron and phosphorus, may increase primary productivity in the high nutrient-low chlorophyll regions or in waters where molecular-nitrogen-fixing, blue-green algae (*cyanobacteria*) bloom. The nitrogen-fixing *Trichodesmium* is abundant in the ECS (Chen et al. 1996b), and nitrogen fixation may account for 20% of the new N input in this upwelling zone (Liu et al. 1996).

The nitrogen budget (Fig. 3.2) is more complicated since denitrification converts nitrate to NH_3, N_2O and N_2 with the latter two degassing at the air-sea interface. On the other hand, nitrogen fixation by plankton utilizes N_2. The input from rivers is still smaller than that from the incoming water masses, but the difference is not as sharp as for phosphorus. Not much nitrogen leaves the ECS with the outflowing seawater. Instead, the largest sinks are the net burial on the shelf, the offshore transport in the form of sediments and that in the form of degassing as N_2. The box model gives the offshore N transport as 0.38 ± 0.19 mmol N $m^{-2} d^{-1}$, and the net denitrification rate as equivalent to 0.103 ± 0.050 mol N $m^{-2} yr^{-1}$. Seitzinger and Giblin (1996) gave an average denitrification rate of 0.252 mol N $m^{-2} yr^{-1}$ for continental shelf sediments in the North Atlantic.

After the completion of the SEEP project, the question as to the sources of N for the shelf to be able to support the measured primary production was considered unresolved. Biscaye et al. (1994) found it difficult to explain the flux of nitrate onto the shelf without imposing an export of flux of water. The above calculations, made by the present authors, however, indicate that the upwelling of nutrient-rich subsurface water would balance the export of nutrient-depleted surface waters and sediments.

It is important to keep in mind that for the ECS, the riverine N/P ratio is 111, a value much higher than the Redfield N/P ratio of 16 for phytoplankton. This makes P more limiting than N in terms of net organic production in the estuaries because the maximum amount of new organic matter that could be produced with land-derived inorganic nutrients is determined by the major nutrients which run out first. The total seawater flux of N and P to the ECS, however, has a ratio of 21 which is much closer to the Redfield ratio. As a result, the phosphorus shortage in the ECS, as a whole, is not as dramatic owing to the large influx from the subsurface Kuroshio waters. Changes in the nutrient structure of small riverine inputs are also not expected to affect the stoichiometric nutrient balance of the phytoplankton ecosystem in the ECS. The Si budget is also given in Fig. 3.2. The riverine input is still smaller than the oceanic input, but the budget is not balanced because it has yet to be determined how to estimate the Q_B, Q_{Re} and Q_{ss} values for Si.

The upwelled Kuroshio Intermediate Water (KIW) actually originates in the South China Sea (SCS) and is high in nutrients, which it contributes more generously to the shelf than do the much smaller riverine fluxes. Furthermore, the PO_4 flux from the Changjiang River can only be identified in the estuary. The NO_3 flux extends farther but can still only be identified in the Yangtze River Plume. This suggests that productivity in much of the ECS is mainly influenced by the upwelled subsurface Kuroshio waters. Any potential change in the upwelling rate would, therefore, have a much larger effect than would a change in the nutrient input from the rivers (Chen and Wang 1999). For instance, cutting back the Yangtze River outflow by 10% would reduce the cross-shelf water exchange by roughly 9%. The nutrient supply would then be reduced by that much as well. This means that new production, primary production and the fish catch in the ECS would be proportionately reduced as well. On the other hand, the passage of typhoons, which enhances vertical mixing, freshwater inflow and the resuspension of bottom sediments are three factors causing the shelf ecosystems to be more productive (Shiah et al. 2000a).

Though it is recognized that the feedback mechanisms are probably complicated, it is nevertheless worth noting that after an El Niño event, the freshwater outflow and the fish catch in the ECS are both reduced. For instance, after the strong 1982–1983 El Niño event, the yield of a major commercial fish, gunther (*Navodon Sepient*), was reduced by over 60% when coastal rainfall was reduced by 50%. This is in agreement with the above estimate (Chen 2000).

The major rivers bring in carbon in the form of dissolved inorganic carbon (DIC), dissolved organic carbon (DOC), particulate inorganic carbon (PIC) and particulate organic carbon (POC). A mass-balance calculation (Table 3.2) gives the downslope contemporary particulate car-

Table 3.2. The fate of organic carbon produced by the primary production on the East China Sea Shelf (mmol m^{-2} of shelf area per year)

In	(mmol m^{-2} yr^{-1})
Primary production	13322
Net POC influx	652
Total in	13974
Out	**(mmol m^{-2} yr^{-1})**
Denitrification	132
Mn reduction	10
Fe reduction	70
Sulfate reduction	730
CH_4 reduction	7
POC deposit on shelf	667
POC transported offshore	772
Net DOC export	1144
Total out	3531
Aerobic regeneration	10443

bon transport rate as 7.72 ±3.9 mmol C m^{-2} d^{-1} of which 27% is organic (Fig. 3.3). The offshelf transport of POC is only 5.7% of the average primary productivity. It should be mentioned that bacteria production in the ECS also constitutes an important part of the carbon cycle. In terms of carbon turnover, bacterial production is on average about 20% of the primary production (F. K. Shiah, pers. comm.). Since the growth efficiency of bacteria is rather low (approximately 30%), the organic carbon consumed during bacterial growth is much greater than that indicated by the bacteria production. In other words, although it is not known how much of the organic carbon is labile, at least three times more organic carbon may be consumed by bacteria in the ECS.

Bacteria production shows considerable spatial and temporal variations (Kemp 1994; Shiah et al. 1999, 2000b). The inner shelf enjoys a rich supply of organic carbon from river runoff, sediment resuspension and in situ production. In cold seasons, namely winter and spring, bacteria production is predominantly temperature-dependent in the inner and mid-shelves, where water temperature is low. In warm seasons, the inner and mid-shelves are not thermally stressed for bacteria nor substrate-limited, but bacteria production does not rise above a certain limit, perhaps due to enhanced grazing. Since the bacteria production is limited by temperature in cold seasons and reaches a plateau in warm seasons, a significant fraction of the CO_2 uptake by phytoplankton seems to escape bacterial decomposition throughout the year. In the outer shelf, where the intruding Kuroshio Surface Water keeps the temperature high and nutrient-depleted throughout the year, bacteria production is substrate-limited. The optimal thermal condition for bacterial growth in the outer shelf

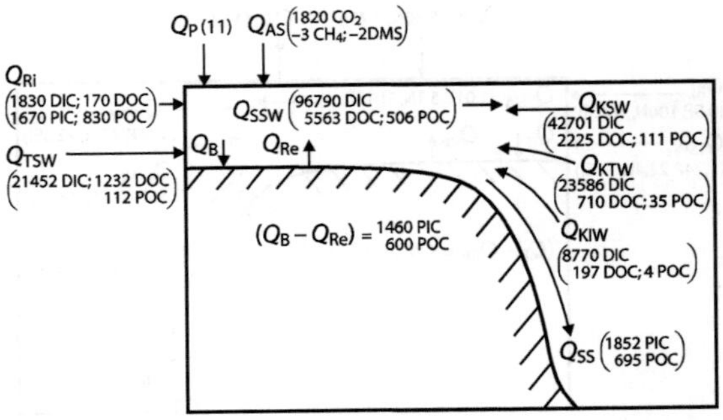

Fig. 3.3. Schematic diagram for the annual carbon budget in the East China Sea (numbers in 10^9 mol C yr^{-1}) (taken from Chen and Wang 1999). Subscripts as in Fig 3.2

results in the rapid oxidation of organic carbon. Little fixed carbon from primary production may escape bacterial degradation, except in areas with strong shelf break upwelling or subsurface water intrusion. A seaward export of organic carbon has been observed in these areas (Liu et al. 1995; Hung et al. 2000).

Continental shelf waters are generally high in total alkalinity (TA) because of river discharge and in situ generation due to the oxidation of organic material. If we formulate the particulate organic materials as $(CH_2O)_{106}(NH_3)_{16}H_3PO_4$, then the aerobic oxidation of organic material in seawater by oxygen can be represented by the Eq. 3.1 which reduces TA by 17 moles for the regeneration of 106 moles of organic carbon. (Chen et al. 1982). Accordingly,

$$(CH_2O)_{106}(NH_3)_{16}(H_3PO_4) + 138O_2 \longrightarrow 106CO_2 + 122H_2O + 16HNO_3 + H_3PO_4 \quad (3.1)$$

When the dissolved oxygen is exhausted, the system turns to the next most abundant source for the oxidation of organic material, NO_3^-. Afterwards, the manganese, iron, sulfate and methane reductions occur.

Based on the water fluxes after taking into account the above reactions, the alkalinity budget is given in Table 3.3 and Fig. 3.4. The order of 3.9 ±3.9 mmol m^{-2} d^{-1} of alkalinity is generated on the shelf, mainly the the result of iron and sulfate reductions (Chen and Wang 1999).

The ECS has a net export of 64 ±32 mg C m^{-2} d^{-1} organic carbon offshelf, which is the equivalent of 15% of primary productivity. Results from the SEEP-II program on the eastern US continental shelf also indicate that most of the biogenic particulate matter is remineralized over the shelf. Only a small proportion (<5%) is exported to the adjacent slope (Biscaye et al. 1994). In the North Sea, only 1.5% of the primary production is accumulated on the shelf as organic carbon, and 2–3% of the primary production is exported over the margin (de Haas et al. 1997). The ECOMARGE and OMEX projects have also led to similar conclusions (Monaco et al. 1990; Antia et al. 1999).

Table 3.3. Processes affecting alkalinity on the East China Sea Shelf (per m² of shelf area per year)

Process	Source	Quantity (mmol in TA)
Input	Ri	2 444 plus 3 711 from PIC
	KSW	56 204
	KTW	30 138
	KIW	10 173
	TSW	27 808
	Aerosol	24 from PIC
Output	SSW	124 214
	Net burial	3 246 from PIC
	S.S.	4 117 from PIC
Water column production	New production	310
	New production that supports anaerobic processes	152
Anaerobic reduction in sediments	Denitrification	104
	Mn reduction	42
	Fe reduction	559
	Sulfate reduction	723

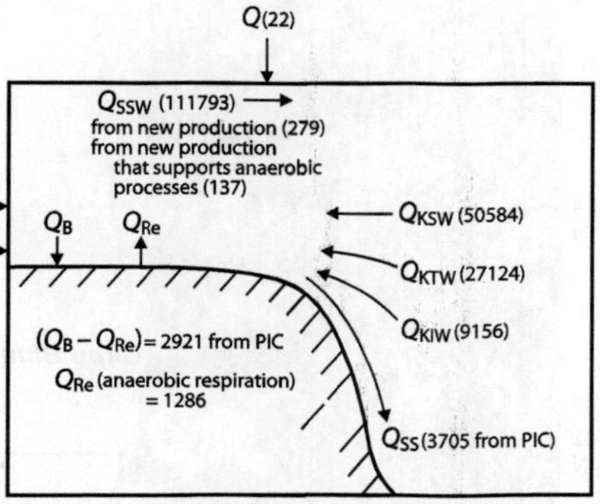

Fig. 3.4. Schematic diagram for the annual alkalinity budget (numbers in 10^9 mol yr^{-1}) in the East China Sea (taken from Chen and Wang 1999). Subscripts as in Fig 3.2

It is also not yet possible to directly measure the offshore transport of particulate matter due to large spatial and temporal variabilities. Again mass-balance calculations reveal an offshore flux at $7.9 \pm 4 \times 10^9$ mol yr^{-1} of particulate organic phosphorus for the ECS. Independent calculations give a downslope flux of contemporary particulate organic and inorganic carbon at 2.12 ± 1.1 and 5.64 ± 2.8 mmol m^{-2} d^{-1}, respectively. At least 20% of this is old, refractory material which accounts for the relatively old ^{14}C ages obtained from core tops (up to 8 000 years old). A similar situation has been found elsewhere (Jahnke and Shimmield 1995).

The ECS shelf is probably saturated with the anthropogenic, excess CO_2. Between the shelf break and Ryukyu Island is a deep trench. The hydrography is such that there is upwelling toward the west, bringing older deep waters upward. As a result, excess CO_2 penetrates to only about 600 m. The entire ECS contained 0.07 (±0.02) Gt C excess carbon in 1992. Since the waters on the ECS shelf are highly supersaturated with respect to calcite and aragonite, sediments on the ECS shelf are not expected to neutralize excess CO_2 in the coming century. The high productivity, nevertheless, causes a large portion of the ECS inner and mid-shelf waters to be undersaturated with respect to CO_2 near the surface all year round. As a result, the flux of CO_2 into the ECS is large. The outer shelf and the Kuroshio region (east of 126° E) are depleted in PO_4 and the surface water pCO_2 is near saturation.

3.3 Export Systems

Export systems are continental margins with effective processes to export materials to the open ocean. For practical reasons, all continental margins with narrow shelves (less than 100 km in width) are considered as export systems. In continental margins with wide shelves, 200 m isobaths may be used to define the outer boundaries of the margins (Gordon et al. 1996). In coastal zones with very narrow shelves, highly active biogeochemical processes are not confined to the shelves but occur in the boundary current systems, which are distinct from the adjacent open ocean and often poorly represented by the coarsely gridded ocean models or databases. Therefore, we need to treat the coastal zone beyond the boundary of the shelf as a part of the continental margin.

Among the 85 continental margins listed in Appendix 3.1, 30 are narrow margins. Many of them are bordered by eastern boundary currents. In fact, most of the eastern coast of the Pacific Ocean, from Vancouver Island to the Central Chilean Shelf, belongs to this category. So does most of the eastern coast of the Atlantic Ocean from the western Iberian Coast to Cape Town. Many of the eastern boundary current systems are characterized by coastal upwelling. Another important coastal upwelling system with narrow shelves is the northwestern coast of the Arabian Sea driven by the monsoons. Essentially all major coastal upwelling systems are export systems.

There are also other types of export systems where the river delta systems extend near the shelf edge; for example, the Mississippi Delta, the East African Coast, the eastern coast of the Bay of Bengal, the southeastern coast of Australia, the northern coast of New Guinea and the eastern coast of Brazil. Yet another type of export system includes many tropical watersheds that discharge a high volume of water and sediments to the ocean due to frequent floods caused by torrential rains. More than half of the total runoff and land-derived sediments are discharged to the oceans from tropical coasts, especially the Indo-Pacific Archipelago (Nittrouer et al. 1995). The narrow shelves and weak Coriolis force favor cross shelf transport of sediments, which may carry a significant amount of carbon, to the deep ocean. This type of continental margin is especially vulnerable to human perturbation, which causes up to ten fold increase of sediment production rate in a typical Oceania watershed (Kao and Liu 1996). The extra sediment transport to the ocean may result in non-negligible change in the carbon cycle, but the available information is not sufficient for quantitative estimation of the associated carbon fluxes.

The most important biogeochemical activities within the export systems are associated with upwelling in the coastal provinces. Therefore, we devote most of this section to describe these systems. One way to define these margins is to use the Rossby deformation radius as the width of the coast zone (Barber and Smith 1981; Hall et al. 1996). Here we adopt the coastal provinces defined by Longhurst (1998) to represent these continental margins. The Coastal Domain is further defined as the system bounded by the continental shelf and the coastline on the landward side and by the vertical surface at the edge of the coastal provinces and the horizontal surface at 200 m depth in regions beyond the shelf.

3.4 Coastal Upwelling Systems

Coastal upwelling is mainly induced by the along shore wind, which drives water across the upwelling front, producing surface divergence near the coast and convergence offshore (Csanady 1990). Upwelling brings subsurface nutrient-rich water to the surface, where it fuels blooming of phytoplankton and enhances primary productivity (Barber and Smith 1981). The upwelling condition has a direct impact upon new production (Kudela and Chavez 2000) and also the air-sea exchange of CO_2 (Torres et al. 1999), which are crucial to the understanding of the marine carbon cycle.

Before the onset of upwelling, the coastal water is usually stratified and low in nutrient. As the wind intensifies, upwelling and offshore surface Ekman transport occurs (Brink 1998). The mass balance in the upper water column is maintained by an indirect circulation cell with upward advection of the subsurface water and the subduction of surface water. The coastal water becomes enriched in nutrients, which is often separated from the warm nutrient-poor offshore water by an upwelling front. When the wind is uniform and sustained, the front tends to be parallel to the coast. When the wind weakens, the front tends to intensify and become unstable. The instability of the current may cause meandering of the along shore flow and produce filaments of upwelled water (Fig. 3.5). These filaments have strong influence over the biological activities and offshore transport of organic carbon (Brink and Cowles 1991).

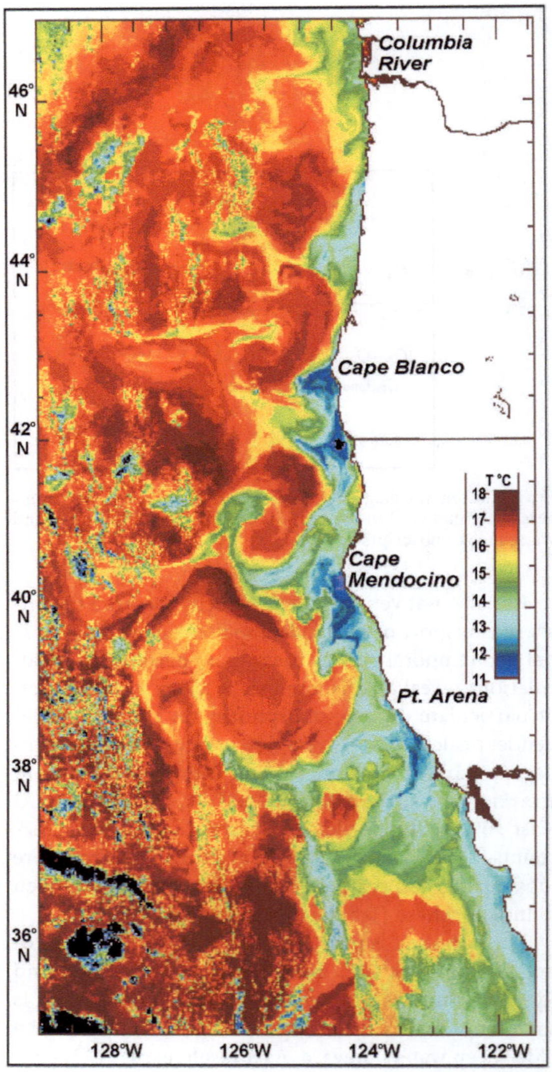

Fig. 3.5. Sea surface temperature along the coast of Oregon and northern California, USA, on September 5, 1994 as measured by satellite-borne AVHRR of NOAA (courtesy of Oregon State University). Upwelling filaments were manifested as offshore-shooting tongues of low temperature waters, which were usually associated with high concentrations of Chl *a*

During intense upwelling, primary production may reach as high as 10 g C m^{-2} d^{-1} (Barber and Smith 1981). However, the spatial and temporal variability is rather high due to the very dynamic nature of the upwelling system. Nitrogen is often the limiting nutrient in upwelling ecosystems due to their juxtaposition with denitrifying zones (e.g., Codispoti and Christensen 1985). Nevertheless, phosphate limiting could also occur in some of the upwelling systems, where the source waters have high N/P ratio, such as in the North Atlantic (Fanning 1992, Michaels et al. 2000). The rich supply of nutrients enhances phytoplankton growth, which in turn may sustain a considerable biomass increase at the higher trophic levels. What fraction of the new production may go into fish harvest is a question of interest.

3.5 California Current System

The west coast of North America from Vancouver Island to the tip of Baja California is bordered by the California Current System (CCS). Major currents in the system include the California Current, the Davison Current and the California Undercurrent (Hickey 1998). The California Current is an equatorward surface current. In winter, there is a poleward surface flow shoreward of the California Current along the coastline north of Southern California Bight, which is the major bend in the otherwise pretty straight coastline. The undercurrent is the poleward return flow. Upwelling is strongest in summer at most places along the northern half of the coast. In the Southern California Bight, upwelling is strongest in winter and early spring. Coastal upwelling, subduction and filaments (Fig. 3.5) make the coastal zone a highly variable and dynamic zone (Brink and Cowles 1991). The primary production and new production are closely related to the upwelling and other dynamic processes in the coastal transition zone (Chavez et al. 1991). Estimates of the transport across the filaments are in the range of 1.5–3.6 Sv (Korso and Huyer 1986)

In the CCS, the sea surface Chl *a* concentrations shows a summer maximum in most locations near the coast (Thomas et al. 1994). The strongest seasonal variation in the CCS within 100 km of the coast occurs at 34–45° N and at 24–29° N. The maximum Chl *a* concentration exceeds 3.0 mg m^{-3} in May–June. The seasonal variation is weaker in the Southern California Bight. In winter, the Chl *a* concentrations in the offshore extension patches reach 0.5–1 mg m^{-3}. The primary production in the CCS was estimated to be 740–1240 mg C m^{-2} d^{-1} based on shipboard observations (Chavez et al. 1991). The estimation based on CZCS pigment concentrations yields a mean PP of 1060 mg C m^{-2} d^{-1} (Longhurst et al. 1995).

Notable inter-annual variation is coupled with El Niño conditions (Fig. 3.6). During ENSO events, the trade wind is weakened in the western Pacific. The Chl *a* in the CCS is reduced during El Niño events. The primary production seems to be less affected by the ENSO condition probably due to a more efficient recycling of nutrients, while the new production is much reduced as the upwelling flux of nutrients is reduced due to the intrusion of nutrient-poor water in the source layer of the upwelling water (Kudela and Chavez 2000).

The pCO_2 off Oregon coast varied in a wide range from 150 to 690 μatm in summer, which is attributable to large spatial variation of upwelling intensity (van

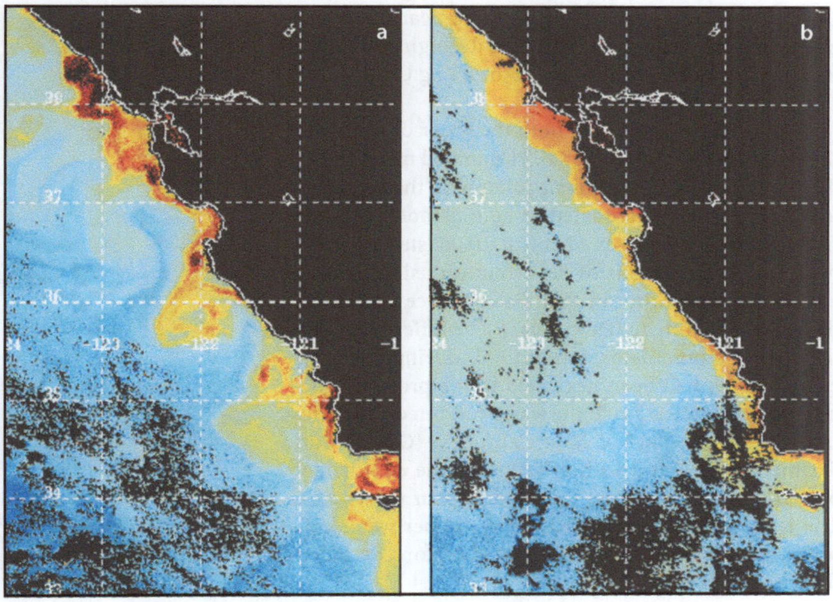

Fig. 3.6.
Imagery of chlorophyll concentration (mg m^{-3}) derived from SeaWiFS data for the central California Current Coastal Province (courtesy of Monterey Bay Aquarium Research Institute, *http://www2.mbari.org/kura/seawifs_share/images/index.htm*). **a** February 9, 1998 during intense El Niño condition; **b** May 17, 1998, when the El Niño condition weakened

Geen et al. 2000). Although TCO_2 accounts for much of the large pCO_2 variation, the temperature and salinity effects may cause a 25% shift of pCO_2. The strong CO_2 uptake is associated with a coastal bloom of phytoplankton dominated by large diatoms (Chavez et al. 1991). Up to 200 mg C kg^{-1} of TCO_2 may be removed by the phytoplankton bloom. One important factor controlling phytoplankton growth and, thereby, pCO_2 in surface seawater is the availability of iron from contact of the upwelled water with the shelf sediments (Johnson et al. 1999). In spite of the active upwelling off California, the observed mean pCO_2 values at the sea surface along an onshore-offshore transect throughout the year in 1993 were below the atmospheric partial pressure of CO_2 in most cases (Friederich et al. 1995), suggesting that the CCS may not be a net source of CO_2.

The VERTEX experiment carried out a direct measurement of the export flux of particulate organic carbon from the euphotic zone off Point Sur, California (Martin et al. 1987). Because of the coupling between offshore advection and high primary production (Chavez et al. 1991), phytodetritus may be transported offshore advectively and settle to the deep water. The POC and PON fluxes depend on the primary production and decreases as a power function of depth (Pace et al. 1987):

$$F_{POC} = 3.523\, Z^{-0.734}\, PP \qquad (3.2)$$

$$F_{PON} = 0.432\, Z^{-0.743}\, PP^{1.123} \qquad (3.3)$$

where F is the flux in mg $m^{-2}\, d^{-1}$, Z is in units of metres and PP is primary production in the same unit as the flux. The observed POC fluxes were 42–85 g C $m^{-2}\, yr^{-1}$ (Martin et al. 1987). The high POC flux near the continental margin is obviously related to the high upwelling-induced primary production (250~420 g C $m^{-2}\, yr^{-1}$). The f-ratio is about 0.17–0.2.

Denitrification occurs in both the water column and the sediments beneath the CCS. The annual mean flux of denitrification in the sediment column on the Washington Shelf has been estimated to be 0.18 mol N $m^{-2}\, yr^{-1}$ from porewater profiles of nitrate (Christensen et al. 1987), but observations of N_2 evolution using benthic chamber suggest a mean flux almost twice as high as the previous estimate (Devol 1991). The difference is attributed to ammonia oxidation to N_2 during denitrification, which had been neglected in the previous estimation. Denitrification in the water column occurs only in the southern part of Baja California (Gruber and Sarmiento 1997) and within basins on the continental borderland off California (Liu and Kaplan 1982, 1989). The latter are much deeper than the upper water column addressed in this section and, therefore, excluded from consideration.

3.6 Humboldt Current System

The Peru-Chile Coast is bordered by the Humboldt Current System between 4° S and 45° S (Longhurst 1998). The upwelling system is determined by the presence of upwelling favorable winds and the pattern of coastal ocean circulation (Strub et al. 1998). The continental shelf along this coast is generally rather narrow and, in some areas practically non-existent. In many places, the shelf width is less than the Rossby radius of deformation, which depends on the Coriolis force, and therefore on latitude. It changes from about 270 km at 4° S, to 46 km at 24° S, and to about 29 km at 40° S (Chavez and Barber 1987).

The Humboldt Current System is the counterpart of the California Current System in the southern hemisphere. It is characterized by a generally equatorward flowing current, but the whole system is quite complex with interleaving currents and countercurrents, which may be separated into the coastal and the oceanic branches (Longhurst 1998). Upwelling favorable winds are seasonal, and coastal divergence and cyclonic wind-stress curl cause extensive upwelling from 4° S latitude to about 40° S (Thomas et al. 1994; Strub et al. 1998). Satellite images show upwelling occurs mainly along the Peruvian Coast from 5–15° S (Fig. 3.7) and the Chilean Coast from 25 to 40° S. The South Pacific high that drives the long-shore winds strongly affects the amount of nutrient upwelled (Strub et al. 1998). The strongest upwelling favorable winds are found during July–August (austral winter) off Peru and during December–January (austral spring and summer) off Chile (Thomas et al. 1994, Strub et al. 1998). It is noteworthy that poleward

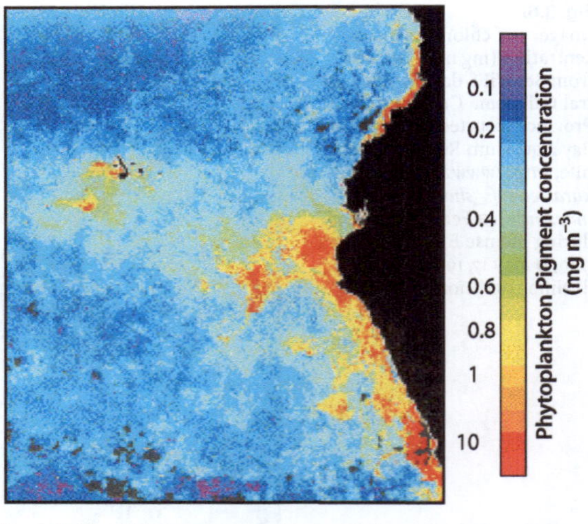

Fig. 3.7. The CZCS composite image of phytoplankton pigment concentration around the Peruvian upwelling centers covering the periods from 16–26 January, 1980 (modified, courtesy of NASA). The equatorward current brought the upwelled nutrient-laden water to the northwest

currents, as undercurrent over the shelf and countercurrent offshore, are dominant off Peru and can continue as far as middle latitudes off Chile. The dynamics responsible for such a feature is not clear, but the wind system and the shelf topography have been suggested to contribute to this special feature (Strub et al. 1998).

Very high concentrations of Chl *a* (10–20 mg Chl m^{-3}) associated with large size phytoplankton (>20 µm) have been observed during intense upwelling in the austral spring (Chavez 1995; Morales et al. 1996). Off Peru between 6 and 17° S, there is a minimum in Chl *a* in fall. Between 17 and 40° S, maximum Chl *a* values occur in austral winter (Thomas et al. 1994). North of 20° S, the maximum Chl concentration reaches 1.5 mg m^{-3} or higher during the austral spring, summer and fall. The primary production for the Peruvian upwelling region was estimated to be as high as 3 200 mg C m^{-2} d^{-1} (Chavez et al. 1989). The estimated *f*-ratio ranges between 0.21 and 0.75 (Minas et al. 1986; Chavez et al. 1989). The upper limit represents the potential export fraction under strong upwelling conditions. The annual average primary production is estimated to be 833 g C m^{-2} yr^{-1} for the coastal zone (Chavez and Barber 1987), but the mean PP drops to 270 g C m^{-2} yr^{-1} for the broad Humboldt Current System (Longhurst et al. 1995). Under El Niño conditions, the sinking flux of foraminifera off the Chilean coast decreased by about 30%, but the flux is still higher than those in other upwelling systems under normal conditions (Marchant et al. 1998). This implies that the export production in this upwelling system could be much higher than other upwelling systems, which is consistent with the fact that the South American fisheries are the most productive among all upwelling systems. Apparently the ecosystem performance is the result of intricate interplay among physical forcing, shelf width, bottom topography and biological responses (Walsh 1977), that is yet to be unveiled.

Very large fluctuations of air-sea CO_2 exchange fluxes have been observed around the upwelling center at 30.7° S off the Chilean coast in January (Torres et al. 1999). During the upwelling period, the surface pCO_2 reached as high as 900 µatm. The maximum outgasssing flux reached 25 mmol C m^{-2} d^{-1}, but there was still uptake of atmospheric CO_2 in the Tongoy Bay where the uptake flux was around −4 mg C m^{-2} d^{-1}. During the relaxation period, the maximum outgasssing flux at the upwelling center dropped to 3 mmol C m^{-2} d^{-1}, while the maximum uptake flux was enhanced slightly to −6 mg C m^{-2} d^{-1}.

Denitrification occurs in the suboxic water off Peru extending from 10° S to 25° S. The suboxic water in the coastal zone off Peru covers an area of 326×10^9 m^2. The average denitrification flux is 3.2 mol N m^{-2} yr^{-1} (Codispoti and Packard 1980). Denitrification in the surface sediments off Peru reaches as high as 1.4 mol N m^{-2} yr^{-1} (Codispoti and Packard 1980), which is apparently caused by the high flux of sinking organic particles.

3.7 Benguela Current System

The Benguela Current System, which forms the eastern limb of the South Atlantic gyre circulation, consists of a series of coastal jets and wind-induced upwelling cells (Shillington 1998). There are four groups of major coastal upwelling centers (Longhurst 1998). In northern and central Namibia, upwelling is strong throughout the year with the highest intensity in the austral winter from June to August. From Luderritz to Walvis Bay, upwelling is most vigorous with intensity peaking in the austral spring (Nelson and Hutchings 1983). In the Hondeklip Bay, maximum upwelling occurs from October to December. In the southern part around Cape Columbine and Cape peninsula, upwelling is largely restricted to the austral summer. The lateral output of the upwelled water is about 1.7 Sv (Shannon 1985). The upwelling water in the offshore region has a mean nitrate concentration around 19 µM and a mean N/P ratio (by atom) around 11 (Chapman and Shannon 1985).

No extensive suboxic water mass has been observed in the Atlantic Ocean (Deuser 1975). However, localized oxygen-deficient waters do exist on the Namibian Shelf around Walvis Bay in the Benguela Current System (Chapman and Shannon 1985). Oxygen concentration in the bottom water on the shelf may be reduced to less than 0.5 ml l^{-1} (Chapman and Shannon 1985). A secondary nitrite maximum often occurs in the bottom water, indicating denitrification in the suboxic bottom layer of the inner shelf (Calvert and Price 1971; Chapman and Shannon 1985). The low N/P ratio and the raised $\delta^{15}N$ values of nitrate (above 6.5‰, K. K. Liu unpublished data) in the upwelling water also indicate the occurrence of denitrification.

3.8 Monsoonal Upwelling Systems

The most famous monsoonal upwelling system is that along the northwest coast of the Arabian Sea (Longhurst 1998). Analogous to the Arabian Sea, the South China Sea also exhibits monsoon-driven upwelling off northwestern Luzon in winter (Shaw et al. 1996) and off the coast of central Vietnam in summer (Wu et al. 1999). Both areas have very narrow shelves. Satellite observations and model predictions suggest enhanced phytoplankton biomass in these upwelling areas, but little direct observations are available to assess the carbon fluxes reliably. Therefore, we focus on the Northwest Arabian Upwelling System below.

The African and the Arabian coasts of the Arabian Sea have very narrow shelves, usually less than 10 km in width (Longhurst 1998). The Arabian Sea experiences two monsoons every year. The more energetic southwest monsoon blows from June to September, and the weaker northeast monsoon blows from December to February (Burkill 1999). During the SW monsoon, the wind maximum is the low level Findlater Jet that peaks in July as it blows along the Somali and Oman coast. Wind stress curl is anti-cyclonic seaward of the jet and cyclonic landward of the jet. The Somali Current occurs shortly following the onset of the jet and upwelling is induced in the northwestern Arabian Sea.

The total volume transport of coastal upwelling in the Arabian Sea is estimated to be 10–15 Sv (Chavez and Toggweiler 1995). The mean nitrate concentration in the source water is probably around 10 μM (Chavez and Toggweiler 1995). The observed N/P ratio is 12.2 (Woodward et al. 1999). In nearshore waters a mixed community of diatoms and *Synechococcus* dominate during the southwest monsoon (Tarran et al. 1999). However, very high concentration of Chl *a* (13 mg m^{-3}) have been found associated with small size (0.2–2 μm) phytoplankton (Savidge and Gilpin 1999), which is quite different from the diatom dominated population in the coastal upwelling systems. Towards the late stage of the SW monsoon, the phytoplankton biomass may reach as high as 69 mg Chl m^{-2} and the production may reach 3 800 mg C m^{-2} d^{-1}.

The total organic carbon (TOC) concentration in the surface water varies between 60 and 100 μM, while the concentration variation in the subsurface water below 100 m is much reduced with a mean concentration around 45 μM (Hansell and Peltzer 1998). The TOC concentration drops abruptly just below the surface mixed layer. The TOC production probably accounts for 6–8% of the total primary production (Hansell and Peltzer 1998). Since a major fraction of the TOC is dissolved organic carbon (DOC), the distribution is probably controlled mostly by the DOC dynamics.

The Arabian Sea has been observed as a persistent CO_2 source to the atmosphere (Fig. 3.8) with the highest fluxes during southwest monsoon in summer (Goyet et al. 1998). The strongest CO_2 outgassing occurs in the coastal zone along the Arabian Peninsula. The monthly mean of ΔpCO_2 in the coastal zone peaks in August (Fig. 3.8), but much higher local maximum up to 400 μatm has been reported (Kortzinger et al. 1997), indicating the high temporal and spatial variability. The outgassing flux of CO_2 shows a much stronger peak during the southwest monsoon than the corresponding ΔpCO_2 peak due to the wind enhanced gas transfer velocity. The estimated annual average of CO_2 outflux in the Arabian Sea is 0.46 mol C m^{-2} yr^{-1} (Goyet et al. 1998). The low N/P ratio in the upwelling water indicates denitrification in the Arabian Sea. The oxygen minimum zone extends from the northern coast of the Arabian Sea towards the central part of the basin (Wyrtki 1971). Along the northwest coastal zone of the Arabian Sea, suboxic condition occurs mainly in the northern and central Arabian Sea and around the northern tip of the Omani coast.

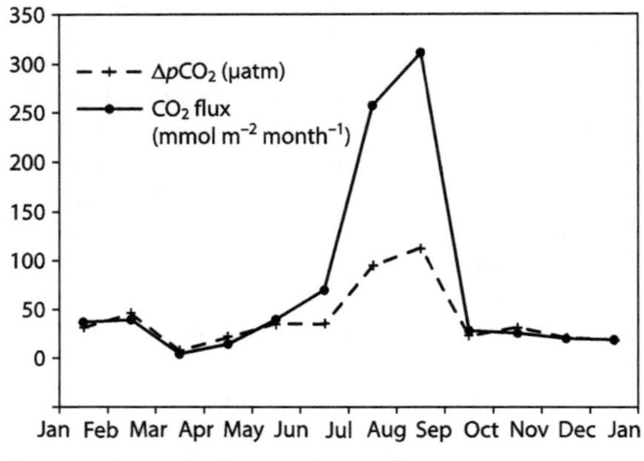

Fig. 3.8. The monthly mean values of ΔpCO_2 (μatm) and the flux of CO_2 (mmol m^{-2} $month^{-1}$) along the Arabian Peninsula (data taken from Goyet et al. 1998)

3.9 Biogeochemical Budgeting

The budget of carbon, nitrogen and phosphorus fluxes in six upwelling systems are given in Table 3.4. As stated above, the domains considered are not restricted to the shelf region. Instead, the domains of study consist of the upper 200 m in the coastal provinces, which are defined following Longhurst et al. (1995) or Longhurst (1998) except one. The one exception is the Portugal-Morocco Coastal Province, which is defined following Walsh (1988) and Barton (1998). Slightly modified is the definition of the Northwest Arabian Upwelling Province, which does not extend as far south as that of Longhurst (1998). Instead, it extends only to the northern Somali Coast where the upwelling is significant (Hill et al. 1998).

The major fluxes are the upwelling of nutrients and the export of organic carbon. Also considered are the river runoff, the burial of biogenic matter and denitrification in shelf sediments and water column. For the calculation of water column processes, the areas of the coastal provinces are used; for that of benthic processes, the areas of the shelves are used. The shelf areas are taken from either Appendix 3.1 or Table III of Walsh (1988).

The upwelling volume transport has been estimated for a few locations by integrating the offshore transport along the coast (e.g., Korso and Huyer 1986). In upwelling regions without such estimates, a mean upwelling index of 100 (in units of m^3 s^{-1} per 100 m or Sv per 1 000 km) of coastline is assumed for calculation (Huthnance 1995). It is cautioned that the upwelling index does show considerable variability, and, therefore, the assumption only represents a ballpark value. The reported mean upwelling index off northwest Africa is as high as 270

(Barton 1998). The length of the coastline for each upwelling system is listed in Table 3.4. The nitrate fluxes are calculated from the volume transports and nitrate concentrations in the source waters. The phosphate fluxes are calculated from the nitrate fluxes and the observed N/P ratios. In four out of the six cases, the N/P ratio is less than the normal Redfield ratio of 16, indicative of denitrification in the upwelling systems. However, in the North Atlantic the N/P ratio in the upwelling source water is slightly above the normal N/P ratio, apparently resulting from the prevalent nitrogen fixation (Michaels et al. 2000).

Most of the upwelling systems receive little runoff from land. Among the few major rivers that discharge into these systems is the Columbia River. The total runoff within the coastal upwelling systems accounts for only 3% of the total global runoff (Walsh 1988). In the six upwelling systems considered here, the total river runoff amounts to 522 $km^3 yr^{-1}$ (0.017 Sv), which is less than 1% of the total upwelling volume transport. The inputs of C, N and P from river discharge are negligible in the upwelling systems.

The primary productivity data are taken from Longhurst et al. (1995) except for the Portugal-Morocco Coastal Province whose PP value is taken from Walsh (1988). The area-integrated primary production is calculated from the PP data and the area of the coastal provinces (Table 3.4). The total primary production amounts to 273×10^{12} mol C yr^{-1}, which represents 6.6–7.3% of the global ocean primary production of $3\,800–4\,200 \times 10^{12}$ mol C yr^{-1} (Longhurst et al. 1995). From the primary productivity several important variables are derived. Most important of all is the export production or new production. By definition, the export systems readily export a significant fraction of primary production to the deep water. Only a small fraction is buried on the shelves. The carbon burial fluxes are listed in Table 3.4. The sum represents less than 0.4% of the total primary production. The export production is derived from the reported *f*-ratios and further constrained by the available nitrate or phosphate, which is provided by upwelling. However, a significant fraction of the upwelled nitrate may be removed during denitrification in the coastal zone.

Since denitrification in the suboxic water column often occurs in the subsurface layer of coastal upwelling systems, the fluxes of nitrate removal during denitrification are derived from primary production data. Liu and Kaplan (1984), based on then available data, suggested that the rate of denitrification was proportional to the availability of organic carbon. This notion was later substantiated with additional data (e.g., Naqvi 1987). The availability of organic carbon may be calculated from the change of the sinking flux of particulate organic carbon within the water column of the denitrifying layer. Using the relationship between the sinking flux of POC and the primary production observed in the California Current upwelling system (Pace et al. 1987), we calculated the available fluxes of sinking POC in the subsurface layer of 100–200 m (Table 3.5). It is noted that the denitrifying zone was assumed to be 150–200 m for the Northwest Arabian Upwelling Province. The relationship reported by Liu and Kaplan (1984) indicates that ¼ to ½ of the available sinking POC flux may be consumed by denitrifying bacteria. For this study, we assume 40% of the available organic carbon flux consumed during denitrification in the water column (0–200 m) of the coastal provinces discussed in this section. We also assume nitrogen gas as the only final product of all nitrogen species reacted. Results of the calculation are listed in Table 3.5. The area-integrated denitrification fluxes are calculated by multiplying the denitrification flux by the area of the denitrifying waters in the coastal provinces. In the California Current and Humboldt Current Coastal Provinces, the suboxic water covers 30% of the coastal provinces, which is indicated by the distribution of negative N* at 200 m depth (Gruber and Sarmiento 1997); in the Northwest Arabian Upwelling Province, the suboxic water covers 60% as indicated by the distribution of the oxygen deficient water (Wyrtki 1971); in the Benguela Current Coastal Province, we assume one half of the shelf regions covered by suboxic waters, based on chemical hydrography (Chapman and Shannon 1985); for the Portugal-Morocco and the Canary Coastal Provinces, we assume no water column denitrification. The total removal of fixed nitrogen sums up to 0.97×10^{12} mol N yr^{-1}.

Since denitrification in the sediment column is directly related to the organic carbon content (Liu and Kaplan 1984), it is assumed for this calculation that the denitrification flux is proportional to the organic carbon burial flux with a C/N ratio of 4:1. The results are listed in Table 3.5. The calculated denitrification flux for the shelf off the western American Coast bordering the California Current is 0.25 mol N $m^{-2} yr^{-1}$, which falls in the range of previous estimates, 0.18–0.37 mol N $m^{-2} yr^{-1}$ (Christensen et al. 1987; Devol 1991). The sum is 0.24×10^{12} mol N yr^{-1}, which is about 24% of the total rate of denitrification in the water column in these domains. Denitrification as a whole removes fixed nitrogen from these coastal domains at a rate equivalent to 10% of the total upwelling flux of nitrate.

The subsurface circulation patterns in the upwelling systems are complicated. It is not clear whether the upwelling waters would pass the suboxic zones, both in the water column and at water-sediment interface, and experience denitrification. To make a conservative estimate, we subtracted the denitrification fluxes from the upwelled fluxes in the calculation of the net inputs of nitrate to the coastal provinces (Table 3.5). For nitrate-limiting waters which have N/P ratios lower than Redfield ratio, the potential new production can be cal-

Table 3.4. Transports of water and biophilic elements (C, N, P) and their transformation in coastal upwelling provinces. All areas are in units of 10^9 m^2; fluxes in mol m^{-2} yr^{-1}; total transports of C, N and P in 10^{12} mol yr^{-1}

Coastal provinces	Shelf area[q]	Coastal zone area[n]	Coastal length (km)	Water transport (Sv)		Primary production[m]		New production	Organic C burial[d]		Source waters		Total upwelled nutrients	
				Upwelling	Runoff[q]	Flux	Total	Total	Flux	Total	NO_3^- (mM)	N/P	NO_3^-	PO_4^{3-}
California Current	270	960	3000	1.5–3.6[m]	0.0080	32.3	31.0	6.2	1.00[c,e]	0.270	20[h]	14.2[k]	0.95–2.27	0.067–0.16
Humboldt Current	100	2610	4000	4.0	0.0035[o]	22.4	58.5	12.3	0.23[i]	0.023	25[g]	12.0[k]	3.15	0.263
Northwest Arabian Upwelling	166	2500	3000	15.0[h]		37.8	94.6	26.5	0.40	0.066	10[h]	12.2[r]	4.73	0.388
Portugal-Morocco	80	600	1300	1.3	0.0038	14.6[q]	8.8	1.9	0.05	0.004	8[a]	17.9[a]	0.33	0.018
Canary Current	100	810	1000	2.7[b]	0.0009	61.0	49.4	8.0	0.05	0.005	15[j]	17.0[k]	1.28	0.075
Benguela Current	144	1130	2700	1.7[p]	0.0004[f]	26.9	30.4	5.5	4.00	0.576	19[f]	11.0[f]	1.02	0.093
Sum	860	8610	15000	27.3	0.017		272.7	60.4		0.94			12.23	0.96

Data sources: [a] Alvarez-Salgado (1997); [b] Barton (1998); [c] Berelson et al. (1996); [d] Berner (1982); [e] Carpenter (1987); [f] Chapman and Shannon (1985); [g] Chavez et al. (1989); [h] Chavez and Toggweiller (1995); [i] Codispoti and Christensen (1985); [j] Codispoti and Friederich (1978); [k] Fanning (1992); [l] Hall et al. (1996); [m] Korso and Huyer (1986); [n] Longhurst et al. (1995); [o] Milliman et al. (1995); [p] Shannon (1985); [q] Walsh (1988); [r] Woodward et al. (1999).

Table 3.5. Nitrogen budget and export fraction in coastal upwelling systems

Coastal provinces	Water column denitrification				Benthic denitrification		Available nitrate (Tmol N y^{-1})	Maximum export fraction	Reported *f*-ratio	Export fraction
	ΔF_{OrgC} (g C m^{-2} y^{-1})	F_N (mol N m^{-2} y^{-1})	Area (10^9 m^2)	Total rate (Tmol N y^{-1})	F_N (mol N m^{-2} y^{-1})	Total rate (Tmol N y^{-1})				
California Current	1.55	0.64	288	0.19	0.250	0.068	1.39	0.30	0.20[b]	0.20
Humboldt Current	1.07	0.45	783	0.35	0.058	0.006	2.80	0.32	0.21[c] 0.75[d]	0.21
Northwest Arabian Upwelling	0.64	0.27	1500	0.40	0.100	0.017	4.31	0.31	0.28 ±0.12[e]	0.28
Portugal-Morocco	0.70	0.29	0	0.00	0.013	0.001	0.33	0.22[a]	0.25[c]	0.22
Canary Current	2.92	1.22	0	0.00	0.013	0.001	1.28	0.16[a]	0.64[c]	0.16
Benguela Current	1.29	0.54	72	0.04	1.000	0.144	0.84	0.18		0.18
Sum			2643	0.97		0.236	10.94			

[a] Value based on phosphate limitation (see text).
Data sources: [b] Martin et al. (1987), [c] Minas et al. (1986), [d] Chavez et al. (1989), [e] Watts and Owens (1999).

culated from the net nitrate input under the assumption of 100% nitrate utilization (Chavez et al. 1989). The maximum export fraction e_{max} may be calculated from the total potential new production and the total primary production (Table 3.5). The potential new production is calculated under the assumption of nitrate limitation for all provinces except the two in the North Atlantic, where the high N/P ratio makes these upwelling systems P-limiting. The calculated e_{max} values for the California Current Coastal Province and the Northwest Arabian Upwelling Province are in reasonable agreement with reported *f*-ratios. The discrepancy is probably attributed to the much bigger area of the provinces defined in this study than the traditional coastal zone. For those provinces where there are reported *f*-ratios, we adopt the minimum among the *f*-ratios and the e_{max} values as the export fraction; for those provinces where there are no available *f*-ratios, we adopt the calculated e_{max} values as the export fraction (Table 3.5). The total new production is 60×10^{12} mol C yr^{-1}, which yields a mean export fraction of 0.22.

If DOC production accounts for 5–7% of the total primary production (based on Hansell and Peltzer 1998) in the upwelling systems, the total DOC production is $14–19 \times 10^{12}$ mol C yr^{-1}. The difference between the DOC concentrations in the surface and the subsurface layers was about 20 µM as observed off the Northwest Arabian Coast. If this represents the mean difference between the input and the output fluxes in coastal upwelling systems, the total export of DOC would be 17×10^{12} mol C yr^{-1}, which is in good agreement with the estimated DOC production.

A major fraction of the fish catch comes from the coastal upwelling systems. The world catch of marine fishes varies from 25.9 to 92.7 Mt yr^{-1} between 1953 and 1993 with a mean of 61 Mt yr^{-1} (FAO 1996). Assuming a carbon content of 6% of wet weight of fish (Walsh 1981), we get an average carbon removal rate of 0.3×10^{12} mol C yr^{-1} by marine fishing. If the coastal provinces discussed in this section provide 50% of the total fish catch in the world, fishing would result in a mean carbon removal rate of 0.15×10^{12} mol C yr^{-1} from these systems, which accounts for only 0.25% of the new production.

Whether the upwelling systems act as a source or sink of atmospheric CO_2 depends on the origin of the upwelling water (Watson 1995). The higher the content of deep ocean water in the upwelling water, the stronger it tends to release CO_2 to the atmosphere; the higher the content of the intermediate water ventilated in the subpolar region, the stronger it tends to absorb atmospheric CO_2. Among the provinces discussed in this section, the Arabian Sea (10–25° N) is a persistent source of CO_2 (Goyet et al. 1998), whereas the eastern North Atlantic Ocean off the Iberian Peninsula (40–43° N) is a sink of CO_2 in spite of coastal upwelling (Perez et al. 1999). It is likely that the closer to the equator the upwelling area, the higher is the tendency of releasing CO_2. However, the air-sea CO_2 exchange fluxes in some of the other upwelling systems have extremely high spatial and temporal variability (e.g., Torres et al. 1999), which prevents any meaningful estimation of the net fluxes. Since strong upwelling occurs mostly at relatively low latitudes, it is reasonable to assume the upwelling systems are a net source of CO_2.

3.10 The Arctic Shelves

3.10.1 Introduction

The Arctic Ocean differs from all others by virtue of the combination of sea ice, large volumes of runoff from land, and enormous continental shelves. Furthermore, the Arctic Ocean is a mediterranean sea surrounded by a drainage basin that exceeds the size of the ocean and has only restricted passages through which it exchanges water with the Pacific and Atlantic Oceans (Fig. 3.9). Arctic shelves undoubtedly behave in many ways the same as other shelves; for example, runoff from the land will stratify shelf surface water and produce estuarine circulation, winds will force exchange and upwelling at the shelf margins, and the shelves will generally have higher primary production and greater terrestrial influence than the interior ocean. However there are also vital differences. Exceptionally strong seasonality in runoff and light intensity occur synchronously around the entire Arctic Basin. Seasonal ice formation has a bipolar impact on the freshwater cycle. In summer, melting ice provides a broadly-distributed source of stratification, but in winter, ice formation destroys stratification by rejecting salt to the water column and thereby enhances mixing or produces convection (Fig. 3.10). Particularly in the flaw leads and polynyas that are widely distributed over the central shelves (Macdonald 2000), intense ice formation can produce convecting water which then flows across the bottom to spread as descending plumes into the interior ocean transporting material with them (see for example, Kämpf et al. 1999; Melling and Lewis 1982). The coupling of shelves to the interior ocean in the Arctic is, therefore, unique because it is so strongly affected by the oscillating cycle of ice formation and melting (Fig. 3.10). On one hand, the ice restricts air-sea interaction reducing wind mixing, sediment resuspension and upwelling. On the other hand, ice produces upwelling at its edge, provides an additional transient habitat for biota, and is an exceptionally powerful medium for material transport either directly in the ice itself or indirectly though thermohaline circulation (Cota et al. 1990; Eicken et al. 2000; Pfirman et al. 1989). The combination of the above seasonally modulated processes has no parallel in any other ocean.

Fig. 3.9. View of the Arctic Ocean showing the drainage basin, major rivers, the shelves and the major passages connecting the Arctic with the Pacific and Atlantic Oceans

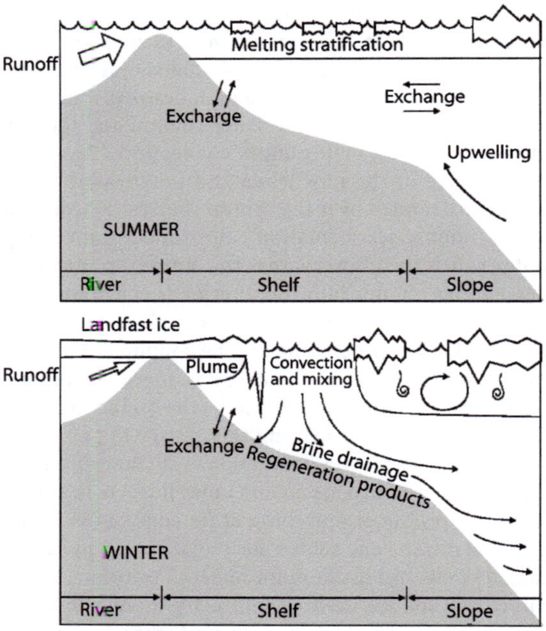

Fig. 3.10. A simplified scheme of the Arctic shelves showing the strong bipolarity produced by the alternation between ice melting and clearing in summer and ice formation during winter. Ice production produces two exchange pathways in ice export and brine export

The Arctic is thought to be particularly sensitive to global climate change (Cuffey et al. 1995; Fyfe et al. 1999; Walsh J. E. 1991a) and recent evidence suggests that, for reasons that remain unclear, the Arctic Ocean is now undergoing significant change in its ice climate and in its water-mass composition and distribution (Dickson 1999; Macdonald 1996; Macdonald et al. 1999; McLaughlin et al. 1996). An alteration in the sea ice climate will inevitably lead to large changes in the biogeochemistry of the shelves where seasonal clearing of ice already occurs. And so, in the world oceans we can expect global change to manifest itself first in the Arctic and, within the Arctic, first in the marginal seas. It has been suggested recently that the Arctic Ocean may be two or three times more important for anthropogenic CO_2 sequestration than its size implies (Bauch et al. 2000). However, the critical question is not how important the Arctic Ocean is presently in global elemental cycles, but, rather, how important it might be in contributing to change in these cycles. It seems likely in view of the size of the shelves and the vulnerability to drastic change from ice cover to open water that the Arctic Ocean will factor well above its relative size in the equation of global change.

Here, we start with a view of the Arctic Ocean as a large, shelf-dominated semi-enclosed sea (Fig. 3.9). We then survey the shelves within the Arctic, outlining their characteristics, and eventually focus on the Mackenzie Shelf where recent material budgets have been constructed for organic carbon, nitrogen and phosphorus. We then construct a preliminary budget for Arctic Ocean treating the shelves as a single box. Finally, we speculate on how global change might affect the marginal Arctic seas.

3.10.2 The Arctic Ocean As a Mediterranean, Shelf-Dominated Sea

The potential to construct budgets for the entire Arctic Ocean by viewing it as a semi-enclosed sea has long been recognized (Aagaard and Carmack 1989; Aagaard and Coachman 1975; Anderson et al. 1983; Anderson et al. 1998; Codispoti and Lowman 1973; Goldner 1999b; Wijffels et al. 1992). Although constraining such budgets is made difficult by recirculation in Fram Strait, which obscures the estimate of net flow, it may be easier to close material budgets for the entire Arctic Ocean than it is for any one of the shelves within the Arctic. All Arctic Ocean budgets have depended first on estimating the volumetric flows of water and sea ice into and out of the Arctic. The assignment of salinity to exchanging water masses together with freshwater runoff and precipitation estimates allow the construction of reasonably accurate freshwater (Fig. 3.11a) and salt budgets (Fig. 3.11b). Once volumetric flows and exchanges have been established, budgets can be constructed for any property for which there are appropriate concentration data. Examples of such budgets include silicate (Anderson et al. 1983; Codispoti and Lowman 1973; Walsh et al. 1989), inorganic carbon (Anderson et al. 1998; Lundberg and Haugan 1996), organic carbon (Anderson et al. 1998; Opsahl et al. 1999), hexachlorocyclohexane (Macdonald et al. 2000), Cd (Macdonald 2000; Yeats and Westerlund 1991) and Al (C. Measures, unpublished). There is considerable variation in the estimates of volumetric flow used in budget constructions and there is considerable variability in the current flows themselves. Despite these variations, reasonable closure on budgets for the entire Arctic Ocean seems presently attainable (Goldner 1999a).

In the context of carbon and nutrients, Anderson and co-workers (Anderson et al. 1998) have undoubtedly produced one of the most detailed budgets to date. They considered the Arctic Ocean as two basins (Canada and Eurasian Basins), amalgamated the shelves into three main boxes (Laptev Sea, Chukchi/East Siberian/Beaufort Seas, Barents and Kara Seas) and, because the vertical stratification is so important in the Arctic Ocean in terms of property exchanges between the shelves and interior oceans, they sectioned the basins vertically into 6 boxes in the Canada Basin and 5 boxes in the Eurasian Basin. Inflows from the Atlantic and Pacific Oceans cross the Barents/Kara Shelves and the Chukchi and East Siberian Shelves before entering the interior ocean. Importantly, the shelves are incorporated into their box

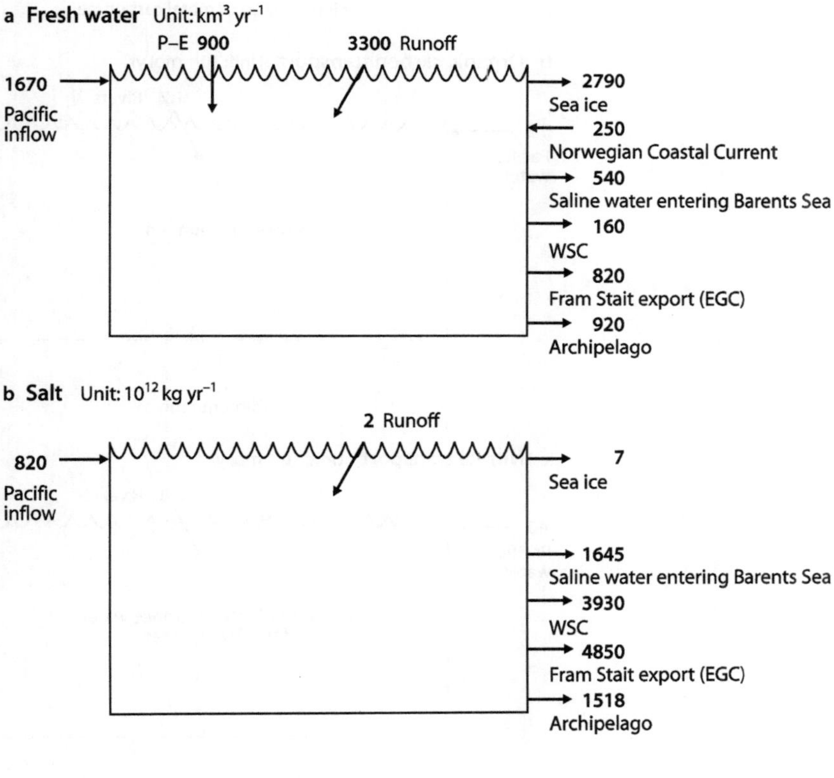

Fig. 3.11.
a A schematic diagram showing the freshwater budget for the Arctic Ocean constructed by Aagaard and Carmack (1989). The freshwater budget has been calculated using a foundation salinity of 34.8 which explains why saline water entering the Barents Sea and in the West Spitzbergen Current (WSC) are entered as freshwater losses to the Arctic Ocean. Each component was calculated independantly with the result that this budget is unbalanced to amount of net accumulation of 890 $km^3 yr^{-1}$ which the authors suggest is well within the uncertainties. **b** A schematic diagram showing the salt budget for the Arctic Ocean adapted from (Goldner 1999b). The original budget, presented relative to a median salinity of 34.4 has been converted and simplified to make it comparable to the freshwater budget in Fig. 3.11a

model so as to allow exchanges with the interior ocean at surface and at depth. This latter exchange process, illustrated in Fig. 3.10, is mediated by the ice formation/brine production over the shelves which nourishes the Arctic halocline (Aagaard et al. 1985). One of the problems identified both by Anderson et al. (1998) and Lundberg and Haugen (1996) is that the Arctic Ocean is not at steady state with respect to CO_2 due to increased atmospheric CO_2 pressure over the past two centuries. They estimate the current vs. the pre-industrial CO_2 budget considering only the altered air-sea exchange (i.e., effects on the biological pump are not considered).

Their budget for inorganic carbon (C_T), simplified in Fig. 3.12a, shows that the exchanges between the Arctic Ocean and the Atlantic dominate. The Pacific provides an important source of C_T for the Arctic Ocean whereas rivers and sea ice play only a minor role. While Anderson et al. (1998) have made significant progress in identifying the major components of the carbon budget and sensitivity to error, Fig. 3.12a underscores the difficulty in using differences between large fluxes to estimate relatively small net fluxes. Furthermore, the error estimates provided by the authors are fairly large and caution is warranted in drawing too many conclusions. As pointed out by Anderson et al. (1998), water that supplies input has been in contact more recently with the atmosphere (younger) than water supplying the output. Since atmospheric CO_2 concentrations have been increasing during that time, the ocean is not in steady state and the output reflects historical conditions whereas the input reflects impacted conditions. In this budget, the present input of C_T to the Arctic Ocean exceeds output

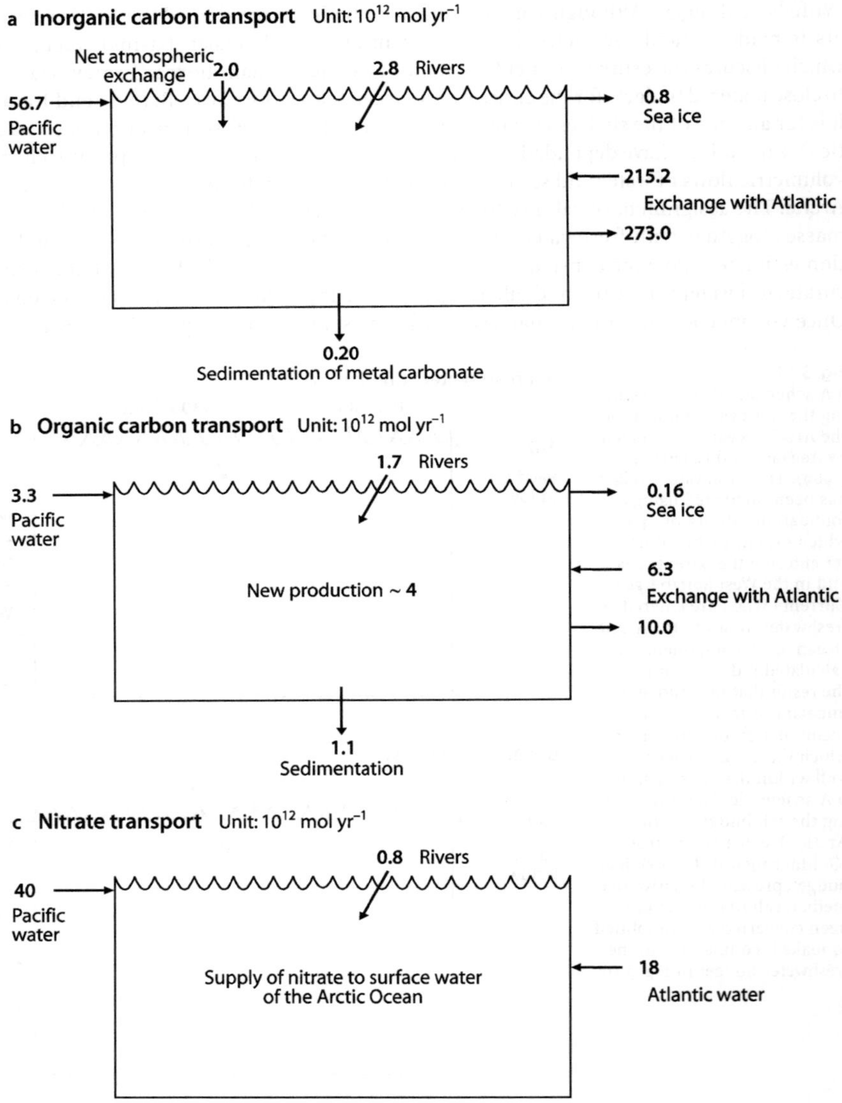

Fig. 3.12. A simplified schematic of the budget for the Arctic Ocean based on (Anderson et al. 1998) for **a** inorganic carbon, **b** organic carbon and **c** nitrate inputs that support new production in surface water

by 0.75 ±0.50 × 10^{12} mol C yr^{-1}. The carbonate burial flux was estimated to be small (0.02 × 10^{12} mol C yr^{-1}) and, considering both the C_T and organic carbon (OC) budgets (Fig. 3.12b), air-sea flux of CO_2 was estimated from the budget imbalance to be 2.0 ±1.4 × 10^{12} mol C yr^{-1} into the Arctic Ocean.

The OC budget (Fig. 3.12b) shows that rivers and sedimentation contribute more significantly than they did in the case of C_T. The burial of OC in sediments is estimated from the Arctic Ocean new production which is supported by nitrate inputs primarily from the Pacific and Atlantic Oceans with only very minor additions from rivers (Fig. 3.12c), and by considering that 80% of the new carbon production is regenerated and therefore does not get captured in sediments. These authors caution that dissolved organic nitrogen may be an important but as yet unconsidered component of the system especially for river waters (and see Gordeev et al. 1996). Based on the denitrification estimate for western Arctic Shelf sediments (0.7 mol m^{-2} yr^{-1}; Devol et al. 1997) the loss of nitrate collectively by the total shelf area would be about 3 × 10^{12} mol yr^{-1}, which would far outweigh any input of nitrate from rivers (Table 3.7).

The first estimation of the silicate budget for the Arctic Ocean (Codispoti and Lowman 1973) concluded that this ocean was not a major source for reactive silicate and could even be a sink. A subsequent budget (Anderson et al. 1983) estimated net outputs (1.89 × 10^{12} mol yr^{-1}) to exceed slightly the net inputs (1.78 × 10^{12} mol yr^{-1}) and, finally, a re-estimation by Walsh et al. (1989) gave a balanced budget but suggested that, depending on the silicate concentration in water entering at Bering Strait, there could be a burial loss of as much as 0.4 × 10^{12} mol yr^{-1}. Unfortunately, the major term in these budgets, the silicate-rich Pacific inflow, was overestimated because flows of 1.3–1.5 Sv were used compared to the more probable 0.83 Sv determined recently by (Roach et al. 1995). The question of whether or not the Arctic Ocean is a sink for silicate has by no means been convincingly demonstrated.

Opaline silicate sequestration in the Arctic Ocean is of more than academic interest. Carbonate oceans (nutrient poor surface water, coccolith dominated) can be contrasted with silicate oceans (nutrient rich surface water, diatom dominated). In the former, little net carbon dioxide uptake by the ocean occurs whereas in the latter, diatoms are more effective in supporting net CO_2 uptake from the atmosphere because they sediment organic carbon decoupled from carbonate (Honjo 1997; Nozaki and Oba 1995). The surface (0–200 m) of the Arctic Ocean comprises two domains – the Pacific-dominated water in the Canada Basin and the Atlantic-dominated water in the Eurasian Basin which have traditionally been separated by a front over the Lomonosov Ridge. Because Pacific water is rich in silicate we might expect regions of the Arctic Ocean where water of Pacific origin dominates to be more productive of diatom assemblages; this would include especially the Chukchi and East Siberian Seas, but also large portions of the Canada Basin and the Canadian Archipelago. Recent observations show that one manifestation of change in the Arctic Ocean is the displacement of the front between Atlantic and Pacific water (McLaughlin et al. 1996). Large-scale change in the size of the Pacific (silicate) Domain could be an effective, but as yet unquantified, way to alter CO_2 sequestration on a large scale for the Arctic Ocean without even necessarily altering the primary production.

3.10.3 The Shelves of the Arctic Ocean

Of the Arctic Ocean total area (~10 × 10^6 km^2), over 30% is shelf (Table 3.6; Fig. 3.13). The predominant shelf area forms a continuous circumpolar band over 7 000 km long and up to 1 000 km wide. Figure 3.13 and Table 3.6 suggest that each of the identified shelves is unique and that there really is no 'representative' Arctic Shelf. Since the last ice age, many of the coastal regions in the Arctic have undergone relative sea-level rise and, consequently, have drowning coastlines. This process which is ongoing suggests that Arctic shelves continue to adjust to the new sea level. Many coastal regions in the Arctic are flat, do not extend much above present sea level, and are composed of poorly bonded soils sensitive to permafrost destruction. Therefore, coastal erosion for many of the regions is an important component of the coastal sediment budget (Are 1999; Kassens et al. 1998; Macdonald et al. 1998). Coastal inputs of freshwater and solids are now reasonably well estimated for most Arctic shelves (Table 3.6) as are their general physical characteristics. However, the exchanges at the shelf edge have not been directly quantified and models do not constrain them particularly well (e.g., see Björk 1990; Goldner 1999b). These latter, which probably dominate C, N and P cycles, are complicated in the Arctic by transport processes mediated by ice and brine production on the shelves (Fig. 3.10).

It is important to clarify how ice effects transport. The ice itself is not a particularly efficient transporter of conservative water properties like salt or nutrients because these tend to be excluded from the ice during formation (e.g., Fig. 3.11b). In contrast, sea ice often provides an important vehicle for the transport of sediment through suspension freezing (Eicken et al. 2000) and organic carbon (Mel'nikov and Pavlov 1978). Ice exerts its greatest influence on conservative property exchange by forcing thermohaline circulation through brine rejection. It is estimated that, for the entire Arctic, as much as 1–1.4 Sv on average is involved in such transport from shelf to basin (Goldner 1999a) which rivals the estimated strength of Arctic shelfbreak upwelling (3.8 Sv; Goldner 1999a).

Table 3.6. Characteristics of the Arctic shelves and inputs from land

Shelf	Shelf area (10^3 km^2)	Freshwater in-flow (km^3 yr^{-1})	Sediment (10^6 t yr^{-1})	Ice export (km^3 yr^{-1})	Residence time (yr)	TOC (10^9 mol yr^{-1})	DIC (10^9 mol yr^{-1})	DIN (10^9 mol yr^{-1})	DIP (10^9 mol yr^{-1})	$Si(OH)_4$ (10^9 mol yr^{-1})	Primary production (g C m^{-2} yr^{-1})
Barents	600	463	22	560		442	510	4.7	0.28	43	150 (nearshore) 40 – 80 (deep)
Kara	880	1 478	33	340	2.5	725	1 180	59	1.9	184	?
Laptev	504	745	25	480–660	3.5 ± 2	433	600	4.1	0.16	47	12 – 27
E. Siberian	890	250	34	Import	3.5 ± 2	150	100	1.4	0.05	13	60 – 360
Chukchi	570	~100	<5	Import	0.2 – 1.2	~15	3	~0.8	~0.1	~8	60 – 360
Beaufort	180	>330	130	450	0.5 – 1	~280	360	1.0	~0.03	~15	20 – 40
Archipelago	300	~60	?	?	~0.5	?	?	<0.5	<0.05	?	12 – 27
Total	3 924	3 426	~250	1 920		~2 000	2 750	~71.5	~2.6	310	

Data sources include Barrie et al. (1998), Dethleff (1995), Gordeev (2000), Gordeev et al. (1996), Hanzlick and Aagaard (1980), Macdonald et al. (1998), Pavlov and Pfirman (1995), Rigor and Colony (1997), Sakshaug et al. (1994), Schlosser et al. (1994), Walsh (1989).

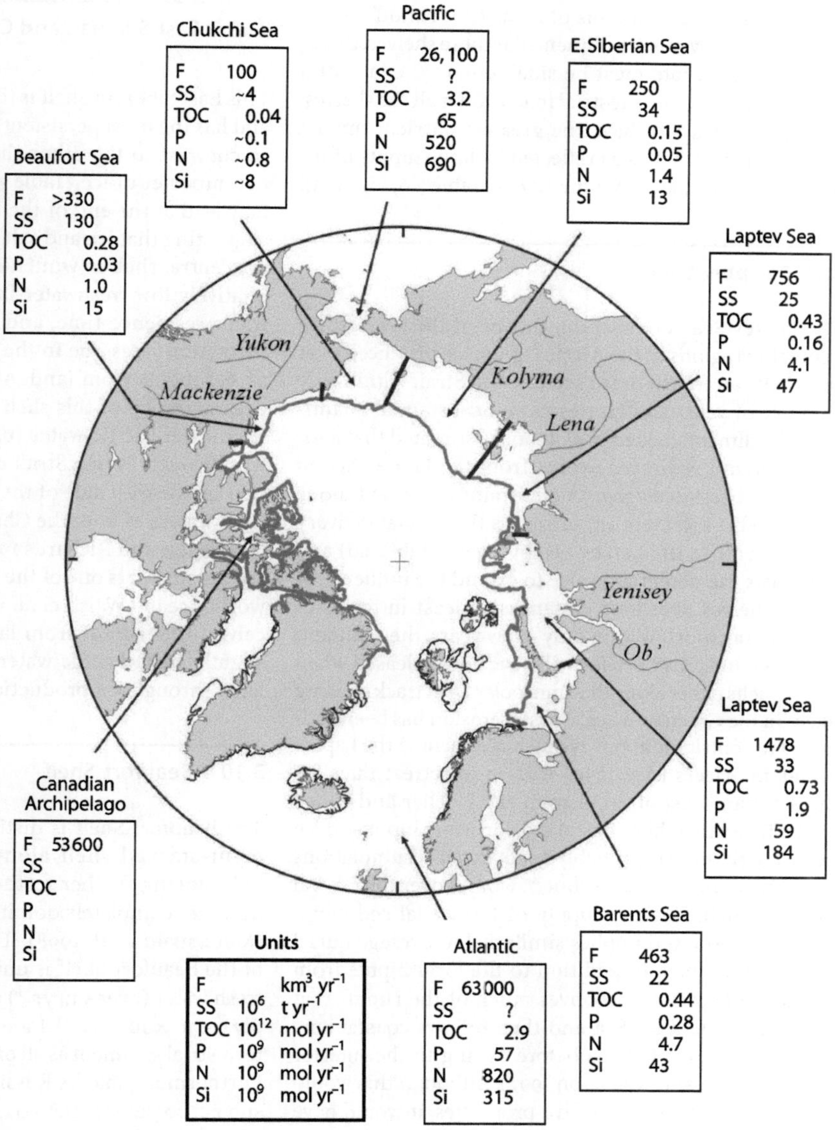

Fig. 3.13.
The Arctic Shelves and coastal inputs. Data have been collated from (Barrie et al. 1998; Gordeev 2000; Macdonald et al. 1998; Sharma 1979; Walsh et al. 1989) and references therein. Shown also is the inflow to the Chukchi from the Pacific Ocean. *F* refers to water and *SS* is the suspended solids. Symbols represent fresh water (*F*), suspended sediment (*SS*), total organic carbon (*TOC*), dissolved inorganic phosphorus (*P*), dissolved inorganic nitrogen (*N*) and reactive silicate (*Si*)

3.10.4 Barents Shelf

This shelf (Fig. 3.13) receives moderate inputs of runoff, small amounts of terrestrial particulate matter and has a relatively high productivity (Table 3.6). The Barents Sea branch of Atlantic water enters over this shelf to undergo seasonal modification by ice formation (Gerdes and Schauer 1997). The polar front results in a migrating and melting ice edge that can be both a contributor to productivity and source of material from the melting ice (cf. Sakshaug and Slagstad 1992). Salinification and cooling is crucial for the production of lower halocline waters which is reflected by the fairly large ice export from this region.

3.10.5 Kara Shelf

The Kara Sea has been relatively well studied because of the radioactive wastes disposed there (Eicken et al. 2000; Pavlov and Pfirman 1995; Pfirman et al. 1997). As an extension of the Barents Sea, the Kara Sea is strongly influenced by Atlantic water which has been cooled and modified while travelling over the Barents Shelf (Loeng et al. 1997). It receives large amounts of runoff from two of the Arctic major rivers, the Yenesey and the Ob, and exports substantial amounts of ice (Pfirman et al. 1997). The circulation and hydrology are highly variable due to winds, runoff and the sea ice cycle (Harms and Karcher 1999). Like all of the shelves on the Russian side of the Arctic, large amounts of freshwater runoff are not accompanied by much sediment. For these shelves, therefore, the burial rate must be small due to the lack of a supply of inorganic terrestrial material. Of all the shelves, the Kara appears to have the greatest nutrient impact from human activities as reflected by large inputs of nitrate and phosphate from the rivers (Table 3.6, Fig. 3.13).

3.10.6 Laptev Shelf

The Laptev Sea is remarkable as one of the greatest ice exporting regions in the Arctic Ocean. Ice produced over this shelf can transit the ocean to Fram Strait within only a couple of years and carries with it substantial quantities of sediment. Eicken et al. (2000) estimated that a total ice-bound sediment export from the Laptev Sea of 18.5×10^6 t resulted from one entrainment event alone in 1994. This sediment export rivals the annual delivery of sediments to the Laptev Sea by rivers (Table 3.6) and illustrates the potential for ice to extend the influence of Arctic shelves over long distances, at least in terms of particle transport. Within only a few years, the sediment entrained in Laptev Sea ice will have been released when the ice melts either along the Transpolar Drift track or, more likely, in the Greenland Sea. Coastal erosion has been estimated to contribute about as much sediment to the Laptev Sea as the rivers (Are 1999) with local retreat rates for the Siberian coast of up to 40 m yr^{-1} (Weller and Lange 1999). Taken together, coastal erosion, river input and ice export of sediments probably result in net accumulations of sediment near the coast but net losses over the central to outer shelf. The net supply of terrestrial sediment, however, is low and implies similarly low average burial rates on this shelf. In addition to the local inputs from rivers, the Laptev Sea receives much of the runoff that first enters the Kara Sea and then follows coastal currents into the Laptev Sea before exiting to the interior ocean (Olsson and Anderson 1997). Although this would transport mainly conservative properties, it would have a strong influence on the freshwater balance in the Laptev Sea. The characteristic formation of a flaw lead in the Laptev Sea at the end of the landfast ice in winter suggests that the substantial ice production and export is mirrored by brine additions to the water with consequent enhancement of mixing and, potentially, convection of dense plumes as illustrated in Fig. 3.10. Despite the low supply of terrestrial solids, the Laptev Sea margin sediment is dominated by terrigenous material as reflected in its organic matter composition (Stein et al. 1999). The terrestrial appearance of Laptev sediments is probably due to the relative stability of terrestrial organic compounds and the rapid remineralization of marine organic carbon in surface sediments (Olsson and Anderson 1997) since marine organic carbon in the form of algal material is abundant at times in the surface water (Peulvé et al. 1996).

3.10.7 East Siberian and Chukchi Shelves

This East Siberian Shelf is the largest in the Arctic Ocean and has the most persistent ice cover (Barrie et al. 1998). In contrast to the other shelves, the East Siberian is a net importer of ice (Table 3.6). However, it does form a flaw lead at the end of the landfast ice (~25 m isobath) suggesting that ice and salt production must occur over the central shelf in winter. The shelf is characterized by relatively low freshwater inputs for its size, has a long water residence time, and can support only low sedimentation rates due to the large area and small supply of sediments from land. A discontinuity occurs in the oceanography of this shelf toward its eastern end. Nutrient rich Pacific water (0.83 Sv; Roach et al. 1995) enters through Bering Strait especially in the Anadyr current on the west side of the strait to leave its imprint on shelf waters of both the Chukchi and East Siberian Seas (Codispoti and Richards 1968). The import of these nutrients supports one of the most productive areas in the world ocean (Walsh et al. 1989). This region, which receives little runoff from land, is instead influenced by slightly fresh Pacific water which is seasonally modulated through ice production and cooling.

3.10.8 Beaufort Shelf

The Beaufort Shelf is distinguished by a narrow, sediment-starved shelf along the Alaskan north slope but, moving further to the east, the shelf widens and becomes completely dominated by the Mackenzie River (Macdonald et al. 1998). The Mackenzie Shelf portion of the Beaufort Shelf is unique by virtue of the yield of fresh water (over 5 m yr^{-1}) and terrestrial sediment supply (~130×10^6 t yr^{-1}). Indeed, this shelf receives as much terrestrial sediment as all of the Russian shelves together. Furthermore, like its Russian counterparts, this shelf is subject to substantial coastal retreat (up to 10 m yr^{-1}) which provides a further 7×10^6 t yr^{-1} of terrestrial sediment. Despite the enormous input of freshwater, which continues at 5 000 $m^3 s^{-1}$ in winter under the landfast ice, this shelf produces convecting water over the middle shelf through salt rejection from ice (Melling and Lewis 1982). Because of its rich supply of inorganic terrestrial sediments, this shelf can sustain reasonable rates of burial.

The shelves of the Archipelago are somewhat disconnected from one another by channels (Fig. 3.9). Many of these shelves, especially the outer ones facing the Arctic Ocean, have not been studied. There is probably little freshwater runoff in the Archipelago, most of it coming from widely distributed small streams that have an exceptionally brief period of flow. Over the outer shelf the Arctic Ocean attains its thickest pack ice (Bourke and Garrett 1987) which probably limits primary production. Within the Archipelago, strong tidal mixing in passages can supply nutrients to support greater production (Cota et al. 1987). The net flow of water is out of the Arctic Ocean through the Archipelago (1.7 Sv) with much of that water originating in the Pacific. As a result the Archipelago forms a mirror image of the Chukchi and net exchange is from the interior Arctic Ocean, across the shelf and, eventually, down into Baffin Bay. Although the Archipelago channels form an important outflow for Arctic Ocean surface water (0–200 m), they cannot be viewed simply as 'pipes'. For much of the Archipelago, the Rossby radius (~10 km) is exceeded by the channel width and the channels, therefore, behave like a series of connected seas.

Although there are common features for all Arctic shelves in terms of seasonal ice cover, inputs from land, and exchange at the shelf edge, this brief survey of the shelves emphasizes how different they all are in terms of freshwater and sediment supply, ice production, source of nutrients and source of saline water (Pacific vs. Atlantic).

3.10.9 The Mackenzie Shelf of the Beaufort Sea as a Case Study

Budgets have been estimated previously for the Mackenzie Shelf for freshwater, sediments and organic carbon (Macdonald et al. 1995; Macdonald et al. 1998). Here, we build on those budgets, adding dissolved inorganic nitrogen (DIN) and dissolved inorganic phosphorus (DIP) using the terms and methods for box-modeling outlined by Hall et al. (1996) and Gordon et al. (1996). With this method, the shelf is viewed as a single box between land and interior ocean.

Within the Arctic Ocean, the Mackenzie Shelf provides one of the best opportunities to construct material budgets because: (1) inputs can be reasonably well constrained; (2) a large amount of data exist with which to estimate water and sediment properties over the shelf; and (3) the important processes are known qualitatively if not quantitatively. Despite many oceanographic studies of this shelf, there remain uncertainties in closing budgets for non-conservative properties including sediments and nutrients. The common feature of all Arctic shelves, the exchange or export of ice (Table 3.6) and brine (Fig. 3.10), creates difficulties not anticipated in Gordon et al. (1996) and Hall et al. (1996).

The Mackenzie Shelf comprises about one third of the Beaufort Shelf area (65 000 km^2, Fig. 3.13, Table 3.6). In the freshwater budget, the Mackenzie inflow by far dominates (Fig. 3.14a) but export of ice, directly or as ice melt, contributes significantly with precipitation minus evaporation only a minor component. The export of fresh water to the interior ocean is shown as two major components – one deriving from meteoric processes (V_R: runoff plus net precipitation), the other from the internal distillation inherent in ice production (V_I). This ice distillation process differs from the meteoric

Fig. 3.14. Budgets for the Mackenzie Shelf for **a** freshwater; subscripts refer to *Q* (inflow), *E* (evaporation), *P* (precipitation) and *R* (meteoric water); **b** salt; subscripts as in Fig. 3.14a and *X* (calculated volume exchange), and the brine component is estimated from the exported ice volume; **c** sediment; *SS* is shelf sedimentation, *CE* is coastal erosion and *DS* is sedimentation in the delta; **d** inorganic phosphorus and; **e** inorganic nitrogen

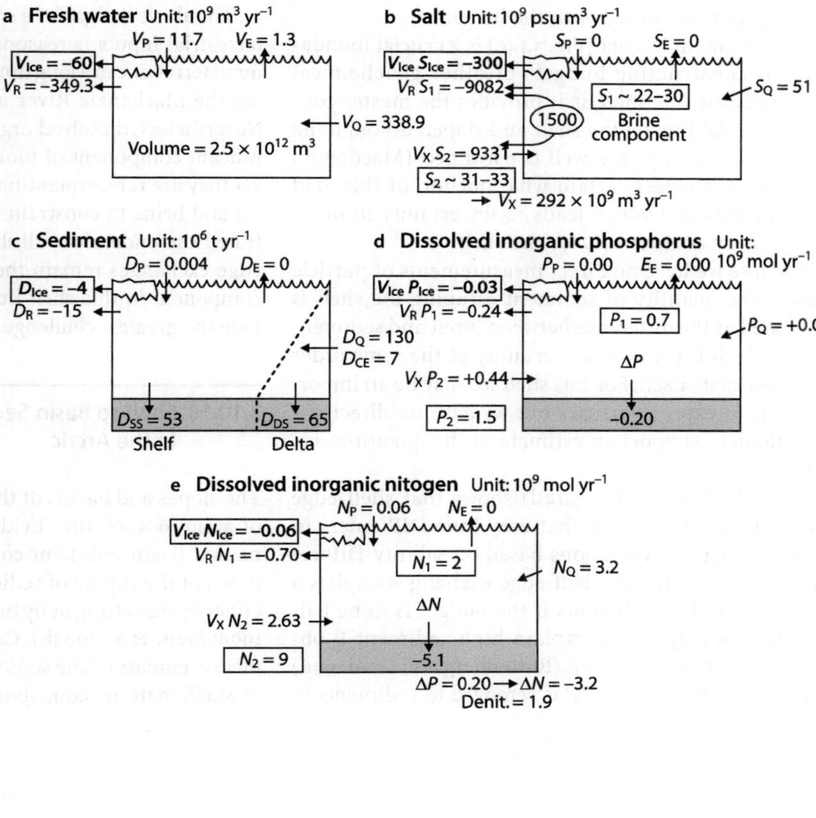

freshwater supply in that conservation of salt requires that it be mirrored elsewhere in the system by the brine rejected during ice growth. How this ice export affects budgets and/or stratification depends crucially on whether the brine is separated from the ice by penetrative convection or whether it remains coupled with the ice in the polar mixed layer (Macdonald 2000).

From the freshwater budget, salt exchanges (Fig. 3.14b) can be estimated using representative salinity distributions. The budget shows that shelf-edge exchange dominates. Although the transport of ice, with salinity of about 5 psu, is only a minor part of the salt budget (~300 × 10^9 kg yr^{-1}), the brine rejected from the ice is likely very significant (~1500 × 10^9 kg yr^{-1}). This brine will be exported near the surface if, for example, the water fails to convect deeply in winter. When sufficiently dense water is formed, the brine may sink and transit the shelf bottom picking up other properties (e.g., regenerated nutrients) to be exported into deeper waters of the interior ocean (Fig. 3.10). Although either route will satisfy salt balance, they differ drastically in where they will transport other components carried by the transport (nutrients, inorganic and organic carbon). If 60 km^3 of ice is exported (Fig. 3.14a), a further 1500 × 10^9 kg of salt must also be exported in surface or deeper water to maintain salt balance. It is difficult to close the ice/brine budget because both ice export and brine export are difficult to measure directly (Macdonald et al. 1995). The lack of certainty in how much salt is involved in convection, discussed above, makes it difficult to estimate other parameters (nitrate, phosphate) transported with the brine.

The sediment budget (Fig. 3.14c) is a crucial foundation for constructing budgets of other geochemical properties for this shelf as it provides the master control for burial flux on the shelf and slope. Although the Mackenzie loadings are well constrained (Macdonald et al. 1998), it is less certain what portion of this load stops in the delta which leads to uncertainty in other sedimentary components for the shelf.

Because we have no direct measurements of particle export, the quantity of sediment escaping the shelf is estimated as the difference between input and sedimentation which leads to an uncertainty of the same order as the estimate itself. For this shelf ice may be an important way to export particles but we have no direct observations to support an estimate of the quantities involved.

The DIP budget (Fig. 3.14d) shows that shelf-edge exchange dominates and that very little DIP enters in runoff. Using concentrations based on salinity-DIP relationships to estimate shelf-edge exchanges implies a loss (–P) to shelf sediments if the budget is to be balanced. This may partly explain high sediment P observed over the central shelf (Ruttenberg and Goñi 1996) but it is also likely that DIP is removed to sediments in the estuary by flocculation and particle settling (cf. Macdonald et al. 1987). In Mackenzie Shelf sediments, there is a close relationship between solid-phase Fe and P and a decrease in solid-phase P with depth (Gobeil et al. 1991). Phosphorus, therefore, is clearly linked to Fe geochemistry and a portion of the P must undergo regeneration back to the water column. It is difficult to use sediment measurements of total P to estimate the fraction of P actively cycling (Fig. 3.12) because the Mackenzie River supplies large amounts of solid phase P (~870 μg g^{-1}), some of which may be bioavailable, carried by a large and variable background sediment flux (Ruttenberg and Goñi 1996).

An important component of the DIN budget is supplied by the Mackenzie River (Fig. 3.14e). There is a significant sink (ΔN) in sediments which, compared to ΔP, exceeds the Redfield ratio (ΔN : ΔP = 26 : 1 – Redfield = 16 : 1) suggesting a preferential loss of DIN.

Porewater profiles clearly show that nitrate and ammonia both diffuse out of sediments into bottom water (Gobeil et al. 1991) and therefore sediments act as a source of regenerated nitrogen. However, the budget suggests a net loss of DIN over the shelf, most likely due to denitrification in sediments (cf. Christensen 1994). This loss (1.9 × 10^9 mol yr^{-1}) implies a denitrification rate of 0.1 mmol m^{-2} d^{-1} which is about a factor of 10 lower than rates measured in Bering and Chukchi sediments (Devol et al. 1997). This discrepancy may be partly explained by organic nitrogen which is often an important component of pristine rivers, equalling or exceeding DIN (Gordeev 2000).

The above budget for a small Arctic shelf dominated by terrestrial inputs is reasonably well constrained for the input terms due to a long time series of measurements both for the Mackenzie River and for the shelf properties. Nevertheless, dissolved organic nutrients could be a significant component of bioavailable nutrient supply and yet they are rarely quantified. The challenge provided by ice and brine to constrain the shelf budget is well illustrated and endemic to all shelves in the Arctic. The shelf-edge exchanges remain the largest and most uncertain component in this shelf budget, and will certainly provide the greatest challenge for Arctic shelves in general.

3.10.10 Shelf to Basin Sediment Transport in the Arctic

The slopes and basins of the Arctic Ocean cover an area of about 6 × 10^6 km^2. Evidence from Eurasian and Canadian Basin sediment cores supports the notion that much of the supply of sediment comes from the shelves either by ice rafting or by turbidity currents (Grantz et al. 1999; Stein et al. 1994b). Compositional and biomarker measurements of the sediments suggest that terrestrial, or shelf, material contributes strongly to both the inorganic and organic components of these basin sediments (Schubert and Stein 1996). As previously discussed, sea ice is recognized as an important transporting agent for sediments and, in some cases may remove sediments from shelves at a rate comparable with riverine supply. Nevertheless, evidence from the basin cores suggests that ice rafting is not the only, and probably not the major, source of sediments to basins (Stein et al. 1994a) and that the dominant transporting mechanism is probably turbidity currents initiated at the margins. It seems likely, therefore, that an important but poorly quantified export process for shelf particulates is slumps or turbidity flows at the shelf edge. Ice rafting, also an important export mechanism, probably delivers most of the sediments it carries to regions where the ice melts – either in the Barents Sea or in the Greenland Sea.

3.10.11 CH_4, DMS (Dimethyl-Sulphide) Production in the Arctic

Methane and DMS, produced in marine environments, are important contributors to greenhouse gases (Houghton et al. 1995) and pathways for organic carbon. In the case of methane, shelf sediments are likely to be the most important source to Arctic surface waters (Kvenvolden et al. 1993), whereas for DMS it will be open water and biological production that supplies the flux to air (Sharma et al. 1999). Based on measurements in the Beaufort Sea, Kvenvolden et al. (1993) estimated that Arctic shelves collectively could contribute a seasonal CH_4 source of about 0.1 Tg yr^{-1} ($\sim 0.006 \times 10^{12}$ mol yr^{-1}). Clearly, in view of the inorganic and organic carbon budgets given in Fig. 3.11a,b, CH_4 is a negligible component. Likewise for DMS, Sharma et al. (1999) estimated a flux equivalent to 0.004×10^{12} mol C yr^{-1}, also clearly a minor component of the OC budget.

3.10.12 A Budget for the Arctic Shelves

Arctic shelves vary so widely that no one shelf – for example, the Mackenzie – can reliably be used as a proxy for the others or by extrapolation to provide estimates for the entire Arctic Shelf. Furthermore, the differences between the shelves imply that they will each respond independently to change and each will have to be monitored to understand how change is manifested over the entire Arctic Shelf. With these cautions, we provide here a budget that builds on previous budgets and models of the Arctic Ocean. A primary difficulty in preparing such a budget, is that various authors use different boxes, assume different flows, and usually treat only selected components of the biogeochemical system (e.g., carbon or silicon). Here, we start with the terrestrial input data listed in Table 3.6 and Fig. 3.13. We build the budget using recent measured flows where available (e.g., Roach et al. 1995, for Bering Sea inflow) or model results where not available (e.g., Goldner 1999b) for upwelling and shelf-edge exchange. For organic and inorganic carbon, we rely heavily on the fluxes estimated by Anderson et al. (1998) recognizing that these authors did not use exactly the same water flows given in Fig. 3.15.

Although the budget (Fig. 3.15) cannot presently be closed, it captures important components of cycling on Arctic Shelves and reveals key weaknesses. The input of nutrients from land to sea is generally not the most important component of shelf cycling: shelf-edge ex-

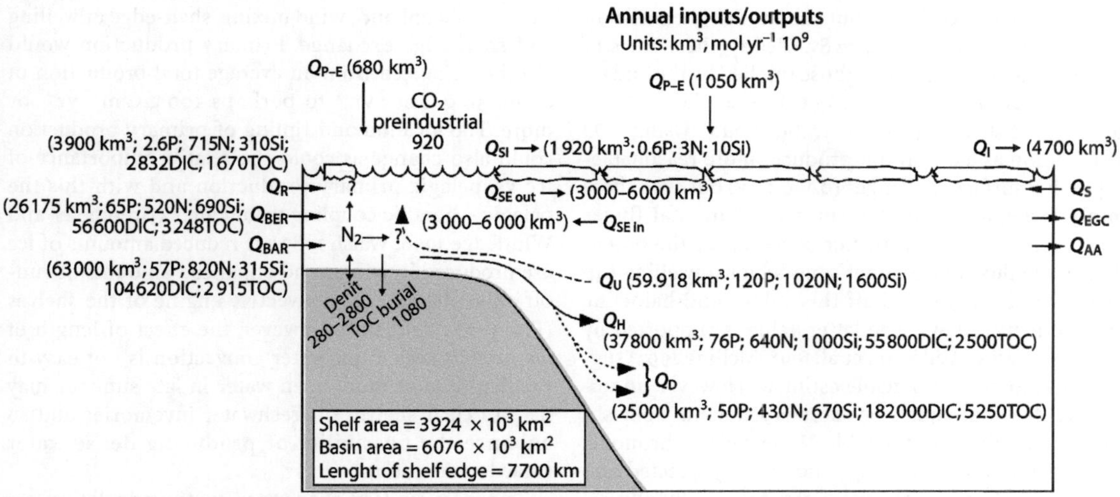

Fig. 3.15. A budget for the Arctic shelves. Q_R (river inflow, Table 3.6); Q_{P-E} (moisture flux – Walsh et al. 1994); Q_{Ber} (Bering Strait inflow – Roach et al. 1995); carbon fluxes (DIC, TOC (Anderson et al. 1998); Q_U, Q_H, Q_D (upwelling, halocline nourishment, deep water formation, Anderson et al. 1998; Goldner 1999b; Melling 1993); Q_{SI} (shelf ice export, Table 3.6). Fluxes of carbon were taken directly from Anderson et al. who used slightly different volumetric flows in their budget

changes appear to have far greater potential fluxes. Two key shelves, the Chukchi and Barents stand out because they are supported by throughflow from the Pacific or Atlantic Oceans respectively. These two regions, therefore, have the greatest potential for export production into the interior ocean. Denitrification over the shelves could, potentially, be a very large component of biogeochemical cycling and well exceeds whatever nitrogen the rivers supply to the shelves. This remains true even if we have underestimated the nitrogen input from rivers by a factor of two by neglecting the organic nitrogen. Clearly, it is crucial to refine our estimate of denitrification either by direct measurements in shelf sediments or by looking for the its 'integrated' signature of N_2O in halocline water. A shelf budget for the Arctic Ocean can be constructed without considering the Arctic Archipelago, as is done in Fig. 3.15, because this latter region is essentially an outflow. However, similar to the Chukchi and Barents Seas, the Archipelago is a flow-through shelf which accepts biogeochemical components from the Arctic Ocean, processes them within the broad channels and basins (mixing, primary production, decay, air-sea exchange, sedimentation), and discharges products into Baffin Bay. The Archipelago, therefore, is another location where strong biogeochemical cycling can occur and which is vulnerable to change especially in its ice climate and it deserves more intensive study.

Shelf-edge exchange remains very problematic in the Arctic Ocean budget. We envision three separate exchange processes (Fig. 3.15) which together with advection of ice and water must balance property fluxes: (1) surface mixed-layer exchange (Q_{SE}); (2) halocline and deep-water formation (Q_H, Q_D); and (3) upwelling (Q_U). Goldner's model (Goldner 1999b) without $\delta^{18}O$ constraint suggests that surface exchanges are of the order of 5 Sv (160 000 $km^3 yr^{-1}$) but the model is consistent with exchanges as high as 72 Sv. Clearly, these rates of exchange are far larger than those implied by the budget for the Mackenzie Shelf (Fig. 3.14; $V_x \sim 300\ km^3 yr^{-1}$ for about 7% of shelf edge in the Arctic Ocean). Using $\delta^{18}O$ distribution as a constraint produces more reasonable-appearing surface exchanges (0.1–0.2 Sv) but even these imply large, as yet undetermined, elemental fluxes (Fig. 3.15). Of equal, or greater concern, are the potentially large fluxes in Goldner's model for upwelling (up to 3.8 Sv; Fig. 3.15 uses half this value) and halocline nourishment (1.4 Sv). The latter value is supported by other estimates (Anderson et al. 1998; Melling 1993) suggesting it to be a reasonable estimate. However, an upwelling flux that equals or exceeds halocline nourishment seems highly improbable. Nevertheless, the model suggests that upwelling ought not to be neglected and measurements of its intensity are needed. Finally, although ice can be an important player in forcing transport through thermohaline circulation, Fig. 3.15 implies that the direct transport of soluble nutrients is small.

Permanent burial of organic carbon in Arctic shelf sediments is estimated to be about $1\,000 \times 10^9\ mol\ yr^{-1}$ by Anderson et al. (1998) which, on an aerial basis, is about ¼ the burial rate estimated for the Mackenzie Shelf. This seems reasonable given that the Mackenzie Shelf is strongly supplied with terrestrial sediments and can therefore support a relatively large burial flux. For the enormous Russian shelves, sediment supply is small and, in some cases, may be less than sediment loss from the shelf making these shelves poor sites for burial but potentially good sites for denitrification. For much of the Arctic Ocean, the shelf slope and basins may provide the most important burial sites.

Recently, Bauch et al. (2000) have inferred from stable isotope ($\delta^{13}C$) measurements that the Arctic Ocean with only 2% of the global ocean area accounts for 4–6% of the global ocean's anthropogenic CO_2 uptake. This anthropogenic carbon (4–7 Gt C or 0.3–0.6 Pmol C) probably entered the ocean through direct ventilation of shelf waters and river inflow, and was subsequently exported to the interior ocean in the halocline either directly or via a growth-decay cycle on the shelf.

3.10.13 Global Change; Speculation on Consequences for Arctic Shelves

The changes most easy to visualize for the Arctic Ocean in the upcoming decades relate to alteration in the Arctic ice cover which has recently shown surprising rapid manifestations of change (Macdonald et al. 1999; Maslanik et al. 1996; Rothrock et al. 1999). With warming, we expect the shelves to have longer periods of little or no ice cover and, therefore, to become more 'temperate' in their behavior. More open water for longer periods would enhance wind mixing, shelf-edge upwelling and shelf-edge exchange. Primary production would also be enhanced from an average total production of about 50 $g\ C\ m^{-2} yr^{-1}$ to perhaps 100 $g\ C\ m^{-2} yr^{-1}$ or more. The location and timing of primary production could also change as could the relative importance of ice vs. pelagic primary production and with this the pelagic – benthic coupling would alter (Grebmeier and Whitledge 1996; Walsh 1989). If reduced amounts of ice are produced in winter and enhanced amounts of run-off enter the sea, the convective engine of the shelves (Fig. 3.10) could stall. However, the effect of length of ice growth season on water convection is not easy to predict because more open water in late summer may help to clear shelves of freshwater inventories and so enhance the possibility of producing dense water (Macdonald 2000).

Longer periods of open water, continued relative sea-level rise and a higher incidence of storms (Everett et al. 1997) will accelerate coastal erosion. In some locations, for example the coasts of the Siberian and western

Canadian Arctic, enhanced coastal erosion could provide significant added supply of terrestrial material to the shelf which would be further augmented by sediment and carbon from rivers which have enhanced flow (Miller and Russell 1992) and which carry more sediments due to breakdown of permafrost in the drainage basins. The delivery of DOC by rivers to the coastal sea could further be affected by changes in the drainage basin including the drying out of peat or increase in fire.

Potentially the most significant changes facing the Arctic could originate from shifts in atmospheric pressure fields (e.g., the Arctic Oscillation, Thompson and Wallace 1998). Such shifts alter the Arctic Ocean surface circulation pattern (Proshutinsky and Johnson 1997) leading to rapid change in the ice climate (Macdonald et al. 1999), water-mass distribution and circulation (Carmack et al. 1997) and thence potentially large-scale change in primary production. In particular the Pacific/Atlantic front is prone to change its location (McLaughlin et al. 1996) which clearly may alter the large-scale distribution of silicate in surface waters. As discussed above, a change like this could alter the way this ocean takes up CO_2 from the atmosphere – smaller areas dominated by $Si(OH)_4$ imply less CO_2 flux into the water due to biological flux of sinking particles (Honjo 1997; Nozaki and Oba 1995). Although polar oceans are not reported to have widespread coccolithophore populations (Tyrrell et al. 1999), recent unprecedented blooms of these carbonate phytoplankton in the Bering Sea (Weller and Lange 1999) should serve as a warning.

Ozone depletion in the upper atmosphere is a concern in the Arctic because it would enhance the incident UV radiation to the surface (Weatherhead and Morseth 1998). DOC – especially highly-coloured terrestrial DOC – strongly attenuates UV in aquatic systems (Schindler et al. 1997) and, in fact, changing its concentration provides a potentially far greater effect on aquatic environments than the projected change to incident UV from ozone depletion. Terrestrial DOC contributes a much larger component of DOC in Arctic surface water than it does in the Atlantic and Pacific Oceans and appears presently to survive transport to Fram Strait which takes between 1–6 years (Opsahl et al. 1999). One of the reasons why allochthonous DOC survives in Arctic surface waters may be that it is protected from photochemical degradation by ice cover (Opsahl et al. 1999). If so, reduced ice cover could not only enhance incident UV to the water surface but, more importantly, it could allow greater destruction of terrestrial DOM further exacerbating the exposure. The above comments underscore the need to construct budgets independently for the terrestrial and marine carbon components as they have different sensitivities to change.

Taken together, the primary forcing for Arctic Ocean change is most likely to come from change in the ice climate which can be effected by increased temperature and by change in the wind fields which provide both thermal and mechanical forcing. Furthermore, the loss of ice cover enhances the penetration of solar radiation leading to further melting – i.e., positive feedback. It is easy to envisage that reduced ice will lead to direct changes in shelf productivity through enhanced upwelling and mixing, for example, and enhanced air-sea exchange (e.g., heat, moisture, DMS). The greatest impact on global climate and carbon, however, may come from displacements or bifurcations in biogeochemical and physical pathways of the Arctic. Examples include: (*1*) The pathways of freshwater and other terrestrial inputs from the Russian Shelves (amount entering the Canadian Basin vs. the amount entering the Eurasian Basin); (*2*) The amount and source of surface water exiting through the Archipelago vs. Fram Strait; (*3*) The amount of primary production from pelagic vs. ice algae (which also may affect the particle flux); (*4*) The amount of primary production supported by silicate phytoplankton vs. carbonate phytoplankton; and (*5*) The amount of organic carbon regenerated in sediments to CO_2 vs. CH_4.

We are presently far from a sufficient understanding of the Arctic Ocean system that would allow us to predict these changes except that the shelves will likely be the first and most important locations for change.

3.11 Marginal Seas

Marginal seas form the linkage between the continents and the oceans, and as such, receive much land runoff, ventilate the deep oceans and exchange much material with the open oceans. The roles of these marginal seas in the context of carbon and nutrient cycles are briefly discussed here, with special emphasis on the CO_2 sink.

3.11.1 High Latitude Marginal Seas

High latitude marginal seas, such as the Bering Sea, the Sea of Okhotsk and the Baltic Sea are relatively nutrient-rich compared with the open oceans, and not surprisingly, enjoy higher productivity. Diatoms often dominate in these waters and because their skeletons are silicious and their soft tissues consume CO_2, the biological pump in these marginal seas continuously removes CO_2 from the atmosphere.

The Bering Sea, with an area of 2.3×10^6 km^2, is the third largest marginal sea in the world. Half of the sea is less than 200 m in depth, and the shelf waters frequently have a pCO_2 of 100 ppm or more below saturation as a result of the cooling effect in winter and primary productivity in spring and summer. Near the shelf break, pCO_2 is frequently supersaturated because of vertical mixing, but overall the shelf ecosystem serves as a sink for atmospheric CO_2 (Codispoti et al. 1986; Chen 1993; Walsh and Dieterle 1994). Primary productivity on the shelf is on the order of 165 (50–300) g C $m^{-2} yr^{-1}$. Although most of it is consumed on the inner shelf, just slightly less than half is exported northward to the Arctic Ocean through the Bering Strait or southward to the Aleutian Basin. The *f*-ratio is 49% on the outer shelf and 17% at mid-shelf (Wollast 1998; Takahashi 1998). In the surface water, dimethylsulfide and N_2O have concentrations of about 50 ng S l^{-1} and 313 ppb, respectively (Nojiri et al. 1997; Uzuka et al. 1997).

The shelf water is probably supersaturated with anthropogenic, or excess CO_2, unlike the deep basin waters which are not. Overall, the Bering Sea contained about 0.21 (±0.05) Gt excess carbon around 1980. Albeit small in value, the carbonate deposits in the vast Bering Sea and the Sea of Okhotsk shelves may well be capable of providing a large sink for excess CO_2 in the near future. It would follow that a doubling of the current CO_2 level in the atmosphere by the latter part of the next century would bring about the dissolution of the calcites on the shelves, which in turn would provide another large sink for CO_2 (Chen 1993).

The surface water in the Sea of Okhotsk is also generally below saturation for pCO_2, except near the Kashevarou Bank and the Boussole Strait where tidal mixing is strong (Rogachev et al. 1997). The dense intermediate water is believed to be an important source of the North Pacific Intermediate Water (NPIW). Kurashina et al. (1967) estimated the gross exchange of water between the North Pacific and the Sea of Okhotsk is about 15 Sv. The present authors have estimated that excess CO_2 totaling 0.18 ±0.08 Gt has already penetrated to at least 1 000 m in the Sea of Okhotsk, and that there is an export value of (0.011–0.18 Gt C yr^{-1}) to the North Pacific ($\sigma_\theta = 27.35–27.5$; Chen and Tsunogai 1998; Andreev et al. 2001).

The Baltic Sea, at 0.42×10^6 km^2, is essentially an estuary with a large runoff; moreover, precipitation exceeds evaporation. As a result, very fresh water flows out of the Baltic Sea through the Kattegat and Danish Straits mainly in the surface layer, while the saltier North Sea water mainly enters near the bottom. With its shallow depth, the entire Baltic Sea is thought to be saturated with respect to excess CO_2, but the total amount is actually only 0.011 ±0.002 Gt C. Most C, N and P are recycled, with little being exported to the North Sea.

3.11.2 Semi-Enclosed Marginal Seas

The South China Sea (SCS), the largest marginal sea in the world (3.5×10^6 km^2), the Mediterranean Sea (2.5×10^6 km^2), the second largest in the world, the Sea of Japan (1×10^6 km^2), the eighth largest and the Red Sea (0.44×10^6 km^2), the 13th largest are all deep basins with narrow shelves. They are characterized by a unique homogeneity in their subsurface waters with only narrow and shallow straits connecting them to the outside. The Red and Mediterranean Seas are unique in the sense that evaporation is greater than precipitation plus runoff. Consequently, there is an inflow of warm, nutrient-poor surface water and an outflow of cooler, nutrient-rich deep waters through the Straits of Gibraltar and Babel Mandeb, respectively.

Since the deep waters in some semi-enclosed marginal seas are formed inside a closed basin, they provide a unique opportunity to estimate the Redfield ratios based on the mass-balance method and a 1-D model (Chen et al. 1996a). For instance, the isolated basins in the Sea of Japan provide a clear indication that the bottom water has become stagnated at least since the 1940s, perhaps because of global warming. Meanwhile, nitrogen and phosphorous concentrations in the deep waters have increased. Similar results have been reported for the Mediterranean (Bethoux et al. 1998a,b; Chen et al. 1999) where the paucity of rivers and upwelling results in low productivity, and P is probably a key factor in the control of the bacterial and phytoplankton growth rates and in the export of carbon (Thierry et al. 2000).

Many marginal seas have deep water formation and, in the process, take up atmospheric CO_2 and transport it to the deep waters. Most phytoplankton also absorbs CO_2 from the surface layer and carries carbon to the deep waters. However, with their coral formations marginal seas in the tropics may release CO_2 to the atmosphere because the formation of coral skeletons results in higher pCO_2. Additionally, the growth of *coccolithophorids* can change the roles of sinks and sources of CO_2 depending on the inorganic vs. organic carbon ratio. The calculated vertical profiles of excess CO_2 indicate that the entire Sea of Japan, Red Sea and the Mediterranean Sea have been penetrated by excess CO_2, amounting to 0.31 ±0.05 Gt C for the Sea of Japan in 1992 (Chen et al. 1995a), 0.07 ±0.02 Gt C for the Red Sea in 1978 (Krumgalz et al. 1990; Chen and Tsunogai 1998) and 1.3 ±0.3 Gt for the Mediterranean in 1990. Owing to its shallow sills, the Sea of Japan does not transport excess CO_2 to the interior of the North Pacific. In contrast, the Mediterranean Sea transports 0.012 ±0.006 Gt yr^{-1} of excess CO_2 to the intermediate waters of the North Atlantic. The *f*-ratio in the Mediterranean Sea ranges from 0.15–0.44, yet the POC flux represents even less than 5% of primary production. This suggests that a large fraction of the carbon export is in the form of DOC, though this is still unquantified (Miquel et al. 2000). There is also deep water formation in the northern Gulf of California in winter, but the newly-formed deep water, along with the excess CO_2 it carries, seems to be confined to only the northern gulf which is also shallow.

The ratio of $CaCO_3$ dissolution and organic carbon decomposition varies from 0.05 at 300 m to about 0.17 below 2 000 m in the Sea of Japan (Chen et al. 1996a). These values are lower than the ratios of 0.14 and 0.36 found in the South and North Pacific, respectively (Chen et al. 1982; Chen 1990), or 0.54 in the Bering Sea (Chen 1993) because diatom and silicoflagellates dominate the phytoplankton in the Sea of Japan (Hong et al., pers. comm.). Another result is that the vertical gradient of SiO_2 is several times higher than that of calcium or alkalinity (Chen et al. 1995a). Carbonates do not seem to be dissolved in the Red Sea because calcite and aragonite are already supersaturated by as much as 300% and 100%, respectively.

The shelves of the SCS occupy an area of about 1.2×10^6 km^2. There is considerable seasonal variation in productivity, with the highest value in winter and the lowest during the inter-monsoon periods. A 3-D numerical model gave an annual mean productivity of 370 mg C m^{-2} d^{-1}, almost identical to the observed value. Further, the model also predicted an *f*-ratio of 0.14 (Liu et al. 2000). Much of the new production is supported by upwelling because the nutrient supply from rivers and that in the surface seawaters through the Luzon and Mindoro Straits and the Sunda Shelf (Fig. 3.16) are only a small fraction of what is required to support new production. In fact, the upwelling of the SCS deep waters conveys nutrients to the intermediate and surface layers. Only a portion of the upwelled nutrients is transported to the ECS shelf through the shallow Taiwan Strait. Further, the outgoing SCS surface water in the Luzon Strait contains many more nutrients than the incoming Kuroshio surface water. The outflowing SCS surface and intermediate waters turn northward, therefore also transporting nutrients to the ECS shelf. It is evident that the SCS acts as a pump sending nutrients away from the deep waters into the euphotic zones in the SCS and the ECS (Chen and Wang 1999; Chen et al. 2001b).

Because of strong upwelling, the deepest excess CO_2 penetration (the depth where the excess CO_2 is at 5 µmol kg^{-1}) is only around 1 500 m, a depth shallower than that in the Northwestern North Pacific. The entire SCS probably contained 0.6 (±0.1) Gt of anthropogenic carbon back in 1994. The saturation horizon of calcite is deeper than 2 000 m in the SCS; thus little enhanced dissolution is expected due to excess CO_2 penetration (Chen and Huang 1995). The saturation horizon of aragonite is 600 m, shallower than the depth of excess CO_2 penetration, but deeper than the saturation horizon found in the Bering Sea (350 m). The upward migration of the saturation horizon (Feely and Chen 1982) would not affect the calcareous deposits on the SCS shelf to the degree that it does the Bering Sea deposits.

There are several other marginal seas, but regrettably, relatively little information is available. The Black Sea, at 0.46×10^6 km^2, connects to the Mediterranean Sea through the very narrow and shallow Straits of Bosphorus and Dardanelles. There is a large excess of precipitation plus runoff relative to evaporation; hence, the fresher surface water flows out of the straits on top of the incoming saltier Mediterranean water. The shallow Persian Gulf (0.24×10^6 km^2), on the other hand, has more evaporation than precipitation plus runoff. As a result, subsurface water flows out at the same time that the outside surface water flows in through the Straits of Hormuz. The oxidation of organic matter is mainly by sulfate (Goyet et al. 1998), and it is expected that the old, anoxic water in the Black Sea below the depth of about 100 m does not contain much excess CO_2. Unlike the Sea of Japan or the Mediterranean Sea that have warmed, the Black Sea has shown a decrease in temperature, perhaps because of decreased freshwater input (Murray et al. 1991). The nutrient supply, however, has actually increased. Indeed, in the last three decades its ecosystem has transformed from a once highly diverse, healthy state in the 1960s to its present eutrophic state charactized by low biodiversity (Oguz et al. 2001).

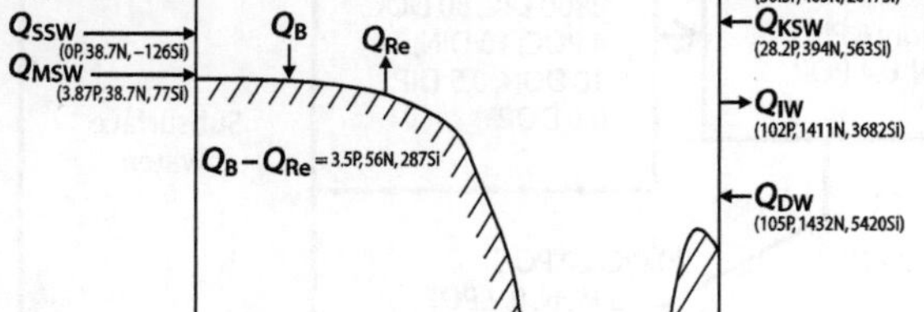

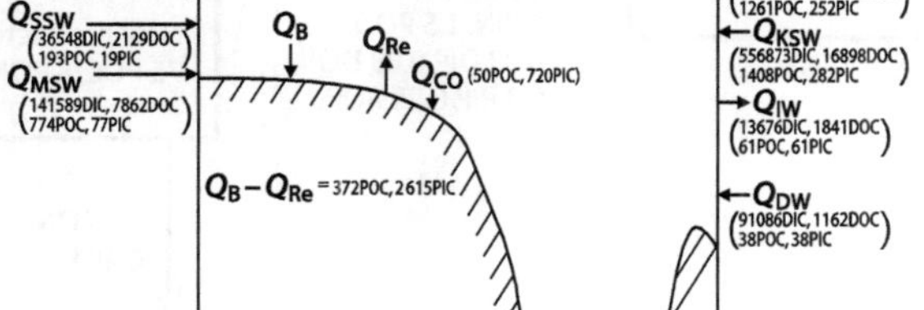

Fig. 3.16. Schematic diagram for the annual (**a**) nutrient and (**b**) carbon budgets (numbers in 10^9 mol yr^{-1}) in the South China Sea. Symbols are the same as Fig. 3.1 except that *SCSW*, *KSW*, *DW*, *MSW*, *SSW* and *CO* denote South China Sea Surface Water, Kuroshio Surface Water, Kuroshio Deep Water, Mindoro Strait Water, Sunda Shelf Water and coral reefs

3.11.3 Initial Synthesis

Almost without exception, the observed data and budget calculations for the continental shelves suggest that although a small fraction of carbon is exported across the shelf-slope break, the principal fate of the carbon which is produced on the shelf is, in fact, oxidation on the shelf. Upwelling seems to support most of the nutrients that are required for new production and denitrification. The global mean dissolved N/P ratio for river outflow is 23 (18–31) according to Bolin and Cook (1983). It is natural to suggest that P is the limiting factor, especially when considering that nitrogen fixation occurs. It is the upwelled seawater, with normally an inorganic N/P ratio close to 16, that supplies DIP to the shelves.

Recently, Nixon et al. (1996) and Galloway et al. (1996) studied the outcome of N and P at the land-sea margin (5.7×10^6 km^2) of the North Atlantic Ocean. They concluded that riverine N is only transported to the open ocean in a few areas, with the flow mostly coming from a few major rivers. This is because net denitrification in most estuaries and continental shelves exceeds the amount of N supplied to the shelves by rivers and requires a supply of nitrate from the open oceans. Their denitrification flux is equal to 12 times the amount deposited on the shelf and slope. Results for the ECS (Fig. 3.2) give a denitrification flux of less than half of the shelf and slope deposit. Another discrepancy is that the P budgets of Nixon et al. and Galloway et al. indicate that there is a net export of DIP to the open oceans in the water column, whereas the opposite holds true in the ECS (Fig. 3.2). The net export from the ECS is mostly supported by upwelling. Extrapolating the ECS value to the global shelves gives an upwelling rate of 4.2×10^{12} mol N yr^{-1}. Brink et al. (1995) gave 7.5×10^{12} mol N yr^{-1}, but J. J. Walsh (1991) gave 40×10^{12} mol N yr^{-1} which seems to be too high as upwelling is probably seldom that strong. The true value is probably on the order of 10×10^{12} mol N yr^{-1} (Fig. 3.17).

J. J. Walsh (1991) and Wollast (1998) gave a net denitrification flux for the world ocean shelf zone (26×10^6 km^2) of 3.6×10^{12} and 2.5×10^{12} mol N yr^{-1}, respectively. Middelburg et al. (1996) gave 7×10^{12} mol N yr^{-1} occurring in shelf sediments. Had the ECS value been extrapolated in this study to cover the world shelves, the flux would be 2.77×10^{12} mol N yr^{-1}, which is in good agreement with the results of Wollast. However, Wollast estimated a net preservation of 0.71×10^{12} mol N yr^{-1} for shelves globally compared with the extrapolated value of 2.17×10^{12} mol N yr^{-1} based on the ECS studies. The offshore transport of organic N was estimated to be 28×10^{12} mol N yr^{-1} by Wollast vs. the extrapolated ECS value of 3.64×10^{12} mol N yr^{-1}. Extrapolating the value of Galloway et al. (1996) would yield a value of only 0.12×10^{12} mol N yr^{-1} for the sum of shelf deposits and offshelf transport of organic N. Despite such a large discrepancy, it is obvious that upwelling is by far the dominant source of nitrogen necessary to sustain organic N production, and hence, burial and offshore transport on the continental shelf. A denitrification value of 2.5×10^{12} mol N yr^{-1} is adopted in this study (Fig. 3.17).

J. J. Walsh (1991) used a primary productivity of 430×10^{12} mol yr^{-1}. He assumed that all upwelled nitrate is taken up by phytoplankton, and he went on to estimate

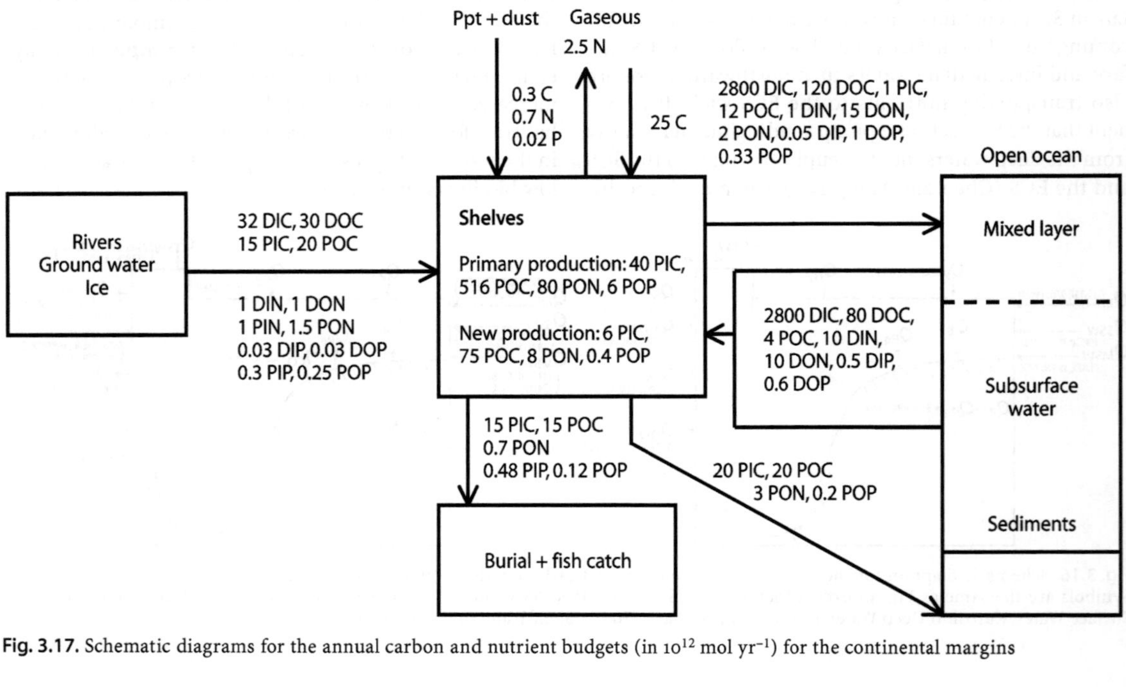

Fig. 3.17. Schematic diagrams for the annual carbon and nutrient budgets (in 10^{12} mol yr^{-1}) for the continental margins

Table 3.7. Non-conservative fluxes in the continental margins

Name	(p–r)	IC air/sea	POC/PIC sed.	DOC/POC offshore	DIC/PIC offshore	IP dust	POP/PIP sed.	DOP/POP offshore	DIP/PIP offshore	IN dust/rain	Net denit	PON/PIN sed.	DON/PON offshore	DIN/PIN offshore	Ref.
Amazon Shelf		–	+/+	–/–	+/–	+	+/+	–/–	+/–	+	–	+/+	–/–	+/–	1,5
Baltic Sea	+	+	–/–	–/0		+	–/–	–/0	0/0	+	–	–/–	–/0	0/0	1,7,11
Barents Sea		+	+/												1,8
Beaufort Sea			+/												
Benguela Current	+	+	–/–	0/0	+/–	+	–/–		+/–	+	–	–/–		+/–	1,2
Bering Sea Shelf, summer	+	+	–/–	–/–	+/–	+	–/	?/–	+/–	?	+	–/	–/–	+/–	1–5
Bering Sea Shelf, winter	?	0	–/–	–/–	+/–	+	–/	?/	+/–	?	?	–/	–/–	+/–	1–6
Brazil Shelf		–				+				+	–				1,9,10,12
Buenos Aires Shelf	+	+				+				+	0				1,9,12
Californian Shelf		+	0	–/–	+/–	+	0/	?/–	+/–	+	–	0/	–/–	+/–	1,5,12
Canadian Archipelago Shelf															
Caribbean Sea						+				+	0				1,12
Chukchi Shelf		+	–/–		+/–		–/–		+/–			–/–		+/–	1,5,8
E. China Sea Shelf	+	+	–/–	–/–	+/–	+	–/	–/	+/–	+	–	–/	–/–	+/–	1,4–6,12
E. Canadian Shelf			–/–								–				2,10
E. Siberian Shelf			+/												
Equatorial W. African Shelf		–				+				+					1,9
Great Barrier Reef	+	–	–/–	?	?	0	–/–	0/0	0/0	0	0	–/–	0/0	0/0	1,2
Gulf of St. Lawrence						+				+	–				1,10
Gulf of Thailand	–	0	–/–	–/–	+/–	0	–/	?/–	+/–	?	–	?/	?	+/–	2
Iberia Shelf		+				+				+					1,9
Irish Sea	+	+	?	?		+	?		+/–	+	–	?		+/–	1,2
Kara Sea		+													1,8
Laptev Sea		+	–/–		+/–		–/?		+/–				–/?	+/–	1,8,13
Mackenzic Shelf	–	–	–/–	+/+	+/–	+	–/–	–/–	+/–	+	–	–/–	–/–	+/–	1,2
Mediterrean Sea		+	–/–		–	+	–/–			+	–	–/–			1
Mid Atlantic Bight			–/–	–/–		+				+	–				2
N. Chile Coast	+	?	?	–/–	+/?	+	?	–/–	+/–		–	?	–/–	+/–	1,2
North Sea	+	I	–/–	–/–	+/–	0	–/–	–/–	+/–	+	–	–/–	–/–	+/–	1,2,5
NW African Shelf		+				+				+	–				1,9,12
NW American Shelf		+	–/–	–/–	+/–	+	–/	?/–	+/–	+	–	–/	–/–	+/–	1,5,12
Patagonia Shelf		+				+				+	–				1,9,12
Red Sea			–		+	+				+	–				1,12
Sea of Japan Basin			–/–			+	–/					–/			1
Sea of Okhotsk Basin			–/–			+	–/					–/			1
S. Atl Bight						+				+	–				1,2
South China Sea Basin	+	+	–/–	?		+	–/	?		+	–	–/	?	–/+	1
Sulu Sea			–/–			+				+		–/–			1
Upper Gulf of Thailand	+	–	–/–	0	0	0	–/	?/–	?/–	+	–	–/	?/–	?/–	2
Weddell Sea		–	–/–												1,12
Western Canadian Shelf			0	–/–	+/–	+	0/	?/–	+/–	+	–	0/	–/–	+/–	1,5

1: This study; *2:* JGOFS (1997); *3:* Walsh (1995); *4:* Walsh et al. (1981, 1989); *5:* de Haas (1997); *6:* Chen (1985, 1988, 1993); *7:* Gordon et al. (1996); *8:* Semiletov (1997); *9:* Bakker (1998); *10:* Seitzinger and Giblin (1996); *11:* Thomas and Schneider (1999); *12:* Weiss et al. (1992); *13:* Stein et al. (1999); *14:* Walsh et al. (1992). "+" means entering the shelf system; "–" means leaving the system.

a global shelf new productivity of 240×10^{12} mol yr^{-1}, which gave an *f*-ratio (new/total production) of 0.54. Wollast (1998) also estimated primary productivity at 500×10^{12} mol yr^{-1} (230 g C m^{-2} yr^{-1}) on the shelf globally, but he claimed that 16.5×10^{12} mol yr^{-1} (7 g C m^{-2} yr^{-1}) is preserved in the sediments and that 180×10^{12} mol yr^{-1} (83 g C m^{-2} yr^{-1}) is exported offshelf in organic form. Mackenzie et al. (1998b) chose to use primary productivity, organic carbon accumulation and offshore transport as 766, 20 and 32×10^{12} mol yr^{-1}, respectively, in their global continental shelf carbon cycle budget calculations. J. J. Walsh (1991) pointed to an offshelf export of 240×10^{12} mol yr^{-1} without shelf deposit although Mackenzie et al. reported only a small offshelf export. In this study, the global shelf productivity is taken to be 516×10^{12} mol yr^{-1} (214 g C m^{-2} yr^{-1}), while the organic C preservation and the downslope offshelf POC export is 15×10^{12} mol yr^{-1} (7.9 g C m^{-2} yr^{-1}) and 20×10^{12} mol yr^{-1} (9.2 g C m^{-2} yr^{-1}), respectively (Fig. 3.17).

In terms of the PIC, Milliman (1993) obtained a preservation value of 7.5×10^{12} mol yr^{-1} for all shelves other than coral reefs in the world. The export is 2.5×10^{12} mol yr^{-1}, again excluding coral reefs. In this study, the preservation value of PIC is taken to be 15×10^{12} mol yr^{-1}, and the export is 20×10^{12} mol yr^{-1}, much higher values than those reported by Milliman. Mackenzie et al. (1998b) chose an accumulation rate of 21×10^{12} mol yr^{-1}. Though they did not report on the PIC export rate, they did use an air-to-sea flux of 0.03×10^{12} mol yr^{-1} which Wollast (1998) ignored. The extrapolated global air-to-sea flux on the shelf based on the ECS data would be 53×10^{12} mol yr^{-1}. Tsunogai et al. (1997) gave an even larger value of 83×10^{12} mol yr^{-1}. In contrast, Holligan and Reiners (1992) and Ver et al. (1999) gave an opposing direction, sea-to-air flux of 33 and 17×10^{12} mol yr^{-1}, respectively. Clearly, large uncertainties and discrepancies exist here, but data for the Sea of Okhotsk, the North Sea and the Arctic Ocean, among others, also show an air-to-sea flux of CO_2 (Table 3.7). It seems that a global air-to-sea transfer rate of 25×10^{12} mol yr^{-1} of CO_2 for the shelves is not unreasonable. In terms of POC, the global net offshelf export data is 28×10^{12} mol yr^{-1}. The net offshelf DOC export is larger, at 40×10^{12} mol yr^{-1} (Fig. 3.17). These values are larger than the riverine inputs thus the shelves are autotrophic.

The above budget calculations, however, are all under the assumption of a steady state. Interannual variations, such as those from an ENSO event, affect the runoff a great deal. As a result, the upwelling rate and all associated fluxes should likely be subject to change. Further, slumping and turbidity currents may wash away the shelf deposit and reduce the reservoir. These possibilities have not been considered. Another point of concern is the reduced freshwater outflow due to irrigation and the damming of rivers. For instance, the Nile, Colorado and the Yellow Rivers now discharge little water. Reduced freshwater input to the shelves reduces the buoyancy effect and diminishes upwelling and the nutrient supply, not to mention productivity as well.

Table 3.7 summarizes the systems studied above to exemplify the similarities and differences in terms of the general typological characteristics explored. These include allocating systems on the basis of positive or negative estimates for a number of processes and fluxes, where the symbol is taken as gains (+) or losses (−) with respect to the shelf. Key processes are ($p - r$), (*nfix* − *denit*), air-to-sea transfer of C, shelf-to-ocean particle transfers and burial, where p is production, r is respiration, *nfix* is nitrogen fixation, and *denit* is denitrification.

One growing concern from these considerations of cross-shelf exchange and regionalization in the construction of some of the models for the various budgets is the constraint of the boundary conditions, particularly in the complexity of physical oceanography which affects upwelling and offshore advection. Another concern is the common use of C and N as currencies of metabolism when P may be a better choice as it exhibits no change in speciation. Phosphorous essentially behaves as a conservative tracer with respect to system-level metabolic processes – except where sediment-loading and riverine input may cause inorganic reactions to dominate P-cycling processes. A case in point is denitrification which is clearly an important process in continental margins, where N/P flux ratios offer the opportunity to explore DIN deficits due to denitrification.

Notwithstanding the uncertainties, the following scenario seems to emerge once all of the available information is summarized and studied (Table 3.8): On the present day shelves, the *f*-ratio is only about 0.15, with most external inorganic nutrients supplied by upwelling. About 97% of the organic matter introduced onto the shelves from terrigenous sources and by primary production remineralize in the water column and in the sediments. Thus, under present day conditions, it can be asserted that shelves do not play an important role in the preservation of organic carbon although the shelves do absorb 0.3 Gt C from the atmosphere each year. Periods with low sea level during the ice ages probably resulted in the exposure and erosion of sediments, thereby reducing the role of the shelves in the preservation of organic carbon and the sequestration of atmospheric CO_2 even more.

3.11.4 Future Research

Global warming is likely to result in increased stratification but decreased productivity in the upper layer of the open ocean. On the other hand, warming is likely to be accompanied by a stronger pressure gradient between the land and the oceans. The stronger pressure gradient will strengthen alongshore geostrophic wind, causing enhanced offshore Ekman transport and amplified coastal upwelling. Whether this may cause the already 'turbulent' upwelling system to become less productive or instead, cause the increased supply of nutrients to generate higher productivity needs to be investigated. The effects of ENSO, known to affect the physical and biological properties of near coast waters, are less known in the marginal seas and the polar regions. The existence or absence of teleconnections must be determined.

Total groundwater discharge and associated chemical fluxes to the coastal zones are also a question of debate, especially over lengthy periods of time and in large areas. The total groundwater discharge is probably about 5 to 10% of the total surface discharge (36 000 to 38 000 $km^3 yr^{-1}$) that reaches the oceans. However, the total flux of dissolved solids in groundwaters may make up a higher proportion of the riverine flux and may even represent as much as 50%. Although the input of DIP via groundwaters is not expected to be higher than that via surface waters because of the poor solubility of mineral phosphates and the tendency of dissolved phosphorus to be adsorbed on solid particles, nitrate loading of groundwaters may result in a substantial input to the coastal zones. DIC concentrations in groundwaters are elevated owing to the dissolution of limestone and the bacterial oxidation of organic matter. Thus, the DIC flux via groundwaters to the coastal oceans may amount to as much as 25% of the riverine flux. Dissolved organic fluxes of carbon, nitrogen and phosphorus associated with groundwater discharge to the coast are poorly known, but they probably do not make up a very substantial portion of the surficial inputs. The global depositional fluxes of nitrogen (NO_y) and reduced nitrogen (NH_3) are known over large areas of the world and, to a first approximation, over certain regions of the oceans. Less well constrained are the depositional fluxes of organic carbon, nitrogen and phosphorus. It is also unknown whether increasing anthropogenic C, N and P inputs from rivers, but maintaining a relatively constant Si supply would affect the relationship between siliceous and calcareous organisms. More siliceous organisms, of course, would lead to larger CO_2 draw down. Si-depleted nutrient loading, however, yields *Phaeocystis* which are neither siliceous nor calcareous, but are major dimethyl sulfide producers in areas such as the North Sea.

It has been hypothesized that the atmospheric delivery of major nutrients (N, P and Si) plus some trace nutrients (e.g., Fe) would affect coastal and ocean biological production, and in some cases, limit production. For instance, inputs of nitrogen via atmospheric deposition to coastal areas may constitute on a local scale as much as 50% of the total river plus atmospheric input. In the North Atlantic, atmospheric delivery of elemental carbon comprises about one third of the total tropospheric anthropogenic carbon flux to the sea surface, a result of biomass burning and fossil fuel combustion. Atmospheric deposition of both organic carbon and nitrogen is substantial, and world-wide, the atmospheric deposition of organic nitrogen is about the same as inorganic nitrogen.

Dissolved organic nitrogen and, for that matter, dissolved organic phosphorus have been little studied. These components are not included in most, if not all, box model calculations. Further, high rates of DON release have been reported for *Trichodesmium*, and DOP may be a major source of P supporting nitrogen fixation. Clearly DON and DOP ought to be measured along with DOC, the latter of course having thus far received more attention. Further, the possibility that excess CO_2 may be stored in an ever increasing DOC pool also requires investigation (Gorshkov 1995). Corals reefs, although covering only 1.12×10^5 km^2 of surface area, are very productive. Whether most material is recycled, however, needs to be evaluated.

In the past few centuries, human activities on land have become an important factor greatly affecting the coastal environment and its exchanges with the atmosphere and the open oceans. A word of caution is that historical records give us only estimates as to the short-term potential responses of ocean margin systems to natural perturbations. Hsu (1991) maintains that if we want to evaluate the consequences of very large anthropogenic perturbations, much longer geological records must be studied so that extreme, catastrophic events can be taken into account. In fact, Hsu postulated that the present rate of species extinction might have already exceeded that which occurred at the end of the Cretaceous. The fluxes of material and energy across ocean margin boundaries will possibly reach just as catastrophic rates as those that prevailed during that critical time in Earth's history. Aeolian flux of material, notably Fe, may have also been much higher during the ice ages because of the exposed shelves. Enhanced Fe influx could increase oceanic productivity but exactly how much of it might be offset by reduced productivity on the shelves is not known. Fortunately, the International Marine Past Global Changes Study (IMAGES) program has obtained cores in several marginal seas. These should broaden our knowledge of cross-shelf exchange processes over a time span of about 300 000 years.

Another issue that deserves further study is the role of the western Bering Sea and the Sea of Okhotsk in the formation of the North Pacific Intermediate Water (NPIW). The accumulation and redistribution of anthropogenic and greenhouse gases, such as CO_2, CH_4 and freons, in the North Pacific are limited and controlled by the lower boundary of the NPIW. It has been shown that the bottom shelf water with density up to 27.05 σ_θ is formed in the coastal polynyas in the Sea of Okhotsk as a result of cooling and brine rejection un-

Table 3.8. Fluxes relevant to continental margins (all values except f are in 10^{12} mol yr^{-1}; numbers in parentheses are reference numbers)

C				N				P			
River plus ground water and ice											
32	(1,3, DIC)	17	(1, DOC)	0.265	(1, DIN)	0.7	(1, DON)	0.012	(1,3, DIP)	0.022	(1, DOP)
14	(1, PIC)	14	(1, POC)	2.4	(1, PON)	4.3	(2, DIN)	0.25	(1, PIP)	0.38	(1, POP)
66.7	(2, DOC)	34	(2a, OC)	3.6	(2, DON)	0.32	(3, DIN)	0.03	(1, excess DOP)	1.32	(3, total)
29.6	(3, POC)	31.8	(3, PIC)	3.43	(3)	4.3	(4)	0.22	(5)	0.053	(13, DIP)
18	(3, DOC)	60	(7, IC)	2.7	(5)	0.5	(1, excess DON)	0.047	(13, PIP)	0.05–0.125	(15, DP)
51	(7, OC)	8	(1, excess OC)	0.5	(14, excess par.)	0.9–2.8	(15)	0.53	(15, par.)	0.14	(24, dis.)
42	(15, DIC)	42	(15, DOC)	5.3	(24)	0.32	(25, DIN)	0.55	(24, par.)	0.025	(25, DIP)
16.7	(15, PIC)	29.2	(15, POC)	0.7	(25, DON)	1	(25, excess dis.)	0.038	(1, DOP)	0.013	(25 excess DIP)
25	(15, glacial)	37.8	(16, DIC)	1.5	(25, par.)	0.42	(25, excess par.)	0.019	(25, excess DOP)	0.25	(25, POP)
10.3	(16, DOC)	5.5	(16, POC)	2.9	(26)			0.38	(25, PIP)		
35	(17, POC)	9.2	(17, DOC)								
39.7	(24, dis.)	15	(24, par.)								
Air-to-sea (gaseous)											
8	(2a)	6	(3)	−3.6	(2)	−1.47	(3)	–			
52.6	(6)	0.03	(7)	−2.5	(4)	−2.74	(6)				
83	(11)	−33	(12)	−3.9	(14)	−0.046	(27, N_2O)				
Precipitation plus dust											
0.3	(6, PIC)	0.008	(16*, DIC)	0.7	(4)	0.59	(5)	0.02	(6)	0.022	(15*)
0.45	(16*)			0.87	(6)	0.64	(14)	0.022	(18*)		
				0.68	(26)						
Net burial plus fish catch											
9	(2a, POC)	15	(4, POC)	40	(2, IN)	0.22	(3)	0.59	(3)	0.109	(5, POP)
24	(3)	15	(4, PIC)	0.71	(4, PON)	0.69	(5, PON)	0.165	(6, POP)		
16.5	(4, POC)	42.2	(6, PIC)	2.17	(6, PON)	1.4	(14)				
17.3	(6, POC)	21	(7, IC)	0.4	(20, PON)						
20	(7, POC)	11	(8, POC)								
7.5	(10, PIC)	4.7	(15)								
2.2	(20, POC)	2.5–5	(22)								
Upwelling plus surface inflow											
467	(3)	2788	(6, DIC)	40	(2)	10.76	(3)	0.44	(3)	0.353	(6, DIP)
126	(6, DOC)	7.6	(6, POC)	26.1	(4, DIN)	3.7	(5, DIN)				
459	(7, IC)	9	(7, OC)	4.93	(6, DIN)	14	(14)				
3330	(15, DIC)			32	(20, DIN)	7.5	(22, DIN)				
				19	(23, DIN)	4.5	(26)				
Down-slope export of particulates											
167	(2)	2	(3, PIC)	2.64	(3, PON)	3.64	(6, PON)	0.15	(3, POP)	0.227	(6, POP)
18	(3, POC)	3.3	(4, PIC)								
53	(6, PIC)	20	(6, POC)								
25	(10, PIC)										
Surface water outflow											
75	(2, DOC)	500	(3)	12.9	(2, DON)	9	(3)	0.38	(3)	0.029	(6, DIP)
2796	(6, DIC)	161	(6, DOC)	0.15	(6, DIN)	15	(20, DIN)				
14.6	(6, POC)	502	(7, IC)								
32	(7, OC)										
Gross offshelf export (down-slope + surface outflow)											
75	(2, DOC)	167	(2, POC)	12.9	(2, DON)	27.8	(2, PON)	0.53	(3)	0.029	(6, DIP)
520	(3)	2796	(6, DIC)	11.64	(3)	0.12	(5, ON)	0.227	(6, POP)		
161	(6, DOC)	53.5	(6, PIC)	3.64	(6, PON)	0.15	(6, DIN)				
34.7	(6, POC)	502	(7, IC)	14	(14)	2.67	(20, DON)				
32	(7, OC)			27	(20, PON)						
Net offshore export (down-slope + surface outflow – upwelling + surface inflow)											
18	(2a, OC)	53	(3)	27.9	(4, ON)	0.7	(2)	0.09	(3)	−0.152	(5)
180	(4, OC)	3.3	(4, PIC)	0.88	(3)	−4.78	(6, DIN)	0.324	(6, DIP)	0.227	(6, DOP)
8	(6, DIC)	35	(6, DOC)	3.64	(6, PON)	−16.9	(20, DIN)				
53.5	(6, PIC)	27.1	(6, POC)	2.67	(20, DON)	27	(20, PON)				
43	(7, IC)	23	(7, OC)								
50	(20, DOC)	71	(20, POC)								

Table 3.8. *Continued*

C				N				P	
Primary productivity									
433	(2, POC)	350	(2a, POC)	90.7	(3)	75.4	(4)	5.66	(3)
600	(3, POC)	25	(4, PIC)						
500	(4, POC)	766	(7, POC)						
750	(19, POC)	658	(20, POC)						
132	(20, DOC)	516	(21)						
New productivity									
242	(2, POC)	195	(4)	48	(2)	51.8	(4)	0.39	(6, POP)
20.5	(6, DOC)	29.8	(6, POC)	7.2	(6, PON)	7.5	(22)		
125	(19, POC)	123	(20)						
75	(21)	50	(22)						
225	(23)								
f-ratio									
0.54	(2)	0.4	(4)	0.54	(2)	0.7	(4)	0.12	(6)
0.15	(6)	0.17	(19)	0.14	(6)	0.12	(20)		
0.16	(20)	0.15	(21)						
N-fixation									
–				2.37	(3)	1.07	(4)	–	
				0.09	(5)				
Denitrification									
–				3.57	(3,4)	6.39	(5)	–	
				7	(9)				
Net denitrification									
–				3.6	(2)	1.2	(3)	–	
				2.5	(4)	6.3	(5)		
				2.74	(6)	2.88	(20)		

Note: *1:* Meybeck 1982, 1993; *2:* Walsh (1991) and references therein; *2a:* Smith and Hollibaugh 1993; *3:* Mackenzie et al. 1998a, total P; *3a:* Mackenzie et al. 1998a, total P, mostly particulates; *4:* Wollast 1998; *5:* extrapolated from Galloway et al. (1996); *6:* extrapolated from Chen et al. (1999) for the ECS; *7:* Mackenzie et al. 1998b; *8:* Berner 1982; *9:* Middelburg et al. 1996; *10:* Milliman 1993; *11:* Tsunogai et al. 1997; *12:* Holligan and Reiners 1992; *13:* Fox 1991, soluble PIP; *14:* Wollast et al. 1993; *15:* Bolin and Cook 1983; *16:* Kempe 1983; *17:* Michaelis et al. 1986; *18:* Mackenzie 1995; *19:* Knauer 1993; *20:* extrapolated from Walsh (1994) for the Atl. Bight; *21:* Liu et al. (2000); *22:* Brink et al. (1995), net burial and upwelling only; *23:* Chavez and Toggweiler (1995); *24:* Martin and Whitfield (1983); *25:* Berner and Berner (1996); *26:* extrapolated from Seitzinger et al. 2000, for the N. Atl.; *27:* Seitzinger and Kroeze (1998).
* Assuming that half of the dust that falls on the oceans deposits on the shelves.

der ice. This water directly ventilates the surface and upper intermediate waters (Talley and Nagata 1995). Another important process is the import of cold and relatively saline water with σ_θ up to 27.25 from the Sea of Japan (Takizawa 1982).

Recent data further indicate that these high density waters underfeed intensive diapycnal mixing induced by strong tidal currents near the Kuril Islands and accelerate the penetration of anthropogenic gases to the lower part of the intermediate water with a σ_θ as high as 27.6 (Riser et al. 1996; Andreev et al. 1999). Strong, cold winds coupled with intensive vertical mixing and interleaving in the winter enhance the oceanic penetration of excess CO_2 in the Bering Sea, as well as the Sea of Okhotsk. Despite the availability of winter carbonate data for the eastern Bering Sea, no corresponding data are available for the western Bering Sea near the Kamchatka Peninsula and in the Sea of Okhotsk, where surface seawater is below −1.5 °C in winter at a time when vertical penetration is enhanced, and at the location where subsurface water is believed to be formed.

Such winter data are essential to obtain complete information on the excess CO_2 penetration in the western Bering Sea and in the Sea of Okhotsk. Similarly, winter data for the Barents and Weddell Sea are also lacking because of their precondition of waters important for the Great Conveyor Belt and deep water formation. Although these seas have only a limited capacity to store excess CO_2, they serve as conveyer belts which transport excess CO_2 to the deep oceans. The excess CO_2 budgets can not be studied adequately without knowledge of source water chemistry (Chen 1993). In view of the large capacity of the NPIW and the North Atlantic Deep Water in storing greenhouse gases, these regions must be studied in much detail.

However, even by pulling all available resources together, it is still not possible to measure all parameters in all seasons everywhere. Simplified, worldwide budgeting is still necessary. One basic requirement for budgets and processes is to clearly define the objectives of the whole budgeting exercise. The LOICZ program has emphasized the following points: extrapolation of budgets and global integration as a validation check; regional distribution of differences between production (p) and respiration (r); and regionalization and global integration of the difference between nitrogen fixation and denitrification.

The optimal exploitation of the existing budgets needs to be related to the rates derived from process studies. Internal and external validations should be performed whenever possible. The first step in the treatment of budgets should be in the selection of the most promising budgets on the basis of certain criteria; these include: a well-defined morphology with either restricted or open boundaries, or with appropriate information on open-boundary exchanges from other sources, such as dynamic models or direct field observations; adequate monitoring for salinity, nutrients, input and output, and; in the presence of horizontal gradients in the concentrations of salt or nutrient, greater care must be taken to multiply the incoming and outgoing water fluxes with the appropriate concentrations.

All budgets should be subjected to a sensitivity analysis in which the input parameters are varied within realistic limits and are ideally set through observation. Sensitivity analyses ought to be part of the presentation of budgets. Finalized budgets will then need as much validation as possible measuring them against independent results of biogeochemical process studies such as those from Lagrangian, time-dependent biological and physical-biological models (Hoffman 1991). Budget results will also need to be compared with process results wherever possible.

3.11.5 Summary

The results of the budget calculations support the conclusions of the SEEP-II study that the hypothesis of the export of a large proportion of the shelf primary productivity is untenable. Only a small fraction of particulate organic carbon, to the order of 5% produced, is exported across the shelf break. The role of DOC is less certain but about 10×10^{12} mol yr^{-1} are exported to the open oceans.

The schematic diagrams of flow patterns of the marginal seas discussed above are given in Fig. 3.18. The Bering Sea and the Sea of Okhotsk do not have very deep water formations but may contribute to the formation of NPIW (Reid 1965, 1973; Talley 1991). The Red and Mediterranean Seas have deep water formations and may contribute to the North Indian Intermediate Water and the dense outflow from the Mediterranean. These processes may thus pump excess CO_2 into the North

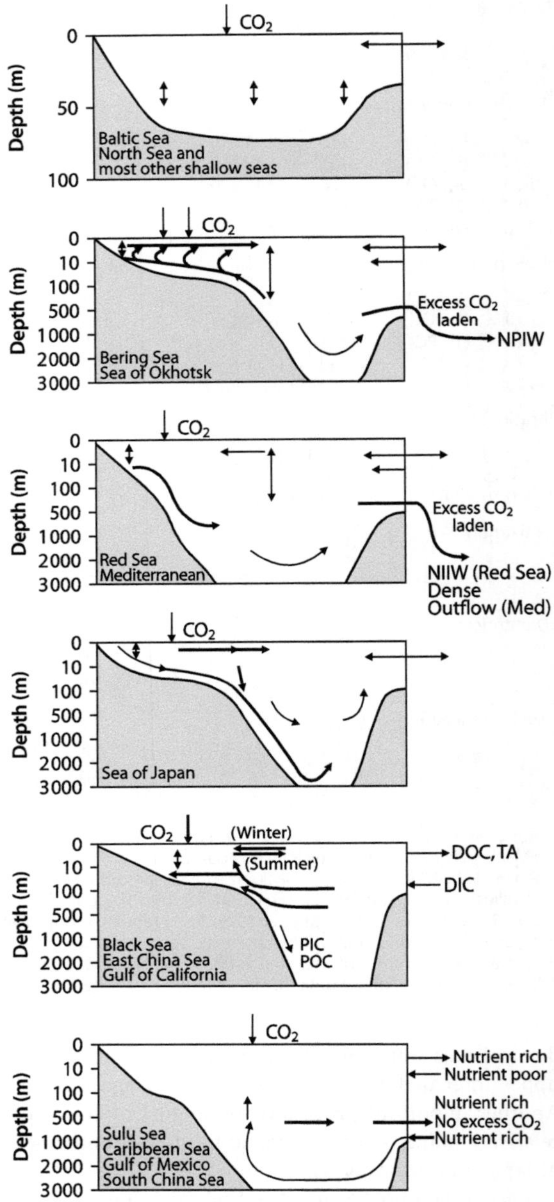

Fig. 3.18. Schematic diagrams of flow patterns in various marginal seas, where *NPIW* denotes the *North Pacific Intermediate Water* and the NIIW the *North Indian Intermediate Water*

Pacific, the Indian and the North Atlantic Oceans. The Sea of Japan has deep and bottom water formation but does not export excess CO_2 due to the shallow sills. Sediments in the shelves of these three seas may neutralize excess CO_2 in the latter part of this century when the shelf waters become undersaturated with respect to calcite and aragonite.

The Bering, East China, and North Seas absorb a large amount of CO_2 because of their high productivity which leads to low pCO_2. Yet the East and South China Seas, and for that matter, the Black Sea, the Gulf of California, the Sulu Sea and the Gulf of Mexico, are not important reservoirs for excess CO_2 owing to their small size and upwelling. It is unlikely the sediments will neutralize excess CO_2 in this century, but carbonate will be regenerated from organic matter at the bottom. This regenerated carbonate will be exported to and stored in the pelagic ocean (Tsunogai et al. 1997).

Another emerging issue will be the large-scale diversion of water and development of hydroelectric power in basins such as the Yangtze and Mekong Rivers. The impact of regulating large river systems is not well understood, and an assessment of the potential impact must be undertaken.

Because of mounting human perturbation in the environment, future coastal zones will become stronger sinks for atmospheric CO_2. Since it is not practical to study each and every estuary or marginal sea in the world, generalizations may suffice. Accordingly, the recommendations made by the CMTT are still valid: there is a need to check, validate and carry out sensitivity analyses on the available budgets. However, all budgets are not equal; some budgets have more precise input data and so have more precise output. This degree of precision must be identified in their descriptions. There is a need to compare budget results with process studies in as many locations as possible. One possibly useful comparison would be with direct measurements of p-r and denitrification, and efforts should be made to explore the finer categorization of different shelf systems based on other globally available parameters, such as shelf width and human impact.

Acknowledgments

The authors wish to thank Prof. S. Smith for his assistance. Professors J. Walsh and M. Fasham provided constructive comments. The National Science Council of the ROC supported the preparation of this manuscript (NSC 89-2611-M-110-001).

References

Aagaard K, Carmack EC (1989) The role of sea ice and other fresh water in the Arctic circulation. J Geophys Res 94, C10:14485–14498

Aagaard K, Coachman LK (1975) Toward an ice-free Arctic Ocean. Eos 56, 7:484–486

Aagaard K, Swift JH, Carmack EC (1985) Thermohaline circulation in the Arctic and Mediterranean Seas. J Geophys Res 90, C3:4833–4846

Alvarez-Salgado XA, Castro CG, Perez FF, Fraga F (1997) Nutrient mineralization patterns in shelf waters of the Western Iberian upwelling. Cont Shelf Res 17:1247–1270

Anderson LG, Dyrssen DW, Jones EP, Lowings MG (1983) Inputs and outputs of salt, fresh water, alkalinity and silica in the Arctic Ocean. Deep-Sea Res 30:87–94

Anderson LG, Olsson K, Chierici M (1998) A carbon budget for the Arctic Ocean. Global Biogeochem Cy 12, 3:455–465

Andreev AG, Bychkov AS, Zhabin IA (1999) Excess CO_2 penetration in the Okhotsk Sea. Extended abstract, 2nd International Symposium on CO_2 in the Oceans. National Institute of Environmental Studies, Tsukuba, January 18–22, 1999, pp 21–05

Andreev A, Honda M, Kumamoto Y, Kusakabe M, Murata A (2001) The excess CO_2 and pH excess in the intermediate water layer of the Northwestern Pacific. J Oceanogr 57:177–188

Antia AN, Bondungen BV, Peinert R (1999) Particle flux across the mid-European continental margin. Deep-Sea Res Pt I 46: 1999–2024

Are FE (1999) The role of coastal retreat for sedimentation in the Laptev Sea. In: Kassens H, Bauch HA, Dmitrenko I, Eicken H, Hubberten HW, Melles M, Thiede J, Tomokhov L (eds) Land-ocean systems in the Siberian Arctic: dynamics and history. Springer-Verlag, Heidelberg, pp 287–295

Bakker DCE (1998) Process studies of the air-sea exchange of carbon dioxide in the Atlantic Ocean. Dissertation, Netherlands Institute of Sea Research, 220 pp

Barber RT, Smith RL (1981) Coastal upwelling ecosystems. In: Longhurst AR (ed) Analysis of marine ecosystem. Academic Press, N.Y, pp 31–68

Barrie L, Falck E, Gregor D, Iverson T, Loeng H, Macdonald R, Pfirman S, Skotvold T, Wartena E (1998) The influence of physical and chemical processes on contaminant transport into and within the Arctic. In: Gregor D, Barrie L, Loeng H (eds) The AMAP assessment. Arctic Monitoring and Assessment Programme, pp 25–116

Barton ED (1998) Eastern boundary of the North Atlantic: Northwest Africa and Iberia. In: Robinson AR, Brink KH (eds) The sea, vol. 11. John Wiley & Sons, New York, pp 633–657

Bauch D, Carstens J, Wefer G, Thiede J (2000) The imprint of anthropognic CO_2 in the Arctic Ocean: evidence from planktic $\delta^{13}C$ data from water column and sediment surfaces. Deep-Sea Res Pt II 47:1791–1808

Bauer JE, Druffel ERM (1998) Ocean margins as a significant source of organic matter to the deep open-ocean. Nature 392:482–485

Berelson WM, McManus J, Coale KH, Johnson KS, Kilgore T, Burdige D, Pilskaln C (1996) Biogenic matter diagensis on the sea floor: a comparison between two continental margin transects. J Mar Res 54:731–762

Berner EK, Berner RA (1996) Global environment. Prentice Hall, Upper Saddle River, N.J., 376 pp

Berner RA (1982) Burial of organic carbon and pyrite sulfur in the modern ocean: its geochemical and environmental significance. Am J Sci 282:451–473

Berner RA (1992) Comments on the role of marine sediment burial as a repository for anthropogenic CO_2. Global Biogeochem Cy 6:1–2

Bethoux JP, Gentili B, Tailliez D (1998a) Warming and freshwater budget change in the Mediterranean since the 1940s, their possible relation to the greenhouse effect. Geophys Res Lett 25: 1023–1026

Bethoux JP, Morin P, Chaumery C, Connan O, Gentili B, Ruiz-Pino D (1998b) Nutrients in the Mediterranean Sea, mass balance and statistical analysis of concentrations with respect to environmental change. Mar Chem 63:155–169

Biscaye PE, Flagg CN, Falkowski PG (1994) The Shelf Edge Exchange Processes experiment, SEEP-II: an introduction to hypotheses, results and conclusions. Deep-Sea Res Pt II 41:231–252

Björk G (1990) The vertical distribution of nutrients and oxygen 18 in the upper Arctic Ocean. J Geophys Res 95, C9:16025–16036

Bolin B (1977) Changes of land biota and their importance for the carbon cycle. Science 196:613–615

Bolin T, Cook RB (1983) The major biogeochemical cycles and their interactions. John Wiley & Sons, New York, 532 pp

Bourke RH, Garrett RP (1987) Sea ice thickness distribution in the Arctic Ocean. Cold Reg Sci Technol 13:259–280

Brink KH (1998) Wind driven currents over the continental shelf. In: Brink K, Robinson A (eds) The sea, vol. 10. John Wiley & Sons, New York, pp 3–20

Brink KH, Cowles TJ (1991) The coastal transition zone program. J Geophys Res 96:14637–14647

Brink KH, Abrantes FFG, Bernal PA, Dugdale RC, Estrada M, Hutchings L, Jahnke RA, Muller PJ, Smith RL (1995) Group Report: How do coastal upwelling systems operate as integrated physical, chemical, and biological systems and influence the geological record? The role of physical processes in defining the spatial structures of biological and chemical variables. In: Summerhayes CP, Emeis KC, Angel MV, Smith RL, Zeitzschel B (eds) Upwelling in the ocean: modern processes and ancient records. John Wiley & Sons, Chichester, pp 103–124
Burkill PH (1999) ARABESQUE: an overview. Deep-Sea Res Pt II 46:529–547
Calvert SE, Price NB (1971) Upwelling and nutrient regeneration in the Benguela Current, October 1968. Deep-Sea Res 18:505–523
Carmack EC, Aagaard K, Swift JH, Macdonald RW, McLaughlin FA, Jones EP, Perkin RD, Smith J, Ellis K, Kilius LK (1997) Rapid changes of water properties and contaminants within the Arctic Ocean. Deep-Sea Res Pt II 44:1487–1502
Carpenter R (1987) Has man altered the cycling of nutrients and organic C on the Washington continental shelf and slope? Deep-Sea Res 34:881–896
Chapman P, Shannon LV (1985) The Benguela ecosystem. Part II. Chemistry and related processes. In: Barnes M (ed) Oceanography and marine biology: an annual review, vol. 23. Aberdeen University Press, Aberdeen, Scotland, pp 183–251
Chavez FP (1995) A comparison of ship and satellite chlorophyll from California and Peru. J Geophys Res 100:24855–24862
Chavez FP, Barber RT (1987) An estimate of new production in the equatorial Pacific. Deep-Sea Res 34:1229–1243
Chavez FP, Toggweiler JR (1995) Physical estimates of global new production. In: Summerhayes CP, Emeis KC, Angel MV, Smith RL, Zeitzschel B (eds) Upwelling in the ocean: modern processes and ancient records. John Wiley & Sons, Chichester, pp 103–124
Chavez FP, Barber RT, Sanderson MP (1989) The potential primary production of the Peruvian upwelling ecosystem, 1953–1984. In: Pauly D, Muck P, Mendo J, Tsukayama I (eds) The Peruvian upwelling system: dynamics and interactions. Instituto del Mar del Peru, Callao, Peru, ICLARM Conference Proceedings 18:50–63
Chavez FP, Barber RT, Kosro PM, Huyer A, Ramp SR, Stanton TP, de Mendiola BR (1991) Horizontal transport and distribution of nutrients in the coastal transition zone off northern California: effect on primary production, phytoplankton biomass and species composition. J Geophys Res 96:14833–14848
Chen CT (1985) Preliminary observations of oxygen and carbon dioxide of the wintertime Bering Sea marginal ice zone. Cont Shelf Res 4:465–483
Chen CT (1988) Summer-winter comparisons of oxygen, nutrients and carbonates in the polar seas. La mer 26:1–11
Chen CTA (1990) Decomposition rate of calcium carbonate and organic carbon in the North Pacific Ocean. Journal of Oceanographical Society of Japan 46:201–210
Chen CTA (1993) Carbonate chemistry of the wintertime Bering Sea marginal ice zone. Cont Shelf Res 13:67–87
Chen CTA (1996) The Kuroshio Intermediate Water is the major source of nutrients on the East China Sea continental shelf. Oceanol Acta 1:523–7
Chen CTA (2000) The Three Gorges Dam: reducing the upwelling and thus productivity of the East China Sea. Geophys Res Lett 27:381–383
Chen CTA, Huang MH (1995) Carbonate chemistry and the anthropogenic CO_2 in the South China Sea. Acta Oceanologica Sinica 14:47–57
Chen CTA, Tsunogai S (1998) Carbon and nutrients in the ocean. In: Galloway JN, Melillo JM (eds) Asian change in the context of global climate change. Cambridge University Press, pp 271–307
Chen CTA, Wang SL (1999) Carbon, alkalinity and nutrient budget on the East China Sea continental shelf. J Geophys Res 104: 20675–20686
Chen CTA, Pytkowicz RM, Olson EJ (1982) Evaluation of the calcium problem in the South Pacific. Geochem J 16:1–10
Chen CTA, Holligan P, Hong HS, Iseki K, Krishnaswami S, Wollast R, Yoder J (1994) Land-ocean interactions in the coastal zone. JGOFS Report No. 15, SCOR, 20 pp
Chen CTA, Wang SL, Bychkov AS (1995a) Carbonate chemistry of the Sea of Japan. J Geophys Res 100:13737–13745
Chen CTA, Ruo R, Pai SC, Liu CT, Wong GTF (1995b) Exchange of water masses between the East China Sea and the Kuroshio off northeastern Taiwan. Cont Shelf Res 15:19–39
Chen CTA, Gong GC, Wang SL, Bychkov AS (1996a) Redfield ratios and regeneration rates of particulate matter in the Sea of Japan as a model of closed system. Geophys Res Lett 23:1785–1788
Chen CTA, Lin CM, Huang BT, Chang LF (1996b) The stoichiometry of carbon, hydrogen, nitrogen, sulfur and oxygen in particular matter of the Western North Pacific marginal seas. Mar Chem 54:179–190
Chen CTA, Bychkov AS, Wang SL, Pavlova GYu (1999) An anoxia Sea of Japan by the year 2200? Mar Chem 67:249–265
Chen CTA, Wann JK, Luo JY (2001a) Aeolian flux of trace metals in Taiwan over the past 2600 years. Chemosphere 43:287–294
Chen CTA, Wang SL, Wang BJ (2001b) Nutrient budgets for the South China Sea basin. Mar Chem 75:281–300
Christensen JP (1994) Carbon export from continental shelves, denitrification and atmospheric carbon dioxide. Cont Shelf Res 14:547–576
Christensen JP, Smethie WM, Devol AH (1987) Benthic nutrient regeneration and denitrification on the Washington continental shelf. Deep-Sea Res 34:1027–1047
Codispoti LA, Christensen JP (1985) Nitrification, denitrification and nitrous oxide cycling in the Eastern Tropical Pacific Ocean. Mar Chem 16:277–300
Codispoti LA, Friederich GE (1978) Local and mesoscale influences on nutrient variability in the northwest African upwelling region near Cabo Corbeiro. Deep-Sea Res 25:751–770
Codispoti LA, Lowman D (1973) A reactive silicate budget for the Arctic Ocean. Limnol Oceanogr 18:448–456
Codispoti LA, Packard TT (1980) Denitrification rates in the eastern tropical South Pacific. J Mar Res 38:453–477
Codispoti LA, Richards FA (1968) Micronutrient distributions in the East Siberian and Laptev Seas during summer, 1963. Arctic 21:61–83
Codispoti LA, Friederich GE, Hood DW (1986) Variability in the inorganic carbon system over the southeastern Bering Sea shelf during spring 1980 and spring-summer 1981. Cont Shelf Res 5:133–160
Cota GF, Prinsenberg SJ, Bennett EB, Loder JW, Lewis MR, Anning JL, Watson NHF, Harris LR (1987) Nutrient fluxes during extended blooms of Arctic ice algae. J Geophys Res 92:C2, 1951–1962
Cota GF, Anning JL, Harris LR, Harrison WG, Smith REH (1990) Impact of ice algae on inorganic nutrients in seawater and sea ice in Barrow Strait, NWT, Canada, during spring. Can J Fish Aquat Sci 47:1402–1415
Csanady GT (1990) Physical basis of coastal productivity – the SEEP and MASAR Experiments. Eos, Transactions, American Geophysical Union, 71:1060–1061, 36:1064–1065
Cuffey KM, Clow GD, Alley RB, Stuiver M, Waddington ED, Saltus RW (1995) Larger Arctic temperature change at the Wisconsin-Holocene Glacial transition. Science 270:455–458
de Haas H (1997) Transport, preservation and accumulation of organic carbon in the North Sea. Dissertation, University of Utrecht. 149 pp
de Haas H, Boer WD, Van Weering TCE (1997) Recent sedimentation and organic carbon burial in a shelf sea: the North Sea. Mar Geol 144:131–146
Degens ET, Kempe S, Richey JE (1991) Summary: biogeochemistry of major world rivers. In: Degens ET, Kempe S, Richey JE (eds) Biogeochemistry of Major World Rivers. SCOPE Report 42, John Wiley & Sons, Chichester, pp 323–347
Dethleff D (1995) Sea ice and sediment export from the Laptev Sea flaw lead during 1991/92 winter season. In: Kassens H, Piepenburg D, Thiede J, Timokhov L, Hubberton H-W, Priamikov SM (eds) Berichte zur Polarforschung. pp 78–93
Deuser WG (1975) Reducing environments. In: Riley JP, Skirrow G (eds) Chemical Oceanography, vol. 3, 2nd Ed. Academic Press, N.Y, pp 1–37
Devol AH (1991) Direct measurement of nitrogen gas fluxes from continental shelf sediments. Nature 349:319–321
Devol AH, Codispoti LA, Christensen JP (1997) Summer and winter denitrification rates in western Arctic shelf sediments. Cont Shelf Res 17:1029–1050

Dickson R (1999) All change in the Arctic. Nature 397:38–40
Eicken H, Kolatschek J, Lindemann F, Dmitrenko I, Freitag J, Kassens H (2000) A key source area and constraints on entrainment for basin scale sediment transport by Arctic sea ice. Geophys Res Lett 26:1919–1922
Everett JT, Fitzharris BB, Maxwell B (1997) The Arctic and the Antarctic. In: Watson RT, Zinyowera MC, Moss RH (eds) The Intergovernmental Panel on Climate Change (IPCC) special report on the regional impacts of climate change. Cambridge University Press, pp 85–103
Fanning KA (1992) Nutrient provinces in the sea: concentration ratios, reaction rate ratios, and ideal conversion. J Geophys Res 97:5693–5712
FAO (1996) Fishery statistics capture production. vol. 82
Fox LE (1991) Phosphorus chemistry in the tidal Hudson River. Geochim Cosmochim Ac 55:1529–1538
Feely RA, Chen CT (1982) The effect of excess CO_2 on the calculated calcite and aragonite saturation horizons in the northeast Pacific. Geophys Res Lett 9:1294–1297
Friederich G, Sakamoto CM, Pennington JT, Chavez FP (1995) On the direction of the air-sea flux of CO_2 in coastal upwelling systems. In: Tsunogai S, Iseki K, Koike I, Oba T (eds) Global fluxes of carbon and its related substances in the coastal sea-ocean-atmosphere system. M & J International, Yokohama, pp 438–445
Fyfe JC, Boer GL, Flato GM (1999) The Arctic and Antarctic oscillations and their projected changes under global warming. Geophys Res Lett 26:1601–1604
Galloway JN, Schlesinger WH, Levy H II, Michaels A, Schnoor JL (1995) Nitrogen fixation: anthropogenic enhancement – environmental response. Global Biogeochem Cy 9:235–52
Galloway JN, Howarth RW, Michaels AF, Nixon SW, Prospero JM, Dentener FJ (1996) Nitrogen and phosphorus budgets of the North Atlantic Ocean and its watershed. Biogeochemistry 35:3–25
Gerdes R, Schauer U (1997) Large-scale circulation and water mass distribution in the Arctic Ocean from model results and observations. J Geophys Res 102, C4:8467–8483
Gobeil C, Paton D, McLaughlin FA, Macdonald RW, Paquette G, Clermont Y, Lebeuf M (1991) Donnés géochimiques sur les eaux interstitielles et les sédiments de la mer de Beaufort. Institut Maurice-Lamontagne, Rapport statistique canadien sur l'hydrographie et les sciences oceaniques, 101, 92 pp
Goldner DR (1999a) On the uncertainty of the mass, heat and salt budgets of the Arctic Ocean. J Geophys Res 104, C12: 29757–29770
Goldner DR (1999b) Steady models of Arctic shelf-basin exchange. J Geophys Res 104:C12, 29733–29755
Gordeev VV (2000) River input of water, sediment, major ions, nutrients and trace metals from Russian territory to the Arctic Ocean. In: Lewis EL, Jones EP, Lemke P, Prowse TD, Wadhams P (eds) The freshwater budget of the Arctic Ocean. NATO Science Series 2. Environmental Security – vol. 70. Klewer Academic Publishers, pp 297–321
Gordeev VV, Martin JM, Sidorov IS, Sidorova MV (1996) A reassessment of the Eurasian river input of water, sediment, major elements, and nutrients to the Arctic Ocean. Am J Sci 296: 664–691
Gordon DC Jr., Boudreau PR, Mann KH, Ong JE, Silvert WL, Smith SV, Wattayakorn G, Wulff F, Yanagi T (1996) LOICZ biogeochemical modelling guidelines. LOICZ Report and Studies, No. 5, 96 pp
Gorshkov VG (1995) Physical and biological bases of life stability. Springer-Verlag, Berlin, 340 pp
Goyet C, Millero FJ, O'Sullivan DW, Eischeid G, McCue SJ, Bellerby RGJ (1998) Temporal variations of pCO_2 in surface seawater of the Arabian Sea in 1995. Deep-Sea Res Pt I 45:609–623
Grantz A, Phillips RL, Jones GA (1999) Holocene pelagic and turbidite sedimentation rates in the Amerasia Basin, Arctic Ocean from radiocarbon age-depth profiles in cores. GeoResearch Forum 5:209–222
Grebmeier JM, Whitledge TE (1996) Arctic system science ocean-atmosphere-ice interactions biological initiative in the Arctic: shelf-basin interactions workshop. ARCSS/OAII Report No. 4
Gruber N, Sarmiento JL (1997) Global patterns of marine nitrogen fixation and denitrification. Global Biogeochem Cy 11:235–266
Hall J, Smith SV, Boudreau PR (eds) (1996) International Workshop on Continental Shelf Fluxes of Carbon, Nitrogen and Phosphorus. LOICZ Reports & Studies, No. 9. Texel, The Netherlands
Hansell DA, Peltzer ET (1998) Spatial and temporal variations of total organic carbon in the Arabian Sea. Deep-Sea Res Pt II 45:2171–2193
Hanzlick D, Aagaard K (1980) Freshwater and Atlantic water in the Kara Sea. J Geophys Res 85:4937–4942
Harms IH, Karcher MJ (1999) Modeling the seasonal variability of hydrography and circulation in the Kara Sea. J Geophys Res 104, C6:13431–13448
Hickey BM (1998) Oceanography of western North America from the tip of Baja California to Vancouver Island. In: Brink K, Robinson A (eds) The Sea, vol. 11. John Wiley & Sons, New York, pp 273–313
Hill AE, Hickey BM, Shillington FA, Strub PT, Brink KH, Barton ED, Thomas AC (1998) Eastern Ocean boundaries. In: Robinson AR, Brink KH (eds) The Sea, vol 11. John Wiley & Sons, New York, pp 29–67
Hoffman EE (1991) How do we generalize coastal models to global scale? In: Mantoura RFC, Martin JM, Wollast R (eds) Ocean margin processes in global change. John Wiley & Sons, Chichester, pp 401–417
Holligan PM, Reiners WA (1992) Predicting the responses of the coastal zone to global change. Adv Ecol Res 22:211–55
Honjo S (1997) The rain of ocean particles and earth's carbon cycle. Oceanus 40, 2:4–7
Houghton JT, Meira Filho LG, Callander BA, Harris N, Kattenberg A, Maskell K (1995) Climate change 1995: the science of climate change. Intergovernmental Panel on Climate Change, Cambridge, 572 pp
Hsu KJ (1991) Fractal theory and time dependency in ocean margin processes. In: Mantoura RFC, Martin JM, Wollast R (eds) Ocean margin processes in global change. John Wiley & Sons, New York, pp 235–250
Hung JJ, Lin PL, Liu KK (2000) Dissolved and particulate organic carbon in the southern East China Sea. Cont Shelf Res 20:545–569
Huthnance JM (1995) Circulation, exchange and water masses at the ocean margin: the role of physical processes at the shelf edge. Prog Oceanogr 35:353–431
Jahnke RA, Shimmield GB (1995) Particle flux and its conversion to the sediment record: coastal upwelling systems. In: Summerhayes CP, Emeis KC, Angel MV, Smith RL, Zeitzschel B (eds) Upwelling in the ocean: modern processes and ancient records. John Wiley & Sons, New York, pp 83–100
JGOFS (1997) Report of the JGOFS/LOICZ Workshop on non-conservative fluxes in the continental margins. JGOFS Report No. 25, LOICZ International Project Office, Texel, The Netherlands, 25 pp
Johnson KS, Chavez FP, Friederich GE (1999) Continental-shelf sediment as a primary source of iron for coastal phytoplankton. Nature 398:697–700
Kämpf J, Backhaus JO, Fohrmann H (1999) Sediment-induced slope convection: Two-dimensional numerical case studies. J Geophys Res 104, C9:20509–20522
Kao SJ, Liu KK (1996) Particulate organic carbon export from the watershed of a subtropical mountainous river (Lanyang Hsi) in Taiwan. Limnol Oceanogr 41:1749–1757
Kashgarian M, Tanaka N (1991) Antarctic intermediate water intrusion into South Atlantic Bight shelf waters. Cont Shelf Res 11:197–201
Kassens H, Dmitrenko I, Rachold V, Thiede J, Timokhov L (1998) Russian and German scientists explore the Arctic's Laptev Sea and its climate system. Eos, Transactions, American Geophysical Union, 79, 27, 317, 322–323
Kemp PF (1994) Microbial carbon utilization on the continental shelf and slope during the SEEP-II experiment. Deep-Sea Res Pt II 41:563–581
Kempe S (1983) Carbon in the freshwater cycle. In: Bolin B, Degens ET, Kempe S, Ketner P (eds) The global carbon cycle. John Wiley & Sons, New York, pp 317–342
Kempe S (1995) Coastal seas: a net source of sink of atmospheric carbon dioxide? LOICZ Report and Studies, No. 1, LOICZ International Project Office, Texel, The Netherlands, 27 pp

Kempe S, Pegler K (1991) Sinks and sources of CO_2 in coastal seas: the North Sea. Tellus 43:224–235
Knauer GA (1993) Productivity and new production of the oceanic system. In: Wollast R, Mackenzie FT, Chou L (eds) Interactions of C, N, P and S biogeochemicaly cycles and global change. Springer-Verlag, Berlin, pp 211–231
Korso PM, Huyer A (1986) CTD and velocity surveys of seaward jets off northern California, July 1981 and 1982. J Geophys Res 93:7680–7690
Kortzinger A, Duinker JC, Mintrop L (1997) Strong CO_2 emissions from the Arabian Sea during south-west monsoon. Geophys Res Lett 24:1763–1766
Krumgalz BS, Erez J, Chen CTA (1990) Anthropogenic CO_2 penetration in the northern Red Sea and in the Gulf of Elat. Oceanol Acta 13:283–290
Kudela RM, Chavez FP (2000) Modeling the impact of the 1992 El Niño on new production in Monterey Bay, California. Deep-Sea Res Pt II 47:1055–1076
Kurashina S, Nishida K, Nakabayashi S (1967) On the open water in the southeastern part of the frozen Okhotsk Sea and the current through the Kuril Islands. Journal of Oceanographical Society of Japan 23:57–71
Kvenvolden KA, Lilley MD, Lorenson TD (1993) The Beaufort Sea continental shelf as a seasonal source of atmospheric methane. Geophys Res Lett 20:2459–2462
Liu KK, Kaplan IR (1982) Nitrous oxide in the sea off Southern California. In: Ernst WG, Morin JG (eds) The environment of the deep sea. Prentice-Hall, pp 73–92
Liu KK, Kaplan IR (1984) Denitrification rates and availability of organic matter in marine environments. Earth Planet Sc Lett 68:88–100
Liu KK, Kaplan IR (1989) Eastern tropical Pacific as a source of ^{15}N-enriched nitrate in seawater off southern California. Limnol Oceanogr 34:820–830
Liu KK, Lai ZL, Gong GC, Shiah FK (1995) Distribution of particulate organic matter in the southern East China Sea: implications in production and transport. Terr Atmos Ocean Sci 6:27–45
Liu KK, Su MJ, Hsueh CR, Gong GC (1996) The nitrogen isotopic composition of nitrate in the Kuroshio Water northeast of Taiwan: evidence for nitrogen fixation as a source of isotopically light nitrate. Mar Chem 54:273–292
Liu KK, Iseki K, Chao SY (2000) Continental margin carbon fluxes. In: Hanson RB, Ducklow HW, Field JG (eds) The changing ocean carbon cycle: a midterm synthesis of the Joint Global Ocean Flux Study. International Geosphere-Biosphere Programme Book Series, Cambridge University Press, Cambridge, pp 187–239
Loeng H, Ozhigin V, Adlandsvik B (1997) Water fluxes through the Barents Sea. Ices J Mar Sci 54:310–317
Longhurst A (1998) Ecological geography of the sea. Academic Press, San Diego, 398 pp
Longhurst AR, Sathyendranath S, Platt T, Caverhill C (1995) An estimation of global primary production in the ocean from satellite radiometer data. J Plankton Res 17:1245–1271
Lundberg L, Haugan PM (1996) A Nordic seas-Arctic Ocean carbon budget from volume flows and inorganic carbon data. Global Biogeochem Cy 10:493–510
Macdonald RW (1996) Awakenings in the Arctic. Nature 380:286–287
Macdonald RW (2000) Arctic estuaries and ice: a positive-negative estuarine couple. In: Lewis EL (ed) The freshwater budget of the Arctic Ocean. NATO Science Series, vol. 70, pp 383–407
Macdonald RW, Wong CS, Erikson PE (1987) The distribution of nutrients in the southeastern Beaufort Sea: implications for water circulation and primary production. J Geophys Res 92:2939–2952
Macdonald RW, Paton DW, Carmack EC, Omstedt A (1995) The freshwater budget and under-ice spreading of Mackenzie River water in the Canadian Beaufort Sea based on salinity and $^{18}O/^{16}O$ measurements in water and ice. J Geophys Res 100:895–919
Macdonald RW, Solomon SM, Cranston RE, Welch HE, Yunker MB, Gobeil C (1998) A sediment and organic carbon budget for the Canadian Beaufort Shelf. Mar Geol 144:255–273
Macdonald RW, Carmack EC, McLaughlin FA, Falkner KK, Swift JH (1999) Connections among ice, runoff and atmospheric forcing in the Beaufort Gyre. Geophys Res Lett 26:2223–2226
Macdonald RW, Barrie LA, Bidleman TF, Diamond ML, Gregor DJ, Semkin RG, Strachan WMJ, Li YF, Wania F, Alaee M, Alexeeva LB, Backus SM, Bailey R, Bewers JM, Gobeil C, Halsall CJ, Harner T, Hoff JT, Jantunen LMM, Lockhart WL, Mackay D, Muir DCG, Pudykiewicz J, Reimer KJ, Smith JN, Stern GA, Schroeder WH, Wagemann R, Yunker MB (2000) Sources, occurrence and pathways of contaminants in the Canadian Arctic: a review. Sci Total Environ 254:93–234
Mackenzie FT (1995) Biogeochemistry. In: Encyclopedia of environmental biology. Academic Press, London, pp 249–276
Mackenzie FT, Lerman A, Ver LMB (1998a) Role of continental margin in the global carbon balance during the past three centuries. Geology 26:423–426
Mackenzie FT, Ver LMB, Lerman A (1998b) Coupled biogeochemical cycles of carbon, nitrogen, phosphrous and sulfur in the land-ocean-atmosphere system. In: Galloway JN, Melillo JM (eds) Asian change in the context of global climate change. Cambridge University Press., pp 42–100
Mantoura RFC, Martin JM, Wollast R (eds) (1991) Ocean margin processes in global change. John Wiley & Sons, New York, 469 pp
Marchant M, Hebbeln D, Wefer G (1998) Seasonal flux patterns of planktonic foraminifera in the Peru-Chile Current. Deep-Sea Res Pt I 45:1161–1185
Martin JM, Whitfield M (1983) The significance of the river input of chemical elements to the ocean. In: Wong CS, Boyle E, Bruland K, Burton JD, Goldberg ED (eds) Trace metals in sea water. Plenum Press, pp 265–296
Martin JH, Knauer GA, Karl DM, Broenkow WW (1987) VERTEX: carbon cycling in the north Pacific. Deep-Sea Res 34:267–285
Maslanik JA, Serreze MC, Barry RG (1996) Recent decreases in Arctic summer ice cover and linkages to atmospheric circulation anomalies. Geophys Res Lett 23:1677–1680
McLaughlin FA, Carmack EC, Macdonald RW, Bishop JKB (1996) Physical and geochemical properties across the Atlantic/Pacific water mass boundary in the southern Canadian Basin. J Geophys Res 101, C1:1183–1197
Melling H (1993) The formation of a haline shelf front in wintertime in an ice-covered Arctic sea. Cont Shelf Res 13:1123–1147
Melling H, Lewis EL (1982) Shelf drainage flows in the Beaufort Sea and their effect on the Arctic Ocean pycnocline. Deep-Sea Res 29:967–985
Mel'nikov IA, Pavlov GL (1978) Characteristics of organic carbon distribution in the waters and ice of the Arctic Basin. Oceanology+ 18:163–167
Meybeck M (1982) Carbon, nitrogen and phosphorus transport by world rivers. Am J Sci 282:401–50
Meybeck M (1993) Natural sources of C, N, P, and S. In: Wollast R (ed) Interactions of C, N, P, and S biogeochemical cycles and global change. Springer-Verlag, Berlin, pp 163–93
Michaelis W, Ittekkot V, Degens ET (1986) River inputs into oceans. In: Lesserre P, Martin JM (eds) Biogeochemical processes at the land-sea boundary. Elsevier, pp 37–52
Michaels AF, Karl DM, Knap AH (2000) Temporal studies of biogeochemical dynamics in oligotrophic oceans. In: Hanson RB, Ducklow HW, Field JG (eds) The changing ocean carbon cycle: a midterm synthesis of the Joint Global Ocean Flux Study. Cambridge University Press, Cambridge, pp 392–416
Middelburg JJ, Soetaert K, Herman DMJ, Heip CHR (1996) Denitrification in marine sediments: a model study. Global Biogeochem Cy 10:661–673
Miller JR, Russell GL (1992) The impact of global warming on river runoff. J Geophys Res 97:2757–2764
Milliman JD (1993) Production and accumulation of calcium-carbonate in the ocean – budget of a nonsteady state. Global Biogeochem Cy 7:927–957
Milliman JD, Rutkowski C, Meybeck M (1995) River discharge to the sea: a global river index (GLORI). LOICZ Reports & Studies, No. 2. LOICZ International Project Office, Texel, The Netherlands, 125 pp
Minas HJ, Minas M, Packard TT (1986) Productivity in upwelling areas deduced from hydrographic and chemical fields. Limnol Oceanogr 31:1182–1206
Miquel JC, Fowler SW, La Rosa J (2000) Seasonal and interannual variations of particle and carbon fluxes in the open NW Mediterranean: 10 years of sediment trap measurements at the Dyfamed station, JGOFS Open Science Meeting, Bergen, 13–18 April, 2000

Monaco A, Biscaye P, Soyer J, Pocklington R, Heussner S (1990) Particle fluxes and ecosystem response on a continental margin: the 1985–1988 Mediterranean ECOMARGE experiment. Cont Shelf Res 10:959–987
Morales CE, Blanco JL, Braun M, Reyes H, Silva N (1996) Chlorophyll-*a* distribution and associated oceanographic conditions in the upwelling regions off northern Chile during the winter and spring 1993. Deep-Sea Res Pt I 43:267–289
Murray JW, Top Z, Ozsoy E (1991) Hydrographic properties and ventilation of the Black Sea. Deep-Sea Res 38, (suppl.) 2:S663–689
Naqvi SWA (1987) Some aspects of the oxygen-deficient conditions and denitrification in the Arabian Sea. J Mar Res 45:1049–1072
Nelson G, Hutchings L (1983) The Benguela upwelling area. Prog Oceanogr 12:333–356
Nittrouer CA, Brunskill GJ, Figueiredo AG (1995) Importance of tropical coastal environments. Geo-Mar Lett 15:121–126
Nixon SW, Ammerman JW, Atkinson LP, Berounsky VM, Billen G, Boicourt WC, Boynton WR, Church TM, Ditoro DM, Elmgren R, Garber JH, Giblin AE, Jahnke RA, Owens NJP, Pilson MEQ, Seitzinger SP (1996) The fate of nitrogen and phosphorus at the land-sea margin of the North Atlantic Ocean. Biogeochemistry 35:141–180
Nojiri Y, Nojiri T, Machida T, Inoue G, Fujinuma M (1997) Meridional distribution and secular trend of atmospheric nitrous oxide concentration over the Western Pacific. In: Tsunogai S (ed) Biogeochemical processes in the North Pacific. Japan Marine Science Foundation, Tokyo, pp 115–118
Nozaki Y, Oba T (1995) Dissolution of calcareous tests in the ocean and atmospheric carbon dioxide. In: Sakai H, Nozaki Y (ed) Biogeochemical processes and ocean flux in the Western Pacific. Terra Scientific Publishing Company, Tokyo, pp 83–92
Oguz T, Ducklow HW, Purcell JE, Malanotte-Rizzoli P (2001) Simulation of recent change in the Black Sea pelagic food web structure due to top-down control by gelatinous carnivores. J Geophy Res 106:4543–4564
Olsson K, Anderson LG (1997) Input and biogeochemical transformation of dissolved carbon in the Siberian shelf seas. Cont Shelf Res, 17, 819–833
Opsahl S, Benner R, Amon RMW (1999) Major flux of terrigenous dissoved organic matter through the Arctic Ocean. Limnol Oceanogr 44:2017–2023
Pace ML, Knauer GA, Karl DM, Martin JH (1987) Primary production, new production and vertical flux in the eastern Pacific Ocean. Nature 325:803–804
Pavlov VK, Pfirman SL (1995) Hydrographic structure and variability of the Kara Sea: implications for pollutant distribution. Deep-Sea Res Pt II 42:1369–1390
Pérez Fiz F, Aida F Ríos, Rosón G (1999) Sea surface carbon dioxide off the Iberian Peninsula (North Eastern Atlantic Ocean). J Marine Syst 19:27–46
Peulvé S, Sicre MA, Saliot A, De Leeuw JW, Baas M (1996) Molecular characterization of suspended and sedimentary organic matter in an Arctic delta. Limnol Oceanogr 41:488–497
Pfirman S, Gascard JC, Wollengurg I, Mudie P, Abelmann A (1989) Particle-laden Eurasian Arctic sea ice: observations from July and August 1987. Polar Res 7:59–66
Pfirman SL, Koegeler JW, Rigor I (1997) Potential for rapid transport of contaminants from the Kara Sea. Sci Total Environ 202:111–122
Proshutinsky AY, Johnson MA (1997) Two circulation regimes of the wind-driven Arctic Ocean. J Geophys Res 102:12493–12514
Reid JL (1965) Intermediate waters of the Pacific Ocean. Johns Hopkins Oceanographic Studies 2, The Johns Hopkins University Press, Baltimore, 85 pp
Reid JL (1973) Northwest Pacific Ocean waters in winter. Johns Hopkins Oceanographic Studies 5, The Johns Hopkins University Press, Baltimore, 96 pp
Rigor I, Colony R (1997) Sea-ice production and transport of pollutants in the Laptev Sea, 1979–1993. Sci Total Environ 202:89–110
Riser SC, Yurasov GI, Warner MJ (1996) Hydrographic and tracer measurements of the water mass structure and transport in the Okhotsk Sea in early spring. PICES Science Report 6:138–143
Roach AT, Aagaard K, Pease CH, Salo SA, Weingartner T, Pavlov V, Kulakov M (1995) Direct measurements of transport and water properties through the Bering Strait. J Geophys Res 100: 18443–18457
Rogachev KA, Bychkov AS, Carmack EC, Tishchenko P Ya, Nedashkovsky AP, Wong CS (1997) Regional carbon dioxide distribution near Kashevarov Bank (Sea of Okhotsk): effect of tidal mixing. In: Tsunogai S (ed) Biogeochemical processes in the North Pacific. Japan Marine Science Foundation, Tokyo, pp 52–69
Rothrock DA, Yu Y, Maykut GA (1999) Thinning of the Arctic sea-ice cover. Geophys Res Lett 26:3469–3472
Rowe GT, Smith S, Falkowski P, Whitledge T, Theroux R, Phoel W, Ducklow H (1996) Do continental shelves export organic matter? Nature 324:559–561
Ruttenberg K, Goñi MA (1996) Phosphorus distribution, C:N:P ratios, and δ^{13}COC in arctic, temperate, and tropical coastal sediments: tools for characterizing bulk sedimentary matter. Mar Geol 139:123–145
Sakshaug E, Bjorge A, Gulliksen B, Loeng H, Mehlum F (1994) Structure, biomass distribution, and energetics of the pelagic ecosystem in the Barents Sea: a synopsis. Polar Biol 14:405–411
Sakshaug E, Slagstad D (1992) Sea ice and wind: effects on primary productivity in the Barents Sea. Atmos Ocean 30:579–591
Savidge G, Gilpin L (1999) Seasonal influences on size-fractionated chlorophyll *a* concentrations and primary production in the north-west Indian Ocean. Deep-Sea Res Pt II 46:701–723
Schindler DW, Curtis PJ, Baylery SE, Parker BR, Beaty KG, Stainton MP (1997) Climate induced changes in the dissolved organic carbon budgets of boreal lakes. Biogeochemistry 36:9–28
Schlesinger WH, Melack JM (1981) Transport of organic carbon in the world's rivers. Tellus 33:172–87
Schlosser P, Bauch D, Fairbanks R, Bönisch G (1994) Arctic river-runoff: mean residence time on the shelves and in the halocline. Deep-Sea Res I 41:1053–1068
Schubert CJ, Stein R (1996) Deposition of organic carbon in Arctic Ocean sediments: terrigenous supply vs. marine productivity. Org Geochem 24:421–436
Seitzinger SP, Giblin AE (1996) Estimating denitrification in North Atlantic continental shelf sediments. Biogeochemistry 35:235–260
Seitzinger SP, Kroeze C (1998) Global distribution of nitrous oxide production and N inputs in freshwater and coastal marine ecosystems. Global Biogeochem Cy 12:93–113
Seitzinger SP, Kroeze C, Styles RV (2000) Global distribution of N_2O emissions from aquatic systems: natural emissions and anthropogenic effects. Chemosphere Global Change Science 2:267–279
Semiletov IP (1997) On global change in the north-east Asia and adjacent seas. Reports of the 7th TEACOM Meeting and International Workshop on Global Change Studies in the Far East Asia, Vladivostok, 10–12, November, 1997, Institute of Marine Biology of Far East Branch of Russian Academy of Sciences, Vladivostok, pp 49–88
Shannon LV (1985) The Benguela ecosystem. Part I. Evolution of the Benguela, physical features and processes. In: Barnes M (ed) Oceanography and marine biology: an annual review, vol. 23. Aberdeen University Press, Aberdeen, Scotland, pp 105–182
Sharma GD (1979) The Alaskan Shelf: hydrographic, sedimentary and geochemical environment. Springer-Verlag, New York, 498 pp
Sharma S, Barrie LA, Plummer D, McConnell JC, Brickell PC, Levasseur M, Gosseliln M, Bates TS (1999) Flux estimation of oceanic dimethyl sulfide around North America. J Geophys Res 104:21327–21342
Shaw PT, Chao SY, Liu KK, Pai SC, Liu CT (1996) Winter upwelling off Luzon in the north-eastern South China Sea. J Geophys Res 101:16435–16448
Shiah FK, Liu KK, Gong GC (1999) Temperature vs. substrate limitation of heterotrophic bacterioplankton production across trophic and temperature gradient in the East China Sea. Aquat Microb Ecol 17:247–264
Shiah FK, Gong GC, Liu KK, Kao SJ (2000a) Biological and hydrographic responses to a typhoon in the Taiwan Strait, JGOFS Open Science Meeting, Japan Marine Science Foundation, Tokyo, 13–18 April, 2000
Shiah FK, Liu KK, Gong GC (2000b) The coupling of bacterial production and hydrography in the southern East China Sea north of Taiwan: spatial patterns in spring and fall. Cont Shelf Res 20:459–477
Shillington FA (1998) The Benguela upwelling system off southwestern Africa. In: Robinson AR, Brink KH (eds) The sea, vol. 11. John Wiley & Sons, New York, pp 583–604.

Smith SV, Hollibaugh JT (1993) Coastal metabolism and the oceanic carbon balance. Rev Geophys 31:75–89
Smith SV, Mackenzie FT (1987) The ocean as a net heterotrophic system: Implications from the carbon biogeochemical cycle. Global Biogeochem Cy 1:187–198
Stein R, Grobe H, Wahsner M (1994a) Organic carbon, carbonate, and clay mineral distributions in eastern central Arctic Ocean surface sediments. Mar Geol 119:269–285
Stein R, Schubert C, Vogt C, Futterer D (1994b) Stable isotope stratigraphy, sedimentation rates, and salinity changes in the Latest Pleistocene to Holocene eastern central Arctic Ocean. Mar Geol 119:333–355
Stein R, Fahl K, Niessen F, Siebold M (1999) Late Quarternary organic carbon and biomarker records from the Laptev Sea continental margin (Arctic Ocean): implications for organic carbon flux and composition. In: Kassens H, Bauch HA, Dmitrenko I, Eicken H, Hubberten HW, Melles M, Thiede J, Tomokhov L (eds) Land-ocean systems in the Siberian Arctic: dynamics and history. Springer-Verlag, Heidelberg, pp 635–655
Strub PT, Mesias JM, Montecino V, Rutllant J (1998) Coastal ocean circulation off western South America. In: Brink K, Robinson A (eds) The sea, vol. 11. John Wiley & Sons, New York, pp 273–313
Takahashi K (1998) The Okhotsk and Bering Seas: critical marginal seas for the land-ocean linkage. In: Saito Y, Ikehara K, Katayama H (eds) Land-sea link in Asia. STA (JISTEC) and Geological Survey of Japan, pp 341–353
Takizawa J (1982) Characteristics of the Soya Warm Current in the Okhotsk Sea. Journal of Oceanographical Society of Japan 38:281–292
Talley LD (1991) A Okhotsk sea water anomaly: implications for ventilation in the North Pacific. Deep-Sea Res 38:171–190
Talley LD, Nagata Y (eds) (1995) The Okhotsk Sea and Oyashio region. PICES Science Report 2, 227 pp
Tarran GA, Burkill PH, Edwards ES, Woodward EMS (1999) Phytoplankton community structure in the Arabian Sea during and after SW monsoon, 1994. Deep-Sea Res Pt II 46:655–676
Thierry MF, Wambeke V, Thingstad FT, Sempere R, Garcia N, Raimbault P, Marie D, Queguiner B, Dolan J, Claustre H (2000) Phosphate as a key factor controlling production and carbon export in the Mediterranean Sea. JGOFS Open Science Meeting, Japan Marine Science Foundation, Tokyo, 13–18 April, 2000
Thomas AC, Strub PT, Huang F, James C (1994) A comparison of the seasonal and interannual variability of phytoplankton pigment concentrations in the Peru and California current system. J Geophys Res 99:7355–7370
Thomas H, Schneider B (1999) The seasonal cycle of carbon dioxide in Baltic Sea surface waters. J Marine Syst 22:53–67
Thompson DWJ, Wallace JM (1998) The Arctic oscillation signature in the wintertime geopotential height and temperature fields. Geophys Res Lett 25:1297–1300
Torres T, Turner DR, Silva N, Rutllant J (1999) High short-term variability of CO_2 fluxes during an upwelling event off the Chilean coast at 30°S. Deep-Sea Res Pt I 46:1161–1179
Tsunogai S, Watanabe S, Nakamura J, Ono T, Sata T (1997) A preliminary study of carbon system in the East China Sea. J Oceanogr 53:9–17
Tyrrell T, Holligan PM, Mobley CD (1999) Optical impacts of oceanic coccolithophore blooms. J Geophys Res 104:3223–3241
Uzuka N, Watanabe S, Tsunogai S (1997) DMS in the North Pacific and its adjacent seas. In: Tsunogai S (ed) Biogeochemical processes in the North Pacific. Japan Marine Science Foundation, Tokyo, pp 127–135
van Geen A, Takesue RK, Goddard J, Takahashi T, Barth JA, Smith RL (2000) Carbon and nutrient dynamics during coastal upwelling off Cape Blanco, Oregon. Deep-Sea Res Pt II 47:975–1002
Ver LMB, Mackenzie FT, Lerman A (1999) Biogeochemical responses of the carbon cycle to natural and human perturbations: past, present and future. Am J Sci 299:762–801
Walsh JE (1991a) The Arctic as a bellwether. Nature 352:19–20
Walsh JE, Zhou X, Portis D, Serreze MC (1994) Atmospheric contribution to hydrologic variations in the Arctic. Atmos Ocean 32:733–755
Walsh JJ (1977) A biological sketchbook for an eastern boundary current. In: Goldberg ED, McCave IN, O'Brien JJ, Steele JH (eds) The sea, vol. 6. John Wiley & Sons, London, pp 923–968
Walsh JJ (1981) A carbon budget for overfishing off Peru. Nature 290:300–304
Walsh JJ (1988) On the nature of continental shelves. San Diego: Academic Press, 520 pp
Walsh JJ (1989) Arctic carbon sinks: present and future. Global Biogeochem Cy 3:393–411
Walsh JJ (1991b) Importance of continental margins in the marine biogeochemical cycling of carbon and nitrogen. Nature 350:53–55
Walsh JJ (1994) Particle export at Cape-Hatteras. Deep-Sea Res Pt II 41:603–628
Walsh JJ (1995) DOC storage in Arctic Seas: the role of continental shelves. Coastal and Estuarine Studies 49:203–230
Walsh JJ, Dieterle DA (1994) CO_2 cycling in the coastal ocean. I – a numerical analysis of the southeastern Bering Sea with applications to the Chukchi Sea and the north Gulf of Mexico. Prog Oceanogr 34:335–392
Walsh JJ, Rowe GT, Iverson RC, McRoy CPM (1981) Biological export of shelf carbon is a sink of the global CO_2 cycle. Nature 291:196–201
Walsh JJ, Premuzic ET, Gaffney JS, Rowe GT, Harbottle G, Stoenner RW, Balsam WL, Betzer PR, Macko SA (1985) Organic storage of CO_2 on the continental slope off the mid-Atlantic bight, the southeastern Bering Sea and the Peru coast. Deep-Sea Res 32:853–883
Walsh JJ, McRoy CP, Coachman LK, Goering JJ, Nihoul JJ, Whitledge TE, Blackburn TH, Parker PL, Wirick CD, Shuert PG, Grebmeier JM, Springer AM, Tripp RD, Hansell DA, Djenidi S, Deleersnijder E, Henriksen K, Lund BA, Andersen P, Muller-Karger FE, Dean K (1989) Carbon and nitrogen cycling within the Bering/Chukchi Seas: source regions for organic matter effecting AOU demands of the Arctic Ocean. Prog Oceanogr 22:277–359
Walsh JJ, Carder KL, Müller-Karger FE (1992) Meridional fluxes of dissolved organic matter in the North Atlantic Ocean. J Geophys Res 97:15625–15637
Watson AJ (1995) Are upwelling zones sources or sinks of CO_2? In: Summerhayes CP, Emeis KC, Angel MV, Smith RL, Zeitzschel B (eds) Upwelling in the oceans: modern processes and ancient records. Dahlem Workshop Reports. Envorinment Science Research Report 18. John Wiley & Sons, New York, pp 321–336
Watts LJ, Owens NJP (1999) Nitrogen assimilation and the *f*-ratio in the northwestern Indian Ocean during an intermonsoon period. Deep-Sea Res Pt II 46:725–743
Weatherhead EC, Morseth CM (1998) Climate change, ozone, and ultraviolet radiation. In: AMAP Assessment report: Arctic pollution issues. Arctic Monitoring and Assessment Programme (AMAP), Oslo, Norway, Chap. 11, pp 717–774
Weller G, Lange M (1999) Impacts of global climate change in the Arctic regions. International Arctic Science Committee, University of Alaska, Fairbanks
Weiss RF, Van Woy FA, Salameh PK (1992) Surface water and atmospheric carbon dioxide and nitrous oxide observations by shipboard automated gas chromatography: results from expeditions between 1977 and 1990. Oak Ridge National Laboratory, NDP-044, 123 pp
Wijffels SE, Schmitt RW, Bryden HL, Stigebrandt A (1992) Transport of freshwater by the oceans. J Phys Oceanogr 22:155–162
Wollast R (1998) Evaluation and comparison of the global carbon cycle in the coastal zone and in the open ocean. In: Brink KH, Robinson AR (eds) The sea, vol. 10. John Wiley & Sons, New York, pp 213–252
Wollast R, Mackenzie FT, Chou L (1993) Interactions of C, N, P and S biogeochemical cycles and global change. Springer-Verlag, Berlin, 507 pp
Woodward EMS, Rees AP, Stephens JA (1999) The influence of the south-west monsoon upon the nutrient biogeochemistry of the Arabian Sea. Deep-Sea Res Pt II 46:571–591
Wu CR, Shaw PT, Chao SY (1999) Assimilating altimetric data into a South China Sea model. J Geophys Res 104:29987–30005
Wyrtki K (1971) Oceanographic atlas of the International Indian Ocean Expedition. National Science Foundation, Washington D.C., 531 pp
Yeats PA, Westerlund S (1991) Trace metal distributions at an Arctic Ocean ice island. Mar Chem 33:261–277

Apendix 3.1 Continental Margins: Site Descriptions

Agulhas Bank
Medium-width shelf influenced by the strong Agulhas Current; modest winds and surface heating but weak riverine outflow; 6×10^4 km²

Amazon Shelf
Medium-to-wide shelves strongly influenced by large rivers, the Brazil Current and tides, but moderate winds; 8.3×10^4 km²

Andaman Sea
Semi-enclosed narrow-to-medium width shelves with weak currents, moderate winds but strong tides and freshwater inputs; 2.4×10^5 km²

Arabian Peninsular Shelf
Narrow shelf with weak currents, tides and freshwater but strong winds; 1.2×10^5 km²

Baltic Sea
Semi-enclosed, shallow sea with strong tides but weak oceanic and moderate freshwater influences; 0.42×10^6 km²

Banda Sea
Semi-enclosed deep basin with narrow shelves, shallow sills, strong tides and freshwater; 0.70×10^6 km²

Barents Sea
Semi-enclosed with wide shelves; large freshwater inputs and winds but exports ice; 6.0×10^5 km²

Bay of Bengal Shelf
Narrow-to-medium width shelves influenced by the moderate E Indian Coastal Current; strong freshwater and winds, but weak tides; 7.2×10^4 km²

Beaufort Sea
Narrow-to-medium shelves with large freshwater inflow; 1.8×10^5 km²

Benguela Current Shelf
Classic wind-forced upwelling system with deep shelf break; influenced by the Agulhas Rings and the Angolan Current; strong winds, but little freshwater or tides; 1.44×10^5 km²

Bering Sea Basin
Deep basin with a maximum depth of 5 121 m; deep connections with the North Pacific; deep winter convection; 1.1×10^6 km²

Bering Sea Shelf
Large, wide, shallow shelf area influenced from shore by large rivers and cyclonic eddies of the Bering Sea; exchange time <6 months; 1.1×10^6 km²

Biscay-French Shelves
Narrow in the south, widening northwards; influenced by moderate winds and tides but weak upwelling; 7.5×10^4 km²

Black Sea
Semi-enclosed sea with narrow-to-wide shelves; strong freshwater but weak ocean and tidal influences; 0.46×10^6 km²

Brazil Shelf
Narrow shelves in the NE and E; strongly influenced by the Brazil and the N Brazil Currents; moderate-to-strong winds but weak tides and freshwater; narrow-to-medium shelves in the south; influenced strongly by the Brazil and Malvinas Currents; moderate-to-strong winds, but weak-to-moderate tides and freshwater; 5×10^5 km²

Buenos Aires Shelves
Medium-width shelves influenced by the strong Malvinas and Brazil Currents but weak-to-moderate tides and freshwater; 1.6×10^5 km²

Canadian Arctic Archipelago Shelves
Semi-enclosed; narrow shelves with moderate freshwater inflow; 3.0×10^5 km²

Caribbean Sea
Semi-enclosed sea with deep basin and wide shelves; strong currents; 1.94×10^6 km²

Central Chilean Shelf
Narrow shelf influenced by the strong Peru-Chile Coastal Current; strong winds but weak freshwater and upwelling; 0.5×10^5 km²

Chukchi Shelf
Medium-to-wide shelves influenced by the Siberian Coastal Current and large freshwater inflow; 5.2×10^5 km²

East China Sea Shelf
Large, wide, shallow shelf strongly influenced from the shore by large river and from the ocean by the Kuroshio; clear seasonal monsoon winds, but moderate tides and surface heating; 0.9×10^6 km²

East Japan Shelf
Narrow shelf strongly influenced by the Oyashio and Kuroshio Currents and other deep-ocean interactions; moderate surface heating and weak tides; freshwater outflow and moderate winds; 1.0×10^5 km²

East African Shelf
Narrow shelf influenced by the strong E African Coastal Current; moderate winds but weak tides and little freshwater; 1.3×10^5 km²

East Madagascar Is. Shelf
Narrow shelf influenced by the strong E Madagascar Current; moderate freshwater and winds but weak tides; 1.0×10^5 km²

East Siberian Sea
Semi-enclosed with wide shelves and large freshwater inflow; 8.9×10^5 km²

Equador-Peru Shelves
Narrow shelf influenced by strong upwelling, winds and the Peru-Chile Count Current, but weak tides and no freshwater; 0.1×10^5 km²

Equatorial W African Shelf
Narrow shelf influenced by the Equatorial Counter Current and upwelling; weak winds, moderate tides, but abundant freshwater; 1.0×10^5 km²

Flores Sea
Semi-enclosed deep basin with narrow shelves, shallow sills, strong tides and freshwater; 1.6×10^5 km²

Great Barrier Reef
Medium-width shelf influenced by the strong Hiro Current, winds and seasonal upwelling, but moderate tides and freshwater; 1.5×10^5 km^2

Gulf of California Shelves
Narrow shelves except in the north; strong tides; little freshwater inflow; 0.16×10^6 km^2

Gulf of Carpentaria
Shallow shelf with strong tides, currents and freshwater; 4×10^5 km^2

Gulf of Maine
Semi-enclosed, with wide shelves weakly influenced by the Labrador Current and the Gulf Stream, but strong tides, modest freshwater and winds; 1.0×10^5 km^2

Gulf of Mexico
Semi-enclosed sea with narrow-to-medium shelves slightly influenced by the Loop Current and tides, but moderate winds; strong freshwater outflow in the north, moderate in the west; 2.3×10^5 km^2

Gulf of St. Lawrence
Nearly closed bay with medium-wide shelf and a weak influence from the Labrador Current; modest winds and tides but strong freshwater outflow; 1.6×10^5 km^2

Gulf of Thailand
Narrow-to-medium width shelf trending N to S hundreds of km with large riverine, tidal and monsoon influences; driven by seasonal upwelling and possibly coastal trapped waves; water exchange time of about 1 year; flushing time with respect to freshwater input ~25 years; 4×10^5 km^2

Halmahera Sea
Semi-enclosed deep basin with narrow shelves, shallow sills, strong tides and freshwater; 1.0×10^5 km^2

Iberian Shelf
Narrow shelf influenced by upwelling and the Portugal Current; moderate winds; small freshwater input; 1.0×10^5 km^2

Indonesian Seas
Open and semi-enclosed seas with narrow-to-wide shelves; strongly influenced by ocean, tides and freshwater.

Irish Sea
Semi-enclosed shelf with through flow, strong tides and winds; 0.18×10^6 km^2

Java Sea
Semi-enclosed shelf with strong tides and freshwater; 0.48×10^6 km^2

Kamchatka Shelf
Narrow shelf influenced by the strong East Kamchatka Current and surface heating; moderate winds and weak deep-ocean influences; tides, freshwater inflow and ice; very high surface nutrient concentration in winter (NO_3 about 20 μM l^{-1}; PO_4 about 2.25 μM l^{-1}; and SiO_2 about 60 μM l^{-1}), but oxygen is 5% below saturation at $T < -1$ °C; 1.8×10^5 km^2

Kara Sea
Semi-enclosed with wide shelves; very large freshwater input but exports ice; 8.8×10^5 km^2

Labrador Shelf
Subpolar shelves of medium width; strongly influenced by the Labrador Current with modest amounts of freshwater; winds and tidal influences; 0.9×10^5 km^2

Laptev Sea
Semi-enclosed; wide shelves with large freshwater inflow but exports ice; 6.6×10^5 km^2

Luzon-Taiwan Mindanao Shelves
Narrow shelves with strong influences from the Kuroshio and other deep-ocean interactions, but weak tides; freshwater inflow, winds and surface heating; 7.3×10^4 km^2

Mackenzie Shelf
Wide shelf dominated by the Mackenzie River that freezes annually; coastal current; exchange time <1 year; 0.06×10^6 km^2

Malacca Straits
Semi-enclosed shallow shelf with weak currents, moderate winds but strong tides and freshwater; 2.4×10^5 km^2

Maluku Sea
Semi-enclosed deep basin with narrow shelves, shallow sills, strong tides and freshwater; 1.6×10^5 km^2

Mediterranean Sea
Semi-enclosed sea with narrow-to-wide shelves; moderate ocean influence but weak tides; 2.51×10^6 km^2

Mid-Atlantic Bight
Narrow-to-medium width shelves moderately influenced by the Gulf Stream and Labrador Current; weak tides, modest freshwater and winds; 1.5×10^5 km^2

Mindanao Shelves
Narrow shelves with strong influences from the Kuroshio and other deep-ocean interactions, but weak tides, freshwater inflow, winds and surface heating; 6.1×10^5 km^2

Mozambique Channel
Semi-enclosed, narrow shelf influenced by the strong Mozambique Current and freshwater, but moderate tides and winds; 9×10^4 km^2

Natal/E. Cape Shelf
Narrow shelf influenced by the strong Agulhas Current; moderate freshwater but weak tides and winds; 4×10^4 km^2

New Zealand Shelves
Open, narrow shelves with a strong ocean influence; moderate tides and freshwater; 2.8×10^5 km^2

Newfoundland Shelf
Medium-width shelf with strong influences from the Labrador Current; weak tides and modest freshwater and winds; 4.5×10^5 km^2

North Sea
Semi-enclosed shallow shelf with strong tides and freshwater; 0.58×10^6 km^2

N Chilean Shelf
Narrow-to-medium width with seasonal wind-forced upwelling; Peru-Chile Counter Current; weak tides but no freshwater; 0.4×10^5 km^2

N Equador-Panama Shelf
Narrow shelf with influences from the Columbia Current, winds and strong precipitation; generally downwelling, but with upwelling north of 4° N in winter; weak tides; 1.0×10^5 km²

North Sea
Shallow shelf sea with water depth ranging from 40 m in the south to 200 m at the shelf edge in the north; strong tides, freshwater inflow and winds; 0.58×10^6 km²

NW African Shelves
Generally narrow shelves influenced by upwelling, the Canary Current and the North and South Atlantic Counter Current; strong winds but weak tides or freshwater inflow; 2×10^5 km²

NW American Shelf
Narrow shelf with strong influences from the California and Davidson Currents, upwelling, seasonal storms, and river water (730 $km^3 yr^{-1}$); buoyancy forced coastal current; little freshwater inflow and weak tides in the south, strong in the north; 0.5×10^5 km²

NW European Shelf
Wide shelf influenced by the North Atlantic Current; strong winds but moderate freshwater and upwelling; 0.9×10^6 km²

Norwegian Shelf
Narrow shelf influenced by the Norwegian Current, strong winds and freshwater, but little upwelling except in fjords; 1.0×10^5 km²

Panama-Mexican Shelf
Narrow shelf with influences from the Costa Rican Coastal Current and offshore winds; little precipitation or freshwater inflow; 1×10^6 km²

Patagonia Shelf
Wide shelf influenced by the strong Malvinas Current, tides and winds, but weak freshwater; 0.8×10^6 km²

Persian Gulf
Shallow semi-enclosed sea with strong tides; evaporation higher than precipitation plus runoff; 0.24×10^6 km²

Red Sea
Deep semi-enclosed sea with narrow shelves; strong evaporation, little precipitation or runoff; incoming surface flow and outgoing subsurface flow; 0.44×10^6 km²

Scotian Shelf
Medium-width shelf with weak influence from the Labrador Current and the Gulf Stream; modest tides, freshwater and winds; 1.5×10^5 km²

Sea of Japan Basin
Deep basin with deep winter convection; straits connecting to the open ocean: all <130 m deep; 7.4×10^5 km²

Sea of Japan Shelves
Narrow shelves; strong surface heating; moderate influences from the Kuroshio and winds; weak tides, freshwater inflow and ice; 2.6×10^5 km²

Sea of Okhotsk Basin
Deep basin with deep winter convection and deep straits connecting the basin to the North Pacific; strong influence from the strong East Kamchatka Current and the Oyashio; 1.01×10^6 km²

Sea of Okhotsk Shelf
Medium-width shelf with strong ice and surface heating influences; moderate tides, freshwater inflow and winds; weak deep-ocean influences; 0.51×10^6 km²

Somalia Shelf
Narrow shelf influenced by strong winds, upwelling and the moderate Somali Current, but little river outflow or tides; 4.6×10^4 km²

S Atlantic Bight
Narrow shelf strongly influenced by the Gulf Stream; weak tides, modest freshwater and winds; 0.3×10^5 km²

S Chilean Shelf
Medium-width shelf influenced by the West Wind Drift and Cape Horn Current; strong winds, tides and freshwater (2.5 $m\,yr^{-1}$); downwelling; 1.2×10^5 km²

South China Sea Basin
Deep basin with moderate monsoon winds, upwelling, river water buoyancy-forced coastal current, and deep connections to the Philippine Sea; weak tides; 1.1×10^6 km²

South China Sea Shelf
Large, wide, shallow shelf moderately influenced from shore by large rivers from the South China Sea Basin by eddies and by surface heating and seasonal monsoon winds; weak tides; exchange time <6 months; 1.1×10^6 km²

SE Australian Shelf
Narrow shelf influenced by the strong E Australia Current and winds, but weak tides and freshwater; 0.4×10^5 km²

S Indonesian Shelf
Narrow shelves with weak currents, weak tides; moderate freshwater but strong winds; 0.67×10^6 km²

Sulawesi (Celebes) Sea
Semi-enclosed deep basin with narrow shelves, shallow sills, strong tides and freshwater; 2.6×10^5 km²

Sulu Sea
Semi-enclosed deep basin with narrow shelves, shallow sills, strong tides and freshwater; 0.35×10^6 km²

W Canadian Shelf
Medium-to-wide shelves with strong influences from winds and the North Pacific and Alaskan Currents; intermittent upwelling in the southern shelf; downwelling in the north; moderate tides and freshwater inflow; 0.1×10^6 km²

W Florida Shelf
Narrow-to-medium width shelves with a moderate influence from the Loop Current; weak tides and freshwater but strong winds; 0.4×10^5 km²

W Indian Shelf
Narrow-to-medium width shelves with weak currents or fresh-water; moderate tides but strong winds; 0.3×10^6 km²

Chapter 4

Phytoplankton and Their Role in Primary, New, and Export Production

Paul G. Falkowski · Edward A. Laws · Richard T. Barber · James W. Murray

4.1 Introduction

Phytoplankton have played key roles in shaping Earth's biogeochemistry and contemporary human economy, yet because the human experience is so closely tied to higher plants as sources of food, fiber, and fuel, the role of phytoplankton in our everyday lives is often overlooked. The most familiar phytoplankton products we consume are petroleum and natural gas. Their uses as fuels, and in its myriad refined forms, as plastics, dyes, and chemical feedstocks are so critical to the industrialized world that wars are fought over the ownership of these fossilized hydrocarbons. Since the beginning of civilization, we have used the remains of calcareous nanoplankton, deposited over millions of years in ancient ocean basins, for building materials. Diatomaceous oozes are mined as additives for reflective paints, polishing materials, abrasives, and for insulation. Phytoplankton provided the original source of oxygen for our planet, without which our very existence would not have been possible. The fossil organic carbon, skeletal remains, and oxygen are the cumulative remains of phytoplankton export production that has occurred uninterrupted for over 3 billion years in the upper ocean (Falkowski et al. 1998). In this chapter we examine what we learned during the JGOFS era about how phytoplankton impact contemporary biogeochemical cycles and their role in shaping Earth's geochemical history.

4.1.1 A Brief Introduction to Phytoplankton

Phytoplankton are a taxonomically diverse group of mostly single celled, photosynthetic aquatic organisms that drift with currents. This group of organisms consists of approximately 20 000 species distributed among at least eight taxonomic divisions or phyla (Table 4.1). In contrast, higher plants are comprised of >250 000 species, almost all of which are contained within one class in one division. Thus, unlike terrestrial plants, phytoplankton are species poor but phylogenetically diverse; this deep taxonomic diversity is reflected in ecological function (Falkowski and Raven 1997).

Within this diverse group of organisms, three basic evolutionary lineages are discernable (Delwiche 2000). The first contains all procaryotic oxygenic phytoplankton, all of which belong to one class of bacteria, namely the cyanobacteria. Numerically these organisms dominate the ocean ecosystems. There are approximately 10^{24} cyanobacterial cells in the oceans. To put that in perspective, the number of cyanobacterial cells in the ocean is 2 orders of magnitude more than all the stars in the sky. Cyanobacteria evolved more than 2.8 billion years ago (Summons et al. 1999) and have played fundamental roles in driving much of the ocean carbon, oxygen and nitrogen cycles from that time to present.

All other oxygen producing organisms in the ocean are eucaryotic, that is they contain internal organelles, including a nucleus, one or more chloroplasts, one or more mitochondria, and, most importantly, in many cases they contain a membrane bound storage compartment, or vacuole. Within the eucaryotes we can distinguish two major groups, both of which have descended from a common ancestor thought to be the endosymbiotic appropriation of a cyanobacterium into a heterotrophic host cell. The appropriated cyanobacterium became a chloroplast.

In one group of eucaryotes, chlorophyll *b* was synthesized as a secondary pigment; this group forms the 'green lineage', from which all higher plants have descended. The green lineage played a major role in oceanic food webs and the carbon cycle from ca. 2.2 billion years ago until the end-Permian extinction, approximately 250 million years ago (Lipps 1993). Since that time however, a second group of eucaryotes has risen to ecological prominence; that group is commonly called the 'red lineage' (Fig. 4.1). The red lineage is comprised of several major phytoplankton divisions and classes, of which the diatoms, dinoflagellates, haptophytes (including the coccolithophorids), and the chrysophytes are the most important. All of these groups are comparatively modern organisms; indeed the rise of dinoflagellates and coccolithophorids approximately parallels the rise of dinosaurs, while the rise of diatoms approximates the rise of mammals in the Cenozoic. The burial and subsequent diagenesis of organic carbon, produced primarily by members of the red lineage in shallow seas in the

Table 4.1. The taxonomic classification and species abundances of oxygenic photosynthetic organisms in aquatic and terrestrial ecosystems. Note that terrestrial ecosystems are domianted by relatively few taxa that are species rich, while aquatic ecosystems contain many taxa but are relatively species poor (from Falkowski and Raven (1997))

Taxonomic Group						Known species	Marine	Freshwater
Empire	Kingdom	Subkingdom	Division	Subdivision	Class			
Bacteria (= Prokaryota)	Eubacteria			Cyanobacteria (sensu strictu) (= Cyanophytes, blue-green algae)		1500	150	1350
				Chloroxybacteria (= Prochlorophyceae)		3	2	1
Eukaryota	Protozoa		Euglenophyta		Euglenophyceae	1050	30	1020
			Dinophyta (Dinoflagellates)		Dinophyceae	2000	1800	200
	Plantae	Biliphyta	Glaucocystophyta		Glaucocystophyceae	13	–	–
			Rhodophyta		Rhodophyceae	6000	5880	120
		Viridiplantae	Chlorophyta		Chlorophyceae	2500	100	2400
					Prasinophyceae	120	100	20
					Ulvophyceae	1100	1000	100
					Charophyceae	12500	100	12400
			Bryophyta (mosses, liverworts)			22000	–	1000
			Lycopsida			1228	–	70
			Filicopsida (ferns)			8400	–	94
			Magnoliophyta (flowering plants)	Monocotyledoneae		52000	55	455
				Dicotyledoneae		188000	–	391
	Chromista	Chlorechnia	Chlorarachniophyta		Chlorarachniophyceae	3–4	3–4	0
		Euchromista	Crytophyta		Crytophyceae	200	100	100
			Haptophyta		Prymensiophyceae	500	100	400
			Heterokonta		Bacillariophyceae (diatoms)	10000	5000	5000
					Chrysophyceae	1000	800	200
					Eustigmatophyceae	12	6	6
					Fucophyceae (brown algae)	1500	1497	3
					Raphidophyceae	27	10	17
					Synurophyceae	250	–	250
					Tribophyceae (Xanthophyceae)	600	50	500
	Fungi		Ascomycontina (lichens)			13000	15	20

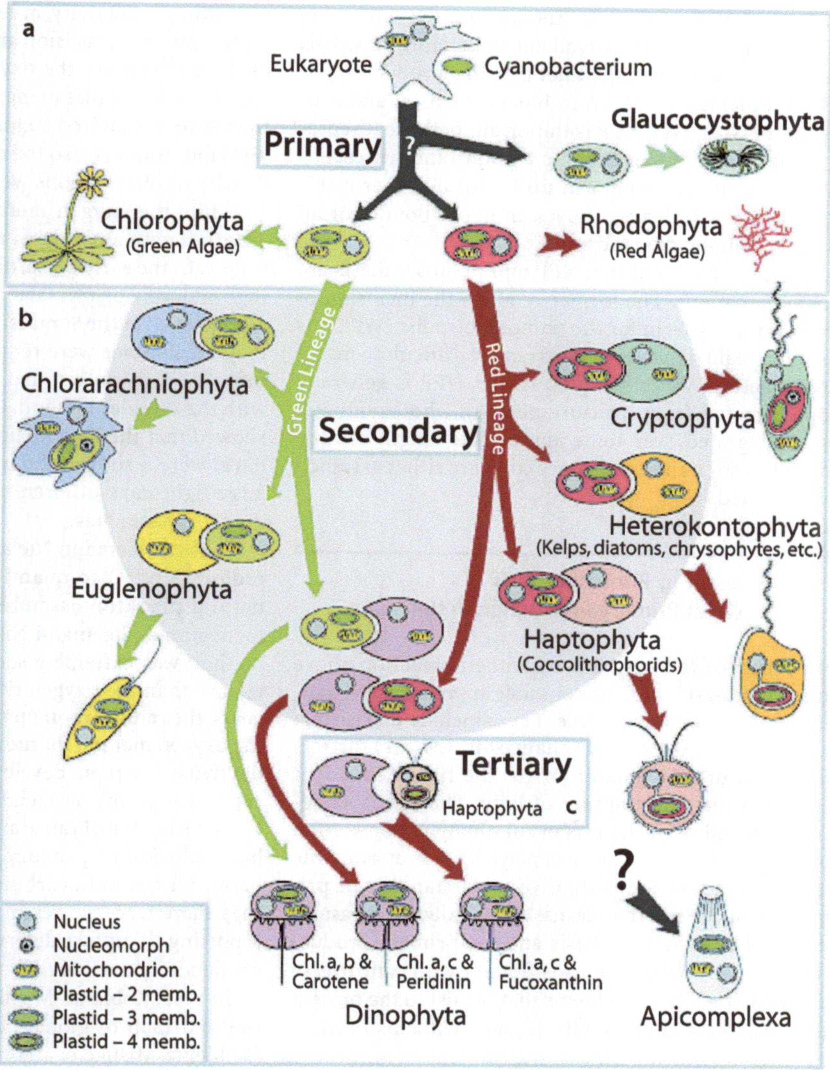

Fig. 4.1. The basic pathway leading the evolution of eucaryotic algae. The primary symbiosis of a cyanobacterium with a apoplastidic host gave rise to both chlorophyte algae and red algae. The chlorophyte line, through secondary symbioses, gave rise to the 'green' line of algae, one division of which was the predecessor of all higher plants. Secondary symbioses in the red line with various host cells gave rise to all the chromophytes, including diatoms, cryptophytes, and haptophytes (modified from Delwiche CF (1999) Tracing the thread of plastid diversity through the tapestry of life. Am Nat 154:164–177

Jurassic period provide the source rocks for most of the petroleum reservoirs that have been exploited for the past century by humans.

4.1.2 Photosynthesis and Primary Production

The evolution of cyanobacteria was a major turning point in biogeochemistry of Earth. Prior to the appearance of these organisms, all photosynthetic organisms were anaerobic bacteria that used light to couple the reduction of carbon dioxide to the oxidation of low free energy molecules, such as H_2S or preformed organics (Blankenship 1992). Cyanobacteria developed a metabolic system that used the energy of visible light (between 400 and 700 nm) to oxidize water and simultaneously reduce CO_2 to organic carbon. Formally oxygenic photosynthesis can be summarized as:

$$CO_2 + H_2O + \text{light} \xrightarrow{\text{Chl } a} (CH_2O)_n + O_2 \qquad (4.1)$$

In Eq. 4.1, light is specified as a substrate, chlorophyll *a* is a requisite catalytic agent, and $(CH_2O)_n$ represents organic matter reduced to the level of carbohydrate.

Like all organisms, phytoplankton provide biogeochemical as well as ecological 'services'; that is they function to link metabolic sequences and properties to form a continuous, self-perpetuating network of elemental fluxes. The fundamental role of phytoplankton is the solar driven conversion of inorganic materials into organic matter via oxygenic photosynthesis. When we subtract the metabolic costs of all other metabolic processes by the phytoplankton themselves, the remaining organic carbon becomes available to heterotrophs. This remaining carbon is called net primary production, or NPP (Lindeman 1942). From biogeochemical and ecological perspectives, NPP is important because it represents the total flux of organic carbon (and hence, energy) that an ecosystem can utilize for all other metabolic processes. Hence, it gives an upper bound for all other metabolic demands.

It should be noted that NPP and photosynthesis are not synonymous. The former requires the inclusion of the respiratory term for the photoautotrophs (Williams 1993). In reality, that term is extremely difficult to measure in natural water samples. Hence, NPP is generally approximated from measurements of photosynthetic rates integrated over some appropriate length of time (usually a day), and respiratory costs are either assumed or neglected.

4.1.3 Measuring Photosynthesis and Net Primary Production in the Sea

Inspection of the basic photosynthetic reaction above suggests at least four major possible approaches to quantifying the process over time. These include measuring: (*a*) changes in oxygen, (*b*) changes in CO_2, (*c*) the formation of organic matter, or (*d*) the time dependent change in the consumption of light. The last is sometimes assayed from changes in chlorophyll fluorescence. Each of these approaches has played a role at one time in helping derive a quantitative understanding of primary production in the oceans. Historically, the measurement of both photosynthesis and net primary productivity in marine ecosystems is fraught with controversy and methodological problems that persist to the present time (Falkowski and Woodhead 1992; Williams 1993).

4.1.4 A Brief History of the Measurement of Primary Productivity in the Oceans

The first quantitative measurements of phytoplankton productivity were made early in the 20th century by Gran and his colleagues. They realized that oxygen could be used as a proxy for the synthesis of new organic material (Gran 1918). Using a chemical titration method developed by a German chemist, Clement Winkler, Gran measured the difference in oxygen concentrations in clear and opaque glass bottles suspended at various depths in the water column. The net O_2 difference in the water column provided a measure of the net synthesis of new organic matter. The oxygen light-dark method resolved the appropriate timescale (hours), measured the productivity of very small phytoplankton (nanophytoplankton) and was sensitive enough to measure plankton productivity, at least in coastal ocean waters, with acceptable precision and accuracy. For the first half of the 20th century the oxygen light-dark method was the method of choice in marine and fresh waters around the world. It required technically skilled personnel and was labor intensive, so the number of replicates and the density of observations were necessarily limited.

While the oxygen method worked well in coastal waters, it gave ambiguous results in oligotrophic ocean waters. In the early 1950s, disagreements arose concerning the levels of primary productivity in oligotrophic regions such as the Sargasso Sea. In oligotrophic regions, long incubations were required to obtain light vs. dark differences in oxygen that could be resolved manually with the Winkler method. Critics of the oxygen method showed that the long incubations (lasting three or four days) were a source of dark-bottle artifacts that led to large light-dark differences and hence very high productivity estimates.

In 1952, Steemann Nielsen introduced the use of the radiotracer, ^{14}C, to quantify the fixation of carbon by natural plankton assemblages in short-term (hours) incubations (Steemann Nielsen 1952). The radiocarbon method was extremely sensitive and far less labor intensive than the oxygen titration approach. Within five years, the radiocarbon method had completely replaced the oxygen method for measuring oceanic primary productivity. The rapid development of large new oceanography programs, especially in the United States, the United Kingdom, France and the USSR, resulted in a global explosion of primary productivity measurements based on the radiocarbon method. Between 1953 and 1973 there were 221 research papers from 16 countries reporting ^{14}C uptake determinations of oceanic primary productivity covering all of the oceans and major seas.

In 1968, Koblentz-Mishke and co-workers published the first map of global oceanic primary productivity (Koblentz-Mishke et al. 1970). The data, compiled from over 7000 stations, were used to derive daily surface primary production estimates, from which a global ocean productivity estimate of ca. 24 Pg C yr^{-1} was made. Still widely reproduced in textbooks, the Koblentz-Mishke et al. (1970) map remains one of the most frequently cited articles dealing with global productivity.

By the mid 1970s it was realized that the radiocarbon method also had problems. Depending upon the length of incubation, the assay could be something closer to gross rather than net photosynthesis. Several attempts were made to develop alternative methods for the application of radiocarbon; these led to short-term incubations (<1 h), in which photosynthesis vs. irradiance curves were derived (Platt et al. 1975). The *P* vs. *I* (later to become *P* vs. *E*) curves were then integrated over time, with varying degrees of complexity, to derive an estimate of 'primary productivity'. The degree to which the assays actually measure NPP remains unclear; however, it is still assumed that the respiratory costs of the phytoplankton themselves are relatively small.

In an effort to help sort out some of the ambiguities of measurements of productivity or photosynthesis, alternative approaches were introduced. The Winkler assay was resurrected with computer-controlled, high precision titration systems. Measurements of total inorganic carbon consumed were made possible by high precision potentiometric titrations with coulombmeters (Williams and Jenkinson 1982). These two approaches still required incubations and were labor intensive; they never replaced the radiocarbon assay, but did help to reveal how difficult it is to interpret radiocarbon measurements as NPP (Grande et al. 1989). In the late 1980s, fluorescence based measurements were introduced (Falkowski et al. 1986; Kolber et al. 1990). These assays, which measure the change in variable chlorophyll fluorescence, were clearly meant to assess photosynthetic activity, and not to replace the radiocarbon method (Kolber and Falkowski 1993). The primary advantages of a variable fluorescence approach are that it requires no incubation and can be done continuously. The fluorescence-based method has proven extremely valuable in mapping processes, such as eddies (Falkowski et al. 1991), fronts, or purposeful ocean fertilization (Behrenfeld et al. 1996), that can influence the efficiency with which light is used in Eq. 4.1.

In the mid-1970s it was recognized that satellite measurements of ocean color could potentially be used to derive global maps of oceanic chlorophyll concentrations (Esaias 1980). The early measurements, by the Coastal Zone Color Scanner, revolutionized our understanding of the global distributions of phytoplankton. The aerial extent and temporal scales of blooms could be seen for the first time in a truly global context. Using productivity models, the pigment retrievals could be used to estimate global primary productivity. This approach forms the basis for the contemporary estimates of oceanic NPP using ocean color sensors on SeaWiFS, MODIS and other Earth observing platforms.

4.1.5 Quantifying Global Net Primary Productivity in the Oceans

From its inception, the JGOFS program realized that a highly standardized, even if imperfect, globally distributed set of measurements of primary productivity was required to help understand sources of variability in NPP in the ocean. To this end JGOFS adopted a set of radiocarbon-based protocols and applied these to measure radiocarbon assimilation in a wide set of oceanographic regimes, including the equatorial, subtropical and subarctic Pacific, the Arabian Sea, the North Atlantic and sub-tropical Atlantic, the Southern Ocean, and from two long term subtropical sites, one north of Oahu in the Hawaiian Islands and the other southeast of Bermuda. Standardization of radiocarbon protocols in the JGOFS program, coupled with high quality phytoplankton pigment analyses (primarily obtained from high performance liquid chromatography), provides a basis for calibrating satellite-based models of primary production. The summarized results of the JGOFS measurements give a relatively consistent perspective of productivity (Fig. 4.2), with the two time series stations, in oligotrophic gyres, showing generally low aerially integrated maximum values relative to the high latitude regions, or regions where major nutrients are available in excess. The seasonal changes in productivity at the two time series stations are also similar, although the phasing of the timing of the maximum productivity periods differs (Fig. 4.3). These data, coupled with satellite im-

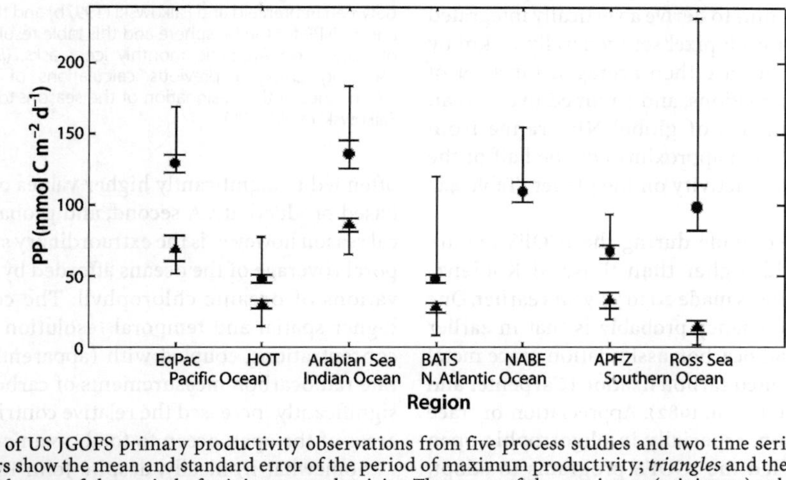

Fig. 4.2. Summary of US JGOFS primary productivity observations from five process studies and two time series studies. The *circles* and their error bars show the mean and standard error of the period of maximum productivity; *triangles* and their error bars show the mean and standard error of the period of minimum productivity. The range of the maximum (minimum) value for both periods is shown by the *thin vertical line*. Data sources are as follows: EqPac (Barber et al. 1996), HOT (Karl et al. 1996), Arabian Sea (Barber et al. 2001), BATS (Steinberg et al. 2000), NABE, APFZ, and Ross Sea (Smith et al. 2000)

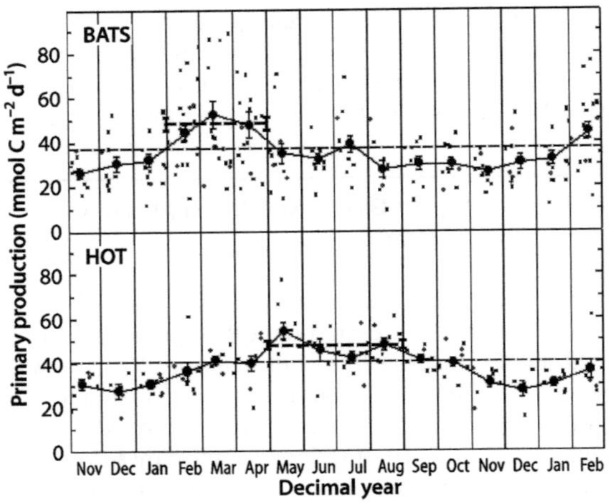

Fig. 4.3. Annual cycle of primary productivity at the two US JGOFS time series studies, the Bermuda Atlantic Time-series Study (BATS) and the Hawaii Ocean Time-series (HOT) from their inception to 2000. *Solid circles* and error bars show the monthly mean and standard error; *small, open circles* represent individual stations; *thin dashed lines* show the annual mean values (BATS = 37 ±1 mmol C m^{-2} d^{-1}; HOT = 40 ±1 mmol C m^{-2} d^{-1}); *thick dashed lines* represent the mean of values during the periods of maximum primary productivity (BATS = 48 ±3 mmol C m^{-2} d^{-1}; HOT = 48 ±2 mmol C m^{-2} d^{-1})

ages of ocean color, knowledge of sea surface temperature, and the incident solar irradiance, are required for model based estimates of net primary productivity for the world oceans (Antoine et al. 1996; Behrenfeld and Falkowski 1997b; Longhurst et al. 1995).

There are several models for estimating global NPP; these have been classified according to the level of integration/differentiation of specific variables (Behrenfeld and Falkowski 1997a), but fundamentally all the models are conceptually similar. The models basically use satellite-based estimates of phytoplankton chlorophyll biomass and incident solar photosynthetically available radiation (400–700 nm) to derive a vertically integrated estimate of NPP for each pixel set (generally 20 km by 20 km). The estimates are then averaged for a set of monthly global observations, and summed over a year. The resulting estimates of global NPP range from ca. 45 to ca. 57 Pg C yr^{-1}; approximately one half of the total net primary productivity on the planet (Table 4.2, Fig. 4.4).

Estimates of NPP made during the JGOFS era are more than two-fold higher than those of Koblenz-Mishke et al., and others made 20 to 30 years earlier. One reason for this discrepancy probably is that in earlier measurements of radiocarbon assimilation, trace metal contamination inhibited carbon fixation (Carpenter and Lively 1980; Fitzwater et al. 1982). Appreciation of trace metal contamination, especially in oligotrophic open ocean ecosystems, led the JGOFS programs to adopt, insofar as possible, trace metal clean techniques (e.g. Sanderson et al. 1995). When so applied, clean techniques

Table 4.2. Annual and seasonal net primary production (NPP) of the major units of the biosphere (after Field et al. 1998)

	Ocean NPP	Land NPP
Seasonal mean		
April–June	10.9	15.7
July–September	13.0	18.0
October–December	12.3	11.5
January–March	11.3	11.2
Annual mean		
Oligotrophic	11.0	
Mesotrophic	27.4	
Eutrophic	9.1	
Macrophytes	1.0	
Biogeographic regions		
Tropical rainforests		17.8
Broadleaf deciduous forests		1.5
Broadleaf and needleleaf forests		3.1
Needleleaf evergreen forests		3.1
Needleleaf deciduous forest		1.4
Savannas		16.8
Perennial grasslands		2.4
Broadleaf shrubs with bare soil		1.0
Tundra		0.8
Desert		0.5
Cultivation		8.0
Total	48.5	56.4

Source: Field et al. (1998). All values in Gt C. Ocean color data are averages from 1978 to 1983. The land vegetation index is from 1982 to 1990. Ocean NPP estimates are binned into three biogeographic categories on the basis of annual average C_{sat} for each satellite pixel, such that oligotrophic = $C_{sat} < 0.1$ mg m^{-3}, mesotrophic = $0.1 < C_{sat} < 1$ mg m^{-3}, and eutrophic = $C_{sat} > 1$ mg m^{-3} (Antoine et al. 1996). This estimate includes a 1 Gt C contribution from macroalgae (Smith 1981). Differences in ocean NPP estimates between Behrenfeld and Falkowski (1997b) and those in the global annual NPP for the biosphere and this table result from (*i*) addition of Arctic and Antarctic monthly ice masks; (*ii*) correction of a rounding error in previous calculations of pixel area; and (*iii*) changes in the designation of the seasons to correspond with Falkowski et al. (1998).

often led to significantly higher values of radiocarbon-based productivity. A second, and probably more critical reason however, is the extraordinary spatial and temporal coverage of the oceans afforded by satellite observations of oceanic chlorophyll. The combination of higher spatial and temporal resolution of chlorophyll concentrations, coupled with (apparently) more accurate radiocarbon measurements of carbon fixation, has significantly increased the relative contribution of large areas of the open ocean to total oceanic NPP.

Somewhat ironically, despite John Ryther's conclusions in 1969 (Ryther 1969) that productivity on continental margins was disproportionate to the aerial ex-

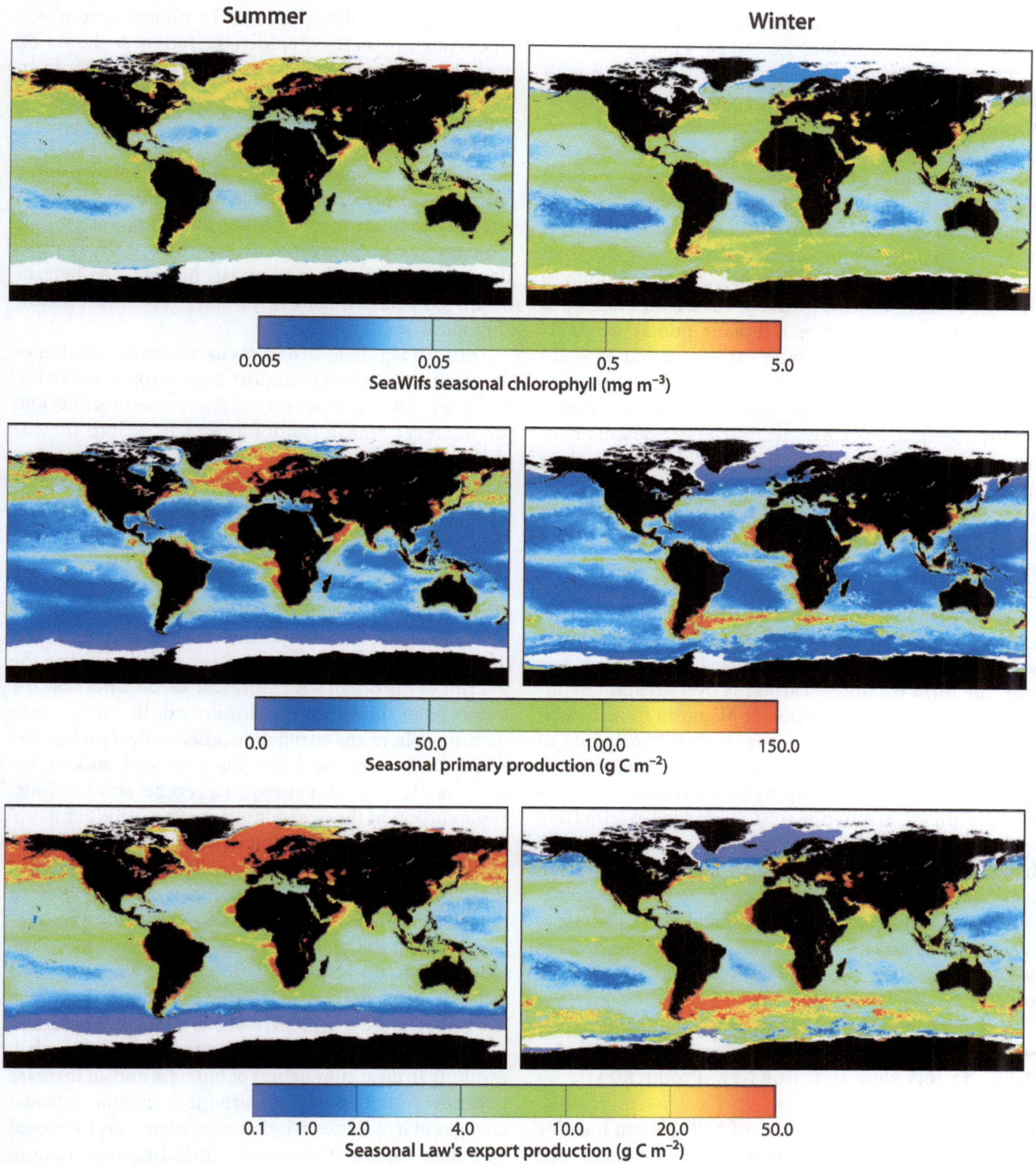

Fig. 4.4. Summer (*left column*) and winter (*right column*) distributions of oceanic chlorophyll, primary production, and export production. Oceanic chlorophyll fields were derived from standard SeaWiFS ocean color algorithms applied to 20 × 20 km pixels for the montly averaged data collected in January, February and March and June, July and August in 1998, 1999, 2000, and 2001. These maps were used to generate monthly averaged net primary production fields using the algorithm described in Behrenfeld and Falkowski (1997a). The net primary production fields were then input into an export model described by Laws et al. 2000 (see Fig. 4.6) to generate global estimates of carbon export

tent of those regions, JGOFS focused almost exclusively on the open ocean (but see the summary of assorted JGOFS studies of continental margin studies by Liu et al. 2000). The resulting extrapolated global NPP emphasizes the open ocean results. Perhaps even more ironically, satellite-based retrievals of chlorophyll on continental margins are often compromised by substances other than phytoplankton that alter the color of the water-leaving radiances. It is entirely conceivable that the estimates of NPP for the ocean provided in Table 4.2

are, in fact, low due to saturation of satellite sensors when chlorophyll concentrations exceed ca. 5 mg m^{-3}.

The results in Table 4.2 compare marine NPP with terrestrial NPP; both are based on satellite compilations of biomass in the respective ecosystems (Field et al. 1998). The striking feature of these calculations is that the contributions of the two ecosystems to the total global NPP are so similar. That is, of the ca. 100 Pg C fixed per annum as NPP on Earth, approximately 45% is due to marine phytoplankton. Given that the total carbon biomass of phytoplankton is <1 Pg, it follows that the average turnover time of the ocean biomass is <1 week. In contrast, terrestrial plant biomass contains approximately 500 Pg of 'living' carbon (including wood in living trees); extending the same logic, the average turnover time of terrestrial plant biomass is approximately 10 years. Moreover, terrestrial ecosystems have much more flexible elemental stoichiometries than marine ecosystems; C/N ratios approach 500 or more. This comparison points out a fundamental difference between the two ecosystems in the context of the global carbon cycle. In terrestrial ecosystems, carbon fixed can be 'stored' in living organic matter (e.g. forests), whereas carbon fixed by marine phytoplankton is rapidly consumed by grazers and transferred from the surface ocean to the ocean interior. In the latter reservoir, the form of 'stored' carbon is overwhelmingly inorganic. Transfer of a fraction of plankton fixed carbon from the euphotic zone to the deep ocean is one of the fundamental controls on the level of atmospheric CO_2. The biological pump helps to reduce the concentration of atmospheric CO_2 by approximately 400 ppm over what it would be if the oceans magically became a sterile reservoir of salty water (see also Watson and Orr, this volume). Thus, elucidating how this transfer occurs, what controls it, how much carbon is transferred via this mechanism, and whether the process is in steady state were primary aims of the JGOFS program.

4.1.6 Export, New and 'True New' Production

Let us now consider some fates of NPP. We can imagine that NPP produced by photoautotrophs in the upper, sunlit regions of the ocean (the euphotic zone) is consumed in the same general region by heterotrophs. In such a case, the basic photosynthetic reaction given above is simply balanced in the reverse direction due to grazing, and no organic matter leaves the ecosystem. This very simple 'balanced state' model, also referred to as the microbial-loop (e.g. Azam 1998), accounts for the fate of most of the organic matter in the oceans (and terrestrial ecosystems as well). In marine ecology, this process is sometimes called 'regenerated production'; that is organic matter produced by photoautotrophs is regenerated by heterotrophic respiration. It should be noted here that with the passage of organic matter from one level of a marine food chain to the next (e.g., from primary producer to heterotrophic consumer), a metabolic 'tax' must be paid in the form of respiration, such that the net metabolic potential of the heterotrophic biomass is always less than that of the primary producers. This does not, a priori, mean that photoautotrophic biomass is always greater, as heterotrophs may grow slowly and accumulate biomass; however, as heterotrophs grow faster, their respiratory rates must invariably increase.

Let us imagine a second scenario. Some fraction of the primary producers and/or heterotrophs sinks below a key physical gradient, such as a thermocline, and for whatever reason cannot ascend back into the euphotic zone. If the water column is very deep, sinking organic matter will most likely be consumed by heterotrophic microbes in the ocean interior. This sinking flux of organic matter is said to be 'exported' from the surface waters, and the fraction of NPP that meets such a fate is called the export production or EP. Export production is of biogeochemical interest because, in the steady state, the oxidation of organic matter in the ocean interior, rather than in the surface waters, enriches the interior of the ocean with inorganic carbon. Because the waters from the ocean interior are not in contact with the atmosphere, the enrichment effectively removes carbon dioxide from the atmosphere. Indeed, sinking organic matter and its subsequent regeneration to inorganic carbon in the ocean interior effectively 'pump up' the inorganic carbon content in the ocean to values that are significantly higher than predicted from equilibration with the atmosphere. This process, where NPP from the ocean surface sinks and is regenerated in the interior is called the biological pump. The intensity of the biological pump is especially important when the food web has been perturbed. Under such conditions (the 'perturbed state'), the mechanisms controlling production and consumption of organic matter are decoupled, resulting in an accumulation of biomass and an increase in export flux. Types of perturbations include seasonal changes in irradiance, wind mixing events, and seasonal or pulsed inputs of nutrients (including iron) due to changes in nutricline depth, upwelling and atmospheric events.

Criteria, which distinguish balanced and perturbed states, are summarized in Table 4.3.

The argument is that the balanced microbial loop is always present, while during non-steady state conditions the perturbed state with larger plankton is added on. A major challenge in the JGOFS synthesis has been to describe these two states mathematically with simple numerical models.

Table 4.3. Criteria, which distinguish balanced and perturbed states

Criteria	'Balanced' state	'Perturbed' state
Temporal forcing	Constant	Episodic
Vertical stratification	High	Low
Phytoplankton	Small – nano/pico	Large – diatom/coccolith
Dominant grazers	Microzooplankton/protists	Mesozooplankton
Food web function	Coupled	Uncoupled
Plankton specific rates	Growth = mortality	Growth > mortality then growth < mortality

4.1.7 Elemental Ratios and Constraints on New Production

All organisms consist of six major 'light' elements, and approximately 54 trace elements (Williams 1981). Let us consider, for the moment, only the former; these are H, C, N, O, P, and S. The amount of S required to satisfy the demands of most organisms is relatively small, and we can ignore it in the following discussion. In 1934, Redfield pointed out that the chemistry of four major elements in the ocean, namely C, N, P, and O, was strongly affected by biological processes (Redfield 1934). Absent biological N_2 fixation on land or in the ocean, there would be virtually no fixed nitrogen in the oceans. In the ocean interior, the ratio of fixed inorganic N to PO_4 in the dissolved phase is remarkably close to the ratios of the two elements in living plankton. Hence, it seemed reasonable to assume that the ratio of the two elements in the dissolved phase was the result of the sinking, grazing and or autocatalyzed cell death, and subsequent remineralization of the elements in the plankton. Further, as carbon and nitrogen in the living organisms is largely reduced, while remineralized forms are virtually all oxidized, the remineralization of organic matter was coupled to the depletion of oxygen. The relationship could be expressed stoichiometrically as:

$$[106(CH_2O)16NH_31PO_4^{3-}] + 138O_2 \longrightarrow 106CO_2 + 122H_2O + 16NO_3^- + PO_4^{2-} + 16H^+ \quad (4.2)$$

Hidden within this balanced chemical formulation are biochemical oxidation-reduction reactions and hydrolytic processes, some of which are contained within specific functional groups of organisms. In the oxidation of organic matter, there is some ambiguity about the stoichiometry of O_2/P. Assuming that the mean oxidation level of organic carbon is that of carbohydrate (as is the case in Eq. 4.1), then the oxidation of that carbon is equimolar with O. On the other hand, some organic matter may be more or less reduced than carbohydrate, and therefore require more or less O for oxidation. Note also that the oxidation of NH_3 to NO_3^- requires 4 atoms of O, and leads to the formation of 1 H_2O and 1 H^+. Let us examine how these sets of metabolic sequences emerged in the ocean on evolutionary and ecological scales. Effectively, Eq. 4.2 is an expanded version of Eq. 4.1, but written in reverse, to emphasize the remineralization of organic matter. When the reactions primarily occur at depth, Eq. 4.2 is driven to the right, while when the reactions primarily occur in the euphotic zone they are driven to the left. Note that in addition to reducing CO_2 to organic matter, formation of organic matter by photoautotrophs requires reduction of nitrate to ammonia. These two forms of nitrogen are critically important in helping to quantify new and export production. Moreover, JGOFS syntheses studies have shown that the oxidation state of plankton carbon is more reduced than carbohydrate; thus more oxygen is required to combust this organic matter during respiration than predicted by Eq. 4.2 (Li and Peng 2002). Hedges et al. (2002) used NMR analyses to estimate that average plankton composition is best represented as 65% protein, 19% lipid and 16% carbohydrate. Complete oxidation of 106 moles of C in this average organism requires 154 rather than 138 moles of O_2.

4.1.8 New Production, Export Production, and Net Community Production

In our previous thought experiment regarding regenerated production, assimilation and metabolism of NPP by heterotrophs leads primarily to excretion of ammonium and other reduced forms of nitrogen (urea and amino acids) into the euphotic zone. In contrast, when waters from the ocean interior are mixed into the euphotic zone, nitrate is the primary form of nitrogen that becomes available. Realizing that the form of nitrogen could be used to identify its source, Dugdale and Goering (1967) developed the concepts of 'new' and 'regenerated' production. They defined new production to be "all primary production associated with newly available nitrogen, for example NO_3-N and N_2-N". In a confined system such as the mixed layer of the ocean, new production is therefore the portion of primary production supported by exogenous nitrogen inputs (Williams et al. 1989). The exogenous nitrogen could include the sources explicitly noted by Dugdale and Goering (1967), but could also include atmospheric inputs of nitrate or ammonium, and in coastal and shelf regions might also include riverine inputs.

The fraction of total production accounted for by new production is referred to as the *f*-ratio, which can be calculated either in terms of new and regenerated nitrogen uptake or in terms of carbon uptake. Note that in Eq. 4.3 nitrogen fixation is not specified in the numerator of the new production equation. This omission is related to methodological limitations on deriving N_2 fixation rates robustly. In high latitude or upwelling regions, where the upward NO_3 flux is high, N_2 fixation is negligible, however, in subtropical gyres, the fraction of new production supported by N_2 fixation may be significant – but remains poorly constrained.

$$f = \frac{\text{new production}}{\text{new + regenerated production}} \cong \frac{NO_3^- \text{ uptake}}{(NO_3^- + NH_4^+ + NO_2^- + DON)\text{uptake}} \cong \frac{^{15}NO_3^- \text{ uptake}}{^{14}CO_2 \text{ uptake}} \times \left(\frac{C}{N}\right)_{\text{plankton}} \quad (4.3)$$

Related to the concept of new production is the term export production, first coined by Berger et al. (1987). Export production refers to the transfer of biogenic material from one region of the ocean to another. Export production is frequently used to describe transfers from the surface waters to the interior of the ocean by processes such as particle sinking, advection or diffusion of dissolved organic matter (DOM), and vertical migration of zooplankton. However, in zones such as the equatorial upwelling system, a substantial amount of export may occur through lateral advection (Landry et al. 1997). "Export production from the primary system, the photosynthetic zone, or on time scales longer than a year will be equal to new production, although they may, or in most cases will, be separated both in space and time" (Williams et al. 1989). The ratio of export production to primary production is sometimes referred to as the *e*-ratio (Murray et al. 1989)

$$e = \frac{\text{export production}}{\text{primary production}} \quad (4.4)$$

Net community production is the difference between net primary production and heterotrophic respiration. It is an old concept that is operationally defined from net changes in oxygen or biomass concentrations over an appropriate period of time.

In a strictly steady state system, new production, export production, and net community production should be equal. However, the ocean is a dynamic system. On a time frame of days or weeks, biomass in the mixed layer may change dramatically. These changes can occur as a result of imbalances in the rates of biological processes or when physical mechanisms transfer biomass across spatial gradients (McGillicuddy et al. 1995). Because new production, export production, and net community production are measured by fundamentally different methods, there is no reason that these measurements should yield identical results unless the system under study is truly in steady state or unless the measurements are averaged over appropriate scales of time and space. All three approaches were used in JGOFS studies but unfortunately never all at the same time. Given this fact, a measurement that targets new production should not be reported as a measure of export production or net community production, and a measure of net community production should not be reported as a measure of new production or export production.

4.1.9 Measurement of New Production

Throughout the mid 1970s several research groups around the world adopted the measurement of assimilation of $^{15}N\text{-}NO_3$ to estimate new production. This approach, originally introduced by Dugdale and Goering (1967), remains conceptually unchanged. ^{15}N labeled NO_3 that is assimilated and converted to PON by phytoplankton over a known incubation period is filtered and analyzed by mass spectrometry or optical emission spectrometry. Incubation rates are kept short to minimize ^{15}N recycling (e.g., McCarthy et al. 1996). Complementary uptake rates of ammonium and urea are also occasionally measured for regenerated production (e.g., Varela and Harrison 1999; Bury et al. 1995). Failure to correct for release of $DO^{15}N$ from cells during incubations may result in underestimation of new production rates (Bronk et al. 1994). The method is difficult to apply when the ambient concentrations of NO_3 are low, such as in the subtropical gyres, because addition of labeled NO_3 frequently violates the concept of a tracer. Due to analytical limitations, for many years ^{15}N-labeled substrates were never added at concentrations less than 30 nM (Glibert and McCarthy 1984). In recent years analytical improvements have allowed investigators to add spikes as small as 3 nM (McCarthy et al. 1999), but even such small additions may amount to significant perturbations at oligotrophic oceanographic sites. At Station ALOHA, off the north coast of Oahu (22.75° N, 158° W), for example, nitrate concentrations in the upper 50 m of the water column average 2 nM, and 95% of the concentrations are less than 7 nM. Hence despite recent improvements in sensitivity, the ^{15}N technique still does not lend itself to truly oligotrophic conditions. Most applications have been in the vast high-nitrate, low-chlorophyll (HNLC) regions – where incomplete utilization of new nutrients leaves surface CO_2 partial pressures elevated with respect to the atmosphere (Chavez and Barber 1987; Dugdale et al. 1992; Kurz and Maier-Reimer 1993). The equatorial Pacific upwelling zone has been one of the most well studied HNLC regions on the globe with respect to ^{15}N-new production (Aufdenkampe et al. 2001). In the JGOFS program, ^{15}N based estimates were broadly applied in all field campaigns, with the exception of the North Atlantic Bloom Experiment (NABE) and the subtropical ocean time series.

4.1.10 Measurement of Net Community Production

Many estimates of net community production are based on changes in oxygen concentration, either in incubation bottles or in situ. In the latter case, corrections must be made for air/sea gas exchange flux. Empirically, in the euphotic zone, organic matter export production and oxygen export are stoichiometrically related and should be equivalent over an annual cycle. The method of estimating net production from O_2 gas exchange fluxes found wide application before and during JGOFS, especially in the subtropical north Atlantic near Bermuda (Jenkins and Goldman 1985; Spitzer and Jenkins 1989); the subarctic Pacific at Station P (Emerson 1987; Emerson et al. 1991) and the subtropical north Pacific at Station ALOHA (Emerson et al. 1995; Emerson et al. 1997). The method hinges on determining the net biologically produced O_2 in the euphotic zone and calculating its flux across the air-sea interface with gas exchange models. The instantaneous rates will reflect average rates during the prior 2–3 weeks, because the residence time of biologically produced O_2 with respect to gas exchange is about that long. The difficult aspects are separating the biologically produced O_2 from air-injection by bubbles and determining the wind speed dependent gas exchange piston velocity (Liss and Merlivat 1986; Wanninkhof 1992). A high-resolution time series is needed to calculate the time rate of change. Estimates of net production by this approach are significantly higher than those derived from satellite color images (e.g. Laws et al. 2000), possibly because surface water chlorophyll is a poor predictor of integrated productivity in many regions (Letelier et al. 1996).

The accuracy of incubation methods is limited by the requirement to keep incubations short enough to avoid bottle confinement artifacts and by the precision of the oxygen measurements. In recent years considerable improvements have been made in analytical methodology (Williams 1993). The ultimate precision of O_2 methods is probably 0.1–0.2 μmole O_2 l^{-1} (Laws et al. 2002). For comparative purposes, this is approximately the amount of net community production at Station ALOHA in one day (Laws et al. 2000). Incubations are typically no longer than 24 h. Regardless of the duration of the incubation, estimates of net community production based on bottle incubations may be biased if the biological community in the bottles is somehow not representative of its in situ counterpart. Bender et al. (1999), for example, noted that in vitro estimates of net community production were 4–20 times greater than estimates from drifting sediment trap and tracer transport studies in the Pacific equatorial upwelling system. They postulated that this difference reflected an anomalous accumulation of POC in incubation bottles due to the exclusion of grazers.

A variation on oxygen-based calculations is to use the observed distributions of oxygen in the deep ocean, below the euphotic zone, to calculate export production. Assuming that the distribution is at steady state, the O_2 distribution reflects the balance between in situ consumption of oxygen by respiration and physical transport of O_2 from high latitudes (ventilation) (Riley 1951; Jenkins 1982). The oxygen utilization rate (OUR) is calculated from the decreasing O_2 content (apparent oxygen utilisation or AOU) as a function of age (determined using ^{3}He/^{3}H and/or chlorofluorocarbon dating; e.g., Doney et al. 1992). This method obviously gives average values over large temporal and spatial scales. By integrating the OUR over a range of density layers below the euphotic zone, areal respiration rates can be calculated. Equating this rate to export production implicitly assumes that the respiration rate below the specified isopycnal surface equals the export of organic matter from the euphotic zone. The respiration rate so calculated is in fact a lower bound on the export flux from the euphotic zone for several reasons. First, there is some accumulation of organic matter in the sediments. Second, no isopycnal surface coincides with the base of the euphotic zone, i.e., there will be some consumption of organic matter below the base of the euphotic zone and the chosen isopycnal surface. Nevertheless, it is reasonable to assume that the consumption of organic matter below an appropriately chosen isopycnal surface is closely correlated with export production from the euphotic zone, and there is evidence that with a suitable time series, decadal scale variability can be detected (Min et al. 2000; Emerson et al. 2002). While this approach was developed prior to JGOFS, the new combined data sets of JGOFS and WOCE have provided excellent opportunities to apply it to different ocean basins (e.g. in the South Atlantic, Warner and Weiss 1992; and North Pacific, Warner et al. 1996).

4.1.11 Measurement of Export Production

Most estimates of export production are in fact estimates of the flux of sinking particulate organic matter. Such calculations need to be augmented to account for other mechanisms of export, including advection of DOM and transport by vertical migrators. Prior to JGOFS free-floating particle interceptor traps (PITS) were the main approach for measuring particle export flux (Martin et al. 1987).

Because of its strong partitioning towards particles, thorium 234 has been widely used as a particle 'scavenging' tracer to determine particle residence times, transformation rates, and sinking rates in the ocean (Santschi et al. 1979; Honeyman and Santschi 1989; Clegg and Whitfield 1990; Dunne et al. 1997). ^{234}Th has a half-life of 24.1 days, which makes it suitable for study of upper ocean processes. The link between ^{234}Th removal from the water and biological processes was demonstrated by Coale and Bruland (1985, 1987) and Bruland and Coale (1986). Based on this evidence, Eppley (1989) suggested that ^{234}Th be used as a tracer for export flux if the particulate Th residence time could be applied to particulate carbon. However, the residence time approach is not used because ^{234}Th and carbon do not always have the same residence times (Murray et al. 1989).

The ^{234}Th method for determining export flux was developed initially by Buesseler et al. (1992) during the JGOFS NABE Study. The basic approach is to use the steady state mass balance for ^{234}Th in the euphotic zone to calculate the export flux of ^{234}Th (P):

$$\partial^{234}Th / \partial t = (^{238}U - {}^{234}Th)\lambda^{234}Th - P + V \qquad (4.5)$$

The method depends on there being a statistically significant deficit of ^{234}Th, a condition that makes application of the method problematic at oligotrophic sites such as BATS (Buesseler et al. 1994) and HOT (Benetiz-Nelson et al. 2001). Various studies have evaluated the importance of the steady state assumption (e.g. Buesseler et al. 1992) and advective and diffusive transport terms (V) in the mass balance (e.g. Buesseler et al. 1995; Bacon et al. 1996; Murray et al. 1996; Dunne and Murray 1999). In general the time rate of change is not significant (e.g. even in the North Atlantic Bloom Experiment; Buesseler et al. 1992), and the transport terms have only been essential in the equatorial and subarctic Pacific and some coastal regions like the Ross Sea (Cochran et al. 2000). The export flux of Th can be converted to carbon (or other elements) knowing the ratio of carbon to ^{234}Th in sinking particles. The relevant equation is:

$$\text{Model POC Flux} = (\text{Model } ^{234}\text{Th Flux})\ (\text{C}/\text{Th}_{\text{sinking particles}}) \qquad (4.6)$$

One of the stumbling blocks has been the fact that different sampling approaches sometimes lead to different C/Th ratios in sinking particles. In EqPac, for example, the C/Th of trap samples was consistently greater than in situ large volume pump samples. In a study conducted at HOT there was much better agreement (Benetiz-Nelson et al. 2001). The ^{234}Th method has been widely applied in all JGOFS studies.

An alternative to the ^{234}Th method is to quantify the rate of accumulation of organic matter in drifting sediment traps. The most commonly used designs of drifting sediment traps used in JGOFS studies are cylinders (e.g. the cylindrical Particle Interceptor Traps (PITs) of Knauer et al. 1979; Martin et al. 1987) and cones (e.g. Honjo and Doherty 1988; Peterson et al. 1993). The most recent development is a neutrally buoyant trap that should reduce shear at the mouth of the trap and improve trap efficiency due to reduction of hydrodynamic bias (Buesseler et al. 2000).

The question of trap accuracy was addressed by Buesseler (1991) who compared ^{234}Th fluxes measured with traps with ^{234}Th fluxes calculated from a scavenging model. He showed that shallow sediment traps typically display either positive or negative collection biases. Most likely this variability in trap efficiency is due to hydrodynamic effects (e.g. Gust et al. 1992) but vertically migrating organisms may also play a role. One approach for eliminating this bias is to compare the ^{234}Th content in trap samples with the modeled flux from the overlying water column. The relevant equation is:

$$\text{Model POC Flux} = (\text{Trap POC Flux}) \times (\text{model } ^{234}\text{Th flux} / \text{trap } ^{234}\text{Th flux}) \qquad (4.7)$$

In this way ^{234}Th can be used as a correction factor for other constituents. However, this approach will only be valid if ^{234}Th and the other components have the same distribution among particles. Note that Eqs. 4.6 and 4.7 are the same when traps are used to collect sinking particles, which means that the model ^{234}Th flux can be viewed alternatively as a tool to correct trap fluxes for hydrodynamic bias or as a primary tool for determining export fluxes if the C/^{234}Th ratio is known from trap samples.

In one study from the equatorial Pacific, Murray et al. (1996) showed a strong correlation between organic carbon and ^{234}Th in PIT type trap samples, supporting the case that ^{234}Th can be used as a tracer for carbon. They then used the comparison of trap and calculated ^{234}Th fluxes as a correction factor for hydrodynamic biases for carbon. Hernes et al. (2001) used the same approach to 'calibrate' traps of a very different design (conical with a large diameter, Peterson et al. 1993), which were deployed at the same times and stations. In the original data the conical POC fluxes were much smaller than the PIT fluxes, in accord with earlier observations reported by Laws et al. (1989). The ^{234}Th correction factors suggest that the conical traps undercollected ^{234}Th while the PIT traps overcollected. When these correction factors are applied to POC, the resulting ^{234}Th corrected POC fluxes are in excellent agreement (Table 4.4). The POC fluxes are about three times larger in the spring (Survey I) than fall (Survey II) and are about 50–70% of the ^{15}N measured new production, which was measured at the same time. The excellent agreement between the ^{234}Th corrected carbon fluxes for these two traps of very different designs verifies the success of this approach.

Table 4.4. Illustration of the ^{234}Th correction approach using conical (105 m) (Hernes et al. 2001) and PIT (100 m) (Murray et al. 1996) data from EqPac, 12° N–12° S at 140° W. Survey I was spring 1992 and Survey II was fall 1992. New production data are from McCarthy et al. (1996). All fluxes given as mmol C m^{-2} d^{-1}

	POC flux		Th correction factor		Th corrected POC flux		New production
	Conical	PIT	Conical	PIT	Conical	PIT	
Survey I	2.2 ±1.5	12.2 ±9.3	1.1 ±0.7	0.3 ±0.2	2.3 ±1.4	2.6 ±0.8	5.7 ±2.2
Survey II	2.4 ±1.3	15.1 ±9.9	4.6 ±1.5	0.6 ±0.2	7.4 ±4.2	8.7 ±5.3	12.4 ±9.2

4.1.12 Summary of Methods

The various methods used to estimate new production, export production, and net community production are clearly not intended to measure the rate of the same process. Although the three rates are equal in a steady state system or when averaged over large enough time and space scales, there is otherwise no a priori reason why the rates should be identical. Comparisons of rates estimated by the various methods should be made with a full awareness of the theoretical differences and practical limitations of the measurements.

In practice by far the most commonly used direct measure of new production is the uptake of nitrate. Traditionally this rate has been estimated using ^{15}N tracer techniques, but with the development of chemiluminescent methods for nitrate analysis (Garside 1985), nitrate uptake can be estimated at low substrate concentrations from the rate of change of nitrate concentration (Allen et al. 1996). Recent studies have suggested that nitrogen fixation may contribute substantially to new production in some parts of the ocean (Karl et al. 1997; Zehr et al. 2001), and nitrification between the 1% and 0.1% light level may confound the interpretation of nitrate as 'newly available nitrogen' (Dore and Karl 1996). Direct estimates of export production are virtually all based on measurements of the downward flux of sinking particles. Very few studies have attempted to directly document the contribution of other mechanisms such as vertical migration (Longhurst et al. 1990; Steinberg et al. 2000) and diapycnal mixing of DOM. The former flux has been estimated to be as much as 20–30% of the sinking POC flux at some locations (Steinberg et al. 2000). These considerations suggest that nitrate uptake and the downward flux of particles are lower bounds on new production and export production if the system in question is defined to be the water column above the 1% light level, i.e., the traditional definition of the euphotic zone. The extent of the bias has often been inferred from estimates of net community production, but comparisons require conversion from an oxygen to a nitrogen or carbon-based budget and implicitly assume that new production, export production, and net community production are equal. Conversions between oxygen, carbon, and nitrogen are usually made assuming Redfield stoichiometry, but this may be a poor assumption in some cases (Sambrotto et al. 1993; Daly et al. 1999). And there is certainly no reason to believe that new production, export production, and net community production are equal when measurements are averaged over different time scales or over equal but short time intervals.

One of the more troubling aspects of all these calculations is the definition of the dimensions of the system being studied. New production and export production are frequently estimated with respect to the euphotic zone, which is conventionally defined to be the water column above the 1% light level. However, this definition is arbitrary. Some photosynthesis certainly occurs between the 1% and 0.1% light level (Hayward and Venrick 1982). During EqPac about 10% of primary production and 20% of new production occurred between the 1% and 0.1% light levels (Murray et al. 1996). Dore and Karl (1996) have reported evidence of nitrification in the same region of the water column. The distinction between new and regenerated production becomes even more confused if one takes physical processes into account. When the mixed layer extends below the depth of the euphotic zone, nutrients regenerated below the depth of the euphotic zone can easily be recycled back into the euphotic zone. This is an important consideration in temperate and high latitudes, where the mixed layer in the winter may be hundreds of meters deep. On an annual basis, recycling of nutrients certainly occurs well below the depth of the euphotic zone in many parts of the ocean. Estimates of net community production based on in situ changes in oxygen concentrations focus on the mixed layer, not the euphotic zone. Comparisons of new production, export production, and net community production will be compromised until a consensus is reached on the vertical dimensions of the system with respect to which these rates are to be calculated.

Subject to these caveats, Table 4.5 summarizes estimates of new production, export production, and net community production based on studies covering a wide range of oceanographic conditions. Much of the work was carried out during the JGOFS program, but results from studies in the Peru upwelling system, at Station P in the subarctic Pacific, and in a Greenland polynya are also included. In most cases the values are estimates of

Table 4.5. Field data and methods of analysis

Area	Z_m	Temperature (°C)	Export or new production (mg N m^{-2} d^{-1})	Method of calculation	Total production (mg N m^{-2} d^{-1})	Method of calculation
BATS	140	21	7.8	Carbon balance assuming Redfield C:N (Michaels et al. 1994)	82	^{14}C production from Michaels et al. (1994) divided by Redfield C:N
HOT	150	25	12.2	Average of O_2 and C mass balances, assuming Redfield ratios (Emerson et al. 1997)	83	^{14}C production from Karl et al. (1996) divided by Redfield C:N
NABE	35	12.5	98	Total production times mean *f* ratio of Bender et al. (1992) and McGillicuddy et al. (1995)	194	^{14}C production from Buesseler et al. (1992) divided by Redfield C:N
EqPac-normal	120	24	32.1	Total production times *f* ratio from McCarthy et al. (1996) based on ^{15}N uptake ratios	260	^{14}C production from Barber et al. (1996) divided by Redfield C:N
EqPac-El Niño	120	27	12.3	Total production times *f* ratio of McCarty et al. (1996) based on ^{15}N uptake ratios	169	^{14}C production from Barber et al. (1996) divided by Redfield C:N
Arabian Sea	65	25	29.2	Total production times *f* ratio of McCarthy et al. (1999) based on ^{15}N uptake ratios	195	^{14}C production from Barber et al. (2001) divided by Redfield C:N
Ross Sea	40	0	165	Total production times *f* ratio of Asper and Smith (1999) from ^{15}N uptake ratios	243	^{14}C production from Asper and Smith (1999) divided by Redfield C:N
Subarctic Pacific-Station P	120	6	40.3	Mean nitrate uptake/utilization from Sambrotto and Lorenzen (1987), Emerson et al. (1993), and Wong et al. (1998)	95	Mean ^{14}C production from Welschmeyer et al. (1991) and Wong et al. (1995) divided by Redfield C:N
Peru-normal	25.5	16.8	339	Nitrate uptake using ^{15}N tracer (Wilkerson et al. 1987)	806	^{14}C production from Wilkerson et al. (1987) divided by Redfield C:N
Peru-El Niño	17.8	17.4	256	Nitrate uptake using ^{15}N tracer (Wilkerson et al. 1987)	867	^{14}C production from Wilkerson et al. (1987) divided by Redfield C:N
Greenland polynya	50	0	35.6	Nitrate uptake using ^{15}N tracer (Smith 1995; Smith et al. 1997)	63.2	^{15}N uptake from Smith (1995) and Smith et al. (1997)

new production based on ^{15}N techniques, but in the absence of ^{15}N results estimates are based on one or a combination of other methods. In all cases the results are based on multiple measurements, and at the time series stations (HOT and BATS) some of the numbers are averages of monthly observations made over a period of several years.

4.2 Synthesis

Following the seminal paper by Eppley and Peterson (1979), it has generally been assumed that a more-or-less hyperbolic relationship existed between primary production and the *f*-ratio, which Eppley and Peterson (1979) defined to be the ratio of new production to primary production. Eppley et al. (1983) suggested that new production and particle sinking are coupled over long time scales and that the residence time of POC in the euphotic zone ranges from 3 to >100 days from shelf to gyre regions. This basic notion was further supported by data from Harrison et al. (1987) that suggested that the *f*-ratio increased hyperbolically with nitrate (i.e., 'new' nitrogen).

The relationship between the ratio of new or export production to primary production, based on the data in Table 4.5, is shown in Fig. 4.5b. The latter ratio is designated the *ef* ratio to indicate that the numerators are in some cases based on estimates of new production and in other cases on export production. There is no significant correlation between primary production and the *ef* ratio in this data set. Similar results have been reported in experimental results summarized by other investigators (e.g., Sarmiento and Armstrong 1997). However, there is excellent agreement (Fig. 4.5a) between the observed *ef* ratios and the *ef* ratios predicted by a model reported by Laws et al. (2000). The model accounts for 97% of the variance in the observed data. The model assumes that primary production is partitioned through both large and small phytoplankton and that the food web adjusts to changes in the rate of exogenous nutrient inputs in a way that maximizes stability,

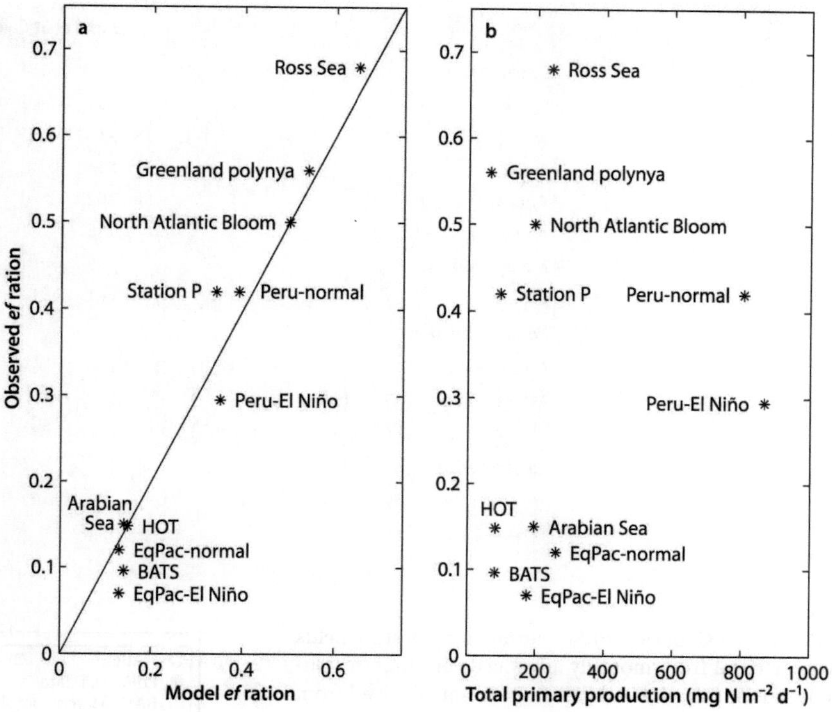

Fig. 4.5.
a Model export ratios vs. observed export ratios at sites summarized by Laws et al. (2000). The *straight line* is the 1:1 line. **b** Total primary production vs. observed export ratios at the same locations (reproduced with permission from Laws et al. (2000) Global Biogeochem Cy 14(4):1231–1246, © 2000 by the American Geophysical Union)

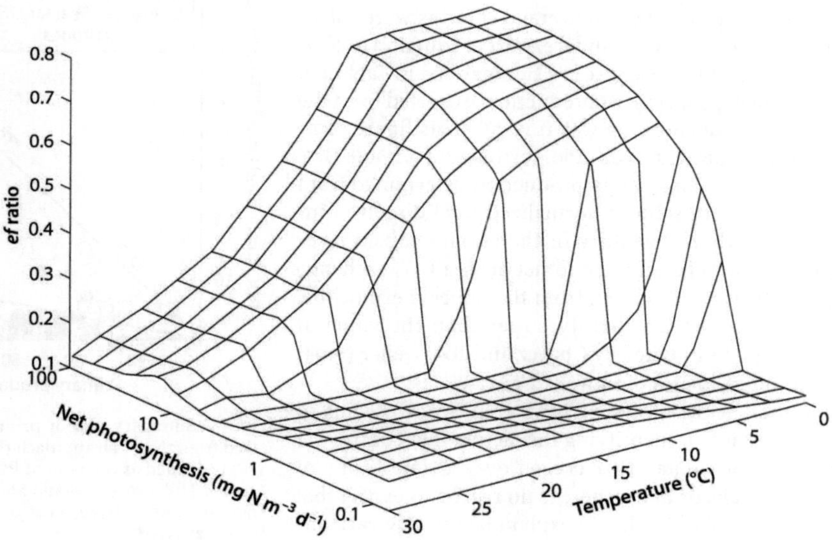

Fig. 4.6.
Calculated export ratios as a function of temperature and net photosynthetic rate (reproduced with permission from Laws et al. (2000) Global Biogeochem Cy 14(4):1231–1246 © 2000 by the American Geophysical Union)

i.e., how rapidly the system returns to steady state following a perturbation. The model predicts that the *ef* ratio will be a function of two variables, primary production per unit volume and temperature. The response curve is shown in Fig. 4.6. At a fixed temperature, the behavior of the response curve is similar to that postulated by Eppley and Peterson (1979). The *ef* ratio is low at low rates of primary production, rises steeply at intermediate rates of primary production, and plateaus at high rates of primary production. The height of the plateau is greatest at low temperatures, and the region of rapid rise shifts to progressively higher rates of primary production as temperature rises.

In order to examine the implications of their model with respect to global export production, Laws et al. (2000) calculated net primary production with the vertically generalized production model of Behrenfeld and Falkowski (1997a,b) using SeaWiFS ocean color data for the 12-month period beginning October 1997 and NASA's derived global chlorophyll fields. The calculated global

Table 4.6.
Export production and *ef* ratios calculated from the model of Laws et al. (2000)

	Export (Gt C yr^{-1})	*ef*
Ocean basin		
Pacific	4.3	0.19
Atlantic	4.3	0.25
Indian	1.5	0.15
Antarctic	0.62	0.28
Arctic	0.15	0.56
Mediterranean	0.19	0.24
Global	11.1	0.21
Total production		
Oligotrophic (chl *a* < 0.1 mg m^{-3})	1.04	0.15
Mesotrophic (0.1 chl *a* < 1.0 mg m^{-3})	6.5	0.18
Eutrophic (1.0 chl *a* mg m^{-3})	3.6	0.36
Ocean depth		
0 – 100 m	2.2	0.31
0.1 – 1 km	1.4	0.33
> 1 km	7.4	0.18

NPP was 52.1 Gt of carbon. Sea surface temperature fields were derived from monthly advanced very high resolution radiometer global data. *ef* ratios were derived from a look-up table generated from the interpolated values shown in Fig. 4.6. Annual average *ef* ratios were calculated from the ratio of annual export production to NPP. The results of this exercise are summarized in Table 4.6. The annual global export production estimated from the model and satellite data was 11.1 Gt C. This figure compares favorably with Lee's (2001) estimate of 9.1–10.8 Gt C for global net community production determined from the decrease in salinity-normalized total dissolved inorganic carbon inventory in the surface mixed layer corrected for changes due to net air-sea CO_2 exchange and diffusive carbon flux from the upper thermocline. Both estimates are slightly larger than the value of 7.2 Gt C yr^{-1} estimated by Chavez and Toggweiler (1995) based on estimates of upwelled nitrate.

Although the model of Laws et al. (2000) seems to do a good job of identifying the factors that control the *ef* ratio when data are averaged over a time frame of several weeks or more, there is no reason to expect that the model could or should explain day-to-day variability in *f*-ratios or *e* ratios. The model assumes a balance between new production and export production, and this assumption may certainly be violated on a short time frame. An alternative approach is needed to describe variability in new or export production on such time scales. Indeed, export production based on ^{234}Th measurements and primary production clearly illustrate the difficulty of extrapolating export from net primary production measurements in highly energetic ecosystems (Fig. 4.7).

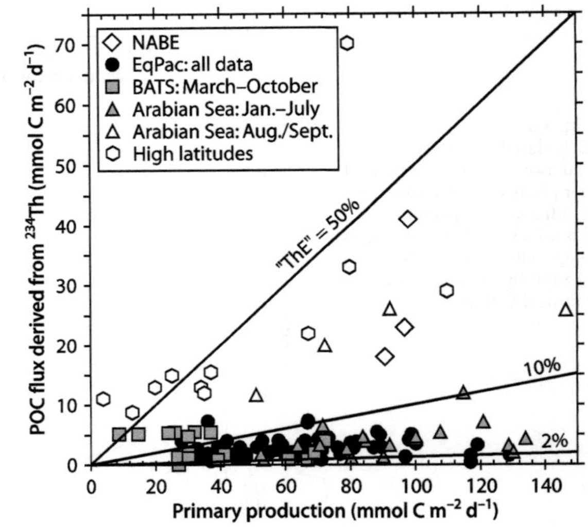

Fig. 4.7. Summary plot of primary production vs. POC flux derived from the ^{234}Th approach (both in terms of mmol C m^{-2} d^{-1}). ThE is defined as the ratio of POC export to primary production. *Lines* of ThE = 50%, 10% and 2% are drawn for comparison to data. Data are shown from most of the JGOFS study regions (after Buesseler (1998))

Aufdenkampe et al. (2001) developed a multiple linear regression (MLR) expression to calculate new production in the tropical Pacific. The model can explain nearly 80% of the variability in new production using four independent variables – rates of primary production (or chlorophyll inventories), inventories of ammonium and nitrate, and temperature. Depth integrated primary production, euphotic zone inventories of ammonium and nitrate, and depth averaged euphotic zone temperature formed the most significant combination of independent variables from more than three dozen tested. The multiple coefficient of determination, R^2, indicated that 79% of the variability in new production could be explained by changes in these four variables alone. Total primary production gives the largest contribution to the regression fit followed by ammonium inventory, nitrate, and finally temperature. Chlorophyll inventory offered a reasonable substitute for primary production with the resulting MLR ($R^2 = 0.63$) receiving similar contributions from nitrate, chlorophyll, and ammonium. MLR analyses of the subsets of stations having data for sediment-trap silica fluxes, and sediment-trap silica-to-nitrogen flux ratios and biogenic silica inventories showed that none of these parameters exhibited stronger partial correlations to new production than did nitrate.

4.2.1 Physical Controls of Export Fluxes: the Importance of Functional Groups

The biologically mediated fluxes of elements between the upper ocean and the ocean interior are critically dependent upon key groups of organisms. Similarly, fluxes between the atmosphere and ocean as well as between the ocean and the lithosphere are mediated by organisms that catalyze phase state transitions from either gas to solute/solid or from solute to solid/gas phases. For example, autotrophic carbon fixation converts gaseous CO_2 to a wide variety of organic carbon molecules, virtually all of which are solid or dissolved solids at physiological temperatures. Respiration accomplishes the reverse. Nitrogen fixation converts gaseous N_2 to ammonium and thence to organic molecules, while denitrification accomplishes the reverse. Calcification converts dissolved inorganic carbon and Ca to solid phase calcite and aragonite whereas silicification converts soluble silicic acid to solid hydrated amorphous opal. Each of these biologically catalyzed processes is dependent upon specific metabolic sequences (i.e., gene families encoding a suite of enzymes) that evolved over hundreds of millions of years of Earth's history, and have, over corresponding periods, led to the massive accumulation of calcite, opal, and organic matter in the lithosphere. Presumably because of parallel evolution as well as lateral gene transfer, these metabolic sequences have frequently co-evolved in several groups of organisms that, more often than not, are not closely related from a phylogenetic standpoint (Falkowski and Raven 1997). Based on their biogeochemical metabolism, these homologously similar sets of organisms can be clustered into 'functional groups' or 'biogeochemical guilds'; i.e., organisms that are related through common biogeochemical processes rather than phylogenetic affiliation.

In the contemporary ocean, the export of particulate organic carbon from the euphotic zone is highly correlated with the flux of particulate silicate. Most of the silicate flux is a consequence of precipitation of dissolved orthosilicic acid by diatoms to form amorphous opal that makes up the cell walls of these organisms. These hard-shelled cell walls presumably help the organisms avoid predation, or if ingested, increase the likelihood of intact gut passage though some metazoans. In precipitating silicate, diatoms simultaneously fix carbon. Upon depleting the euphotic zone of nutrients, the organisms frequently sink en masse, and while some are grazed en route, many sink as intact cells. Ultimately, either fate leads to the gravitationally driven export flux of particulate organic carbon into the ocean interior.

Silica is supplied to the oceans from the weathering of continental rocks. Because of the precipitation of silicious organisms however, the ocean is relatively depleted in dissolved silica. Although diatom frustules (i.e., their silicified cell walls) tend to dissolve and are relatively poorly preserved in marine sediments, enough silica is buried to keep the ocean undersaturated. As the residence time of silica in the oceans is about 10 000 years (i.e., about an order of magnitude longer than the mean deep water circulation), one can get an appreciation for the silicate demands and regeneration rates by following the concentration gradients of dissolved silica along isopycnals. While these demands are generally attributed to diatoms, radiolarians (a group of non-photosynthetic, heterotrophic protists with silicious tests, that are totally unrelated to diatoms) are not uncommon, and radiolarian shells are abundant in the sediments of Southern Ocean. Silica is also precipitated by various sponges, and other protists. As a functional group, the silicate precipitators are identified by their geochemical signatures in the sediments and in the silica chemistry of the oceans. Diatoms can be elucidated by photosynthetic pigment analyses in situ, but cannot be uniquely identified from satellite imagery. Because of their importance in mediating carbon export, a significant effort was spent in JGOFS attempting to understand the factors controlling the distribution of diatoms in the world oceans. Interestingly, diatoms appear to have evolved relatively recently; the first clear evidence of their presence in the oceans is recorded in fossils from cherts dating to 120 Mybp (i.e. in the early Cretaceous). Thus, although export carbon fluxes are associated with this group of organisms in the contemporary ocean, they usurped that role from an unknown group(s) of phytoplankton that dominated earlier in the Mesozoic. What organisms mediated export of carbon in the Proterozoic (prior to the emergence of eucaryotes) and Paleozoic (prior to the dominance of the chromophyte algae) remains totally unknown.

A major strategy of diatoms is to acquire nutrients rapidly under highly physically dynamic conditions, and to store the nutrients in vacuoles for later cell growth. This strategy simultaneously deprives competing groups of phytoplankton of essential nutrients while allowing diatoms to grow rapidly, forming blooms. In this strategy nitrate, but not ammonium can be stored. Nevertheless, Aufdenkampe et al. (2001) showed that silicate itself does not appear to exert a strong direct control on new production during typical conditions. Diatom-associated chlorophyll did indeed seem to be important in the equatorial Pacific, but only a few stations during uncommonly high new production (NP) dictated much of the NP-Chl_{dia} relationship. These findings are consistent with the conclusion of Dunne et al. (1999) that diatoms regulate new production fluxes only during highly dynamic, non-steady state conditions such as the passage of tropical instability waves (referred to earlier as the 'perturbed' state), but that diatoms do not dominate new production during more common and less dynamic conditions.

4.2.2 Calcium Carbonate Precipitation

Like silica precipitation, calcium carbonate is not confined to a specific phylogenetically distinct group of organisms, but evolved (apparently independently) several times in marine organisms. Carbonate sediments blanket much of the Atlantic Basin, and are formed from the shells of both coccolithophorids and foraminifera. (In the Pacific, the carbon compensation depth is generally higher than the bottom, and hence, in that basin carbonates tend to dissolve rather than become buried.) As the crystal structures of the carbonates in both groups is calcite (as opposed to the more diagenically susceptible aragonite), the preservation of these minerals and their co-precipitating trace elements, provides an invaluable record of ocean history. Although on geological time scales, huge amounts of carbon are stored in the lithosphere as carbonates, on ecological time scales, carbonate formation depletes the ocean of Ca^{2+}, and in so doing, potentiates the efflux of CO_2 from the oceans to the atmosphere. This sequence can be summarized by the following:

$$2HCO_3^- + Ca^{2+} \longrightarrow CaCO_3 + CO_2 + H_2O \qquad (4.8)$$

Unlike silicate precipitation, calcium carbonate precipitation leads to strong optical signatures that can be detected both in situ and remotely. The basic principles of detection are the large, broad band (i.e., 'white') scattering cross sections. The high scattering cross sections are detected by satellites observing the upper ocean as relatively highly reflective properties (i.e., a 'bright' ocean). Using this detection scheme, one can reconstruct global maps of planktonic calcium carbonate precipitating organisms in the upper ocean. In situ analysis can be accompanied by optical rotation properties (polarization) to discriminate calcite from other scattering particles. In situ profiles of calcite can be used to construct the vertical distribution of calcium-carbonate-precipitating planktonic organisms that would otherwise not be detected by satellite remote sensing because they are too deep in the water column.

Unlike diatoms, coccolithophores do not store nutrients very efficiently, but can bloom when nutrients are supplied at slow rates. Hence, coccolithophorids are primarily found at low abundance in tropical and subtropical seas, and at higher concentrations at high latitudes in midsummer, following diatom blooms. Hence, export of inorganic carbon by diatoms in spring at high latitudes can be offset by an efflux of carbon to the atmosphere with the formation of coccolithophore blooms later in the year.

4.2.3 Primary, New and Export Production and the Global Carbon Cycle on Longer Time Scales

Now we come to a feature at the intersection of oceanic biology and geochemistry that sometimes causes some confusion, namely, do changes in export production or the biological pump affect the atmospheric concentration of CO_2? The answer is, "Not necessarily, but they can; it depends how the changes come about."

Let us imagine that we could stir the oceans with a large spoon or egg beater, such that we mixed the ocean interior up and it was exposed to the atmosphere. Initially CO_2 would evade from the oceans to the atmosphere, but the stirring process would also bring up essential nutrients that help stimulate NPP. The two major nutrients are fixed inorganic N (primarily in the form of nitrate) and P (primarily in the form of phosphate). As these nutrients are consumed to make new cells in the upper ocean, inorganic carbon would also be fixed back to organic matter, and hence returned to the oceanic reservoir. Unlike any other ecosystem that we know of, the elemental stoichiometry specified in Eq. 4.2 between C, N, P, and O appears to be relatively constrained in the oceans.

On geological time scales, there is one final fate for NPP we should consider, namely burial in the sediments. By far, the largest reservoir of organic matter on Earth is locked up in rocks (Table 4.7). Virtually all of this organic carbon was the result of the burial of the export production in the oceans over literally billions of years of Earth's history. Indeed, without burial, there would be no net evolution of oxygen in Earth's atmosphere, as the production of oxygen would have been

Table 4.7. Carbon pools in the major reservoirs on Earth (from Falkowski and Raven 1997)

Pools	Quantity ($\times 10^{15}$ g)
Atmosphere	720
Oceans	38 400
Total inorganic	37 400
Surface layer	670
Deep layer	36 730
Dissolved inorganic	600
Total organic	1 000
Lithosphere	
Sedimentary carbonates	>60 000 000
Kerogens	15 000 000
Terrestrial biosphere (total)	2 000
Living biomass	600–1 000
Dead biomass	1 200
Aquatic biosphere	1–2
Fossil fuels	4 130
Coal	3 510
Oil	230
Gas	140
Other (peat)	250

balanced by its consumption (e.g. Walker 1974). On geological time scales, the burial of marine NPP effectively removes carbon from biological cycles, and places that carbon into another, much slower, carbon cycle. The latter is driven by tectonics and weathering (discussed by Watson and Orr, this volume), in which the organic (and inorganic; i.e., carbonate) sedimentary rocks are subducted into the upper mantle along tectonic boundaries, and heated in the Earth's interior. The resulting carbon-containing products of this heating process (primarily CO_2 and CH_4) enter the atmosphere through volcanic outgassing. The CH_4 is rapidly oxidized to CO_2 in the upper atmosphere. CO_2 reacts with Ca and Mg silicates to form carbonates through the weathering reactions. On time scales of millions of years, this cycle determines the concentration of O_2 and CO_2 in the atmosphere and oceans.

A very small fraction of the carbon buried in sediments undergoes transformation to become polymers of petroleum (Table 4.7) or exploitable methane. These organic carbon pools are part of the 'slow' carbon cycle; they are effectively removed from the atmosphere for hundreds of millions of years unless they are extracted from the lithosphere and burned. The burial of organic carbon in the modern oceans is primarily confined to a few regions where the supply of sediments from terrestrial sources is extremely high. Such regions include the Amazon outfall and Indonesian mud belts. In contrast, the oxidation of organic matter in the interior of the contemporary ocean is extremely efficient; virtually no carbon is buried in the deep sea. Similarly, on most continental margins, organic carbon that reaches the sediments is consumed by microbes within the sediments, such that very little is actually buried.

As discussed earlier, in the contemporary ocean export of organic carbon to the interior is often associated with diatom blooms. This group has only risen to prominence over the past 40 million years.

Over geological time, the distributions of key functional groups change. For example, relative coccolithophorid abundance generally increases through the Mesozoic, and undergoes a culling at the K/T boundary, followed by numerous alterations in the Cenozoic. The changes in the coccolithophorid abundances appear to trace eustatic sea level variations, suggesting that transgressions lead to higher calcium carbonate fluxes. In contrast, diatom sedimentation increases with regressions and since the K/T impact, diatoms have generally replaced coccolithophorids as ecologically important eucaryotic phytoplankton. On much finer time scales, during the Pleistocene, it would appear that interglacial periods favor coccolithorphorid abundance, while glacial periods favor diatoms. The factors that lead to glacial-interglacial variations between these two functional groups are relevant to elucidating their distributions in the contemporary ecological setting of the ocean.

We shall assert, based on first principles, that the distribution of all planktonic organisms in the oceans obeys a rule of *universal distribution and local selection*. This rule states that in any given body of water, there is a finite probability of finding any organism at any time, but that the local environment will be more conducive to the growth of some organisms than others. This rule implies that, within reason, the rules of 'selection' can be largely elucidated for major functional groups. If, for example, the water column is cold and nutrient rich we may assume that nitrogen fixers will be present but not highly selected while diatoms are more likely to be abundant. Conversely, in warm, stratified, oligotrophic seas, it is unlikely that diatoms will emerge as dominant organisms, while nitrogen fixers are more likely to be abundant.

Over the past ca. 150 years, human energy-related activities have led to a large and rapid injection of CO_2 into Earth's atmosphere, largely through the combustion of fossil fuels. In this process, organic carbon has been extracted from the slow carbon cycle of the lithosphere and placed into the rapid carbon cycle that includes the atmosphere/ocean abiotic exchange, and biologically mediated exchanges that include carbon fixation, respiration, and calcification. To influence the anthropogenic CO_2 emissions, the latter processes must not only deviate from the steady state, but must become imbalanced over time scales of decades. The JGOFS program forced a discussion of whether or if primary production plays or will play a significant role in sequestering the anthropogenic carbon dioxide.

Inspection of Eq. 4.2 suggests that, like Eq. 4.1, the production of organic matter at the surface and its remineralization at depth is normally in a steady state. Thus, moving nitrate and phosphate (at a fixed ratio of 16:1 by moles) from the right hand side to the left would result in a predictable formation of organic matter. To first order however, there are four mechanisms by which the biological pump can deviate from the steady state and impact the net exchange of CO_2 between the atmosphere and ocean. Given that at the present time there is very little burial of organic matter in ocean sediments, we shall ignore burial. The remaining four options operate on time scales of decades to centuries. First, if a limiting nutrient(s) is added to the ocean from an external source, the utilization of that nutrient by primary producers would increase the net formation of organic matter. Some of that organic material would be exported to the ocean interior and remineralized. Alternatively, there are some regions in the ocean where nitrate and phosphate were brought from depth to the surface but not consumed before these nutrients returned to depth. If the efficiency of nutrient utilization were to be somehow enhanced, in principle more carbon would be stored in the ocean interior. Third, if the elemental ratio of the organic matter were to deviate from that given in Eq. 4.2, then in principle the net new flux of carbon to depth would also be altered. Finally, if the ratio of the fluxes of particulate organic/particulate inorganic carbon were to change, the net flux of carbon could change. These scenarios are not mutually exclusive, but require external perturbation. One perturbation will be climate change forced by the combustion of fossil fuels. While it is presently not possible to determine quantitatively whether this perturbation will have any significant effect on the net rate of uptake of anthropogenic carbon by the ocean, or whether any of the four aforementioned processes that are critical to the biological pump will be altered, it is clear that ocean circulation and stratification almost certainly will change.

Synthesis and modeling of the extraordinary data set was the culmination of the formal JGOFS program. These efforts have led to a greater appreciation for the interactions between physical and biological processes in determining which specific functional group dominates, and how that outcome affects the fluxes of organic and inorganic carbon. While still in its infancy, the modeling efforts have further revealed that changes in ocean circulation and stratification can lead to significant changes in export production. In an increasingly stratified ocean of the future, one might predict that diatom blooms occur less frequently or with lower amplitude, and that coccolothophores might become relatively more abundant. Indeed, a major challenge for the next few decades will be to identify the major rules by which key functional groups are distributed in the ocean, both past and present, and to develop predictive models of their distributions in this first century of the third millennium of the common era. Our work is just beginning and the JGOFS data sets will continue to provide a framework in to explore these issues for years to come.

References

Allen CB, Kanda J, Laws EA (1996) New production and photosynthetic rates within and outside a cyclonic mesoscale eddy in the North Pacific subtropical gyre. Deep-Sea Res 43:917–936

Antoine D, Andre JM, Morel A (1996) Oceanic primary production 2. Estimation at global-scale from satellite (coastal zone color scanner) chlorophyll. Global Biogeochem Cy 10:57–69

Asper VL, Smith WO (1999) Particle fluxes during austral spring and summer in the southern Ross Sea (Antarctica). J Geophys Res 104:5345–5360

Aufdenkampe AK, McCarthy JJ, Rodier M, Navarette C, Dunne J, Murray JW (2001) Estimation of new production in the tropical Pacific. Global Biogeochem Cy 15:101–112

Azam F (1998) Microbial control of oceanic carbon flux: the plot thickens. Science 280:694–696

Bacon MP, Cochran JK, Hirschberg D, Hammer TR, Fleer AP (1996) Export flux of carbon at the equator during the EqPac time series cruises estimated from ^{234}Th measurements. Deep-Sea Res Pt II 43:1133–1154

Barber RT, Sanderson MP, Lindley ST, Chai F, Newton J, Trees CC, Foley DG, Chavez FP (1996) Primary productivity and its regulation in the equatorial Pacific during and following the 1991–92 El Niño. Deep-Sea Res Pt II 43:933–969

Barber RT, Marra J, Bidigare RC, Codispoti LA, Halpern D, Johnson Z, Latasa M, Goericke R, Smith SL (2001) Primary productivity and its regulation in the Arabian Sea during 1995. Deep-Sea Res Pt II 48:1127–1172

Behrenfeld M, Falkowski P (1997a) A consumer's guide to phytoplankton productivity models. Limnol Oceanogr 42:1479–1491

Behrenfeld MJ, Falkowski PG (1997b) Photosynthetic rates derived from satellite-based chlorophyll concentration. Limnol Oceanogr 42:1–20

Behrenfeld M, Bale A, Kolber Z, Aiken J, Falkowski P (1996) Confirmation of iron limitation of phytoplankton photosynthesis in the equatorial Pacific. Nature 383:508–511

Bender M, Ducklow H, Kiddon J, Marra J, Martin J (1992) The carbon balance during the 1989 spring bloom in the North Atlantic Ocean, 47° N, 20° W. Deep-Sea Res 39:1707–1725

Bender ML, Orchardo J, Dickson M-L, Barber R, Lindley S (1999) In vitro O_2 fluxes compared with ^{14}C production and other rate terms during the JGOFS Equatorial Pacific experiment. Deep-Sea Res Pt II 46:637–654

Benitez-Nelson C, Buessler KO, Karl DM, Andrews J (2001) A times-series study of particulate matter export in the North Pacific subtropical gyre based on ^{234}Th and ^{238}U disequilibrium. Deep-Sea Res Pt I 48:2595–2611

Berger WH, Fisher K, Lai C, Wu G (1987) Oceanic productivity and organic carbon flux. Scripps Institute of Oceanography Reference 87(30):1–67

Blankenship RE (1992) Origin and early evolution of photosynthesis. Photosynth Res 33:91–111

Bronk DA, Glibert PM, Ward BB (1994) Nitrogen uptake, dissolved organic nitrogen release and new production. Science 265:1843–1846

Bruland KW, Coale KH (1986) Surface water ^{234}Th:^{238}U disequilibria: spatial and temporal variations of scavening rates within the Pacific Ocean. In: Burton JD, Brewer PG, Chesselet R (eds) Dynmanic processes in the chemistry of the upper ocean. Plenum, New York, pp 159–172

Buesseler KO (1991) Do upper-ocean sediment trap studies prive an accurate estimate of sediment trap flux? Nature 353:420–423
Buesseler KO, Bacon MP, Cochran JK, Livingston HD (1992) Carbon and nitrogen export during the JGOFS North Atlantic bloom experiment estimated from ^{234}Th:^{238}U disequilibria. Deep-Sea Res 39:1115–1137
Buesseler KO, Michaels AF, Siegel DA, Knap AH (1994) A three-dimensional time-dependent approach to calibrating sediment trap fluxes. Global Biogeochem Cy 8:179–193
Beussler KO, Andrews JE, Hartman MC, Belastock R, Chai F (1995) Regional estimates of the export flux of particulate organic carbon derived from thorium-234 during the JGOFD EQPAC program. Deep-Sea Res Pt II 42:777–804
Buesseler KO, Steinberg DK, Michaels AF, Johnson RJ, Andrews JE, Valdes JR, Price JF (2000) A comparison of the quantity and composition of material caught in a neutrally buoyant versus surface-tethered sediment trap. Deep-Sea Res Pt I 47:277–294
Bury SJ, Owens NJP, Preston T (1995) ^{13}C and ^{15}N uptake by phytoplankton in the marginal ice zone of the Bellingshausen Sea. Deep-Sea Res Pt II 42:1225–1252
Carpenter EJ, Lively JS (1980) Review of estimates of algal growth using ^{14}C tracer techniques. In: Falkowski PG (ed) Primary productivity in the sea. Plenum Press, New York, pp 161–178
Chavez FP, Barber RT (1987) An estimate of new production in the equatorial Pacific. Deep-Sea Res 34:1229–1243
Chavez FP, Toggweiler JR (1995) Physical estimates of global new production: the upwelling contribution. In: Summerhayes CP et al. (eds) Upwelling in the ocean: modern processes and ancient records. John Wiley & Sons, Chichester, pp 313–320
Clegg SL, Whitfield M (1990) A generalized model for the scavenging of trace metals in the open ocean, I. Particle cycling. Deep-Sea Res 38:91–120
Coale KH, Bruland KW (1985) ^{234}Th:^{238}U disequilibria within the California current. Limnol Oceanogr 30:22–33
Coale KH, Bruland KW (1987) Oceanic stratified euphotic zone as elucidated by ^{234}Th:^{238}U disequilibria. Limnol Oceanogr 32: 189–200
Cochran JK, Bueseler KO, Bacon MP, Wang HW, Hirschberg DJ, Ball L, Andrews J, Crossin G, Fleer A (2000) Short-lived isotopes (^{234}Th, ^{228}Th) as indicators of POC export and particle cycling in the Ross Sea, Southern Ocean. Deep-Sea Res Pt II 47:3451–3490
Daly KL, Wallace DWR, Smith WO Jr., Skoog A, Lara R, Gosselin M, Falck E, Yager PL (1999) Non-Redfield carbon and nitrogen cycling in the Arctic: effects of ecosystem structure and dynamics. J Geophys Res 104:3158–3199
Delwiche C (2000) Tracing the thread of plastid diversity through the tapestry of life. Am Nat 154:164–177
Doney SC, Bullister JL (1992) A chlorofluorocarbon section in the eastern North Atlantic. Deep-Sea Res 39:1857–1883
Dore JE, Karl DM (1996) Nitrification in the euphotic zone as a source for nitrite, nitrate, and nitrous oxide at Station ALOHA. Limnol Oceanogr 41:1619–1628
Downs J (1989) Export of production in oceanic systems: information from phaeopigment carbon and nitrogen analyses. PhD Dissertation, University of Washington, Seattle
Dugdale RC, Goering JJ (1967) Uptake of new and regenerated forms of nitrogen in primary productivity. Limnol Oceanogr 12:196–206
Dugdale RC, Wilkerson FP, Barber RT, Chavez FP (1992) Estimating new production in the equatorial Pacific Ocean at 150° W. J Geophys Res 97:681–686
Dunne JP, Murray JW (1999) Sensitivity of ^{234}Th export to physical processes in the equatorial Pacific. Deep-Sea Res Pt I 46:831–854
Dunne JP, Murray JW, Young J, Balistrieri LS, Bishop J (1997) ^{234}Th and particle cycling in the central equatorial Pacific. Deep-Sea Res Pt II 44:2049–2084
Dunne JP, Murray JW, Aufdenkampe A, Blain S, Rodier M (1999) Silica: nitrogen coupling in the equatorial Pacific upwelling zone. Global Biogeochem Cy 13:715–726
Emerson S (1987) Seasonal oxygen cycles and biological new production in surface waters of the subarctic Pacific Ocean. J Geophys Res 100:15873–15887
Emerson S, Quay PD, Stump C, Wilber D, Knox M (1991) O_2, Ar, N_2 and ^{222}Rn in surface waters of the subarctic ocean: net biological O_2 production. Global Biogeochem Cy 5:49–60
Emerson S, Quay P, Wheeler PA (1993) Biological productivity determined from oxygen mass balance and incubation experiments. Deep-Sea Res 40:2351–2358
Emerson S, Quay PD, Stump C, Wilber D, Schudlich R (1995) Chemical tracers of productivity and respiration in the subtropical Pacific Ocean. J Geophys Res 100:15873–15887
Emerson S, Quay P, Karl D, Winn C, Tupas L, Landry M (1997) Experimental determination of the organic carbon flux from open-ocean surface waters. Nature 389:951–954
Emerson S, Mecking S, Abell J (2002) The biological pump in the North Pacific Ocean: nutrient sources, Redfield ratios and recent changes. Global Biogeochem Cy 15:535–554
Eppley RW (1989) New production: history, methods, problems. In: Berger WH, Smetacek VS, Wefer G (eds) Productivity of the ocean: present and past. John Wiley & Sons, New York, pp 85–97
Eppley RW, Peterson BJ (1979) Particulate organic matter flux and planktonic new production in the deep ocean. Nature 282: 677–680
Eppley RW, Renger EH, Betzer RR (1983) The residence time of particulate organic carbon in the surface layer of the ocean. Deep-Sea Res 30:311–323
Esaias WE (1980) Remote sensing of oceanic phytoplankton: present capabilities and future goals. In: Falkowski PG (ed) Primary productivity in the sea. Plenum Press, New York, pp 321–337
Falkowski PG, Raven JA (1997) Aquatic photosynthesis. Blackwell Scientific Publishers, Oxford, 375 pp
Falkowski PG, Woodhead AD (1992) Primary productivity and biogeochemical cycles in the sea. Plenum Press, New York, 550 pp
Falkowski PG, Wyman K, Ley AC, Mauzerall D (1986) Relationship of steady state photosynthesis to fluorescence in eucaryotic algae. Biochim Biophys Acta 849:183–192
Falkowski PG, Ziemann D, Kolber Z, Bienfang PK (1991) Role of eddy pumping in enhancing primary production in the ocean. Nature 352:55–58
Falkowski P, Barber R, Smetacek V (1998) Biogeochemical controls and feedbacks on ocean primary productivity. Science 281:200–206
Field C, Behrenfeld M, Randerson J, Falkowski P (1998) Primary production of the biosphere: integrating terrestrial and oceanic components. Science 281:237–240
Fitzwater SE, Knauer GA, Martin JH (1982) Metal contamination and its effects on primary production measurements. Limnol Oceanogr 27:544–551
Garside C (1985) The vertical distribution of nitrate in open ocean surface water. Deep-Sea Res 32:723–732
Glibert PM, McCarthy JJ (1984) Uptake and assimilation of ammonium and nitrate by phytoplankton: Indices of nutritional status for natural assemblages. J Plankton Res 6:677–697
Gran H (1918) Kulturforok med planktonalger, Fordhandlinger Skand. Naturforskeres 16 de mo Kristiana (1916), 391 pp
Grande KD, Williams PJL, Marra J, Purdie DA, Heinemann K, Eppley RW, Bender ML (1989) Primary production in the North Pacific Gyre: a comparison of rates determined by the ^{14}C, O_2 concentration and ^{18}O methods. Deep-Sea Res 36:1621–1634
Gust G, Byrne RH, Bernstein RE, Betzer PR, Bowles W (1992) Particle fluxes and moving fluids: experience from synchronous trap collections in the Sargasso Sea. Deep-Sea Res Pt I 41: 831–857
Harrison WG, Platt T, Lewis MR (1987) *f*-ratio and its relationship to ambient nitrate concentration in coastal waters. J Plankton Res 9:235–248
Hayward TL, Venrick EL (1982) Relation between surface chlorophyll, integrated chlorophyll and integrated primary production. Mar Biol 69:247–252
Hedges JI, Baldock JA, Gelinas Y, Lee C, Peterson ML, Wakeham SG (2002) The biochemical and elemental composition of marine plankton: a NMR perspective. Mar Chem 78:47–63
Hernes PJ, Peterson ML, Murray JW, Wakeham SG, Lee C, Hedges JI (2001) Particulate carbon and nitrogen fluxes and compositions in the central equatorial Pacific. Deep-Sea Res Pt I 48: 1999–2023
Honeyman BD, Santschi PH (1989) A Brownian-pumping model for oceanic trace metal scavenging: Evidence from Th isotopes. J Mar Res 47:951–992

Honjo S, Doherty KW (1988) Large aperture time-series sediment traps: design objectives, construction and application. Deep-Sea Res 35:133–149
Jenkins WJ (1982) Oxygen utilization rates in the north Atlantic Subtropical Gyre and primary production in oligotrophic systems. Nature 300:246–248
Jenkins WJ, Goldman J (1985) Seasonal oxygen cycling and primary production in the Sargasso Sea. J Mar Res 43:465–491
Karl DM, Christian JR, Dore JE, Hebel DV, Letelier RM, Tupas LM, Winn CD (1996) Seasonal and interannual variability in primary production and particle flux at station ALOHA. Deep-Sea Res Pt II 43:539–568
Karl DM, Letelier R, Tupas L, Dore J, Christian J, Hebel D (1997) The role of nitrogen fixation in biogeochemical cycling in the subtropical North Pacific Ocean. Nature 388:533–538
Knauer GA, Martin JH, Bruland KW (1979) Fluxes of particulate carbon, nitrogen and phosphorus in the upper water column of the northeast Pacific. Deep-Sea Res 26:97–108
Koblentz-Mishke OJ, Volkovinsky VV, Kabanova JG (1970) Plankton primary production of the world ocean. In: Wooster WS (ed) Scientific exploration of the South Pacific. U.S. National Academy of Science, Washington, pp 183–193
Kolber Z, Falkowski PG (1993) Use of active fluorescence to estimate phytoplankton photosynthesis in situ. Limnol Oceanogr 38:1646–1665
Kolber Z, Wyman KD, Falkowski PG (1990) Natural variability in photosynthetic energy conversion efficiency: a field study in the Gulf of Maine. Limnol Oceanogr 35:72–79
Kurz KD, Maier-Reimer E (1993) Iron fertilization of the Austral Ocean: the Hamburg model assessment. Global Biogeochem Cy 7:229–244
Landry MR, Barber RT, Bidigare R, Chai F, Coale KH, Dam HG, Lewis MR, Lindley ST, McCarthy JJ, Roman MR, Stoecker DK, Verity PG, White JR (1997) Iron and grazing constraints on primary production in the central equatorial Pacific: an EqPac synthesis. Limnol Oceanogr 42:405–418
Laws EA, DiTullio GR, Betzer PR, Karl DM, Carder KL (1989) Autotrophic production and elemental fluxes at 26° N, 155° W in the North Pacific subtropical gyre. Deep-Sea Res 36:103–120
Laws EA, Falkowski PG, Smith WO, Ducklow H, McCarthy JJ (2000) Temperature effects on export production in the open ocean. Global Biogeochem Cy 14:1231–1246
Laws EA, Sakshaug E, Babin M, Dandonneau Y, Falkowski P, Geider R, Legendre L, Morel A, Sondergaard M, Takahashi M, Williams PJ leB (2002) Photosynthesis and primary productivity in marine ecosystems: practical aspects and application of techniques. JGOFS special publication (in press)
Lee K (2001) Global net community production estimated from the annual cycle of surface water total dissolved inorganic carbon. Limnol Oceanogr 46:1287–1297
Letelier RM, Karl DM (1996) Role of *Trichodesmium* spp. in the productivity of the subtropical North Pacific Ocean. Mar Ecol Prog Ser 133:263–273
Li Y-H, Peng T-H (2002) Latitudinal change of remineralization ratios in the oceans and its implications for nutrient cycles. Global Biogeochem Cy (in press)
Lindeman R (1942) The trophic-dynamic aspect of ecology. Ecology 23:399–418
Lipps JH (1993) Fossil prokaryotes and protists. Blackwell, Oxford, 342 pp
Liss PS, Merlivat L (1986) Air-sea gas exchange rates: Introduction and synthesis. In: Buat-Menard P (ed) The role of air-sea exchange in geochemical cycling. D. Reidel, Hingham, Ma, pp 113–129
Liu KK, Atkinson L, Chen CTA, Gao S, Hall J, Macdonald RW, Talaue McManus L, Quinones R (2000) Exploring continental margin carbon fluxes on a global scale. EOS 81:641–644,
Longhurst A, Sathyendranath S, Platt T, Caverhill C (1995) An estimate of global primary production in the ocean from satellite radiometer data. J Plankton Res 17:01245–01271
Longhurst AR, Bedo AW, Harrison WG, Head EJH, Sameoto DD (1990) Vertical flux of respiratory carbon by oceanic diel migrant biota. Deep-Sea Res 37:685–694
Martin JH, Knauer GA, Karl DM, Broenkow WW (1987) VERTEX: carbon cycling in the northeast Pacific. Deep-Sea Res 32:267–286
McCarthy JJ, Garside C, Nevins JL, Barber RT (1996) New production along 140° W in the equatorial Pacific during and following the 1992 El Niño event. Deep-Sea Res Pt II 43:1065–1093
McCarthy JJ, Garside C, Nevins JL (1999) Nitrogen dynamics during the Arabian Sea northeast monsoon. Deep-Sea Res Pt II 46:1623–1664
McGillicuddy DJ Jr., McCarthy JJ, Robinson AR (1995) Coupled physical and biological modeling of the spring bloom in the North Atlantic (I): model formulation and one dimensional bloom processes. Deep-Sea Res Pt I 42:1313–1357
Michaels AF, Bates NR, Buesseler KO, Carlson CA, Knap AH (1994) Carbon-cycle imbalances in the Sargasso Sea. Nature 372: 537–540
Min D-H, Bullister JL, Weiss RF (2000) Constant ventilation age of thermocline water in the eastern subtropical North Pacific Ocean from chlorofluorocarbon measurements over a 12-year period. Geophys Res Lett 27:3909–3912
Murray JW, Downs JN, Strom S, Wei C-L, Jannasch HW (1989) Nutrient assimilation, export production and ^{234}Th scavenging in the eastern equatorial Pacific. Deep-Sea Res 36: 1471–1489
Murray JW, Young J, Newton J, Dunne J, Chapin T, Paul B (1996) Export flux of particulate organic carbon from the central Equatorial Pacific using a combined drifting trap – ^{234}Th approach. Deep-Sea Res Pt II 43:1095–1132
Peterson ML, Hernes PJ, Thoreson DS, Hedges JI, Lee C, Wakeham SG (1993) Field evaluation of a valved sediment trap. Limnol Oceanogr 38:1741–1761
Platt T, Denman KL, Jassby AD (1975) The mathematical representation and prediction of phytoplankton productivity. Fisheries and Marine Services Technical Report 523
Redfield AC (1934) On the proportions of organic derivatives in sea water and their relation to the composition of plankton. James Johnstone Memorial Volume, Liverpool, 176 pp
Riley GA (1951) Oxygen, phosphate and nitrate in the Atlantic Ocean. Bulletin Bingham Oceanographic College 13:1–126
Ryther JH (1969) Photosynthesis and fish production in the sea. Science 166:72–77
Sambrotto RN, Lorenzen CJ (1987) Phytoplankton, phytoplankton production in the coastal, oceanic areas of the Gulf of Alaska. In: Hood DW, Zimerman ST (eds) The Gulf of Alaska: physical environment, biological resources. U.S. Department of Commerce, Washington, D.C., pp 249–282
Sambrotto RN, Savidge G, Robinson C, Boyd P, Takahashi T, Karl DM, Langdon C, Chipman D, Marra J, Codispoti L (1993) Elevated consumption of carbon relative to nitrogen in the surface ocean. Nature 363:248–250
Sanderson MP, Hunter CN, Fitzwater SE, Gordon RM, Barber RT (1995) Primary productivity and trace metal contamination measurements from a clean rosette system versus ultra clean Go-Flo bottles. Deep-Sea Res Pt II 42:431–440
Santschi PH, Li Y-H, Bell J (1979) Natural radionuclides in the water of Narragansett Bay. Earth Planet Sc Lett 45:201–213
Sarmiento JL, Armstrong RA (1997) U.S. JGOFS synthesis and modeling project implementation plan: the role of oceanic processes in the global carbon cycle. AOS Program, Princeton University, 67 pp
Smith SV (1981) Marine macrophytes as a global carbon sink. Science 211:838–840
Smith WO Jr. (1995) Primary productivity and new production in the Northeast Water (Greenland) Polynya during summer 1992. J Geophys Res 100:4357–4370
Smith WO Jr., Gosselin M, Legendre L, Wallace D, Daly K, Kattner G (1997) New production in the Northeast Water Polynya. Journal of Marine Systems 10:199–209
Smith WO Jr., Barber RT, Hiscock MR, Marra J (2000) The seasonal cycle of phytoplankton biomass and primary productivity in the Ross Sea, Antarctica. Deep-Sea Res Pt II 47:3119–3140
Spitzer WS, Jenkins WJ (1989) Raters of vertical mixing, gas exchange and new production: estimates from seasonal gas cycles in the upper ocean near Bermuda. J Mar Res 47:169–196
Steemann-Nielsen E (1952) The use of radio-active carbon (^{14}C) for measuring organic production in the sea. Journal du Conseil International pour Exploration de la Mer 18:117–140

Steinberg DK, Carlson CA, Bates NR, Goldthwait SA, Madin LP, Michaels AF (2000) Zooplankton vertical migration and the active transport of dissolved organic and inorganic carbon in the Sargasso Sea. Deep-Sea Res Pt I 47:137–158

Summons R, Jahnke L, Hope J, Logan G (1999) 2-Methanhopanoids as biomarkers for cyanobacterial oxygenic photosynthesis. Nature 400:554–557

Varela DE, Harrison PJ (1999) Seasonal variability in nitrogenous nutrition of phytoplankton assemblages in the northeastern subarctic Pacific Ocean. Deep-Sea Res Pt II 46:2505–2538

Walker JCG (1974) Stability of atmospheric oxygen. Am J Sci 274:193–214

Wanninkhof R (1992) Relationship between wind speed and gas exchange over the sea. J Geophys Res 97:7373–7382

Warner MJ, Weiss RF (1992) Chlorofluoromethanes in South Atlantic intermediate water. Deep-Sea Res 39:2053–2075

Warner MJ, Bullister JI, Wisegarver DP, Gammon RH, Weiss RF (1996) Basin-wide distributions of chlorocarbons CFC-11 and CFC-12 in the North Pacific: 1985–1989. J Geophys Res 101: 20525–20542

Welschmeyer N, Goericke R, Strom S, Peterson W (1991) Phytoplankton growth and herbivory in the subarctic Pacific: A chemotaxonomic analysis. Limnol Oceanogr 36:1631–1649

Wilkerson FP, Dugdale RC, Barber RT (1987) Effects of El Niño on new, regenerated, and total production in eastern boundary upwelling systems. J Geophys Res 92:14347–14353

Williams RJP (1981) Natural selection of the chemical elements. P Roy Soc Lond 213:361–397

Williams PJ leB (1993a) Chemical and tracer methods of measuring plankton production. ICES Marine Science Symposium 197: 20–36

Williams PJ leB (1993b) On the definition of plankton production terms. ICES Marine Science Symposium 197:9–19

Williams PJ leB, von Bodungen B, Bathmann U, Berger WH, Eppley RW, Feldman GC, Fischer G, Legendre L, Minster J-F, Reynolds CS, Smetacek VS, Toggweiler JR (1989) Group report: export productivity from the photic zone. In: Berger WH, Smetacek VS, Wefer G (eds) Productivity of the ocean: present and past. John Wiley & Sons, New York, pp 99–115

Williams PJL, Jenkinson NW (1982) A transportable microprocessor controlled Winkler titration suitable for field and shipboard use. Limnol Oceanogr 27:576–584

Wong CS, Whitney FA, Iseki K, Page JS, Zeng J (1995) Analysis of trends in primary productivity and chlorophyll a over two decades at Ocean Station P. (50° N, 145° W) in the subarctic Northeast Pacific Ocean. Can J Fish Aquat Sci 121:107–117

Wong CS, Whitney FA, Matear RJ, Iseki K (1998) Enhancement of new production in the northeast subarctic Pacific Ocean during negative North Pacific index events. Limnol Oceanogr 43:1418–1426

Zehr JP, Waterbury JB, Turner PJ, Montoya JP, Omoregie E, Steward GF, Hansen A, Karl DM (2001) Unicellular cyanobacteria fix N_2 in the subtropical North Pacific Ocean. Nature 412:635–638

Chapter 5

Carbon Dioxide Fluxes in the Global Ocean

Andrew J. Watson · James C. Orr[1]

5.1 Introduction

Atmospheric carbon dioxide concentration is one of the key variables of the 'Earth system' – the web of interactions between the atmosphere, oceans, soils and living things that determines conditions at the Earth surface. Atmospheric CO_2 plays several roles in this system. For example, it is the carbon source for nearly all terrestrial green plants, and the source of carbonic acid to weather rocks. It is also an important greenhouse gas, with a central role to play in modulating the climate of the planet. During the five thousand years prior to the industrial revolution, we know (from measurements of air trapped in firn ice and ice cores) that atmospheric CO_2 varied globally by less than 10 ppm from a concentration of 280 ppm (Indermuhle et al. 1999). During the late Quaternary glaciations, the regular advance and retreat of the ice was accompanied by, and to some extent at least driven by (Li et al. 1998; Shackleton 2000), an oscillation in atmospheric CO_2 of about 80 ppm. Evidence from the geologically recent past indicates, therefore, that quite small changes in atmospheric carbon dioxide have big effects on planetary climate. Conversely, a stable concentration of CO_2 is necessary for a stable climate. By this reasoning, we can be fairly certain that human activities will have a major effect on the climate of the planet in the near future, given that we have raised CO_2 by 90 ppm in the last 150 years and it is projected to double from the pre-industrial concentration during the coming century. This gives our investigations into sources and sinks of carbon dioxide a special urgency.

For reasons that are made clear below, the oceans occupy a central role in the global carbon cycle and the processes influencing the concentration of CO_2 in the atmosphere. The JGOFS program represented the first sustained global effort to document the present state of the oceanic carbon cycle, and to test our understanding by comparing that state with the predictions of increasingly sophisticated numerical models. In the subsequent sections, we first discuss the role of the oceans in setting global atmospheric CO_2. JGOFS has made a major contribution to our estimate of the size of the present net flux of CO_2 from atmosphere to ocean, and we next review our estimates of this 'sink' and how it may change in the future. This leads us to a discussion of the advances made in understanding the processes involved in setting that flux. Finally we summarize what we have learned during JGOFS and what the major topics of research are likely to be in the next 10 years.

5.2 The Oceans' Influence on Atmospheric CO_2

In this section, we review some basic facts and figures about the oceanic component of the carbon cycle. Some of this knowledge was well established prior to JGOFS, but much of the detail is recent, and comes from JGOFS and related activities.

5.2.1 The Ocean Sets the Steady-State Atmospheric CO_2 Concentration

Figure 5.1, modified from Houghton et al. (1990) and Watson and Liss (1998), is a summary of the global carbon cycle in reservoir-flux form. One-way fluxes between the atmosphere and ocean are about 70 Pg C yr^{-1}, of the same order as the exchange between the atmosphere and the terrestrial biota. The present sink for atmospheric CO_2 in the oceans is the net imbalance between the ingoing and outgoing fluxes and is of order 2 Pg C yr^{-1}, just ~2% of the one-way fluxes. This net uptake has occurred as a response to increasing CO_2 in the atmosphere. In the pre-industrial era the oceans were a small net source to the atmosphere, as explained below in the section on 'The Pre-Industrial Steady State'.

[1] Ocean carbon model intercomparison results were contributed by the following authors: O. Aumont, K. G. Caldeira, J.-M. Campin, S. C. Doney, H. Drange, M. J. Follows, Y. Gao, A. Gnanadesikan, N. Gruber, A. Ishida, F. Joos, R. M. Key, K. Lindsay, F. Louanchi, E. Maier-Reimer, R. J. Matear, P. Monfray, A. Mouchet, R. G. Najjar, G.-K. Plattner, C. L. Sabine, J. L. Sarmiento, R. Schlitzer, R. D. Slater, I. Totterdell, M.-F. Weirig, M. E. Wickett, Y. Yamanaka and A. Yool.

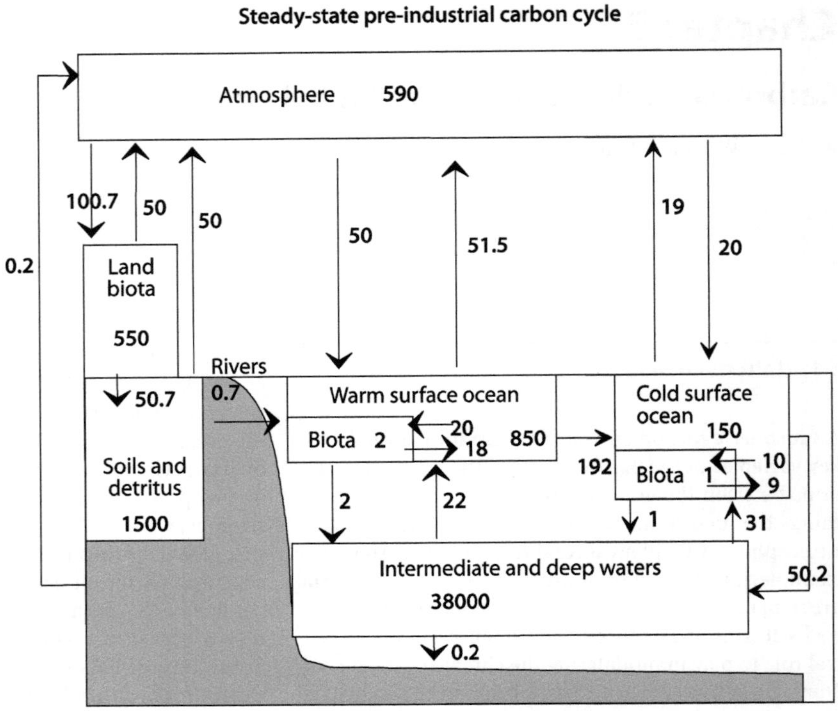

Fig. 5.1.
Reservoir-flux representation of the steady-state, pre-industrial global carbon cycle. Reservoir sizes are in Pg C, and fluxes in Pg C yr^{-1} (1 Pg = one petagram ≡ 1 Gt = one gigatonne = 10^9 tonnes). The figure is adapted from Houghton et al. (1990) with additional information from Sarmiento and Sundquist (1992), Sarmiento and Toggweiler (1984) and Sundquist (1985). Fluxes are adjusted so that the net flux into each reservoir is zero, which sometimes involves specifying fluxes to a higher accuracy than that to which they are in reality known

As the figure shows, most of the carbon that is not locked up in carbonate rocks resides in the ocean, which contains 15–20 times as much carbon as the atmosphere, land vegetation and soils combined. About 90% of the carbon in the ocean is in the form of bicarbonate ion, mostly in the deep sea. The atmosphere exchanges CO_2 rapidly with both the ocean surface and with the land vegetation, such that the residence time of a CO_2 molecule in the atmosphere is only about 10 years with respect to these reservoirs.

The 'one-way' fluxes between ocean and atmosphere are proportional to the partial pressures of CO_2 in the atmosphere and at the ocean surface. (Strictly we should use fugacity rather than partial pressure, to account for the slight departure from ideal gas behaviour of CO_2. In practice, the difference is about 1%. We have chosen to ignore it to make the discussion more accessible to non-chemists who may understand the concept of partial pressure but be unfamiliar with fugacity.)

$$F_{air \rightarrow sea} = KpCO_{2air}$$
$$F_{sea \rightarrow air} = KpCO_{2sea} \quad (5.1)$$

where the constant of proportionality is K, the CO_2 gas exchange coefficient (and is a 'constant' only in the sense that it is independent of CO_2 concentrations). Since the ingoing and outgoing fluxes balance within 2% when integrated over the ocean surface as a whole and periods of a year, (and provided variations in K do not correlate with variations in pCO_2),

$$\overline{\overline{pCO_{2air}}} \approx \overline{\overline{pCO_{2sea}}}$$

where the double overbars represent the time and area averages. If some disturbance were to force atmospheric CO_2 away from near-equilibrium with the ocean, the resulting imbalance of fluxes across the air-sea interface would, on a time scale of a decade, tend to adjust atmospheric and sea surface CO_2 until the steady state was re-established. On a time scale of many centuries, when steady state was restored between deep ocean and the surface, any anomaly in CO_2 would be diluted into the large deep ocean reservoir, so that the final departure from the original steady state would be comparatively small. However, altering the parameters that set the partial pressure of CO_2 at the sea surface, such as the amount of biologically driven flux out of the surface layers or the temperature of the sea surface, will have a direct effect on the steady-state CO_2 content of the atmosphere, within a few decades.

Examples of this behaviour are given in Fig. 5.2a and b using calculations from a very simple box model of the ocean and atmosphere (Sarmiento and Toggweiler 1984), and ignoring changes in the land vegetation and interaction with the sediments. Figure 5.2a shows the result of disturbing the system by an exponentially increasing release of CO_2 into the atmosphere, similar to

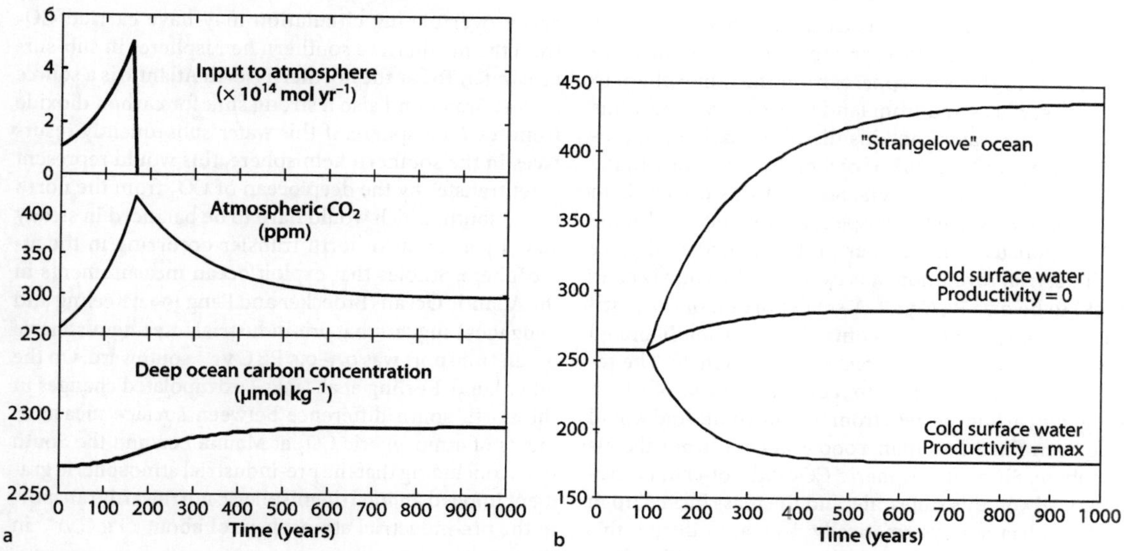

Fig. 5.2. Examples of the behavior of a simple model of the atmosphere-ocean carbon system, in which the ocean is modeled as three boxes (Sarmiento and Toggweiler 1984). **a** Perturbation by an exponentially growing release of CO_2 into the atmosphere that starts at 10^{14} mol yr^{-1} (1.2 Pg C yr^{-1}) and grows at 1% per year for 150 years, then ceases. Atmospheric CO_2 is forced up while the release is growing, but begins to decline by ocean uptake as soon as it ceases. The deep sea is the ultimate repository for most of the released carbon, but concentrations rise only slightly there because the reservoir is so large. **b** Effect of changing the biological pump; 'Strangelove' Ocean (all biological activity ceases), or cold surface ocean productivity ceases, or cold surface ocean productivity raised to the maximum consistent with phosphate concentrations. Changes in biological activity can shift the steady state pre-industrial atmospheric CO_2 concentration by a substantial amount

the effect of the industrial revolution on the carbon cycle. In the model scenario, a release is imposed that grows at 1% per year for 150 years, at which point it abruptly ceases. Atmospheric CO_2 is initially forced up rapidly by the release because the ocean cannot take up CO_2 from the atmosphere as rapidly as it is being added but, as soon as the release is stopped, the concentrations start to decline by uptake to the ocean. After a few hundred years equilibrium is re-established at concentrations only about 30 ppm above the pre-industrial level. The majority of the carbon released ends in the deep sea, but concentrations here increase by only ~1% because this reservoir is so big. Steady state atmospheric concentrations are ~10% higher than previously, and this is because a 1% increase in total carbon in the ocean has a proportionately larger effect on the steady state pCO_2 at the ocean surface. This effect is quantified by the Revelle Buffer factor (β), defined as the fractional change in pCO_2 for a unit fractional change in total carbon content. For average seawater $\beta \sim 10$ (at constant temperature and alkalinity), meaning that a 1% change in total carbon leads to ~10% change in pCO_2 at equilibrium.

In Fig. 5.2b, the effect of changing the efficiency of the 'biological pump' in the oceans is investigated. Increasing net productivity in the 'cold surface water' box of the model (notionally corresponding to the Southern Ocean) removes some carbon from this surface box and rains it down into the deep ocean. The removal of carbon from the surface decreases the carbon content there by a few percent, but this is sufficient to change the equilibrium CO_2 substantially, again due to the high value of β. Atmospheric CO_2 decreases by up to 80 ppm below the previous steady state. Setting the biological pump to zero everywhere (the 'Strangelove' ocean) causes it to increase by about 180 ppm. In these cases no carbon is added to the system in total, but atmospheric CO_2 is affected by modulating processes at the surface ocean so as to change the distribution between ocean and atmosphere.

5.2.2 The Pre-Industrial Steady State

Records obtained from ice core and ice-firn (Indermuhle et al. 1999) show that, before the industrial revolution, the CO_2 concentration in the atmosphere was in the range 280 ±10 ppm for five thousand years. By contrast, from Fig. 5.1 we can see that the residence time for exchange of atmospheric CO_2 to the surface ocean or terrestrial biosphere (the size of the reservoir divided by the flux out of it) is only ~10 years, while the time to adjust to redistribution of carbon between atmosphere and the deep ocean is a few hundred years (Fig. 5.2). It is apparent therefore that the atmosphere was closely in steady state with regard to the net inputs and outputs to the land and oceans before the industrial era. This does not imply that air-sea flux or the air-land flux were each identically zero, only that the sum of these two fluxes was zero. In fact it is clear that there was a net flux of about 0.6 ±0.2 Pg C yr^{-1} that ran in a circuit from the oceans to the atmosphere, from the atmosphere to the land surfaces and from land to ocean. We know this because river water contains dissolved carbon, so globally there is today a flux of carbon from the land to the sea down the world's rivers. Some of this carbon is in the form of dissolved or suspended organics, and some as bicarbonate. This flux has probably increased since the industrial revolution. It is estimated by Ludwig et al. (1998) to be 0.72 Pg C yr^{-1}. A small portion of this, estimated at 0.096 Pg C yr^{-1}, comes from carbon liberated from the rocks by the dissolution of carbonate. The remainder of this 'riverine' flux, about two thirds of which is organic carbon, derives from the land biota and would have been fixed less than 1 000 years ago from the atmosphere. Since atmospheric CO_2 did not change systematically on this timescale, and the mass locked up in the biota has not systematically decreased during this time, it follows that there must be an equivalent flux from oceans to atmosphere to keep the system in steady state. Whether this flux mostly returns to the atmosphere in river estuaries and the coastal ocean, or whether it is distributed over the global ocean, is unknown at present. It depends in particular on the lifetime against oxidation of the organic carbon being carried by the rivers. Depending on the distribution of this flux, direct estimates of the global atmosphere-ocean CO_2 flux may need to be increased by up to ~0.6 Pg C yr^{-1} before comparison with estimates of how the flux has changed since pre-industrial times (see Table 5.2).

5.2.3 Pre-Industrial North-South Transports

Another aspect of the pre-industrial steady state that is important for our present understanding is the issue of whether there was a north-south asymmetry in the distributions of sources and sinks of CO_2 before the industrial era. This would have given rise to a slight gradient in the atmosphere at that time. There are two possible reasons for thinking that there would be such an asymmetry. First, because most of the land surfaces are in the northern hemisphere and the land was a net sink, whereas most of the oceans are in the southern hemisphere and the oceans were a net source, we might expect an enhanced sink in the north and corresponding source in the south. How large such an asymmetry would be depends on the residence time of the riverine carbon against air-sea transfer once it reaches the sea. If this is measured in decades or longer, then we might expect the net source it engenders to be evenly spread over the surface of the global ocean, in other words partly in the southern hemisphere.

A second reason for believing that there might be a north-south imbalance in pre-industrial time is that the ocean overturning circulation may have carried CO_2 from the northern to southern hemispheres in sub-surface water. Today the northern North Atlantic is a source of deep water and also a strong sink for carbon dioxide from the atmosphere. If this water subsequently resurfaces in the southern hemisphere, this would represent a net transfer by the deep ocean of CO_2 from the north to the south, which would have to be balanced in steady state by a south-to-north transfer occurring in the atmosphere. Studies that exploit ocean measurements in the Atlantic Ocean (Broecker and Peng 1992; Keeling and Peng 1995) suggest that pre-industrial inter-hemispheric ocean transport was 0.3–0.5 Pg C yr^{-1} southward. On the other hand, Keeling et al. (1989) extrapolated changes in the north-south difference between surface measurements of atmospheric CO_2 at Mauna Loa and the South Pole, concluding that the pre-industrial atmospheric gradient implied an inter-hemispheric northward transport in the pre-industrial atmosphere of about 1 Pg C yr^{-1} in the ocean.

These inferred transports can be compared with direct estimates from ocean carbon cycle models. Three ocean carbon cycle models were used to investigate this pre-industrial carbon transport during the first Ocean Carbon Models Intercomparison Program (OCMIP-1) (Sarmiento et al. 2000). None of these models produced a southward, pre-industrial, transport of more than 0.1 Pg C yr^{-1}. At the other extreme was one model that actually produced a northward transport of 0.1 Pg C yr^{-1}. All three OCMIP-1 models revealed a global-ocean carbon loop, where 0.6 Pg C yr^{-1} is transported from the Bering Strait, across the Arctic Ocean, southwards through the Atlantic, across the Southern Ocean, northwards again through the Indian and Pacific Oceans, and finally back to Bering Strait. This loop is entirely oceanic however, and barely affects the atmospheric gradients. During OCMIP-2, there was an effort to test the robustness of the OCMIP-1 conclusions by performing similar analysis on a more diverse group of 12 models. OCMIP-2 found a much larger range of total southward pre-industrial transport, from 0.0 up to 0.7 Pg C yr^{-1}. However, excluding one outlier, southward cross-equatorial transport was less than 0.35 Pg C yr^{-1}.

Furthermore, a separate simulation with one of the OCMIP-1 models has highlighted that the ocean models are missing a 'riverine' carbon flux, and that this affects the pre-industrial north-south transport. Aumont et al. (2001) showed with the IPSL model that simulated southward carbon transport increased by 0.1 to 0.3 Pg C yr^{-1} when riverine carbon, from (mostly northern hemisphere) rivers was added, assuming subsequent transport of this carbon within the ocean and outgassing over the whole ocean surface. The Aumont et al. transport is thus consistent with the transports derived from ocean measurements. On the other hand, there remains a discrepancy between Aumont et al.'s results and the

Keeling et al. (1989) estimate of the pre-industrial difference based on atmospheric measurements. The latter study estimates the pre-industrial difference to be −0.82 ppm based on extrapolation of the trend in the difference of atmospheric CO_2 between both stations vs. fossil emissions. For comparison, Aumont et al. use their ocean model's air-sea CO_2 fluxes as boundary conditions in a 3-D atmospheric model, ('TM2', due to M. Heimann). Their pre-industrial air-sea fluxes, including ocean and riverine carbon components, could explain at most −0.3 ppm of the −0.82 ppm estimate from Keeling et al. (1989): the river effect caused a reduction of −0.4 ±0.1 ppm, relative to the ocean-only Mauna Loa-South Pole difference of +0.1 ppm. In OCMIP-2, the ocean-only components of the air-sea fluxes were also transported in the TM2 atmospheric transport model. The resulting Mauna Loa-South Pole difference ranged from −0.28 to +0.22 ppm. Summing both effects suggests that some carbon cycle models may now be able to simulate up to a −0.7 ppm Mauna Loa-South Pole pre-industrial difference. This preliminary calculation indicates the need for more ocean carbon models to explicitly include the river loop. Meanwhile, the debate concerning the magnitude of pre-industrial ocean carbon transport remains open.

Two recent data-analysis studies confirm the inter-hemispheric difference in atmospheric CO_2 derived by Keeling et al., while offering more spatial detail. These new studies find similar results. However, they differ substantially in their interpretation of the cause of the difference. To explain the difference, Conway and Tans (1999) invoke a large anthropogenic sink in the northern hemisphere, that has been essentially constant over the last 40 years. Conversely, Taylor and Orr (2000) suggest that the cause must be natural. They invoke the 'rectifier effect' – the name given to the gradient set up by covariance between seasonal changes in atmospheric transport and the seasonal variability of atmospheric CO_2 concentrations (Pearman and Hyson 1986).

5.3 How Big is the Global Ocean Sink?

During JGOFS we have been able to estimate the global ocean sink by several different independent or quasi-independent methods. Broadly, these are based on three different approaches, models, atmospheric observations and ocean observations, which we describe below.

5.3.1 1-D Models Calibrated with ^{14}C

This approach has been used since the 1970s and was first made possible by the GEOSECS measurements made by Östlund and colleagues of the distribution of ^{14}C in the oceans (Östlund et al. 1987). There is a natural background of ^{14}C in the oceans, derived from cosmic rays colliding with nitrogen atoms in the atmosphere, that has itself been of great value in oceanography. However, the ^{14}C derived from the global pollution caused by the atmospheric nuclear tests of the 1950s and early 1960s overwhelmed the natural signal in surface waters. The intensity of the release rose rapidly to a peak in the early sixties, and then fell off abruptly following the 1963 test ban treaty between the Soviet Union and the USA. The incidental result of this was a 'spike' of radiotracers injected into the atmosphere and surface ocean that could be used to trace the penetration of surface waters into the deep sea. The GEOSECS program of cruises in the early 1970s was ideally timed to take advantage of this signal, and the resulting set of data is probably still the single most powerful constraint on the rate of penetration of CO_2 into the oceans.

Simple models for the overturning circulation of the ocean could be set up and parameters adjusted to reproduce the penetration of ^{14}C into the oceans. When these models were first used to calculate the uptake of fossil fuel CO_2 in response to increasing atmospheric concentrations, answers of the order of 2 Pg C yr^{-1} were obtained. These were compatible with the paradigm of the time, that the land biosphere was neutral with respect to the uptake and release of carbon, so that the difference between the rate of release by fossil fuel burning and the rate of accumulation in the atmosphere was all due to ocean uptake. Since that time, ideas about the sink or source nature of the land biosphere have changed substantially, but the estimate of the ocean sink has remained quite stable.

The success of even simple models in constraining ocean uptake is due to the parallel between the uptake of fossil fuel CO_2 and that of bomb radiocarbon, but this parallel is not exact. For example, the rise in atmospheric CO_2 has taken place over a period of considerably more than a century whereas the bomb signal was over in less than a decade. Furthermore, the time constant for exchange with the atmosphere of injected ^{14}C is an order of magnitude longer than that of ^{12}C (Broecker et al. 1985). The reliability of the models will decrease over longer time scales, although performance over longer times can be constrained by requiring that natural ^{14}C as well as bomb-^{14}C concentrations are reproduced by the models (Siegenthaler and Joos 1992).

5.3.2 3-D Models of the Ocean Carbon Cycle

In the late eighties global circulation models were first combined with simple carbon models of the ocean to produce ocean carbon models capable of diagnosing the fossil fuel sink (Sarmiento et al. 1988). This approach has been refined and developed over the past decade, and there are now many such models in existence. A major initiative under international JGOFS has been the evaluation of the many different such models in the Ocean Carbon Models Intercomparison Programs, OCMIP-1 and 2. These models are one of the best means to test current ideas of the ocean carbon cycle, and as such we repeatedly refer to them in this chapter. Of their many uses, perhaps the most important has been to improve our estimates of the ocean sink for anthropogenic CO_2.

Compared to the earlier generation of models, these have three-dimensional circulations derived from finite-difference solutions of the equations for conservation of mass and momentum. The boundary conditions used to initiate the equations that govern the circulation are typically surface ocean wind stress, and fluxes of heat and fresh water. The resolution is typically two to four degrees in the horizontal and 12 to 30 levels in the vertical, and the forcing is usually climatological and resolved to monthly or finer time scales. These models reproduce the main features of the ocean circulation including the major wind-driven currents and zones of surface convergence and divergence. Early estimates of the anthropogenic uptake during the 1980s made with these models were consistent with the results of simpler models, for example 2.0 ±0.6 Pg C yr^{-1} (Siegenthaler and Sarmiento 1993), and 2.0 ±0.5 Pg C yr^{-1} (Orr 1993). Such estimates were the main basis of the widely-quoted IPCC value for the ocean sink of 2.0 ±0.8 Pg C yr^{-1} (Houghton et al. 1990, 1995).

To help evaluate model uncertainties, and to set the stage for improving predictability, we show here results from OCMIP-2 of simulations of anthropogenic CO_2 in thirteen climatologically forced, coarse-resolution ocean models (Table 5.1). The simulated global uptake for the 1980s was in the range 1.65 to 2.51 Pg C yr^{-1}, with a mean of 1.99 Pg C yr^{-1}. This range falls within the spread of flux estimates from previous compilations of 1-, 2-, and 3-D ocean model results as given above. The wider uncertainties of the earlier studies reflect mainly the additional uncertainties due to our imperfect understanding of the global distribution of bomb ^{14}C, which is used to calibrate ocean models. Additionally, the OCMIP-2 range is similar to that found with four models during OCMIP-1, i.e., 1.6 to 2.1 Pg C yr^{-1} (Orr et al. 2001).

Table 5.1. Change in global annual air-to-sea flux of CO_2 (Pg C yr^{-1}) from pre-industrial value (1765)

Model	1980–1989	1990–1999
PRINCE	1.65	1.98
IPSL.DM1 (HOR)	1.67	1.98
LLNL	1.78	2.08
CSIRO	1.78	2.11
MIT	1.91	2.29
NCAR	1.93	2.30
PRINC2	1.93	2.32
IPSL (GM)	1.97	2.36
MPIM	2.01	2.43
SOC	2.01	2.39
IPSL.DM1 (GM)	2.03	2.43
IGCR	2.05	2.47
PIUB	2.11	2.52
AWI	2.14	2.58
NERSC	2.38	2.84
UL	2.51	3.04
Mean of models	1.99	2.38

Acronyms are: *AWI* (Alfred Wegener Institute for Polar and Marine Research), Bremerhaven, Germany; *CSIRO*, Hobart, Australia; *IGCR/* CCSR, Tokyo, Japan; *IPSL* (Institute Pierre Simon Laplace), Paris, France; *LLNL* (Lawrence Livermore National Laboratories), Livermore, California, USA; *MIT*, Boston, MA, USA; *MPIM* (Max Planck Institut für Meteorologie), Hamburg, Germany; *NCAR* (National Center for Atmospheric Research), Boulder, Colorado, USA; *NERSC* (Nansen Environmental and Remote Sensing Center), Bergen, Norway; *PIUB* (Physics Institute, University of Bern), Switzerland; *PRINCEton* (Princeton University [AOS, OTL]/GFDL), Princeton, NJ, USA; *SOC* (Southampton Oceanography Centre) / SOES / Hadley Center, UK.

5.3.3 ^{13}C Changes with Time in the Ocean

The $^{13}C/^{12}C$ ratio of carbon dissolved in the ocean is changing as fossil fuel carbon invades the surface – termed the ^{13}C Suess effect. Surveys of $\delta^{13}C$ spaced many years apart can therefore be used to calculate the invasion rate independently of other methods. Quay et al. (1992) and Sonnerup et al. (1999) report on this technique, with the most recent estimate giving 1.9 ±0.9 Pg C yr^{-1} as the rate of carbon uptake over the period 1970–1990. This method is potentially valuable as an independent validation of other methods, but the error bars quoted are comparatively large. In addition, there are some problems with possible offsets in some of the early ^{13}C data sets, as reported by Lerperger et al. (2000), who present a set of Pacific data suggesting substantially lower values for ocean uptake.

5.3.4 Atmospheric Observations

Below the stratosphere, the global atmosphere has a time constant for mixing on the order of one year, much faster than that of the ocean, where even the surface requires decades to homogenise. Observations at a few representative sites in the atmosphere can therefore be extrapolated to the entire globe with an accuracy that would require orders of magnitude more effort if it were to be achieved from marine observations alone. As a result, comparatively sparse atmospheric observations impose powerful constraints on the global ocean uptake of CO_2.

High precision observations of atmospheric oxygen (actually the ratio of O_2/N_2) of the kind initiated by R. F. Keeling (Keeling and Shertz 1992) can be combined with atmospheric CO_2 observations to give a direct estimate of the marine and terrestrial sinks of anthropogenic CO_2. In Keeling's method, the rate of decline of atmospheric oxygen and increase in CO_2 in the atmosphere are estimated from atmospheric time series. Burning of fossil fuel releases CO_2 and consumes oxygen in a known molar ratio of about $O_2:CO_2 \approx 1.4$, while net biological uptake or release of carbon and oxygen by the land vegetation occurs in a ratio $O_2:CO_2 \approx 1.1$. Uptake of fossil fuel carbon by the ocean is not accompanied by any net exchange of oxygen, i.e., $O_2:CO_2 \approx 0$, and it is assumed that the ocean is neither a source nor sink of oxygen when averaged over a year. Two equations, one for the CO_2 and one for the O_2 change, can then be written and solved for the size of the land and ocean sinks. This method has since been applied to more extensive data sets (Battle et al. 2000).

Atmospheric oxygen is only one example of measurements in the atmosphere that can be used to help separate the oceanic and terrestrial components of the sink for anthropogenic CO_2. Another is the interpretation of atmospheric ^{13}C variations. Fractionation of ^{13}C relative to ^{12}C during photosynthesis is substantial, while ocean-atmosphere exchange results in little fractionation. Thus seasonal and inter-annual changes in the $\delta^{13}C$ of the atmosphere can similarly be related to the strength of these sinks. Using data from a network of atmospheric sampling stations in an atmospheric transport model, information can be obtained about the distribution around the world of the sources and sinks. Addition of measurements besides CO_2 and O_2 that discriminate between ocean and land surface also improves the estimate. This 'inversion' technique can potentially synthesise many different measurements that provide constraints on the sources and sinks to give an objective best estimate of the ocean sink. Bousquet et al. (2000), Rayner et al. (1999) and Langenfelds et al. (1999) are recent examples of estimates made by the atmospheric inversion technique.

As always, caution is needed in the application of these methods. An accurate assessment of the uncertainties in the data has to be incorporated into the analysis. If noisy or biased data are used without allowance being made, the resulting estimates of sinks are similarly noisy and biased. Similarly, the assumptions going into the model will bias the outcome if they are incorrect. For example, Keeling's method is sensitive to the assumption that oxygen in the ocean is in steady state, which is questionable if the oceans have begun to warm or change their degree of stratification, (as further discussed below). It also assumes specific values for $O_2:CO_2$ ratios during fossil fuel burning and vegetation uptake, which are not invariant or precisely known.

5.3.5 Observations of the Air-Sea Flux

Conceptually, the simplest way to specify the global flux of CO_2 from the atmosphere into the oceans is to measure it (over all the oceans, and all the time). Unfortunately, until very recently (see section on 'Gas Transfer Velocity') the direct measurement of CO_2 flux has been too insensitive to give useful results even at one place and one time. The approach followed over the past several decades has therefore been to split the problem into two parts, for one of which at least we do have an accurate means of measurement. Subtracting the two equations (5.1) to obtain the net flux, we have:

$$F_{net} = F_{air\text{-}sea} - F_{sea\text{-}air} = K\Delta pCO_2$$

where ΔpCO_2 is the difference between the atmospheric and surface ocean partial pressures of CO_2. The gas transfer coefficient K is normally further divided into the product $k_v\alpha$, where k_v is the gas transfer velocity (or piston velocity) and α is the solubility of carbon dioxide, which is accurately known as a function of temperature and salinity (Weiss 1974).

$$F_{net} = k_v\alpha\Delta pCO_2$$

This equation expresses the air-sea flux as a product of a readily measurable chemical gradient across the sea surface ($\alpha\Delta pCO_2$), and a variable gas transfer velocity k_v that expresses the ease with which a molecule of gas can pass from the gaseous to the dissolved phase or vice versa. The gas transfer velocity is not so easily measured, but progress has been made in parameterising it. We discuss the transfer velocity below.

Measurements of surface pCO_2 or ΔpCO_2 have been made since the early 1970s from research vessels, and more recently from commercial ships of opportunity and moored or drifting buoys. JGOFS activities have contributed to this database enormously, and in truly international fashion, during the nineties. Major contributions came both from the global CO_2 survey sponsored by JGOFS and WOCE, and from the many process and monitoring studies occurring in different parts of the world at a national and international level.

Syntheses of these data have been published (Takahashi et al. 1997, 1999; Lefèvre et al. 1999) and the synthesis program is ongoing. Globally the database now exceeds one million measurements. The syntheses to date have been in the form of monthly or seasonal climatologies, in which it is assumed that the major variability in pCO_2 at a given site is on a seasonal cycle, so that data from different years but the same time of year can be pooled to give a meaningful average. A major uncertainty in such analyses is how to deal with the continuous increase in atmospheric CO_2 that has occurred during the period of the measurements. Though the majority of the data are from the 1990s, data have been collected over a period spanning thirty years. During this time atmospheric pCO_2 has increased by around 40 ppm, about five times larger than the mean difference between atmosphere and ocean, so this is not a small correction. In regions where the surface waters have a long residence time (years to decades) we can expect sea-surface pCO_2 to tend to increase towards the atmosphere, whereas in regions where the surface is being rapidly replaced by deeper water this tendency would be less. Takahashi et al. therefore assume that for all waters equatorward of 50° N and 50° S, surface ocean pCO_2 has increased at the same rate as in the atmosphere. This corresponds to the subtropical gyres where we expect water to remain for a long time at the surface. Poleward of these latitudes they assume either no increase in pCO_{2sea} or an increase at half the rate of the atmosphere. These assumptions are of necessity arbitrary and introduce uncertainties into derived fluxes.

The global distribution of the air-sea flux of CO_2 over the world ocean derived by Takahashi et al. (2000) is shown in Fig. 5.3. By integrating over the ocean surface of Fig. 5.3, we can arrive at an estimate of the size of the global air-to-sea flux. Takahashi et al. quote 2.17 Pg C yr^{-1} for this flux, corrected to 1995. This is an estimate of the actual air-sea flux at that time. To allow for direct comparison with estimates by other methods (see Table 5.2) a 'riverine' component of up to 0.6 Pg C yr^{-1} should probably be added to this, corresponding to the steady-state sea-to-air flux in pre-industrial time. However, it is not at present known how much of the riverine flux arises from the open ocean, and how much from wetlands, estuaries and the near-shore ocean.

The error bars on this observational estimate are large. Major sources of uncertainty are the interpolation to account for the rise in atmospheric CO_2 described above, the uncertainty on parameterization of the gas transfer velocity (described more fully below) and the fact that there remain some areas of the world ocean very sparsely covered, particularly the South Pacific and Southern Ocean. (The substantial difference between Takahashi et al.'s 1997 and 1999 estimates derives in part from the inclusion of a data set, due to A. Poisson and colleagues, covering the Southern Indian Ocean.) Because of these uncertainties, the integration to obtain the global net flux is presently not a particularly useful constraint on the global ocean sink. The value of the global pCO_2 survey is immense however, for it gives us a detailed picture of the relative importance of source and sink regions and how these vary seasonally. In the large areas where the data coverage is good, this description is an important test for any model (either conceptual or numerical) of the processes governing air-sea flux of carbon. The global pCO_2 survey shows how the ocean 'breathes', and as it is further refined this information grows ever more valuable.

5.3.6 Preformed Total Carbon Methods and the Ocean Inventory of CO_2

The global survey of CO_2 sponsored by WOCE and JGOFS, consisted of full-depth sections of inorganic CO_2 over all the major oceans. These measurements were made to an accuracy of order 1 μmol kg^{-1} (0.05%) in total carbon, with intercalibration maintained across the many groups and long time interval by the production of thousands of standards to this high accuracy (A. G. Dickson, ms in preparation). This effort has given us a three-dimensional picture of the distribution of inorganic carbon in the ocean of unprecendented detail and accuracy. It will likely serve as the basis of the description of the CO_2 distribution in the oceans for decades to come.

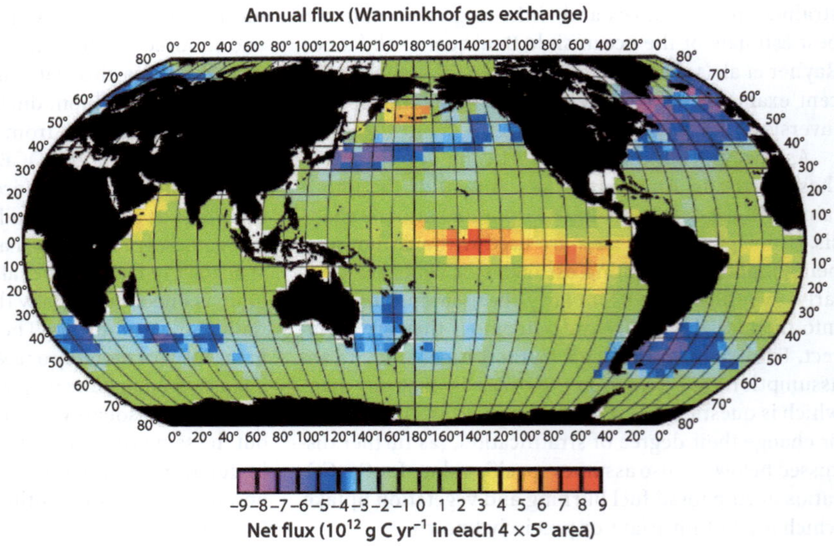

Fig. 5.3. The global, annual distribution of air-sea flux of CO_2 derived from pCO_2 measurements in the surface ocean (reprinted from Takahashi et al. (2002)). These estimates are based on the air-sea gas exchange equation, with ocean pCO_2 data collected over several decades, but corrected to 1995 under assumptions described in the text. The Wanninkhof (1992) gas exchange-wind speed relationship was used

In using this resource to calculate the amount of anthropogenic CO_2 taken up by the oceans, the difficulty is to separate the man-made signal from the natural processes, which are larger and normally dominate the variations in carbon content. A method for making this separation was introduced in the 1970s (Brewer 1978; Chen and Millero 1979) and an improved technique due to Gruber et al. (1996) has now been applied by several authors to parts of the global data base (Gruber 1998; Feely et al. 1999; Sabine et al. 1999).

In detail the method now being used is fairly complex, but in outline what is done is to estimate what the total carbon content of a given sample would have been pre-industrially and to subtract this from the presently observed value to give the excess carbon. To do this, Gruber et al. make use of a quasi-conservative tracer that is invariant with respect to biological processes. Soft-tissue remineralisation is accounted for by tracking the oxygen content and carbonate remineralisation is accounted for by corrections to alkalinity. The tracer is defined by:

$$\Delta C^* = C_m - R_{C:O}[(O_2)_{eq} - (O_2)_m] - 0.5\{TA_m - TA_0 - R_{N:O}[(O_2)_{eq} - (O_2)_m]\} - C_{eq}$$

where C is total inorganic carbon, and O_2 dissolved oxygen. The subscript 'm' signifies measured quantities, while 'eq' refers to equilibrium with the pre-industrial atmosphere, $R_{C:O}$ and $R_{N:O}$ are Redfield ratios of carbon and nitrate to oxygen utilization (both negative in sign), TA is the total alkalinity and TA_0 the preformed total alkalinity, determined as a function of temperature, salinity and nutrient content.

If remineralization of organic and carbonate carbon were the only in situ processes that could cause carbon content to change once water leaves the surface layer, this tracer for change in 'preformed total carbon' would be approximately conservative. Anthropogenic carbon content is calculated as the difference between the preformed total carbon calculated for a given sample, and that estimated for the sample pre-industrially. The method of calculating this pre-industrial value varies. Mixing is assumed to occur only along density surfaces (isopycnals). If an end member can be found on such a surface at sufficient distance from the surface, the anthropogenic influence there can be assumed to be negligible, and this allows an estimate of the pre-industrial preformed total carbon. In other situations, transient tracers such as the CFCs or tritium-helium can be used to estimate a 'ventilation age' for the sample. This can be related to the degree to which surface pCO_2 would have been increased at the time the water mass was ventilated.

The information retrieved by the inventory technique is invaluable for global carbon studies because it relates specifically to the total uptake of the ocean since the industrial revolution, a variable that is difficult to ascertain by any other observational method. The uncertainty in the sink for individual ocean regions has been estimated as ~20% (Gruber 1998). The total amount of carbon burned as fossil fuel, and the amount by which atmospheric CO_2 has increased, are well known (to better than 10%), so the net uptake or release of the land biosphere can be calculated by difference from these figures with reasonable accuracy. The approximations involved in back-calculating pre-industrial pre-formed carbon for water masses are the chief uncertainty of the method.

Table 5.2 Recent estimates of the ocean sink for CO_2

Reference	Method	Mean ($Pg\ C\ yr^{-1}$)	Uncertainty ($Pg\ C\ yr^{-1}$)	Period covered
	Model estimates			
Thirteen OCMIP-2 models	Ocean carbon cycle models	2.0	0.3 (2-σ between models)	1980–1989
Thirteen OCMIP-2 models	Ocean carbon cycle models	2.4	0.5 (2-σ between models)	1990–1999
Takahashi et al. (1999)	Surface ocean pCO_2	1.5[a]	0.7	Corrected to 1990
Takahashi et al. (1999)	Surface ocean pCO_2	2.2[a]	1.1	Corrected to 1995
Sonnerup et al. (1999)	Oceanic ^{13}C change	1.9	0.9	1970–1990
	Atmospheric inversions			
Battle et al. (2000)	O_2/N_2 and CO_2, ^{13}C used as check	2.0	0.6 (1-σ variability)	1991–1997
Keeling et al. (1996)	O_2/N_2 and CO_2	1.9		1991–1994
IPCC (Hougton et al. 2001)	O_2/N_2 and CO_2	1.7	0.5 (1-σ variability)	1991–2000
Ciais et al. (2000)	$^{12}CO_2$ distributions	1.5	0.5 (1-σ variability)	1985–1995
Kaminski et al. (1999)	$^{12}CO_2$ distributions	1.5	0.4 (1-σ variability)	1980–1989
Kheshgi et al. (1999)	Carbon cycle model (^{12}C, ^{13}C, ^{14}C)	1.7	0.7 (90% confidence int.)	1980–1989
Kheshgi et al. (1999)	O_2/N_2, ^{12}C, ^{13}C	2.1	0.3 (1-σ variability)	1980–1996

[a] These figures refer to the actual, observed flux across the air-sea interface, whereas all the other figures in the table refer to the change from the pre-industrial steady state. As such, for comparison with the other figures an (unknown) proportion of the riverine flux of 0.6 $Pg\ C\ yr^{-1}$ should be added to them, as explained further in the text.

5.3.7 Summary of Recent Estimates of the Ocean Sink

In Table 5.2 we have drawn together recent estimates of the ocean sink for the periods of the 1980s and 1990s. Overall, at the present time there is little reason to update the 'canonical' estimate of 2.0 Pg C yr^{-1}. However, it is notable that, taking all the methods into account, we are still unable to be confident in the size of the sink to much greater than about 30%. In particular, the models consistently predict that the sink should have increased by about 20% between these two decades, as the atmospheric concentration continues to rise, and that this trend might be expected to continue in the coming decades. The O_2/N_2 method meanwhile, especially as quoted in the recent IPCC third assessment report, gives a clearly lower value for the decade of the 1990s than do the model estimates. At the time of writing, the cause of this discrepancy is unknown. One possibility is that it is due to a breakdown in the assumption that the ocean is a zero net source of oxygen. Studies have suggested there should be inter-annual variation in the net source of oxygen from regions subject to seasonal convection such as the North Atlantic (McKinley et al. 2000) and it may be that convection is becoming globally less vigorous and deep as a result of surface warming and increased fresh water runoff. If so, this would be one of the first indications of that global change is beginning to affect the overturning properties of the ocean.

5.4 What Processes Control Air-Sea CO_2 Flux?

The ocean sink, integrated over the whole ocean and periods of a decade, is controlled mainly by the rate of vertical mixing and overturning of the ocean – how quickly surface waters penetrate into the interior. On shorter time and space scales a more complex pattern has emerged from the global survey. In detail we cannot as yet perfectly explain this pattern, but in broad outline, it is readily understood as an interplay between thermodynamic, biological and hydrodynamic forcing of surface pCO_2. (The terms 'solubility pump', and 'biological pump' are frequently used to refer to the thermodynamic and biological forcings, but the important role of circulation and mixing is not obvious using this terminology). The biological effect is due to the local influence of biological activity as it varies from place to place and seasonally. The hydrodynamic effect is the influence of circulation and convection in changing surface water carbon content by mixing it with higher concentrations from greater depth. The thermodynamic effect is the influence of temperature on the solubility of CO_2 and its distribution between carbonate, bicarbonate and dissolved gas. This last is the easiest to quantify: other factors being constant, heating seawater causes pCO_2 to increase by an accurately known amount, about 4% per °C, and cooling does the reverse (Takahashi et al. 1993). Surface water that is warming will therefore tend to have high pCO_2 and be a source to the atmosphere, while cooling water will be a sink. Since the time for equilibration of the CO_2 system with the atmosphere is long (~1 year) compared to the time for thermal equilibration of the mixed layer (a few months), this thermal CO_2 signal may remain measurable even after the thermal forcing has ceased.

5.4.1 Patterns in the Global Survey

The most notable feature on the air-sea flux map (Fig. 5.1) is the equatorial Pacific CO_2 'bulge'. In non-El-Niño years, the Eastern equatorial Pacific is a strong and continuous source of CO_2 to the atmosphere – the strongest oceanic source on Earth. This is due to the upwelling of water into the surface, forced by the equatorial divergence. Subsurface waters have a relatively higher carbon content than the surface as a result of the downward fractionation due to the biological and solubility pumps. More importantly, as the water enters the surface layer it is heated strongly, causing it to begin degassing CO_2 to the atmosphere. The biological effect tends to offset this flux by fixing nutrients and CO_2 out of the newly upwelled water. However, biological CO_2 drawdown in the equatorial Pacific is strongly iron-limited, as shown during JGOFS by bottle incubations (Fitzwater et al. 1996) and in situ fertilizations (Cooper et al. 1996). The biological pump is therefore comparatively inefficient in this region (Murray et al. 1994), and this increases the net CO_2 efflux. All regions of upwelling in the tropics or subtropics may be similarly expected to be sources of atmospheric CO_2, for example the Arabian Sea during the Southwest Monsoon (Sabine et al. 2000), or the Coast of Peru and Chile (Torres et al. 1999). In coastal upwelling systems, net biological drawdown is greater than in the equatorial Pacific, presumably because of a greater supply of available iron.

Warm, poleward-travelling water currents lose heat to the atmosphere and therefore tend to be sinks for atmospheric CO_2. The regions where cooling is strongest are the western boundary currents, particularly the Gulf Stream and the Kuroshio in the Pacific. In the North Atlantic, there is an overall northward drift and cooling of the water as a result of the Atlantic meridional overturning circulation (MOC) and this contributes to a net undersaturation in CO_2 throughout the region north of the subtropical gyre. Furthermore, net biological export is also high. The biological pump in the North Atlantic is efficient, removing most of the nitrate and phosphate in the surface waters, as a result of the ready supply of iron in the form of atmospheric dust from the surrounding land, particularly the Sahara desert. Both the thermodynamic and the biological effects therefore tend to make the region a net sink for atmospheric CO_2, and the North Atlantic and Nordic Seas are the strongest consistent sink regions in the world ocean.

Other things being equal, strong net sink regions of the ocean should tend to be coincident with regions where water that is cooling or has recently been cooled is at the surface. Because this implies increasing density, such regions are likely to be formation zones for subsurface water. By contrast, the strongest net source regions are warm-water upwelling zones. Again, other things being equal, regions where the biological utilization of nutrients is efficient should be stronger sinks, while 'High nutrient low chlorophyll' (HNLC) regions, where biological utilization of CO_2 is inefficient, should be stronger sources. This reasoning is qualitatively consistent with the major features of the pCO_2 survey. Thus for example, the sink region in the Southern Indian Ocean and in the South Atlantic in the Brazil-Malvinas regions (Fig. 5.3) are located in the formation zones for Antarctic Intermediate Water, which is the major source of thermocline waters for the world ocean. Biological activity is also particularly strong in the Southwest Atlantic, due perhaps to relatively abundant iron supply from the nearby Patagonian Shelf or from windblown dust. The Southern Ocean south of the Polar front is a site of strong upwelling, and also an HNLC zone, so might be expected to be a source zone by this reasoning, though much less marked than in warm-water upwellings because in these cold waters the contribution of heating to raising surface pCO_2 will be absent. In fact, in winter at least, the upwelling water is probably being cooled rather than heated. Measurements south of the Polar front (of which there are very few) suggest the region today is approximately neutral with respect to the atmosphere.

5.4.2 Comparison Using Models

Ocean carbon models enable us to compare quantitatively our expectations with the data. Here, we compare the outputs of models both with the data and with each other. Furthermore, in the models we can decompose the fluxes into the pre-industrial steady-state and anthropogenic components, and the pre-industrial component into solubility-driven and biologically driven parts – impossible to do with the observations, which represent the sum of all the processes occurring simultaneously. Anthropogenic uptake was determined in the OCMIP models as the difference between the model uptake in 1765 and that in 1995. To separate the biological and solubility effects, the models were run for two equilibrium simulations:

a Abiotic – the solubility component which includes carbon chemistry and realistic gas exchange between surface ocean and atmospheric CO_2 (held to 278 ppm for the preindustrial case);
b Biotic – the solubility plus biological components, together.

A measure of the steady-state biological fluxes can be found from the difference between the two simulations.

Sea-to-air fluxes decomposed into these components are displayed as zonal integrals for the global ocean in Fig. 5.4. All the solubility simulations exhibit ocean outgassing in the tropics and uptake in the high latitudes, as expected if high-latitude waters are cooling and sinking while the tropical upwellings are warming. Additionally, the biological component counteracts the solubility component either by bringing respired CO_2 (produced by bacterial degradation of organic matter) to the surface via upwelling and deep convection (mostly in the high latitudes), or by consuming CO_2 at the surface via photosynthesis (mostly in the subtropical gyres and the tropics). Anthropogenic fluxes are generally smaller than natural fluxes, but the contribution from the modern anthropogenic component is not negligible.

Figure 5.5 compares the distribution of the total (anthropogenic plus pre-industrial) fluxes predicted by the models with the climatology of Takahashi et al. (Fig. 5.3, redrawn as the first two panels of Fig. 5.5). The zonal integrals of these distributions are given in Fig. 5.6. All the models predict the equatorial CO_2 bulge, and the strong sinks in the North Atlantic and associated with the Kuroshio in the Northwest Pacific, as also seen in the data. The models show greatest agreement with each other and with the data in lower latitudes (40° S to 40° N).

But in detail the model distributions differ quite substantially, the disagreement being greatest in the Southern Ocean. It is important also to remember that the climatological fluxes represented by the data may be in error for the reasons described above. In particular, Takahashi et al. assumed that at high latitudes, the ocean pCO_2 has not increased, or has increased at only half the rate of the atmosphere, over recent decades. At low latitudes they assumed it has tracked the atmosphere. The models suggest that the low-latitude assumption is probably reasonable, but that at high latitudes the ocean pCO_2 does increase. Hopefully, it will soon be possible to use the models to examine the uncertainties introduced by the assumptions used in constructing the data-based climatologies.

The distribution of the anthropogenic carbon dioxide fluxes in isolation from the natural signals are examined in Fig. 5.7a, while Fig. 5.7b shows the storage of anthropogenic CO_2. As discussed above the models agree well on global mean anthropogenic uptake. General patterns of regional uptake are also grossly similar between

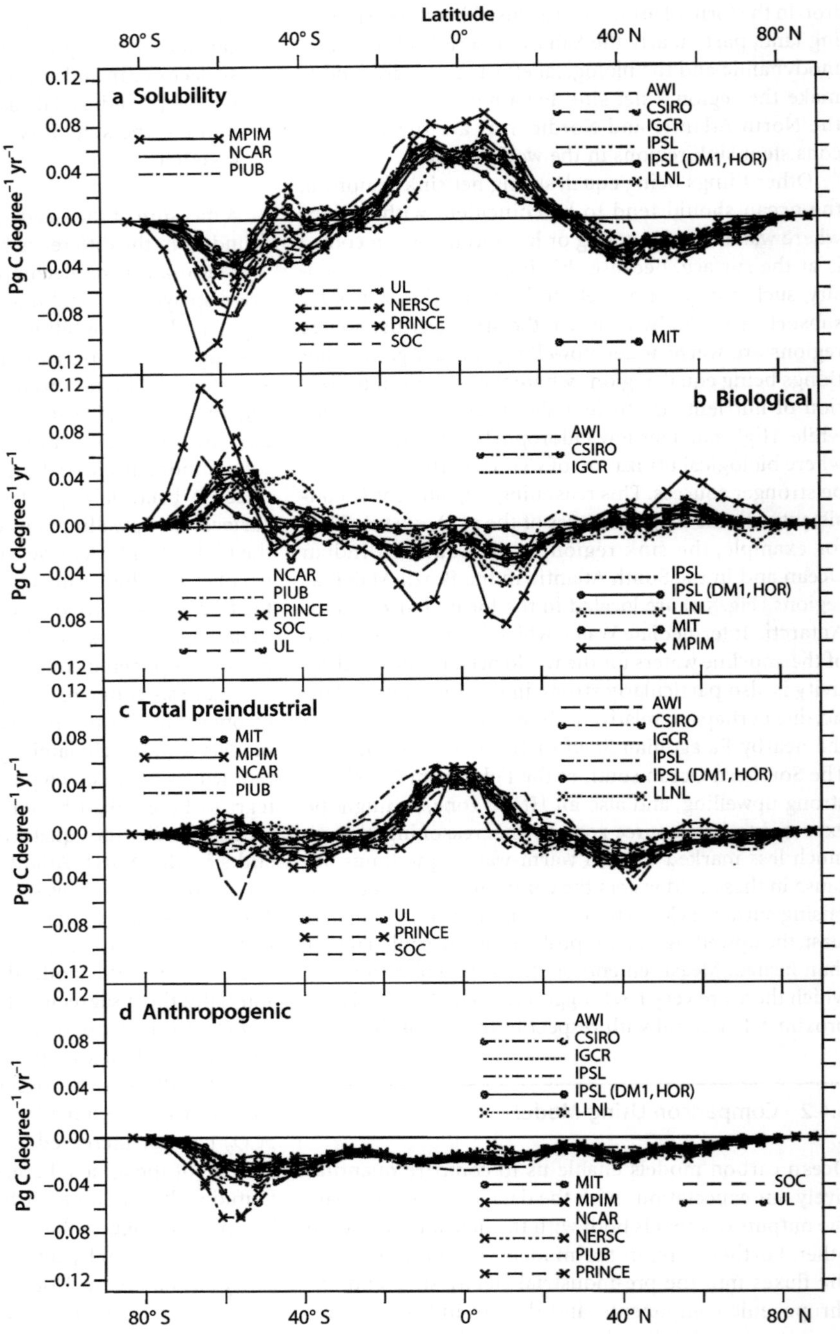

Fig. 5.4. The zonally integrated sea-to-air flux of CO_2 for the global ocean as predicted by the OCMIP models. The flux is separated into **a** the solubility component, with realistic gas exchange, **b** the biological component only, **c** the sum of $a + b$, i.e., the total natural flux, and **d** the component due to anthropogenic change during 1765 to 1995. The biological component **b** is determined by difference between the 'biotic' and 'abiotic' simulations, described in the text

the OCMIP-2 models, with uptake highest in the high latitudes and at the equator, i.e., in zones where deep waters uncontaminated with anthropogenic CO_2 communicate readily with the surface via upwelling and convection. Low fluxes are evident in the subtropics, where surface waters have had longer to equilibrate with the atmosphere. However, the models disagree substantially about local patterns of anthropogenic carbon uptake. Large differences in model uptake are found in the Southern Ocean, in the tropics, and in the North Atlantic. When ocean surface area is taken into account, the largest differences between models are found south of 30° S, which occupies about one third of the surface of the entire ocean.

As opposed to the large difference in uptake of anthropogenic CO_2, there is surprising agreement among

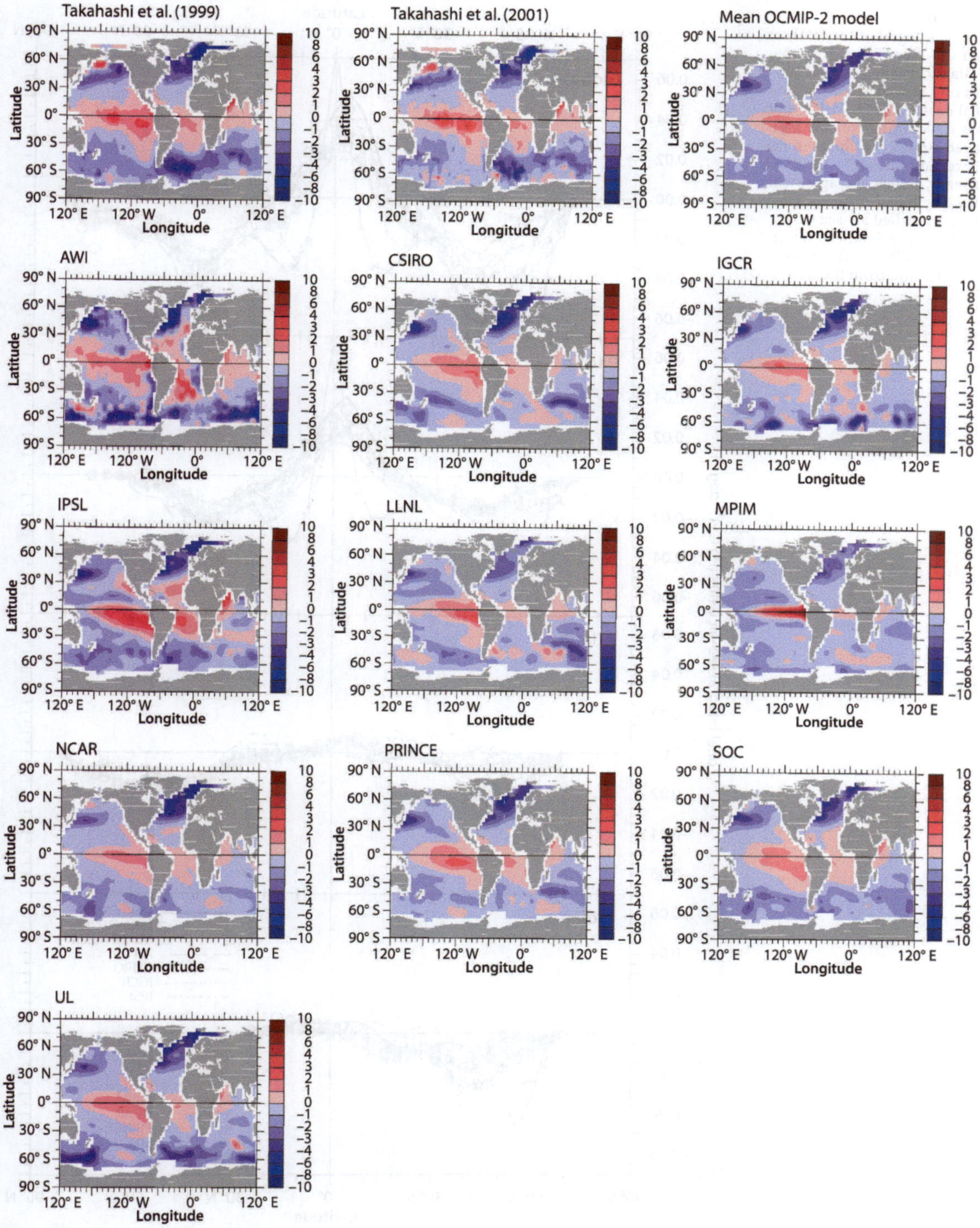

Fig. 5.5. Model and data-based estimates of the annual mean, sea-to-air CO_2 flux in 1995. Observed fluxes are based on pCO_2 observations that were interpolated in space and time to a climatological grid (Takahashi et al. 1999; Takahashi et al. 2002). For consistency, these gridded ΔpCO_2 data fields were then multiplied by the same gas exchange fields used for the OCMIP-2 simulations. The monthly OCMIP-2 gas exchange field is based on satellite observed winds and the gas exchange formulation of Wanninkhof (1992) (see *http://www.ipsl.jussieu.fr/OCMIP*). The model flux fields represent the total sea-air flux, obtained by summing the pre-industrial state (Biotic Equilibrium run) plus the anthropogenic perturbation (abiotic historical minus abiotic equilibrium or control run). For consistency, all model fields were interpolated to the same 4 × 5 data grid

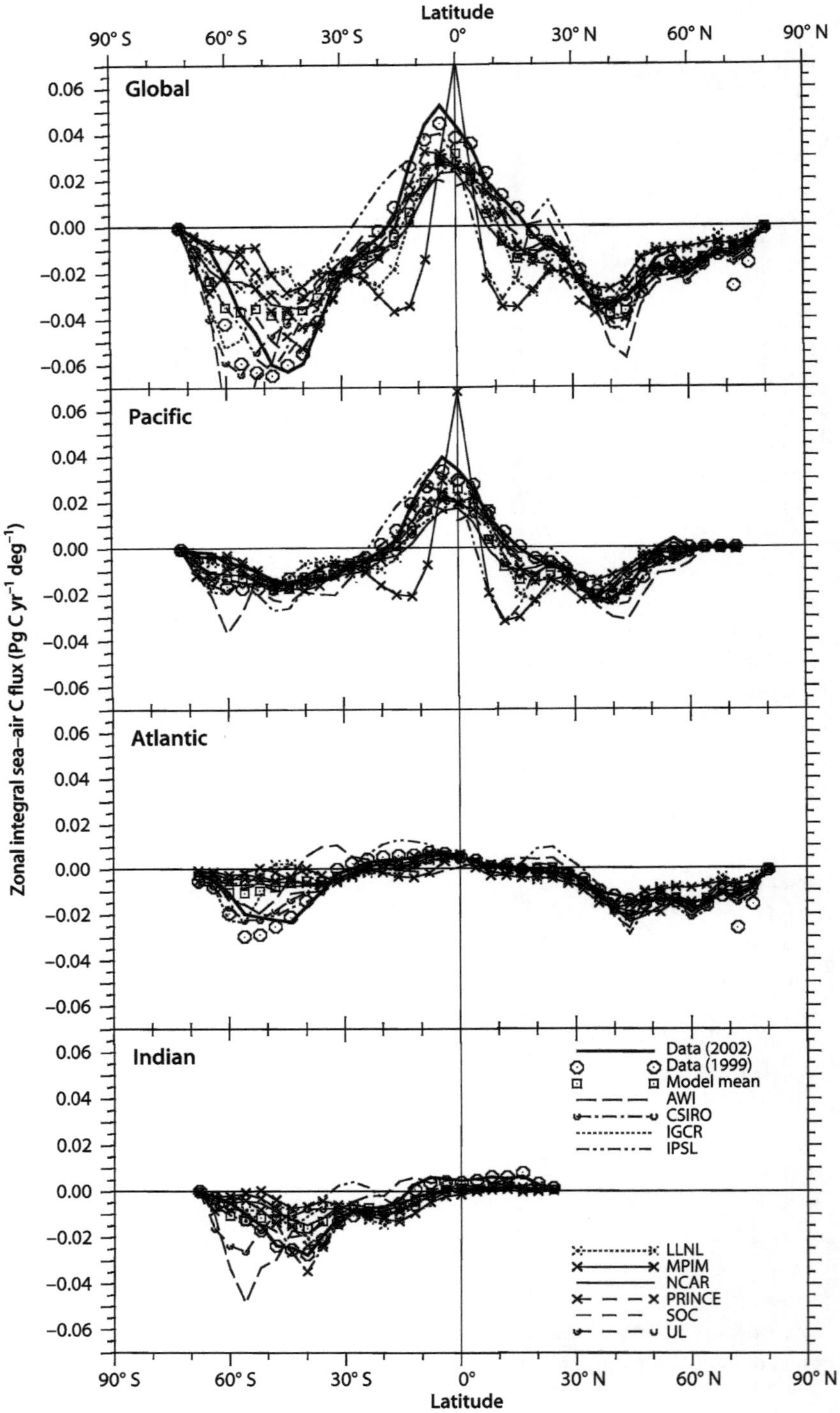

Fig. 5.6.
Zonal integral of the total sea-air CO_2 flux (Fig. 5.5) for the Atlantic, Pacific, and Indian Basins, as well as for the global ocean. The OCMIP-2 models show similar zonally averaged air-sea CO_2 fluxes in the lower latitudes (40° S to 40° N); zonally, models also agree with the data-based estimates. However, modelled air-sea CO_2 fluxes in the Southern Ocean are very different from one another and from the data

models concerning average meridional patterns of storage, as represented by the zonally integrated column inventory. All models store most of the anthropogenic CO_2 in the subtropics (where uptake is lowest), particularly in the southern hemisphere, and they store the least anthropogenic CO_2 in the high latitudes and in the tropics (where uptake is highest). The low storage in these regions results from the surface layer being flushed by subsurface water, low in anthropogenic CO_2, newly entering the mixed layer from below.

Fig. 5.7.
a Zonally integrated, global, cumulative *uptake*, i.e., air-sea flux of anthropogenic CO_2; **b** *storage*, i.e., the inventory of anthropogenic carbon in the ocean (vertical column integral of the concentration), from 1765 to 1995, **c** *convergence* of anthropogenic CO_2 (i.e., storage minus uptake), and **d** northward transport of anthropogenic CO_2

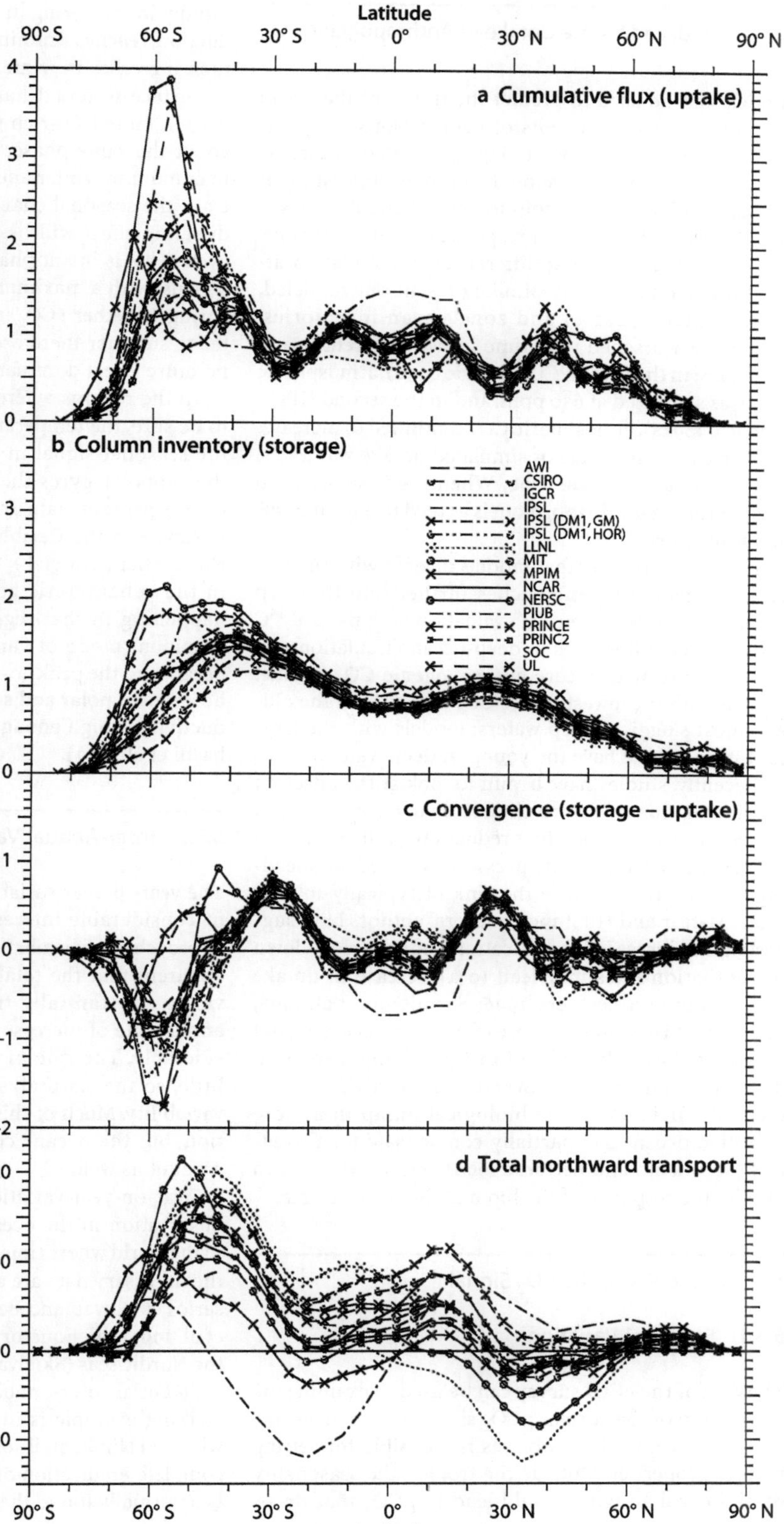

5.4.3 Modelled Future Uptake of Anthropogenic CO_2

Models can be used to predict the future of the ocean CO_2 sink, and some results of comparisons using the OCMIP models are shown in Fig. 5.8. In all of these, the biological pump was assumed to remain operating at the same efficiency as it does today and circulation was not altered in the future in response to climate change, so that only pre-existing differences in circulation affect the outcomes. Despite similar present-day, modeled, global uptake (±22%) and zonal mean inventories, agreement worsened with time in two different future scenarios: in the first (IPCC scenario S650), atmospheric CO_2 was stabilized at 650 ppm, and in the second (IPCC scenario IS92a), atmospheric CO_2 continued to increase. In both cases, the range of simulated uptake was ±30% about the mean in year 2100. When the S650 scenario was continued until 2300, models agreed to within ±33% about the mean.

It is not surprising that models diverge with time, as anthropogenic CO_2 encroaches further into the deep ocean. The OCMIP-2 model evaluation with natural ^{14}C reveals large differences in deep-ocean circulation and a logical link with future anthropogenic CO_2 uptake: models with the lowest future CO_2 uptake have the oldest (most sluggish) deep waters; models with the largest future uptake have the youngest deep waters.

Recently, studies have begun to look at the effect on the air-sea flux of future climate change. Warming of the ocean surface leads to a reduction in the uptake of CO_2 as the surface partial pressure is raised, and leading to a net reduction in the sink of typically 10% by 2100 (Matear and Hirst 1999). Several authors have suggested that increased stratification and/or a slow-down in ventilation rates may lead to a decrease in uptake (Sarmiento and LeQueré 1996; Sarmiento et al. 1998; Matear and Hirst 1999). These effects are uncertain, but might also be on the order of 10% providing there is no dramatic change in the overturning circulation. Increased efficiency of the biological pump in a more stratified ocean may partially compensate for this effect (Sarmiento and LeQueré 1996), but it is difficult to predict the response of the biota at the present time.

5.5 Variability in the CO_2 Signal

5.5.1 Seasonal Variation

The data of the pCO_2 survey can be used to examine the seasonality of the surface pCO_2 signal, which gives further insight into the processes responsible for setting surface properties. Outside the tropics, the seasonality of biological activity should lead to pCO_2 that drops rapidly in the spring in response to maximum net production, reaches a minimum in summer and the fall and then increases in winter as net respiration takes place in surface waters. Changes in the upward mixing of deeper, more CO_2-rich water will produce a cycle with about the same phase, since mixing rates will be at a maximum in winter and minimum in summer. By contrast, the seasonal change in pCO_2 due to the thermodynamic effect will be in phase with the temperature cycle. This is in antiphase to the biological and mixing signals, with a maximum in the fall and a minimum in spring. Whether pCO_2 increases or decreases from winter to summer therefore depends on whether the temperature effect dominates the other two factors.

In the regions where data coverage is good enough to be sure, the temperature effect in general dominates the seasonal signal in the subtropical gyres while in the subpolar gyres the opposite occurs. Cooper et al. (1998) present data from a shipping route between Europe and the Caribbean, showing this effect in the North Atlantic. Metzl et al. (1999) discuss the seasonality in the subantarctic zone and show a similar effect happening in that region. Seasonal variability is the dominant mode of variation over much of the world ocean, and the peak-to-peak magnitude of the cycle can be large. In polar and sub-polar regions where the productivity is high enough it can exceed 100 μatm (Takahashi et al. 1993).

5.5.2 Inter-Annual Variation

The year-to-year variability of the ocean sink is a topic of considerable interest at present, because different approaches give substantially different estimates. It is apparent that the total natural sink, land-plus-ocean, varies substantially from year to year, because the annual rate of increase of atmospheric CO_2 varies considerably. The rate of release from fossil fuel varied little, so the 'natural' sinks must be the cause of this variability. Much of this is caused by terrestrial vegetation, but the oceans contribute an as-yet-unspecified amount as well.

Year-on-year variation is difficult to study by direct observation in the ocean because there are few places in the world where time series have been run. Sites where the necessary data are available include the JGOFS time series at Hawaii and Bermuda (Bates et al. 1996; Sabine et al. 1995) the Equatorial Pacific (Feely et al. 1997) and the Nordic seas (Skjelvan et al. 1999). In all of these sites, substantial inter-annual variability is seen. The most striking example is in the Eastern equatorial Pacific, where El Niño conditions are accompanied by the near-complete eradication of the above-saturation pCO_2 values normally found there. Feely et al. (1997) calculate an

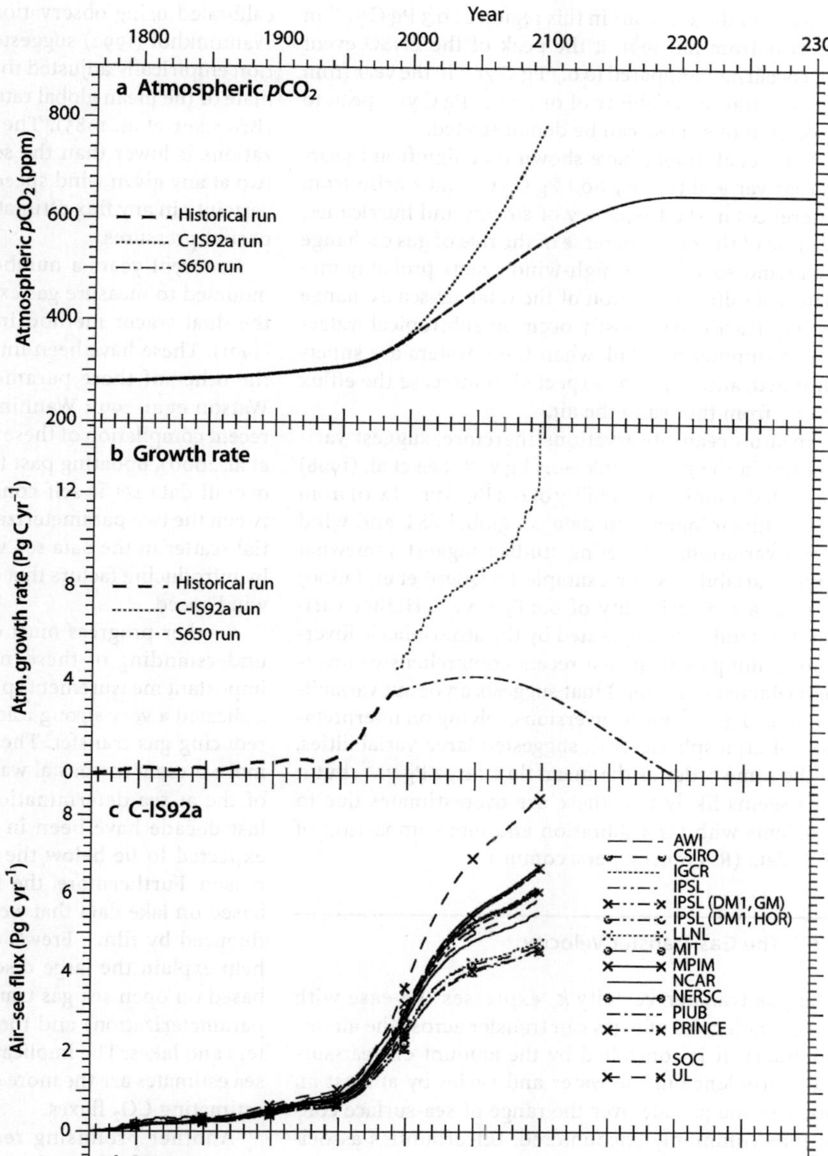

Fig. 5.8. History of a atmospheric CO_2, b its growth rate, and the ocean uptake of anthropogenic carbon, under the historical and two future scenarios CIS92a (c) and S650 (d). All models were forced by observed atmospheric pCO_2 from 1765 to 1990 and subsequently by a concentration scenario chosen by the IPCC for its Third Assessment Report (TAR). The latter scenario is based on the IS92a emissions scenario

efflux from the sea to air in this region of 0.3 Pg C yr^{-1} in the year from fall 1991 at the peak of the ENSO event then occuring, compared to 0.7 Pg C yr^{-1} in the year from fall 1993, thus a variability of order 0.4 Pg C yr^{-1} peak to peak from this cause can be demonstrated.

Bates et al. (1998) have shown that significant year-to-year variability (i.e., >0.1 Pg C yr^{-1}) may arise from differences in the frequency of storms and hurricanes. Because of the rapid increase of the rate of gas exchange with wind speed, rare, high-wind events probably mediate a significant portion of the total air-sea exchange of CO_2. Hurricanes mostly occur in subtropical waters in late summer and fall, when these waters are supersaturated, and might be expected to increase the efflux of CO_2 from the sea to the air.

In situ ocean observations therefore, suggest variabilities in the global sink ~0.5 Pg yr^{-1}. Lee et al. (1998) estimated a similar variability of 0.4 Pg C yr^{-1} (2-σ) from $p$$CO_2$ climatologies and data on global SST and wind speed variations. Modeling studies suggest somewhat larger variabilities, for example Le Queré et al. (2000) suggest a 1-σ variability of 0.4 Pg C yr^{-1}. Higher variabilities tend to be suggested by the atmospheric inversion techniques such as a recent comprehensive inversion (Rayner et al. 1999) that suggests an ocean variability of 0.8 Pg yr^{-1}. Early inversions, relying on interpretation of atmospheric $\delta^{13}C$, suggested large variabilities, of the same order as the mean flux, i.e., 2 Pg yr^{-1}, but it now seems likely that these are overestimates due to problems with the calibration and intercomparison of $\delta^{13}C$ data (R. Francey, pers. comm.).

5.6 The Gas Transfer Velocity

The gas transfer velocity k_v expresses the ease with which molecules of a gas can transfer across the air-sea boundary. It is controlled by the amount of near-surface turbulence in the water and varies by at least an order of magnitude over the range of sea-surface conditions commonly encountered. Uncertainties associated with the gas transfer are one of the chief impediments to an accurate specification of air-sea CO_2 fluxes from oceanic measurements. k_v has traditionally been expressed as an empirical function of wind speed only, but we know this to be a crude simplification. It would be expected to be a function also of variables such as fetch, atmospheric stability (Erickson 1993), rainfall, (Ho et al. 1997), surface films (Frew 1997), and degree of whitecapping (Monhan and Spillane 1984) that are not solely a function of wind speed.

Over the last decade, two different parametrizations of k_v with wind speed have have been widely used. The Liss-Merlivat formulation (Liss and Merlivat 1986) is based on a synthesis of wind-wave tunnel experiments calibrated using observations from a small lake, while Wanninkhof (1992) suggested a simple quadratic function empirically adjusted to be compatible with an estimate of the mean global rate of (^{14}C-based) gas transfer (Broecker et al. 1985). The first of these parameterizations is lower than the second by nearly a factor of two at any given wind speed, creating a substantial uncertainty in any flux estimate based on measurement of partial pressures.

In recent years a number of campaigns have been mounted to measure gas exchange at sea, mostly using the dual tracer method introduced by Watson et al. (1991). These have been interpreted as favoring one or the other of these parameterizations (e.g., compare Watson et al. 1991; Wanninkhof et al. 1993). The most recent compilation of these measurements (Nightingale et al. 2000), updating past treatments, suggests that the overall data set is self-consistent and falls largely between the two parameterizations. There is still substantial scatter in the data set, which can only be explained by introducing factors that do not correlate closely with wind speed.

Further progress must depend on gaining a better understanding of these 'non-wind' factors. Recently, important measurements published by Frew (1997) have indicated a very strong role for natural organic films in reducing gas transfer. These films appear to be much more present in coastal waters than the deep sea. Most of the at-sea determinations of gas transfer over the last decade have been in shelf waters and might be expected to be below the open-ocean values for this reason. Furthermore, the Liss-Merlivat formulation is based on lake data that we might also expect to be influenced by films. Frew's observations may therefore help explain the large discrepancy between estimates based on open sea gas transfer (such as Wanninkhof's parameterization) and those calibrated in coastal waters and lakes. The implication is that the higher, open-sea estimates are the more correct ones to employ when estimating CO_2 fluxes.

Another promising recent development is in the measurement of CO_2 fluxes. Direct measurement using the eddy correlation technique has been used for CO_2 fluxes over land vegetation for decades, but has been too imprecise to be useful for the much lower net fluxes between air and sea. Recently, a careful re-engineering of this technique has improved the precision sufficiently to make at-sea eddy correlation measurements (McGillis et al. 2001a,b). This advance has already enabled new parameterizations of k_v to be suggested (Wanninkhof and McGillis 1999). If it can be reduced to routine, the method offers the possibility of open sea measurements of CO_2 flux that would enable direct testing of the existing theory, and rapid improvements to our CO_2 flux estimates.

5.7 Conclusion: the Next Ten Years

At the conclusion of the JGOFS programme, we have a hugely improved picture of the present state of global ocean fluxes of CO_2 compared to the situation in 1990. We understand in broad terms the processes that give rise to the observed air-sea fluxes, but that understanding is as yet insufficient to make us very confident about predicting how these fluxes will change in the future under a given forcing.

To become confident in our predictions we require a deeper understanding of both the physics and biology of the oceans. In particular, our knowledge of the biological aspects of carbon cycling is largely 'static'; we can describe what happens today, but we are only beginning to understand the processes at the deeper level required to predict how they will change in the future when forced by global change. How, for example, will changes in temperature, pH, circulation and nutrient availability affect the marine biological communities? Will this lead to changes in the efficiency of the biological CO_2 pump? Few models have as yet tried to tackle these issues, and the basic assumption in current use is that the biological parts of the system will not change in the future. To make progress here, we should adopt more laboratory and field investigations targeted at particular processes.

It is clear too that we need ongoing observations, and new types of observations, in order to monitor those processes. The knowledge obtained during JGOFS required an enormous effort by a substantial fraction of the world's marine scientists and marine research resources. Such an effort is unlikely to be repeated for decades to come, so we must develop new methods of observation if we are to maintain and update our understanding in the future.

Fortunately, such techniques are becoming available. For much of the JGOFS decade, no adequate satellite ocean color sensor was flying, but since 1997 high resolution global ocean color coverage has been available. New instruments and techniques for the remote measurement of nutrients, carbon dioxide and fluorescence from drifting and moored buoys and ships of opportunity have been developed (Cooper et al. 1998; Merlivat and Brault 1995; DeGrandpré et al. 1999). We can look forward in the next ten years to a global ocean observing system that will include not only the important physical measurements such as temperature and salinity, but also these vital biogeochemical variables.

Acknowledgements

This work was supported by contract no. EVK2-2000-22058 (CAVASSOO) from the European commission.

References

Aumont O, Orr JC, Monfray P, Ludwig W, Amiotte-Suchet P, Probst J-L (2001) Riverine-driven interhemispheric transport of carbon. Global Biogeochem Cy 15:393–405

Bates NR, Michaels AF, Knap AH (1996) Seasonal and interannual variability of oceanic carbon dioxide species at the US JGOFS Bermuda Atlantic Time-series Study (BATS) site. Deep-Sea Res Pt II 43:347–383

Bates NR, Knap AH, Michaels AF (1998) Contribution of hurricanes to local and global estimates of air-sea exchange of CO_2. Nature 395:58–61

Battle M, Bender ML, Tans PP, White JWC, Ellis JT, Conway T, et al. (2000) Global carbon sinks and their variability inferred from atmospheric O_2 and $\delta^{13}C$. Science 287:2467–2470

Bousquet P, Peylin P, Ciais P, Le Queré C, Friedlingstein P, Tans PP (2000) Regional changes in carbon dioxide fluxes of land and oceans since 1980. Science 290:1342–1346

Brewer PG (1978) Direct measurement of the oceanic CO_2 increase. Geophys Res Lett 5:997–1000

Broecker WS, Peng TH (1992) Interhemispheric transport of carbon-dioxide by ocean circulation. Nature 356:587–589

Broecker WS, Peng TH, Ostlund G, Stuiver M (1985) The distribution of bomb radiocarbon in the ocean. J Geophys Res 90: 6953–6970

Chen C-T, Millero FJ (1979) Gradual increase of oceanic CO_2. Nature 277:205–206

Ciais P, Peylin P, Bousquet P (2000) Regional biospheric carbon fluxes as inferred from atmospheric CO_2 measurements. Ecol Appl 10:1574–1589

Conway TJ, Tans PP (1999) Development of the CO_2 latitude gradient in recent decades. Global Biogeochem Cy 13: 821–826

Cooper DJ, Watson AJ, Nightingale PD (1996) Large decrease in ocean-surface CO_2 fugacity in response to in-situ iron fertilization. Nature 383:511–513

Cooper DJ, Watson AJ, Ling RD (1998) Variation of pCO_2 along a North Atlantic shipping route (UK to Caribbean): a year of automated observations. Mar Chem 60:147–164

DeGrandpré MD, Baehr MM, Hammar TR (1999) Calibration-free opitcal chemical sensors. Anal Chem 71:1152–1159

Erickson DJ (1993) A stability dependent theory for air-sea gas-exchange. J Geophys Res 98:8471–8488

Feely RA, Wanninkhof R, Goyet C, Archer DE, Takahashi T (1997) Variability of CO_2 distributions and sea-air fluxes in the central and eastern equatorial Pacific during the 1991–1994 El Niño. Deep-Sea Res Pt II 44:1851–1867

Feely RA, Sabine CL, Keys RM, Peng T-H (1999) CO_2 synthesis results: estimating the anthropgenic carbon dioxide sink in the Pacific Ocean. US JGOFS News 9:1–5

Fitzwater SE, Coale KH, Gordon RM, Johnson KS, Ondrusek ME (1996) Iron-deficiency and phytoplankton growth in the Equatorial Pacific. Deep-Sea Res Pt II 43:995–1015

Frew NM (1997) The role of organic films in air-sea gas exchange. In: Liss PS, Duce RA (eds) The sea surface and global change. Cambridge University Press, Cambridge, pp 121–171

Gruber N (1998) Anthropogenic CO_2 in the Atlantic Ocean. Global Biogeochem Cy 12:165–191

Gruber N, Sarmiento JL, Stocker TF (1996) An improved method for detecting anthropogenic CO_2 in the oceans. Global Biogeochem Cy 10:809–837

Ho DT, Bliven LF, Wanninkhof R, Schlosser P (1997) The effect of rain on air-water gas exchange. Tellus B 49:149–158

Houghton JT, Jenkins GJ, Ephraums JJ (eds) (1990) Climate change: the IPCC scientific assessment. Cambridge University Press, Cambridge, 364 pp

Houghton JT, Meira Filho LG, Callendar BA, Harris N, Kattenberg A, Maskell K (eds) (1995) The science of climate change: contribution of Working Group I to the second assessment of the Intergovernmental Panel on Climate Change. Cambridge University Press, Cambridge, 572 pp

Houghton JT, Ding Y, Griggs DJ, Noguer M, van der Linden PJ, Xiaosu D (eds) (2001) Climate change 2001: the scientific basis contribution of Working Group I to the third assessment report of the Intergovernmental Panel on Climate Change. Cambridge University Press, Cambridge, 944 pp

Indermuhle A, Stocker TF, Joos F, Fischer H, Smith HJ, Wahlen M, et al. (1999) Holocene carbon-cycle dynamics based on CO_2 trapped in ice at Taylor Dome, Antarctica. Nature 398:121–126

Kaminski T, Heimann M, Giering R (1999) A coarse grid three-dimensional global inverse model of the atmospheric transport – 2. Inversion of the transport of CO_2 in the 1980s. J Geophys Res 104:18555–18581

Keeling CD, Bacastow RB, Carter AF, Piper SC, Whorf TP, Heimann M, et al. (1989) A three-dimensional model of atmospheric CO_2 transport based on observed winds. 1. Analysis of observational data. In: Peterson DH (ed) Aspects of climate variability in the Pacific and the Western Americas Geophysical Monograph. American Geophysical Union, 55, 165–235

Keeling RF, Peng TH (1995) Transport of heat, CO_2 and O_2 by the Atlantic's thermohaline circulation. Philos T Roy Soc B 348: 133–142

Keeling RF, Shertz SR (1992) Seasonal and interannual variations in atmospheric oxygen and implications for the global carbon-cycle. Nature 358:723–727

Keeling RF, Piper SC, Heimann M (1996) Global and hemispheric CO_2 sinks deduced from changes in atmospheric O_2 concentration. Nature 381:218–221

Kheshgi HS, Jain AK, Wuebbles DJ (1999) Model-based estimation of the global carbon budget and its uncertainty from carbon dioxide and carbon isotope records. J Geophys Res 104: 31127–31143

Langenfelds RL, Francey RJ, Steele LP, Battle M, Keeling RF, Budd WF (1999) Partitioning of the global fossil CO_2 sink using a 19-year trend in atmospheric O_2. Geophys Res Lett 26:1897–1900

Le Quere C, Orr JC, Monfray P, Aumont O, Madec G (2000) Interannual variability of the oceanic sink of CO_2 from 1979 through 1997. Global Biogeochem Cy 14:1247–1265

Lee K, Wanninkhof R, Takahashi T, Doney SC, Feely RA (1998) Low interannual variability in recent oceanic uptake of atmospheric carbon dioxide. Nature 396:155–159

Lefevre N, Watson AJ, Cooper DJ, Weiss RF, Takahashi T, Sutherland SC (1999) Assessing the seasonality of the oceanic sink for CO_2 in the northern hemisphere. Global Biogeochem Cy 13:273–286

Lerperger M, McNichol AP, Peden J, Gagnon AR, Elder KL, Kutschera W, et al. (2000) Oceanic uptake of CO_2 re-estimated through $\delta^{13}C$ in WOCE samples. Nucl Instrum Meth B 172: 501–512

Li XS, Berger A, Loutre MF (1998) CO_2 and northern hemisphere ice volume variations over the middle and late quaternary. Clim Dynam 14:537–544

Liss PS, Merlivat L (1986) Air-sea gas exchange rates: introduction and synthesis. In: Buat-Menard P (ed) The role of air-sea exchange in geochemical cycling. D. Reidel, Dordrecht, pp 113–127

Ludwig W, Amiotte-Suchet P, Munhoven G, Probst JL (1998) Atmospheric CO_2 consumption by continental erosion: present-day controls and implications for the last glacial maximum. Global Planet Change 17:107–120

Matear RJ, Hirst AC (1999) Climate change feeback on the future oceanic CO_2 uptake. Tellus B 51:722–733

McGillis WR, Edson JE, Hare JE, Fairall CW (2001a) Direct covariance air-sea CO_2 fluxes. J Geophys Res 106:16729–16745

McGillis WR, Edson JB, Ware JD, Dacey JWH, Hare JE, Fairall CW, Wanninkhof R (2001b) Carbon dioxide flux techniques performed during GasEx 98. Mar Chem 75:267–280

McKinley GA, Follows MJ, Marshall J (2000) Interannual variability in the air-sea flux of oxygen in the North Atlantic. Geophys Res Lett 27:2933–2936

Merlivat L, Brault P (1995) CARIOCA buoy, carbon dioxide monitor. Sea Technol 10:23–30

Metzl N, Tilbrook B, Poisson A (1999) The annual fCO_2 cycle and the air-sea CO_2 flux in the sub-Antarctic Ocean. Tellus B 51: 849–861

Monahan EC, Spillane MC (1984) The role of oceanic whitecaps in air-sea gas exchange. In: Brutaert W, Jirka GH (eds) Gas transfer at water surfaces. D. Reidel, Dordrecht, pp 495–504

Murray JW, Barber RT, Roman MR, Bacon MP, Feely RA (1994) Physical and biological-controls on carbon cycling in the Equatorial Pacific. Science 266:58–65

Nightingale PD, Malin G, Law CS, Watson AJ, Liss PS, Liddicoat MI, et al. (2000) In situ evaluation of air-sea gas exchange parameterizations using novel conservative and volatile tracers. Global Biogeochem Cy 14:373–387

Orr JC (1993) Accord between ocean models predicting uptake of anthropogenic CO_2. Water Air Soil Poll 70:465–481

Orr JC, Maier-Reimer E, Mikolajewicz U, Monfray P, Sarmiento JL, Toggweiler JR, et al. (2001) Estimates of anthropogenic carbon uptake from four three-dimensional global ocean models. Global Biogeochem Cy 15:43–60

Ostlund HG, Possnert G, Swift JH (1987) Ventilation rate of the deep Arctic Ocean from ^{14}C-data. J Geophys Res 92:3769–3777

Pearman GI, Hyson P (1986) Global transport and inter-reservoir exchange of carbon dioxide with particular reference to stable isotope distributions. J Atmos Chem 4:81–124

Quay PD, Tilbrook B, Wong CS (1992) Oceanic uptake of fossil-fuel CO_2–^{13}C evidence. Science 256:74–79

Rayner PJ, Law RM, Dargaville R (1999) The relationship between tropical CO_2 fluxes and the El Niño-Southern Oscillation. Geophys Res Lett 26:493–496

Sabine CL, Mackenzie FT, Winn C, Karl DM (1995) Geochemistry of carbon dioxide in seawater at the Hawaii Ocean Time-Series Station, Aloha. Global Biogeochem Cy 9:637–651

Sabine CL, Key RM, Johnson KM, Millero FJ, Poisson A, Sarmiento JL, et al. (1999) Anthropogenic CO_2 inventory in the Indian Ocean. Global Biogeochem Cy 13:179–198

Sabine CL, Wanninkhof R, Key RM, Goyet C, Millero FJ (2000) Seasonal CO_2 fluxes in the tropical and subtropical Indian Ocean. Mar Chem 72:33–53

Sarmiento JL, Le Quere C (1996) Oceanic carbon dioxide uptake in a model of century-scale global warming. Science 274: 1346–1350

Sarmiento JL, Sundquist ET (1992) Revised budget for the oceanic uptake of anthropogenic carbon-dioxide. Nature 356:589–593

Sarmiento JL, Toggweiler JR (1984) A new model for the role of the oceans in determining atmospheric pCO_2. Nature 308:621–624

Sarmiento JL, Toggweiler JR, Najjar R (1988) Ocean carbon-cycle dynamics and atmospheric pCO_2. Philos Tr R Soc S-A 325:3–21

Sarmiento JL, Hughes TMC, Stouffer RJ, Manabe S (1998) Simulated response of the ocean carbon cycle to anthropogenic climate warming. Nature 393:245–249,

Sarmiento JL, Monfray P, Maier-Reimer E, Aumont O, Murnane RJ, Orr JC (2000) Sea-air CO_2 fluxes and carbon transport: a comparison of three ocean general circulation models. Global Biogeochem Cy 14:1267–1281

Shackleton NJ (2000) The 100 000-year ice-age cycle identified and found to lag temperature, carbon dioxide, and orbital eccentricity. Science 289:1897–1902

Siegenthaler U, Joos F (1992) Use of a simple model for studying oceanic tracer distributions and the global carbon cycle. Tellus B 44:186–207

Siegenthaler U, Sarmiento JL (1993) Atmospheric carbon-dioxide and the ocean. Nature 365:119–125

Skjelvan I, Johannessen T, Miller LA (1999) Interannual variability of fCO_2 in the Greenland and Norwegian Seas. Tellus B 5:477–489

Sonnerup RE, Quay PD, McNichol AP, Bullister JL, Westby TA, Anderson HL (1999) Reconstructing the oceanic ^{13}C Suess effect. Global Biogeochem Cy 13:857–872

Sundquist ET (1985) Geological perspectives on carbon dioxide and the carbon cycle. In: Sundquist ET, Broecker WS (eds) The carbon cycle and atmospheric CO_2: natural variations archean to present. American Geophysical Union, Washington D.C., pp 5–69

Takahashi T, Olafsson J, Goddard JG, Chipman DW, Sutherland SC (1993) Seasonal-variation of CO_2 and nutrients in the high-latitude surface oceans – a comparative study. Global Biogeochem Cy 7:843–878

Takahashi T, Feely RA, Weiss RF, Wanninkhof RH, Chipman DW, Sutherland SC, et al. (1997) Global air-sea flux of CO_2: an estimate based on measurements of sea-air pCO_2 difference. P Natl Acad Sci Usa 94:8292–8299

Takahashi T, Wanninkhof RH, Feely RA, Weiss RF, Chipman DW, Bates N, et al. (1999) Net sea-air CO_2 flux over the global oceans: an improved estimate based on the sea-air pCO_2 difference. In: Nojiri Y (ed) Proceedings of the 2nd Internations Symposium on CO_2 in the Oceans, Tsukuba, January 1999. National Institute for Environmental studies, Tsukuba, Japan, pp 9–15

Takahashi T, Sutherland SC, Sweeney C, Poisson A, Metzl N, Tilbrook B, Bates N, Wanninkhof R, Feely RA, Sabine C, Olafsson J, Nojiri Y (2002) Global sea-air CO_2 flux based on climatological surface ocean pCO_2, and seasonal biological and temperature effects. Deep-Sea Res Pt II 49:1601–1622

Taylor JA, Orr JC (2000) The natural latitudinal distribution of atmospheric CO_2. Global Planet Change 26:375–386

Torres R, Turner DR, Silva N, Rutllant J (1999) High short-term variability of CO_2 fluxes during an upwelling event off the Chilean coast at 30° S. Deep-Sea Res Pt I 46:1161–1179

Wanninkhof R (1992) Relationship between wind speed and gas exchange over the ocean. J Geophys Res 97:7373–7382

Wanninkhof R, McGillis WR (1999) A cubic relationship between air-sea CO_2 exchange and wind speed. Geophys Res Lett 26:1889–1892

Wanninkhof R, Asher W, Weppernig R, Chen H, Schlosser P, Langdon C, et al. (1993) Gas transfer experiment on Georges Bank using 2 volatile deliberate tracers. J Geophys Res 98:20237–20248

Watson AJ, Liss PS (1998) Marine biological controls on climate via the carbon and sulphur geochemical cycles. Philos T Roy Soc B 353:41–51

Watson AJ, Upstill-Goddard RC, Liss PS (1991) Air sea gas-exchange in rough and stormy seas measured by a dual-tracer technique. Nature 349:145–147

Weiss RF (1974) Carbon dioxide in water and sea water; the solubility of a non-ideal gas. Mar Chem 2:203–215

Chapter 6

Water Column Biogeochemistry below the Euphotic Zone

Paul Tréguer · Louis Legendre · Richard T. Rivkin · Olivier Ragueneau · Nicolas Dittert

6.1 Introduction

The main focus of the International JGOFS research inititiatives was on the cycling of carbon and of associated elements within the surface layer, and their downward export from the upper ocean. Relatively few coordinated measurements and experiments were made below the photic zone so our understanding and modeling of the biogeochemistry of the ocean's interior is still in its infancy. However from the numerous data acquired in the 1990s during JGOFS and JGOFS-like process studies it is possible to extract sufficient information to make preliminary statements about the biogeochemistry of the water column below the euphotic zone. An important preliminary result of these studies is that we now are beginning to realize that the biogeochemistry of the surface ocean, of the ocean's interior, and of the surface sediments appears to be more coupled than was thought fifteen years ago. Moreover, the rate of sedimentation of particulate biogenic carbon into the ocean's interior can be very fast; particles can travel from the surface waters to the abysses in only a few days or weeks, and once the organic matter that is produced in the photic zone (Falkowski et al. 2003, this book) reaches the ocean's interior it is subjected to intense processes of biodegradation and recycling that release nutrients and dissolved CO_2 into subsurface waters. The subsequent upward transport of nutrients and dissolved CO_2 that have been produced and stored in the ocean's interior occurs at a slow rate; the dissolved CO_2 stored below the main ocean thermocline (the 'ventilation depth') will not return to the surface waters and to the atmosphere for centuries to millenia (Fig. 6.1). Whatever the exact time scale, it is clear that the biogeochemistry of the water column is largely driven by the biological pumping of CO_2 (and of other associated essential nutrients), a major process for exporting matter and energy from the photic zone towards the deep layers, and ultimately towards the sediments (see Lochte et al, this book).

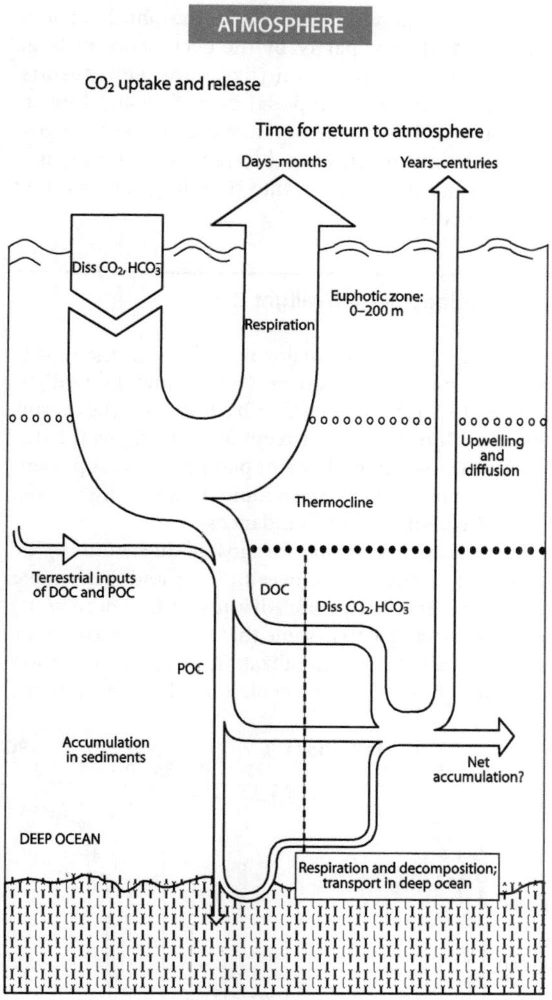

Fig. 6.1. A 1-D vision of the world ocean: the classical JGOFS scheme of the biological pump of CO_2 in the ocean with time scaling in the different layers of the ocean and in the sediment. This scheme first shows that the focus of JGOFS was really on the surface layer, which makes sense because about 90% of the cycle of carbon occurs in the surface layer. It also shows the biological pump is responsible for the transfer of carbon (and of associated elements) to the ocean's interior and ultimately to the sediments

6.2 The Twilight Zone: Biology, Biogeochemical Processes and Fluxes

The 'twilight' zone, immediately below the photic zone between 100 and 1 000 m, is a key layer for remineralization/respiration and biotic processes resulting in an intensive oxygen consumption and in a dramatic decrease of the vertical distributions of the concentrations in dissolved and particulate biogenic matter (Fig. 6.2). As demonstrated by the bifurcation model (Fig. 6.3, Legendre and Le Fèvre 1989; Le Fèvre et al. 1998), the transfer of biogenic matter out of the photic zone is channelled, at least partly, by the occurrence of large microphagous zooplankton (like salps, euphausiids, large copepods and pteropods) in surface and subsurface waters. These large organisms are able to ingest microbial-sized particles and to repackage small, usually non-sinking particles, into both large and rapidly sinking faeces.

6.2.1 Biology of the Twilight Zone

Living organisms have major roles in both the downward transport of carbon into and below the twilight zone, and its remineralization back to CO_2 (i.e., respiration) within that zone. Except for autotrophs (phytoplankton), essentially all major plankton groups present in the euphotic zone are also found in the twilight zone, albeit often with lower abundances.

Bacteria are responsible for most of the heterotrophic respiration in the surface ocean (Sherr and Sherr 1996; del Giorgio and Cole 1998; Rivkin and Legendre 2001) and it is reasonable to assume that they also account for essentially all the remineralization of biogenic carbon in the twilight zone (Nagata et al. 2000; Harris et al. 2001). Bacterial abundances tend to decrease from ca 5 to 10×10^8 cells l^{-1} in the upper ocean to ca. 5 to 10×10^7 at 1 000 m (Ducklow 1993; Turley and Mackie 1994; Patching and Eardly 1997; Nagata et al. 2000). Areal abundances and biomass however can be 2- to 4-fold greater in the 100 to 1 000 m depth interval than the upper 100 m. The few studies on the production of twilight-zone bacteria show that the rates tend to be ca. 1 to 10 ng C l^{-1} d^{-1}, or about 10^2- to 10^3-fold lower than in the upper ocean (Hoppe et al. 1993; Biddanda and Benner 1997; Patching and Eardly 1997; Hoppe and Ullrich 1999; Dixon and Turley 2000; Turley and Stutt 2000; Nagata et al. 2000 and references cited therein). Areal production is typically 2- to 10-fold lower in the 100 to 1 000 m depth interval than in the upper 100 m. Karner et al. (2001) reported that two groups of archaea (pelagic euryarchaeota and crenarchaeota) were very abundant below the euphotic zone at the Hawaii Ocean Time-Series station. The fraction of pelagic crenarchaeota increased with depth, reaching ca. 40% of the total picoplankton. The high proportion of cells containing significant amounts of rRNA suggests that most pelagic deep-sea microorganisms are metabolically active. This strongly supports the premise that the heterotrophic microbial community in the twilight zone is active, with high rates of autochthonous production that is fuelled by the input of surface-derived dissolved and particulate organic material.

Compared to bacteria, relatively little is known about microzooplankton in the twilight zone (Turley et al. 1988; Turley and Carstens 1991; Patterson et al. 1993). Studies have described the distributions of testate foraminiferia or radiolaria (Swanberg et al. 1986) and tintinnids (Krsinic 1988), and a recent study reported that twilight-zone heterotrophic nanoflagellates actively ingest bacterial-size prey (Cho et al. 2000). The exponential decrease in bacterial biomass with depth (e.g., Nagata et al. 2000, 2001) suggest that mortality is due to a combination of

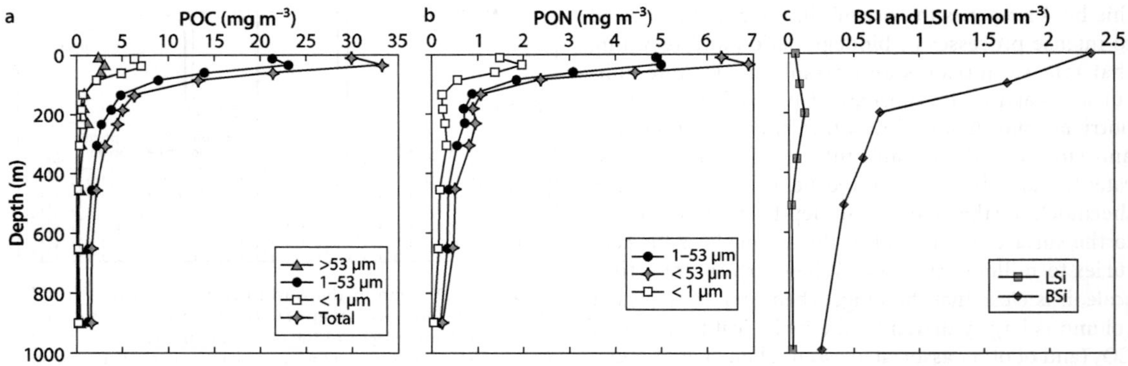

Fig. 6.2. Typical vertical profiles in the 0–1 000 m layer of the ocean. **a** Particulate organic carbon (POC) and **b** particulate organic nitrogen (PON) with size-fractionation, in the 0–1 000 m layer in the North Pacific, station PAPA (reprinted from Deep-Sea Research II 46, Boyd et al. (1999) Transformations of biogenic particulates from the pelagic to deep ocean realm. pp 2761–2792, © 1999, with permission from Elsevier Science). **c** Biogenic silica (BSi) and lithogenic silica (LSi) in the Southern Ocean (reprinted from Deep-Sea Research I 37, Tréguer et al. (1990) The distribution of biogenic and lithogenic silica and the composition of particulate organic matter in the Scotia Sea and the Drake Passage during autumn 1987. pp 833–851, © 1990, with permission from Elsevier Science)

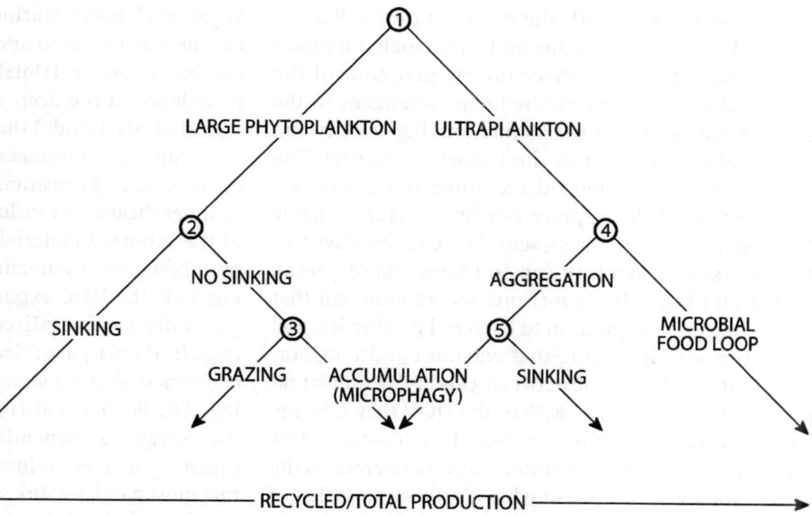

Fig. 6.3. The 'bifurcation model' of Legendre et Le Fèvre (1989) for the export production in the ocean. Although a pre-JGOFS concept this model has proven to be useful in JGOFS process studies, to correctly understand the major pathways of the export of biogenic matter under physical and biological forcing (reproduced with permission from W. H. Berger et al. (1989) (ed.) Productivity in the ocean: present and past. © John Wiley & Sons)

active ingestion of bacteria by heterotrophic protists (Cho et al. 2000) and viral lysis (Hara et al. 1996). Twilight-zone zooplankton and fish have been well studied relative to microheterotrophic components of the food web (Gjosaeter and Kawaguchi 1980; Barnett 1975; Gasser et al. 1998; Tseitlin and Rudjakov 1999; Silguero and Robinson 2000; Yamaguchi and Ikeda 2000; Luo et al. 2000).

Several groups of metazoans, mostly crustacean zooplankton but also some fish and other taxa, migrate upwards, generally to feed in the richer surface waters (Longhurst and Harrison 1988; Hay et al. 1997; Al-Mutari and Landry 2001; Hidaka et al. 2001). There are two general types of vertical migrations, i.e., nycthemeral (or daily) migrations, which occur on a circadian or 24-h cycle, and onthogenic (or seasonal) migrations, that take place over the annual cycle. The same organisms may undergo the two types of migrations during different phases of their life cycle. The typical pattern of nycthemeral migrations is upward movement of the population around sunset and return to depth around sunrise, but the reverse pattern also exists. Depending on the taxa, their size and life stage, nycthemeral migrations vertically range over a few tens to hundreds m, and the upper migration depth ranges from surface to hundreds of meters. During the nycthemeral migrations, the organisms actively transport carbon downwards, mostly from the euphotic into the twilight zones, where they remineralise organic carbon, release (respire) CO_2 and excrete DOC. This downward transport of organic and inorganic carbon may be a significant contribution to carbon export to the twilight zone in some areas (Longhurst and Harrison 1988; Longhurst et al. 1990; Hay et al. 1997; Steinberg 2000; Al-Mutari and Landry 2001; Hidaka et al. 2001) and may account for a large fraction of the bacterial demand for carbon substrates (Steinberg 2000; Nagata et al. 2000; Hidaka et al. 2001). Seasonal migrations generally include a downward movement of the population at the end of the phytoplankton production season (autumn), followed by overwintering at depth (down to 1 000 m), where the eggs of some species are released, followed by a slow movement toward the surface of juvenile stages or gravid females near the beginning of the productive season (end of winter or spring). The downward component of the migration transports carbon below the twilight zone, where there is both respiration by the organisms and often mass mortality followed by the oxidation of the dead body masses. Morales (1999) estimates that this process transports into the deep waters of the North Atlantic Ocean as much carbon as the sinking of organic particles.

6.2.2 Nature of the Exported Material and Processes

For convenience, the size continuum of organic carbon in oceans is divided by filtration into DOC and POC, which are both exported out of the photic zone. The distinction between dissolved and particulate is often functionally defined, based upon filter type and porosity. Even within the functionally 'dissolved' component, a large proportion (i.e., 30 to 50%) of the dissolved organic matter may be colloidal (Kepkay 2000). Using new (but now non-controversial) techniques (Hedges and Lee 1993; Cauwet 1994; Williams 2000) the role of dissolved organic matter in the carbon cycle of the world ocean is in the process of being reassessed.

The concentration of dissolved organic carbon decreases from the surface values of ca 60 to 80 μM (depending on the region; Williams 2000) to a relatively constant 38 to 40 μM below 1 000 m (Hansell and Carlson 1998; Williams 2000). The shape of the profile and the turnover rate of organic carbon in the twilight zone depend upon the nature of the exported material, and its interactions with the communities of heterotrophic bacteria, protozoans and (metazoans) animals (see above). The main mechanisms that control the remineralization of carbon by living organisms in the twilight zone are the size structure, sinking velocity and elemental composition of the exported material. The DOC is often characterised according to its turnover time (inverse of their uptake rate by bacteria) as labile (a few hours to a few days), semi-labile (a few days to a few weeks) or refractory (up to thousands of years). Rivkin and Legendre (2001) provide an equation that relates bacterial respiration to bacterial production and water temperature. Using that equation and temperatures in the twilight zone between 3 and 15 °C shows that bacteria respire ca. 65 to 85% of the DOC they take up, and incorporate ca. 15 to 35% into their biomass. Over long-term steady state, the biomass is transferred to the pelagic food web, the fate of which is the same as that of other forms of particulate matter (see below). The organic particles (POC) may be solubilised by diffusible exoenzymes liberated by their attached bacteria (Alldredge and Youngbluth 1985; Hoppe et al. 1993), or they may alternatively be fragmented into smaller particles by various processes and/or consumed by metazoans or protozoans. Using an empirical equation given by Rivkin and Legendre (2001) and the same range of temperatures as above shows that marine protozooplankton (phagotrophic flagellates and ciliates) in the twilight zone respire ca. 40 to 55% of the carbon they consume, so that ca. 45 to 60% are eventually transferred to larger pelagic organisms or exported below the twilight zone. Hence, respiration in the twilight zone is an inverse function of the size of the material exported from the euphotic zone.

Sediment traps are classically used for collecting the sinking particulate organic carbon including large size material named 'marine snow' by Suzuki and Kato (1953). This sticky and fibrous material, >500 µm in length, is able to scavenge particles that easily adhere to it, forming aggregates. Polysaccharide-rich transparent exopolymeric (TEP) material acts as a 'biological glue' and greatly enhances the formation of these aggregates (Alldredge and Gotschalk 1988; Kiorboe et al. 1998; Mari 1999; Mari et al. 2001). Colonization of marine snow aggregates by a wide diversity of marine herterotrophs, ranging from bacteria to zooplankton, has been reported (Alldredge and Youngbluth 1985; Smith et al. 1992; Silver et al. 1998; Azam 1998; Kiorboe 2000). Microbial communities within and upon the marine snow can undergo complex successive changes on time scales of hours to days, with significant alteration of the chemical and biological properties of the particles (Alldredge and Silver 1988; Steinberg et al. 1997; Silver et al. 1998). Aggregates may also break apart, spilling their organic content into the water, so that modeling the fate of marine snow in the water column becomes a real challenge (Jackson 1995; Logan et al. 1995). Marine snow is distributed throughout the water column and has been recently observed at depths of 1 000 m (Gorsky et al. 2000). Nonetheless its abundance in the deep sea is invariably lower (several orders of magnitude) than in the surface waters.

Because the remineralization of organic matter increases with its residence time in the twilight zone, remineralization is influenced by the sinking velocity of the exported material. DOC is exported downwards by mixing and pycnocline ventilation. Because it does not sink, the DOC exported from the euphotic zone is generally remineralised in the upper twilight zone (e.g., in the Sargasso Sea, the exported DOC is mostly consumed above 400 m; Carlson et al. 1994). Concerning POC, Fortier et al. (1994) provide a figure that shows the overall size dependency of the sinking velocity of organic particles (volumes from 10^1 to $>10^{11}$ µm^3), and that most particles sink slowly (about 100 m d^{-1}), including the faecal pellets of most crustacean zooplankton. Because of this, only a few types of exported particles sink fast enough to escape remineralization within the twilight zone (e.g., phytoplankton aggregates, faecal pellets of Antarctic krill, salps and houses of twilight-zone appendicularians; Fortier et al. 1994).

So, the elemental composition of organisms (expressed as the ratio of carbon to various elements, e.g., N, P, Si) differs for different food web components. For example, when bacteria, protists and zooplankton use resources with elemental ratios higher than their own (e.g., $C:N_{Bacteria} < C:N_{DOM}$), they remineralise and release into the environment the excess C as inorganic (CO_2) and/or organic compounds. Because the C:N of dissolved and particulate organic matter increases with depth (Loh and Bauer 2000; Kaehler and Koeve 2001), the proportion of the ingested or assimilated organic material that is respired also increases with depth.

The structure and activity of the food webs in both the euphotic and twilight zones will influence the composition, and the rates and patterns of degradation and remineralization of particulate and dissolved biogenic carbon. The combined effect of the above described processes of organic matter transformation and partitioning leads to the remineralization, within the twilight zone, of most of the organic carbon exported from the euphotic zone. It can be estimated that almost none the exported DOC and only ca. 10% of the POC exported from the euphotic zone reach the bottom of the twilight zone at about 1 000 m.

6.2.3 Microbial Production of Nitrous Oxide

As soon as sub-oxic conditions prevail in subsurface and deep waters, active microbial production of nitrous oxide (N_2O), a green-house gas, occurs either by nitrification or by denitrification (Naqvi et al. 1998). This is especially the case for ecosystems with high export fluxes of organic matter (e.g., the Somali upwelling and the Gulf of Aden, De Wilde and Helder 1997). Very intense production of N_2O occurred in the N_2O peak layers that can be supersaturated by as much as 600–800% (Lal et al. 1996). Dissolved N_2O is then transferred to the surface layer, and ultimately out-gassed into the atmosphere. N_2O is a long-lived atmospheric trace gas, about 20 times more effective in radiative forcing than CO_2 on a per mole basis, and significantly contributes to global warming (Houghton et al. 1996). During the JGOFS Netherlands Indian Ocean Program (NIOP) within the upwelling areas of the NW Indian Ocean, De Wilde and Helder (1997) documented strong N_2O emissions into the atmosphere (they were 3 orders of magnitude above the global mean oceanic N_2O flux). According to Lal and Patra (1998) the total annual emission of N_2O from the Arabian Sea accounts for 13–17% of the net global oceanic source, estimated at about 4.4 Tg N_2O yr^{-1} by Nevison et al. (1995).

6.3 The Fluxes of Biogenic Matter vs. Depth

6.3.1 The Export Flux out of the Euphotic Zone

Siegenthaler and Sarmiento (1993) estimated that the export flux of dissolved organic carbon out of the photic zone might represent as much as 60% of the total export flux, but the exact contribution of dissolved organic carbon to the the total export flux of carbon towards the ocean's interior is still a matter of discussion (Williams 1995; Emerson et al. 1997).

The export flux of particulate organic carbon can be estimated using proxies like ^{234}Th (Murray et al. 1996; Buesseler 1998) and barite (Dymond and Collier 1996; Dehairs et al. 1997). It can also be simulated by coupled physical-biogeochemical models (Sarmiento et al. 1998; Pondaven et al. 1999). Buesseler (1998) showed that the export flux of organic carbon represents more than 50% of the annual primary productivity in high latitude systems, differing from the oligotrophic gyres of the Atlantic or of the Pacific oceans, where it is usually less than 10% (Fig. 6.4). Vertical fluxes of particulate matter are not necessarily unidirectional: zooplankters and micronektoners are responsible for active upward or downward migrations that can bypass sediment traps (Angel 1989; Zhang and Dam 1997). Most models that simulate the vertical distributions and the fluxes of biogenic particulate matter in the water column rely on a curve fit of particle flux as an exponential function of depth, avoiding the difficulty of simulating the internal dynamic of the deep ecosystem. New and promising modeling approaches are now emerging (Jackson and Burd 2002) that consider the large diversity of the organic matter that populates the 'twilight zone', taking into account falling particles, particle feeding animals, predatory animals, vertical migrating animals that feed near the surface, and bacteria. Jackson and Burd's model simulates large oscillations in the deep vertical particle flux, not synchronized to the changes in the patterns of surface productivity, but resulting from the autochtonous activity of the deep ocean biota.

As described above, the vertical flux of particulate biogenic matter out of the photic zone is largely size-dependent and highly variable on seasonal and annual time scales. Buesseler (1998) demonstrated all the high-export events appear to be related to the composition of the phytoplankton assemblages and more specifically to the occurrence of diatoms blooms and large cells in the photic zone (Fig. 6.4; Boyd and Newton 1995, 1999; Bory et al. 2001). Recent works referring to specific environmental conditions (e.g., Kemp et al. 2000) point out the importance of giant diatoms (2–5 mm in diameter or length) to export fluxes. These algae are part of the shade flora (Smetacek 2000) that grow slowly in the less well lit waters at the base of the euphotic zone or in frontal systems

The export production of organic carbon out of the photic zone has to be balanced by an equivalent flux of limiting nutrients (i.e., new production) into the photic zone over the long-term steady state (i.e., annual or longer time periods). This paradigm, proposed by Eppley and Peterson (1979) over two decades ago, is a well known and important concept for biogeochemists. It is important to recall that Eppley and Peterson's original concept was based on the nitrogen cycle, and that the polar oceans and, specifically the Southern Ocean, "where growing seasons are short while ambient nutrients are high", were excluded from their conceptual approach. Since the 1980s this concept has been system-

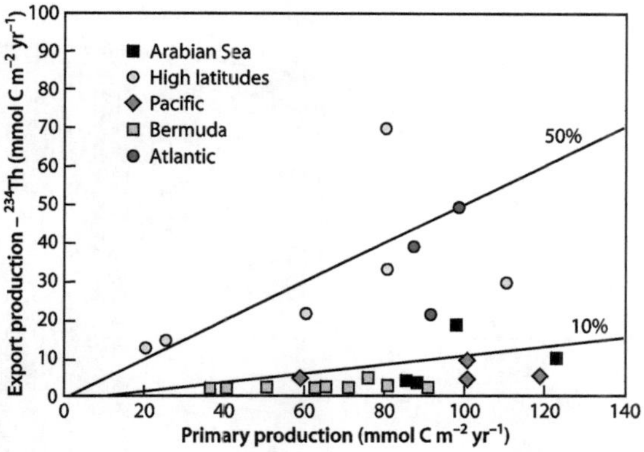

Fig. 6.4. The relative importance of export flux of particulate organic carbon, POC (estimated by the ^{234}Th method) vs. primary productivity, PP. Much of the ocean is characterized by low export of POC relative to PP, except for the high latitudes ecosystems. Diatoms and large cells are responsible for large export events out of the euphotic layer (modified by permission of American Geophysical Union from Buessler KO (1998) The decoupling of production and particulate export in the surface ocean. Global Biogeochem Cy 12:297–310, © 1998 American Geophysical Union)

atically but simplistically extrapolated by a number of authors to the carbon cycle. Its validity has been tested during the numerous JGOFS process studies. As shown by Garside and Garside (1993) for the North Atlantic Bloom Experiment, although the concept is valid at appropriate time scales (a year), the balance between export and new production is not matched at short time scales (days to months). It is also clear that extrapolation from nitrogen to carbon is not straitghtforward because the cycles of carbon and of nitrogen in the euphotic zone (as well as those of carbon and silicon) can be decoupled (see Falkowski et al., this book). This has profound consequences for the assumption about elemental ratios of the exported biogenic material, which can vary from the Redfield ratios (Sambrotto et al. 1993; Maier-Raimer 1996; Karl et al. 2001; Kaehler and Koeve 2001; Koertzinger et al. 2001). Indeed, the structure of the food web can influence both the quantity (Boyd and Newton 1995; Rivkin et al. 1996) and quality (i.e., C:N ratios) of the exported material (Walsh 1996; Daly et al. 1999; Agusti et al. 2001), and thus predicting that carbon export from the cycling or flux of other elements may lead to large systematic errors. This also has important major consequences for paleoproductivity reconstructions (Ragueneau et al. 2000).

Based on monthly mean total production maps (estimated from CZCS chlorophyll and *f*-ratios), Behrenfeld and Falkowski (1997) and Falkowski and Raven (1997) estimated that the total export flux was 16 Gt C yr^{-1} (1300 Tmol C yr^{-1}). A revised estimate of about 11 Gt C yr^{-1} (916 Tmol C yr^{-1}) is now proposed by Falkowski et al. (2003, this book) using recent SeaWIFs data. Thus, export production might represent 20–40% of the recent estimates of the global marine primary production (Antoine et al. 1997; Falkowski et al. 1998; Laws et al. 2000). Based on inverse modeling of the distributions of physical and chemical parameters for the world ocean, including a large amount of available historical nutrient data (>20 000 stations, *http://www.awi-bremerhaven.de/GEO/Flux/model.html*), Schlitzer (2002) estimates the global marine export flux of biogenic carbon to be 10 Gt C yr^{-1} (or 833 Tmol C yr^{-1}), a value that is very similar to Falkowski's most recent estimate. However, as regards the regional distribution of this export production, large discrepancies exist between the two approaches. Compared to Schlitzer's method the satellite-derived estimates give a much lower export flux in the Southern Ocean and a higher estimate in the oligotrophic open ocean areas (Fig. 6.5). The discrepancies between the two approaches remain unsolved to date. The export of $CaCO_3$ in Schlitzer's model roughly follows that of organic carbon and amounts to about 16% of the organic carbon flux. Schlitzer's estimate for the export flux of opal is 162.5 Tmol yr^{-1}, i.e., 58–81% of the global marine production of biogenic silica according to Nelson et al. (1995) and Tréguer et al. (1995).

6.3.2 The Export Flux towards the Ocean's Interior (>1 000 m)

Although they are still criticised (review in Gardner 1995; also see Yu et al. 2001), sediment traps have proven to be very useful tools to study the spatial and temporal variability of sinking biogenic matter in the deep ocean. Sediment traps have been deployed on moorings many times in the various provinces of the world ocean (Honjo and Manganini 1993; Honjo 1997; also see Ragueneau et al. 2000) and the regional distribution of the seasonal and annual fluxes of organic carbon mat-

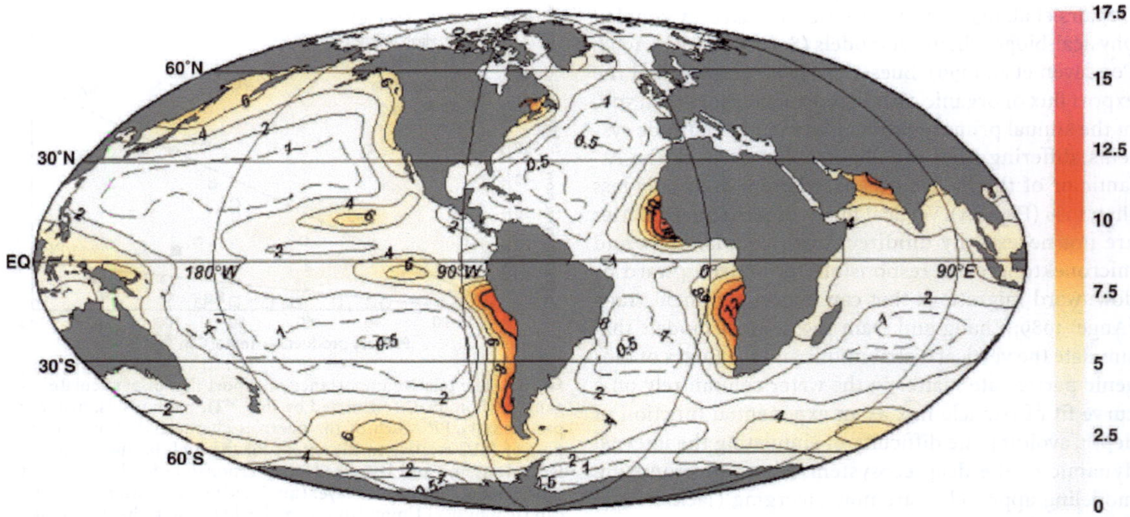

Fig. 6.5. Export production of particulate organic matter (in mol C $m^{-2} yr^{-1}$) out of the surface layer for the world ocean, calculated by inverse modeling of the nutrient fields in the ocean's interior (courtesy of R. Schlitzer)

ter below 1 000–2 000 m (i.e., below the layer of fast remineralization and dissolution rates, see above) is now relatively well documented (Lampitt and Antia 1997). The polar seas are however still poorly sampled (Pondaven et al. 2000; Honjo et al. 2000). The proportion of the primary production that reaches the deep sea does not vary much with latitude (Jahnke 1996). Globally, the open ocean flux of organic carbon at 2 000 m, normalised using the empirical relationship due to Martin et al. (1987), is 0.34 Gt C yr^{-1} that is <1% of the total net primary production (Lampitt and Antia 1997). Lampitt and Antia showed the organic carbon flux at 2 000 m (outside the Polar Domains) ranges between 0.38 and 4.2 g C m^{-2} yr^{-1} (Fig. 6.6); the monsoonal environments show the highest (1.7%) export ratio (calculated as the ratio of export flux of carbon at 2 000 m: the primary productivity in the surface layer). These authors found the lowest percentages (about 0.1%) in the Southern Ocean. Recently however (Pondaven et al. 2000; Ragueneau et al. 2000) values as high as 0.6% have been reported for the Southern Ocean, indicating a smaller range than previously thought. The global mean rain rate of opal at 3 000 m is estimated at 0.08 mol Si m^{-2} yr^{-1} (Tréguer et al. 1995). This rain rate ranges between 0.02–0.09 mol Si m^{-2} yr^{-1} and 0.05–0.51 mol Si m^{-2} yr^{-1}, respectively for the North Atlantic Ocean on the one hand and for the two major High Nutrient Low Chlorophyll (HNLC) systems, the Southern Ocean and the Equatorial Pacific, on the other hand. Honjo (1997) showed that the ratio of silicon to calcium in biogenic particles collected by deep ocean sediment traps can be used as a proxy for the rate of removal of carbon-dioxide carbon from the upper ocean to the deep-ocean 'sink'. Based on this ratio Honjo (1997) has defined two types of biogeochemical oceans: the 'silicic acid ocean' (mostly the HNLC areas) corresponds to about 40% of the world ocean with nutrient rich surface waters and opal rich particle rains and sediments. The rest of the oceanic domain is the 'carbonate ocean' (mostly the central gyres) with nutrient poor surface waters and particle rains and sediments rich in calcium carbonate.

6.4 The Variable Composition of the World Ocean Waters along the Conveyor Belt

The JGOFS strategy was largely inspired by a 1-D (vertical) vision of the ocean biogeochemistry (Fig. 6.1). This vision has proven to be very fruitful in understanding the cycling of biogenic matter in the ocean; for a given site of the world ocean the export ratio of the flux of biogenic silica to that of organic carbon exported from the surface layer (Ragueneau et al. 2000) increases with depth, demonstrating preferential remineralization of organic carbon (the same is true for PON) in the twilight zone. Nonetheless large variations in the Si/C molar ratio (Fig. 6.7) are now reported at regional scales; the ratio ranges from 0.1 in the North Atlantic, to 0.3–0.4 in the Indian Ocean, to 0.7 in the Equatorial Pacific, reaching values >1 in the North Pacific. In other words, this Si/C ratio increases along the track of the conveyor belt. So, one important JGOFS output is that the world ocean biogeochemistry experiences drastic differences at regional scales. According to Broecker and Peng (1982), biogenic silica recycles more slowly than organic carbon (or organic nitrogen) causing silicic acid accumulation in subsurface and deep waters along the track of the conveyor belt. More specifically the chemical composition of the subsurface layers of the ocean also varies, with an increase in the silicic acid/nitrate ratio from the North

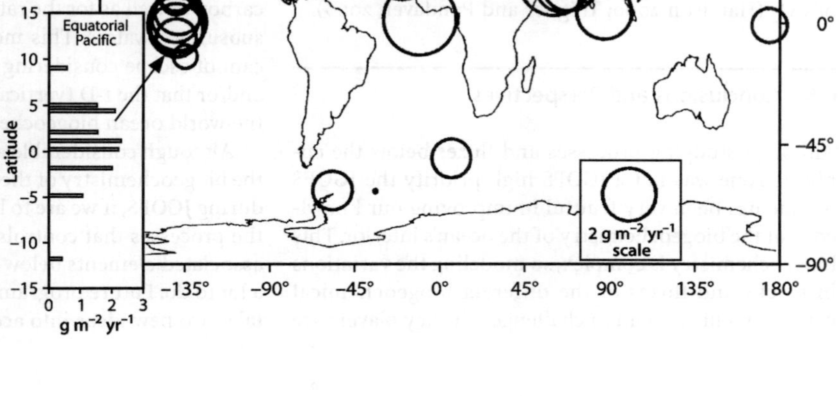

Fig. 6.6. Export flux of organic carbon in the deep ocean as measured from sediment traps. The flux is proportional to the area of each circle. Organic carbon flux normalized to a depth of 2 000 m exhibits a range of an order of magnitude in areas outside the Polar Domains (0.38 to 4.2 g C m^{-2} yr^{-1}). Globally the average oceanic flux at 2 000 m deep is about 1% of the total net primary production (reprinted from Deep-Sea Res II 8, Lampitt RS, Antia AN (1997) Particle flux in deep seas: regional characteristics and temporal variability. pp 1377–1403, © 1997, with permission from Elsevier Science)

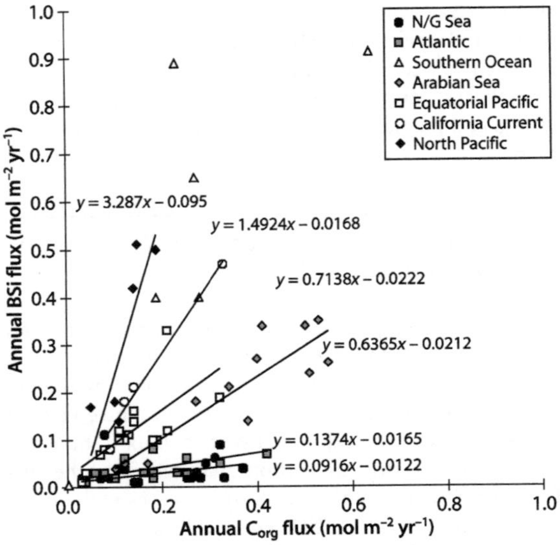

Fig. 6.7. Export fluxes (in mol $m^{-2} yr^{-1}$) of biogenic silica (BSi) and of particulate organic carbon (all depths below the euphotic zone) for various regions of the world ocean. Note that the Si/C ratio increases going from the Atlantic Ocean (the 'carbonate ocean' to the North Pacific and the Southern Ocean (the 'silicate ocean'), i.e., following the track of the Conveyor Belt (reproduced with permission from Ragueneau et al. 2000)

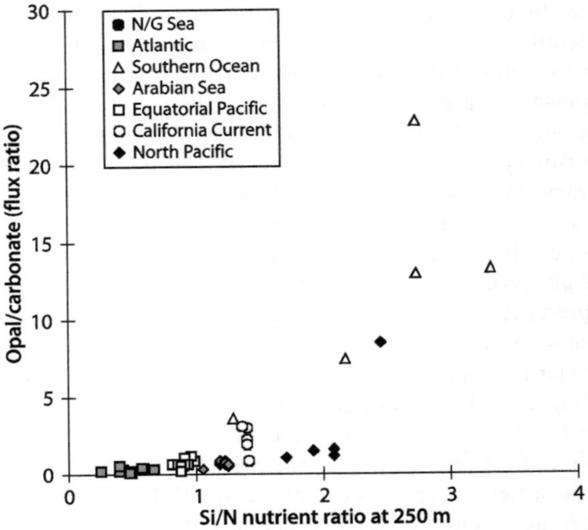

Fig. 6.8. Variations of the silicic acid/nitrate ratio in subsurface waters (250 m depth) at stations where the export fluxes have been measured (see Fig. 6.7). This Si/N ratio increases as the silicic acid concentrations in subsurface water increases with the age of the water masses (reproduced with permission from Ragueneau et al. 2000)

Atlantic to the Pacific Ocean and to the Southern Ocean. On average the nutrient source waters of the coastal upwelling of the Pacific Ocean have higher Si/N ratios than those of the Atlantic (Fig. 6.8). Inter-basin silicon fractionation has been described already by Codispoti (1983) through comparisons of nutrients in the two major oceans. Taking into account the JGOFS data for particulate matter we are now able to complete the explanation. These differences in the biogeochemistry of the ocean's interior have important consequences for the surface ocean ecosystems. Indeed a higher silicic acid/nitrate ratio for the Pacific or for the Southern Ocean (compared to the Atlantic) could favour the predominance of diatoms over that of non-siliceous species (Ragueneau et al. 2000). It is important to note that these conditions prevailing in the modern ocean can vary dramatically, depending on the external sources of silicic acid in surface waters, with important consequences for the biological pump of CO_2 (Harrison 2000; Tréguer and Pondaven 2000).

6.5 Conclusions and Perspectives

Although studying processes and fluxes below the euphotic zone was not a JGOFS high priority the JGOFS decade has been very fruitful in improving our knowledge of the biogeochemistry of the ocean's interior. This biogeochemistry is complex, so modeling the variations in stocks and fluxes of the different biogeochemical compartments remains a challenge. The key players are the dissolved and colloidal organic matter, the falling particles, the particle feeding animals, the predatory animals, the vertically migrating animals that feed near the surface, and the bacteria. It is clear that the biogeochemistry of the ocean's interior is much more linked to that of the surface layer than we thought in the 80s. The twilight zone is the key layer for remineralization/respiration processes that result in a drastic decrease of stocks and fluxes of biogenic matter vs. depth. The export flux of organic carbon out of the surface mixed layer is estimated at about 10 Gt C yr^{-1}, i.e., about 20% of the global marine primary production; at 2 000 m depth the export flux of organic carbon decreases to about 1% of the primary production. Contrary to what was expected from Eppley and Peterson (1979), the export production is usually not balanced by new production at short time scales. The cycles of carbon, nitrogen and biogenic silica can be decoupled both in the surface layer and in the ocean's interior. Following the track of the conveyor belt, large variations have been evidenced in the ratios of the export fluxes of biogenic silica to particulate organic carbon as well as for the ratios of silicic acid to nitrate in subsurface waters. This means marine biogeochemists cannot escape considering the ocean at regional scales, and/or that the 1-D (vertical) vision typical of JGOFS for the world ocean biogeochemistry is no longer valid.

Although considerable advances in the knowledge of the biogeochemistry of the water column were achieved during JGOFS, if we are to better understand and model the processes that controls the fluxes of carbon and of associated elements below the photic zone we have still a lot to do. Future programs in biogeochemistry should take two new ideas into account.

6.5.1 The Ventilation Depth and the *v*-Ratio

The dissolved CO_2 stored below the main ocean thermocline will not return to the surface waters or to the atmosphere for centuries to millenia. So the export flux of carbon which is really relevant to the JGOFS perspective is definitely not the flux which escapes the photic zone (i.e., the export flux of carbon as defined by Eppley and Peterson 1979), but the flux which survives to degradation by mineralization and recycling within the twilight zone, and which goes down beyond the ventilation depth. This depth is variable regionally and seasonally, ranging from 50 m in low latitudes to 800 m in high latitudes where deep convection occurs especially during winter (see Fig. 10.2 in Falkowski and Raven 1997). For future studies the ventilation depth of a given system is a characteristic feature that should be systematically determined. In parallel to the classical Eppley and Peterson's *f*-ratio for the euphotic zone, determining the relative export flux for the ventilation depth, so called the *v*-ratio ($v = F_A / P_D$, F_A being the annual flux of biogenic element at the ventilation depth, and P_D the time-integrated annual gross photosynthetic production in the surface layer) is of particular interest. For example in the Pacific sector of the Southern Ocean, where very deep convection occurs in the Southern Antarctic Circumpolar region during winter, the *v*-ratio for organic carbon is estimated at 0.04, which is about 10 times lower that the annual average *f*-ratio (Nelson et al. 2002).

6.5.2 The Role of Mineral Ballasts in the Export of Carbon to the Ocean Interior

The idea that the rates of C, N and P remineralization (as well as that of oxygen consumption) in the deep waters of the ocean at a given depth can be used to calculate estimates of the flux of carbon exported from the surface layer emerged in the beginings of the chemical oceanography (Riley 1954). Among others, Suess (1980), Betzer et al. (1984), and Martin et al. (1987) suggested that the remineralization of particulate organic carbon (POC) can be specified as a function of depth alone, and did not depend on the location of the study site.

But we now have numerous evidences of decoupling between surface production and deep remineralization (Armstrong and Jahnke 2001). This is because mineral ballasts, either of biogenic and/or of lithogenic origin, can play a major role in the sinking of organic carbon (Armstrong and Jahnke 2001). In other words the export of organic carbon towards the ocean interior and the sediment is ecosystem dependent. Hence diatoms are key players in high latitude regions typified with high export flux systems (Buesseler 1998) and high *f*-ratios, i.e., with low recycling of carbon in the surface layer. In other words this means that the organic matter exported from the surface layer in association with biogenic silica is very labile, so it is rapidly degraded in the upper layer of the twilight zone, giving low *v*-ratios at the ventilation depth. The contrary is true for low latitude systems typified with low *f*-ratios, where phytoplankters like coccolithophorids predominate. François et al. (2001) now give support to the idea that, in carbonate-dominated systems with low *f*-ratios, *v*-ratios are high, i.e., a higher fraction of the exported organic carbon sinks to the deep ocean. If this is true calcium carbonate might be more efficient a transporter of organic carbon to the ocean interior than opal.

References

Agusti S, Duarte CM, Vaque D, Hein M, Gasol JM, Vidal M (2001) Food web structure and elemental (C, N and P) fluxes in the eastern tropical North Atlantic. Deep-Sea Res Pt I 48:2295–2321

Alldredge AL, Gotschalk C (1988) In situ behavior of marine snow. Limnol Oceanogr 33:339–351

Alldredge AL, Silver MW (1988) Characteristics, dynamics and significance of marine snow. Prog Oceanogr 20:41–82

Alldredge AL, Youngbluth MJ (1985) The significance of macroscopic aggregates (marine snow) as sites for heterotrophic bacterial production in the mesopelagic zone of the Subtropical Atlantic. Deep-Sea Res Pt I 32:1445–1456

Al-Mutari H, Landry MR (2001) Active export of carbon and nitrogen at Station ALOHA by diel migrat zoolankton. Deep-Sea Res Pt I 48:2083–2103

Angel MV (1989) Does mesopelagic biology affect the vertical flux? In: Berger WH, Smetacek VS, Wefer G (eds) Productivity of the ocean: present and past. John Wiley & Sons, Chichester New York, pp 155–173

Antoine D, André JM, Morel A (1997) Oceanic primary production 2 – Estimation at global scale from satellite. Global Biogeochem Cy 10:57–69

Armstrong RA, Jahnke RA (2001) Decoupling surface production from remineralization and benthic deposition: the role of mineral ballasts. US JGOFS News 11:1–2

Azam F (1998) Microbial control of ocean carbon flux. The plot thickens. Science 280:694–696

Barnett MA (1975) Studies on the patterns of distribution of mesopelagic fish faunal assemblages in the central Pacific and their temporal persistence in the gyres. PhD Dissertation University of California at San Diego. 145 pp

Behrenfeld MJ, Falkowski PG (1997) A consumer's guide to phytoplankton primary productivity models. Limnol Oceanogr 42: 1479–1491

Berger WH, et al. (ed) (1989) Productivity in the ocean: present and past. John Wiley & Sons

Betzer PR, Showers WJ, Laws EA, Winn CD, DiTullio GR, Kroopnick PM (1984) Primary productivity and particles fluxes on a transect of the equator at 153° W in the Pacific Ocean. Deep-Sea Res Pt I 31:1–11

Biddanda B, Benner R (1997) Major contribution from mesopelagic plankton to heterotrophic metabolism in the upper ocean. Deep-Sea Res Pt I 44:2069–2085

Bory A, Jeandel C, Leblond N, Vangriesheim A, Khripounoff A, Beaufort L, Rabouille C, Nicolas E, Tachikawa K, Etcheber H, Buat-Ménard P (2001) Downward particle fluxes in different productivity regimes off the Mauritanian upwelling zone (EUMELI program). Deep-Sea Res Pt I 48:2251–2282

Boyd P, Newton P (1995) Evidence of the potential influence of planktonic community structure on the interannual variability of particulate organic carbon flux. Deep-Sea Res Pt I 42: 619–639

Boyd P, Newton P (1999) Does planktonic community structure determine downward particulate organic carbon flux in different oceanic provinces? Deep-Sea Res Pt I 46:63–91

Boyd PW, Sherry ND, Berges JA, Bishop JKB, Calvert SE, Charrette MA, Giovannoni SJ, Goldblatt R, Harrison PJ, Moran SB, Roy S, Soon M, Strom S, Thibault D, Vergin KL, Whitney FA, Wong CS (1999) Transformations of biogenic particulates from the pelagic to deep ocean realm. Deep-Sea Res Pt II 46:2761–2792

Broecker WS, Peng TH (1982) Tracers in the sea. Eldigio Press, Columbia University, Palisades, New York, 690 pp

Buesseler KO (1998) The decoupling of production and particulate export in the surface ocean. Global Biogeochem Cy 12:297–310

Carlson CA, Ducklow HW, Michaels AF (1994) Annual flux of dissolved organic carbon from the euphotic zone in the northwestern Sargasso Sea. Nature 371:405–408

Cauwet G (1994) HTCO method for DOC analysis in seawater: influence of catalyst on blank estimation. Mar Chem 47:55–64

Cho BC, SC Na, Choi DH (2000) Active ingestion of fluorescently labeled bacteria by mesopelagic heterotrophic nanoflagellates in the East Sea, Korea. Mar Ecol Prog Ser 206:23–32

Codispoti LA (1983) On nutrient availibility and sediments in upwellings areas. In: Suess E, Thiede J (eds) Coastal upwellings. Part A. Plenum Press, New York, pp 125–145

Daly KL, Wallace DWR, Smith WO, Skoog A, Lara R, Gosselin M, Falck E, Yager PL (1999) Non-Redfield carbon and nitrogen cycling in the Arctic: effects of ecosystem structure and dynamics. J Geophys Res 104:3185–3199

De Wilde HPJ, Helder W (1997) Nitrous oxide in the Somali Basin: the role of upwelling. Deep-Sea Res Pt II 44:1319–1340

Dehairs F, Shopova D, Ober S, Veth C, Goeyens L (1997) Particulate barium stocks and oxygen consumption in the Southern Ocean mesopelagic water column during spring and early summer. Deep-Sea Res Pt II 44:497–516

del Giorgio PA, JJ Cole (1998) Bacterial growth efficiency in natural systems. Annu Rev Ecol Syst 29:503–541

Dixon JL, Turley CM (2000) The effect of water depth on bacterial numbers, thymidine incorporation rates and C:N ratios in northeast Atlantic surficial sediments. Hydrobiology 440:217–225

Ducklow HW (1993) Bacterioplankton distribution and production in the northwestern Indian Ocean and Gulf of Oman, September 1986. Deep-Sea Res 40:753–771

Dymond J, Collier R (1996) Particulate barium fluxes and their relationships to biological productivity. Deep-Sea Res Pt II 43: 1283–1308

Emerson S, Quay P, Karl D, Winn C, Tupas L, Landry M (1997) Experimental determination of the organic carbon flux from open-ocean surface waters. Nature 389:951–954

Eppley RW, Peterson BJ (1979) Particulate organic flux and planktonic new production in the deep ocean. Nature 282:677–678

Falkowski PG, Raven JA (1997) Aquatic photosynthesis. Blackwell Science Pub, New York, 375 pp

Falkowski PG, Barber RT, Smetacek VS (1998) Biogeochemical controls and feedbacks on ocean primary production. Science 281:200–206

Falkowski PG, Laws EA, Barber RT, Murray JW (2003) Phytoplankton and their role in primary, new, and export production. In: Fasham MJR (ed) Ocean biogeochemistry, Chap. 4. Springer-Verlag, Heidelberg, (this volume)

Fortier L, Le Fèvre J, Legendre L (1994) Export of biogenic carbon to fish and to the deep ocean: the role of large planktonic microphages. J Plankton Res 16:809–839

François R, Honjo S, Krishfield R, Manganini S (2001) Factors controlling the flux of organic carbon to the bathypelagic zone of the ocean? US JGOFS SMP workshop (abstract), 16–10 July 2001, WHOI

Gardner WD (1995) Sediment trap technology and sampling in surface waters. Report on the JGOFS Symposium in Villefranche sur mer, France, in May 1995, available on the TAMU web site (*http://www.tamu.edu*)

Garside C, Garside JC (1993) The '*f*-ratio' on 20° W during the North Atlantic Bloom Experiment. Deep-Sea Res Pt II 40:75–90

Gasser B, Payet G, Sardou J, Nival P (1998) Community structure of mesopelagic copepods (>500 μm) in the Ligurian Sea (Western Mediterranean). J Marine Syst 15:511–522

Gjosaeter J, Kawaguchi K (1980) A review of the world resources of mesopelagic fish. Food and Agriculture Organization of the United Nations Series: (FAO fisheries technical paper. No. 193). Rome, Italy

Gorsky G, Picheral M, Stemmann L (2000) Use of the underwater video profiler for the study of aggregate dynamics in the northern Mediterranean. Estuar Coast Shelf S 50:121–128

Hansell DA, Carlson CA (1998) Deep ocean gradients in the concentration of dissolved organic carbon. Nature 395:263–266

Hara S, Koike I, Terauchi K, Kamiya H, Tanoue E (1996) Abundance of viruses in deep oceanic waters. Mar Ecol Prog Ser 145:29–277

Harris JRW, Stutt ED, Turley CM (2001) Carbon flux in the northwest Mediterranean estimated from microbial production. Deep-Sea Res Pt I 48:231–248

Harrison KG (2000) Role of increased marine silica input on paleo-pCO_2 levels. Paleoceanography 15:292–298

Hay GC, Harris RP, Head RN, Kennedy H (1997) A technique for the in situ assessment of vertical nitrogen flux caused by the diel vertical migration of zooplankton. Deep-Sea Res Pt I 44: 1085–1089

Hedges JI, Lee C (1993) Measurements of DOC and DON in natural waters. Proceedings of NSF/NOAA/DOE Workshop, Seattle, WA, 15–16 July 1991. Mar Chem 41:290 pp

Hidaka K, Kawaguchi K, Murakami M, Takahashi M (2001) Downward transport of organic carbon by diel migratory micronecton in the western Pacific: its quantitative and qualitative importance. Deep-Sea Res Pt I 48:1923–1939

Honjo S (1997) The rain of ocean particles and Earth's carbon cycle. Oceanus 40:4–7

Honjo S, Manganini SJ (1993) Annual biogenic particles fluxes to the interior of the North Atlantic Ocean: studied at 34° N, 21° W and 48° N, 21° W. Deep-Sea Res Pt I 40:587–607

Honjo S, François R, Manganini S, Dymond J, Collier R (2000) Particle fluxes to the interior of the Southern Ocean in the western Pacific Ocean sector along 170° W. Deep-Sea Res Pt II 47:3073–3093

Hoppe HG, Ullrich S (1999) Profiles of ectoenzymes in the Indian Ocean: phenomena of phosphatase activity in the mesopelagic zone. Aquat Microb Ecol 19:139–148

Hoppe HG, Ducklow H, Karrasch B (1993) Evidence for dependency of bacterial growth on enzymatic hydrolysis of particulate organic matter in the mesopelagic ocean. Mar Ecol Prog Ser 93:277–283

Houghton JT, Meira Filho LG, Callander BA, Harris N, Kattenberg A, Maskell K (1996) Climate change 1995. Cambridge University Press, Cambridge, 572 pp

Jackson G (1995) TEP and coagulation during a mesocosm experiment. Deep-Sea Res Pt I 42:215–222

Jackson GA, Burd AB (2002) A model for the distribution of particle flux in the mid-water column controlled by subsurface biotics interactions. Deep-Sea Res Pt II 49:193–217

Jahnke RA (1996) The global ocean flux of particulate organic carbon: areal distribution and magnitude. Global Biogeochem Cy 10:71–88

Kaehler P, Koeve W (2001) Marine dissolved organic matter: can its C:N ratio explain carbon overconsumption? Deep-Sea Res Pt I 48:49–62

Karl DM, Björkman KM, Dore JE, Fujieki L, Hebel V, Houlihan T, Letelier M, Tupas LM (2001) Ecological nitrogen-to-phosphorus stoichiometry at station ALOHA. Deep-Sea Res Pt II 48: 1529–1566

Karner MB, DeLong EF, Karl DM (2001) Archaeal dominance in the mesopelagic zone of the Pacific Ocean. Nature 409:507–510

Kemp AES, Pike J, Lange RB (2000) The 'fall dump' – a new perspective on the role of a 'shade flora' in the annual cycle of diatom production and export flux. Deep-Sea Res Pt II 47:2129–2154

Kepkay P (2000) Colloids and the ocean carbon cycle. In: Wangersky PJ (ed) The handbook for environmental chemistry. Marine Chemistry. Springer-Verlag, New York, pp 35–56

Kiorboe T (2000) Colonization of marine snow aggregates by invertebrates zooplankton: abundance, scaling, and possible role. Limnol Oceanogr 45:479–484

Kiorboe T, Tiselius P, Mitchell-Ines B, Hansen JLS, Visser A, Mari X (1998) Intensive aggregate formation with low vertical flux during an upwelling-induced diatom bloom. Limnol Oceanogr 43:104–106

Koertzinger A, Koeve W, Kaehler P, Mintrop L (2001) C: N ratios in the mixed layer during the productive season in the northeast Atlantic Ocean. Deep-Sea Res Pt I 48:661–688

Krsinic F (1988) The family *Xystonellidae* (*Ciliophora*, *Tintinnina*) in the Adriatic Sea. J Plankton Res 10:413–429

Lal S, Patra PK (1998) Variabilities in the fluxes of annual emissions of nitrous oxide from the Arabian sea. Global Biogeochem Cy 12:321–327

Lal S, Patra PK, Venkataramani S, Sarin MM (1996) Distributions of nitrous oxide and methane in the Arabian Sea. Curr Sci India 71:894–899

Lampitt RS, Antia AN (1997) Particle flux in deep seas: regional characteristics and temporal variability. Deep-Sea Res Pt II 8: 1377–1403

Laws EA, Falkowski PG, Smith WO, Ducklow H, McCarthy JM (2000) Temperature effect on export production in the open ocean. Global Biogeochem Cy 14:1231–1246

Le Fèvre J, Legendre L, Rivkin RB (1998) Fluxes of biogenic carbon in the Southern ocean: the role of large mircrophagous zooplankton. J Marine Syst 17:325–346

Legendre L, Le Fèvre J (1989) Hydrodynamical singularities as controls of recycled vs. export production in oceans. In: Berger WH, Smetacek VS, and Wefer G (eds) Productivity in the ocean: present and past. John Wiley & Sons, S. Bernard, Dahlem Konferenzen, 49–63

Lochte K, Anderson R, Francois R, Jahnke R, Shimmield G, Vetrov A (2003) Benthic processes and the burial of carbon. In: Fasham MJR (ed) Ocean biogeochemistry, Chap. 8. Springer-Verlag, Heidelberg, (this volume)

Logan BE, Passow U, Alldredge AL, Grossart P, Simon M (1995) Rapid formation and sedimentation of large aggregates is predicted from coagulation rates (half-lives) of transparent exopolymer particles. Deep-Sea Res Pt I 42:203–214

Loh AN, Bauer JE (2000) Distribution, partitioning and fluxes of dissolved and particulate organic C, N and P in the eastern North Pacific and Southern Oceans. Deep-Sea Res Pt I 47:2287–2316

Longhurst AR, Harrison WG (1988) Vertical nitrogen flux from the photic zone by diel migrant zooplankton and necton. Deep-Sea Res 35:881–889

Longhurst AR, Bedo AW, Harrison WG, Head EJH, Sameoto DD (1990) Vertical flux of respiratory carbon by oceanic diel migrant biota. Deep-Sea Res Pt I 37:685–694

Luo J, Ortner PB, Forcucci D, Cummings SR (2000) Diel vertical migration of zooplankton and mesopelagic fish in the Arabian Sea. Deep-Sea Res Pt I 47:1451–1473

Maier-Reimer E (1996) Dynamic vs. apparent Redfield ratio in the oceans: a case for 3-D-models. J Marine Syst 9:113–120

Mari X (1999) Carbon and C/N ratio of transparent exopolymeric particles (TEP) produced by bubbling exudates of diatoms. Mar Ecol Prog Ser 183:59–71

Mari X, Beauvais S, Lemée R, Luizia Pedrotti M (2001) Non-Redfield C:N ratio of transparent exopolymeric particles in the northwestern Mediterranean Sea. Limnol Oceanogr 46:1831–1836

Martin JH, Knauer GA, Karl DM, Broenkow WW (1987) VERTEX: carbon cycling in the northeast Pacific. Deep-Sea Res Pt I 34: 267–285

Morales CE (1999) Carbon and nitrogen fluxes in the oceans: the contribution by zooplankton migrants to active transport in the North Atlantic during the Joint Global Ocean Flux Study. J Plankton Res 21:1799–1808

Murray JW, Young J, Newton J, Dunne J, Chapin T, Paul B, McCarthy J (1996) Export flux of particulate organic carbon from the central equatorial Pacific determined using a combined drifting trap-^{234}Th approach. Deep-Sea Res Pt II 4–6:1095–1132

Nagata T, Fukuda H, Fukuda R, Koike I (2000) Bacterioplankton distribution and production in deep Pacific waters: large-scale geographic variations and possible coupling with sinking particle fluxes. Limnol Oceanogr 45:426–435

Nagata T, Fukuda H, Fukuda R, Koike I (2001) Basin scale geographic patterns of bacterioplankton biomass and production in the subarctic Pacific, July–September 1997. J Oceanogr 57:301–313

Naqvi SWA, Yoshinari T, Jayakumar DA, Altabet MA, Narvekar PV, Devol AH, Brandes JA, Codispoti LA (1998) Budgetary and biogeochemical implications of N_2O isotope signatures in the Arabian Sea. Nature 394:462–464

Nelson DM, Tréguer P, Brzezinski MA, Leynaert A, Quéguiner B (1995) Production and dissolution of biogenic silica in the ocean: revised global estimates, comparison with regional date and relationship to biogenic sedimentation. Global Biogeochem Cy 9:359–372

Nelson DM, Anderson RF, Barber RT, Brzezinski MA, Buesseler KO, Chase Z, Collier RW, Dickson ML, François R, Hiscock MR, Honjo S, Marra J, Martin WR, Sambrotto RN, Sayles FL, Sigmon DE (2002) Vertical budgets for organic carbon and biogenic silica in the Pacific sector of the Southern Ocean, 1996–1998. Deep-Sea Res Pt II 49:1645–1674

Nevison CD, Weiss RF, Erickson DJ III (1995) Global oceanic emissions of nitrous oxide. J Geophys Res 100:809–820

Patching J, Eardly D (1997) Bacterial biomass and activity on the deep water of the eastern Atlantic – evidence of a barophilic community. Deep-Sea Res Pt I 44:155–170

Patterson DJ, Nygaard K, Steinberg G, Turley CM (1993) Heterotrophic flagellates and other protists associated with oceanic detritus throughout the water column in the mid North Atlantic. J Mar Biol Assoc Uk 73:67–95

Pondaven P, Ruiz-Pino D, Druon JN, Fravalo C, Tréguer P (1999) Factors controlling silicon and nitrogen biogeochemical cycles in HNLC systems (the Southern Ocean and the North Pacific) compared to a mesotroph system (the North Atlantic) Deep-Sea Res Pt I 46:1923–1968

Pondaven P, Ragueneau O, Tréguer P, Hauvespre A, Dezileau L, Reyss JL (2000) Resolving the 'opal paradox' in the Southern Ocean. Nature 405:168–172

Ragueneau O, Tréguer P, Anderson RF, Brzezinski MA, DeMaster DJ, Dugdale RC, Dymond J, Fischer G, François R, Heinze C, Meir-Reimer E, Martin-Jézéquel V, Nelson DM, Quéguiner B (2000) A review of the Si cycle in the modern ocean: recent progress and missing gaps in the application of biogenic opal as a paleoproductivity proxy. Global Planet Change 26:317–365

Riley CA (1954) Oxygen, phosphate and nitrate in the Atlantic Ocean. B Bingham Oceanogr C 13–14:1–26

Rivkin RB, Legendre L (2001) Biogenic carbon cycling in the upper ocean: effects of microbial respiration. Science 291:2398–2400

Rivkin RB, Legendre L, Deibel D, Tremblay J-E, Klein B, Crocker K, Roy S, Silverberg S, Lovejoy C, Mesplé F, Romero N, Anderson MR, Matthews P, Savenkoff C, Vézina AF, Therriault J-C, Wesson J, Bérubé C, Ingram RG (1996) Vertical flux of biogenic carbon in the ocean: is there food web control? Science 272:1163–1166

Sambrotto RN, Savidge G, Robinson C, Boyd P, Takahashi T, Karl D, Langdon C, Chipman D, Marra J, Codispotti L (1993) Elevated consumption of carbon relative to nitrogen in the surface ocean. Nature 363:248–250

Sarmiento JL, Hughes TMC, Stouffer R, Manabe S (1998) Simulated response of the ocean carbon cycle to anthropogenic climate warming. Nature 393:245–249

Schlitzer R (2002) Carbon export fluxes in the Southern Ocean: results from inverse modeling and comparison with satellite based estimates. Deep-Sea Res Pt II 49:1623–1644

Sherr EB, Sherr BF (1996) Temporal offset in the oceanic production and respiration processes implied by seasonal changes in atmospheric oxygen: the role of heterotrophic microbes. Aquat Microb Ecol 11:91–100

Siegenthaler U, Sarmiento JL (1993) Atmospheric carbon dioxide and the ocean. Nature 365:119–125

Silguero JMB, Robinson BH (2000) Seasonal abundance and vertical distribution of mesopelagic calycophoran siphonophores in Monterey Bay, CA. J Plankton Res 22:1139–1153

Silver MW, Coale SL, Pilskaln CH, Steinberg DR (1998) Giant aggregates: importance as microbial centers and agents of material flux in the mesopelagic zone. Limnol Oceanogr 43:498–507

Smetacek V (2000) Oceanography: the giant diatom dump. Nature 406:574–575

Smith DC, Simon M, Alldredge AL, Azam F (1992) Intense hydrolytic enzyme activity on marine aggregates and implications for rapid particle dissolution. Nature 359:139–142

Steinberg DK (2000) Zooplankton vertical migration and the active transport of dissolved organic and inorganic carbon in the Sargasso Sea. Deep-Sea Res Pt I 47:137–158

Steinberg DK, Silver MW, Pilskaln CH (1997) Role of mesopelagic zooplankton in the community metabolism of giant larvacean house detritus in Monterey Bay, California, USA. Mar Ecol Prog Ser 147:167–179

Suess E (1980) Particulate organic carbon flux in the oceans: surface productivity and oxygen utilisation. Nature 288: 260–263

Suzuki N, Kato K (1953) Studies on suspended material. Marine snow in the sea I – Sources of marine snow. Bulletin of the Faculty of Fisheries Hokkaido University 4:132–135

Swanberg N, Bennett P, Lindsey JL, Anderson OR (1986) The biology of a coelodendrid: a mesopelagic phaeodarian radiolarian. Deep-Sea Res 33:15–25

Tréguer P, Pondaven P (2000) Global change: silica control of carbon dioxide. Nature 406:358–359

Tréguer P, Nelson DM, Gueneley S, Zeyons C, Morvan J, Buma A (1990) The distribution of biogenic and lithogenic silica and the composition of particulate organic matter in the Scotia Sea and the Drake Passage during autumn 1987. Deep-Sea Res Pt I 37:833–851

Tréguer P, Nelson DM, van Bennekom AJ, DeMaster DJ, Leynaert A, Quéguiner B (1995) The balance of silica in the world ocean: a re-estimate. Science 268:375–379

Tseitlin VB, Rudjakov JA (1999) Seasonal variations of zooplankton biomass vertical distributions in the mesopelagial of the Indian Ocean. Oceanology+ 39:573–580

Turley CM, Carstens M (1991) Pressure tolerance of oceanic flagellates: implications for remineralization of organic matter. Deep-Sea Res 38:403–413

Turley CM, Mackie PJ (1994) Biogeochemical significance of attached and free-living bacteria and the flux of particles in the NE Atlantic Ocean. Mar Ecol Prog Ser 115:191–203

Turley CM, Stutt ED (2000) Depth-related cell-specific bacterial leucine incorporation rates on particles and its biogeochemical significance in the Northwest Mediterranean. Limnol Oceanogr 45:419–425

Turley CM, Lochte K, Patterson DJ (1988) A barophilic flagellate isolated from 4 500 m in the mid-North Atlantic. Deep-Sea Res Pt I 35:1079–1092

Walsh JJ (1996) Nitrogen fixation within a tropical upwelling ecosystem: evidence for a Redfield budget of carbon/nitrogen cycling by the total phytoplankton community. J Geophys Res 101:20607–20616

Williams PJ leB (1995) Evidence for the seasonal accumulation of carbon-rich dissolved material, its scale in comparison with changes in particulate material and the consequential effect on net C/N assimilation ratios. Mar Chem 51:17–29

Williams PJ leB (2000) Heterotrophic bacteria and the dynamics of dissolved organic material. In: Kirchman DL (ed) Microbial ecology of the oceans. John Wiley & Sons, New York, pp 153–200

Yamaguchi A, Ikeda T (2000) Vertical distribution, life cycle, and developmental characteristics of the mesopelagic calanoid copepod *Gaidius variabilis* (Aetideidae) in the Oyashio region, western North Pacific Ocean. Mar Biol 137:99–109

Yu EF, François R, Bacon MP, Honjo S, Flee AP, Manganini SJ, van der Loeff MM Rutgers, Ittekkot V (2001) Trapping efficiency of bottom-tethered sediment traps estimated from the intercepted fluxes of ^{230}Th and ^{231}Pa. Deep-Sea Res Pt I 48:865–889

Zhang X, Dam HG (1997) Downward export of carbon by diel migrant mesozooplankton in the central equatorial Pacific. Deep-Sea Res Pt II 9–10:2191–2202

Chapter 7

The Impact of Climate Change and Feedback Processes on the Ocean Carbon Cycle

Philip W. Boyd · Scott C. Doney

7.1 Introduction

7.1.1 Climate and Change – Present Status

We have been aware of the concept of global climate change since the advent of modern science in the 17th Century and the emergence of disciplines such as geology. However, it is only in the last century that a putative link, termed the 'the Greenhouse Effect' (Wood 1909), has been suggested between the atmospheric concentrations of particular gases and climate. The composition of the atmosphere has been studied routinely since the late fifties/early sixties with the establishment of monitoring sites for atmospheric CO_2 (such as Mauna Loa) where the 40-year dataset clearly demonstrates the rise of atmospheric CO_2 (Keeling et al. 1995). Similar anthropogenically-mediated increases in the atmospheric concentrations of other gases such as nitrous oxide and methane have also been recorded in the last 40 years (Bigg 1996; IPCC 2001). Such increases in the concentrations of these gases in the atmosphere alter the radiative forcing globally by decreasing the long-wave radiative flux leaving the trophosphere (Houghton et al. 1990) which is thought to lead to climatic effects. Alongside the global monitoring of atmospheric concentrations and distributions of greenhouse gases, there have been concerted efforts to use mathematical models to better understand the nature of the relationship between the observed changes in gas concentrations, subsequent alteration of radiative forcing and climate.

While it is now well established that anthropogenic activities are responsible for observed increases in atmospheric gas concentrations such as CO_2 (IPCC 2001), the relationship between altered atmospheric composition and climate change is less well established. Nevertheless "there is new and stronger evidence that most of the warming observed over the last fifty years is attributable to human activities" (IPCC 2001), and that "emissions of greenhouse gases and aerosols due to human activites continue to alter the atmosphere in ways that are expected to affect climate" (IPCC 2001).

Recently, a synthesis of many observational studies indicates that the global average surface temperature has increased by ca. 0.6 °C in the 20th Century (IPCC 2001), and there is also strong evidence that the oceans are exhibiting a warming trend (Levitus et al. 2000). Indeed, the present rate of change in atmospheric CO_2 is without precedent, necessitating the development of a mechanistic understanding of the processes controlling climate change (IPCC 2001). In addition to contemporary changes in atmospheric CO_2, there is evidence of substantial climate change from the geological past with, for example, glacial core records showing large variations in temperature and greenhouse gas concentrations associated with glacial/interglacial oscillations over the last 400 000 years (the Vostok core record; Petit et al. 1999). Throughout this period, there are consistent trends of lower atmospheric CO_2 levels during the glacials, and higher levels during the interglacials (Woodwell et al. 1998). There has been much debate and many theories proposed as to what factors caused these shifts from interglacial to glacial periods (Archer and Johnson 2000; Schrag 2000; Sigman and Boyle 2000).

What is the role of the ocean in climate change? The ocean is inextricably linked with the atmosphere and the land, carries 50-fold more CO_2 (mostly in the form of Dissolve Inorganic Carbon, DIC) than the atmosphere, and thus dominates the global carbon cycle, playing a major role in climate change on a range of scales (Siegenthaler and Sarmiento 1993; Sigman and Boyle 2000). Current model and data-based estimates indicate that the ocean presently takes up approximately 2 Pg C yr^{-1} of anthropogenic carbon, or roughly a third of the atmospheric fossil fuel emissions (Battle et al. 2000; Doney et al., this volume). The potential influence of marine biota on atmospheric CO_2 levels suggest that a fully effective oceanic biological pump would result in an atmospheric CO_2 level of around 160 ppm (the unperturbed interglacial concentration of atmospheric CO_2 is 280 ppm) while an abiotic ocean would yield atmospheric CO_2 levels of ca. 450 ppm (Shaffer 1993). One of the goals of the Joint Global Ocean Flux Study (JGOFS) project is to "assess and understand the sensitivity to climate change of the ocean carbon cycle" and to "improve our ability to predict future climate-related changes" (JGOFS Science Plan 1991). JGOFS has produced comprehensive biogeochemical datasets from all major oceanographic regions. These datasets, which provide an unprecedented level of detail, enable us to define further biogeochemical provinces, and to understand better the functioning of the oceanic carbon cycle. Present coupled atmosphere-ocean climate or general circulation models (GCMs) are useful tools but have limited predictive abilities due to many uncertainties, in particular with regard to the response of the biota (Sarmiento et al. 1998; Doney 1999). A further aim of the JGOFS programme is to reduce these uncertainties by improving our understanding of the role of the biota in climate change.

The global climatic system is driven by a series of external and internal processes that influence global temperature and other climatic factors (Woodwell et al. 1998). External perturbations (such as variations in solar radiation, emissions of CO_2 from fossil fuel burning, human-induced alterations of surface albedo) elicit changes in the Earth's radiation balance and subsequently the climate (e.g., temperature, cloud cover, water vapour). These climate responses may in turn feedback on the initial perturbation. An example is the water vapour feedback where increased temperatures lead to higher atmospheric concentrations of water vapour, which is also a greenhouse gas leading to an amplification of the original temperature rise. Feedbacks may be classified as positive or negative i.e., in the case of climate change they may enhance or diminish the initial effects of anthropogenic perturbations on climate, respectively (Woodwell et al. 1998). Also, as feedbacks may be interactive (Lashof 1989) they can be viewed as the 'gears' of climate and as such may result in the non-linear effects that are increasingly receiving attention in climate research (Rahmstorf 1999).

This chapter provides examples of feedbacks, discusses the nature and dynamics of feedback mechanisms, identifies the main mechanisms reported to be relevant to climate change, and reviews aspects of feedbacks such as classification, forcing, magnitude and 'degree of belief' (after Pate-Cornell 1996). To be comprehensive both physically- and biogeochemically-mediated feedbacks have been included, but here only the latter are considered in detail. This is followed by an examination of case histories, from the JGOFS pro-

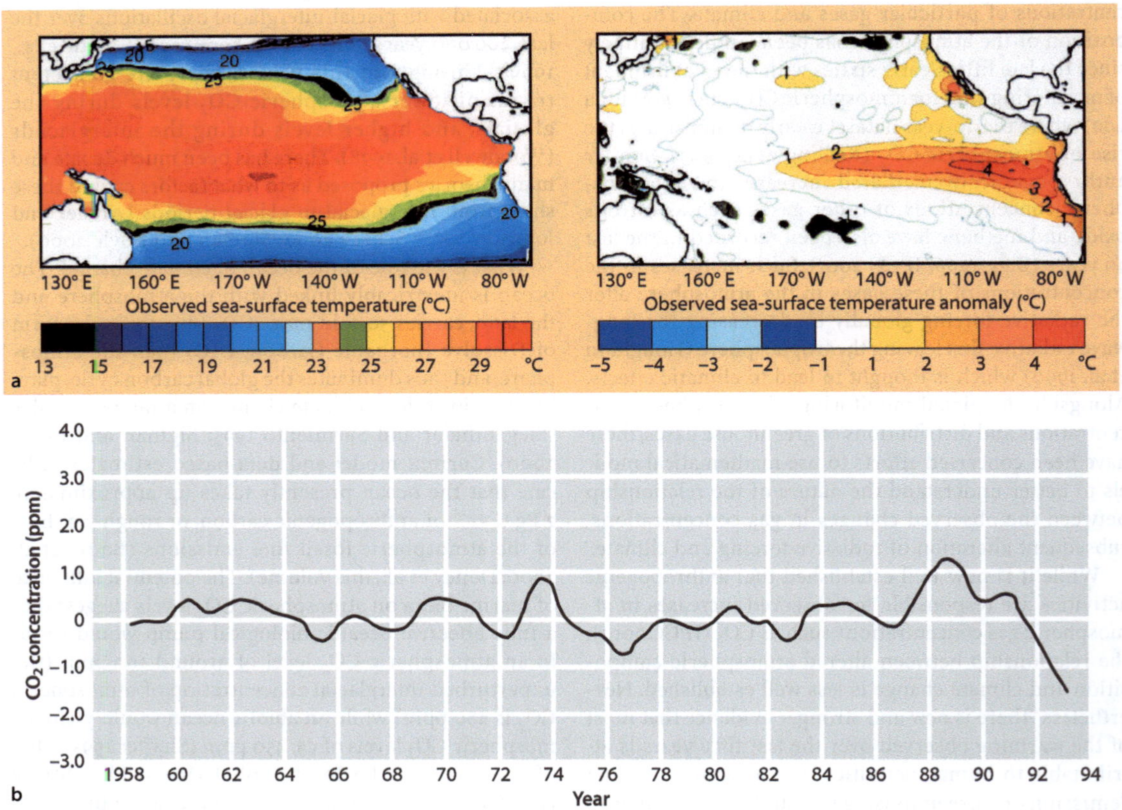

Fig. 7.1. Evidence of climate variability from **a** the 1997 El Niño event. The observed mean sea surface temperature (October 1997), and its deviation from the long-term (15 year) mean temperature for October are presented in the left and right panels respectively (reprinted by permission from Nature (Webster and Palmer 1997), © 1997 Macmillan Publishers Ltd.); and **b** the Mauna Loa CO_2 anomaly which displays the effects of El Niño events (1982–1983; 1986–1987) and the Pinatubo eruption (1991) (Sarmiento 1993). For details of the removal of the seasonal signal and de-trending of the remaining signal see Sarmiento (1993)

gramme, that shed light on biotic feedbacks that were previously poorly constrained at the time of the second IPCC (International Panel on Climate Change) review (Denman et al. 1996). Finally, our present understanding of each feedback is appraised, the recent emergence of any new mechanisms is discussed, and the current scientific understanding is related to projection (modeling simulations) and detection (ocean observations) of feedbacks in the future.

7.1.2 Examples of Feedbacks in the Present and the Geological Past

Recent examples demonstrate that large amplitude climate variability can occur on short timescales, tens of years or less, both in the present (Fig. 7.1a), and in the geological past (the 'North Atlantic Trigger', Fig. 7.2). Each case illustrates the short timescales and large magnitude of the response by ocean-atmosphere feedbacks, and more importantly indicates how poor our understanding is of the complex interplay between feedbacks and the resulting shifts in the functioning of the global carbon cycle. Sarmiento (1993) demonstrated how the removal of the seasonal signal and long-term trends from the Mauna Loa atmospheric CO_2

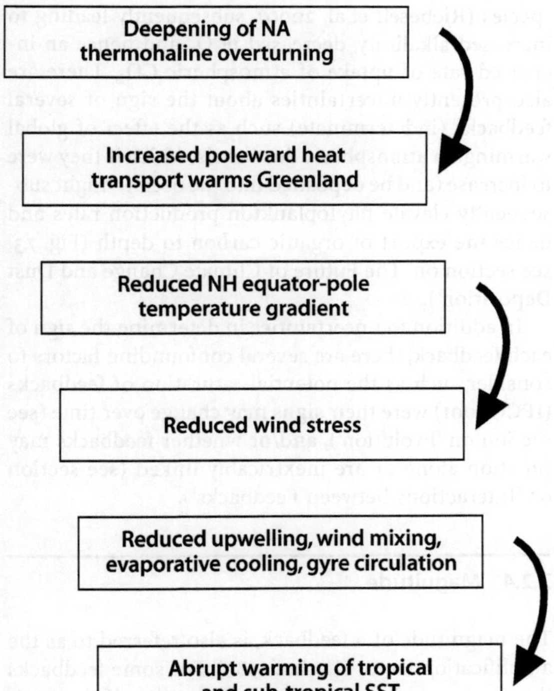

Fig. 7.2. A schematic of interactive feedbacks, chronologically arranged from the top of the panel downwards, hypothesised to drive the putative North Atlantic Trigger around 15 000 years ago (from Severinghaus and Brook 1999)

record provides evidence of the effects of climate variability due to El Niño and other perturbations such as the Pinatubo eruption (Fig. 7.1b) on CO_2 levels. In the latter case, the eruption has been attributed to cause a −1.5 ppm change in CO_2. Sarmiento assesses the plausibility of potential candidate mechanisms to explain which feedbacks might have caused this marked shift in CO_2. These include both terrestrial (cooling/rainfall altering the balance of respiration and photosynthesis on land) and oceanic (iron-rich ejected rocks elevating ocean productivity; Watson 1997). Sarmiento concluded that the nature of the feedback mechanism(s) effecting atmospheric CO_2 were not clear, and that this analysis was restricted by our lack of understanding of the functioning of the global carbon cycle. However, the magnitude of such climatic variability – as in the case of Pinatubo – shows that the oceanic biota may potentially have a pronounced and rapid effect on atmospheric CO_2 concentrations. Severinghaus and Brook (1999) report on a 9 °C change in temperature over several decades around 15 000 years ago (i.e., the last glacial termination). This warming may have led to an abrupt increase in methane concentrations, and the authors put forward a causal mechanism comprising a suite of interactive feedbacks called the 'North Atlantic Trigger' (Fig. 7.2). Again in this example it is evident how interactions between feedbacks result in a marked amplification of the climatic signal over a relatively short time period.

7.2 Feedbacks

7.2.1 Definition

In the field of climate change, the term feedback has often been used widely and loosely (Lashof 1989). Lashof refers to the term in a manner analogous to an electronic amplifier with the output (W) determined by the input signal (L) and a feedback signal proportional to W (such that the resultant (G), or gain is defined as $G = (W - L) / W$. Thus, if $G > 0$ the feedback is positive, and if $G < 0$ it is negative. In the case of the amplifier analogy, for clarity, care must be taken to use commensurate inputs and outputs. For example if the input is a 100 Pg increase in atmospheric CO_2, then the output will be the resulting change in atmospheric CO_2 due to terrestrial and oceanic carbon sources/sinks and climatic warming feedbacks.

7.2.2 Identification

Lashof (1989) divided feedbacks into those which are geophysically- and biogeochemically-mediated, examples of which include sea-ice/solar albedo and CO_2 solubility. It cannot be ruled out that there are other feedbacks that have yet to be identified. Geophysical feedbacks were defined by Lashof (1989) as those due to physical forcing (as opposed to chemical or biochemical processes) in the geosphere-biosphere that alter the radiative characteristics of the system in response to the initial radiative or temperature perturbation. Physical feedbacks, including the effect of water vapour, clouds, polar ice, and ocean heat storage, are usually included in GCMs (Woodwell et al. 1998). Biogeochemical feedbacks were defined by Lashof (1989) as those that involve the response of the biosphere and biogeochemical components of the geosphere, and at that time had seldom been considered in typical climate models (Lashof 1989). Examples of biogeochemical feedback mechanisms include short-term biological responses to warming and atmospheric CO_2 increases (e.g., changes in leaf area and terrestrial photosynthesis that in turn impact surface albedo and surface water vapour exchange), and longer-term effects due to the re-organisation of ecosystems (e.g., a shift in vegetation, forest to grassland; Lashof 1989). The biogeochemical feedbacks that have received most attention are those that impact biotic exchanges of greenhouse gases which are often sensitive to changes in temperature, radiation and moisture (Woodwell et al. 1998). Until recently, the latter have not been generally included in GCMs which have been generally driven by prescribed CO_2 concentrations rather than by prescribed emissions. Such biogeochemical feedbacks are complicated but potentially large and have the potential for either damping or accelerating global warming once begun (Woodwell et al. 1998). In the last few years there have been major efforts by climate modellers to include such biogeochemical feedbacks into coupled GCMs (e.g., Cox et al. 2000).

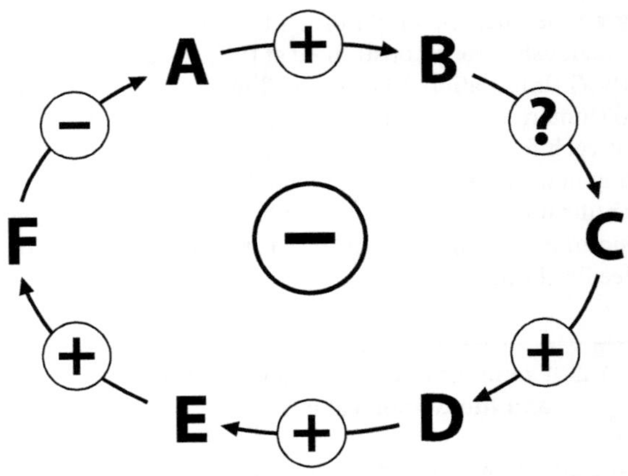

Fig. 7.3. A means by which to represent feedback loops (after Kellogg 1983). For example, the link between *A* increased atmospheric carbon dioxide concentrations, *B* warming, *C* enhanced aridification, *D* increased dust supply, *E* iron-elevated rates of primary production, and *F* higher downward particulate export fluxes. Positive (exacerbation) and negative (diminution) denote the signs of the feedbacks, note the relationship between B and C remains controversial/unproven (see Fig. 7.13). Although global warming is thought to increase aridity and hence the supply of iron-rich dust to iron-depleted phytoplankton (Falkowski et al. 1998), global atmospheric moisture content is predicted to increase as a result of global warming (Trenberth 1998), and thus it is not known if warming will result in an increase or decrease in iron-mediated phytoplankton production in the future

7.2.3 Classification

Feedbacks may be classified as positive (exacerbation), negative (diminution) (after Kellogg 1983), or in some cases indeterminate. Examples of those considered to be positive include the CO_2 buffering capacity feedback (where as the ocean takes up more CO_2 and becomes more acidic, the thermodynamic capacity of seawater to take up CO_2 (δDIC per unit CO_2) becomes smaller (Siegenthaler and Sarmiento 1993; IPCC 2001). Changes in ice and snow cover also constitute a positive feedback because warming the Earth reduces the planetary albedo and increases the surface shortwave heating by reducing the extent and persistence of sea-ice and snow cover (Lashof 1989). An example of a feedback viewed to be negative is the predicted decrease in pH, due to increased pCO_2, which in laboratory studies result in decreased calcification rate in several coccolithophorid species (Riebesell et al. 2000), subsequently leading to increased alkalinity, decreased pCO_2 and hence an increased rate of uptake of atmospheric CO_2. There are also presently uncertainties about the sign of several feedbacks (indeterminate) such as the effect of global warming on atmospheric dust levels, which if they were to increase (and be deposited into the ocean) might subsequently elevate phytoplankton production rates and hence the export of organic carbon to depth (Fig. 7.3, see section on 'The Future of Climate Change and Dust Deposition').

In addition to uncertainties in determining the sign of each feedback, there are several confounding factors to consider such as the potential saturation of feedbacks (IPCC 2001) were their signs may change over time (see section on 'Evolution'), and/or whether feedbacks may function alone or are inextricably linked (see section on 'Interactions between Feedbacks').

7.2.4 Magnitude

The magnitude of a feedback, is also referred to as the amplification factor (Lashof 1989). For some feedbacks this factor can be taken at face value (i.e., if the feedback effect is global), whereas for others it must be scaled to take into account the regional nature of a particular mechanism (see Scales and Response Times). Although

Kellogg (1983) discussed several biogeochemical feedback scenarios, this was done in a qualitative manner. Lashof (1989) provided some of the first estimates of the magnitude of both geophysical and biogeochemical feedbacks. In general, based on modeling outputs, Lashof reported that geophysical feedbacks were all positive and had the largest magnitude (water vapour followed by ice/snow albedo and clouds). In contrast, although biogeochemical feedbacks were more numerous they had relatively small magnitudes, with some positive (ocean eddy diffusion) and negative (trophospheric chemistry). The cumulative effect of biogeochemical feedbacks was positive but only 25% of the magnitude (although of the same sign) of the geophysical feedbacks. Recently, Joos et al. (1999) performed a modeling experiment to assess the effects of various feedbacks on the global carbon cycle including sea surface temperature (SST), oceanic circulation, and some general biotic effects. They also included model runs employing different scenarios for the marine biota (such as more marine calcifiers). Joos et al. reported that the projected changes in the ocean carbon cycle produced modest changes in atmospheric CO_2 concentrations (a 4% and 20% increase by 2100 and 2500). The feedback with the greatest magnitude appeared to be a reduction in oceanic CO_2 uptake due to sea-surface warming. The projected changes in the biota due to climate change compensated in part for some of the projected physico-chemical changes (Joos et al. 1999). See also Denman and Pena (2000) for a sensitivity analysis of how the various biogeochemical feedbacks considered by two models (Heinze et al. 1991; Shaffer 1993) might alter atmospheric CO_2 concentrations.

7.2.5 Evolution

Will the effects of climate change result in the spatial and/or temporal evolution of feedbacks, and does the magnitude of each mechanism therefore alter with time or in space? The IPCC (2001) report discusses the possibility of 'sink saturation' whereby the effectiveness (and hence the magnitude and possibly sign) of a particular feedback mechanism – in this case the magnitude of the CO_2 buffering capacity feedback – may decrease over time. It is also possible that the nature of the forcing on climate and subsequently on feedbacks may alter over decadal to centennial timescales (IPCC 2001); for example temperature presently exerts a major role in forcing climate. However the influence of shifts in physical circulation may become increasingly important in the future. Furthermore, the areal extent over which a feedback is influential may presently be regional (uptake and storage of atmospheric CO_2 in the Southern Ocean, Calderia and Duffy 2000), rather than global (such as CO_2 buffering capacity, Siegenthaler and Sarmiento 1993), but may alter as climate changes. Parmesan (1996) examined the validity of the assumption that warming trends would result in a poleward expansion of the area occupied by certain terrestrial species and concluded that the existing small datasets had been overly upscaled. More recently, Lipschultz (1999) pointed to a poleward expansion of the oceanic biome for nitrogen fixers that might result from global warming. Future field/modeling experiments might include an assessment of whether there is a temporal sequence of feedbacks. It is unlikely that feedback mechanisms will remain as static features in the future, and such evolution needs to be taken into account when attempting to rank feedbacks.

7.2.6 Interactions between Feedbacks

Kellogg (1983) states that "in actuality, these various feedback (loops) cannot be considered by themselves – they all interact with each other – something that is well recognized by climate system modellers". The interactions between feedbacks is thought to be largely responsible for the non-linearity in response to climate (Rahmstorf 1999) observed both in the geological record (Broecker and Henderson 1998; Severinghaus and Brook 1999) and in the present day (El Niño, Webster and Palmer 1997; Chavez et al. 1999; Pinatubo, Sarmiento 1993). As such interactions usually result in enhanced system complexity, they will be difficult to assess, even qualitatively.

7.2.7 Scales and Response Times

At present, most climate model predictions with full three-dimensional GCMs only extend for 100 years or so into the future, and interest therefore has focussed on those feedbacks that will take place on decadal timescales (or less). To assess the contribution of each feedback mechanism, all of the relevant, and often wide-ranging time and space scales must be considered (such as algal physiology, Wolf-Gladrow et al. 1999; or ocean circulation, Calderia and Duffy 2000). Such temporal and spatial scales will also reflect the response time for each mechanism. Algal physiological responses are likely to be rapid and localised (hours/days Behrenfeld et al. 1996; Boyd et al. 2000) while those for foodweb shifts might take longer – from weeks (Boyd et al. 2000) to months (Chavez et al. 1999, see section on 'Regime Shifts'). Basin-scale might operate on a decadal/centennial time horizon (shutdown of North Atlantic Deep Water and subsequent alteration of thermohaline circulation, Broecker 1997). These time and space scales must be placed in context in the structuring of the coupled atmosphere/ocean models.

7.2.8 Degree of Confidence – Understanding Feedbacks

Prior to any ranking of feedbacks as is required to assign future research priorities, the degree of confidence in our understanding of their many aspects (magnitude, complexity – isolated vs. interactive) must be assessed. Such a synthesis must be made despite many uncertainties; it may be possible to adopt an approach analogous to that of Pate-Cornell (1996) who combined existing information on global climate change estimates with Bayesian probabilities that serve as a proxy for a 'degree of belief'. Climate scientists have recently begun to assign 'level of scientific understanding' or 'confidence' estimates for the expectations of forcing changes, responses and feedbacks (IPCC 2001). Similar exercises can be conducted for biogeochemistry. For example, the general consensus is that a decrease in pH is highly probable (99%) (Kleypas et al. 1999) while a decrease in calcification is perhaps only moderately likely (Riebesell et al. 2000). Here, the expected changes in the physicochemical environment of the future may be put forward as follows (in order of confidence): the most probable changes will be increased ocean DIC; decreased pH and $[CO_3^-]$concentration; probable changes include an increase in surface temperature; somewhat probable might be decreased solar irradiance because of increased cloudiness; coastal eutrophication; decrease in high latitude surface salinity (sea ice meltback and enhanced hydrological cycle); increased vertical stratification (temperature at low latitudes and salt at high latitudes); and increased storminess and hurricanes (Royer et al. 1998; Bates et al. 1998). The processes/changes that there is least confidence in our understanding of include changes in gyre shape/amplitude, dust deposition (see section on 'The Future – Climate Change and Dust Deposition'). On the basis of the unknowns raised so far in this chapter it is not possible to presently rank feedbacks in order of importance for climate change but a scheme by which to do so is presented in Fig. 7.4.

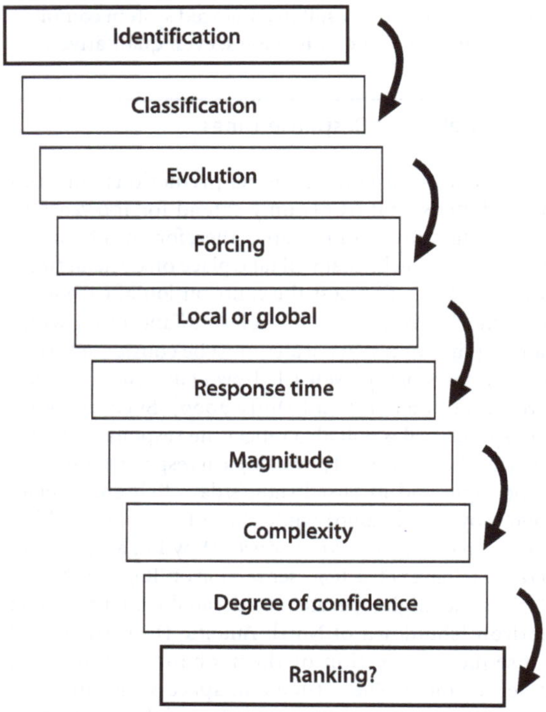

Fig. 7.4. Aspects of feedbacks and the order (from *top* to *bottom*) in which they should be considered to appraise our understanding/confidence in estimates of each feedback mechanism

7.3 What do Current Models Predict?

Current model estimates concur with data-based estimates indicating that the ocean takes up around a third of the atmospheric fossil fuel emissions (Schimel et al. 1995; Doney et al., this volume). Model projections suggest that the ocean will continue to play a major role in drawing down excess atmospheric CO_2 over the next several centuries (Fig. 7.5). As the magnitude of the anthropogenic impact on the climate system grows, significant changes in ocean physics and other external forcing factors will undoubtedly alter the uptake of an-

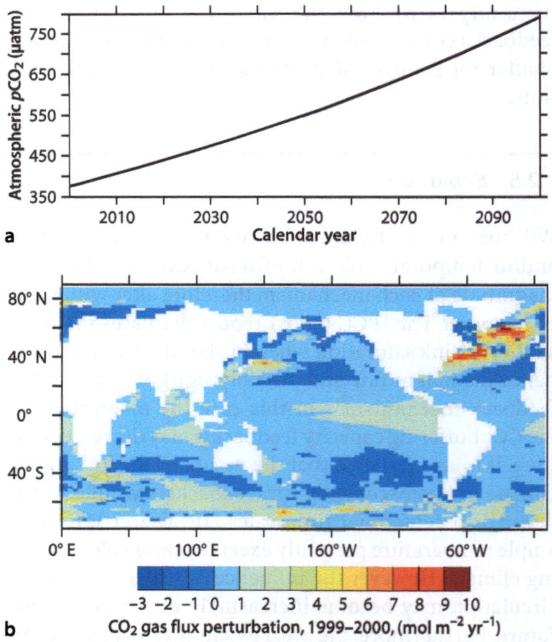

Fig. 7.5. Modelled projections for the next century of the rise in atmospheric CO_2 from the IPCC IS92A scenario (**a**) and the resulting magnitude of the global oceanic uptake of CO_2 (**b**) as a function of time from the NCAR (National Center for Atmospheric Research, Boulder, Colorado, USA) ocean model

thropogenic carbon and perturb the 'natural' biogeochemical carbon cycle, and these potential ocean feedbacks are a key topic of current research.

Recent advances in coupled ocean-atmosphere models (e.g., Manabe and Stouffer 1993; Boville and Gent 1998) have resulted in a series of reasonably credible physical scenarios for the next 100 years. Fossil fuel consumption and land-use change since the Industrial Revolution have led to substantial increases in the atmospheric levels of greenhouse trace gases such as CO_2, N_2O, CFCs and CH_4. The resulting net heating of the atmosphere is partially offset by cooling due to atmospheric sulphate aerosols over Northern Hemisphere industrial regions (Schimel et al. 1996). Calculating the full climatic impact of these perturbations, including feedbacks from clouds, sea-ice, and water vapour, requires the use of three-dimensional, coupled ocean-atmosphere models. The present generation of coupled simulations suggest continued surface warming of between 1.5–5.0 °C (above pre-industrial) and a strengthening of the hydrological cycle over the next century (e.g., Kattenberg et al. 1996).

From the perspective of ocean biogeochemistry, perhaps the most relevant features observed in almost all coupled model projections are a general warming of the upper ocean and thermocline and increased vertical stratification in both low latitude (warming) and high latitude (freshening) surface waters (Fig. 7.6). The regional patterns of climate change can differ considerably among coupled models (model intercomparisons provide useful tests to look at such regional variations, see Orr 1999) and should be viewed with some scepticism. However, many coupled model simulations show trends such as: reduction of polar sea-ice cover; shift toward more El Niño like conditions in the equatorial Pacific; and decreased deepwater production in the North Atlantic and Southern Ocean. Other potentially important physical climate forcings are changes in the pattern and magnitude of river runoff and thus nutrient input to the coastal ocean, a possible decrease in surface solar insolation under a more humid and cloudy atmosphere, and modified terrestrial dust iron deposition to the surface ocean (Mahowald et al. 1999). Terrestrial dust production and subsequent oceanic deposition depends on a complex mixture of physical climatic factors (e.g., soil moisture, see section on 'The Future Climate Change and Dust Deposition') and it is as yet unclear whether future dust depostion will increase or decrease. Finally, the basic ocean chemical state, and in particular the pH and carbonate system, will be dra-

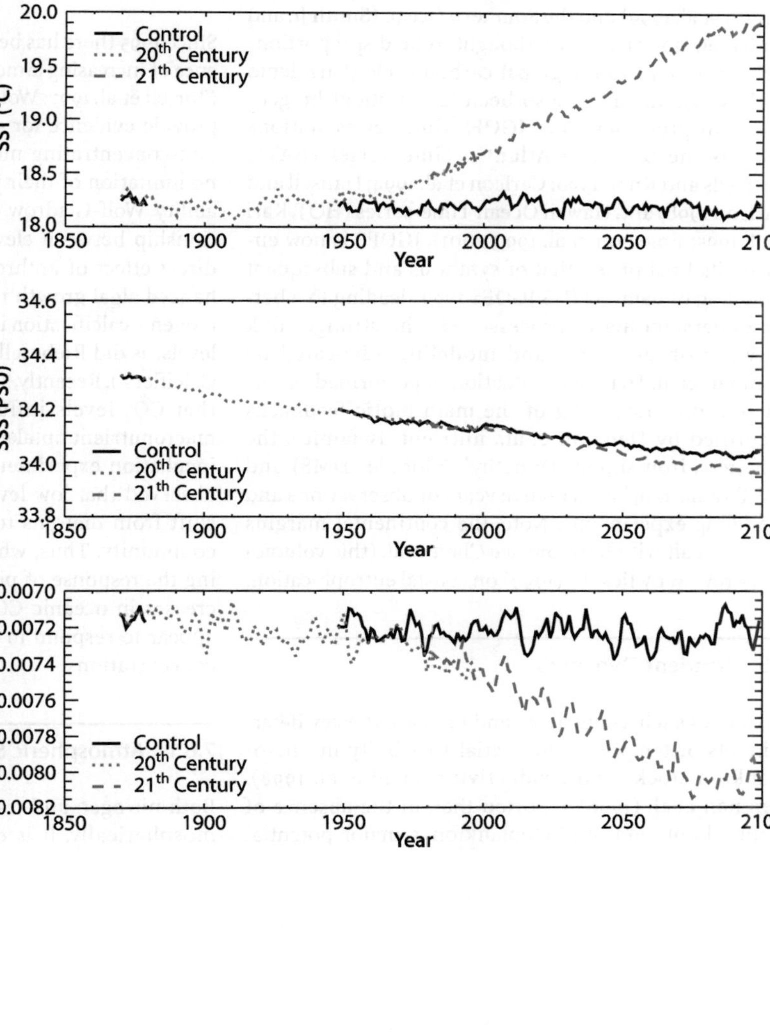

Fig. 7.6. Projections of the increases in global average sea surface temperature and salinity, and upper ocean (50 m) vertical density gradient (a measure of stratification) over the next century from the NCAR Climate System Model (CSM)

matically altered due to the uptake of anthropogenic CO_2. By the middle of the next century, the addition of CO_2 to the surface ocean will shift the acid-base chemistry by about −0.25 pH and reduce carbonate ion concentrations by about 30%, with the potential for significantly decreasing marine calcification (e.g., Kleypas et al. 1999). For future directions in ocean-atmosphere modeling see section on 'Modeling – Future Goals'.

7.4 Status of Our Understanding of Feedbacks

The most comprehensive review of marine biotic feedbacks was conducted by Denman et al. (1996) as part of the IPCC second scientific assessment of climate change. Their review identified and discussed several potential feedbacks, but concluded that their sign and magnitude were relatively poorly constrained. Moreover, Denman et al. stated that our most powerful tools to investigate feedbacks – climate models – could not adequately simulate complex ecosystem interactions, and that modeling and observational studies must be used in tandem to address these complex climate-feedback-responses.

Since 1995, the JGOFS programme has conducted further field studies in regions, such as the Arabian Sea (Smith et al. 1999) and the Southern Ocean (Smith Jr. and Anderson. 2000), that are thought to be disproportionately important to the global carbon cycle (Sarmiento et al. 1998). Much has also been learnt about biogeochemical processes from JGOFS time-series stations such as the Bermuda Atlantic Time Series (BATS, Michaels and Knap 1996; Carlson et al. 1994; Hansell and Carlson 1998) and Hawaii Ocean Time-series (HOT, Karl et al. 1995; Emerson et al. 1997, 2001). JGOFS is now entering its final phase, that of synthesis and subsequent modeling of datasets (US JGOFS 1999) leading to a better understanding of processes via the stronger link between observations and modeling advocated by Denman et al. Here, an evaluation is performed of our current understanding of the main biotic feedbacks identified by Denman et al., nutrient dynamics, the calcifiers, iron supply, Dimethyl Sulphide (DMS) and UV-B radiation, based on five years of observations and modeling experiments. Note, the continental margins are not dealt with here, but see Chen et al. (this volume) and a review by Jickells (1999) on coastal eutrophication.

7.5 Nutrient Dynamics

Resources such as nutrients and light are the key determinants of temporal and spatial variability in phytoplankton stocks and productivity (Field et al. 1998). Denman et al. (1996) reported that, in the absence of carbon limitation of phytoplankton, a major potential feedback resulting from climate change would be any alteration in the relationship between carbon and nutrients – via shifts in the supply of new nutrients to the open ocean, or to changes in the stoichiometry of nutrient uptake. Such shifts might result in an enhanced role of the biological pump in influencing climate, for example by fixing more carbon per unit of nutrient taken up. Much of their discussion of feedbacks was based on the decoupling of the carbon/macronutrient relationship. For new nutrients, Denman et al. focussed on atmospheric inputs (Michaels et al. 1993), nitrogen fixation (Carpenter 1973; Carpenter and Romans 1991) and coastal/riverine inputs (Schindler and Bayley 1993), while for altered nutrient uptake they discussed studies by Sambrotto et al. (1993), Banse (1994), and Karl et al. (1995). Denman et al. also reviewed the potential for uptake of presently unused nutrients (such as in High Nitrate Low Chlorophyll (HNLC) regions) due to changes in the supply rate of trace metals (such as iron), and the role of the continental margins in supplying new nutrients to the open ocean. Here, our present understanding of macronutrients is reported on, see section on 'Atmospheric Deposition of Iron' for details on changes in iron supply.

7.6 Phytoplankton and Carbon Limitation

Since 1995 there has been further research into the influence of increasing atmospheric CO_2 levels on algal growth (Tortell et al. 1997; Wolf-Gladrow et al. 1999). Tortell et al. provide evidence for active uptake of HCO_3^-, via a carbon-concentrating mechanism, in diatoms and hence no limitation of their photosynthetic rate by CO_2 availability. Wolf-Gladrow et al. reported evidence of a relationship between elevated CO_2 levels (mimicking the direct effect of anthropogenic CO_2 increases) and enhanced algal growth rates. They also observed reduced biogenic calcification in coccolithophorids at higher CO_2 levels, as did Riebesell et al. (2000) (see section on 'The Calcifiers'). Recently, Tortell et al. (2002) report evidence that CO_2 levels influence species composition and macronutrient uptake stoichiometry. During shipboard incubation experiments in the Equatorial Pacific they observed that low levels of CO_2 resulted in a floristic shift from diatoms towards a *Phaeocystis*-dominated community. Thus, while uncertainties remain regarding the response of phytoplankton to the projected increases in oceanic CO_2 levels, algal functional groups appear to respond in different ways to changes in CO_2 concentrations.

7.6.1 Atmospheric Supply of Nutrients

Both nitrogen and trace elements may be supplied atmospherically. It is estimated that there has been a

5–10-fold increase in sources of atmospheric pollution since the Industrial Revolution (Paerl and Whitall 1999). Model projections suggest that this source of nutrients might be 3- to 4-fold higher in 2020 (relative to 1980) in the coastal ocean (Galloway et al. 1994). However, increases in atmospheric inputs to offshore waters were predicted by these authors to be somewhat lower. Since 1995, research has focussed on several themes including the contribution of atmospheric nitrogen to ocean basin nitrogen budgets (Spokes et al. 2001), and the bioavailability of atmospheric nitrogen (Paerl et al. 1999; Peierls and Paerl 1997). Moreover, other atmospheric nitrogen sources such as dissolved organic nitrogen (DON) (Peierls and Paerl 1997) and urea (Cornell et al. 1998) have gained attention with respect to nitrogen loading of the oceans. Spokes et al. (2001) reported that up to 30% of new production in NE Atlantic waters could be supported by atmospheric inputs in spring. They identified that most nitrogen deposition occurred during short-lived, high concentration, SE transport events (i.e., when an airmass passes over Northern Europe first). In the SE Mediterranean Sea, estimates suggest that nitrogen inputs represent a 8–20% contribution to new production and point to possible phosphate limitation of algal growth in these waters (Herut et al. 1999).

Paerl and Whitall (1999) discuss the possibility of links between anthropogenically-derived atmospheric nitrogen, coastal eutrophication and the expansion of harmful algal blooms (see section on 'Regime Shifts'). Moreover, Paerl (1999) points out that nitrogen loading may uniquely mediate coastal eutrophication via the bypassing of estuarine filters of terrigenous nitrogen inputs. In the open ocean, enhanced atmospheric transport, via pollutants, of nitrate has mainly been reported in the oligotrophic Atlantic (Galloway et al. 1995). Paerl et al. (1999) point out that natural rainfall and Dissolved Inorganic Nitrogen (DIN) additions most often stimulated CO_2 fixation and elevated chlorophyll levels in western Atlantic waters. They also suggest that the observed high levels of stimulation may be due to the supply of both DIN and iron, and that this may contribute to the eutrophication potential of oceanic waters downwind of urban, industrial and agricultural emissions. Atmospheric DON has also gained attention as a significant new source of nitrogen loading of the oceans (Peierls and Paerl 1997). Seitzinger and Sanders (1999) report that DON in rainwater can be an important source of nitrogen to coastal ecosystems – diatoms and dinoflagellates dominated the assemblage in experimental treatments receiving rainwater DON, whereas small monads dominated in ammonium-enriched treatments. The effect of atmospheric nitrogen on marine productivity depends on the biological availability of both inorganic and organic nitrogen forms, and there is a need to include all nitrogen inputs in atmospheric loading estimates.

The effects of new nitrogen would appear to mainly impact the coastal ocean in the near future (see projections for 2020 in Galloway et al. 1994), but nonetheless provides interesting insights into ecological shifts 'not in the textbooks' that may be analogous to such shifts, in response to perturbations, in the open ocean (see section on 'Regime Shifts'). Also such nitrogen inputs appear to be regional rather than global, and include parts of the N Atlantic, Mediterranean Sea, Baltic Sea, North and Yellow Seas (Paerl 1999). Human alterations of the nitrogen cycle – rather than climate change per se – have approximately doubled the rate of nitrogen input into terrestrial systems globally (Vitousek et al. 1997). Thus, anthropogenic inputs rather than climate change will likely be the key determinants of the magnitude and impact of atmospheric nitrogen.

7.6.2 Nitrogen Fixation

The magnitude of the oceanic nitrate inventory is due to the balance between rates of nitrogen fixation and denitrification (Codispoti 1989, 1995). Nitrogen fixation by diazotrophs is observed over much of the subtropical and tropical oligotrophic oceans and is likely a major input to the marine and global nitrogen cycle (Karl et al. 1995; Michaels et al. 1996; Capone et al. 1997; Emerson et al. 2001). Of particular significance has been further evidence from the HOT site of both increased abundances of nitrogen fixers and fixation rates during El Niño 'years' (Karl et al. 1995, 1997; Karl 1999; Karl et al., this volume, see section on 'Regime Shifts', this chapter). Karl (1999) concludes that the enhancement of nitrogen fixation is in response to increased stratification, and that this elevated activity drives the resident phytoplankton in the North Pacific Subtropical Gyre (NPSG) into phosphate limitation. Emerson et al. (2001) conducted an analysis on 11 years of data from the HOT site and report that, based on changes in apparent oxygen utilisation and organic C:N:P ratios in the upper ocean, a large proportion of the nitrogen supply to surface waters is probably from nitrogen fixation.

Recently, considerable progress has been made on two aspects of research into nitrogen fixers – identification of algal groups (in addition to *Trichodesmium*) that can fix nitrogen, and secondly on the iron requirements of nitrogen fixers. One of the puzzles of studies into nitrogen fixation in oligotrophic waters has been a perceived imbalance between estimates of nitrogen fixation derived from biogeochemical models/studies (Lipschultz and Owens 1996; Gruber and Sarmiento 1997) and that from observed *Trichodesmium* abundances and measured fixation rates. Zehr et al. (2001) suggest they have partially reconciled this enigma through their discovery, using molecular tools, of unicellular cyanobacteria capable of fixing substantial amounts of nitrogen in oligotrophic regions such as the NPSG.

Theoretical studies of algal iron requirements (Raven 1990) suggested that nitrogen fixers such as *Trichodesmium* had relatively high iron requirements compared to diatoms. It was subsequently speculated that an upper limit on nitrogen fixation rates would be imposed by such high iron requirements (Falkowski 1997) and/or by phosphate limitation (Jickells and Spokes 2001). Thus, elevated iron supply to oligotrophic waters in the geological past might have increased the activity of nitrogen fixers and therefore the availability of new nutrients (Falkowski 1997), and has been invoked to explain the timing and magnitude of glacial/interglacial CO_2 changes (Broecker and Henderson 1998).

Recent experimental evidence suggests that phosphate rather than iron limitation of nitrogen fixers is evident in the subtropical Atlantic (Sanudo-Wilhelmy et al. 2001), which may be explained by the relatively high ambient dissolved iron concentrations (Martin et al. 1993; Powell et al. 1995). This mode of control may not be applicable to other oligotrophic waters as iron levels are lower in the NPSG (based on limited data, Karl 1999), and in the tropical Pacific there is evidence that the dominant cells (picophytoplankton) are iron-stressed (Behrenfeld and Kolber 1999). Furthermore, in the subtropical coastal waters off Australia, a laboratory study using *Trichodesmium* provides evidence that under iron-limiting conditions there is a marked decline in physiological rates such as nitrogen fixation (Berman-Frank et al. 2001). In the W Atlantic Shelf region, Lenes et al. (2001) add important observational data on this subject, via monitoring the response of *Trichodesmium* in these waters to a Sahara-derived dust deposition event. They report on a dust-mediated 100-fold increase in *Trichodesmium* stocks, the concomitant reduction in phosphate levels to below the limit of detection, and that the increased organic nitrogen levels associated with the bloom had the potential to support toxic algal blooms.

7.6.3 Changes in Nutrient Uptake Stoichiometry – the Redfield Ratio

The possibility that the uptake of nutrients might depart from fixed Redfieldian stoichiometry has been an overriding theme in discussions of the role and magnitude of biotic feedbacks (Sambrotto et al. 1993; Banse 1994; Anderson and Sarmiento 1994; Kortzinger et al. 2001), and provides a counter argument to the geochemical standpoint put forward by Broecker (1990). Denman et al. (1996) suggested that model simulations (Shaffer 1993) of ocean carbon cycle changes were particularly sensitive to changes in the Redfield ratio between carbon and the limiting nutrient. Denman and Pena (2000) provided evidence, via a sensitivity analysis of biogeochemical models (Heinze et al. 1991; Shaffer 1993), that a two-fold shift in the carbon:nutrient Redfield ratio might indeed result in the greatest change in atmospheric CO_2 levels (relative to two-fold shifts in other properties such as productivity, or ventilation rates).

Sambrotto et al. (1993) presented evidence – based on estimates of nutrient utilisation ratios i.e., the net community drawdown of nitrate and DIC – of elevated carbon consumption relative to nitrate during the spring bloom in the NE Atlantic, Bering Shelf and Gerlache Strait, and asked whether this was evidence of widespread inconsistencies in the Redfield ratio. These variations in nutrient stoichiometry (with C:N ratios of up to 14 along the 20° W meridian in the NE Atlantic) were up to double that of estimates made by Takahashi et al. (1993) at a station (47° N) along this meridian in the same year. Mechanisms invoked for such elevated ratios include preferential recycling of nitrogen relative to carbon in the upper ocean (Sambrotto et al. 1993) or a high release of dissolved organic carbon (DOC) during an algal bloom (Toggweiler 1993). Furthermore, Lancelot and Billen (1985) had previously reported similarly elevated nutrient utilisation ratios during algal blooms and considered them due to a temporary imbalance between uptake of nitrogen and carbon consumption.

The study of Sambrotto et al. (1993) did not correct the observed macronutrient concentrations for abiotic factors such as evaporation minus precipitation, concurrent air-sea CO_2 flux, or to consider in their stoichiometric calculations the influence of winter reserve carbon and nutrient concentrations.

Since 1995 there has been only one published study to investigate further such aspects of the relationship between nutrient and carbon consumption (Kortzinger et al. 2001). They examined the C:N relationships in the mixed layer in the NE Atlantic in June/July 1996 using a transect (the 20° W meridian) as a proxy for the seasonal progression of biological events. Kortzinger et al. reported that in spring the system was dominated by nitrogen over-consumption, whereas during summer oligotrophy there was a marked carbon over-consumption (Fig. 7.7). Their study highlights the need to examine nutrient utilisation rates and stoichiometry over the entire algal growth season, and thus it is not presently known what the impact of such transient changes in C:N uptake stoichiometry have on the ocean carbon cycle over longer timescales.

Recently, there has been evidence of shifts in carbon/nutrient uptake stoichiometry due to changes in phytoplankton community structure (Arrigo et al. 1999), and in iron supply (Hutchins and Bruland 1998; Takeda 1998; Franck et al. 2000). Karl et al. (1995) have also observed shifts in the ratios of nitrogen to phosphorus at the HOT site between years due to pronounced changes in nitrogen fixation rates. In the Ross Sea, *Phaeocystis antarctica*

Fig. 7.7.
Changes in the C:N ratio of mixed layer Particulate Organic Matter, new production and exported production along the 20° W meridian during late spring/summer in the NE Atlantic Ocean (reprinted from Deep-Sea Res I 48, Kortzinger et al. (2001) C:N ratios in the mixed layer during the productive season in the Northeast Atlantic Ocean. pp 661–688, © 2001, with permission Elsevier Science)

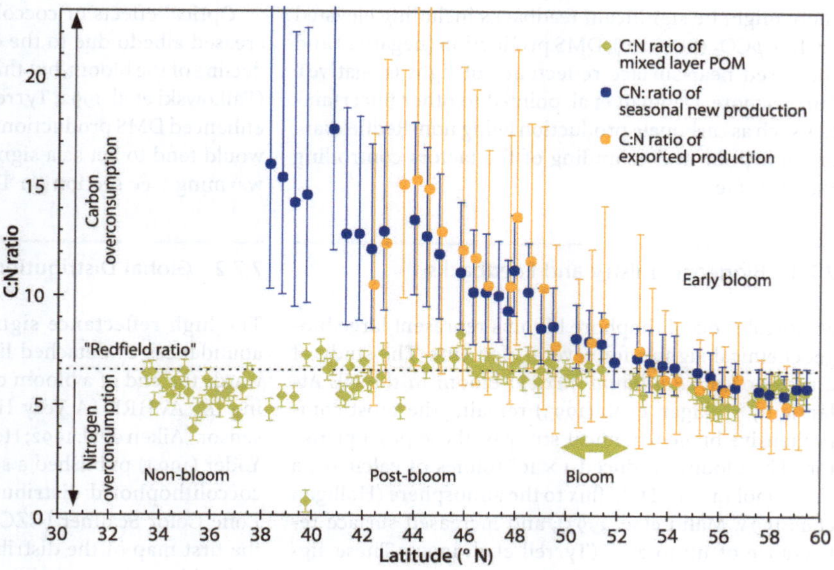

blooms under conditions of low light (Moisan and Mitchell 1999) due to deeper mixed layers (Arrigo et al. 1999), and are characterised by more carbon fixed per unit phosphate compared with diatom blooms. The latter are observed in summer (blooming under high light, shallow mixed layers, Arrigo et al. 1999). Evidence of the influence of iron supply on macronutrient uptake ratios comes from reports of decreased algal silicic acid: carbon and silicic acid: nitrate uptake ratios upon iron supply to phytoplankton from open (Takeda 1998) and coastal ocean HNLC waters (Hutchins and Bruland 1998). Such iron-mediated shifts in uptake stoichiometry, if sustainable over long timescales, have been shown to have important implications for increased drawdown of atmospheric CO_2 during glacial maxima (Watson et al. 2000).

7.6.4 Export Production and Remineralisation in the Deep Ocean

In a review of the ocean carbon cycle Siegenthaler and Sarmiento (1993) suggested that climate change might also influence the relationship between the relative magnitude of the export of carbon via either particulate or dissolved forms by causing a more rapid remineralisation of sinking organic particles. Moreover, Denman and Pena (2000) from a sensitivity analysis of biogeochemical models (Heinze et al. 1991; Shaffer 1993), suggested that shifts in remineralisation depth scales could have a pronounced effect on atmospheric CO_2 levels. Levitus et al. (2000) have reported that the surface waters of the World Ocean have warmed significantly during the last 40 years, with the largest warming occurring in the upper 300 metres (on average by 0.56 °C). This may have implications for bacterial ecto-enzyme activity (Smith et al. 1992; Christian and Karl 1995) which if it increased might reduce the depth scales for remineralisation for each major element (Christian et al. 1997).

Since 1995 considerable progress has been made in better understanding the role of DOC in carbon biogeochemistry, its relative contribution to export production (Carlson et al. 1994), and in its global and seasonal distributions and budgets (Hansell and Carlson 1998). However, in the latter study, the authors point to the need for more knowledge on the controls on the production and consumption of the semi-labile DOC pool. These are essential before the relative contribution of this pool to the downward export of biogenic carbon, and to the inter-hemispheric transport of DOC can be accurately assessed. Recent estimates that considered both DOC and POC fluxes at the HOT site suggest that the subtropical oceans may be responsible for a significantly higher contribution to the global biological pump than previously thought (Emerson et al. 1997).

7.7 The Calcifiers

The calcifiers include coralline algae, calcareous zooplankton (such as foraminifera) and coccolithophorids, but only the latter is considered here. Denman et al. (1996) focussed on this group as they are carbonate producers, influence the rain ratio (the ratio of organic carbon to carbonate in settling particles (see Denman and Pena 2000), help set the pCO_2 of surface waters via calcification, and have been linked to large changes in surface ocean DMS levels (Holligan et al. 1993; Westbroeck et al. 1993). Denman et al. suggested that if climate change resulted in a shift towards the dominance of this group, there might be significant feedbacks including elevated surface pCO_2 (positive), DMS production (negative) and increased near-surface reflectance of light (negative). Furthermore, Denman et al. pointed to other uncertainties such as carbonate production being non-Redfieldian, and our poor understanding of the factors controlling the rain rate.

Optical effects of coccolithophore blooms cause increased albedo due to the detached liths that mark the decline of the bloom, but this effect is thought to be small (Falkowski et al. 1992; Tyrrell et al. 1999). In contrast, the enhanced DMS production associated with such blooms would tend to act as a significant negative feedback on warming (see section on 'DMS and the Biota').

7.7.1 Biogeochemistry and Feedbacks

Mesoscale coccolithophore blooms represent large biogeochemical signals for several elements. The study of a 250 000 km^2 *Emiliania huxleyi* bloom in the NE Atlantic by Holligan et al. (1993) remains the most comprehensive biogeochemical study of these phytoplankton. The bloom produced 1×10^6 tonnes of calcite-C, a 1122 nmol m^{-2} h^{-1} DMS flux to the atmosphere (Holligan et al. 1993; Malin et al. 1993), and increased surface reflectance of up to 25% (Tyrrell et al. 1999). These signals represent a relative increase of up to 50 µatm in surface pCO_2 in association with alkalinity and water temperature changes (Holligan et al. 1993), and were calculated to cause a 17% reduction in CO_2 uptake over the spring/summer growth period in the NE Atlantic (Robertson et al. 1994).

Model projections of the effect of global warming predict that pCO_2 in surface waters will increase, in line with increases in atmospheric CO_2 (IPCC 2001), leading to a corresponding decrease in the pH of these waters (Kleypas et al. 1999), and potentially a decrease in calcification rate. In laboratory culture studies, Wolf-Gladrow et al. (1999) and Riebesell et al. (2000) have examined the effects on coccolithophorids of increased CO_2 concentrations over a range predicted from IPCC GCM climate change scenarios (such as the 'business as usual' scenario IS 92a, which will cause carbonate ion concentrations and seawater pH to drop by ca. 50% and 0.35 units, respectively, relative to pre-industrial values). Both studies point to reduced biogenic calcification. Riebesell et al. report that, over a range of mean surface ocean conditions between pre-industrial times and that expected by the year 2100, the ratio of calcification to organic matter production of *E. huxleyi* and *Gephyrocapsa oceanica* is predicted to have decreased by 23% and 50%, respectively. Scanning electron microscopy revealed that malformed coccoliths and incomplete coccospheres increased in relative numbers with increasing CO_2, suggesting that predicted shifts in seawater carbonate chemistry on a decadal timescale may reduce the production of calcium carbonate in the surface ocean and its subsequent transport to depth. The predicted increases in algal growth rate, enhanced algal C : P ratios and reduced calcification reported by Wolf-Gladrow et al. (1999) are predicted to all increase the ocean's capacity to store atmospheric CO_2 and thus act as a negative feedback.

7.7.2 Global Distributions

The high reflectance signature associated with high abundances of detached liths of coccolithophorids towards the end of a bloom can be viewed from space using the AVHRR (A Very High Resolution Radiometer) sensor (Aiken et al. 1992; Holligan et al. 1993). Brown and Yoder (1994) published a series of global composites of coccolithophorid distributions using satellite Coastal Zone Color Scanner (CZCS) composites that provided the first map of the distributions of *E. huxleyi* blooms. Indeed, by comparing observed biogeographical patterns with global maps of other oceanographic properties Brown and Yoder suggested that low silicic acid concentrations might be related to the distribution of coccolithophorids (see section on 'Controlling Factors and Modeling'). The launch of the SeaWiFS sensor in late 1997 has provided a means to remotely-sense cocolithophore calcite concentrations from space (Gordon et al. 2001) and provided further detailed global coverage of coccolithophorid distributions (Fig. 7.8). Recently observed distributional trends, such as the appearance of high coccolithophorid abundances in the subarctic NE Atlantic, are similar to those from CZCS during the 1976–1984 (Brown and Yoder 1994). Such distributions suggest that coccolithophorids are dominant in distinct regions and that their response to climate change or variability might be on a regional, as opposed to a global scale.

Recently, Iglesiais-Rodreguez et al. (2002) have used SeaWiFS data on coccolithophorid distributions to construct a detailed monthly time-series, and by using probability analysis and modeling to assess what oceanographic properties best fit with these distributions. They

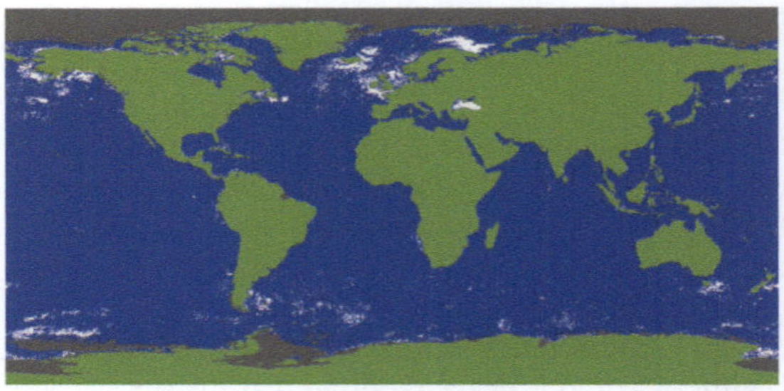

Fig. 7.8. An annual composite (1998/1999) of global coccolithophore distributions (denoted by *white areas*) derived from the SeaWiFS (Sea-viewing Wide-Field-of-View Sensor) in conjunction with an algorithm devised by Chris Brown

suggest that a composite proxy incorporating stratification, insolation and nitrate supply is most closely associated with the distribution of coccolithophorid blooms. Thus, satellite-derived coccolithophorid maps provide a powerful tool to monitor any future changes in coccolithophorid distributions, and may be compared with any concurrent shifts in maps of other oceanographic properties (see section on 'Regime Shifts').

7.7.3 Controlling Factors, Forcing and Modeling

Coccolithophorids are a major contributor to both carbonate and DMS production (see section on 'DMS and the Biota'), yet until recently comparatively little research has been devoted to the mode(s) of control on their distributions. What will be the effects of the projected higher ocean temperatures and stratification, will they alter the global distributions of coccolithophorids (Fig. 7.9)? Despite biogeochemical studies of coccolithophorids during JGOFS (Holligan et al. 1993) uncertainties remain regarding what prompts the onset of such blooms, with inconclusive results from modeling studies (Tyrrell and Taylor 1995). Also little is known about the factors leading to the decline and collapse of such blooms (Boyd et al. 1998). Causative factors thought to control coccolithophorid blooms include conditions of high insolation/shallow mixed layer depths (<30 m) thought to favour the photophysiology of coccolithophorids (Nanninga and Tyrrell 1996), high irradiances and low phosphate levels, (cells using dissolved organic rather than dissolved inorganic phosphate) (Tyrrell and Taylor 1995), low silicic acid concentrations (Brown and Yoder 1994) and/or viral control (Bratbak et al. 1996; Malin et al. 1998).

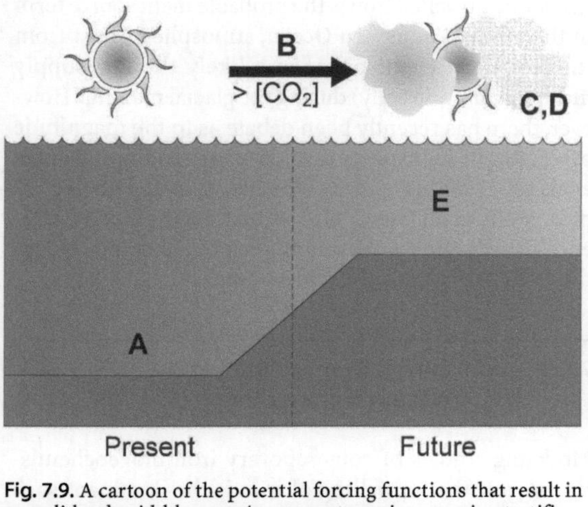

Fig. 7.9. A cartoon of the potential forcing functions that result in coccolithophorid blooms. *A* represents an increase in stratification/shoaling of the mixed layer depth (elevated mean light levels, Tyrrell et al. 1999), *B* an increase in atmospheric and hence upper ocean carbon dioxide levels (and its effect of calcification, Riebesell et al. 2000); *C* the degree of *cloudiness* (alteration of incident irradiances, Tyrrell and Taylor 1995); *D* changes in the aerosol dust supply (cobalt – Ellwood and van den berg 2001); and *E* increased *water temperatures* (Tyrrell et al. 1999)

Few studies have examined the relationship between trace element supply and coccolithophorid activity. Laboratory culture work on *E. huxleyi* isolates from HNLC NE Pacific waters indicate that they have low iron quotas relative to diatoms (Muggli et al. 1996). Sunda and Huntsman (1995) have shown that higher Co^{2+} concentrations in laboratory experiments promoted coccolithophorid growth, relative to that of diatoms, when both algal groups were grown under zinc-limiting conditions (i.e., cobalt may have been substituting for zinc in the enzyme carbonic anhydrase, Morel et al. 1994). Recently, Ellwood and Van den berg (2001) reported extremely low (<5 fM) free Co^{2+} concentrations in the NE Atlantic suggesting that coccolithophorids may be cobalt-limited in this province.

Climate modeling experiments, such as by Sarmiento et al. (1998), predict increased near surface/thermocline stratification due to warming at low latitudes. Thus, it would appear that temperature, indirectly by decreasing mixed layer depths, might alter the biogeography of coccolithophorids. Furthermore, the effects of altering temperature and/or mixed layer depth are likely to reduce upper ocean nitrate, or phosphate levels. Therefore, if the general consensus, that high mean irradiances and/or low macronutrient levels (Nanninga and Tyrrell 1996; Tyrrell and Taylor 1995; Iglesiais-Rodreguez et al. 2002), promote coccolithophorid blooms then this conclusion in tandem with model predictions points to increased incidences of such blooms in the future. The impact of such forcing (i.e., increased stratification etc.) on the distributions of coccolithophorids must be reconciled with that of increased pCO_2 on their physiology (Riebesell et al. 2000). Although calcification rates may decrease significantly in response elevated atmospheric CO_2 levels, these changes may be offset if the incidence of coccolithophore blooms was to increase significantly due to impact of other oceanic forcing.

7.7.4 A Case Study – the Bering Sea

There is recent evidence in these waters of unprecedented incidences of coccolithophorid blooms over the period 1997–2000 (Napp and Hunt Jr. 2001; see section on 'Regime Shifts'). A bloom (>200 km length-scale) was initially observed in July–August 1997. A further bloom was recorded in 1998, and by summer 2000 a bloom was observed here for the fourth consecutive year suggesting a potential regime shift (Napp and Hunt Jr. 2001). The blooms were visible in SeaWiFS imagery and the duration of the 1998 bloom was 9 months – much greater than previously observed in the NE Atlantic (Holligan et al. 1993). Reports of the bloom in each year indicate that upper ocean temperatures were 2 °C above normal.

Napp and Hunt Jr. (2001) report that anomalous regional weather over the SE Bering Sea in spring and summer 1997 resulted in significant changes in the chemical and biological oceanography of this region. That these observations may be related to climatic shifts such as elevated temperatures and changes in sea-ice melt patterns, strongly suggests that we do not yet fully understand what controls the development of coccolithophorid blooms.

7.8 Iron Supply to the Oceans

The Iron hypothesis (Martin 1990) was based on the relationship between the magnitude of dust (a proxy for iron) and atmospheric CO_2 concentrations in the geological record, with dust peaks coinciding with CO_2 minima during each of the glacial maxima. Martin proposed that alleviation of iron stress in phytoplankton would enhance rates of both primary and export production, and increase carbon sequestration to depth during the glacial maxima. Such a mechanism may exist today, such that any climate-induced alteration of iron supply to the ocean might alter the magnitude of global primary production. Hence, this relationship has received particular attention over the last decade (reviewed by de Baar and Boyd 2000). Denman et al. (1996) summarised that there was mounting evidence of the role of iron in enhancing phytoplankton production when macronutrients were available in excess. Since then, advances in this field have included debate on what sets global oceanic iron concentrations (Johnson et al. 1997); the first detailed models of iron biogeochemistry (Fung et al. 2000; Archer and Johnson 2000) and the response of the biota to iron supply (Lancelot et al. 2000; Hannon et al. 2001); comprehensive upper ocean iron budgets (Price and Morel 1999; Bowie et al. 2001); in situ confirmation of iron limitation of algal growth in the Southern Ocean (Boyd et al. 2000); remotely-sensing global maps of aerosol dust distributions (Stegmann and Tindale 1999); revised estimates of dust-borne iron flux to the oceans (Jickells and Spokes 2001); and new interpretations, from modeling studies, on the magnitude of iron-enhanced carbon sequestration during the glacial terminations (LeFevre and Watson 1999; Watson et al. 2000; Maher and Dennis 2001).

7.8.1 How Much of the Ocean Is Iron-Poor?

The areal extent of iron-poor waters is now thought to be more widespread than previously acknowledged (Mullineaux 1999). In addition to open ocean HNLC regions (equatorial and north-east Pacific Ocean, Southern Ocean) which comprise around 30% of the World Ocean (de Baar et al. 1999), HNLC waters have been reported in coastal regions (Hutchins et al. 1998) where, due to the complex interplay of geomorphology, sediment distributions and riverine input, waters are iron-poor. Large expanses of Low Nitrate Low Chlorophyll (LNLC) waters such as the tropical South Pacific are reported to have iron-limited resident cells (mainly picophytoplankton, Behrenfeld and Kolber 1999).

7.8.2 The Supply of Iron to the Ocean

Denman et al. (1996) discussed the need to obtain estimates of current natural and anthropogenic iron inputs to the atmosphere and the ocean, which are critical to improved understanding of the control of iron supply on primary production rates in HNLC regions. At present, sources of iron to the open ocean include atmospheric (Duce and Tindale 1991; Piketh et al. 2000), upwelling (Martin et al. 1989), ice melt (Sedwick and DiTullio 1997, Sedwick et al. 2000), volcanic eruptions (Watson 1997; Boyd et al. 1998), and extra-terrestrial dust (Johnson 2001). The relative rate of supply by each mechanism varies regionally; Martin et al. (1989) concluded that aerosol iron was the greatest supply term in the NE subarctic Pacific, whereas de Baar et al. (1995) reported that upwelling was most important in the Southern Ocean. In coastal waters, iron from resuspended shelf sediments (and subsequent upwelling), rather than from riverine sources, is thought to be the main supply term (Johnson et al. 1999).

In the geological past, there is evidence that the main modes of iron supply may have altered; based on a modeling study, LeFevre and Watson (1999) concluded that while upwelled iron is the probable main source term in the modern Southern Ocean, atmospheric dust from the Northern Hemisphere was likely the key supply mechanism (indirectly) during the glacial maxima. However, there has recently been debate as to the magnitude of Northern Hemisphere dust fluxes at this time (Watson et al. 2000; Maher and Dennis 2001). In the geological past, Wells et al. (1999) also report on the likely influence of tectonic processes on periodic iron inputs into the eastern equatorial Pacific Ocean.

7.8.3 Atmospheric Deposition of Iron vs. Upwelling Supply

Modeling studies of contemporary iron biogeochemistry suggest that upwelling of iron is the dominant supply term (Archer and Johnson 2000), whereas Fung et al. (2000) in a model using longer timescales conclude that aeolian supply of iron is dominant. Both agree upon the large uncertainties in the present models. Climate change may result in shifts in the rate of upwelling to the surface ocean due to the projected changes in the degree of stratification (Sarmiento et al. 1998), and/or changes in aerosol deposition of iron (temperature/aridification, Falkowski et al. 1998; Berman-Frank et al. 2001), which will alter the magnitude of global productivity. Here the question of how climate change might influence the atmospheric deposition of iron is considered in detail.

7.8.4 Dust Supply – Global Maps and Fluxes

Duce and Tindale (1991) published the first global map of the supply of aerosol iron to the oceans, which evinced low inputs to HNLC waters, and that most of the global supply is to Northern Hemisphere waters. Their study, as they point out, was characterised by a paucity of data. This, along with the episodic nature of dust supply (Prospero 2000), introduces large uncertainties as to the accuracy of dust flux estimates. Recently, some of these uncertainties have been overcome by using by-products of satellite sensors for aerosols (CZCS, Stegmann and Tindale 1999; AVHRR, Husar et al. 1997; SeaWiFS, Husar et al. 2001; TOMS, Fig. 7.10). Thus, the detection of interannual variability in the global dust input to the ocean may be estimated indirectly from satellite observations. Such datasets also provide useful information on the seasonality of dust input (Stegmann and Tindale 1999) the intermittency of dust storms, and on the main dust routes (Husar et al. 1997, 2001; Fig. 7.10).

Dust supply can be converted into iron supply using the crustal abundance of iron and by assuming a dust solubility factor (Duce and Tindale 1991). Recently, Jickells and Spokes (2001) conducted laboratory experiments on the solubility of aerosol particles, and these presently provide the most reliable estimates (overall solubility of atmospherically-transported iron at seawater pH is 0.8–2.1% of the total Fe deposited) of how much aerosol iron is available to the biota, and in what regions the main depositional fluxes would occur. Jickells and Spokes (2001) also present revised estimates of total dust flux ($400–1\,000 \times 10^{12}$ g yr^{-1}), total iron input ($0.25–0.63 \times 10^{12}$ mol yr^{-1}), and total soluble iron input ($0.2–1.32 \times 10^{10}$ mol yr^{-1}) to the oceans.

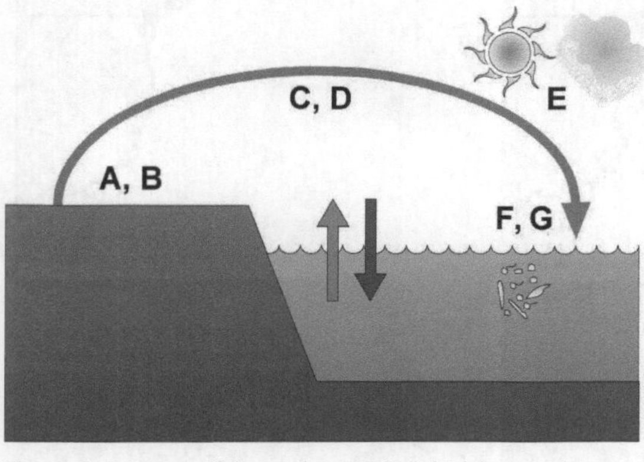

Fig. 7.10. A cartoon of the mechanisms by which iron can be transported from terrestrial sources to the open ocean. *A* Denotes changes in aridity, vegetation cover in arid environments (e.g., sensitivity to CO_2 fertilization and hydrology) and soil moisture (see also Fig. 7.11); *B* changes in land use (human activity) and/or in the dust size spectra; *C* alteration of atmospheric dust transport routes; *D* changes to the dust load/source-atmospheric carrying capacity, iron content and solubility; *E* nature (wet or dry) of deposition; *F* the iron status of the resident algal community (replete or deplete); *G* the iron requirements of the dominant species (diatoms vs. nitrogen fixers)

7.8.5 Dust Transport – from Soil to Phytoplankton

Many factors may influence the passage of dust from arid regions to the utilization by phytoplankton in the ocean (Fig. 7.10). In arid and semi-arid regions complex relationships between soil, micro-meteorology and dust production are apparent (detailed in Pye 1987; Nickovic and Dobricic 1996). Anthropogenic disturbances of soil (agriculture) and land use practices (Tegen and Fung 1995) will also influence significantly this transition from soil to dust. The uplifting and transport of dust depends on the size spectra of dust particles produced from the soil. Such transport is often episodic – via dust storms (Uematsu et al. 1983; Merrill 1989; Merrill et al. 1994), and both aridity and a rapid uplifting of the dust to altitudes above 1 km is needed to ensure long-range dust transport. Dust may then be deposited into the ocean by either dry or wet deposition which has implications for the solubility of aerosol particles (Jickells and Spokes 2001). The distance over which 50% of the dust load is lost from high altitude is thought to be around 500 km (Prospero et al. 1989). Finally, upon deposition into the ocean where a small proportion of the aerosol iron will be soluble, there are additional uncertainties, such as if and for how long the iron deposited will be 'bio-available' (Rue and Bruland 1997; Wu et al. 2001; Barbeau et al. 2001), and what are the iron requirements of the resident biota (nitrogen fixers have higher iron requirements than diatoms (Raven 1990; Berman-Frank et al. 2001).

7.8.6 Response by the Biota – Detection

What are the likely effects of dust inputs on the biota? DiTullio and Laws (1991) and Young et al. (1991) opportunistically sampled surface waters during a dust storm in the North Central Subtropical Pacific Gyre (NSPG), and observed transient increases in rates of primary production. Boyd et al. (1998) provided indirect evidence from the HNLC waters of the NE Pacific that episodic increases (of up to ten-fold) in phytoplankton stocks might have been driven by such episodic dust supply (Fig. 7.11). They also reported that the fate of such episodic algal blooms might be sedimentation to the deep ocean, resulting in up to 5-fold increases in POC export

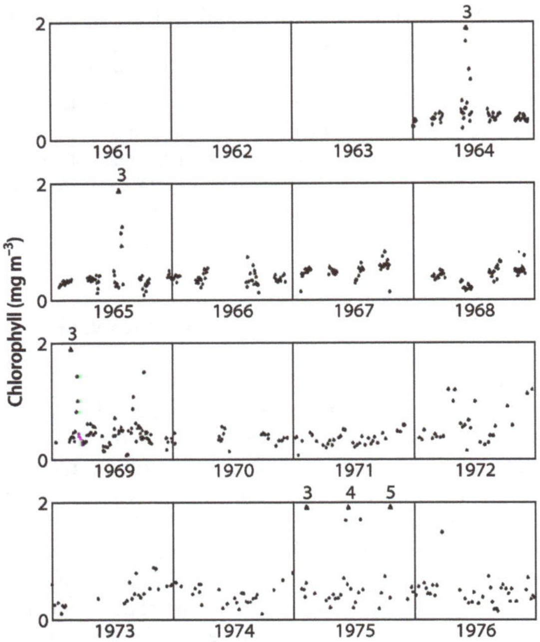

Fig. 7.11. A time-series of chlorophyll concentrations from the NE subarctic Pacific (Ocean Station Papa) in which episodic blooms in High Nitrate Low Chlorophyll (HNLC) waters were observed in 1964, 1965, 1969 and 1975 (numbers in each of these panels denote the highest chlorophyll levels (in mg m^{-3}) attained (data are from Parslow (1981))

(relative to ambient fluxes) over a month – equivalent to a 15% increase in POC export over the annual cycle. Clearly, the data of Boyd et al. (1998) indicate that the effects of dust supply in the present day may be episodic but are pronounced. Any increase in dust supply – due to climate change – might result in a relatively large response by the biota, but how can such a response be detected? Recent monitoring of HNLC waters has employed moored bio-optical sensors (Boyd et al. 1999) which record chlorophyll concentrations and other associated properties (temperature/irradiance) on an hourly basis. These platforms by providing the opportunity to monitor the expected biological responses to dust – such as increased levels of chlorophyll. Such mooring records, in conjunction with remotely-sensed evidence of dust storms (Fig. 7.12; Husar et al. 2001) may provide a means to both detect changes in dust supply to the ocean and the subsequent biological response.

7.8.7 The Future – Climate Change and Dust Deposition

In the future, if climate change results in elevated temperatures, it has been suggested that subsequent increases in aridification will result in enhanced dust input and increased primary productivity (i.e., a negative

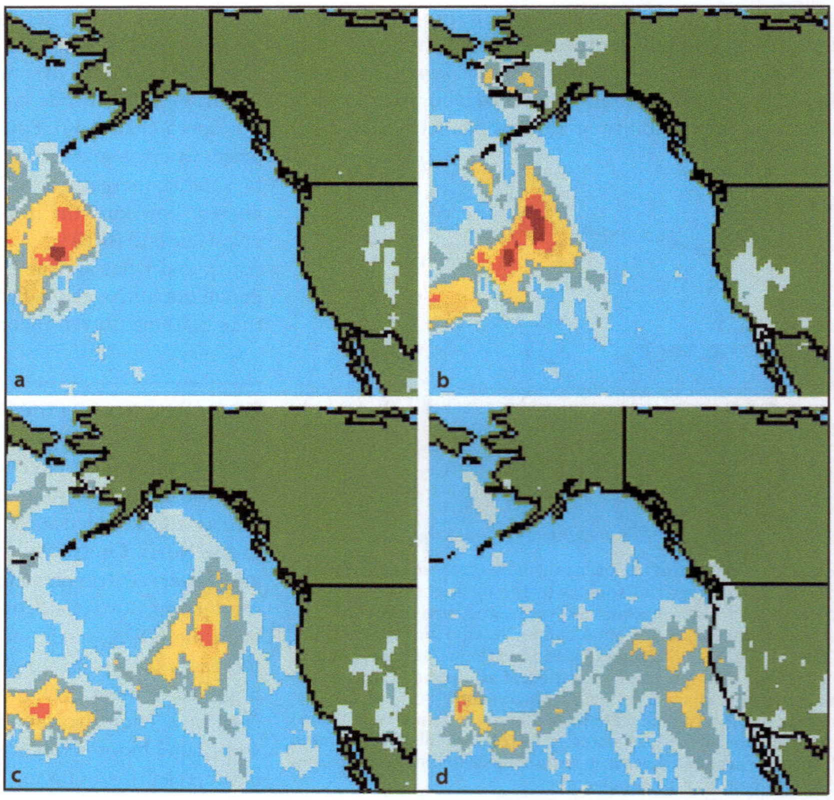

Fig. 7.12. Four consecutive daily plots of the aerosol index from the TOMS satellite in the North Pacific during the period 22–25 April 1998 (**a** April 22; **b** April 23; **c** April 24; **d** April 25;), showing the movement of a large dust storm from west to east. *Red* represents the highest dust concentrations, *blue* background levels. This was the largest dust storm to pass over the N Pacific in 10 years (Husar et al. 2001) but it did not appear to have any effect on the biota at Ocean Station Papa (P. Boyd, pers. comm.). This indicates that dust passing over the ocean may not always be deposited and thus such satellite images must be interpreted with due caution (images courtesy of TOMS/NASA)

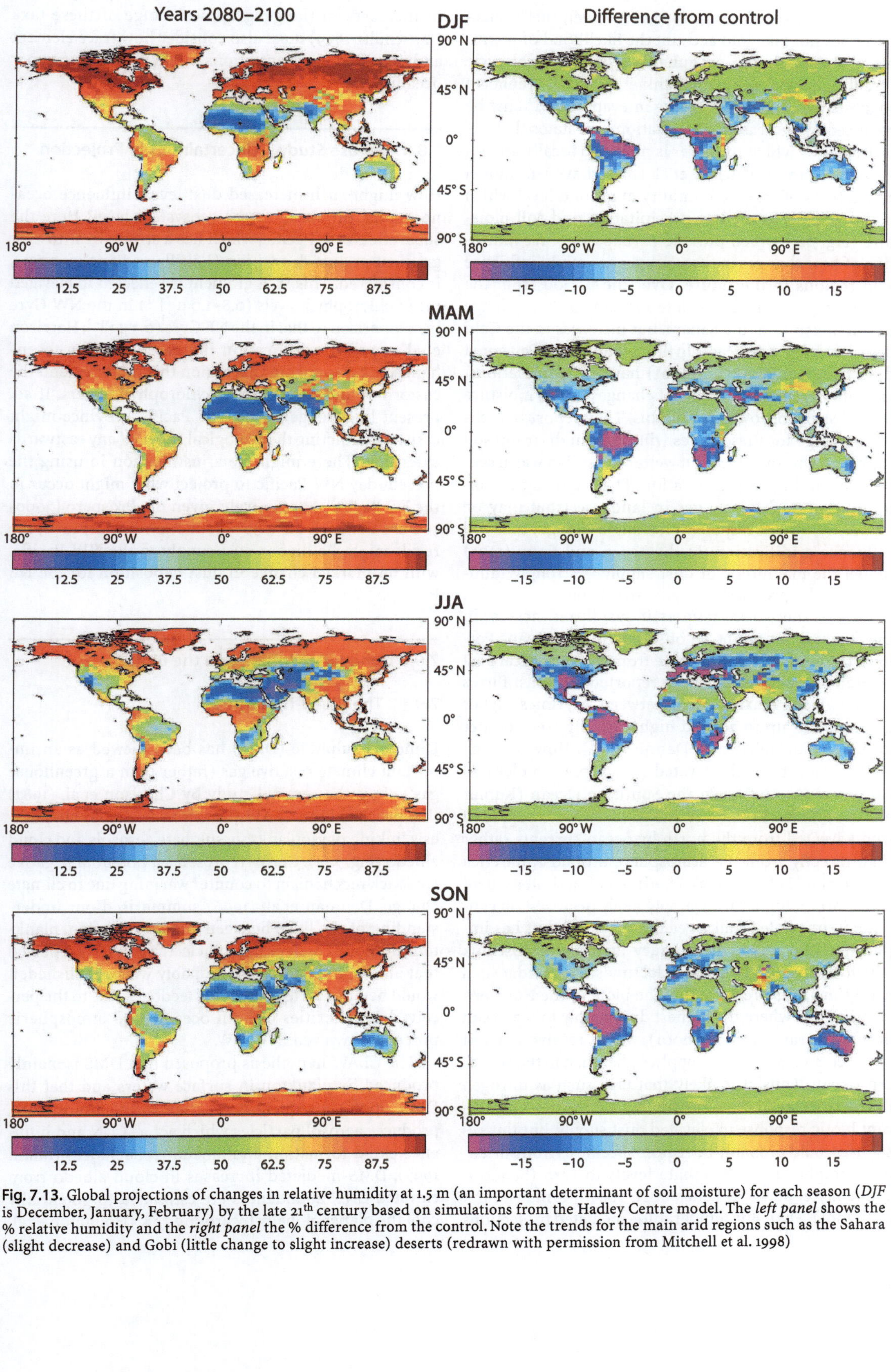

Fig. 7.13. Global projections of changes in relative humidity at 1.5 m (an important determinant of soil moisture) for each season (*DJF* is December, January, February) by the late 21th century based on simulations from the Hadley Centre model. The *left panel* shows the % relative humidity and the *right panel* the % difference from the control. Note the trends for the main arid regions such as the Sahara (slight decrease) and Gobi (little change to slight increase) deserts (redrawn with permission from Mitchell et al. 1998)

feedback; Falkowski et al. 1998). However, this projection does not take into account the likelihood of warming also resulting in an enhanced hydrological cycle (Trenberth 1998; Berman-Frank et al. 2001). Trenberth suggests that a global increase in evaporation must be balanced by increased precipitation, but states that the processes by which the latter is modified locally are not well understood. Mitchell et al. (1998) have run model simulations of relative humidity at ground level which will be closely related to precipitation and soil moisture (Fig. 7.13) which provide perhaps the first indication of how changes in these properties might influence arid regions in the future. Over the period from the present day to 2069–2099 there appear to be little change in soil moisture in the Sahara, but increases in the Gobi desert and slight decreases in the vicinity of the Patagonia desert region. Pan et al. (2001) have run simulations, validations, and projections of changes in soil moisture for the states of Iowa and Illinois. They report that climate change for these States will result in drier top soil (upper 10 cm) in winter, but wetter top soil in warm seasons due to higher precipitation. There is also the confounding issue of changes in land use (that may be caused by, or as a response to, climate change) that will influence aridification: in the Tegen and Fung (1995) model the magnitude of dust supply is strongly influenced by the proportion of disturbed soils.

Despite these uncertainties, if dust fluxes increase in the future what might be observed in the ocean? Perhaps the best indications come from the geological past where Mahowald et al. (1999) reported that dust fluxes during glacial maxima were between 2–3 times higher globally, and up to 20-fold higher locally (see counter argument in Maher and Dennis 2001). However, any contribution by such elevated iron supply to elevated carbon sequestration in the Southern Ocean (Kumar et al. 1995) may have been due primarily to a supply of iron advected from the north by ocean currents rather than directly from the atmosphere around Antarctica (Watson and Lefevre 1999). This input of, and subsequent transport of, higher iron supply likely occurred on centennial timescale (Watson et al. 2000), and thus has implications for any contemporary increase in dust flux which are likely to be on decadal timescales. Under such conditions, it is probable that the biota in the Northern Hemisphere, where the highest dust supply to the ocean occurs (Husar et al. 1997, 2001), would receive most of any such increase in iron supply, rather than in the Southern Ocean. Thus, it is likely that taxa such as nitrogen fixers, rather than diatoms, might yield the most significant biotic response to elevated dust supply, but this response would probably be region-specific as iron enrichment might drive phosphate levels to zero (Sanudo-Wilhelmy et al. 2001; Lenes et al. 2001; Berman-Frank et al. 2001). Thus, any temperature-mediated projections of increases in the geographical range of these taxa (Lipschultz 1999) must also consider how trace element and phosphate concentrations might alter in the upper ocean.

7.8.8 A Case Study – Uncertainties in Projection

How might such increased dust levels influence oceanographic provinces at the ecosystem level? Here the subarctic North Pacific, which has a strong west-to-east gradient in dust deposition (Jickells and Spokes 2001), is considered. This W-E gradient is reflected in elevated mean chlorophyll levels (0.8–1.0 $\mu g\ l^{-1}$) in the NW Gyre in comparison to that in the NE Gyre (0.3 $\mu g\ l^{-1}$; Harrison et al. 1999). If aridification increased, leading to enhanced dust supply in this region this might extend the eastern boundary of higher chlorophyll waters. If so, present knowledge of the NW Pacific Province might assist in predicting the ecological effect, of any eastwards extension. There might be a justification in using the present day NW Pacific to project what might occur in the NE Pacific. Alternatively, given the Prospero (2000) 500 km distance for a 50% reduction in dust load, the result might be higher dust supply to the NW Pacific, with no marked change in dust deposition for the NE Pacific.

7.9 Dimethyl Sulphide and the Biota

7.9.1 The CLAW Hypothesis

Dimethyl sulphide (DMS) has been viewed as an important climate reactive gas (rather than a 'greenhouse gas') since the seminal study by Charlson et al. (1987) that proposed the CLAW (Charlson et al. 1987) hypothesis linking phytoplankton, sulphate aerosols and cloud albedo (Fig. 7.14). CLAW represents a potential negative feedback mechanism to counter warming due to climate change. Denman et al. (1996) summarised our understanding of the links between DMS production, plankton and cloud condensation nuclei (CCN). They reported that although links were reasonably well established, it would be difficult to model this feedback due to the perceived complexities of both oceanic and atmospheric processes involved in CLAW.

The CLAW hypothesis proposed that DMS is mainly produced by plankton in surface waters and that this DMS, upon subsequent ventilation to the atmosphere, produces aerosol particles which act as CCN and influence cloud formation and albedo (reflectance) (Malin 1997). DMS-mediated increases in cloud albedo from elevated CNN abundances are estimated to potentially cool the planet globally by around 1 °K (Houghton et al.

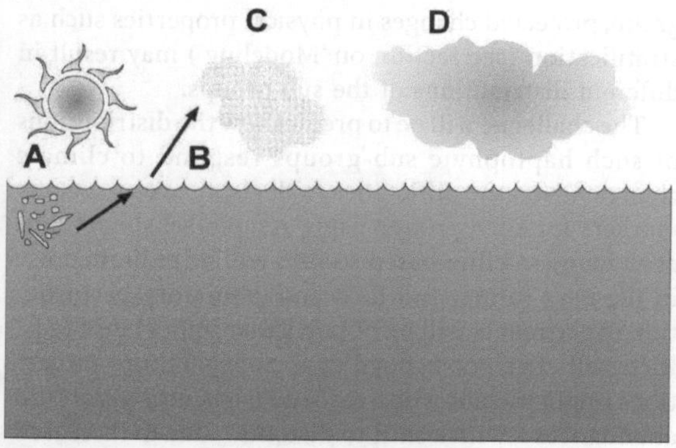

Fig. 7.14. A cartoon of the marine and atmospheric factors that may alter cloud cover via the sulphur cycle as outlined in the CLAW hypothesis. *A* Denotes haptophyte algae (coccolithophores, *Phaeocystis*, or 'HNLC' nanoflagellates) that produce dimethylsulfoniopropionate (DMSP); the distributions of these algal groups appears to be controlled by different environmental factors; *B* represents processes that mediate the conversion of DMSP to dimethyl sulphide (DMS) such as heterotrophic bacterial activity and micro-zooplankton herbivory; *C* and *D* denote atmospheric components of the CLAW hypothesis (many aspects of which have yet to be quantified), including the conversion of DMS to sulphur dioxide, and then to sulphuric acid. The sulphuric acid may subsequently form a new particle 'a cloud condensation nuclei' but this stage is dependent on the background abundances of condensation nuclei. As the Northern Hemisphere has higher background levels than the Southern, then DMS production in the latter will probably have a disproportionate effect on cloud formation and the resulting feedbacks (such as albedo)

1996). Furthermore, paleo-oceanographic evidence from the Vostok ice-core points to pronounced changes in Methanesulphonate (MSA, an atmospheric DMS oxidation product) between the glacials and interglacials (Legrande et al. 1991; Turner et al. 1996). Here, only the oceanic components of the CLAW hypothesis – which in February 2003 remains untested – are considered in detail.

7.9.2 What Produces DMSP/DMS?

Since 1995 there have been significant advances in understanding DMS production mechanisms (Malin 1997). Early studies investigated whether links existed between DMS production and bulk biological signals such as phytoplankton stocks and/or production. It was concluded that DMS levels were poorly related to such bulk signals (Holligan et al. 1987; reviewed by Kettle et al. 1999), and that the production of the precursor DMSPp appeared to be more closely related to specific aspects of phytoplankton processes. Furthermore, large changes in upper ocean DMSP and DMS concentrations have been observed during phytoplankton blooms (Holligan et al. 1993; DiTullio and Smith 1996), and mesoscale iron enrichments, such as IronEx II (Turner et al. 1996) in the equatorial Pacific, and SOIREE (Boyd et al. 2000) in the Southern Ocean. Therefore, algal community structure, in particular the haptophytes – such as coccolithophorids (Malin et al. 1993), *Phaeocystis antarctica* (DiTullio and Smith 1996), or 'HNLC' haptophytes (see section on 'The Haptophyte Connection') were reported to play a disproportionately important role in the production of DMSP.

Research into the mechanisms to convert DMSP to DMS have mainly focussed on grazer- or bacterially-mediated pathways. Wolfe et al. (1997) have consolidated their previous work (Wolfe et al. 1994) that proposed an important linkage between microzooplankton grazing and haptophytes (such as coccolithophorids) upon the lysis of the algae by DMSP lyase. Wolfe et al. (1997) further elucidated this mechanism suggesting that the algal production of DMSP is a grazer-activated chemical defence mechanism. Other studies have provided a new level of biochemical detail on the pathways/intermediates formed (such as DMSHB) involved in the synthesis of DMSP (Gage et al. 1997; reviewed by Simo 2001). It is also now acknowledged that much of the DMS released (up to 90%) is turned over by bacteria, or photo-oxidised to form non-volatile compounds (Malin 1997; Wolfe et al. 1999).

7.9.3 Global Distributions of DMS

Distributions of DMS levels in the surface ocean (15 000 observations) have been collated by Kettle et al. (1999). They produced maps of DMS concentrations based on the biogeochemical provinces of Longhurst (1995) and corrected following mapping procedures outlined in Conkwright et al. (1994). The maps (plate 5 in Kettle et al.) for mid-summer in each of the Northern and Southern Hemisphere point to high DMS concentrations in Southern Ocean polar waters (although the dataset is relatively small), and to high levels in the subarctic NE Atlantic and NE Pacific. Intermediate DMS concentrations were evident in both January and July in the HNLC waters of the Equatorial Pacific. In particular, the elevated DMS levels in the Southern Ocean may be disproportionately important for sulphur emissions/CLAW hypothesis as this region is characterized by generally low CNN/aerosol levels. Kettle et al. investigated whether global DMS distributions were correlated with other physical (temperature and salinity), chemical (DMSP) or biological (chlorophyll) factors but found none. Furthermore, they reported that no simple algorithm could be used to reproduce the observed monthly fields of DMS concentrations. Recently, Anderson et al. (2001) used the database of Kettle et al. and report success, upon expanding this database with highly resolved nutrient and irradiance datasets, in predicting the global fields of DMS using a combination of chlorophyll, irradiance and nutrients.

7.9.4 The Haptophyte Connection

Anderson et al. (2001) have produced reliable simulations of the DMS global fields, based on a relatively simple relationship amenable to incorporation into a GCM. However, they caution that it is unlikely that DMS feedbacks (such as shifts in DMSP-producing species) on the climate system could be represented and thus realistically simulated using this approach. In addition, reports of inconclusive attempts to relate DMSP/DMS to bulk phytoplankton signals (Watanabe et al. 1995; Curran et al. 1998) are not surprising given the specificity of the mechanisms for DMSP/DMS (Wolfe et al. 1997, 1999). From field studies the main producers of DMSP are haptophytes, but different taxa may be particularly important in specific oceanic provinces. There are at least three groups in the open ocean (Fig. 7.14) that may responsible for the trends in global DMS maps (Kettle et al. 1999). In the NE Atlantic the impact of coccolithophorid (Holligan et al. 1993), and to a lesser extent *Phaeocystis pouchetti* blooms (Owens et al. 1989; Marra et al. 1995; Liss et al. 1994) on DMSP/DMS levels has been reported. Another haptophyte, *Phaeocystis antarctica*, dominates blooms in the Ross Sea and in the vicinity of the marginal ice zone (MIZ), regions which are subsequently characterised by elevated DMSP/DMS levels (DiTullio and Smith 1996). The third group – the so-called 'HNLC' haptophytes (<20 µm autotrophic flagellates) have been reported to be important during mesoscale in situ iron enrichments in HNLC waters. During both the IronEx II (Turner et al. 1996) and SOIREE (Boyd et al. 2000) studies a tripling of surface DMSP concentrations was recorded, which followed the evolution of the algal bloom, and although no specific candidates as to the biological source of DMSP were proposed for the former, the timing of increases in haptophyte abundances was similar to that for increases in DMSP during SOIREE (Boyd et al. 2000).

To predict how such DMSP producers might respond to climate change, more information on the enviromental control of each haptophyte group is required. Iron supply plays a role in the increases in DMSP, via increases in HNLC haptophyte stocks observed in the Southern Ocean (Boyd et al. 2000), and equatorial Pacific waters (Turner et al. 1996). In contrast, for coccolithophorids several factors such as stratification, insolation, nutrients and/or cobalt concentrations may determine the frequency and location of blooms (see section on 'The Calcifiers'). Another mode of control may best explain the dominance of the Ross Sea and MIZ waters waters by *Phaeocystis* – that of low light environments characterised by deep mixed layers with weak stratification (Arrigo et al. 1999). Moisan and Mitchell (1999) investigated the photophysiology of this taxa, and their laboratory culture results support the field observations of Arrigo et al. (1999). Thus, even within the haptophyte group, projected changes in physical properties such as stratification (see section on 'Modeling') may result in different distributions of the sub-groups.

The challenge will be to predict how the distributions of such haptophyte sub-groups respond to climate change. Given the difficulties in devising taxon-specific markers for such groups using remote-sensing (Ciotti et al. 1999), satellite-based studies will be of limited use in the near future, and field and laboratory perturbation experiments will be of particular importance (e.g., Riebesell et al. 2000; Boyd et al. 2000). Future projections might include a decrease in *Phaeocystis antarctica* abundances – with a shift to diatoms – due to increased stratification as predicted by GCMs and mediated by climate change (Sarmiento et al. 1998). For projected changes in coccolithophorid distributions see section on 'The Calcifiers', while for 'HNLC' haptophytes see the section on 'Atmospheric Deposition of Iron'.

The main pathway for the conversion of DMSP to DMS may be grazer-mediated (Wolfe et al. 1994, 1997, 1999). During SOIREE, the timing and magnitude of an increase in heterotrophic ciliate abundances was invoked as the most likley mechanism to explain the observed increase in DMS (and concurrent decrease in DMSP) (Boyd et al. 2000). However, it remains unclear whether such an iron-mediated increase in haptophyte stocks followed by an increase in herbivore stocks is a transient response on a time-scale of weeks, or part of a longer term and ongoing series of predator-prey oscillations. Thus, it is presently unknown whether persistently elevated DMS concentrations would result from a sustained and higher iron supply to the ocean (such as during the last glacial maxima, or due to increased iron levels from aridification).

7.10 UV-B and Ozone Depletion

The erosion of the protective ozone layer by depleting compounds (such as CFCs) has resulted in oceanic biota being exposed to elevated UV-B (280–320 nm) solar radiation. Denman et al. (1996) concluded that phytoplankton species-specific sensitivity to UV-B could result in shifts in community structure, particularly at high latitudes. They also acknowledged UV-B effects on heterotrophs and DOC. Here findings and observations on UV-B since 1995 are assessed (Fig. 7.15).

7.10.1 Present Status of Ozone Depletion

A report on ozone depletion by the World Meteorological Organisation (WMO Report No. 44, 1999) concluded that the total combined abundance of ozone-depleting compounds in the lower atmosphere peaked around 1994 and is slowly declining. The WMO estimated that, based on projections on the maximum allowances un-

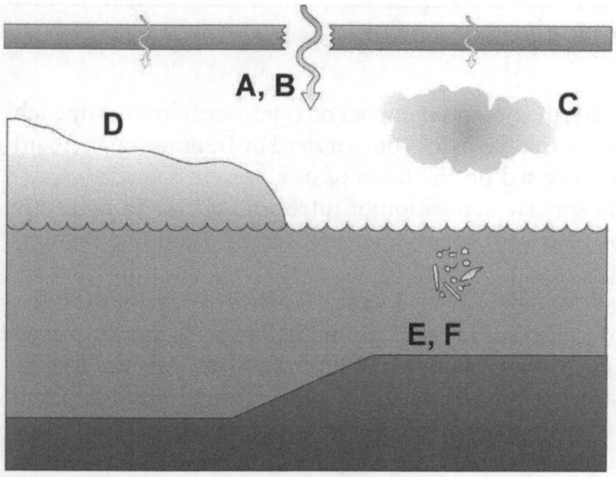

Fig. 7.15. A cartoon summarizing the factors controlling levels of UV-B reaching the upper ocean and its effect on the resident biota. *A* Denotes the influence of human activities resulting in the release of ozone-depleting compounds in the lower atmosphere (presently in decline under restrictions imposed by the Montreal Convention); *B* represents interactions between ozone depletion and the (cooling) effect of global warming on the stratosphere (Shindell et al. 1998); *C* the influence of cloud cover in the lower atmosphere (albedo) which can reduce UV-B penetration; *D* sea-ice cover which alters albedo (reflectance); *E* mixed layer depth, driven by wind stress and upper ocean stratification, which will also influence algal photoinhibition; *F* subsequent floristic and faunistic shifts than are driven by the degree of UV tolerance, in which organisms show differential sensitivity

der the Montreal Protocol, the period of peak ozone depletion is predicted to lie within at least the next 20 years. However, both identification of this maximum and evidence for the recovery of the ozone layer will require more than two decades. McKenzie et al. (1999) recently confirmed long-term decreases in ozone in the New Zealand region, and increases in the UV index. The authors point out that although stratospheric ozone loading is close to the minimum expected under the present control regime, there are concerns about the possible interactions between ozone depletion and the effects of global warming on the stratosphere. In this stratum, warming of the lower atmosphere results in radiative cooling which may delay the recovery of the ozone layer by decades (Shindell et al. 1998).

7.10.2 Phytoplankton and Primary Production

Previous studies established that elevated environmental UV-B reduces primary production rates in surface waters (Smith and Baker 1984; Fig. 7.15 this study), and led to the development of models on the spectral effects of UV on photosynthesis. Laboratory culture experiments, in which UV-B levels were manipulated, suggested that cells were able to counter the effects of UV-B inhibition within hours (Cullen and Lesser 1991). Since 1995, there has been progress in understanding the effects of UV-B on algal physiology, with more emphasis on field studies and refining predictive models. Studies in the Weddell-Scotia Confluence (WSC) by Neale et al. (1998a) reported that the relationship between UV-B exposure and inhibition of photosynthesis was more complex that previously thought. In contrast to previous laboratory experiments (Cullen and Lesser 1991), resident cells demonstrated no ability to counter UV-induced inhibition of photosynthesis over <4 h periods, implying that UV inhibition was a non-linear function of cumulative exposure (Neale et al. 1998a). Furthermore, studies of algal assemblages in WSC waters with contrasting hydrographic conditions, indicated that both photo-acclimatory ability and species-selection against less UV-tolerant species were important determinants of the sensitivity of the phytoplankton assemblage to UV radiation. Thus, algal UV tolerance was highest in shallower mixed layers (with a proviso – the mixed layer must have been established for sufficient time to allow photoacclimation) whereas low light adapted cells in deeper mixed layers had the lowest UV-tolerances.

Models to predict the relative biological effects of UV-B dosage must incorporate a Biological Weighing Function (BWF), i.e., a UV wavelength-dependent function (Cullen et al. 1992). Their model successfully described the majority of the spectrally-dependent experimental variations in photosynthetic rate in the WSC region. However, as the model yielded 6 specific BWFs for the resident cells, Neale et al. concluded that no single BWF was applicable to the WSC, and that the development of a suite of BWFs was required for Southern Ocean waters. A further modeling study (Neale et al. 1998b) of the interactive effects of ozone depletion and vertical mixing indicate that inhibition of photosynthesis can be enhanced or decreased by vertical mixing, and was dependent on both mixed layer depth and mixing rate. They conducted a sensitivity analysis of various factors to a worst case scenario in which ozone levels were suddenly reduced by half, and reported that column-integrated production could decrease by up to 10%. However, their analysis points to more pronounced inhibitory effects on primary production rates due to alteration of other factors that may be influenced by climate change such as vertical mixing rates, or the degree of cloudiness.

7.10.3 Dissolved Organic Matter and Heterotrophic Bacteria

UV-B radiation may also alter rates of marine bacterial and viral activity, and rates of photodegradation of DOM (Dissolved Organic Matter) from different sources (Moran and Hodson 1994). Recent studies indicate the relationship between UV radiation, DOM source, DOM reactivity, and subsequent utilisation by bacteria are complex (Pausz and Herndl 1999). They report that algal exudates of DOM become more resistant to bacterial utilisation upon UV exposure in the water column. This contrasts with the findings of Amon and Benner (1996) who observed that refractory DOC after UV exposure was more labile for bacterial utilisation. Other recent studies (Visser et al. 1999) have provided evidence of UV damage to heterotrophic bacterial DNA, and inhibition of protein and DNA synthesis in microbial communities in the Caribbean Sea.

7.10.4 Pelagic Community Response

In addition to specific UV damage or inhibitory effects on plankton at the molecular or physiological level (reviewed by Vincent and Neale 2000), the resulting interactive effects of UV damage at the ecosystem level must also be considered. Bothwell et al. (1994) observed that differential sensitivities to UV-B for freshwater algae and herbivores led to a counter-intuitive increase in algal biomass in habitats exposed to UV-B. Bothwell et al. cautioned that predictions of ecosystem response to elevated UV-B should not therefore be made using assessments of a single trophic level. A further plea for careful interpretation of experimental results was also made by Cabrera et al. (1997) who provided evidence of species-specific differences in UV sensitivity of herbivores in high altitude Andean lakes. Cabrera et al. advocated the careful definition of timescales for such manipulative experiments since different trophic groups may have different response times. A mesoscosm experiment by Mostajir et al. (1999) in the lower St Lawrence estuary indicates that UV-B exposure may alter the structure and dynamics of the pelagic foodweb, from large to small organisms, and thus favour a microbial rather than a herbivorous foodweb (sensu Legendre and Rassoulzadegan 1995). Thus, the carbon pool of the foodweb may primarily be within small organisms with implications for the proportion of carbon (or other elements) that are recycled or exported from the system.

7.10.5 The Future

Uncertainties remain as to how such inhibitory effects of elevated UV-B on the biota will interact with other projected shifts due to climate change. For example, the implications of predicted increases in upper ocean stratification, particularly in the Southern Ocean, due to increased freshening (Sarmiento et al. 1998) must be carefully related to the complex relationship between vertical mixing and UV-B dosage (Neale et al. 1998b). Superimposed on these changes, are the often counter-intuitive effects of UV-B radiation at the ecosystem level, and the potential changes in other factors controlling UV-B levels, such as cloud cover, aerosol extinctions, stratospheric temperatures, or albedo (McKenzie et al. 1999).

7.11 Summary of Biotic Feedbacks

Here, the potential impact on open ocean waters by each of the five feedbacks put forward by Denman et al. (1996) are revised on the basis of our present knowledge. Atmospheric deposition of nitrogen appears to presently be restricted to the coastal waters and continental margins, and projected changes will be greatest in these regions (Galloway et al. 1994). Further to previous reports of variations in the C:N Redfield ratio (Sambrotto et al. 1993) other examples of alterations of nutrient uptake stoichiometry have been put forward (Hutchins and Bruland 1998; Arrigo et al. 1999). However, caution is urged as to how robust and significant such trends in carbon: nitrogen uptake stoichiometry might be over longer timescales (Kortzinger et al. 2001). In contrast, iron-mediated decreases in the silicic-acid:carbon uptake ratio, if sustained over centennial timescales, may play a significant role in modulating atmospheric CO_2 levels during the glacial maxima (Watson et al. 2000).

Laboratory studies on coccolithophorids have demonstrated that elevated atmospheric CO_2 concentrations result in decreased calcification (Riebesell et al. 2000), yet little is known about what controls the onset or decline of these blooms of calcifiers. This points to the dangers of making projections based on only one set of perturbed conditions, and to the need for a holistic approach whereby the effects of climate-mediated shifts at all levels (physiological to biogeographical) are taken into account when making such projections. There is now evidence that the main DMSP producers reside within three distinct sub-groups of the haptophytes. However, blooms by each group may be triggered by different environmental factors; much may be learnt from the approaches developed to determine what factors control phytoplankton processes in HNLC regions (reviewed by de Baar and Boyd 2000). Iron limitation of phytoplankton growth is now more widespread that previously thought and has been observed in both coastal HNLC and oceanic LNLC regions. Defining the controls on iron supply in the present, geological past, and future will be central to understanding the biogeochemical functioning of much of the surface waters of the World Ocean. Although the Montreal Convention appears to be slowly redressing the 'balance' with respect to ozone depletion which is forecast to decrease on a timescale of decades, uncertainties remain on the interactive effects of UV-B and global warming.

Denman et al. (1996) in their IPCC review considered these main potential marine feedback mechanisms discretely. However, a point may have been reached whereby it is simplistic to think in terms of discrete feedbacks, since there are strong links between them, such as the role of iron supply on the algal uptake ratio of silicic acid to carbon, or that of elevated UV-B levels on the photo-

Table 7.1. Examples of recently reported regime and domain shifts in phytoplankton community structure, at higher trophic levels, and potential impacts on the magnitude of the biological pump. NSPG denotes North Subtropical Pacific Gyre; *MIZ* denotes Marginal Ice Zone. Earlier reports of potential regime shifts include Southward (1963, English Channel); Venrick (1971), Venrick et al. (1987) (NPSG); Aebischer et al. (1990) and Colebrook et al. (1984) and Colebrook (1986) (NE Atlantic)

Biota	Region	Reference
Phytoplankton – nitrogen fixers	NPSG	Karl et al. (1995, 1999); Karl (1999)
Phytoplankton	NE subarctic Pacific	Whitney et al. (1998)
Phytoplankton – pico to large diatoms	Equatorial Pacific	Chavez et al. (1999)
Small to large phytoplankton[a]	NE subarctic Pacific	Parslow (1981)
Coccolithophores	Bering Sea	Napp and Hunt Jr. (2001)
Phytoplankton	NE Atlantic and North Sea	Reid et al. (1998)
Euphausids	Bering Sea	Napp and Hunt Jr. (2001)
Zooplankton and salmon	NE subarctic Pacific	Beamish et al. (1999)
Calanoid copepods	North Sea	Heath et al (1999)
Krill and Salps	Polar Southern Ocean	Loeb et al. (1997)
Krill and Salps	Southern Ocean – East Antarctica	Nicol et al. (2000)
Penguin stocks	MIZ – Antarctic Peninsula	Smith et al. (1999)
Cod stocks	North Sea	O'Brien et al. (2000)
Elevated export to depth	NE subarctic Pacific	Boyd et al. (1998)
Fulmar populations	N Atlantic	Thompson and Ollason (2001)

[a] By inference – see Boyd et al. (1998).

chemistry of iron or DOM. Denman et al. acknowledged the significance of observations by Karl et al. (1995) of higher abundances of nitrogen fixers at the HOT site as evidence of a marked change in the composition of a pelagic community (often termed a regime or domain shifts, Karl et al. 1997). However, Denman et al., while also citing previous evidence of interdecadal variations in planktonic stocks in the N Pacific (e.g., Venrick 1971, 1982; Venrick et al. 1987), cautioned that potentially confounding effects of methodological changes over time might account for much of this apparent variability.

One striking trend since 1995 has been the number of studies reporting evidence of regime or domain shifts (summarised in Table 7.1), and how they may provide the best clues as to the likely response of the biota to climate change (see below).

7.12 Climate – Variability vs. Change

7.12.1 Climate Change

There is evidence of pronounced shifts in the magnitude of many properties over the geological record – for example in the Vostok ice core there are gradients of 6–7 °C in temperature, and accompanying shifts of 100 ppm in atmospheric CO_2, tenfold changes in dust (iron) levels, and five-fold shifts in both MSA concentrations and methane levels. This contrasts with a 20 ppm change in atmospheric CO_2 between 1958 and 1990 (Keeling et al. 1996; Archer and Johnson 2000). Research has focussed on the relative timing of these events to assist in the identification of forcing functions for shifts in temperature and CO_2, and the phasing of these changes such that the mechanisms for abrupt climate change may be elucidated. Plausible biotic mechanisms considered so far include elevated iron supply and nitrogen fixation (Broecker and Henderson 1998). In a recent modeling study, Watson et al. (2000) conclude that increased iron supply during the last glacial periods may account for up to 30% of the observed 100 ppm change in atmospheric CO_2, and that multiple feedbacks are probably responsible for pronounced biogeochemical gradients at the glacial terminations. These paleo-oceanographic studies provide estimates of the magnitude of change due in part to feedbacks, and offer compelling evidence for relatively abrupt changes in properties brought upon by non-linear responses (Hasselmann 1999; Rahmstorf 1999). Yet such changes are mainly on centennial or millennial timescales, and thus tell us little about the response time in the present day of feedbacks to climate change predicted to occur on annual to decadal timescales (Sarmiento et al. 1998).

In only a few cases from the geological past are there records of changes occurring on decadal timescales as are predicted from present models. One example is derived from changes in nitrogen and argon isotopes in trapped air from the Greenland ice record which indicates that the Greenland summit warmed 9 ±3 °C over a period of several decades around 15 000 years ago. Furthermore, atmospheric methane concentrations (possibly from wetlands) rose abruptly over a 50 yr period not long after the onset of this abrupt temperature change (Severinghaus and Brook 1999; see Introduction to this chapter). The mechanisms behind other climatic anomalies such as the Medieval Warming Period or the Little Ice Age may also assist in explaining the magnitude of such events (Reid 1997). However, to obtain a mechanistic understanding of the role of such feedbacks, more comprehensive datasets than can currently be provided by the paleo-oceanographic record are required.

7.12.2 Climate Variability

A close examination of trends in atmospheric CO_2 concentrations over the last three decades suggest that large shifts, due to interannual climate variability such as the effects of El Niño events, are superimposed onto the pronounced rate of increase in CO_2 concentrations (Sarmiento 1993). Furthermore, there is strong evidence of links between such transient events and modes of atmospheric variability – indices of climate such as El Niño Southern Oscillation (ENSO), or NAO (North Atlantic Oscillation). For example in the North Atlantic, Sutton and Allen (1997) analysed patterns of sea surface temperature during the winter and reported that the trends appeared to be 'remarkably organised' over a decadal time-scale. McCartney (1997) proposed that the long-term variability in westerly winds (as evidenced from the NAO) may be driven by temperature variations in the ocean (see counter-argument in Seagar et al. 2000). While such climatic variability potentially confounds the detection and interpretation of any such climate change, it does provide information on the response times of oceanic processes to climatic variability (and vice-versa). Furthermore, with the improved resolution of ocean time-series observations (such as BATS) there are increasing opportunities to compare observed trends in biogeochemical properties with such climatic indices.

Although the dataset is small, several studies suggest indirectly that oceanic biota may have a pronounced and rapid response to climate variability, with a concomitant effect on atmospheric CO_2 concentrations (Fig. 7.1). The biota may respond to climate variability or change, by alteration of rates of bulk properties, such as primary production, and/or via shifts in ecosystem structure (Falkowski et al. 1998). Recently, Behrenfeld et al. (2001) provided the most comprehensive estimate of how rates of primary production might alter in response to marked climate variability. They report an increase of around 10% in global primary production (derived from SeaWiFS via a productivity algorithm) over the 3 yr duration of an El Niño event, i.e., a negative feedback.

In contrast, because ecological processes are highly dynamic (Field et al. 1998) little is known about the timing or magnitude of climate-mediated floristic and/or faunistic shifts, and how they might in turn impact climate via feedbacks. However, detailed oceanographic datasets, from SeaWiFS, time-series sites, mooring arrays and shipboard surveys/experiments have provided clues as to how ecosystem structure might respond to climate variability (Table 7.1). These studies suggest that ecological shifts (often termed regime or domain shifts) may subsequently influence the biogeochemical cycles of several elements via feedback mechanisms. As such, these detailed observations on climate-mediated regime shifts may be viewed as fortuitous experiments under which conditions for an entire ecosystem were altered for a sustained period of time (months), such as shoaling the mixed layer depth (Karl et al. 1995), or in-

Fig. 7.16. An example of a rapid (six months) shift in algal stocks and community structure in the waters of the E Equatorial Pacific as evidenced by a ribbon like filament of high chlorophyll waters (*green*) surrounded by HNLC waters (*blue*). Shipboard sampling of these waters indicated that they were diatom-dominated (rather than the usual picophytoplankton dominated HNLC community), with signi-ficantly high rates of primary and export production being driven by elevated upwelling of iron-rich waters during the La Nina condition (SeaWIFS data reprinted with permission from Chavez et al. (1999) Biological and chemical response of the equatorial Pacific Ocean to the 1997–1998 El Niño. Science 286: 2126–2131, © 1999, American Association for the Advancement of Science)

creasing the rate of upwelling of nutrients (Chavez et al. 1999; Fig. 7.16) in a way that could not be mimicked in the laboratory or in mesocscale field experiments (Coale et al. 1996). This approach offers the best means to investigate the response time of the biota to climatic variability and hence will likely provide our best indicators of how biotic feedbacks might respond to climate change.

Such evidence of regime shifts may represent a departure in how climate-mediated alteration of pelagic marine systems are perceived. Previously, researchers examined datasets with a view to detecting changes in the phytoplankton signals. Several studies have suggested little evidence in the archives of change in algal stocks etc (Falkowski and Wilson 1992), whereas most of the reported regime shifts are characterised by floristic changes which may have marked implications for the strength of feedback mechanisms (nitrogen fixers, *Phaeocystis*, and also large vs. small cells and their influence on particulate export flux, Fig. 7.17). Thus, it has led to perhaps the largest single shift in the emphasis of research since Denman et al. (1996), the investigation of how feedbacks are controlled or forced at the taxon-level rather than solely via alteration of bulk signals.

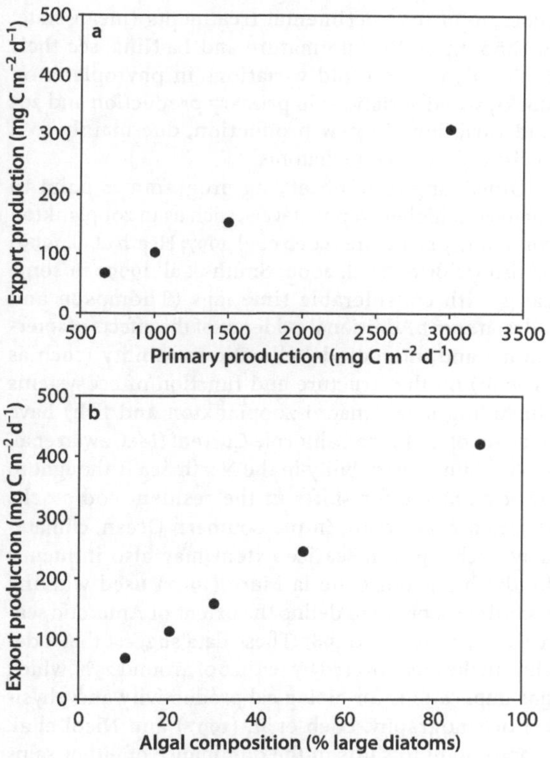

Fig. 7.17. A comparison of the effect of increasing rates of primary production without changing community structure (in a population dominated by picophytoplankton, with few large cells), and altering community structure from small to large cells without increasing rates of primary production (mean rate is 660 mg C m^{-2} d^{-1}) on the potential export production from the base of the mixed layer to depth. Projections are based on the approach of Boyd and Newton (1999) and using data from the HNLC NE subarctic Pacific (Harrison et al. 1999)

7.12.3 Regime Shifts

There is increasing evidence for strong links between climate variability and changes in biological populations (Smith et al. 1999; Wuethrich 2000). Such variations in climate may result in a shift in algal taxa (e.g., Karl et al. 1995, 1997) and may also have implications at higher trophic levels (Heath et al. 1999, Table 7.1). In terrestrial systems climate variability may result in complex biological responses resulting from the triggering of several chains of ecological events that cascade through an ecosystem (Wuethrich 2000; Smith et al. 2000). Such shifts are referred to in the literature as regime or domain shifts but care is required in the usage of this term, and a distinction should be made between isolated events or curios, and shifts that are repeatedly observed over several years (summary in Table 7.1).

In the ocean, there is compelling evidence of regime shifts in response to climatic variability (such as based on SOI and El Niño events) in both the NPSG (Karl et al. 1995, 1997, this volume; Karl 1999) and in the NE subarctic Pacific (Whitney et al. 1998). Both regions are characterised by long time-series records, suggesting that regime shifts of a similar magnitude may also be occurring in other oceanic regions but have so far not been detected due to a paucity of data. Several other studies provide evidence of a shift in phytoplankton community structure in response to climatic changes (Reid et al. 1998; Chavez et al. 1999; Napp and Hunt Jr. 2001). Here three examples of such shifts are considered in detail.

In the NPSG (HOT site) Karl et al. (1995) reported ecosystem changes in response to the 1991–1992 El Niño. They present evidence of a shift from a nitrogen- to a phosphorus-limited system caused by a floristic shift towards greater abundances and activity of nitrogen-fixers. Karl et al. summarise this domain shift as resulting in changes in the proportion of nitrogen fixers during El Niño and non El Niño years, with enhanced nitrogen fixation in response to increased stratification in El Niño year (which has major implications for nutrient cycling in this region and for future shifts due to climate change), and a suggestion that such a shift in the degree of nitrogen fixation is driven by nitrogen limitation. The increased rates of nitrogen fixation results in a phosphate-limited ecosystem (for details see Karl et al., this volume). In the Western Tropical Pacific, Hansell and Feely (2000) hypothesise that water column stratification, forced by high net precipitation, favours enhanced rates of nitrogen fixation in these waters. Such changes may be driven by a more active hydrological cycle in this region due to increased tropical ocean temperatures. Projected increases in upper ocean stratification due to future climate change (Sarmiento et al. 1998) might result in a shift towards increased nitrogen fixation and hence a negative feedback. Furthermore, Lipschultz (1999) has suggested that with the projected warming of the ocean there might be a potential polewards increase in the geographical range of nitrogen fixers. The ability to remotely-sense blooms of nitrogen fixers is required to answer these contentions (Subramaniam et al. 1999a,b; Dupouny et al. 2000).

Whitney et al. (1998) provide evidence of interannual variability in nitrate supply to surface waters of the NE subarctic Pacific (Fig. 7.18), and report shifts between the El Niño/La Nina cycles in salinity, temperature and nitrate. During the El Niño period the reduced nitrate supply was estimated to have reduced new production over the growth season by 40% (2 million t C) over a 290 000 km^2 area. This resulted in a shift to an oligotrophic system (characterised by a subsurface chlorophyll maximum) as opposed to the previously observed mesotrophic system akin to the NE Atlantic. There is anecdotal evidence of subsequent changes in the magnitude of production and the composition of the higher trophic levels due to this climatic variability (Beamish et al. 1999; Mantua et al. 1997).

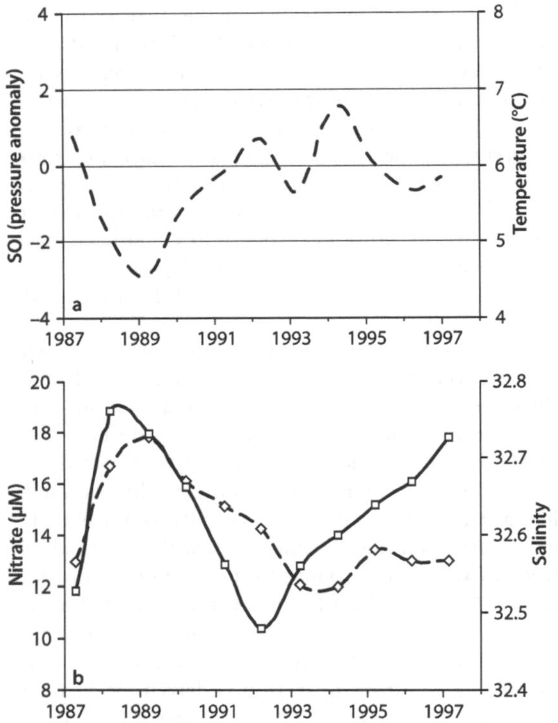

Fig. 7.18. a Changes in the Southern Oscillation Index (SOI, derived from the monthly mean pressure anomalies between Tahiti and Darwin, Australia) for the N Pacific in relation to **b** the observed shifts in winter reserve nitrate and salinity for these waters between 1993 and 1997. The nitrate changes resulted in massive fluctuations in the productivity of this region between these years (redrawn with permission from Whitney et al. (1998) Interannual variability in nitrate supply to surface waters of the Northeast Pacific Ocean. Marine Ecology Progress Series 170:15–23)

Other regime shifts in the NE subarctic Pacific include transient events in which phytoplankton processes were altered in the HNLC waters (Ocean Station Papa, OSP). In three of ten years there was evidence of a temporary shift from HNLC conditions to those characterised by transient bloom events (Parslow 1981; Fig. 7.17 this study). Such events, may be due to diatoms as evidenced by occasional exceptionally large drawdown of surface silicic acid levels at OSP (Wong and Matear 1999; see section on 'Atmospheric Deposition of Iron').

In another province, the equatorial Pacific, Chavez et al. (1999) recently reported marked biological and chemical changes driven by the 1997–1998 El Niño event. There were shifts in the upwelling rate of macronutrients, CO_2 and iron, resulting in subsequent changes in community structure, increases in phytoplankton stocks, and in the proportion of new production during La Niña compared to the mean, El Niño (onset and mature) conditions. Thus, variations in conditions resulted in four states, akin to experimental treatments (mean state, El Niño onset, El Niño mature and La Niña, see their Table 7.1), with 20-fold variations in phytoplankton stocks, 5-fold variations in primary production and 20-fold variations in new production, due mainly to a floristic shift to large diatoms.

Other long-term observing programmes point to changes at higher trophic levels, such as in zooplankton community structure (Loeb et al. 1997; Heath et al. 1999) or fish (O'Brien et al. 2000; Smith et al. 1999) in some cases with considerable time lags (Thompson and Ollason 2001). Additional evidence of the effects of interannual and interdecadal climatic variability (such as El Niño) on the structure and function of ecosystems (including meso, macro-zooplankton and fish) have been reported in the California Current (McGowan et al. 1998). Climate variability in the North Sea is thought to be responsible for shifts in the resident cod stocks (O'Brien et al. 2000). In the Southern Ocean, climate-driven changes in sea-ice extent may also influence foodweb structure. de la Mare (1997) used whaling records as a proxy to define the extent of Antarctic sea-ice between 1931 and 1987. These data suggest that a decline in the area covered by sea-ice of around 25%, which has implications for biological productivity and physical oceanography. Loeb et al. (1997) and Nicol et al. (2000) point to shifts in the dominance of either salps or krill in response to shifts near the ice-edge.

The reportage of a wide range of regime shifts points to a change in the dominant flora or fauna which may have marked effects on biogeochemical cycles due to both alterations in community structure and rates of production (Karl et al. 1997). Several studies evince a rapid response time for such shifts, in some cases within

6 months (Chavez et al. 1999). Due to the wide range of these shifts, they may be less amenable to model or predict than that, for example, of a direct (temperature-driven) increase in the rate of growth or primary production with no concurrent floristic shift. Moreover, these domain shifts appear to regional in scale (Karl et al. 1995; Whitney et al. 1998), and indicate the need to assess such changes or feedbacks on a regional basis (sensu Longhurst 1995, 1998) before attempting to determine the cumulative or global effects of biotic feedbacks due to climate change (see section on 'The Future'). Thus, to comprehensively estimate the influence of such regime shifts, scales of biotic response from physiological (Beardall et al. 1998) to shifts in the areal extent of biogeographical provinces must be taken into account.

7.12.4 Unexpected Biological Responses to Climate Change

Although the significance of regime shifts, and their use as proxies of biotic responses to climate change, has only recently been acknowledged, the environmental control of such shifts may be amenable to modeling. For example, in the Equatorial Pacific study by Chavez et al. (1999) during the La Niña state the enhanced upwelling of macronutrients and iron resulted in a diatom-dominated assemblage, a similar response as observed in equatorial Pacific iron enrichment experiments (Price et al. 1994; Coale et al. 1996). However, our ability to model and hence predict regime shifts in the future will be limited if such shifts result in the dominance of species that are unexpected. Valuable lessons may be learnt from the terrestrial biosphere (Pounds et al. 1999) where

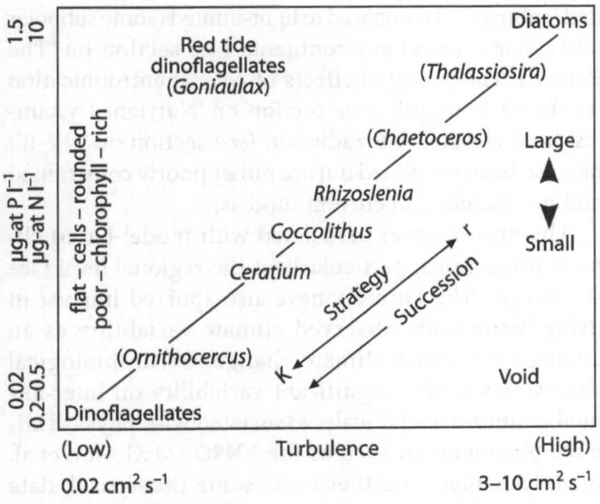

Fig. 7.19. The phytoplankton mandala of Margalef that summarises conceptually the role of forcing (gradients in nutrient supply and turbulence) in determining phytoplankton species and size composition (reprinted with permission from Margalef R (1997) Our biosphere. In: Kinne O (ed) Excellence in Ecology. Ecology Institute, Oldendorf)

"widespread amphibian extinctions in seemingly undisturbed highland forests may attest to how profound and unpredictable the outcome can be when climate change alters ecological systems". In the ocean the proliferation of toxic algal blooms in the nearshore eutrophic environment (Paerl 1999) or the recent sustained presence and overwintering of a cocolithophore bloom in the Bering Sea (Napp and Hunt Jr. 2001) may be examples of such unpredictable changes. Several terrestrial studies have investigated the ecological impact of species invasions (either natural or anthropogenic) and conclude that there is considerable difficulty in predicting the outcome, due to the transient nature of such systems (Petchley et al. 1999; Sait et al. 2000). In the ocean, Margalef (1997) has come closest to exploring such ecological concepts. His phytoplankton mandala (nutrient concentrations vs. turbulence, Fig. 7.19) is an example of the type of predictive conceptual frameworks that will be required to tackle these issues.

The evidence presented for regime shifts suggests that future climate change is unlikely to result solely in the alteration of one trophic level, that such shifts may be abrupt, and have subsequent ramifications for higher trophic levels, the export of carbon to depth, and for pathways of nutrient recycling (Karl et al. 1995). Thus, a multi-stranded approach comprising laboratory culture (Riebesell et al. 2000) and mesoscale (Coale et al. 1996) experiments, remote-sensing of algal distributions (Fig. 7.8), and time-series observations (Fig. 7.11) are needed to predict the nature and geographic extent of regime shifts in the ocean in response to climate change.

7.13 Modeling – Future Goals

Numerical models play an important dual role in climate change research with respect to ocean biogeochemistry. Firstly, physical climate models are required to estimate the manner and degree to which the environment may change, and secondly ocean biogeochemical models are needed to predict the subsequent response to and possible feedbacks of biota and biogeochemistry to these forcings. The two are of course linked because one of the main driving factors of climate change, atmospheric CO_2 levels, is modulated by the ocean carbon cycle (Siegenthaler and Sarmiento 1993). Indeed, the future levels of atmospheric greenhouse gases such as CO_2 have become one of the major uncertainties associated with climate predictions through the next few centuries (Hansen et al. 1995). Robust ocean biogeochemical models are required for hind-casting past historical climate change and projecting future effects and feedbacks. Particular focus should be on the nature, times-scale and magnitude of changes that might be observed over the next few decades, providing observationalists attempting to detect climate change with some suggestions for future fieldwork.

A summary of state of biogeochemical ocean modeling at the time of the last IPCC report (Denman et al. 1996) commented on the limitations of the available coupled atmosphere-ocean GCMs. They reported that the coupled GCMs included only the physical climate system with no explicit inclusion of the chemical or biological processes related to climate-reactive gases and atmospheric aerosols. The only consideration of the relationships between climate change and biological processes was via box models (Shaffer and Sarmiento 1995). At that time Denman et al. pointed to the lack of data to formulate, calibrate and evaluate marine climate effect models as being the greatest limitation to modeling efforts. Recently, there have been substantial improvements with coupled GCMs now including physical and biogeochemical feedbacks in a simplified form.

The results from current coupled ocean-atmosphere models on the extent of secular climate change predicted for the next century under greenhouse warming scenarios are at present based on a series of four coupled ocean-atmosphere experiments looking at the effects of climate change on ocean sequestration (Sarmiento et al. 1998; Joos et al. 1999; Matear and Hirst 1999; Cox et al. 2000). These simple ocean carbon cycle models have been run with either intermediate complexity (e.g., Joos et al. 1999) or fully coupled 3-D atmosphere-ocean GCM simulations (Sarmiento et al. 1998; Matear and Hirst 1999; Cox et al. 2000). The physical effects of warming of the surface ocean and reduction in meridional overturning, convective mixing, and isopycnal mixing combined lead to a 30–40 µatm decrease in CO_2 over the next century. The biogeochemical response is governed by two opposing factors, a reduction in the upward nutrient supply due to the increased stratification, which leads to decreased export production and CO_2 uptake, and a decrease in the upward vertical flux of dissolved inorganic carbon. The latter factor generally dominates in the present (biologically-unsophisticated) simulations, so that the effect of altered biogeochemistry is a net positive CO_2 uptake partially compensating the physical effects.

Numerical models are being improved dramatically as better, data-based mechanistic parameterizations are developed and evaluated based in part on the JGOFS process study, time-series and global survey data (Doney 1999; Doney et al., this volume). With respect to climate change, processes that need further observational and modeling attention include (Doney and Sarmiento 1999): direct physiological effects due to alterations in ocean temperature, pH etc.; sensitivity of HNLC regions, particularly in the Southern Ocean and subtropical waters (with resident nitrogen fixers) to changing dust deposition; and response of algal community structure, export flux and subsequently air-sea carbon exchange to projected increased water column vertical stability. As shown in Fig. 7.20, a number of potential feedbacks are possible, both local (e.g., wind speed or buoyancy forcing altering mixed layer depth) and non-local associated with changes in large-scale circulation patterns and the depth of the nutricline etc. Observational (Polovina et al. 1995) and preliminary modeling studies (Bopp et al. 2001) suggest that enhanced stratification can have different impacts on rates of primary production in nutrient-limited (subtropics) compared to light-limited (some subpolar and polar oceans) environments (see section on 'The Future'). The potential effects of coastal eutrophication on the carbon cycle (see section on 'Nutrient Dynamics') and elevated UV radiation (see section on 'UV-B') has also been suggested but are rather poorly constrained and not included in current models.

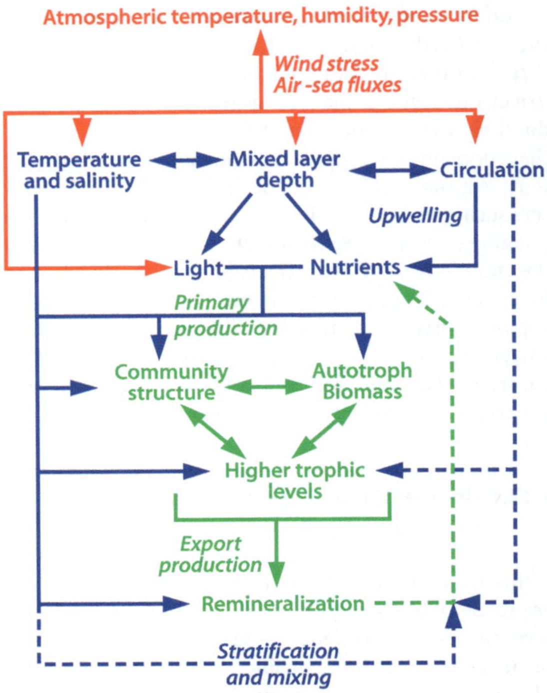

Fig. 7.20. Schematic of ecosystem level responses to physical factors of climate change including both rapid local effects (e.g., sea surface temperature, mixed layer depth) and longer term, non-local effects (e.g., sub-surface nutrient distributions)

The uncertainties associated with model-based climate projections, particularly at the regional level (see section on 'The Future'), have also spurred interest in using historically observed climate variability as an analogy for future climate change. Ocean biological time-series exhibit significant variability on inter-annual to inter-decadal scales associated with physical climate phenomenon such as the ENSO (McGowan et al. 1998; Karl 1999). Furthermore, some theoretical, data analysis, and modeling studies argue that climate change may project significantly onto these existing modes of natural variability (Corti et al. 1999). Thus, one could infer the biogeochemical climate change response by extrapolating present distributions during climate mode extrema. However, the projected physical climate change for the next century greatly exceeds observed natural variability, and the ecosystem response to physical forcing may be time-scale dependent and quite non-linear (Fig. 7.20). Most climate variability studies focus on bottom-up control of autotrophic production (e.g., nutrient supply, irradiance), but top-down control from higher trophic levels are also possible. The feedbacks associated with gyre scale circulation, large-scale nutrient distribution, planktonic community structure, and top-down controls have inherently longer time-scales and may not be well resolved by interannual climate variability. The paleoclimate record offers another valuable resource (see section on 'Climate Change') yielding information about the state of ocean biogeochemistry under climate perturbations of similar magnitude (though not rate of change) to projected anthropogenic effects and to a variety of climatic conditions both colder and warmer than today.

7.14 The Future

A central goal of JGOFS was to assess and understand the sensitivity to climate change of the ocean carbon cycle, and to "improve our ability to predict future climate-related changes". The decade long JGOFS programme has enabled us to reduce uncertainties in our understanding of the carbon cycle but, as reported here, to make smaller advances in our predictive abilities with respect to climate change. The legacy of JGOFS will be to provide datasets, directions and design criteria for future programmes – the progeny of JGOFS – to supply a suite of projections for the biogeochemical shifts induced by climate change, and recommendations on how to approach the detection of such changes, and on monitoring approaches to recognise any 'unexpected' ecological shifts (Pounds et al. 1999).

7.14.1 Detection and Projection

Are there preferred modes (amplitudes of patterns) of natural variability (Corti et al. 1999) and how can they be detected (Fedorov and Philander 2000)? The time-series data obtained during JGOFS and the launch of the SeaWiFS sensor in 1997 have provided powerful tools to detect, with confidence, any dramatic shifts in the composition of the biota. Observations of such shifts in several cases have been linked with indices of climatic variability. Furthermore, such observations are at the ecosystem or domain level and provide an invaluable 'domain wide' set of trends resulting from climatic variability. In the future the present suite of ocean observatories must be supplemented in selected locations that represent key biogeochemical provinces. These observatories will provide the detection of shifts and/or feedbacks to climate change. In tandem with detection will be projection, derived from increasingly sophisticated coupled ocean atmosphere models embedded with sub-models with higher resolution biological detail. These two strands together offer 'Earth system' analysis – a macroscope to look at global change (Schellnhuber 1999).

7.14.2 Does the 'Initial' Condition Still Exist?

Climate-mediated shifts in ocean properties may already be underway (Sarmiento et al. 1998; Broecker and Henderson 1998; Calderia and Duffy 2000) such as shutting down of midwater convective processes in the Southern Ocean. This in conjunction with the relatively large oscillations around the mean state due to climatic variability (IPCC 2001) may make specifying the 'baseline' condition, i.e., prior to any marked change in climate, difficult. Thus, a further requirement of field programmes, in addition to ocean observatories (McGowan 1990; Macilwain 2000), will be additional surveys and process voyages to obtain the 'baseline' condition in oceanographic provinces. Leading examples of such comprehensive studies include the JGOFS time-series sites, the CALCOFI study (McGowan 1999), the Gulf of Mexico observatory (Blaha et al. 2000), and in the Ross Sea by the ROVVERS group (which also includes a modeling component; Arrigo et al. 1999, 2000).

7.14.3 The Need for a Regional Approach

The siting of additional observatories should reflect one of the main conclusions of this chapter – the regional nature of biotic feedbacks – in contrast to physico-chemical feedbacks, which tend mainly to be global in their impact. Although the modeling study of Joos et al. (1999) suggests that the feedbacks with the greatest magnitude are physico-chemical, the impact of increased stratification on ocean ecosystems is likely to be pronounced (Falkowski et al. 2000). At present little is known globally about the effects of stratification on floristic or faunistic shifts (Margalef 1997). Shifts in the degree of stratification and depth of the mixed layer will probably have different effects on the dominant biota in each oceanic province (Fig. 7.21). For example, in the NPSG it may result in a shift towards nitrogen fixers (Karl et al. 1995, 1997), or a shift from a *Phaeocystis* to a diatom-dominated community in the Ross Sea (Arrigo et al. 2000; Moisan and Mitchell 1999), whereas in the NE Atlantic altered stratification may lead to the predominance of coccolithophores (Tyrrell and Taylor 1995), or diatoms in the Equatorial Pacific (Chavez et al. 1999). Each floristic shift will have associated feedbacks with different signs and magnitude.

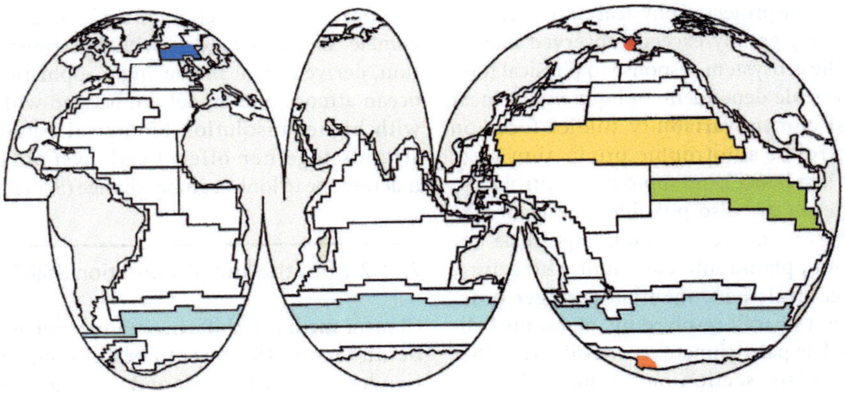

Fig. 7.21. A summary of the likely effects of increased stratification and decreased mixed layer depths on flora in various ocean provinces, based on the scheme of Longhurst (1995, 1998). Note the forcing will result in different outcomes in each region. *Yellow* denotes increased abundances of nitrogen fixers (Karl et al. 2001; a negative feedback); *brown* represents fewer *Phaeocystis* but more diatoms (Arrigo et al. 1999; positive feedback). *Green* denotes a shift to more diatoms (see Fig. 7.17, Chavez et al. 1999; a positive and negative feedback – as waters with higher carbon dioxide concentrations will also be upwelled along with iron). *Red* denotes the persistence of coccolithophorid blooms in the Bering Sea that may have been initiated by the warming of surface waters and increased stratification (Napp and Hunt Jr 2001); *blue* represents subantarctic waters where a decrease in mixed layer depth – in the presence of iron enrichment – results in elevated diatom stocks (Boyd et al. 1999; Blain et al. 2001); *purple* denotes changes in phytoplankton stocks and the length of the growth season in response to climate variability (Reid et al. 1998)

7.14.4 A New Definition of Biogeochemical Provinces?

JGOFS has produced comprehensive biogeochemical datasets from all major oceanic regions. These datasets, which provide an unprecedented level of detail, enable us to refine further the definition of biogeochemical provinces. They also, for the first time, permit us to focus in detail on the complex physicochemical and biological feedback mechanisms that may result from climate change. The observed regional nature of regime shifts and hence biotic feedbacks requires that we adopt a regional-based approach akin to the 'ecological geography' provinces put forward by Longhurst (1995, 1998). Existing criteria for dividing the ocean into provinces has been criticised and referred to as arbitrary (McGowan 1999). Perhaps it is now time to better define ecological boundaries with new goals in mind, to relate them to our improved understanding of ocean physics from the WOCE (World Ocean Circulation Study), to reformulate them so that they are not static and rigid but can be dynamic and alter in time in step with projected changes in ocean circulation. This will be a considerable task beyond the scope and efforts of one individual (Banse 1998). However, this approach will yield a more reliable assessment of the magnitude, evolution, ranking and degree of confidence that presently available for each biotic feedback. It would also be helpful to relate such site-specific observations regionally using remote sensing (Hansen et al. 1995; Chavez et al. 1999). By linking this regional approach with remote-sensing may reduce the dangers of upscaling – 'the process of extrapolating from the site-specific scale at which observations are usually made to the smallest scale that is resolved in global scale models' (Parmesan 1996; Harvey 2000).

7.15 Summary

The most dramatic shift in our perception of marine biotic feedbacks over the last five years has been the widespread occurrence of regime shifts in response to climate variability. Their detection has been possible by using ocean time-series datasets, the first step towards an ocean-observing system. Observations indicate that such regime shifts are relatively rapid, may be forced by a number of different environmental factors, and influence many aspects of biogeochemistry, such as nutrient inventories and the fate of fixed carbon. Furthermore, these regime shifts and their associated feedbacks encompass and overlap the original five feedbacks identified by Denman et al. (1996), suggesting that such biological responses to climate change are likely to be interlinked and complex. In contrast to many of the physicochemical feedbacks, such as stratification or gas solubility, which are amenable to global modeling (Kleypas et al. 1999), biotic feedbacks as manifested by regime shifts appear to be regional in nature. Thus, any modeling attempt must take place initially at this regional level. Datasets from both WOCE and JGOFS will be useful in understanding better how each feedback and/or shift is mediated and forced. Perturbation experiments can provide data needed on the physiological and ecological responses, and this information together with data from satellites on global distributions and geographical extent may be used to define the environmental mode(s) of control on taxa. In addition, a series of ocean-observing systems will be required in the main biogeochemical provinces to permit further assessment of these shifts over time. Such observatories, in tandem with remote-sensing and mathematical models will be needed to obtain 'baseline' conditions for each biogeochemical province. The final goal will be to assess the cumulative impact of such biotic feedbacks, in response to climate change, on the global ocean.

Acknowledgements

We are grateful to C. Brown, F. Chavez, A. Kortzinger, and J. Mitchell for the provision of figures and are indebted to the IPCC and US-JGOFS, respectively for providing unpublished manuscripts (including the relevant chapters of the IPCC 2001 draft report, and a draft copy of the US-JGOFS SMP 1999 report) that assisted with the preparation of this chapter. We thank M. Fasham and an anonymous reviewer for the insights and comments which helped improve this chapter. We acknowledge the help of Martin Manning and David Wratt (NIWA, NZ) for sign-posts into the vast literature on climate change. This chapter is dedicated to the vision of Alan Longhurst and his 'ecological geography' provinces, and to the foresight of the people who put JGOFS together back in Paris in the mid-eighties.

References

Aebischer NJ, Coulson JC, Colebrook JM (1990) Parallel long-term trends across 4 marine trophic levels and weather. Nature 347: 753–755

Aiken J, Moore GF, Holligan PM (1992) Remote sensing of oceanic biology in relation to global climate change. J Phycol 28(5):579–590

Amon RMW, Benner R (1996) Bacterial utilization of different size classes of dissolved organic matter. Limnol Oceanogr 41(1):41–51

Anderson LA, Sarmiento JL (1994) Redfield ratios of remineralization determined by nutrient data analysis. Global Biogeochem Cy 8:65–80

Anderson TR, Spall SA, Yool A, Cipollini P, Challenor PG, Fasham MJR (2001) Global fields of sea surface dimethylsufide predicted from chlorophyll nutrients and light. J Mar Res 30:1–20

Archer DE, Johnson K (2000) A model of the iron cycle in the ocean. Global Biogeochem Cy 14:269–279

Arrigo KR, Robinson DH, Worthen DL, Dunbar RB, DiTullio GR, VanWoert M, Lizotte MP (1999) Phytoplankton community structure and the drawdown of nutrients and CO_2 in the Southern Ocean. Science 283:365–365

Arrigo KR, DiTullio GR, Dunbar RB, Robinson DH, VanWoert M, Worthen DL, Lizotte MP (2000) Phytoplankton taxonomic variability in nutrient utilization and primary production in the Ross Sea. J Geophys Res 105:8827–8846

Banse K (1994) Uptake of inorganic carbon and nitrate by marine plankton and the Redfield ratio. Global Biogeochem Cy 8:81–84

Banse K (1998) Review of Longhurst's 'Ecological geography of the sea'. Limnol Oceanogr 43:1763

Barbeau K, Rue EL, Bruland KW, Butler A (2001) Photochemical cycling of iron in the surface ocean mediated by microbial iron (III) – binding ligands. Nature 413:409–413

Bates NR, Knap AH, Michaels AF (1998) Contribution of hurricanes to local and global estimates of air-sea exchange of CO_2. Nature 395:58–61

Battle M, Bender ML, Tans PP, White JWC, Ellis JT, Conway T, Francey RJ (2000) Global carbon sinks and their variability inferred from atmospheric O_2 and $\delta^{13}C$. Science 287:2467–2470

Beamish RJ, Noakes DJ, McFarlane GA, Klyashtorin L, Ivanov VV, Kurashov V (1999) The regime concept and natural trends in the production of Pacific salmon. Can J Fish Aquat Sci 56:516–526

Beardall J, Beer S, Raven JA (1998) Biodiversity of marine plants in an era of climate change: some predictions based on physiological performance. Bot Mar 41:113–123

Behrenfeld MJ, Kolber ZS (1999) Widespread iron limitation of phytoplankton in the South Pacific Ocean. Science 283:840–843

Behrenfeld MJ, Bale AJ, Kolber ZS, Aiken J, Falkowski PG (1996) Confirmation of iron limitation of phytoplankton photosynthesis in the equatorial Pacific Ocean. Nature 383:508–511

Behrenfeld MJ, Randerson JT, McClain CR, Feldman GC, Los SO, Tucker CJ, Falkowski PG, Field CB, Frouin R, Esaias WE, Kolber DD, Pollack NH (2001) Biospheric primary production during an ENSO transition. Science 291:2594–2597

Berman-Frank I, Cullen JT, Shaked Y, Sherrell RM, Falkowski PG (2001) Iron availability, cellular iron quotas, and nitrogen fixation in *Trichodesmium*. Limnol Oceanogr 46(6):1249–1260

Bigg GR (1996) The oceans and climate. University Press, Cambridge, UK

Blaha JP, Born GH, Guinasso NL, Herring HJ, Jacobs GA, Kelly FJ, Leben RR, Martin RD, Mellor GL, Niiler PP, Parker ME, Patchen RC, Schaudt K, Scheffner NW, Shum CK, Ohlmann C, Sturges W, Weatherly GL, Webb D, White HJ (2000) Gulf of Mexico ocean monitoring system. Oceanography 13:10–15

Blain S, et al. (2001) A biogeochemical study of the island mass effect in the context of the iron hypothesis: Kerguelen Islands, Southern Ocean. Deep-Sea Res Pt I 48:163–187

Bopp L, Monfray P, Aumont O, Dufresne J –L, Le Treut H, Madec G, Terray L, Orr JC (2001) Potential impact of climate change on marine export production. Global Biogeochem Cy 15:81–99

Bothwell ML, Sherbot DMJ, Pollock CM (1994) Ecosystem response to solar ultraviolet-B radiation: influence of trophic-level interactions. Science 265:97–100

Boville BA, Gent PR (1998) The NCAR Climate System Model, version one. J Climate 11:1115–1130

Bowie AR, Maldonado MT, Frew RD, Croot PL, Achterberg EP, Mantoura RFC, Worsfold PJ, Law CS, Boyd PW (2001) The fate of added iron during a mesoscale fertilisation experiment in the Southern Ocean. Deep-Sea Res Pt II 48:2703–2745

Boyd PW, Newton PP (1999) Does planktonic community structure determine downward particulate organic carbon flux in different oceanic provinces. Deep-Sea Res Pt I 46:63–91

Boyd PW, Wong CS, Merrill J, Whitney F, Snow J, Harrison PJ, Gower J (1998) Atmospheric iron supply and enhance vertical carbon flux in the NE subarctic Pacific: Is there a connection? Global Biogeochem Cy 12:429–441

Boyd PW, Sherry ND, Berges JA, Bishop JKB, Calvert SE, Charette MA, Giovannoni SJ, Goldblatt R, Harrison PJ, Moran SB, Roy S, Soon M, Strom S, Thibault D, Vergin KL, Whitney FA, Wong CS (1999) Transformations of biogenic particulates from the pelagic to the deep ocean realm. Deep-Sea Res Pt II 46:2761–2792

Boyd PW, Watson A, Law CS, Abraham E, Trull T, Murdoch R, Bakker DCE, Bowie AR, Buesseler K, Chang H, Charette M, Croot P, Downing K, Frew R, Gall M, Hadfield M, Hall J, Harvey M, Jameson G, La Roche J, Liddicoat M, Ling R, Maldonado M, McKay RM, Nodder S, Pickmere S, Pridmore R, Rintoul S, Safi K, Sutton P, Strzepek R, Tanneberger K, Turner S, Waite A, Zeldis J (2000) A mesoscale phytoplankton bloom in the polar Southern Ocean stimulated by iron fertilisation of waters. Nature 407:695–702

Bratbak G, Wilson W, Heldal M (1996) Viral control of *Emiliania huxleyi* blooms. J Marine Syst 9(1–2):75–81

Broecker WS (1990) Comment on 'Iron deficiency limits phytoplankton growth in Antarctic waters' by John H. Martin et al. Global Biogeochem Cy 4:3–4

Broecker WS (1997) Thermohaline circulation, the Achilles heel of our climate system: Will man-made CO_2 upset the current balance? Science 278:1582–1594

Broecker WS, Henderson GM (1998) The sequence of events surrounding Termination II and their implications for the cause of glacial-interglacial CO_2 changes. Paleoceanography 13: 352–364

Brown CW, Yoder JA (1994) Coccolithophore blooms in the global ocean. J Geophys Res 104C:1541–1558

Cabrera S, López M, Tartarotti B (1997) Phytoplankton and zooplankton response to ultraviolet radiation in a high-altitude Andean lake: short- versus long-term effects. J Plankton Res 19:1565–1582

Caldeira K, Duffy PB (2000) The role of the Southern ocean in uptake and storage of anthropogenic carbon dioxide. Science 287:620–622

Capone DG, Zehr JP, Paerl HW, Bergman B, Carpenter EJ (1997) *Trichodesmium*: a globally significant cyanobacterium. Science 276:1221–1229

Carlson CA, Ducklow HW, Michaels AF (1994) Annual flux of dissolved organic carbon from the euphotic zone in the northwestern Sargasso Sea. Nature 371:405–408

Carpenter EJ (1973) Nitrogen fixation by *Oscillatoria* (*Trichodesmium*) *thiebautii* in the southwestern Sargasso Sea. Deep-Sea Res 20:285–288

Carpenter EJ, Romans K (1991) Major role of the cyanobacterium *Trichodesmium* in nutrient cycling in the North Atlantic Ocean. Science 254:1356–1358

Charlson RJ, Lovelock JE, Andreae MO, Warren SG (1987) Oceanic phytoplankton, atmospheric sulphur, cloud albedo and climate. Nature 326:655–661

Chavez FP, Strutton PG, Friederich GE, Feely RA, Feldman GC, Foley DG, McPhaden MJ (1999) Biological and chemical response of the equatorial Pacific Ocean to the 1997–98 El Nino. Science 286:2126–2131

Chen C-TA, Liu KK, McDonald R (2003) Continental margin exchanges. In: Fasham MJR (ed) Ocean biogeochemistry, Chap. 3. Springer-Verlag, Heidelberg (this volume)

Christian JR, Karl DM (1995) Bacterial ectoenzymes in marine waters: activity ratios and temperature responses in three oceanographic provinces. Limnol Oceanogr 40:1042–1049

Christian JR, Lewis MR, Karl DM (1997) Vertical fluxes of carbon, nitrogen and phosphorus in the North Pacific subtropical gyre near Hawaii. J Geophys Res 102:15667–15677

Ciotti A-M, Cullen JJ, Lewis MR (1999) A semi-analytical model of the influence of phytoplankton community structure on the relationship between light attenuation and ocean color. J Geophys Res 104(C1):1559–1578

Coale KH, Johnson KS, Fitzwater SE, Gordon RM, Tanner S, Chavez FP, Ferioli L, Sakamoto C, Rodgers P, Millero F, Steinberg P, Nightingale P, Cooper D, Cachlan WP, Landry MR, Constantinou J, Rollwagen G, Trasvina A, Kudela R (1996) A massive phytoplankton bloom induced by an ecosystem-scale iron fertilization experiment in the equatorial Pacific Ocean. Nature 383:495–501

Codispoti LA (1989) Phosphorus vs. nitrogen limitation of new and export production. In: Berger WH, Smetacek VS, Wefer G (eds) Productivity of the ocean: present and past. pp 377–394, John Wiley & Sons, New York

Codispoti LA (1995) Is the ocean losing nitrate? Nature 376:724

Colebrook JM (1986) Environmental influences on long-term variability in marine plankton. Hydrobiology 142:309–325

Colebrook JM, Robinson GA, Hunt HG, Roskell J, John AWG, Bottrell HH, Lindley JA, Collins NR, Halliday NC (1984) Continuous plankton records: a possible reversal in the downward trend in the abundance of the plankton of the North Sea and the Northeast Atlantic. Journal du Conseil International pour l'Exploration de la Mer 41:304–306

Conkwright M, Levitus S, Boyer TP (1994) NOAA atlas NESDIS 1, world ocean atlas 1994, vol. 1: nutrients. Technical report, Natural Environment Satellite, Data and Information Service, Nat. Oceanic and Atmospheric Administration, U.S. Department of Commerce, Washington D.C

Cornell SE, Jickells TD, Thornton CA (1998) Urea in rainwater and atmospheric aerosol. Atmos Environ 32:1903–1910

Corti S, Molteni F, Palmer TN (1999) Signature of recent climate change in frequencies of natural atmospheric circulation regimes. Nature 398:799–802

Cox PM, Betts RA, Jones CD, Spall SA, Totterdell IJ (2000) Acceleration of global warming due to carbon-cycle feedbacks in a coupled climate model. Nature 408:184–187

Cullen JJ, Lesser MP (1991) Inhibition of photosynthesis by ultraviolet-radiation as a function of dose and dosage rate – results for a marine diatom. Mar Biol 111:183–190

Cullen JJ, Neale PJ, Lesser MP (1992) Biological weighting function for the inhibition of phytoplankton photosynthesis by ultraviolet radiation. Science 258:646–650

Curran MAJ, Jones GB, Burton H (1998) The spatial distribution of DMS and DMSP in the Australasian sector of the Southern Ocean. J Geophys Res 103:16677–16698

de Baar HJW, Boyd PW (2000) The role of iron in plankton ecology and carbon dioxide transfer of the global oceans. In: Hanson RB, Ducklow HW, Field JG (eds) The dynamic ocean carbon cycle: a midterm synthesis of the Joint Global Ocean Flux Study, Chap. 4. International Geosphere Biosphere Programme Book Series, pp 61–140, Cambridge University Press

de Baar HJW, de Jong JTM, Bakker DC, Löscher BM, Veth C, Bathman U, Smetacek V (1995) Importance of iron for plankton blooms and carbon dioxide drawdown in the Southern Ocean. Nature 373:412–415

de Baar HJW, de Jong JTM, Nolting RF, Timmermans KR, van Leeuwe MA, Bathmann U, van der Loeff MR, Sildam J (1999) Low dissolved Fe and the absence of diatom blooms in remote Pacific waters of the Southern Ocean. Mar Chem 66:1–34

de la Mare WK (1997) Abrupt mid-twentieth century decline in Antarctic sea-ice extent from whaling records. Nature 389:57–60

Denman K, Hofmann E, Marchant H (1996) Marine biotic responses and feedbacks to environmental change and feedbacks to climate. In: Houghton JT, Meira Filho LG, Callander BA, Harris N, Kattenberg A, Maskell K (eds) Climate change 1995, The science of climate change. Cambridge University Press, pp 483–516

Denman KL, Pena MA (2000) Beyond JGOFS. In: Hanson RB, Ducklow HW, Field JG (eds) The dynamic ocean carbon cycle: a midterm synthesis of the Joint Global Ocean Flux Study, Chap. 4. International Geosphere Biosphere Programme Book Series, Cambridge Univerity Press, pp 469–490

DiTullio GR, Laws EA (1991) Impact of an atmospheric-oceanic disturbance on phytoplankton community dynamics in the North Pacific Central Gyre. Deep-Sea Res 38:1305–1329

DiTullio GR, Smith WO Jr. (1996) Spatial patterns in phytoplankton biomass and pigment distributions in the Ross Sea. J Geophys Res 101:467–477

Doney SC (1999) Major challenges confronting marine biogeochemical modeling. Global Biogeochem Cy 13:705–714

Doney SC, Sarmiento JL (1999) Ocean biogeochemical response to climate change. U.S. JGOFS, Synthesis and Modeling Project Planning Workshop Report

Doney SC, Lindsay K, Moore JK (2003) Global ocean carbon cycle modeling. In: Fasham MJR (ed) Ocean biogeochemistry, Chap. 9. Springer-Verlag, Heidelberg (this volume)

Duce RA, Tindale NW (1991) Chemistry and biology of iron and other trace metals. Limnol Oceanogr 36:1715–1726

Dupouny C, Neveux J, Subramaniam A, Mulholland MR, Montoya JP, Campbell L, Carpenter EJ, Capone DG (2000) Satellite captures *Trichodesmium* blooms in the southwestern tropical Pacific. EOS 81:13–16

Ellwood MJ, van den Berg CMG (2001) Determination of organic complexation of cobalt in seawater by cathodic stripping voltammetry. Mar Chem 75(1–2):33–47

Emerson S, Quay P, Karl D, Winn C, Tupas L, Landry M (1997) Experimental determination of the organic carbon flux from open-ocean surface waters. Nature 389:951–954

Emerson S, Mecking S, Abell J (2001) The biological pump in the subtropical North Pacific Ocean: nutrient sources, redfield ratios and recent changes. Global Biogeochem Cy 15:535–554

Falkowski PG (1997) Evolution of the nitrogen cycle and its influence on the biological sequestration of CO_2 in the ocean. Nature 387:272–275

Falkowski PG, Wilson C (1992) Phytoplankton productivity in the North Pacific ocean since 1900 and implications for absorption of anthropogenic CO_2. Nature 358:741–743

Falkowski PG, Kim Y, Kolber Z, Wilson C, Wirick C, Cess R (1992) Natural versus anthropogenic factors affecting low-level cloud albedo over the North-Atlantic. Science 256:1311–1313

Falkowski PG, Barber RT, Smetacek V (1998) Biogeochemical controls and feedbacks on primary production. Science 281: 200–206

Falkowski PG, Scholes RJ, Boyle E, Canadell J, Canfield D, Elser J, Gruber N, Hibbard K, Högberg P, Linder S, Mackenzie FT, Moore B III, Pedersen T, Rosenthal Y, Seitzinger S, Smetacek V, Steffen W (2000) The global carbon cycle: a test of our knowledge of earth as a system. Science 290:291–296

Fedorov AV, Philander SG (2000) Is El Niño changing? Science 288:1997–2002

Field CB, Behrenfeld MJ, Randerson JT, Falkowski P (1998) Primary production of the biosphere: integrating terrestrial and oceanic components. Science 281:237–240

Franck VM, Brzezinski MA, Coale KH, Nelson DM (2000) Iron and silicic acid concentrations regulate Si uptake north and south of the polar frontal zone in the Pacific Sector of the Southern Ocean. Deep-Sea Res Pt II 47:3315–3338

Fung I, Meyn SK, Tegen I, Doney SC, John JG, Bishop KB (2000) Iron supply and demand in the upper ocean. Global Biogeochem Cy 14:281–295

Gage DA, Rhodes D, Nolte KD, Hicks WA, Leustek T, Cooper AJL, Hanson AD (1997) A new route for synthesis of dimethylsulphoniopropionate in marine algae. Nature 387:891–894

Galloway JN, Levy H II, Kasibhatla PS (1994) Year 2020: consequences of population growth and development on deposition of oxidised nitrogen. Ambio 23:120–123

Galloway JN, Schlesinger WH, Levy H, Michaels A, Schnoor JL (1995) Nitrogen fixation: anthropogenic enhancement-environmental response. Global Biogeochem Cy 9(2):235–25

Gordon HR, Boynton GC, Balch WM, Groom SB, Harbour DS, Smyth TJ (2001) Retrieval of coccolithophore calcite concentration from seaWiFS imagery. Geophys Res Lett 28(8): 1587–1590

Gruber N, Sarmiento JL (1997) Global patterns of marine nitrogen fixation and denitrification. Global Biogeochem Cy 11: 235–266

Hannon E, Boyd PW, Silvoso M, Lancelot C (2001) Modeling the bloom evolution and carbon flows during SOIREE: implications for future in situ iron-enrichments in the Southern Ocean. Deep-Sea Res Pt II 48:2745–2774

Hansell DA, Carlson CA (1998) Net community production of dissolved organic carbon. Global Biogeochem Cy 12:443–453

Hansell DA, Feely RA (2000) Atmospheric intertropical convergence impacts surface ocean carbon and nitrogen biogeochemistry in the western tropical Pacific. Geophys Res Lett 27: 1013–1016

Hansen J, Rossow W, Carlson B, Lacis A, Travis L, Del Genio A, Fung I, Cairns B, Mishchenko M, Sato M (1995) Low-cost long-term monitoring of global climate forcings and feedbacks. Climatic Change 31:247–271

Harrison PJ, Boyd PW, Varela DE, Takeda S, Shiomoto A, Odate T (1999) Comparison of factors controlling phytoplankton productivity in the NE and NW Subarctic Pacific gyres. Prog Oceanogr 42:205–234

Harvey LDD (2000) Upscaling in global change research. Climatic Change 44:225–263

Hasselmann K (1999) Linear and nonlinear signatures. Nature 398:755–756

Heath MR, Backhaus JO, Richardson K, McKenzie E, Slagstad D, Beare D, Dunn J, Fraser JG, Gallego A, Hainbucher D, Hay S, Jónasdóttir S, Madden H, Mardaljevic J, Schacht A (1999) Climate fluctuations and the spring invasion of the North Sea by *Calanus finmarchicus*. Fish Oceanogr 8:163–176

Heinze C, Maier-Reimer E, Winn K (1991) Glacial rCO_2 reduction by the world ocean: experiments with the Hamburg carbon cycle model. Paleoceanography 6:395–430

Herut B, Krom MD, Pan G, Mortimer R (1999) Atmospheric input of nitrogen and phosphorus to the southeast Mediterranean: sources, fluxes and possible impact. Limnol Oceanogr 44: 1683–1692

Holligan PM, Turner SM, Liss PS (1987) Measurements of dimethyl sulphide in frontal regions. Cont Shelf Res 7:213–224

Holligan PM, Fernández E, Aiken J, Balch WM, Boyd P, Burkill PH, Finch M, Groom SB, Malin G, Muller K, Purdie DA, Robinson C, Trees CC, Turner SM, van der Wal P (1993) A biochemical study of the coccolithophore, *Emiliania huxleyi*, in the North Atlantic. Global Biogeochem Cy 7:879–900

Houghton JT, Jenkins GJ, Ephraums JJ (1990) Climate change – the IPCC scientific assessment. Cambridge University Press, Cambridge

Houghton JT, Filho LGM, Harris CN, Kattenberg A, Maskell K (eds), Lakeman JA (production ed) (1996) Climate change 1995: the science of climate change. Contribution of WGI to the Second Assessment Report of the Intergovernmental Panel on Climate Change, Press Syndicate of the University of Cambridge, USA

Husar RB, Prospero JM, Stowe LL (1997) Characterization of tropospheric aerosols over the oceans with the NOAA advanced very high resolution radiometer optical thickness operational product. J Geophys Res 102:16889–16909

Husar RB, Tratt DM, Schichtel BA, Falke SR, Li F, Jaffe D, Gasso S, Gill T, Laulainen NS, Lu F, Reheis MC, Chun Y, Westphal D, Holben BN, Gueymard C, McKendry I, Kuring N, Feldman GC, McClain C, Frouin RJ, Merrill J, DuBois D, Vignola F, Murayama T, Nickovic S, Wilson WE, Sassen K, Sugimoto N, Malm WC (2001) Asian dust events of April 1998. J Geophys Res 106(D16):18317–18330

Hutchins DA, Bruland KW (1998) Iron-limited diatom growth and Si:N uptake ratios in a coastal upwelling regime. Nature 393: 561–564

Hutchins DA, DiTullio GR, Zhang Y, Bruland KW (1998) An iron limitation mosaic in the California upwelling regime. Limnol Oceanogr 43:1037–1054

Iglesias-Rodriguez MD, Brown CW, Doney SC, Kleypas J, Kolber D, Kolber Z, Hayes PK, Falkowski PG (2002) Representing key phytoplankton functional groups in ocean carbon cycle models: coccolithophores. Global Biogeochem Cy (in press)

IPCC (2001) Climate change 2001: the scientific basis. Contribution of Working Group I to the third assessment report of the Intergovernmental Panel on Climate Change, edited by J.T. Houghton et al., Cambridge University Press, New York

JGOFS (1991) The Joint Global Ocean Flux Study Science Plan. Scientific Committee on Ocean Research, JGOFS, Bergen, Norway, 61 pp

JGOFS (1999) Synthesis and modeling project. In: Doney SC, Sarmiento JL (ed) Ocean biogeochemical response to climate change workshop. US JGOFS Planning Office Report No. 22, Woods Hole, MA, 106 pp

Jickells TD (1999) The inputs of dust-derived elements to the Sargasso Sea: a synthesis. Mar Chem 68:5–14

Jickells TD, Spokes LJ (2001) Atmospheric iron inputs to the oceans. In: Turner D, Hunter KA (eds) The Biogeochemistry of iron in seawater. John Wiley & Sons, New York Chichester, pp 85–122

Johnson KS (2001) Iron supply and demand in the upper ocean: is extraterrestrial dust a significant source of bioavailable iron? Glob Biogeochem Cy 15:61–63

Johnson KS, Gordon RM, Coale KH (1997) What controls dissolved iron concentrations in the world ocean? Mar Chem 57:137–161

Johnson KS, Chavez FP, Friederich GE (1999) Continental-shelf sediment as a primary source of iron for coastal phytoplankton. Nature 398:697–699

Joos F, Plattner G -K, Stocker TF, Marchal O, Schmittner A (1999) Global warming and marine carbon cycle feedbacks on future atmospheric CO_2. Science 284:464–467

Karl DM (1999) A sea of change: biogeochemical variability in the north Pacific subtropical gyre. Ecosystems 2:181–214

Karl DM, Letelier R, Hebel D, Tupas L, Dore J, Christian J, Winn C (1995) Ecosystem changes in the north Pacific subtropical gyre attributed to the 1991–92 El Niño. Nature 373:230–234

Karl DM, Bidigare RR, Letelier RM (2001) Long-term changes in plankton community structure and productivity in the North Pacific Subtropical Gyre: the domain shift hypothesis. Deep-Sea Res II 48:1449–1470

Karl D, Letelier R, Tupas L, Dore J, Christian J, Hebel D (1997) The role of nitrogen fixation in biogeochemical cycling in the subtropical North Pacific Ocean. Nature 388:533–538

Karl DM (1999) A sea of change: biogeochemical variability in the North Pacific Subtropical Gyre. Ecosystems 2:181–214

Karl DM, Bates N, Emerson S, Harrison PJ, Jeandel C, Llinas O, Liu KK, Marty J-C, Michaels AF, Miguel JC, Neuer S, Nojiri Y, Wong CS (2003) Temporal studies of biogeochemical processes determined from ocena time-series observations during the JGOFS era. In: Fasham MJR (ed) Ocean biogeochemistry. Springer-Verlag, Heidelberg, (this volume)

Kattenberg A, et al. (1996) Climate models – projections of future climate. In: Houghton JT, Filho LGM, Harris CN, Kattenberg A, Maskell K (eds) Lakeman JA (production ed) Climate change 1995: the science of climate change. Contribution of WGI to the Second Assessment Report of the Intergovernmental Panel on Climate Change. Press Syndicate of the University of Cambridge, USA

Keeling CD, Whorf T, Wahlen M, Van Der Plicht J (1995) Interannual extremes in the rate of rise of atmospheric carbon dioxide since 1980. Nature 375:666–670

Keeling RF, Piper SC, Heimann M (1996) Global and hemispheric CO_2 sinks deduced from changes in atmospheric O_2 concentration. Nature 381:218–221

Kellogg WW (1983) Feedback mechanisms in the climate system affecting future levels of carbon dioxide. J Geophys Res 88: 1263–1269

Kettle AJ, Andreae MO, Amouroux D, Andreae TW, Bates TS, Berresheim H, Bingemer H, Boniforti R, Curran MAJ, DiTullio GR, Helas G, Jones GB, Keller MD, Kiene RP, Leck C, Levasseur M, Malin G, Maspero M, Matrai P, McTaggart AR, Mihalopoulos N, Nguyen BC, Novo A, Putaud JP, Rapsomanikis S, Roberts G, Schebeske G, Sharma S, Simó R, Staubes R, Turner S, Uher G (1999) A global database of sea surface dimethylsulfide (DMS) measurements and a procedure to predict sea surface DMS as a function of latitude, longitude, and month. Global Biogeochem Cy 13:399–444

Kleypas JA, Buddemeier RW, Archer D, Gattuso J –P, Langdon C, Opdyke BN (1999) Geochemical consequences of increased atmospheric carbon dioxide on coral reefs. Science 284:118–120

Körtzinger A, Koeve W, Kähler P, Mintrop L (2001) C:N ratios in the mixed layer during the productive season in the Northeast Atlantic Ocean. Deep-Sea Res Pt I 48:661–688

Kumar N, Anderson RF, Mortlock RA, Froelich PN, Kubik P, Dittrich-Hannen B, Suter M (1995) Increased biological productivity and export production in the glacial Southern Ocean. Nature 378:675–680

Lancelot C, Billen G (1985) Carbon-nitrogen relationship in nutrient metabolism of coastal marine ecosystems. In: Jannash HW, Leb Williams PJ (eds) Advance in aquatic microbiology, vol. 3. Academic Press, London, pp 263–321

Lancelot C, Hannon E, Becquevort S, Veth C, de Baar HJW (2000) Modeling phytoplankton blooms and related carbon export production in the Southern Ocean: application to the Atlantic sector in austral spring 1992. Deep-Sea Res Pt I 47:1621–1662

Lashof DA (1989) The dynamic greenhouse: feedback processes that may influence future concentrations of atmospheric trace gases and climatic change. Climatic Change 14:213–242

Lefèvre N, Watson AJ (1999) Modeling the geochemical cycle of iron in the oceans and its impact on atmospheric CO_2 concentrations. Global Biogeochem Cy 13:727–736

Legendre L, Rassoulzadegan F (1995) Ice core record of oceanic emissions of dimethylsulphide during the last climate cycle. Nature 350:144–146

Legrand M, Feniet-Saigne C, Saltzman ES, Germain C, Barkov NI, Petrov VN (1991) Plankton and nutrient dynamics in marine waters. Ophelia 41:153–172

Lenes JM, Darrow BP, Cattrall C, Heil CA, Callahan M, Vargo GA, Byrne RH, Prospero JM, Bates DE, Fanning KA, Walsh JJ (2001) Iron fertilization and the *Trichodesmium* response on the West Florida Shelf. Limnol Oceanogr 46(6):1261–1277

Levitus S, Antonov JI, Boyer TP, Stephens C (2000) Warming of the world ocean. Science 287:2225–2229

Lipschultz F (1999) Climate change and the areal extent of nitrogen fixers. U.S. JGOFS Report 22

Lipschultz F, Owens NJP (1996) An assessment of nitrogen fixation as a source of nitrogen to the North Atlantic. Biogeochemistry 35:261–274

Liss PS, Malin G, Turner SM, Holligan PM (1994) Dimethyl sulphide and *Phaeocystis*: a review. Workshop on the ecology of *Phaeocystis*-dominated ecosystems, January 1991, Brussels (Belgium). J Marine Syst 5(1):41–53

Loeb V, Siegel V, Holm-Hansen O, Hewitt R, Fraser W, Trivelpiece W, Trivelpiece S (1997) Effects of sea-ice extent and krill or salp dominance on the Antarctic food web. Nature 387:897–900

Longhurst A (1995) Seasonal cycles of pelagic production and consumption. Prog Oceanogr 36:77–167

Longhurst A (1998) Ecological geography of the sea. Academic Press, San Diego

Macilwain C (2000) Ocean researchers dive to deep-sea stations. Nature 406:449–449

Maher BA, Dennis PF (2001) Northern hemisphere dust fluxes: significance for Fe fertilisation in the Southern Ocean? Nature 411:176–180

Mahowald N, Kohfeld K, Hansson M, Balkanski Y, Harrison SP, Prentice IC, Schulz M, Rodhe H (1999) Dust sources and deposition during the last glacial maximum and current climate: a comparison of model results with paleodata from ice cores and marine sediments. J Geophys Res 104:15,895–15916

Malin G (1997) Sulphur, climate and the microbial maze. Nature 387:857–859

Malin G, Turner S, Liss P, Holligan P, Harbour D (1993) Dimethylsulphide and dimethylsulphoniopropionate in the Northeast Atlantic during the summer coccolithophore bloom. Deep-Sea Res Pt I 40(7):1487–1508

Malin G, Wilson WH, Bratbak G, Liss PS, Mann NH (1998) Elevated production of dimethylsulfide resulting from viral infection of cultures of *Phaeocystis pouchetii*. Limnol Oceanogr 43(6):1389–1393

Manabe S, Stouffer RJ (1993) Century-scale effects of increased atmospheric CO_2 on the ocean-atmosphere system. Nature 364(6434):215–218

Mantua NJ, Hare SR, Zhang Y, Wallace JM, Francis RC (1997) A Pacific interdecadal climate oscillation with impacts on salmon production. B Am Meteorol Soc 78:1069–1079

Margalef R (1997) Our biosphere. In: Kinne O (ed) Excellence in ecology, Chap. 10. Ecology Institute, Oldendorf/Luhe, Germany

Marra J, Langdon C, Knudson CA (1995) Primary production, water column changes, and the demise of a *Phaeocystis* bloom at the marine light-mixed layers site (59° N, 21° W) in the Northeast Atlantic Ocean. J Geophys Res 100:6633–6644

Martin JH (1990) Glacial-interglacial CO_2 change: the iron hypothesis. Paleoceanography 5(1):1–13

Martin JH, Gordon RM, Fitzwater SE, Broenkow WW (1989) VERTEX: phytoplankton/iron studies in the Gulf of Alaska. Deep-Sea Res 36: 649–680

Martin JH, Fitzwater SE, Gordon RM, Hunter CN, Tanner SJ (1993) Iron, primary production and carbon-nitrogen fluxes during the JGOFS North Atlantic Bloom Experiment. Deep-Sea Res Pt II 40:115–134

Matear RJ, Hirst AC (1999) Climate change feedback on the future oceanic CO_2 uptake. Tellus B 51:722–733

McCartney M (1997) Climate change: is the ocean at the helm? Nature 377:521–522

McGowan JA (1990) Climate and and change in oceanic ecosystems: the value of time-series data. Tree 5(9):293–295

McGowan JA (1999) A biological WOCE. Oceanography 12:33–35

McGowan JA, Cayan RC, Dorman LM (1998) Climate-ocean variability and ecosystem response in the Northeast Pacific. Science 281:210–217

McKenzie RJ, Conner B, Bodeker G (1999) Increased summertime UV radiation in New Zealand in response to ozone loss. Science 285:1709–1711

Merrill J, Arnold E, Leinen M, Weaver C (1994) Mineralogy of aeolian dust reaching the North Pacific Ocean 2. Relationship of mineral assemblages to atmospheric transport patterns. J Geophys Res 99:21025–21032

Merrill JT (1989) Atmospheric long-range transport to the Pacific Ocean. In: Riley JP, Chester R, Duce RA (eds) Chemical oceanography, vol. 10. Academic Press, London

Michaels AF, Knap AH (1996) Overview of the U.S. JGOFS Bermuda Atlantic Time-series Study and the Hydrostation S program. Deep-Sea Res Pt II 43(2–3):157–198

Michaels AF, Siegel DA, Johnson RJ, Knap AH, Galloway JN (1993) Episodic inputs of atmospheric nitrogen to the Sargasso Sea – contributions to new production and phytoplankton blooms. Global Biogeochem Cy 7:339–351

Michaels AF, Olson D, Sarmiento JL, Ammerman JW, Fanning K, Jahnke R, Knap AH, Lipschultz F, Prospero JM (1996) Inputs, losses and transformations of nitrogen and phosphorus in the pelagic North Atlantic Ocean. Biogeochemistry 35(1):181–226

Mitchell JFB, Johns TC, Senior CA (1998) Transient response to increasing greenhouse gases using models with and without flux adjustment. Hadley Centre Technical Note 2, Hadley Centre for Climate Prediction and Research, Meteorological Office, Bracknell, RG12 2SY, UK, 26 pp

Moisan TA, Mitchell BG (1999) Photophysiological acclimation of *Phaeocystis antartica* Karsten under light limitation. Limnol Oceanogr 44:247–258

Moran MA, Hodson RE (1994) Support of bacterioplankton production by dissolved humic substances from three marine environments. Mar Ecol Prog Ser 110(2–3):241–247

Morel FMM, Reinfelder JR, Roberts SB, Chamberlain CP, Lee JG, Yee D (1994) Zinc and carbon co-limitation of marine phytoplankton. Nature 369:740–742

Mostajir B, Demers S, de Mora S, Belzile C, Chanut J-P, Gosselin M, Roy S, Villegas PZ, Fauchot J, Bouchard J, Bird D, Monfort P, Levasseur M (1999) Experimental test of the effect of ultraviolet-B radiation in a planktonic community. Limnol Oceanogr 44:586–596

Muggli DL, Lecourt M, Harrison PJ (1996) Effects of iron and nitrogen source on the sinking rate, physiology and metal composition of an oceanic diatom from the subarctic Pacific. Mar Ecol Prog Ser 132:215–227

Mullineaux CW (1999) The plankton and the planet. Science 28:801–802

Nanninga HJ, Tyrrell T (1996) Importance of light for the formation of algal blooms by *Emiliania huxleyi*. Mar Ecol Prog Ser 136:195–203

Napp JM, Hunt GL Jr. (2001) Anomalous conditions in the southeastern Bering Sea 1997: linkages among climate, weather, ocean and biology. Fish Oceanogr 10:61–69

Neale PJ, Cullen JJ, Davis RF (1998a) Inhibition of marine photosynthesis by ultraviolet radiation: variable sensitivity of phytoplankton in the Weddell-Scotia confluence during the Austral spring. Limnol Oceanogr 43:433–488

Neale PJ, Cullen JJ, Davis RF (1998b) Interactive effects of ozone depletion and vertical mixing on photosynthesis of Antarctic phytoplankton. Nature 392:585–589

Nickovic S, Dobricic S (1996) A model for long-range transport of desert dust. Mon Weather Rev 124:2537–2544

Nicol S, Pauly T, Bindoff NL, Wright S, Thiele D, Hosie GW, Strutton PG, Woehler E (2000) Ocean circulation off east Antarctica affects ecosystem structure and sea-ice extent. Nature 406:504–507

O'Brien CM, Fox CJ, Planque B, Casey J (2000) Climate variability and North Sea cod. Nature 404:142–143

Orr JC (1999) On ocean carbon-cycle modal comparison. Tellus B 51:509–510

Owens NJP, Cook D, Colebrook M, Hunt H, Reid PC (1989) Long term trends in the occurrence of *Phaeocystis* sp. in the northeast Atlantic. J Mar Biol Assoc Uk 69(4):813–821

Paerl HW (1999) Cultural eutrophication of shallow coastal waters: coupling changing anthropogenic nutrient inputs to regional management approaches. Limnologica 29(3):249–254

Paerl HW, Whitall DR (1999) Anthropogenically-derived atmospheric nitrogen deposition, marine eutrophication and harmful algal bloom expansion: Is there a link? Ambio 28:307–311

Paerl HW, Willey JD, Go M, Peierls BL, Pinckney JL, Fogel ML (1999) Rainfall stimulation of primary production in western Atlantic Ocean waters: roles of different nitrogen sources and co-limiting nutrients. Mar Ecol Prog Ser 176:205–214

Pan Z, Arritt RW, Gutowski WJ Jr., Takle ES (2001) Soil moisture in a regional climate model: simulation and projection. Geophys Res Lett 28:2947–2950

Parmesan C (1996) Climate and species' range. Nature 382: 765–766

Parslow JS (1981) Phytoplankton-zooplankton interactions: data analysis and modelling (with particular reference to ocean station Papa [50° N, 145° W] and controlled ecosystem experiments). Deep-Sea Res Pt I 41:1617–1642

Pate-Cornell ME (1996) Uncertainties in global change estimates. Climatic Change 33:145–149

Pausz C, Herndl GJ (1999) Role of ultraviolet radiation on phytoplankton extracellular release and its subsequent utilization by marine bacterioplankton. Aquat Microb Ecol 18:85–93

Peierls BL, Paerl HW (1997) Bioavailability of atmospheric organic nitrogen deposition to coastal phytoplankton. Limnol Oceanogr 42:1819–1823

Petchey OL, McPhearson PT, Casey TM, Morin PJ (1999) Environmental warming alters food-web structure and ecosystem function. Nature 402:69–72

Petit JR, Jouzel J, Raynaud D, Barkov NI, Barnola JM, Basile I, Bender M, Chappellaz J, Davis M, Delaygue G, Delmotte M, Kotlyakov VM, Legrand M, Lipenkov VY, Lorius C, Pépin L, Ritz C, Saltzman E, Stievenard M (1999) Climate and atmospheric history of the past 420 000 years from the Vostok ice core, Antartica. Nature 399:429–436

Piketh SJ, Tyson PD, Steffen W (2000) Aeolian transport from southern Africa and iron fertilization of marine biota in the South Indian Ocean. S Afr J Sci 96:244–246

Polovina JJ, Mitchum GT, Evans GT (1995) Decadal and basin-scale variation in mixed layer depth and the impact on biological production in the Central and North Pacific, 1960–88. Deep-Sea Res Pt I 42(10):1701–1716

Pounds JA, Fogden MPL, Campbell JH (1999) Biological response to climate change on a tropical mountain. Nature 398:611–615

Powell RT, King DW, Landing WM (1995) Iron distributions in surface waters of the south Atlantic. Mar Chem 50:13–20

Price NM, Morel FMM (1999) Biological cycling of iron in the ocean. In: Sigel A, Sigel H (eds) Metal ions in biological systems, Chap. 1, iron transport and storage in microorganisms, plants and animals, vol. 35. Marcel Dekker Inc., New York, pp 1–36

Price NM, Ahner BA, Morel FMM (1994) The Equatorial Pacific Ocean: grazer-controlled phytoplankton populations in an iron-limited ecosystem. Limnol Oceanogr 3(3):520–534

Prospero JM (2000) Global aerosol report website. *http://www.nrlmry.navy.mil/aerosol/globaer_world_loop.htm*

Prospero JM, Uematsu M, Savoie DL (1989) Mineral aerosol transport to the Pacific Ocean. In: Riley JP, Chester R, Duce RA (eds) Chemical oceanography, vol. 10. Academic Press, London

Pye K (1987) Aeolian dust and dust deposits, Academic, San Diego, California

Rahmstorf S (1999) Shifting seas in the greenhouse. Nature 399:523–524

Raven JA (1990) Predictions of Mn and Fe use efficiencies of phototrophic growth as a function of light availability for growth and C assimilation pathway. New Phytol 116:1–17

Reid GC (1997) Solar forcing of global climate change since the mid 17th century. Climatic Change 37:391–405

Reid PC, Edwards M, Hunt HG, Warner AJ (1998) Phytoplankton change in the North Atlantic. Nature 391:546

Riebesell U, Zondervan I, Rost B, Tortell PD, Zeebe RE, Morel FMM (2000) Reduced calcification of marine plankton in response to increased atmospheric CO_2. Nature 407:364–367

Robertson JE, Robinson C, Turner DR, Holligan P, Watson AJ, Boyd P, Fernandez E, Finch M (1994) The impact of a coccolithophore bloom on oceanic carbon uptake in the northeast Atlantic during summer 1991. Deep-Sea Res Pt I 41:297–314

Royer J-F, Chauvin F, Timbal B, Araspin P, Grimal D (1998) A GCM study of the impact of greenhouse gas increase on the frequency of occurrence of tropical cyclones. Climate Change 38:307–343

Rue EL, Bruland KW (1997) The role of organic complexation on ambient iron chemistry in the equatorial Pacific Ocean and the response of a mesoscale iron addition experiment. Liminol Oceanogr 42:901–910

Sait SM, WC Liu, Thompson DJ, Godfray HCJ, Begon M (2000) Invasion sequence affects predator-prey dynamics in a multi-species interaction. Nature 405:448–450

Sambrotto RN, Savidge G, Robinson C, Boyd P, Takahashi T, Karl DM, Langdon C, Chipman D, Marra J, Codispoti L (1993) Elevated consumption of carbon relative to nitrogen in the surface ocean. Nature 363:248–250

Sanudo-Wilhelmy SA, Kustka AB, Gobler CJ, Hutchins DA, Yang M, Lwiza K, Burns J, Capone DG, Raven JA, Carpenter EJ (2001) Phosphorus limitation of nitrogen fixation by *Trichodesmium* in the central Atlantic Ocean. Nature 411:66–69

Sarmiento JL (1993) Atmospheric CO_2 stalled. Nature 365:697–698
Sarmiento JL, Hughes TMC, Stouffer RJ, Manabe S (1998) Simulated response of the ocean carbon cycle to anthropogenic climate warming. Nature 393:245–249
Schellnhuber HJ (1999) 'Earth system' analysis and the second Copernican revolution. Nature 402, (suppl.) C19–C23
Schimel D, Enting IG, Heimann M, Wigely TML, Raynaud D, Alves D, Siegenthaler U (1995) CO_2 and the carbon cycle. In: Houghton JT, Meira Filho LG, Bruce J, Lee H, Callander BA, Haites E, Harris N, Maskell K (eds) Climate change 1994. IPPC. Cambridge University Press, Cambridge, pp 35–71
Schimel D, Alves D, Enting I, Heimann M, Joos F, Raynaud D, Wigley T (1996) CO_2 and the carbon cycle. In: Climate change 1995: the science of climate change: contribution of WGI in the second assessment Report of the IPCC. Cambridge University Press, Cambridge, pp 65–86
Schindler DW, Bayley SE (1993) The biosphere as an increasing sink for atmospheric carbon: estimates from increased nitrogen deposition. Global Biogeochem Cy 7:717–733
Schrag DP (2000) Of ice and elephants. Nature 404:23–69
Seager R, Kushnir Y, Visbeck M, Naik N, Miller J, Krahmann G, Cullen H (2000) Causes of Atlantic Ocean climate variability between 1958 and 1998. J Climate 13(16):2845–2862
Sedwick PN, DiTullio GR (1997) Regulation of algal blooms in Antarctic shelf waters by the release of iron from melting sea ice. Geophys Res Lett 24:2515–2518
Sedwick PN, DiTullio GR, Mackey DJ (2000) Iron and manganese in the Ross Sea, Antarctica: seasonal iron limitation in Antarctic shelf waters. J Geophys Res 105, C5:11321–11336
Seitzinger SP, Sanders RW (1999) Atmospheric inputs of dissolved organic nitrogen stimulate estuarine bacteria and phytoplankton. Limnol Oceanogr 44:721–730
Severinghaus JP, Brook EJ (1999) Abrupt climate change at the end of the last glacial period inferred from trapped air in polar ice. Science 286:930–934
Shaffer G (1993) Effects of the marine biota on global carbon cycling. In: Heinmann M (ed) The global carbon cycle. Springer-Verlag, Berlin, pp 431–455
Shaffer G, Sarmiento JL (1995) Biogeochemical cycling in the global ocean I: a new, analytical model with continuous vertical resolution and high-latitude dynamics. J Geophys Res 100: 2659–2672
Schindell DT, Rind D, Lonergan P (1998) Increased polar stratospheric ozone losses and delayed eventual recovery owing to increasing greenhouse-gas concentrations. Nature 392:589–592
Siegenthaler U, Sarmiento JL (1993) Atmospheric carbon dioxide and the ocean. Nature 365:119–125
Sigman DM, Boyle EA (2000) Glacial/interglacial variations in atmospheric carbon dioxide. Nature 407:859–869
Simo R (2001) Production of atmospheric sulfur by oceanic plankton: biogeochemical, ecological and evolutionary links. Tree 16:287–294
Smith DC, Simon M, Alldredge AL, Azam F (1992) Intense hydrolytic enzyme activity on marine aggregates and implications for rapid particle dissolution. Nature 359:139–142
Smith RC, Baker KS (1984) The analysis of ocean optical data. Proceedings of the Society of Photo-Optical Instrumentation and Engineering 489, Available as: Ocean Optics VII, pp 119–126
Smith RC, Ainley D, Baker K, Domack E, Emslie S, Fraser B, Kennett J, Leventer A, Mosley-Thompson E, Stammerjohn S, Vernet M (1999) Marine ecosystem sensitivity to climate change: historical observations and paleoecological records reveal ecological transitions in the Antarctic Peninsula region. Bioscience 49:393–404
Smith SD, Huxman TW, Zitzer SF, Charlet TN, Housman DC, Coleman JS, Fenstermaker LK, Seemann JR, Nowak RS (2000) Elevated CO_2 increases productivity and invasive species success in an arid ecosystem. Nature 408:79–82
Smith WO Jr., Anderson RF (2000) US Southern Ocean Joint Global Ocean Flux Study Program (AESOPS). Deep-Sea Res Pt II 47:3073–3548
Southward AJ (1963) The distribution of some plankton animals in the English Channel and approaches. III. Theories about long term biological changes including fish. J Mar Biol Assoc Uk 43:1–29
Spokes LJ, Yeatman SG, Cornell SE, Jickells TD (2001) The input of nitrogen species to the North East Atlantic: the importance of south easterly flow. Tellus 52:37–49
Stegmann PM, Tindale NW (1999) Global distribution of aerosols over the open ocean as derived from the coastal zone color scanner. Global Biogeochem Cy 13:383–397
Subramaniam A, Carpenter EJ, Falkowski PG (1999a) Bio-optical properties of the marine diazotrophic cyanobacteria *Trichodesmium* spp. II. A reflectance model for remote sensing. Limnol Oceanogr 44:618–627
Subramaniam A, Carpenter EJ, Karentz D, Falkowski PG (1999b) Bio-optical properties of the marine diazotrophic cyanobacteria *Trichodesmium* spp. I. Absorption and photosynthetic action spectra. Limnol Oceanogr 44:608–617
Sunda WG, Huntsman SA (1995) Interrelated influence of iron, light and cell size on marine phytoplankton growth. Nature 390:389–392
Sutton RT, Allen MR (1997) Decadal predictability of North Atlantic sea surface temperature and climate. Nature 388:563–567
Takahashi T, Olafsson J, Goddard JG, Chipman DW, Sutherland SC (1993) Seasonal variation of CO_2 and nutrients in the high-latitude surface oceans: a comparative study. Global Biogeochem Cy 7:843–878
Takeda S (1998) Influence of iron availability on nutrient consumption ration of diatoms in oceanic waters. Nature 3, 93:774–777
Tegen I, Fung I (1995) Contribution to the atmospheric mineral aerosol load from land surface modification. J Geophys Res 100:707–726
Thompson PM, Ollason JC (2001) Lagged effects of ocean climate change on fulmar population dynamics. Nature 413:417–420
Toggweiler JR (1993) Oceanography – carbon overconsumption. Nature 363:210–211
Tortell PD, Reinfelder JR, Morel FMD (1997) Active uptake of bicarbonate by diatoms. Nature 390:243–244
Tortell PD, DiTullio GR, Sigman DM, Morel FMM (2002) CO_2 effects on species composition and nutrient utilization in an Equatorial Pacific phytoplankton assemblage. Mar Ecol Prog Ser 236:37–43
Trenberth KE (1998) Atmospheric moisture residence times and cycling: implications for rainfall rates and climate change. In: Climate change 39. Kluwer Academic Publishers, The Netherlands, pp 667–694
Turner SM, Nightingale PD, Spokes LJ, Liddicoat MI, Liss PS (1996) Increased dimethyl sulphide concentrations in sea water from in situ iron enrichment. Nature 383:513–517
Tyrrell T, Taylor AH (1995) A modelling study of *Emiliania huxleyi* in the NE Atlantic. J Marine Syst 9:83–112
Tyrrell T, Holligan PM, Mobley CD (1999) Optical impacts of oceanic coccolithophore blooms. J Geophys Res 102:3223–3241
Uematsu M, Duce RA, Prospero JM, Chen L, Merrill JT, McDonald RL (1983) Transport of mineral aerosol from Asia over the North Pacific Ocean. J Geophys Res 88:5343–5352
Venrick EL (1971) Recurrent groups of diatom species in the North Pacific. Ecology 52:614–625
Venrick EL (1982) Phytoplankton in an oligotrophic ocean: observations and questions. Ecol Monogr 52:129–154
Venrick EL, McGowan JA, Cayan DR, Hayward TL (1987) Climate and chlorophyll *a*: long-term trends in central North Pacific Ocean. Science 238:70–72
Vincent WF, Neale PJ (2000) Mechanisms of UV damage to aquatic organisms. In: de Ora S, Demers S, Vernet M (eds) The effects of UV radiation in the marine environment, Chap. 6. Cambridge University Press, U.K, pp 149–176
Visser PM, Snelder E, Kop AJ, Boelen P, Buma AGJ, van Duyl FC (1999) Effects of UV radiation on DNA photodamage and production in bacterioplankton in the coastal Caribbean Sea. Aquat Microb Ecol 20:49–58
Vitousek PM, Aber JD, Howarth RW, Likens GE, Matson PA, Schindler DW, Schlesinger WH, Tilman DG (1997) Human alteration of the global nitrogen cycle: sources and consequences. Ecol Appl 7:737–750
Watanabe S, Yamamoto H, Tsunogai S (1995) Dimethyl sulphide widely varying in surface water of the eastern North Pacific. Mar Chem 512:253–259

Watson AJ (1997) Volcanic iron, CO_2, ocean productivity and climate. Nature 385:587–588
Watson AJ, Lefèvre N (1999) The sensitivity of atmospheric CO_2 concentrations to input of iron to the oceans. Tellus B 51: 453–460
Watson AJ, Bakker DCE, Ridgwell AJ, Boyd PW, Law CS (2000) Effect of iron supply on Southern Ocean CO_2 uptake and implications for glacial atmospheric CO_2. Nature 407:730–733
Webster PJ, Palmer TN (1997) The past and the future of El Niño. Nature 390:562–564
Wells ML, Vallis GK, Silver EA (1999) Tectonic processes in Papua New Guinea and past productivity in the eastern equatorial Pacific Ocean. Nature 398:601–604
Westbroek P, Brown CW, Van Bleijswijk J, Brownlee C, Brummer GJ, Conte M, Egge J, Fernandez E, Jordan R, Knappertsbusch M, Stefels J, Veldhuis M, Van der Wal P, Young J (1993) A model system approach to biological climate forcing. The example of *Emiliania huxleyi*. Global Planet Change 8(1–2):27–46
Whitney FA, Wong CS, Boyd PW (1998) Interannual variability in nitrate supply to surface waters of the Northeast Pacific Ocean. Mar Ecol Prog Ser 170:15–23
WMO (1999) Scientific assessment of ozone depletion 1998. In: Albritton DL, Aucamp PJ, Megie G, Watson RT (eds) World Meteorological Organization Global Ozone Research and Monitoring Project, Report No. 44. Geneva
Wolf-Gladrow DA, Riebesell U, Burkhardt S, Bijma J (1999) Direct effects of CO_2 concentration on growth and isotopic composition of marine phytoplankton. Tellus B 51:461–476
Wolfe GV, Sherr EB, Sherr BF (1994) Release and consumption of DMSP from *Emiliania huxleyi* during grazing by *Oxyrrhis marina*. Mar Ecol Prog Ser 111(1–2):111–119
Wolfe GV, Steinke M, Kirst GO (1997) Grazing-activated chemical defence in a unicellular marine alga. Nature 387:894–897
Wolfe GV, Levasseur M, Cantin G, Michaud S (1999) Microbial consumption and production of dimethyl sulfide (DMS) in the Labrador Sea. Aquat Microb Ecol 18:197–205
Wong CS, Matear RJ (1999) Sporadic silicate limitation of phytoplankton productivity in the subarctic NE Pacific. Deep-Sea Res Pt II 46:2539–2555
Wood RW (1909) Note on theory of the greenhouse. Philosophy Magazine (series 6) 17:319–320
Woodwell GM, Mackenzie FT, Houghton RA, Apps M, Gorham E, Davidson E (1998) Biotic feedbacks in the warming of the earth. Climatic Change 40:495–518
Wu J, Boyle E, Sunda W, Wen L-S (2001) Soluble and colloidal iron in the oligotrophic North Atlantic and North Pacific. Science 293:847–849
Wuethrich B (2000) How climate change alters rhythms of the wild. Science 287:793–795
Young RW, Carder KL, Betzer PR, Costello DK, Duce RA, DiTullio GR, Tindale NW, Laws EA, Uematsu M, Merrill JT, Feely RA (1991) Atmospheric iron inputs and primary productivity: phytoplankton responses in the North Pacific. Global Biogeochem Cy 5:119–134
Zehr JP, Waterbury JB, Turner PJ, Montoya JP, Omoregle E, Steward GF, Hansen A, Karl DM (2001) Unicellular cyanobacteria fix N_2 in the subtropical North Pacific ocean. Nature 412:635–638

Chapter 8

Benthic Processes and the Burial of Carbon

Karin Lochte · Robert Anderson · Roger Francois · Richard A. Jahnke · Graham Shimmield · Alexander Vetrov

8.1 Introduction

A major goal of the Joint Global Ocean Flux Study (JGOFS) has been to understand the export of carbon from the surface ocean to the deep sea, a process which removes carbon from the active exchange with the atmosphere for long periods of time. Deep-sea sediments are the final sink of organic matter which is not degraded in the water column nor at the water-sediment interface. This interface is a physical boundary collecting and concentrating sinking particulate organic matter from fine debris to dead whales and consequently supports a fairly active benthic community. The level of biotic activity, the rates of remineralization, and last, but not least, the amount of material buried and preserved in the sediment, all depend on the mass flux and the composition of the material reaching the sea floor. As will be shown in this chapter, the connection between the surface ocean and the seafloor is not a simple one. However, the integration of signals at the sea floor allows conclusions to be drawn about upper ocean processes which go beyond the period of direct observation.

The remnants of surface water productivity deposited at the sea floor are not only the drivers of benthic biological activity and biogeochemical processes, but they also carry signals of productivity which are eventually buried in the sediments and retain the information of past ocean productivity. These signals undergo major changes not only in the water column but also at the water-sediment interface. Thus, this biologically active interface acts as a final filter which determines in which chemical form and quantity the settling material is preserved in the deeper layers of the sediment. Although much of the orginal material is lost in these degradation processes, empirical relationships between the signals stored in the sediments and various aspects of ocean productivity have been detected and used to asses levels of productivity in the past. One uncertainty of this approach is the alteration of the material by biological and chemical processes at the sea floor prior to burial. Hence, there is a common interest of biogeochemists, biologists and paleoceanographers to understand the link between productivity, benthic processes and preservation of signals in deep-sea sediments.

Paleoproductivity studies have made large efforts to reconstruct past changes in environmental and climatic conditions by following variations in signals of ocean biology, in particular of the 'biological pump', that are preserved in the sediments. The knowledge of changes in paleoproductivity is a key to understanding how external factors affect biogeochemical cycles and why the ocean is today as it is. Unfortunately, the intuitively-obvious approach to reconstruct past changes in ocean productivity from the profiles of organic carbon in the sediments is not valid. There is no simple, direct relationship between the burial rate of organic carbon and the biological productivity of the overlying waters. Preservation of organic matter is low and very variable. Furthermore, while there is strong evidence that preservation efficiency depends on bottom water oxygen content, sediment exposure time and mass sedimentation rates there is as yet no valid algorithm for assessing preservation efficiency. A further complication involves the supply of refractory 'old' organic matter originating from reworked ancient sediments and redeposited on top of new material, particularly at sites near ocean margins. Due to these and further problems paleoceanographers have turned to 'proxies' to determine past changes in ocean productivity. Proxies are indirect indicators of past processes which are, ideally, better preserved in marine sediments than bulk organic matter. Studies in the modern ocean by JGOFS and other projects have helped to establish a linkage between the proxy and either export flux, concentration of nutrients in the surface water or nutrient utilization efficiency. With this set of tools, which still presents us with many questions and unresolved problems, changes of paleoproductivity can be elucidated and related to environmental change.

This chapter deals with two aspects of deep-ocean fluxes: firstly, the removal of carbon from the surface waters by export to the deep ocean, deposition and burial in the deep-sea sediments as far as we understand it today; secondly, the use of proxies preserved in the deep-sea sediments as indicators of past ocean productivity. The study of benthic processes and burial of carbon in the sediments has not been pursued by all national JGOFS programmes, hence this information is compiled from a number of different studies.

8.2 Processes of Transport and Turnover of Material in the Deep Ocean

8.2.1 Transfer of Organic Material from the Surface to the Deep Ocean

The export of biogenic particles from the productive upper layer of the ocean removes a small and variable fraction of algal biomass from the euphotic zone transferring biologically bound carbon and associated elements into the deeper water layers. Mass occurrences of phytoplankton often lead to agglomeration of individual algal cells forming relatively large and rapidly sinking aggregates, which also scavenge other particles from the water column (Alldredge and Silver 1988). Oceanic regions characterized by seasonal phytoplankton blooms are known for high and episodic sedimentation while oligotrophic oceanic gyres are characterized by low and more constant particle fluxes (Antia et al. 2001). During settling in the water column a large part of labile organic matter is lost from the particles due to zooplankton grazing, microbial degradation and leaching. These losses are most pronounced in the zone between the upper mixed layer and about 1500 m water depth (e.g., Berger et al. 1987; Martin et al. 1987; Louanchi and Najjar 2000). Hence, what reaches the sea floor is the net result of production in the surface ocean and subsequent alteration in the deeper water column. This statement is true for the organic material, which feeds the deep-sea organisms, as well as for the proxies which are used to reconstruct past productivity levels from the sediment record.

Sinking particles can be collected by moored sediment traps deployed in different depths in the water column. Such particle interceptor traps collect and preserve the sinking material for later analysis of its constituents. This method has provided valuable long term measurements of the vertical flux of matter in selected regions of the ocean (e.g., Haake et al. 1993; Deuser 1996; Honjo et al. 2000). It has demonstrated the temporal and spatial variability of both, the flux rates and the composition of the sedimenting material. Furthermore, trap material provides information about the major groups of phytoplankton contributing to the export from the upper ocean, and particularly those with siliceous or calcareous shells can be well identified. A major problem associated with traps is the uncertainty of the efficiency by which they collect sinking particles. Sediment traps moored in shallow water depth are subjected to stronger currents and turbulence and are, therefore, prone to major errors (Buesseler 1991; Buesseler et al. 2000; Gust et al. 1994; Yu et al. 2001a). Shallow traps also frequently capture live zooplankton when they search for food, called 'swimmers', which can be separated from the sinking dead debris only with difficulty. A further complication involves release of dissolved organic matter from particles which can amount to a large fraction of the vertical organic carbon flux in shallow traps and which is usually not accounted for in flux measurements (Kähler and Bauerfeind 2001). In contrast, flux measurements from deep moored traps agree reasonably well with data from other methods (Bacon et al. 1985; Emerson et al. 1987; Buesseler 1991; Yu et al. 2001a; Scholten et al. 2001). The uncertainty in the vertical flux at shallow depth severely limits attempts to estimate the total export from the upper ocean to the deep sea.

The organic carbon flux recorded using sediment traps generally resembles primary productivity patterns in time and space. Hence, sediment trap data were used to develop algorithms which link the primary production to the flux of particulate organic carbon (POC) and water depth (e.g., Suess 1980; Betzer et al. 1984; Martin et al. 1987; Pace et al. 1987; Berger et al. 1987). Such algorithms are frequently applied to estimate the fraction of primary production that is exported out of the upper productive layer and reaches the deep ocean. This approach is tempting and convenient, since basin-wide or global export can be estimated from surface water primary productivity derived from satellite images (e.g., Antoine et al. 1996; Behrenfeld and Falkowski 1997), but it is also subject to a number of uncertainties. Many intermediate steps are required to infer export production from satellite chlorophyll data and the various algorithms describing the vertical flux differ considerably. Trap data indicate that between 10% and 40% of the primary production as derived from satellite data is being exported out of the upper mixed layer at 125 m water depth (Antia et al. 2001) and between 0.4% and 3% at 1000 m (Fischer et al. 2000) with highest export fractions generally associated with high primary production and lowest ones in oligotrophic regions. Particularly high and highly variable percentages of primary production are exported at 100 m water depth in polar regions and in the Arabian Sea at certain periods of the monsoon cycle, while export in subtropical and equatorial regions is lower and less variable (Buesseler 1998). Furthermore, the seasonality of production seems to influence export. Regions with highly variable export production tend to export more organic carbon to the deep ocean than more stable ones (Lampitt and Antia 1997; Antia et al. 2001). Algorithms with an exponential relationship between primary production and export flux seem to describe this relationship best (e.g., Betzer et al. 1984; Antia et al. 2001).

The speed of sinking of particles is an important factor determining the export flux. A range of different types of particles with different sedimentation rates ranging from 1 m to ca. 300 m d^{-1} (Alldredge and Silver 1988) are produced in the euphotic zone and fast-sinking particles are the first ones to reach the deep waters preceding the bulk of slower particles. Among fast sinking particles are diatom aggregates and large fecal pellets such as those of salps, pteropods or krill which may sediment with a speed of several hundred up to thousand meters per day (e.g., Madin 1982; Pfannkuche and Lochte 1993; Noji et al. 1997). These particles arrive in deep waters only a few days after their production in the upper water column and are important vehicles for transfer of organic matter into the deep ocean. For instance, shallow (500 m) and deep (3 200 m) moored traps at the long-time series station near Bermuda recorded the seasonal peaks of sedimentation almost without a significant time lag indicting that below the upper mixed water layer fast transfer may prevail (Conte et al. 2001), which may be caused by repackaged particles produced by zooplankton grazing in mid-water. The importance of fast sinking diatoms, or rather aggregates of diatoms, for export fluxes has been emphasized in a modeling study by Boyd and Newton (1999) highlighting diatom blooms as the plankton community with the highest export potential. Biogenic minerals, such as calcite, aragonite and opal, increase the ballast of settling particles and, hence, their sinking speed (Armstrong et al. 2002). Additional ballast from lithogenic particles introduced to the surface waters by dust, river plumes or melting of 'dirty' ice and scavenged from the water by the organic particles also increases the sinking speed (Ittekkot 1993). Large food falls to the deep-sea floor, as for instance observed from a mass mortality of swimming crabs (Christiansen and Boetius 2000), may also introduce a sizeable amount of fresh organic matter to the sea floor. Such stochastic fluxes may be large in certain regions, but cannot be assessed by conventional means and remain generally unaccounted for.

Another factor determining the flux of organic matter is the degree to which it is mineralized in the water column. Most of the organic material is lost in water depths between 500 m to 1 500 m while at greater depths only small losses occur (e.g., Berger et al. 1987; Martin et al. 1987; Louanchi and Najjar 2000). In regions with strong seasonal phytoplankton 'blooms' forming fast sinking aggregates higher losses in the water column were found, while in systems with less pulsed export (e.g., subtropical and tropical regions) less material is degraded below the upper mixed layer on the way to the deep ocean (Antia et al. 2001). This difference in loss of organic material may be caused by the types of particles being exported. Particles sedimenting after phytoplankton blooms are composed of relatively fresh organic matter from senescent algal cells and are remineralised rapidly within days (Turley and Lochte 1990). In contrast, physically more stable systems seem to recycle most material in the upper water column and loss in deeper waters is small. Exceptionally high exports and little degradation were associated with melting ice probably due to very fast sinking aggregates (Peinert et al. 2001). In contrast to shallow traps recording local differences in productivity, deep moored traps in around 3 000 m water depth show less regional variability (Antia et al. 2001). This implies that there is an effective biological filter removing organic material in mid water. These degradation processes in the mesopelagic zone beneath the upper mixed water layer are still poorly known and represent a major uncertainty in our understanding of carbon transport processes in the ocean. In particular, the influence of different types of zooplankton and their different feeding strategies on particle dynamics in the deeper water column is still an open question.

8.2.2 Benthic Carbon Turnover Processes

The sedimenting material accumulates at the sea floor and at times of high sedimentation, e.g., after the phytoplankton spring bloom, it can form a fluffy layer of organic-rich particles on the sediment surface altering the chemical and biological conditions in this layer (Billett et al. 1983; Rice et al. 1986; Thiel et al. 1989). Although in the deep water column below 1 500 m the organic material seems largely unaffected by biotic consumption, there is a substantial remineralization of this material once it settles on the sea floor (Lochte and Turley 1988; review by Turley et al. 1995). Generally >90% of the organic carbon reaching the sea floor is remineralized and only a small residual fraction is permanently buried in the sediments (Bender and Heggie 1984; Emerson et al. 1985). The community inhabiting the deep-sea floor seems well adapted to low food supply of poor quality and possesses specific capabilities to deal with this material (e.g., Lochte and Turley 1988; Boetius and Lochte 1996). As a result of high biological activity and changing physico-chemical conditions in the water-sediment interface, both sedimenting organic material and proxies experience major modifications.

The water-sediment interface comprises the lower few meters of water above the sediment with increased concentrations of resuspended sediment particles and elevated bacterial activity (called 'the benthic boundary layer', Ritzrau 1996), and the upper few centimeters of the sediment with high biological activity (Ritzrau et al. 2001a). In deep-sea sediments, highest biological activity is found in the upper two centimeters declining down to 5 or 10 cm, below which activity is very low (e.g., Boetius et al. 2000a). However, active bacteria and archaea are still present in great sediment depths, but except for specific geological settings their impact on cycling of material in sediments is assumed to be small (Parkes et al. 2000).

Gradients of oxygen in pore water indicate the rate of mineralization of organic matter in the sediment and allow one to model the degradation rates of this partly recalcitrant material. Various organic fractions with different degradation rates can be identified (Soetaert et al. 1996; Hedges et al. 1999) ranging from mean life times of a few months in regions with high vertical flux (e.g., Smith et al. 1997; Luff et al. 2000) up to 0.7 to 3.5 years (Sayles et al. 1994) or 1.7 to 33 years (e.g., Smith et al. 1997; Sauter et al. 2001). Life times ranging from 0.3 to 80 years have been reported for Southern Ocean sediments (Sayles et al. 2001). The considerable differences in the degradation rates of organic material depend on its chemical composition and shortest degradation rates are observed in relatively 'fresh' organic matter which has settled quickly to the sea floor. If material is turned over slowly, benthic remineralization does not follow directly the temporal pulses of sedimentation, but the signal is spread out in time. According to diagenetic models, the amplitude in seasonal variations in benthic remineralization depends critically on the reactivity of the deposited material and can only show a direct response to a seasonal pulse of sedimentation when mean lifetimes of less than 0.25 years prevail (Martin and Bender 1988; Sayles et al. 2001). Therefore, seasonal changes in vertical flux as recorded by sediment traps are not always reflected by similar seasonal changes in sediment oxygen consumption. Absence of seasonal shifts in oxygen consumption in response to sedimentation events has been observed by direct measurements of benthic respiration in some investigations (e.g., Sayles et al. 1994), while other studies indicate seasonal fluctations in respiration linked to sedimentation events (Fig. 8.1) (e.g., Pfannkuche 1993; Drazen et al. 1998). Thus, the benthic response depends on the deposition of relatively 'fresh' organic matter which has the strongest impact on benthic metabolic rates.

In particular the small sized organisms, such as bacteria, foraminifera, nematodes and other meiofauna, react to sedimentation of particulate organic matter (reviews by Gooday and Turley 1990; Lochte and Pfannkuche 2002; see also Pfannkuche et al. 1999). On a long-term station in the North Atlantic, the sediments showed a seasonal increase of concentrations of algal pigments, adenylates as indicator of biomass of small organisms, bacterial biomass and respiration (Fig. 8.1) indicating that there is seasonal growth of small sized organisms triggered by the sedimentation of the phytoplankton spring bloom. Increased metabolism, rising enzyme activity, increasing growth rates or migration towards the source of food were observed in different groups of these small organisms (Gooday and Turley 1990; Lochte 1992; Pfannkuche 1993). Owing to the poor nutritional quality of the organic material, it seems that bacteria and archaea are best adapted to degrade this material as they possess the greatest diversity of hydrolytic enzymes (e.g., Boetius and Lochte 1996). The largest share of organic carbon in deep-sea sediments is primarily utilized by microorganisms, but larger organisms feeding on the organic particles and the attached microbes also benefit from the sedimenting material (Rice et al. 1986; Gooday and Turley 1990; Heip et al. 2001).

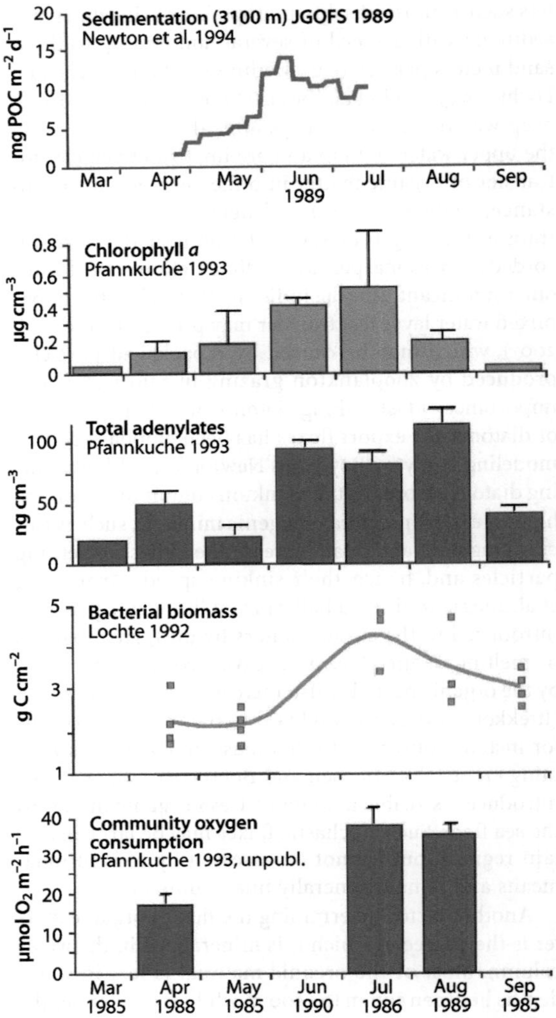

Fig. 8.1. Seasonal pattern of sedimentation of particulate organic carbon (POC) in the Northeast Atlantic in 1989, and seasonal fluctuations in the sediment at the same station of chlorophyll *a*, total adenylates, prokaryotic biomass and oxygen consumption of the sediment community. The sediment data are compiled from several years. This compilation indicates that approximately 4–6 weeks after the surface phytoplankton bloom sedimenting phytodetritus reaches the sea floor (4500 m depth) and gives rise to increased benthic microbial biomass and respiration. The sources of data are indicated with each variable (modified after Lochte et al. 1993, Lochte and Pfannkuche 2002)

From the more productive continental slopes towards the mid-oceanic regions with low sedimentation rates the proportion of large benthic organisms (macrofauna) within the total benthic community, both in respect to biomass and contribution to organic carbon degradation, decreases (Heip et al. 2001). This shift in community composition has an additional impact on the degradation process as the burrowing and pumping activity of larger organisms facilitates rapid incorporation of freshly deposited particles into the upper sediment layers.

Bioturbation and bioirrigation are positively related to the level of food input (e.g., Smith et al. 1997; Shimmield and Jahnke 1995; Pope et al. 1996), but they also depend on the type of organisms involved (Turnewitsch et al. 2000). Long-term studies in the Porcupine Abyssal Plain indicate that major shifts in the composition and abundance of benthic macrofauna may occur. For instance, increases of a species of sea cucumber (*Amperima rosea*) were observed over a period of >10 years but the reasons remain obscure (Bett et al. 2001; Billett et al. 2001). Long-term observations in the eastern North Pacific revealed a decline in sedimentation of POC to the sea floor and, hence, indicate declining food supply for the benthos (Smith and Kaufmann 1999; Smith et al. 2001). Such interannual or decadal variations are likely to have profound effects on the structure of deep-sea communities and on the turnover of material at the sea floor. Understanding the extent of these changes is critical to any interpretation of biogeochemical cycling in the deep sea. However, we are just beginning to appreciate such changes, we are not aware to which extent they are happening at present and in the past, and we are uncertain about their consequences.

Since most deep-sea sediments are well oxygenated, the organic compounds are primarily remineralized via oxygen (Emerson et al. 1985). In areas with high deposition of organic matter the increased respiration in the sediments can not be balanced by diffusion of oxygen from the overlying bottom water. In these sediments oxygen becomes depleted and the pathways of microbial degradation processes change. Such conditions are found in the deep-sea sediments adjacent to intense upwelling regions and at continental margins. For instance, the upwelling regions off Namibia, Peru and Chile are characterized by extensive anoxic zones in the sediments occasionally even extending into the water column. Anoxic conditions are also observed in upper sediment layers of the Arabian Sea where the oxygen minimum zone of the water column impinges on the continental margins and in regions where high sedimentation from monsoonal upwelling occurs. In these regions, sulfate reduction is found in the upper sediment layers, a rare occurrence in deep-sea sediments (Ferdelman et al. 1999; Boetius et al. 2000a). Furthermore, anoxic layers may also develop when turbidites cover the sea floor and prevent oxygen exchange with the overlying water. Under these conditions higher proportions of organic carbon are buried, since oxygen exposure time has been identified as the key factor influencing organic carbon preservation in sediments (Hartnett et al. 1998; review by Canfield 1994). Also the preservation of proxies may change under low oxygen or anoxic conditions (see below).

Productive continental margins also contain large amounts of methane in form of methane hydrates which represents a very large store of organically bound carbon. This ice-like material is kept stable under low temperature and high pressure, but it disintegrates when temperature rises or pressure is reduced (Zatsepina and Buffett 1998). Methane hydrate deposits close to their stability limits can be released in large amounts under changing environmental conditions allowing methane to escape to the atmosphere. Such events may be initiated by switches in thermohaline circulation and evidence of massive methane release from the sea floor is found in geological records (Kennett et al. 2000). Under present day conditions, however, a very effective benthic filter consisting of methane oxidizing bacteria normally consumes this climatically active gas within the sediment when it is released gradually (Hinrichs et al. 1999; Boetius et al. 2000b) and prevents an efflux to the atmosphere. These sedimentary sources of methane also sustain very active benthic communities which are not directly dependent on the organic matter supply from the upper ocean (Sibuet and Olu 1998). The quantitative role of this additional energy source for organic matter cycling in the deep sea is probably limited to local effects only.

8.3 Quantitative Estimates of Carbon Deposition and Carbon Turnover

8.3.1 Strategies for Quantification of Benthic Fluxes

The total flux of organic matter to the sea floor can be estimated as the sum of the burial rate and remineralization rate at any specific location with the remineralization rate estimate dominating the calculation (Jahnke 1996). Deep-moored sediment trap results have indicated that particle fluxes in the water column generally decrease with depth by only a factor of 2–3 between 1 km and 4 km water depth. Thus, quantifying deposition fluxes to the sea floor also provides an important constraint on the magnitude and distribution of the water column fluxes below the main thermocline. As mentioned above, the processes within the sediments that control solute fluxes, remineralization and burial tend to dampen the effects of short-term variability of the particle input fluxes and facilitate estimation of mean fluxes. The reduced temporal variability also facilitates the combining of individual results into basin-wide compilations of flux distributions.

Two basic strategies have been employed to quantify recent fluxes to the deep-sea floor. Reaction/transport models of organic carbon distributions within the sediment column have been developed to estimate remineralization and burial (e.g., Rabouille and Gaillard 1991; Soetaert et al. 1996; Boudreau 1997). Since sediment cores can be routinely recovered from the sea floor and analyzed, there is a large data set from which to work. It is of disadvantage, however, that since the majority of the deposited organic matter has already been remineralized, this method attempts to predict the total initial flux from a small residual. Inaccuracies in the model could lead to large inaccuracies in the estimated flux. Furthermore, the upper few centimeters of the sediments in the deep sea are generally comprised of materials deposited over the preceding few to tens of thousands of years. Recent variations in the particle flux would, therefore, not be reflected in sediment characteristics. Thus, while fluxes derived from sediment studies alone may reflect the average deposition occurring over the last few thousands of years, there is considerable uncertainty when extending these estimates to the present.

The other principal strategy for estimating sea-floor fluxes is to compute the benthic solute flux of metabolic oxidants (mainly oxygen and nitrate) or carbon through pore water or benthic flux chamber studies (e.g., Glud et al. 1994; Jahnke et al. 1990). In principle, solute fluxes only provide an estimate of remineralization, and burial must be estimated independently. However, only a few percent of the deposited organic carbon survive remineralization in deep-sea sediments and a rough estimate of burial is sufficient to account for this term. Limitations are that accurate flux estimations require (*a*) high resolution measurements of mm scale near the sediment surface, (*b*) in situ measurement to avoid pressure and temperature related sampling artifacts and (*c*) molecular diffusive transport to dominate solute exchange processes (Glud et al. 1994).

In the last decade, methods for determining benthic solute fluxes (mainly used for oxygen fluxes) have become more sophisticated due to the development of free falling lander systems (Tengberg et al. 1995). They are equipped either with benthic chambers in which the rate of loss of oxygen in the water above the sediment can be measured or with oxygen microsensors which record oxygen profiles in the sediment. These in situ measurements have added new data to the traditional shipboard measurements of oxygen consumption and have improved our understanding of the biological degradation of organic matter at the deep-sea floor. In situ benthic flux chamber measurements are generally thought to be the most accurate technique for evaluating benthic fluxes in deep-sea sediments, but are limited in that they require sophisticated instruments and are time consuming, restricting the number of measurements that can be obtained. In some cases such in situ benthic chamber measurements indicate higher sediment carbon turnover than the vertical organic carbon flux measured by sediment traps (e.g., Smith 1987; Smith and Kaufmann 1999; Witte and Pfannkuche 2000; Ritzrau et al. 2001b). This discrepancy can not yet be fully resolved. Possible explanations are lateral input of organic matter from higher productive regions not captured by sediment traps, which is particularly obvious near continental margins (see below), or systematic underestimation of total vertical flux by sediment traps. Good agreement between carbon regeneration estimated from benthic studies and vertical organic carbon fluxes measured by sediment traps in certain mid-oceanic environments (e.g., equatorial Pacific Ocean: Berelson et al. 1997; Southern Ocean: Sayles et al. 2001) supports the view that discrepancies are greatest in ocean-margin environments.

8.3.2 Regional Assessments of Deep-Ocean Fluxes

A correlation between benthic fluxes and surface water productivity, estimated from satellite images, has been found in several well-studied regions. For instance, in the Arabian Sea biogeochemical processes in the benthos match the pattern of monsoon-driven primary productivity (Pfannkuche and Lochte 2000). In the northern North Atlantic, a consistent relationship between primary production, water depth and benthic oxygen fluxes was found (Schlüter and Sauter 2000). The global assessment of Jahnke (1996) shows a general agreement between benthic oxygen fluxes and export measured by sediment traps as well as primary productivity patterns estimated from satellite images, but in some regions deviations are observed.

Notable exceptions to this general pattern are found at stations close to continental margins. Their benthic fluxes exceed by far the rates measured by sediment traps or estimated from surface water productivity. This was observed, for example, at stations in the western Arabian Sea (Witte and Pfannkuche 2000), in the Argentine Basin (Hensen et al. 2000) or on the Iceland Plateau (Schlüter et al. 2001; Ritzrau et al. 2001b). Based on a nutrient and carbon budget Chen and Wang (1999) estimated for the East China Sea an export of organic matter from the shelf via downslope transport of POC of 0.7×10^{12} mol C yr^{-1}. This has to be compared to a total primary production of ca. 12×10^{12} mol C yr^{-1} indicating that most of the primary produced material is recycled on the shelf and that <6% is exported to the adjacent deep sea. During the SEEP study at the northeastern American Atlantic Margin and the Mid-Atlantic bight, <5% of primary production was estimated to be exported from the shelf (Anderson et al. 1994) and studies at the European continental margin (OMEX) indicate an export of ca. 10% (McCave et al. 2001; Wollast and Chou 2001). Such estimations are hampered by many poorly constrained factors, but they are an example of the magnitudes of downslope export to be expected. Although this is only a small fraction of total production on the shelves, it represents a large influx to the sediments of the continental margins. It has to be noted that under different hydrographic and topographic conditions the export from a specific shelf system may be very different. Most of the transport from the shelf occurs in the bottom turbidity layer, which occupies a few tens of meters above the sediment. This process is difficult to observe and has been largely neglected in the past. Major conduits for export are canyons of the shelf slope (Biscaye and Anderson 1994) through which most of the export is funnelled. Upwelling and frontal systems at the shelf edge give rise to high productivity and exceptionally high sedimentation contributing additional material to the flux at continental margins. Although there is an increasing recognition that export of organic matter from the productive shelf areas is a quantitatively important flux to the deep ocean, this export is still very difficult to quantify (see also Chen et al., this volume).

Another region of possible discrepancies is the Southern Ocean. Relatively high benthic fluxes of oxygen and nitrate, an indicator of benthic remineralization, are found in the upwelling regions off Namibia and Chile and in the Atlantic part of the Antarctic Circumpolar Current zone (Hensen et al. 2000; Grandel 2000) which are not matched by estimates of primary production from satellite images. The relatively high benthic mineralisation rates in parts of the Southern Ocean may be attributed to fast sinking large diatom aggregates and krill fecal pellets. These observations are supported by recent analyses from the ANTARES and AESOPS (Southern Ocean – JGOFS) programmes which showed that the fraction of exported production in some regions of the Southern Ocean is higher than in lower latitudes (Pondaven et al. 2000; Honjo et al. 2000; Nelson et al. 2002). Organic carbon fluxes recorded in sediment traps at 1 000 m depth in the Polar Frontal Zone and in the Antarctic Zone were about twice as high as the global average (Honjo et al. 2000). This is in contrast to other sediment trap studies which indicate that fluxes in the north and south polar regions were similar to the global average and that around 1.2% of primary production reaches the sea floor (Schlüter et al. 2001). These different observations indicate that we do not yet understand sufficiently production and export processes in the Southern Ocean and that we may underestimate the vertical flux in this region. Another important aspect is extensive sediment focusing in some regions of the Southern Ocean (Francois et al. 1993; Kumar 1994; Frank et al. 1999; Dezileau et al. 2000; Sayles et al. 2001) which produces local rates of sediment deposition far in excess (by as much as a factor of 20) of regionally-averaged particle rain rates. Sediment focusing creates high benthic fluxes in regions of modest productivity, thereby accounting for some of the discrepancies noted above.

8.3.3 Global Estimates of Deep Ocean Carbon Deposition and Remineralization

The major obstacle in obtaining large scale estimates of sea-floor fluxes is the paucity of data. There are some well studied regions, some examples are given above, but for wide areas of the deep sea no measurements exist. The best strategy for assessing the distribution of sea floor fluxes at the basin scale is to combine benthic flux and sediment modeling strategies (see above). Point measurements of benthic fluxes provide the most accurate data and constrain the overall magnitude of the fluxes. Sediment models provide the greatest spatial coverage and could be used to interpolate between and extrapolate from point flux measurements to basin scale estimates. This, of course, implies that there is a reasonable relationship between burial rates of organic carbon, which span a long period of time, and benthic fluxes, which record the instantaneous rate of organic matter turnover. A compilation of benthic remineralization rates (from in situ benthic flux chamber deployments or in situ microelectrode oxygen profile measurements) and burial rates (from organic carbon contents of the sediments and sediment mass accumulation rates estimated primarily through ^{14}C chronology) shows a significant linear correlation (Jahnke 1996). Thus, despite the time scale difference between sediment accumulation and benthic flux response to varying inputs, there is an overall correlation between sediment burial and benthic fluxes. Burial efficiencies (100 × burial rate / deposition rate) in the deep ocean gyre regions average about 1%, in higher flux areas of open ocean and coastal upwelling regions they range from a few percent to 10%, and on the upper continental slope very high efficiencies >10% are estimated. Thus, although burial efficiencies are not constant, a regular progression from low to high deposition locations is observed.

Utilizing these techniques, global estimates of the distributions of organic carbon fluxes to the deep-sea floor and global benthic oxygen fluxes have been estimated between 60° S and 60° N (Jahnke 1996) (Fig. 8.2). Overall, the distribution of the sea floor fluxes displays the same general features as previously published distributions of primary and new production (e.g., Berger

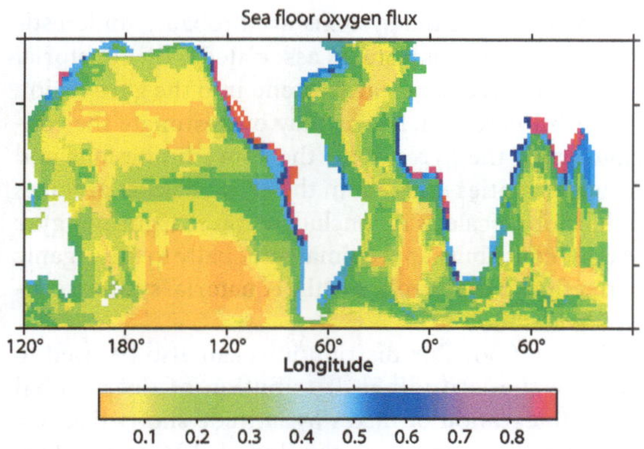

Fig. 8.2. Global distribution of sea floor oxygen flux (mol m^{-2} yr^{-1}) estimated from benthic flux estimates. Values are extrapolated to basin scales using empirical correlations between benthic oxygen flux and organic carbon burial rates (redrawn from Jahnke 1996)

et al. 1987; Antoine et al. 1996). On average, low fluxes are estimated for the deep, central gyre regions and high fluxes are calculated for continental boundary regions, especially those areas adjacent to coastal upwelling regions such as the Pacific margins of the Americas, the Atlantic African margin, and the monsoonally influenced regions of the Arabian Sea. Elevated fluxes are also estimated for the equatorial Pacific region and the polar-front region of the southern Atlantic and Indian Oceans.

The sea-floor fluxes imply a global organic carbon deposition rate to the deep-sea floor of 33×10^{12} mol C yr^{-1}, excluding areas shallower than 1 000 m (Jahnke 1996). Of this flux approximately 3% are permanently buried. The resulting total remineralization rate in the sediments accounts for 45% of the total oxygen utilization of the deep ocean estimated from Apparent Oxygen Utilisation-^{14}C relationships. This estimate of global organic carbon flux agrees well with estimates from deep moored sediment traps, which indicate a flux at 2 000 m of 0.34 Gt C yr^{-1} (28×10^{12} mol C yr^{-1}). Thus, the magnitude of the deep fluxes estimated from the sediment studies is consistent with other measures of deep-ocean metabolism.

Based on Jahnke's (1996) results from benthic oxygen fluxes, the relative importance of continental margin, equatorial and gyre regions in POC transport to the deep ocean can be assessed. For this purpose, the margin regions are defined as any location within 6° of latitude or longitude of the continental shelf, the equatorial region is defined as those areas between 5° N and 5° S and the gyre region is all of the sea floor area remaining. Respectively, these areas represent 24.5%, 7.3% and 68.2% of the total deep-sea floor area. In general, despite comprising the majority of ocean area, the gyre region contributes only approximately 50% of the estimated total deep flux. Margin areas contribute 40% or more and equatorial areas contribute less than 10%. Since some marginal seas and basins such as the Caribbean Sea and Gulf of Mexico are not included, contributions from margin systems are probably underestimated. Some of the features associated with equatorial or margin processes clearly extend into the gyre region as defined here and, hence, may overestimate the contribution of the gyres. With the above limitations and the uncertainties inherent in the extrapolation of fluxes to the global scale, it is concluded that margin and gyre systems contribute approximately equally to the organic matter flux to the deep sea while equatorial systems contribute less than 10%.

The sea floor flux distributions can also be used to examine the latitudinal distribution of deep global fluxes. Integration of fluxes by latitude shows that low latitude regions dominate the flux of POC to the deep ocean and suggests that 2/3 of the total POC flux to the sea floor occurs within 30° of the equator. This low latitude dominance follows very closely the latitudinal distribution of deep-ocean surface area as approximately 60% of the total ocean area lies within 30° of the equator (Menard and Smith 1966). Given this distribution, high latitude regions could contribute equally to the deep POC flux only if deep POC fluxes there were at least a factor of 5 larger than low-latitude fluxes. Such high deep POC fluxes are not indicated by sediment trap data (Fischer et al. 2000; Antia et al. 2001; Schlüter et al. 2001), even if the recently reported higher fluxes in the Southern Ocean are considered (see above).

The above described benthic oxygen flux distribution based on sea-floor results can also be compared to a distribution estimated from surface water productivities based on CZCS satellite data and direct productivity measurements, vertical transfer relationships and water depth (Romankevich et al. 1999; Tseitlin 1993; Vinogradov et al. 1996) (Fig. 8.3). Overall, there is a good agreement in the estimated total carbon and oxygen fluxes to the deep-sea floor (sea floor greater than 1 000 m water depth) between the two approaches. A value of 53×10^{12} mol O_2 yr^{-1} ($= 32 \times 10^{12}$ mol C yr^{-1}) is estimated from the surface productivity and vertical carbon flux relationship for the global deep-ocean floor. A value of 33×10^{12} mol C yr^{-1} resulted from benthic measurements (see above, Jahnke 1996). Approximately 72% of the total sea floor flux derived from the primary production-depth relationship occurs within 30° of the equator. Considering that these results are obtained from completely different approaches and data sets, such agreement may be to some extent fortuitous since regional differences are apparent between both approaches.

A totally different approach to deep-ocean fluxes is made by inverse models. The large data base of deep ocean dissolved nutrients, oxygen and hydrographic data is used to fit a coupled circulation-biogeochemistry global ocean model until the observed properties are realistically reproduced (Schlitzer 2002). This inverse model yields total integrated export fluxes which are necessary for a realistic reproduction of nutrient and oxygen data in the deep water colum. This model indicates that the tropical and subtropical areas between 30° N and 30° S contribute the largest export flux of organic carbon (ca. 53% of global export flux). This is a consequence of the very large area covered by this zonal belt. The second largest contribution comes from the Southern Ocean south of 30° S (ca. 33%) while the export of the oceans north of 30° N is smaller (ca. 14%). Most of the Southern Ocean export is found in a zonal belt along the Antarctic Circumpolar Current, associated with the high productivity at the Polar Front, and in the coastal upwelling regions off Chile and Namibia. While in most areas of the ocean the model estimates of export flux are in agreement with estimates based on satellite images, they are higher by factors of two to five in the Southern Ocean. These model results of high export relative to satellite estimates support the observa-

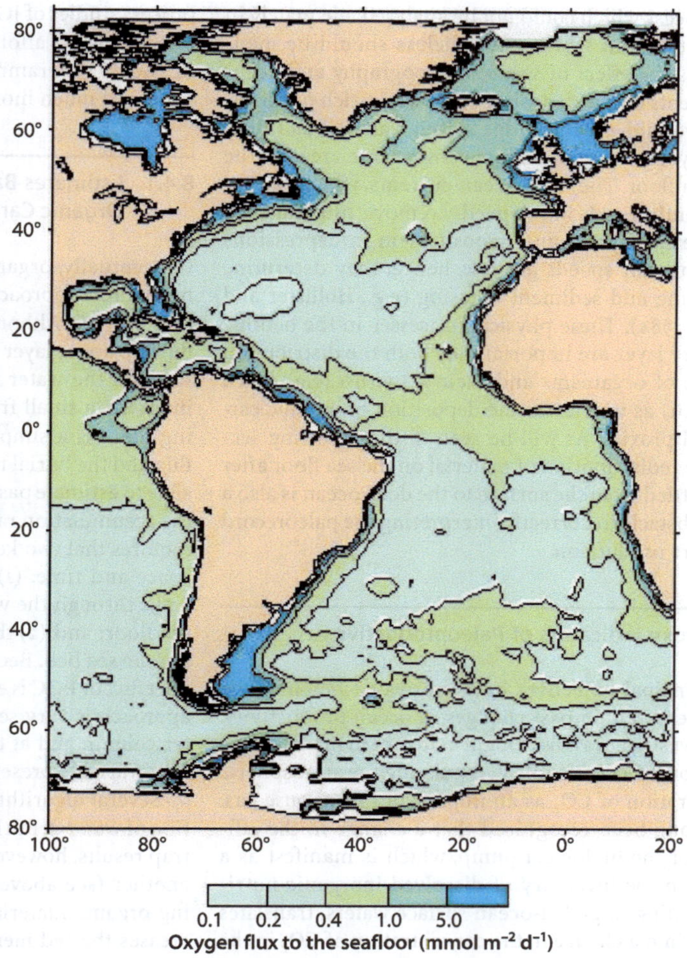

Fig. 8.3. Distribution of sea floor oxygen flux (mmol m^{-2} d^{-1}) in the Atlantic Ocean based on estimates of surface primary production rates, vertical transfer relationships and water depth (Romankevich et al. 1999)

tions of high benthic fluxes in parts of the Southern Ocean (Grandel 2000), AESOPS sediment trap studies (Honjo et al. 2000) and analyses of nutrient budgets at ANTARES (Pondaven et al. 2000) as well as throughout the Southern Ocean (Louanchi and Najjar 2000). All these studies also indicate that the organic carbon fluxes are accompanied by exceptionally high silicate fluxes in the Southern Ocean stressing the importance of diatoms for efficient export fluxes (see also Tréguer, this volume). Since the ratio of silicate to organic carbon varies strongly in settling material (e.g., Honjo et al. 2000; Pondaven et al. 2000), care has to be taken when using sedimentary opal accumulation rates as indicators of paleoproductivity.

When the sea floor fluxes are compared to distributions of primary productivity reported by Behrenfeld and Falkowski (1997) differences are obvious in the high latitudes. Unlike benthic assessments that were dominated by low latitude regions (within 30° of the equator), 55% of the reported primary productivity occurs at latitudes greater than 30°. While the estimates of the global deep-ocean carbon fluxes converge to values around 0.3–0.4 Gt C yr^{-1} (25–30 × 10^{12} mol C yr^{-1}), the discrepancies between regional estimates offer a chance to refine our perception of the export processes. It seems that in particular the productivity and the export associated with the polar front zone of the Southern Ocean needs to be critically analysed. Part of the discrepancies may be caused by poor data coverage in the southern hemisphere for benthic processes and, hence, benthic fluxes are more difficult to assess. Satellite data in high latitudes may also misjudge total productivity due to very deep mixing of the water column and frequent cloud cover. Furthermore, the observations imply that the efficiency with which organic material is transported from the photic zone to the sea floor varies with latitude and may be particularly high in plankton communities dominated by diatoms and large zooplankton. Export fraction and degradation in the mesopelagic zone beneath the euphotic zone seem to be influenced by the plankton community, but as yet there are few hard facts which allow estimates of the regionally variable transfer efficiencies from the upper to the deep ocean to be made.

A process which could not be analysed adequately in this context, but which nevertheless should be mentioned, is the effect of sea-floor topography and deep-sea currents on the deposition of organic-rich particles. Extensive mid-oceanic ridge systems as well as ridges and canyons at ocean margins cover large areas of the deep-sea floor. The deep-ocean currents, which can intermittently reach high speeds, remove fine material from elevated parts and deposit them in depressions where current speeds are low, hence, they determine winnowing and sediment focusing (e.g., Hollister and McCave 1984). These physical processes in the benthic boundary layer are important for both the distribution patterns of organisms and their activities (e.g., Flach et al. 2001) as well as for the deposition of organic carbon and proxies. As will be seen in the following section, the redistribution of material on the sea floor after it has settled from the surface to the deep ocean is also a major obstacle in correctly interpreting the paleorecord of export production.

8.4 Proxy Indicators of Paleoproductivity

Two principal objectives have motivated much of the recent research on past changes in ocean productivity. First, investigators have sought evidence in marine sediments for the ocean's role in regulating the atmospheric concentration of CO_2 as an important greenhouse gas. It has long been recognized that a change in the efficiency of the biological pump, which is manifest as a change in the inventory of dissolved inorganic nutrients residing in global-ocean surface waters, translates directly into a change in the concentration of CO_2 in the atmosphere (e.g., Broecker 1982; Sarmiento and Toggweiler 1984). Various factors, ranging from changes in wind-driven upwelling (Pedersen and Bertrand 2000), ocean nutrient inventory (Falkowski 1997; Ganeshram et al. 2000) and to fertilization by iron (Martin 1990) have been hypothesized to induce climate-related changes in ocean productivity. Much of the recent paleoproductivity research has been designed to test these hypotheses. The second, and no less important, objective motivating paleoproductivity research has been the desire to understand the response of ocean ecosystems to changing environmental boundary conditions associated with climate change. A sound understanding of the sensitivity of ocean ecology to perturbation by climate change in the past will help guide efforts to predict the response of ocean ecosystems to anticipated global warming. A number of methods have been developed to reconstruct past changes in ocean productivity. Careful calibration of such methods is essential in order to test the assumptions and resolve the limitations inherent in these methods. Progress has been made in calibrating productivity proxies through the comprehensive biogeochemical process studies of JGOFS. However, because this research was not always afforded high priority during the design of JGOFS programmes, progress occurred at a limited pace, and much more remains to be done.

8.4.1 Estimates Based on Organic Carbon Burial Rates

Conceptually, organic carbon burial rates provide the most direct approach for reconstructing export production of POC. Although organic matter leaving the productive upper layer is largely oxidized during its transit through the water column and at the water-sediment interface, a small fraction is eventually buried. Assuming there is a simple relationship between the export flux and the burial rate of carbon, then it should be possible to estimate past changes in export production from the accumulation rate of organic carbon. This approach requires that two key parameters do not vary greatly in space and time: (*1*) the remineralization of POC as it sinks through the water column and as it settles on the sea floor; and (*2*) the preservation of POC after reaching the sea bed. Because only a small fraction of the export flux of POC is eventually preserved and buried, this approach is very sensitive to variable losses in the water column and at the water-sediment interface as well as to variable preservation during sediment diagenesis.

Several algorithms for POC regeneration as a function of water depth have been developed using sediment trap results, however, they differ substantially from one another (see above). Furthermore, 'ballasting' of sinking organic material by denser inorganic particles increases the sedimentation speed and, hence, the depth of remineralization (Ittekkot 1993; Armstrong et al. 2002). An algorithm relating carbon preservation to sediment accumulation rate has been derived empirically at continental margin sites (Mueller and Suess 1979). Subsequently, it has been learned that much of the POC of continental margin sediments originates from lateral transport down the continental slope. Much of this POC is old (^{14}C age of several hundred years, Anderson et al. 1994) and refractory, thereby creating artificially high apparent POC preservation efficiencies in continental margin sediments (e.g., Jahnke 1990; Anderson et al. 1994). For this reason, the preservation algorithm overestimates the sensitivity to sediment accumulation rate, and should not be used in paleoproductivity reconstructions.

As the supply of POC by lateral transport is often large at ocean margins, sometimes exceeding the vertical rain from surface waters (Anderson et al. 1994, see above), the burial rate of POC is largely decoupled from the export flux sinking from overlying waters. In some cases, POC of terrestrial origin dominates the organic matter preserved and buried in ocean-margin sediments (e.g., Lyle et al. 1992; Villaneuva et al. 1997), even in regions far removed from the mouths of major rivers. Where terrestrial POC constitutes a significant portion of buried organic matter, the carbon accumulation rate obviously cannot be used to reconstruct past changes in ocean productivity.

The sensitivity of sedimentary POC preservation to bottom water oxygen concentration has long been debated (see review by Canfield 1994), but there is now increasing evidence that POC preservation is sensitive to multiple factors, including bottom water oxygen concentration (Keil et al. 1999), oxygen exposure time (Hartnett et al. 1998), abundance of bacterial grazers (Lee 1992), bioturbation rate (Kristensen et al. 1992; Andersen and Kristensen 1992), and protection by sorption to minerals (Mayer 1994; Keil et al. 1994). This introduces substantial uncertainty into paleoproductivity reconstructions based on the accumulation rate of organic carbon. For this reason, together with the other sources of uncertainty identified above, paleoceanographers have sought independent methods free of these problems to estimate past levels of ocean productivity.

8.4.2 Estimates Based on Biomarker Accumulation Rates

Specific organic biomarker compounds, known to be produced by marine phytoplankton, are used as proxies alternative to organic carbon burial rates for paleoproductivity studies. Biomarkers may be specific to selected groups of phytoplankton (e.g., alkenones from coccolithophorids, dinosterol from dinoflagelates, brassicasterol from diatoms), or they may include the full suite of the pigment transformation products of chlorophyll, known as chlorins. Use of these biomarkers eliminates the error associated with input of terrestrial POC, and also supplies information about the dominant groups of phytoplankton preserved in the sediments (e.g., Schubert et al. 1998).

Like any method that relies on the absolute accumulation rate to reconstruct changes in export production, the use of biomarkers is sensitive to errors introduced by sediment focusing; i.e., the redistribution of particles by deep-sea currents, and by variable rates of preservation of the biomarkers. While sediment focusing is well known in environments influenced by bottom currents (e.g., the Southern Ocean, see above), recent evidence suggests that sediment focusing is also prevalent in other regions, such as the central equatorial Pacific Ocean, where its effects were previously not recognized (Marcantonio et al. 1996, 2001). Fortunately, it is possible to correct for sediment focusing by normalizing concentrations of biomarkers to ^{230}Th, a radiogenic product of ^{234}U dissolved in seawater, which is highly particle reactive and whose flux to the sediments in most oceanic regions nearly equals its known rate of production in the overlying water column (Suman and Bacon 1989; Francois et al. 1990). Analysis of sediment trap samples collected world-wide is contributing to the further calibration of ^{230}Th as a constant flux proxy by evaluating the degree to which its flux deviates from its known production rate and defining the conditions under which these deviations occur (Yu et al. 2001a; Scholten et al. 2001). This approach only provides estimates of 'preserved rain rates', i.e., rain rate minus dissolution or remineralization before burial. Variable losses of biomarkers due to biological action or changes in redox conditions are still a major problem. As yet, no method has been proposed to correct for such variation in the preservation of biomarkers, although research conducted during JGOFS process studies may help to improve the understanding of the factors regulating biomarker preservation.

8.4.3 Estimates Based on Barium Accumulation Rates

High concentrations of barium, in excess of average concentrations in aluminosilicate minerals, have long been known to occur in marine sediments underlying regions of high productivity. This has led to the suggestion that accumulation rates of excess barium can be used to estimate export production. Support for this view came from sediment trap data which showed a tight, but non-linear, relationship between the flux of organic carbon and that of Ba (Dymond et al. 1992; Francois et al. 1995) and from empirical correlations between barite accumulation rate and primary productivity (Paytan et al. 1996). These relationships provide a means of quantifying the export flux of carbon from the accumulation rate of excess Ba in sediments, if the preservation of excess Ba can be constrained. The main advantage of Ba is that it is much better preserved in sediments than is organic carbon. An empirical algorithm for excess Ba preservation as a function of sediment accumulation rate has been developed (Dymond et al. 1992). It has also been suggested that barite, the common carrier of excess Ba, would only derive from biogenic material freshly produced in the overlying water, so that the Ba record would not be affected by lateral transport of more refractory POC from shelves and continents (Francois et al. 1995). Because of these advantages, the use of excess Ba for paleoproductivity reconstructions has become extremely popular.

While initial results suggested that Ba offers many advantages as a paleoproductivity proxy, more recent research has discovered that it also suffers from some serious limitations. Calibration studies using sediment trap samples have shown that the Ba/POC rain ratio in sinking particles varies substantially from one location to another, even among sites where lateral supply of reworked POC is expected to be negligible (Dymond and Collier 1996; Dehairs et al. 2000). Preservation of barite in sediments is poor where pore water sulfate concentrations are lowered by sulfate reduction (e.g., Brumsack 1986; von Breymann et al. 1992). It was discovered more recently that preservation of excess Ba plummets even under modestly reducing (sometimes referred to as suboxic) conditions (Kumar et al. 1996; McManus et al. 1998), such as occur commonly at ocean margins and in other regions of high export production. Furthermore, barite is subject to extensive reworking and export from continental shelves (Fagel et al. 1999), similar to organic carbon. An example from the JGOFS Equatorial Pacific Study shows the absolute accumulation rates of barite in sediment cores derived from ^{18}O-based stratigraphy and in comparison to the accumulation rate corrected for sediment focusing by normalisation to ^{230}Th (Fig. 8.4). These two traces indicate very different pictures of the sensitivity of export production to climate change over the last 200 kyr in the equatorial Pacific Ocean. The absolute accumulation rate (^{18}O) implies much greater export during glacial climate intervals, whereas the accumulation rate corrected for sediment focusing (^{230}Th) indicates very little climate sensitivity. Supply of excess barite from sediment focusing obviously increased the barite accumulation rate during glacials (Marcantonio et al. 2001). Therefore, this temporal variability may be interpreted as changes in climate-related deep-sea currents responsible for sediment focusing rather than changes in export production. Another example from the Atlantic sector of the Southern Ocean shows ^{230}Th normalized accumulation rates of organic carbon (C_{org}) and excess Ba in two sediment cores (Fig. 8.5). Through the time interval from the last glacial maximum (ca. 20 kyr) to the present, the carbon accumulation rates have decreased more than 5-fold, yet there has been no significant change in the accumulation rate of Ba. This exemplifies that one obtains entirely different views of the climate sensitivity of export production depending on which tracer is used. Either the initial Ba/C_{org} ratio has changed through time or the relative preservation of C_{org} or excess Ba is the factor that has changed. These two examples illustrate that one must be cautious in using Ba as a productivity proxy in regions where supply of barite is decoupled from export production by winnowing from shelves, or where preservation of Ba is low and variable in suboxic sediments. Suboxic sediments exist not only in many productive regions (e.g., Kumar et al. 1995), but also in areas where sediment focusing is pronounced (e.g., Francois et al. 1993), precluding the use of Ba accumulation in these regions as well.

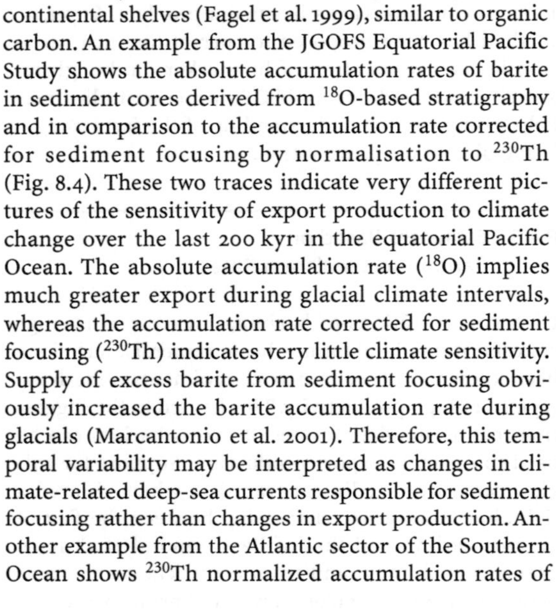

Fig. 8.4. Absolute accumulation rate of barite derived from ^{18}O-based stratigraphy (*diamonds*) and accumulation rate normalised to ^{230}Th to correct for sediment focusing (*squares*). The absolute accumulation rate implies much greater export production during glacial climate intervals, whereas normalization to ^{230}Th indicates very little climate sensitivity. The difference is caused by supply of excess barite due to sediment focusing. *Scale bar* shows marine isotope stages. *Roman numericals III* and *V* represent interglacial periods; *filled intervals* represent glacial periods (data from US JGOFS Equatorial Pacific study at 0° N and 140° W; barite data from Paytan (1995); this figure has been redrafted from results presented by Marcantonio et al. 2001)

In pelagic regions, where sediments remain well oxygenated to great depth, it was believed until recently that the principal interference influencing the accumulation rate of excess Ba is sediment focusing, which can be corrected for by normalizing to ^{230}Th. However, first-order budgets for Ba in the central equatorial Pacific Ocean, constructed using benthic incubation chambers, sediment trap samples and cores collected during the US JGOFS EqPac programme, show that the present rates of Ba rain and benthic remobilization are nearly in balance, indicating that the net rate of Ba accumulation is negligible (McManus et al. 1999). Thus, there seems to be a minimum threshold in carbon flux that needs to be reached to leave a biogenic Ba signal in the sediment. The usefulness of Ba-based paleoproductivity algorithms may thus be limited to a relatively narrow window of productivity, whose exact limits still need to be established.

8.4.4 Estimates Based on Radionuclide Ratios

Radionuclide ratios as productivity proxies exploit the systematic relationship between mass particle flux and the adsorption of certain radionuclides to sinking particles. Here, the fact that ^{230}Th is sufficiently particle-reactive that it is removed from the ocean at a rate equivalent to its known rate of production (see above) is exploited. Thus, this proxy indicates changes in the rate of mass particle flux and indirectly past changes in productivity. In contrast to this ultra-reactive behavior of Th, less-reactive tracers are scavenged from the ocean on time scales comparable to, or greater than, the mixing time of an ocean basin, and their flux from the water column increases with increasing particle flux. Tracers that fall into this category include ^{231}Pa and ^{10}Be. ^{231}Pa, like ^{230}Th, is produced uniformly throughout the

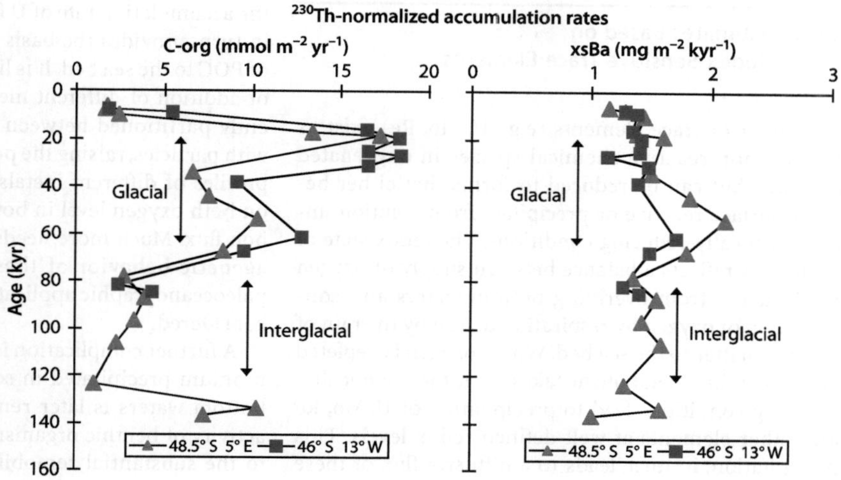

Fig. 8.5. Accumulation rates of (*left*) organic carbon and (*right*) excess Ba in two cores from the Atlantic sector of the Southern Ocean from a sampling site between the present positions of the Antarctic Polar Front and the Subantarctic Front. The accumulation rates have been normalized to ^{230}Th to account for sediment focusing. The last glacial and interglacial periods as recorded in the sediments are indicated. Note the major changes of export production indicated by C_{org} accumulation rate between interglacial and glacial, but not by excess Ba (R. Anderson, unpubl.)

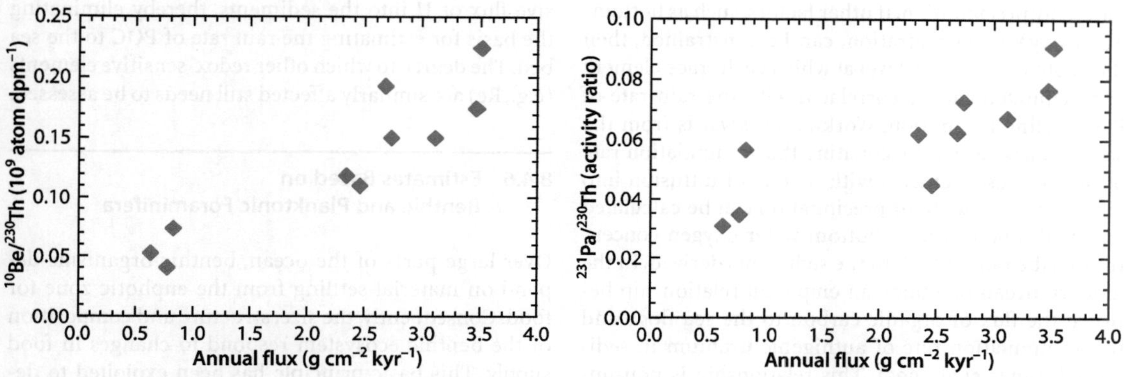

Fig. 8.6. Flux-weighted mean radionuclide ratios are plotted against annual average mass flux for sediment trap samples recovered during the US JGOFS study along a transect normal to the equator at 140° W from 9° N to 12° S. Radionuclide ratios correlate with mass flux, supporting their use as proxies for vertical flux. Sediment traps used in this calibration were deployed at depths greater than 2000 m (R. Anderson, unpubl.)

ocean by U decay (^{235}U in this case), while ^{10}Be is produced in the atmosphere and supplied to the ocean primarily in precipitation. The tendency for ^{231}Pa/^{230}Th and ^{10}Be/^{230}Th ratios to increase with increasing sediment accumulation rate was proposed as a paleoproductivity proxy (Kumar et al. 1995; Anderson et al. 1998). A recent calibration of these proxies using JGOFS sediment trap samples from the equatorial Pacific Ocean shows good correlations between ^{231}Pa/^{230}Th and ^{10}Be/^{230}Th ratios and the mean annual flux of particles, supporting the original contention that these radionuclide ratios can be used as proxies of particle flux (Fig. 8.6) (see also Marcantonio et al. 2001; Yu et al. 2001b).

There are some shortcomings with this approach, however. To the extent that these nuclide ratios reflect particle flux, rather than export production per se, they may be influenced by changes in the flux of detrital particles eroded from continents, e.g., riverine inputs or aeolian dust input, over the considered time scales. In addition, scavenged fluxes of these tracers at any one site in the ocean also depend on the intensity of scavenging in surrounding regions, which determines whether the surrounding regions serve primarily as a source (i.e., surrounding regions have lower particle fluxes and scavenging intensity) or a sink (higher particle fluxes) for the laterally-transported tracers (^{231}Pa and ^{10}Be) relative to the site being studied. Furthermore, scavenged elemental ratios are controlled not only by particle flux, but also by the chemical composition of particles (e.g., Walter et al. 1997; Yu et al. 2001b; Chase 2001). Finally, deep water thermohaline circulation also affects radionuclide ratios independently of particle flux (Yu et al. 1996; Marchal et al. 2000; Yu et al. 2001b; Chase 2001). In order to exploit these tracers reliably, it will be necessary to fully characterize the effects of particle composition-dependent elemental fractionation, and to take into account possible change in deep water circulation. The greatest potential for these ratios is in providing synoptic maps reflecting the combined effects of deep water circulation and relative particle fluxes between oceanic regions.

8.4.5 Estimates Based on Redox-Sensitive Trace Elements

A number of trace elements (e.g., U, Mo, Re) exist as soluble, non-reactive chemical species in oxygenated seawater, but can be reduced to forms that either become surface reactive or precipitate from solution under chemically-reducing conditions. The redox state of sediments reflects a balance between supply of oxygen by diffusion from overlying bottom waters and consumption of oxygen by respiration driven by the rain of organic matter to the sea bed. Where oxygen is depleted and anaerobic metabolism takes over, the chemically-reducing conditions lead to precipitation of U, Mo, Re and other elements at well-defined redox levels. This precipitation, in turn, leads to a diffusive flux of these elements into the sediments to replace the atoms removed by precipitation. If other factors, such as bottom-water oxygen concentration, can be constrained, then the depth of the redox level at which each trace element is precipitated can be correlated with the rain rate of POC fueling respiration. Working backwards from the sedimentary record, by equating the accumulation rate of one of these elements with its rate of diffusion into sediments, the depth of precipitation can be calculated and, with knowledge of bottom water oxygen concentration, the rain of POC to the sediments derived. In the modern ocean one finds an empirical relationship between the flux of organic carbon to the sea floor and the accumulation rate of authigenic uranium in sediment (Kumar et al. 1995). This relationship is non-linear, and involves a threshold carbon flux below which no uranium is buried. If other parameters can be constrained, then the change in accumulation rate of authigenic uranium, or other redox-sensitive elements, could be used to reconstruct past changes in export production.

Several problems limit the use of this proxy. First, the redox conditions of sediments depend on bottom-water oxygen concentration as well as on POC supply. If changes in bottom-water oxygen concentration cannot be excluded, or quantified, then it is not possible to derive a carbon flux uniquely from the accumulation rate of uranium and other redox-sensitive metals. Also, the interpretation of authigenic metal accumulation delineated above relies on the principle that metal accumulation is primarily due to diffusive fluxes from bottom waters into the sediment. It was recently discovered, however, that particulate authigenic U formed in surface waters, although very labile, can contribute a substantial fraction of authigenic U buried in sediments under suboxic or anoxic conditions (Zheng et al. 2002a). If particulate authigenic U formed in surface waters is preserved and buried, then this flux partially decouples the accumulation rate of U from its diffusive flux which, in turn, provides the basis for estimating the rain rate of POC to the sea bed. It is likely, however, that the mode of addition of different metals into sediment is differently partitioned between diffusive flux and addition with particles, raising the possibility that combining the profiles of different metals could provide information on both oxygen level in bottom water and organic carbon flux. Much more needs to be known about the diagenetic behavior of these metals, however, before paleoceanographic applications of this approach can be considered.

A further complication is that much of the authigenic uranium precipitated in sediments under oxygenated bottom waters is later remobilized by the burrowing activity of benthic organisms (Zheng et al. 2002b). Due to the substantial remobilization of authigenic U, its accumulation rate is again decoupled from the diffusive flux of U into the sediments, thereby eliminating the basis for estimating the rain rate of POC to the sea bed. The degree to which other redox-sensitive elements (e.g., Re) are similarly affected still needs to be assessed.

8.4.6 Estimates Based on Benthic and Planktonic Foraminifera

Over large parts of the ocean, benthic organisms depend on material settling from the euphotic zone for food. Consequently, the size, structure and composition of the benthic ecosystem respond to changes in food supply. This basic principle has been exploited to derive empirical relationships between carbon flux and the abundance, accumulation rate, and species composition of benthic foraminifera (e.g., Herguera and Berger 1991, 1994; McCorkle et al. 1994; Thomas et al. 1995). These relationships, in turn, applied to down core records have been used to reconstruct past changes in surface productivity. While a response to food supply is easily understood, one would also expect the composition of the benthic community to respond to other environmental factors, such as food quality, seasonality of food supply, and bottom water oxygen concentration. Recent expansion of calibration studies is beginning to address these multiple factors, and offers hope of extracting information concerning carbon flux as well as its seasonality (Loubere and Fariduddin 1999). The carbon isotopic composition of benthic foraminiferan $CaCO_3$ shells also contains information about the environment in which they grew (Woodruff and Savin 1985; Zahn et al. 1986; McCorkle et al. 1990).

Species assemblage of planktonic foraminifera deposited in sediments are also used to estimate surface water productivity and statistical transfer functions have been derived (e.g., Mix 1989a,b). However, planktonic foraminifera species assemblages are known to respond to other environmental parameters as well, including temperature and mixed layer depth, so caution is warranted in using transfer functions to predict multiple properties from the sedimentary record. Many of the properties (for example, productivity and mixed layer depth) are correlated in the modern ocean. This means that transfer functions calibrated with modern samples may confuse one ocean property for another. Furthermore, if the relationships between properties change through time, then the transfer function estimates could be in error. JGOFS process studies have provided a valuable opportunity to relate foraminiferal species assemblages to sea surface temperature, integrated primary productivity, and mixed layer depth. The calibration study by Watkins and Mix (1998), conducted during the US JGOFS Equatorial Pacific study, further benefitted from the opportunity to collect samples under contrasting El Niño and non-El Niño conditions. Further studies are needed to produce more robust global algorithms for productivity reconstruction based on this principle.

8.4.7 Estimates Based on Coccolithophorids and Diatoms

At low latitudes, coccolithophore communities of the lower euphotic zone (~60 to ~180 m) are dominated by *Florisphaera profunda*. Most other coccolithophore species live in the upper euphotic zone (0 to ~60 m). This vertical zonation has been used for paleoproductivity studies (Molfino and McIntyre 1990; McIntyre and Molfino 1996) where the abundance of *F. profunda* in fossil assemblages was used to monitor the depth of the nutricline. The relative abundance of *F. profunda* increases when the upper euphotic zone is impoverished in nutrients and the nutricline is deep. Conversely, the relative abundance of *F. profunda* decreases when wind stress induces a rise of the nutricline and an increase in productivity of the upper euphotic zone. Beaufort et al. (1997) correlated the relative abundance of *F. profunda* in surface sediments from the Indian Ocean with primary productivity estimated from satellite information. They further used principal component analysis to correlate species assemblages of planktonic foraminifera in Indian Ocean core top samples to primary productivity estimated from satellite information. Primary productivity estimates derived from foraminifera were similar to those derived from the abundance of *F. profunda*, and both proxies showed similar patterns of variability in a 270-kyr record from a site near the Maldives Islands.

Because diatoms form a major component of phytoplankton in upwelling regions of high productivity, and because diatoms contribute a substantial portion of the export flux even in less productive regions (Goldman 1993; Buesseler 1998), much effort has been devoted to correlating diatom species assemblages (e.g., Sancetta 1992; Abrantes 2000) and flux of diatom shell material (opal) (Ragueneau et al. 2000) to primary productivity. A complication in the exploitation of these approaches to reconstruct productivity is the poor preservation of diatoms throughout most of the world's oceans. On average, only 3–5% of the opal produced by diatoms is preserved in sediments (Nelson et al. 1995; Tréguer et al. 1995; Ragueneau et al. 2000). While significantly better than the average preservation of organic matter, opal preservation is still low enough for both diatom assemblages and opal accumulation in sediment to be sensitive to relatively small changes of the degree of preservation, obscuring their relationship to paleoproductivity. Opal preservation on the seafloor varies spatially and temporally (Shemesh et al. 1989; Pichon et al. 1992; Abrantes 2000; Pondaven et al. 2000; Sayles et al. 2001) and must be accurately constrained to estimate past changes in opal productivity from opal accumulation (e.g., Charles et al. 1991; Mortlock et al. 1991). Empirical algorithms have been developed to estimate opal preservation from diatom species assemblages (e.g., Pichon et al. 1992), but they only provide relative preservation estimates and are still associated with relatively large margins of error. While refinement of these algorithms is warranted, we also need to better understand the factors that control opal dissolution both in the water column and sediments. Furthermore, opal accumulation rates can be profoundly influenced by sediment focusing. This was shown to be particularly true in the Southern Ocean where focusing factors in excess of ten are not uncommon (e.g., Kumar 1994; Francois et al. 1993; Frank et al. 1999; Dezileau et al. 2000). In order to utilize opal accumulation rates to reconstruct productivity, better preservation algorithms will need to be developed (Sayles et al. 2001), and it will be necessary to correct for sediment focusing using ^{230}Th, or an alternative proxy of known flux. Recent work has also emphasised that the relationships between fluxes of organic carbon and opal vary strongly geographically depending on the plankton community (e.g., Honjo et al. 2000; Pondaven et al. 2000; Schlitzer 2002; see above). Biogeochemical process studies, such as those conducted during the international JGOFS programmes, have provided opportunities to start evaluating those factors. Much more remains to be done, however, particularly in terms of alteration of the species assemblage preserved in the sediments.

8.4.8 Proxies of Surface Nutrient Concentration

The export production is equal to the new production relying on the supply of nutrients by upwelling and mixing. If this supply can be assumed constant, or constrained by independent methods, then a change in surface nutrient concentration can be interpreted in terms of a change in export flux. Consequently, substantial effort has been devoted to developing proxies of surface nutrient concentration. The two proxies used most commonly, the carbon isotopic composition and the Cadmium/Calcium (Cd/Ca) ratio of planktonic foraminifera, both rely on the modern empirical relationship between nutrient (e.g., PO_4) concentration and the relevant parameter recorded in foraminifera shells (i.e., the carbon isotopic composition of dissolved inorganic carbon, or the Cd concentration of seawater).

Carbon isotopic composition of planktonic foraminifera have been used widely to estimate past changes in surface nutrient concentration (e.g., Labeyrie and Duplessey 1985; Charles and Fairbanks 1990; Ninnemann and Charles 1997). Extending this approach, the difference between the carbon isotopic composition of planktonic foraminifera and that of benthic foraminifera has been used to infer past changes in the overall strength of the biological pump (e.g., Curry and Crowley 1987). Recent studies have shown, however, that the carbon isotopic composition of foraminifera is offset from that of dissolved inorganic carbon due to the influence of other environmental parameters. For example, the carbon isotopic composition of planktonic foraminifera varies systematically with the carbonate ion concentration in seawater (Spero et al. 1997) and depends, as well, on the carbon isotopic composition of ingested prey and on temperature (Kohfeld et al. 2000). Isotopic fractionation during air-sea exchange of CO_2 also affects the $\delta^{13}C$ of surface water DIC, independently of nutrient concentration (Charles et al. 1993). In addition, planktonic foraminifera species grow over a range of water depths that include not only surface but also pycnocline waters (Fairbanks et al. 1980). When all present uncertainties are taken into account, the propagated error in the final estimate of past nutrient concentration can exceed the largest climate related changes recorded in the sediments, at least over the interval from the Last Glacial Maximum (LGM) to the present (Kohfeld et al. 2000). It is thus essential that these uncertainties be reduced before reliably interpreting the $\delta^{13}C$ of planktonic foraminifera.

Cadmium/calcium ratios of planktonic foraminifera have not been widely exploited to reconstruct surface nutrient concentrations because the low concentrations of Cd in low-latitude surface waters leave little Cd to be acquired by foraminifera. Foraminiferal Cd/Ca ratios hold potential for reconstructing nutrient concentrations in high-latitude waters, however, where elevated nutrient concentrations persist throughout the year, provided that other factors affecting the Cd/Ca ratio can be constrained (Elderfield and Rickaby 2000). Early work in the Southern Ocean found no change in nutrient concentration from the LGM to the present (Keigwin and Boyle 1989). More recently, Rosenthal et al. (1997) interpreted planktonic foraminiferal Cd/Ca ratios from the Subantarctic Indian Ocean to indicate that surface nutrient concentrations had been lower, and productivity higher, than today during the LGM. This interpretation has been challenged, recently, following the calibration of the temperature dependence of the partition coefficient for uptake of Cd by planktonic foraminifera (Rickaby and Elderfield 1999). When this temperature dependence is taken into account, for example by using the Mg/Ca ratio or the oxygen isotopic composition of the foraminifera shells, Rickaby and Elderfield find little change since the LGM in the surface nutrient concentration of Subantarctic waters. Further tests are needed to assess the general applicability of the temperature dependence of metal uptake by foraminiferal carbonate shells, as well as to constrain the preferential loss of trace metals during partial dissolution of foraminifera shells (McCorkle et al. 1995).

8.4.9 Proxies of Surface Nutrient Utilization Efficiency

The efficiency of the biological pump, expressed as the fraction of new nutrient added to the euphotic zone which is utilized by phytoplankton, can be evaluated from N isotopes for nitrogen and, possibly, Si isotopes for silicate. The isotope systematics of both elements follow the same principle, in that the light isotope is taken up preferentially and, as a result, the residual nutrient and planktonic material produced from it become increasingly heavy with nutrient depletion. While the study of Si isotopes is just beginning (De la Rocha et al. 1998), the possibilities afforded by N isotopes have been investigated for several years (Calvert et al. 1992; Francois and Altabet 1992; Altabet and Francois 1994; Francois et al. 1997). These studies have shown that the $^{15}N/^{14}N$ ratio of particulate organic matter in the water column and surface sediments reflects the fraction of surface nitrate utilized by phytoplankton, with heavier isotope ratios found in areas of greater nitrate depletion. The isotopic composition of organic nitrogen preserved in sediments has been used as a means of assessing changes in this important variable through time, although its interpretation is complicated by a diagenetic increase in the $^{15}N/^{14}N$ ratio at the water-sediment interface. Recent findings indicate, however, that this latter problem could be resolved by analyzing organic nitrogen associated with the protein template locked within diatom frustules (Sigman et al. 1999). This pool of sedimentary nitrogen appears to be unaffected by diagenesis, while recording the primary signal produced in surface waters.

Calibration work conducted during JGOFS process studies, and elsewhere, has found remarkable spatial uniformity in the fractionation factor associated with the uptake of nitrate by phytoplankton (Altabet and Francois 2001), although unique species assemblages (e.g., blooms of *Phaeocystis antarctica* common in the Southern Ocean) and phytoplankton growing in certain unique habitats (e.g., sea ice algae) remain to be tested. Thus, the original isotopic composition of nitrate taken up by phytoplankton determines their N isotope signal. For instance, nitrogen fixation reduces ^{15}N in the nitrate pool and, hence, causes light N isotopic composition in phytoplankton. Major changes in the level of nitrogen fixation will, therefore, also affect the $^{15}N/^{14}N$ signal in the sediments. Denitrification, on the other hand, enriches the ^{15}N isotope in nitrate, hence, nutrient sources affected by denitrification will lead to a heavy isotopic composition in phytoplankton. If the supply of nitrate to the surface waters from different source water masses with distinct $^{15}N/^{14}N$ characteristics changes, this will affect the isotopic composition of phytoplankton and finally of the sediment record. Therefore, the interpretation of the N isotope signals in the sediments will require a sound understanding of the sources of nitrate being utilized by phytoplankton. This introduces a complicating factor into the interpretation of the paleorecord, since it cannot be assumed that the ^{15}N isotope signal in surface water nitrate has remained constant.

8.5 Conclusions

Sea floor studies provide a powerful strategy for estimating the magnitude and the global distribution of organic matter fluxes in the oceans. The general coherence between primary productivity estimates from satellites and benthic fluxes is, in fact, quite astounding considering the many intermediate steps required to link both processes, all of which are subject to errors. Although regional deviations are observed, general correlations are found when considering longer periods of time. This strengthens the view that there is a systematic link between upper and deep ocean processes. This systematic link is also the ultimate justification for using proxies from deep-sea sediments to reconstruct paleoproductivity. The reason for this correlation has to be sought in a fairly consistent relationship between nutrient supply and phytoplankton export production, in the relatively uniform chemical composition of sedimenting material, and in only small deviations from average settling speeds of the majority of sinking particles. On shorter time scales, like seasonal cycles, the coherence between primary production and benthic fluxes is relaxed due to the different speeds on which both systems function.

Patterns of fluxes at the deep-sea floor do not always follow the distributions of surface water primary production as derived from satellites. In particular, discrepancies are observed along continental margins and in some areas of the Southern Ocean. Additional supply of organic material enters the deep ocean from productive shelf areas and it is important to better quantify this flux. In highly productive regions, such as in upwelling regions and in the polar front zone of the Southern Ocean, the transfer efficiency of organic material to the deep ocean may be greater than elsewhere. Hence, the role of plankton communities in regulating the transfer efficiency in differently productive regions of the ocean needs to be elucidated. This also concerns the removal and modification of organic particles in the mesopelagic zone of the ocean beneath the productive upper layers which probably has a noticeable influence on the amount and composition of particles reaching the deep-sea floor.

The biological processes that modulate vertical flux and modify material at the sea floor also create substantial uncertainties when using proxies for paleoproductivity studies. The inherent uncertainties are partly due to the fact that each productivity proxy is affected by different secondary processes which can act independently of productivity. These uncertainties also stem in part from our lack of understanding of the processes affecting the links between productivity and proxy. To remedy the first problem we need a to use a multiproxy approach. Biases induced by secondary processes are more likely to be identified in such an approach, and consistency between proxies may also provide important information on some of these secondary processes. In particular, a direct comparison of geochemical proxies with ecological proxies has yet to be undertaken. Addressing the second issue requires comprehensive biogeochemical process studies, in which expression of the proxy signal can be related directly to the biogeochemical parameter of interest, and in which the preservation of the proxy during sediment diagenesis can be evaluated. Future biogeochemical process studies, along the lines of those conducted by JGOFS, provide ideal opportunities for proxy calibration, and should include a proxy calibration component. In return, improved understanding of the carbon cycle under forcing conditions (radiations, sea level, circulation etc.) very different from those prevailing today will highlight important aspects of the fundamental processes at play that may have been missed by studying the modern ocean alone. The synergy between paleoceanographic and modern process studies is evident and should mold future comprehensive research programmes on the carbon cycle.

Acknowledgement

We thank the Paleo-JGOFS Task Team for support for this review. The collection of benthic data was supported by the EU project ADEPD.

References

Abrantes F (2000) 200 000 yr diatom records from Atlantic upwelling sites reveal maximum productivity during LGM and a shift in phytoplankton community structure at 185 000 yr. Earth Planet Sc Lett 176:7–16

Alldredge AL, Silver MW (1988) Characteristics, dynamics and significance of marine snow. Prog Oceanogr 20:41–82

Altabet MA, Francois R (1994) Sedimentary nitrogen isotope ratio as a recorder for surface ocean nitrate utilization. Global Biogeochem Cy 8:103–116

Altabet MA, Francois R (2001) Nitrogen isotope biogeochemistry of the Antarctic Polar Front Zone at 170° W. Deep-Sea Res Pt II 48:4247–4273

Andersen FØ, Kristensen E (1992) The importance of benthic macrofauna in decomposition of microalgae in coastal marine sediment. Limnol Oceanogr 37:1392–1403

Anderson RF, Rowe GT, Kemp PF, Trumbore S, Biscaye PE (1994) Carbon budget for the mid-slope depocenter of the Middle Atlantic Bight. Deep-Sea Res Pt II 41:669–703

Anderson RF, Kumar N, Mortlock RA, Froelich PN, Kubik P, Dittrich-Hannen B, Suter M (1998) Late-Quaternary changes in productivity of the Southern Ocean. J Marine Syst 17:497–514

Antia AN, Koeve W, Fischer G, Blanz T, Schulz-Bull D, Scholten J, Neuer S, Kremling K, Kuss J, Hebbeln D, Bathmann U, Fehner U, Zeitzschel B (2001) Basin-wide particulate carbon flux in the Atlantic Ocean: regional export patterns and potential for atmospheric CO_2 sequestration. Global Biogeochem Cy 15:845–862

Antoine D, Andre J-M, Morel A (1996) Oceanic primary production 2. Estimation at global scale from satellite (coastal zone color scanner) chlorophyll. Global Biogeochem Cy 10:57–69

Armstrong RA, Lee C, Hedges JI, Honjo S, Wakeham SG (2002) A new, mechanistic model for organic carbon fluxes in the ocean: a quantitative role for the association of POC with ballast minerals. Deep-Sea Res Pt II 49:219–236

Bacon MP, Huh CA, Fleer AP, Deuser W (1985) Seasonality in the flux of natural radionuclides and plutonium in the deep Sargasso Sea. Deep-Sea Res 32:273–286

Beaufort L, Lancelot Y, Camberlin P, Cayre O, Vincent E, Bassinot F, Labeyrie LD (1997) Insolation cycles as a major control of equatorial Indian Ocean primary productivity. Science 278:1451–1454

Behrenfeld MJ, Falkowski PG (1997) Photosynthetic rates derived from satellite-based chlorophyll concentration. Limnol Oceanogr 42:1–20

Bender ML, Heggie DT (1984) Fate of organic carbon reaching the sea floor: a status report. Geochim Cosmochim Ac 48:977–986

Berelson WM, Anderson RF, Dymond J, Demaster D, Hammond D, Collier ER, Honjo S, Leinen M, McManus J, Pope R, Smith C, Stephens M (1997) Biogenic budgets of particle rain, benthic remineralization and sediment accumulation in the equatorial Pacific. Deep-Sea Res Pt II 44:2251–2282

Berger WH, Fischer K, Lai C, Wu G (1987) Ocean carbon flux: global maps of primary production and export production, In: Agegian CR (ed) Biogeochemical cycling and fluxes between the deep euphotic zone and other realms. Research Rep. 88-1, NOAA Undersea Research Programme, Silver Spring, Md., pp 131–176

Bett BJ, Malzone MG, Narayanaswamy BE, Wigham BD (2001) Temporal variability in phytodetritus and megabenthic activity at the sea bed in the deep Northeast Atlantic. Prog Oceanogr 50: 349–368

Betzer PR, Showers WJ, Laws EA, Winn CD, DiTullio GR, Kroopnick PM (1984) Primary productivity and particle fluxes on a transect of the equator at 153° W in the Pacific Ocean. Deep-Sea Res 31:1–11

Billett DSM, Lampitt RS, Rice AL, Mantoura RFC (1983) Seasonal sedimentation of phytoplankton to the deep-sea benthos. Nature 302:520–522

Billett DSM, Bett BJ, Rice AL, Thurston MH, Galéron L, Sibuet M, Wolff GA (2001) Long-term changes in the megabenthos of the Porcupine Abyssal Plain (NE Atlantic). Prog Oceanogr 50: 325–348

Biscaye PE, Anderson RF (1994) Fluxes of particulate matter on the slope of the southern Middle Atlantic Bight – Seep-II. Deep-Sea Res Pt II 41:459–509

Boetius A, Lochte K (1996) Effect of organic enrichments on hydrolytic potentials and growth of bacteria in deep-sea sediments. Mar Ecol Prog Ser 140:239–250

Boetius A, Ferdelman T, Lochte K (2000a) Bacterial activity in sediments of the deep Arabian Sea in relation to vertical flux. Deep-Sea Res Pt II 47:2835–2875

Boetius A, Ravenschlag K, Schubert C, Rickert D, Widdel F, Gieseke A, Amann R, Joergensen BB, Witte U, Pfannkuche O (2000b) A marine microbial consortium apparently mediating anaerobic oxidation of methane. Nature 407:623–626

Boudreau BP (1997) Diagenetic models and their implementation. Springer-Verlag, Berlin 414 pp

Boyd PW, Newton PP (1999) Does planktonic community structure determine downward particulate organic carbon flux in different oceanic provinces? Deep-Sea Res Pt I 46:63–91

Breymann M von, Emeis K-C, Suess E (1992) Water depth and diagenetic constraints on the use of barium as a paleoproductivity indicator. In: Upwelling systems: evolution since the early Miocene. Geological Society Special Publication No. 64, pp 273–284

Broecker WS (1982) Glacial to interglacial changes in ocean chemistry. Prog Oceanogr 2:151–197

Brumsack HJ (1986) The inorganic geochemistry of cretaceous black shales (DSDP Leg 41) in comparison to modern upwelling sediments from the Gulf of California. In: Summerhayes CP, Shackleton NJ (eds) North Atlantic Paleoceanography. Geological Society Special Publication, No. 21, pp 447–462

Buesseler KO (1991) Do upper-ocean sediment traps provide an accurate record of the particle flux? Nature 353:420–423

Buesseler KO (1998) The decoupling of production and particle export in the surface ocean. Global Biogeochem Cy 12:297–310

Buesseler KO, Steinberg DK, Michaels AF, Johnson RJ, Andrews JE, Valdes JR, Price JF (2000) A comparison of the quantity and composition of material caught in a neutrally buoyant versus surface-tehered sediment trap. Deep-Sea Res Pt I 47:277–294

Calvert SE, Nielsen B, Fontugne MR (1992) Evidence from nitrogen isotope ratios for enhanced productivity during formation of eastern Mediterranean sapropels. Nature 359:223–225

Canfield DE (1994) Factors influencing organic carbon preservation in marine sediments. Chem Geol 114:315–329

Charles CD, Fairbanks RG (1990) Glacial to interglacial changes in the isotopic gradients of Southern Ocean surface waters. In: Bleil U, Thiede J (eds) Geological history of the Polar Oceans: Arctic versus Antarctic. Kluwer Academic Publishers, The Netherlands, pp 519–538

Charles CD, Froelich PN, Zibello MA, Mortlock RA, Morley JJ (1991) Biogenic opal in Southern Ocean sediments over the last 450 000 years: implication for surface water chemistry and circulation. Paleoceanography 6:697–728

Charles CD, Wright JD, Fairbanks RG (1993) Thermodynamic influences on the marine carbon-isotope record. Paleoceanography 8:691–697

Chase Z (2001) Trace elements as regulators (Fe) and recorders (U, Pa, Th, Be) of biological productiviy in the ocean. Ph.D. Dissertation, Columbia University, New York, 292 pp

Chen C-TA, Wang S-L (1999) Carbon, alkalinity and nutrient budgets on the East China Sea continetal shelf. J Geophys Res 104: 20675–20686

Chen C-TA, Liu KK, MacDonald R (2003) Continental margin exchanges. In: Fasham M, Field J, Platt T, Zeitzschel B. Springer-Verlag, (this volume)

Christiansen B, Boetius A (2000) Mass sedimentation of the swimming crab *Charybdis smithii* (Crustacea: Decapoda) in the deep Arabian Sea. Deep-Sea Res Pt II 47:2687–2706

Conte MH, Ralph N, Ross EH (2001) Seasonal and interannual variability in deep-ocean particle fluxes at the Oceanic Flux Program (OFP)/Bermuda Atlantic Time Series (BATS) site in the western Sargasso Sea near Bermuda. Deep-Sea Res Pt II 48:1471–1505

Curry WB, Crowley TJ (1987) The $\partial^{13}C$ of equatorial Atlantic surface waters: implications for ice age pCO_2 levels. Paleoceanography 2:489–517

De La Rocha CL, Brzezinski MA, DeNiro MJ, Shemesh A (1998) Silicon-isotope composition of diatoms as an indicator of past oceanic change. Nature 395:680–683
Dehairs F, Fagel N, Antia AN, Peinert R, Elskens M, Goeyens L (2000) Export production in the Bay of Biscay as estimated from barium-barite in settling material, a comparison with new production. Deep-Sea Res Pt I 47:583–601
Deuser WG (1996) Temporal variability of particle flux in the deep Sargasso Sea. In: Ittekkot V, et al. (ed) Particle flux in the Ocean. John Wilex, New York, pp 185–198
Dezileau L, Bareille G, Reyss J-L, Lemoine F (2000) Evidence for strong sediment redistribution by bottom currents along the southeast Indian ridge. Deep-Sea Res Pt I 47:1899–1936
Drazen JV, Baldwin RJ, Smith KL Jr. (1998) Sediment community response to a temporarally varying food supply at an abyssal station in the NE Pacific. Deep-Sea Res Pt II 45:893–913
Dymond J, Collier R (1996) Particulate Ba fluxes and their relationships to biological productivity. Deep-Sea Res Pt II 43:1283–1308
Dymond J, Suess E, Lyle M (1992) Barium in deep-sea sediment: a geochemical proxy for paleoproductivity. Paleoceanography 7:163–181
Elderfield H, Rickaby REM (2000) Oceanic Cd/P ratio and nutrient utilization in the glacial Southern Ocean. Nature 405:305–310
Emerson S, Fischer K, Reimers C, Heggie D (1985) Organic carbon dynamics and preservation in deep-sea sediments. Deep-Sea Res 32:1–21
Emerson S, Stump C, Grootes PM, Stuiver M, Farwell GW, Schmidt FH (1987) Estimates of degradable organic carbon in deep-sea surface sediments from ^{14}C concentrations. Nature 329:51–53
Fagel N, Andre L, Dehairs F (1999) Advective excess Ba transport as shown from sediment and trap geochemical signatures. Geochim Cosmochim Ac 63:2353–2367
Fairbanks RG, Wiebe PH, Be AWH (1980) Vertical-distribution and isotopic composition of living planktonic-foraminifera in the Western North-Atlantic. Science 207:61–63
Falkowski PG (1997) Evolution of the nitrogen cycle and its influence on the biological sequestration of CO_2 in the ocean. Nature 387:272–275
Ferdelman T, Fossing H, Neumann K, Schulz HD (1999) Sulfate reduction in surface sediments of the South-East Atlantic continental margin between 15°38' S and 27°57' S (Angola and Namibia). Limnol Oceanogr 44:650–661
Fischer G, Ratmeyer V, Wefer G (2000) Organic carbon fluxes in the Atlantic and the Southern Ocean: relationship to primary production compiled from satellite radiometer data. Deep-Sea Res Pt II 47:1961–1997
Flach E, Lavaleye M, de Stigter H, Thomsen L (2001) Feeding types of the benthic community and particle transport across the slope of the N.W. European continental margin (Goban Spur). Prog Oceanogr 42:209–231
Francois R, Altabet MA (1992) Glacial to interglacial changes in surface nitrate utilization in the Indian sector of the Southern Ocean as recorded by $\partial^{15}N$. Paleoceanography 7:589–606
Francois R, Bacon MP, Suman DO (1990) Th-230 profiling in deep-sea sediments: high-resolution records of flux and dissolution of carbonate in the equatorial Atlantic during the last 24000 years. Paleoceanography 5:761–787
Francois R, Bacon MP, Altabet MA, Labeyrie LD (1993) Glacial/interglacial changes in sediment rain rate in the S.W. Indian sector of subantarctic waters as recorded by ^{230}Th, ^{231}Pa, U and $\delta^{15}N$. Paleoceanography 8:611–629
Francois R, Honjo S, Manganini SJ, Ravizza GE (1995) Biogenic barium fluxes to the deep sea: Implications for paleoproductivity reconstruction. Global Biogeochem Cy 9:289–303
Francois R, Altabet MA, Yu E-F, Sigman D, Bacon MP, Frank M, Bohrmann G, Bareille G, Labeyrie LD (1997) Contribution of southern ocean surface water stratification to low atmospheric CO_2 concentrations during the last glacial period. Nature 389: 929–935
Frank M, Gersonde R, Mangini A (1999) Sediment redistribution, ^{230}Th ex-normalization and implications for reconstruction of particle flux and export productivity. In: Fischer G, Wefer G (eds) Use of proxies in paleoceanography: examples from the South Atlantic. Springer-Verlag, New York, pp 409–426
Ganeshram RS, Pedersen TF, Calvert SE, McNeill GW, Fontugne MR (2000) Glacial-interglacial variability in denitrification in the world's oceans: causes and consequences. Paleoceanography 15:361–376
Glud RN, Gundersen JK, Jorgensen BB, Revsbech NP, Schulz HD (1994) Diffusive and total oxygen uptake of deep-sea sediments in the eastern South Atlantic ocean: in situ and laboratory measurements. Deep-Sea Res Pt I 41:1767–1788
Goldman JC (1993) Potential role of large oceanic diatoms in new primary production. Deep-Sea Res Pt I 40:159–168
Gooday AJ, Turley CM (1990) Responses by benthic organisms to inputs of organic matter to the ocean floor: a review. Philos Tr R Soc S-A 331:119–138
Grandel S (2000) Untersuchungen zum regionalen Verteilungsmuster benthischer Stoffflüsse unter Berücksichtigung biogeographischer Provinzen im Arabischen Meer und im Atlantik. Ph.D. Dissertation, Christian-Albrechts-University, Kiel, 170 pp
Gust G, Michaels AF, Johnson R, Deuser WG, Bowles W (1994) Mooring line motions and sediment trap hydromechanics: in situ intercomparison of three common deployment designs. Deep-Sea Res Pt I 41:831–857
Haake B, Ittekkot V, Rixen T, Ramaswamy V, Nair RR, Curry WB (1993) Seasonality and interannual variability of particle fluxes to the deep Arabian Sea. Deep-Sea Res Pt I 40:1323–1344
Hartnett HE, Keil RG, Hedges JI, Devol AH (1998) Influence of oxygen exposure time on organic carbon preservation in continental margin sediments. Nature 391:572–574
Hedges JI, Hu FS, Devol AH, Hartnett HE, Tsamakis E, Keil RG (1999) Sedimentary organic matter preservation: a test for selective degradation under oxic conditions. Am J Sci 299:529–555
Heip CHR, Duineveld G, Flach E, Graf G, Helder W, Herman PMJ, Lavaleye M, Middelburg JJ, Pfannkuche O, Soetaert K, Soltwedel T, de Stigter H, Thomsen L, Vanaverbeke J, de Wilde P (2001) The role of the benthic biota in sedimentary metabolism and sediment-water exchange processes in the Goban Spur area (NE Atlantic). Deep-Sea Res Pt II 48:3223–3243
Hensen C, Zabel M, Schulz HD (2000) A comparison of benthic nutrient fluxes from deep-sea sediments off Namibia and Argentina. Deep-Sea Res Pt II 47:2029–2050
Herguera JC, Berger WH (1991) Paleoproductivity from benthic foraminifera abundance: glacial to postglacial change in the West-Equatorial Pacific. Geology 19:1173–1176
Herguera JC, Berger WH (1994) Glacial to postglacial drop in productivity in the western Equatorial Pacific: Mixing rate vs. nutrient concentrations. Geology 22:629–632
Hinrichs K-U, Hayes JM, Sylva SP, Brewer PG, DeLong EF (1999) Methane-consuming archaebacteria in marine sediments. Nature 398:802–805
Hollister CD, McCave IN (1984) Sedimentation under deep-sea storms. Nature 309:220–225
Honjo S, Francois R, Manganini S, Dymond J, Collier R (2000) Particle fluxes to the interior of the Southern Ocean in the Western Pacific sector along 170° W. Deep-Sea Res Pt II 47:3521–3548
Ittekkot V (1993) The abiotically driven biological pump in the ocean and short-term fluctuations in atmospheric CO_2 contents. Global Planet Change 8:17–25
Jahnke RA (1990) Early diagenesis and recycling of biogenic debris at the sea floor, Santa Monica Basin, California. J Mar Res 48:413–436
Jahnke RA (1996) The global ocean flux of particulate organic carbon: Areal distribution and magnitude. Global Biogeochem Cy 10:71–88
Jahnke RA, Reimers CE, Craven DB (1990) Intensification of recycling of organic matter at the sea floor near ocean margins. Nature 348:50–54
Kähler P, Bauerfeind E (2001) Organic particles in a shallow sediment trap: substantial loss to the dissolved phase. Limnol Oceanogr 46:719–723
Keigwin LD, Boyle EA (1989) Late Quaternary paleochemistry of high-latitude surface waters. Palaeogeogr Palaeocl 73:85–106
Keil RG, Montluçon DB, Prahl FG, Hedges JI (1994) Sorptive preservation of labile organic matter in marine sediments. Nature 370:549–552

Keil RG, Tsamakis E, Devol A (1999) Amino acid composition and OC:SA ratios indicate enhanced preservation of organic matter in Pacific Mexican margins ediments. In: Transactions of the American Geophysical Union, pp OS189. Ocean Sciences Meeting Supplement

Kennett JP, Cannariato KG, Hendy LL, Behl RJ (2000) Carbon isotopic evidence for methane hydrate instability during Quaternary interstadials. Science 288:128–133

Kohfeld KE, Anderson RF, Lynch-Stieglitz J (2000) Carbon isotopic disequilibrium in polar planktonic foraminifera and its impact on modern and Last Glacial Maximum reconstructions. Paleoceanography 15:53–64

Kristensen E, Andersen FØ, Blackburn TH (1992) Effects of benthic macrofauna and temperature on degradation of macroalgal detritus: the fate of organic carbon. Limnol Oceanogr 37:1404–1419

Kumar N (1994) Trace metals and natural radionuclides as tracers of ocean productivity. Ph.D. Dissertation, Columbia University, New York, 317 pp

Kumar N, Anderson RF, Mortlock RA, Froelich PN, Kubik PW, Dittrich-Hannen B, Suter M (1995) Iron fertilization of glacial-age Southern Ocean productivity. Nature 378:675–680

Kumar N, Anderson RF, Biscaye PE (1996) Remineralization of particulate authigenic trace metals in the Middle Atlantic Bight: implications for proxies of export production. Geochim Cosmochim Ac 60:3383–3397

Labeyrie LD, Duplessy JC (1985) Changes in the oceanic $^{13}C/^{12}C$ ratio during the last 140 000 years – high-latitude surface-water records. Palaeogeogr Palaeocl 50:217–240

Lampitt RS, Antia A (1997) Particle flux in deep seas: regional characteristics and temporal variability. Deep-Sea Res Pt I 44:1377–1403

Lee C (1992) Controls on organic carbon preservation: the use of stratified water bodies to compare intrinsic rates of decomposition in oxic and anoxic systems. Geochim Cosmochim Ac 56: 3323–3335

Lochte K (1992) Bacterial standing stock and consumption of organic carbon in the benthic boundary layer of the abyssal North Atlantic. In: Rowe GT, Pariente V (eds) Deep-sea food chains and the global carbon cycle. Kluwer Academic Publishers, Netherland, pp 1–10

Lochte K, Pfannkuche O (2002) Processes driven by the small sized organisms at the water-sediment interface. In: Hebbeln D, Wefer G (eds) Ocean margin systems. Springer-Verlag, Heidelberg

Lochte K, Turley CM (1988) Bacteria and cyanobacteria associated with phytodetritus in the deep sea. Nature 333:67–69

Lochte K, Ducklow H, Fasham MJR, Stienen C (1993) Plankton succession and carbon cycling at 47° N 20° W during the JGOFS North Atlantic Bloom Experiment. Deep-Sea Res Pt II 40:91–114

Louanchi F, Najjar RG (2000) A global monthly climatology of phosphate, nitrate, and silicate in the upper ocean: spring-summer export production and shallow remineralization. Global Biogeochem Cy 14:957–977

Loubere P, Fariduddin M (1999) Quantitative estimation of global patterns of surface ocean biological productivity and its seasonal variation on time scales from centuries to millennia. Global Biogeochem Cy 13:115–133

Luff R, Wallmann K, Grandel S, Schlüter M (2000) Numerical modelling of benthic processes in the deep Arabian Sea. Deep-Sea Res Pt II 47:3039–3072

Lyle M, Zahn R, Prahl FG, Dymond J, Collier R, Pisias NG, Suess E (1992) Paleoproductivity and carbon burial across the California Current: the MULTITRACERS transect, 42° N. Paleoceanography 7:251–272

Madin LP (1982) Production, composition and sedimentation of salp fecal pellets in oceanic waters. Mar Biol 67:39–45

Marcantonio F, Anderson RF, Stute M, Kumar N, Schlosser P, Mix A (1996) Extraterrestrial ^{3}He as a tracer of marine sediment transport and accumulation. Nature 383:705–707

Marcantonio F, Anderson RF, Higgins S, Stute M, Schlosser P, Kubik P (2001) Sediment focusing in the central equatorial Pacific Ocean. Paleoceanography 16:260–267

Marchal O, Francois R, Stocker TF, Joos F (2000) Ocean thermohaline circulation and sedimentary $^{231}Pa/^{230}Th$ ratio. Paleoceanography 15:625–641

Martin JH (1990) Glacial-interglacial CO_2 change: the iron hypothesis. Paleoceanography 5:1–13

Martin JH, Knauer GA, Karl DM, Broenkow WW (1987) VERTEX: carbon cycling in the northeast Pacific. Deep-Sea Res 34:267–285

Martin WR, Bender ML (1988) The variability of benthic fluxes and sedimentary reminerlization in response to seasonally variable organic carbon rain rates in the deep-sea: a modeling study. Am J Sci 288:561–574

Mayer LM (1994) Surface area control of organic carbon accumulation in continental shelf sediments. Geochim Cosmochim Ac 58:1271–1284

McCave IN, Hall IR, Antia AN, Chou L, Dehairs F, Lampitt RS, Thomsen L, van Weering TC, Wollast R (2001) Distribution, composition and flux of particulate material over the European margin at 47°50' N. Deep-Sea Res Pt II 48:3107–3139

McCorkle DC, Keigwin L, Corliss BH, Emerson SR (1990) The influence of microhabitats on the carbon isotopic composition of deep sea benthic foraminifera. Paleoceanography 5:161–185

McCorkle D, Veeh HH, Heggie DT (1994) Glacial-holocene paleoproductivity off western Australia: a comparison of proxy records. In: Zahn R, Kaminski M, Pedersen TF (eds) Carbon cycling in the glacial ocean: constraints of the ocean's role in global change. NATO ASI Series, vol. 117. Springer-Verlag, Berlin, pp 443–479

McCorkle DC, Martin PA, Lea DW, Klinkhammer GP (1995) Evidence of a dissolution effect on benthic foraminiferal shell chemistry – $\partial^{13}C$, Cd/Ca, Ba/Ca, and Sr/Ca: results from the Ontong Java Plateau. Paleoceanography 10:699–714

McIntyre A, Molfino B (1996) Forcing of Atlantic equatorial and subpolar millennial cycles by precession. Science 274:1867–1870

McManus J, Berelson WM, Klinkhammer GP, Johnson KS, Coale KH, Anderson RF, Kumar N, Burdige DJ, Hammond DE, Brumsack HJ, McCorkle DC, Rushdi A (1998) Geochemistry of barium in marine sediments: implications for its use as a paleoproxy. Geochim Cosmochim Ac 62:3453–3473

McManus J, Berelson WM, Hammond DE, Klinkhammer GP (1999) Barium cycling in the North Pacific: implications for the utility of Ba as a paleoproductivity and paleoalkalinity proxy. Paleoceanography 14:53–61

Menard HW, Smith SM (1966) Hypsometry of ocean basin provinces. J Geophys Res 71:4305–4325

Mix AC (1989a) Influence of productivity variations on long-term atmospheric CO_2. Nature 337:541–544

Mix AC (1989b) Pleistocene paleoproductivity: evidence from organic carbon and foraminiferal species. In: Berger WH, Smetacek VS, Wefer G (eds) Productivity of the ocean: present and past. Wiley Interscience, Chichester, pp 313–340

Molfino B, McIntyre A (1990) Precessional forcing of nutricline dynamics in the Equatorial Atlantic. Science 249:766–769

Mortlock RA, Charles CD, Froelich PN, Zibello MA, Saltzman J, Hays JD, Burckle LH (1991) Evidence for lower productivity in the Antarctic Ocean during the last glaciation. Nature 351:220–223

Mueller PJ, Suess E (1979) Productivity, sedimentation rate, and sedimentary organic matter in the oceans – I. Organic C on preservation. Deep-Sea Res 26A:1347–1362

Nelson DM, Tréguer P, Brzezinski MA, Leynaert A, Queguiner B (1995) Production and dissolution of biogenic silica in the ocean: revised global estimates, comparison with regional data, and relationship to biogenic sedimentation. Global Biogeochem Cy 9:359–372

Nelson DM, Anderson RF, Barber RT, Brzezinski MA, Buesseler KO, Chase Z, Collier RW, Dickson M-L, François R, Hiscock MR, Honjo S, Marra J, Martin WR, Sambrotto RN, Sayles FL, Sigmon DE (2002) Vertical budgets for organic carbon and biogenic silica in the Pacific sector of the Southern Ocean, 1996–1998. Deep-Sea Res Pt II 49:1645–1674

Newton PP, Lampitt RS, Jickells TD, King P, Boutle C (1994) Temporal and spatial variability of biogenic particle fluxes during the JGOFS northeast Atlantic process studies at 47° N, 20° W (1989–1990). Deep-Sea Res Pt II 41:1617–1642

Ninnemann US, Charles CD (1997) Regional differences in Quaternary Subantarctic nutrient cycling: link to intermediate and deep water ventilation. Paleoceanography 12:560–567

Noji TT, Bathmann UV, von Bodungen B, Voss M, Antia A, Krumbholz M, Klein B, Peeken I, Noji CI-M, Rey F (1997) Clearance of picoplankton-sized particles and formation of rapidly sinking aggregates by the pteropod Limacina retroversa. J Plankton Res 19:863–875

Pace ML, Knauer GA, Karl DM, Martin JH (1987) Primary production, new production and vertical flux in the Eastern Pacific. Nature 325:803–804
Parkes RJ, Cragg BA, Wellsbury P (2000) Recent studies on bacterial populations and processes in subseafloor sediments: a review. Hydrogeol J 8:11–28
Paytan A (1995) Marine barite, a recorder of oceanic chemistry, productivity, and circulation. Ph.D. Dissertation, University of California, San Diego, 111 p
Paytan A, Kastner M, Chavez FP (1996) Glacial to interglacial fluctuations in productivity in the equatorial Pacific as indicated by marine barite. Science 274:1355–1357
Pedersen TF, Bertrand P (2000) Influences of oceanic rheostats and amplifiers on atmospheric CO_2 content during the Late Quaternary. Quat Sci Reviews 19:273–283
Peinert R, Bauerfeind E, Gradinger R, Haupt O, Krumbholz M, Peeken I, Ramseier RO, Werner I, Zeitzschel B (2001) Biogenic particle sources and vertical flux patterns in the seasonally ice-covered Greenland Sea. In: Schäfer P, Ritzrau W, Schlüter M, Thiede J (eds) The Northern North Atlantic. Springer-Verlag, Berlin Heidelberg, pp 69–79
Pfannkuche O (1993) Benthic response to the sedimentation of particulate organic matter at the BIOTRANS station, 47° N, 20° W. Deep-Sea Res Pt I 40:135–149
Pfannkuche O, Lochte K (1993) Open ocean pelago-benthic coupling: cyanobacteria as tracers of sedimenting salp faeces. Deep-Sea Res Pt I 40:727–737
Pfannkuche O, Lochte K (2000) The biogeochemistry of the deep Arabian Sea: overview. Deep-Sea Res Pt II 47:2615–2628
Pfannkuche O, Boetius A, Lochte K, Lundgreen U, Thiel H (1999) Responses of deep-sea benthos to sedimentation patterns in the North-East Atlantic in 1992. Deep-Sea Res Pt I 46:573–596
Pichon JJ, Bareille G, Labracherie M, Labeyrie LD, Baudrimont A, Turon JL (1992) Quantification of the biogenic silica dissolution in Southern-Ocean sediments. Quaternary Res 37:361–378
Pondaven P, Ragueneau O, Tréguer P, Hauvesre A, Dezileau L, Reyss JL (2000) Resolving the 'opal paradox' in the Southern Ocean. Nature 405:168–172
Pope RH, Demaster DJ, Smith CR, Seltmann H (1996) Rapid bioturbation in equatorial Pacific sediments: evidence from excess ^{234}Th measurements. Deep-Sea Res Pt II 43:1339–1364
Rabouille C, Gaillard J-F (1991) Towards the EDGE: early diagenetic global explanation: a model depicting the the early diagenesis of organic matter, O_2, NO_3, Mn, and PO_4. Geochim Cosmochim Ac 55:2511–2525
Ragueneau O, Tréguer P, Leynaert A, Anderson RF, Brzezinski MA, DeMaster DJ, Dugdale RC, Dymond J, Fischer G, François R, Heinze C, Maier-Reimer E, Martin-Jézéquel V, Nelson DM, Quéguiner B (2000) A review of the Si cycle in the modern ocean: recent progress and missing gaps in the application of biogenic opal as a paleoproductivity proxy. Global Planet Change 26:317–365
Rice AL, Billet DSM, Fry J, John AWG, Lampitt RS, Mantoura RFC, Morris RJ (1986) Seasonal deposition of phytodetritus to the deep-sea floor. P Roy Soc Edinb B 88:265–279
Rickaby REM, Elderfield H (1999) Planktonic foraminiferal Cd/Ca: paleonutrients or paleotemperature? Paleoceanography 14: 293–303
Ritzrau W (1996) Microbial activity in the benthic boundary layer (BBL): small scale distribution and its relationship to the hydrodynamic regime. J Sea Res 36:171–180
Ritzrau W, Graf G, Schlüter M (2001a) Exchange processes across the sediment water interface. In: Schäfer P, Ritzrau W, Schlüter M, Thiede J (eds) The Northern North Atlantic. Springer-Verlag, Berlin Heidelberg, pp 199–206
Ritzrau W, Graf G, Scheltz A, Queisser W (2001b) Bentho-pelagic coupling and carbon dynamics in the northern North Atlantic. In: Schäfer P, Ritzrau W, Schlüter M, Thiede J (eds) The Northern North Atlantic. Springer-Verlag, Berlin Heidelberg, pp 207–224
Romankevich EA, Vetrov AA, Korneeva GA (1999) Geochemistry of organic carbon in the ocean. In: Gray JS, Ambrose W Jr., Szaniawska A (eds) Biogeochemical cycling and sediment ecology. NATO ASI Series, Kluwer Academic Publishers, Dordrecht Boston London, p 1–27
Rosenthal Y, Boyle EA, Labeyrie L (1997) Last glacial maximum paleochemistry and deepwater circulation in the Southern Ocean: evidence from foraminiferal cadmium. Paleoceanography 12:787–796
Sancetta C (1992) Primary production in the glacial North Atlantic and North Pacific Oceans. Nature 360:249–251
Sarmiento JL, Toggweiler JR (1984) A new model for the role of the oceans in determining atmospheric pCO_2. Nature 308:621–624
Sauter EJ, Schlüter M, Suess E (2001) Organic carbon flux and reminerlization in surface sediments from the northern North Atlantic derived from pore-water oxygen microprofiles. Deep-Sea Res Pt I 48:529–553
Sayles FL, Martin WR, Deuser WD (1994) Response of benthic oxygen demand to particulate organic carbon supply in the deep-sea near Bermuda. Nature 371:686–689
Sayles FL, Martin WR, Chase Z, Anderson RF (2001) Benthic remineralization and burial of biogenic SiO_2, $CaCO_3$, organic carbon and detrital material in the Southern Ocean along a transect at 170° W. Deep-Sea Res Pt II 48:4323–4383
Schlitzer R (2002) Carbon export fluxes in the Southern Ocean: results from inverse modeling and comparison with satellite based estimates. Deep-Sea Res Pt II 49:1623–1644
Schlüter M, Sauter EJ (2000) Spatial budget of organic carbon flux to the seafloor of the northern North Atlantic (60° N–80° N). Global Biogeochem Cy 14:329–340
Schlüter M, Sauter E, Schulz-Bull D, Balzer W, Suess E (2001) Fluxes of organic carbon and biogenic silica reaching the seafloor: a comparison of high northern and southern latitudes of the Atlantic Ocean. In: Schäfer P, Ritzrau W, Schlüter M, Thiede J (eds) The Northern North Atlantic. Springer-Verlag, Berlin Heidelberg, pp 225–240
Scholten JC, Fietzke J, Vogler S, van der Loeff MM Rutgers, Mangini A, Koeve W, Waniek J, Stoffers P, Antia A, Kuss J (2001) Trapping efficiencies of sediment traps from the deep eastern North Atlantic: the ^{230}Th calibration. Deep-Sea Res Pt II 48:2383–2408
Schubert CJ, Villaneuva J, Calvert SE, Cowie GL, von Rad U, Schulz H, Berner U, Erlenkeuser H (1998) Stable phytoplankton community structure in the Arabian Sea over the past 200 000 years. Nature 394:563–566
Shemesh A, Burckle LH, Froelich PN (1989) Dissolution and preservation of antarctic diatoms and the effect on sediment thanatocoenoses. Quaternary Res 31:288–308
Shimmield GB, Jahnke RA (1995) Particle flux and its conversion to the sediment record: open ocean upwelling systems. In: Summerhayes CP, Emeis K-C, Angel MV, Smith RL, Zeitzschel B (eds) Upwelling in the ocean: modern processes and ancient records. John Wiley & Sons, New York, pp 171–192
Sibuet M, Olu K (1998) Biogeography, biodiversity and fluid dependence of deep-sea cold-seep communities at active and passive margins. Deep-Sea Res Pt II 45:517–567
Sigman DM, Altabet MA, Francois R, McCorkle DC, Gaillard JF (1999) The isotopic composition of diatom-bound nitrogen in Southern Ocean sediments. Paleoceanography 14:118–134
Smith CR, Berelson W, Demaster DJ, Dobbs FC, Hammond D, Hoover DJ, Pope RH, Stephen M (1997) Latitudinal variations in benthic processes in the abyssal equatorial Pacific: control by biogenic particle flux. Deep-Sea Res Pt II 44:2295–2317
Smith KL Jr. (1987) Food energy supply and demand: a discrepancy between particulate organic carbon flux and sediment community oxygen consumption in the deep ocean. Limnol Oceanogr 32:201–220
Smith KL Jr., Kaufmann RS (1999) Long-term discrepancy between food supply and demand in the deep eastern North Pacific. Science 284:1174–1177
Smith KL Jr., Kaufmann RS, Baldwin RJ, Carlucci AF (2001) Pelagic-benthic coupling in the abyssal eastern North Pacific: an 8-year time-series study of food supply and demand. Limnol Oceanogr 46:543–556
Soetaert K, Herman PMJ, Middelburg JJ (1996) Dynamic response of deep sea sediments to seasonal variations: a model. Limnol Oceanogr 41:1651–1668
Spero HJ, Bijma J, Lea DW, Bemis BE (1997) Effect of seawater carbonate concentration on foraminiferal carbon and oxygen isotopes. Nature 390:497–500

Suess E (1980) Particulate organic carbon flux in the oceans – surface productivity and oxygen utilisation. Nature 288:260–263

Suman DO, Bacon MP (1989) Variations in holocene sedimentation in the North American basin determined from ^{230}Th measurements. Deep-Sea Res 36:869–878

Tengberg A, de Bovee F, Hall P, Berelson W, Cicceri G, Crassous P, Devol A, Emerson S, Glud R, Graziottin F, Gundersen J, Hammond D, Helder W, Jahnke R, Khripounoff A, Nuppenau V, Pfannkuche O, Reimers C, Rowe G, Sahami A, Sayles F, Schuster M, Wehrli B, de Wilde P (1995) Benthic chamber and profile landers in oceanography – a review of design, technical solutions and functioning. Prog Oceanogr 35:253–294

Thiel H, Pfannkuche O, Schriever G, Lochte K, Gooday AJ, Hemleben C, Mantoura RFC, Turley CM, Patching JW, Riemann F (1989) Phytodetritus on the deep-sea floor in a central oceanic region of the Northeast Atlantic. Biol Oceanogr 6:203–239

Thomas E, Booth L, Maslin M, Shackleton NJ (1995) Northeastern Atlantic benthic foraminifera during the last 45000 years: Changes in productivity seen from the bottom up. Paleoceanography 10:545–562

Tréguer P, Legendre L, Rivkin RT, Ragueneau O, Dittert N (2003) Water column biogeochemistry below the euphotic zone. In: Fasham M (ed) Ocean biogeochemistry. Springer-Verlag, Heidelberg, (this volume)

Tréguer P, Nelson DM, Vanbennekom AJ, Demaster DJ, Leynaert A, Queguiner B (1995) The silica balance in the world ocean – a reestimate. Science 268:375–379

Tseitlin VB (1993) The relationship between primary production and vertical organic carbon flux in the ocean mesopelagial. Okeanologya 33:224–228

Turley CM, Lochte K (1990) Microbial response to the input of fresh detritus to the deep-sea bed. Palaeogeogr Palaeocl 89:3–23

Turley CM, Lochte K, Lampitt RS (1995) Transformations of biogenic particles during sedimentation in the northeastern Atlantic. Philos T Roy Soc B 348:179–189

Turnewitsch R, Witte U, Graf G (2000) Bioturbation in the abyssal Arabian Sea: influence of fauna and food supply. Deep-Sea Res Pt II 47:2877–2911

Villaneuva J, Grimalt JO, Cortijo E, Vidal L, Labeyrie LD (1997) A biomarker approach to the organic matter deposited in the North Atlantic during the last climatic cycle. Geochim Cosmochim Ac 61:4633–4646

Vinogradov ME, Shushkina EA, Kopelevitch OV, Sheberstov SV (1996) Photosynthetic primary production in the ocean, based on expedition's and satellite data. Okeanologiya+ 36:566–575

Walter H-J, van der Loeff MMR, Hoeltzen H (1997) Enhanced scavenging of ^{231}Pa relative to ^{230}Th in the south Atlantic south of the polar front: implications for the use of the ^{231}Pa/^{230}Th ratio as a paleoproductivity proxy. Earth Planet Sc Lett 149:85–100

Watkins JM, Mix AC (1998) Testing the effects of tropical temperature, productivity, and mixed-layer depth on foraminiferal transfer functions. Paleoceanography 13:96–105

Witte U, Pfannkuche O (2000) High rates of benthic carbon remineralisation in the abyssal Arabian Sea. Deep-Sea Res Pt II 47:2785–2804

Wollast R, Chou L (2001) The carbon cycle at the ocean margin in the Northern Gulf of Biscay. Deep-Sea Res Pt II 48:3265–3293

Woodruff F, Savin SM (1985) ∂^{13}C values of Miocene Pacific benthic foraminifera: correlations with sea level and biological productivity. Geology 212:119–122

Yu E-F, Francois R, Bacon MP (1996) Similar rates of modern and last-glacial ocean thermohaline circulation inferred from radiochemical data. Nature 379:689–694

Yu E-F, Francois R, Bacon MP, Honjo S, Fleer AP, Manganini SJ, van der Loeff MM Rutgers, Ittekkot V (2001a) Trapping efficiency of bottom-tethered sediment traps estimated from the intercepted fluxes of ^{230}Th and ^{231}Pa. Deep-Sea Res Pt I 48:865–889

Yu E-F, Francois R, Bacon MP, Fleer AP (2001b) Fluxes of ^{230}Th and ^{231}Pa to the deep sea: Implications for the interpretation of excess ^{230}Th and ^{231}Pa/^{230}Th profiles in sediments. Earth Planetetary Science Letters 191: 219–230.

Zahn R, Winn K, Sarnthein M (1986) Benthic foraminiferal ∂^{13}C and accumulation rates of organic carbon: *Uvigeina peregrina* group and *Cibicidoides wuellerstorfi*. Paleoceanography 1: 27–42

Zatsepina OY, Buffett BA (1998) Thermodynamic conditions for the stability of gas hydrates in the seafloor. J Geophys Res B 103:24127–24139

Zheng Y, Anderson RF, van Geen A, Fleisher MQ (2002a) Preservation of particulate non-lithogenic uranium in marine sediments. Geochim Cosmochim Ac 66:3085–3092

Zheng Y, Anderson RF, van Geen A, Fleisher MQ (2002b) Remobilization of authigenic uranium in marine sediments by bioturbation. Geochim Cosmochim Ac 66:1759–1772

Chapter 9

Global Ocean Carbon Cycle Modeling

Scott C. Doney · Keith Lindsay · J. Keith Moore

9.1 Introduction

One of the central objectives of the Joint Global Ocean Flux Study (JGOFS) is to use data from the extensive field effort to improve and evaluate numerical ocean carbon cycle models. Substantial improvements are required in the current suite of numerical models if we are to understand better the present ocean biogeochemical state, hindcast historical and paleoclimate variability, and predict potential future responses to anthropogenic perturbations. Significant progress has been made in this regard, and even greater strides are expected over the next decade as the synthesis of the JGOFS data sets are completed and disseminated to the scientific community. The goals of this chapter are to outline the role of modeling in ocean carbon cycle research, review the status of basin to global-scale modeling, and highlight major problems, challenges, and future directions.

Marine biogeochemical models are quite diverse, covering a wide range of complexities and applications from simple box models to global 4-D (space and time) coupled physical-biogeochemical simulations, and from strict research tools to climate change projections with direct societal implications. Model development and usage are strongly shaped by the motivating scientific or policy problems as well as the dynamics and time/space scales considered. A common theme, however, is that models allow us to ask questions about the ocean we could not address using data alone. In particular, models help researchers quantify the interactions among multiple processes, synthesize diverse observations, test hypotheses, extrapolate across time and space scales, and predict future behavior.

A well posed model encapsulates our understanding of the ocean in a mathematically consistent form. Many, though not all, models can be cast in general form as a coupled set of time-dependent advection, diffusion, reaction equations:

$$\frac{\partial X}{\partial t} + \vec{u}\nabla X - \nabla(K\nabla X) = \text{sources/sinks} \qquad (9.1)$$

where X refers to a set of prognostic or predicted variables (e.g., temperature, phytoplankton biomass, dissolved inorganic carbon). The second and third terms on the left hand side of the equation describe the physical processes of advection and mixing, respectively. All of the chemical and biological interactions are subsumed into the final source/sink term(s) on the right hand side, which often involve complex interactions among a number of prognostic variables. In addition, the model may require external boundary conditions (e.g., solar radiation, wind stress, dust deposition) and, for time varying problems, initial conditions. The model equations are then solved numerically by integrating forward in time for X.

Numerical models cannot capture all of the complexity of the real world. Part of the art of modeling is to abstract the essence of a particular problem, balancing model complexity with insight. Many processes must be either parameterized in a simple fashion or neglected altogether. For example, the biophysical details of photosynthesis, though quite well known, may not necessarily be crucial and certainly not sufficient for simulating the seasonal bloom in the North Atlantic. On the other hand, a number of key processes (e.g., phytoplankton mortality, the controls on community structure) are not well characterized and are often used as model tuning parameters.

As opposed to much of ocean physics, fundamental relationships either are not known or may not exist at all for much of marine biogeochemistry. Therefore, ocean biogeochemical modeling is inherently data driven. The JGOFS field data are invaluable in this regard, providing the basis for highlighting model deficiencies, developing improved parameterizations, and evaluating overall model performance. The desire for increasing model realism and sophistication must be tempered by the realization that models can quickly outstrip the ability to parameterize the appropriate processes or evaluate the overall simulation. Inverse methods and data assimilation will certainly help in this regard, but the true benefits will only be gained when the underlying models rest on a sound, mechanistic basis.

Broadly speaking, much of current ocean carbon cycle modeling can be condensed into a few overarching scientific questions that match well with the other individual chapters of this book. These include: What are the physical and biological controls on primary, new and export production? What are the roles of multiple limiting nutrients, mesoscale variability and trophic structure? How are organic and inorganic carbon transported, transformed and remineralized below the surface layer? How much anthropogenic carbon does the ocean take up and where? How does ocean biogeochemistry respond to climate variability and are there feedbacks on climate change? Ocean carbon modeling is a diverse and growing field and can not be covered comprehensively in a single chapter. Rather, we present an overview of the current state and major issues involving ocean biogeochemical and ecosystem modeling drawing mostly on specific examples from the NCAR modeling program.

Historically, global ocean biological and chemical modeling has evolved along three related, though often distinct, paths. First, a number of early efforts were directed toward improving oceanic anthropogenic carbon uptake estimates, building on simple box models and coarse resolution ocean physical general circulation models (GCMs). Transient tracer simulations (radiocarbon, tritium, chlorofluorocarbons) developed in conjunction as a way to assess model physical circulation and mixing. Second, biogeochemical carbon cycle models, while often relying on the same physical model frameworks, were developed to improve our understanding of the dynamics controlling large-scale biogeochemical fields (e.g., surface pCO_2, subsurface nutrient, oxygen and dissolved inorganic carbon distributions) and their responses to climate variability and secular change (e.g., glacial-interglacial transition and greenhouse warming). The treatment of biological processes in this class of models has been rather rudimentary in most cases. Third, marine ecosystem models have been focused much more on the details of biological interactions within the upper ocean, tracking the controls on upper ocean primary and export production as well as the flow of mass and energy though the marine food web. These models often are created for specific biogeographical regions commonly based on local surface or 1-D time-series data sets. More recently, ecosystem models have been extended to basin and global scale. One of the most important trends in the field is the unification of these three approaches, leading ultimately to a coherent modeling framework linking ocean physics, biology and chemistry over a range of time and space scales.

9.2 Anthropogenic Carbon Uptake, Transient Tracers, and Physics

An initial and ongoing focus of ocean biogeochemical modeling research is to quantify the rate at which the ocean takes up transient tracers and excess anthropogenic CO_2. The water column and upper few meters of marine sediments contain the largest mobile, natural reservoir of carbon on time-scales of 10^2 to 10^5 years. With about 50 times more carbon than that stored in the atmosphere (Fig. 9.1) (Sarmiento and Sundquist

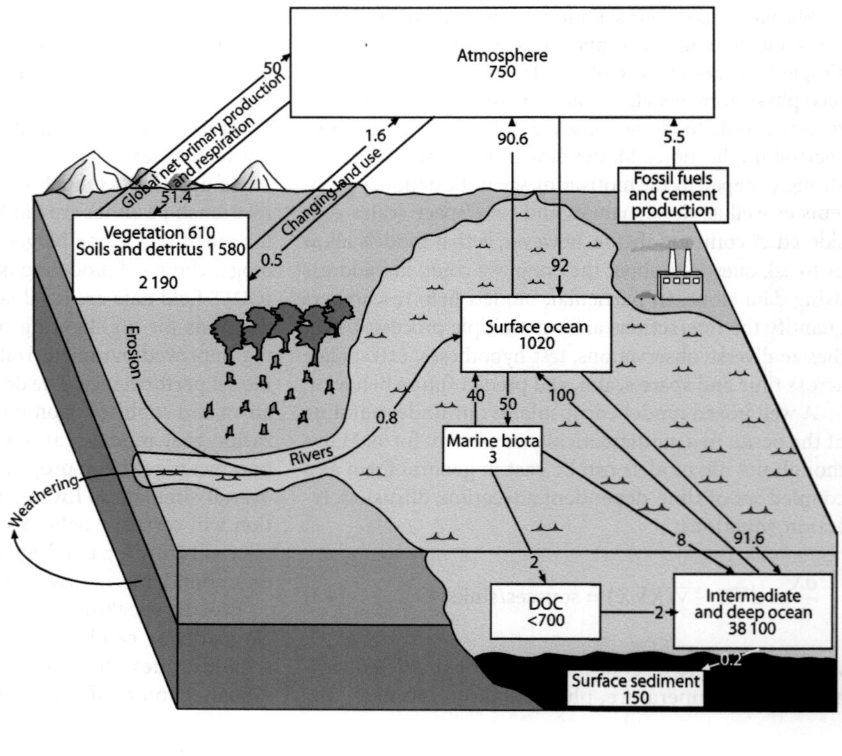

Fig. 9.1. Schematic of present global carbon cycle budget. The budget includes the natural background cycle as well as anthropogenic perturbations. Reservoir sizes are given in units of Pg C (1 Pg equals 10^{15} g), while fluxes are given in Pg C yr^{-1} (adapted from Schimel et al. (1995) and US CCSP (1999))

1992; Siegenthaler and Sarmiento 1993), the ocean will serve as the ultimate sink for about 90% of human fossil fuel emissions (Archer et al. 1998). Anthropogenic carbon uptake is often computed as a passive perturbation to the natural dissolved inorganic carbon (DIC) field (Sarmiento et al. 1992), a fairly reasonable assumption for the pre-industrial to the present time period. Under these conditions (i.e., fixed circulation and background biogeochemical cycles), net carbon uptake is simply a matter of ocean physics, primarily determined by the ventilation time-scales exposing deep water to the surface and, to a much lesser degree, air-sea gas exchange. The invasion into the ocean of transient tracers such as radiocarbon, tritium, and the chlorofluorocarbons provides a direct, often quite dramatic illustration of ocean ventilation and is commonly used either to calibrate/evaluate ocean physical models or as proxies for anthropogenic CO_2 uptake.

Early attempts to calculate ocean CO_2 uptake in the 1970s and 1980s relied heavily on ocean box and 1-D advection diffusion models of varying complexity (Oeschger et al. 1975; Siegenthaler and Oeschger 1978; Siegenthaler and Joos 1992). This class of models represents, in a fairly crude, schematic form, the basics of ocean thermocline ventilation and thermohaline circulation. The crucial model advection and mixing parameters are typically set by calibrating simulated transient tracer distributions (tritium, natural and bomb radiocarbon) to observations. More recently, such models have mostly been supplanted by full 3-D general circulation models for the anthropogenic CO_2 question. But because they are simple to construct (and interpret) and computationally inexpensive, box models and a related derivative the 2-D, zonally averaged basin model (Stocker et al. 1994) continue to be used today for a number of applications requiring long temporal integrations including paleoceanography (Toggweiler 1999; Stephens and Keeling 2000) and climate change (Joos et al. 1999). Some caution is advised, however, as recent studies (Broecker et al. 1999; Archer et al. 2000) clearly demonstrate that box model predictions for key carbon cycle attributes can differ considerably from the corresponding GCM results.

Ocean general circulation model studies of anthropogenic carbon uptake date back to the work of Maier-Reimer and Hasselmann (1987) and Sarmiento et al. (1992), and the number of model estimates (and modeling groups) for CO_2 uptake has increased significantly over the 1990s. For example, more than a dozen international groups are participating in the IGBP/GAIM Ocean Carbon Model Intercomparison Project (OCMIP; *http://www.ipsl.jussieu.fr/OCMIP/*). These numerical experiments are closely tied to and greatly benefit from efforts to evaluate ocean GCMs using hydrographic (Large et al. 1997; Gent et al. 1998) and transient tracer data (Toggweiler et al. 1989a,b; Maier-Reimer 1993; England 1995; Heinze et al. 1998; England and Maier-Reimer 2001). More recently, empirically based methods have been developed for estimating anthropogenic carbon distributions directly from ocean carbon and hydrographic observations (Gruber et al. 1996; Gruber 1998; Wanninkhof et al. 1999; Watson this volume). With the completion of the high quality, JGOFS/WOCE global CO_2 survey (Wallace 1995, 2001), a baseline can be constructed for the world ocean for the pre-industrial DIC field and the anthropogenic carbon perturbation as of the mid-1990s, an invaluable measure for testing numerical model skill and monitoring future evolution.

As an example of this class of carbon uptake simulations, the large-scale patterns of anthropogenic CO_2 air-sea flux and integrated water column inventory from the NCAR CSM Ocean Model (Large et al. 1997; Gent et al. 1998) are shown in Fig. 9.2. The regions of highest anthropogenic carbon uptake – equatorial upwelling bands, western boundary currents, high latitude intermediate and deep water formation regions – are associated with the transport of older subsurface waters to the air-sea interface (Doney 1999). Although the maximum specific uptake rates are found in the subpolar North Atlantic, the area is relatively small, and the integrated uptake of the Southern Ocean and Equatorial band are larger. The anthropogenic DIC water column anomaly is stored primarily in the thermocline and intermediate waters of the subtropical convergence zones and the lower limb of the North Atlantic thermohaline circulation as illustrated by the second panel of Fig. 9.2 and Fig. 9.3, a depth vs. latitude comparison of field data derived and model simulated anthropogenic DIC. The two meridional sections follow the thermohaline overturning circulation from the northern North Atlantic to the Southern Ocean and then back to the northern North Pacific. The model simulates in a reasonable fashion the patterns from empirical estimates except perhaps in the subpolar and intermediate depth North Atlantic, which may reflect problems with the model formation of North Atlantic Deep Water (Large et al. 1997).

At present, most numerical models predict a similar net uptake of anthropogenic CO_2 for the 1990s of approximately 2 Pg C yr^{-1} (1 Pg C equals 10^{15} g C) (Orr et al. 2001), a result supported by atmospheric biogeochemical monitoring and a variety of other techniques (Schimel et al. 1995; Keeling et al. 1996; Rayner et al. 1999). The models, however, show considerable regional differences, particularly in the Southern Ocean (Orr et al. 2001). The agreement of the NCAR model with empirical basin inventories is quite good (Table 9.1), suggesting that at least at this scale the NCAR model transport is relatively skillful.

While based on a more complete description of ocean physics, the coarse resolution, global GCMs used for these carbon studies still require significant parameterization of sub-gridscale phenomenon such as deep wa-

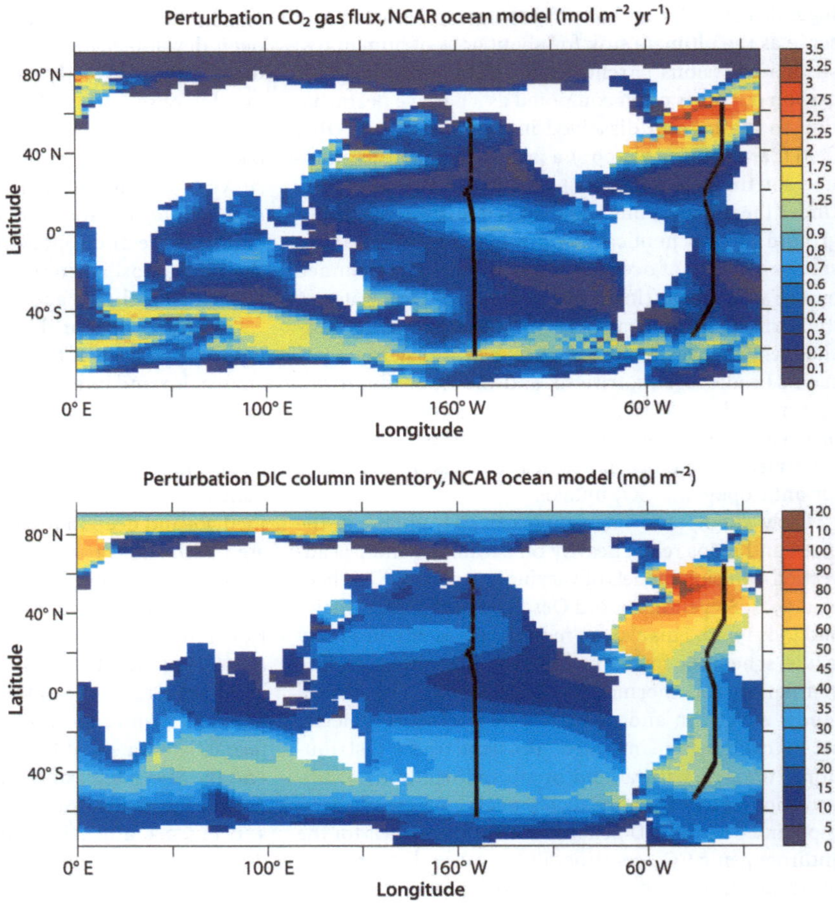

Fig. 9.2. Spatial distributions of model simulated ocean anthropogenic (perturbation) carbon. Simulated fields are shown for (*top*) air-sea flux (mol C m^{-2} yr^{-1}) and (*bottom*) water column inventory (mol C m^{-2}) for 1990 from the NCAR CSM Ocean Model. The two *lines* indicate the Atlantic and Pacific transects used for the horizontal sections in Fig. 9.3 and 9.7

ter formation, surface and bottom boundary layer physics, and mixing rates along and across density surfaces (isopycnal and diapycnal diffusion). The ongoing OCMIP effort is comparing about a dozen current generation global ocean carbon models among themselves and against ocean observations. Completed analysis of OCMIP Phase 1 and early results from Phase 2 demonstrate significant differences among the models in the physical circulation and simulated chlorofluorocarbon (Dutay et al. 2001), radiocarbon, and current and projected future anthropogenic CO_2 fields (Orr et al. 2001). The largest model-model differences and model-data discrepancies are found in the Southern Ocean, reflecting differences in the relative strength and spatial patterns of Antarctic Mode (Intermediate) Waters and Antarctic Bottom Water (AABW) (Dutay et al. 2001). Models using horizontal mixing rather than an isopycnal scheme (Gent and McWilliams 1990) tend to overestimate convective mixing in the region of the Antarctic Circumpolar Current (Danabasoglu et al. 1994). Not surprisingly, the formation of AABW appears quite sensitive to the under-ice, surface freshwater fluxes in the deep water formation zones (Doney and Hecht 2002); ocean models without active sea ice components appear to have weak AABW formation while many of the interactive ocean-sea ice models tend to have way too much bottom water production.

These known deficiencies in ocean GCM physics hamper quantitative model-data comparisons of biogeochemical and ecosystem dynamical models as well. Uncertainties in the physical flow field, particularly vertical velocity (Harrison 1996), mixing and convection, affect a variety of biogeochemical processes – nutrient supply, boundary layer stability and mean light levels, downward transport of transient tracers, anthropogenic carbon and semi-labile dissolved organic matter – and thus obscure the validation of tracer and biogeochemical components. The refinement of global ocean GCMs is an on-going process, and substantial progress will likely arise from improved treatments of surface boundary forcing and subgrid-scale physics (McWilliams 1996; Haidvogel and Beckmann 1999; Griffes et al. 2000). Transient tracers and biogeochemistry can contribute in this regard by providing additional, often orthogonal, constraints on model performance to traditional physical measures (Gnanadesikan 1999; Gnanadesikan and

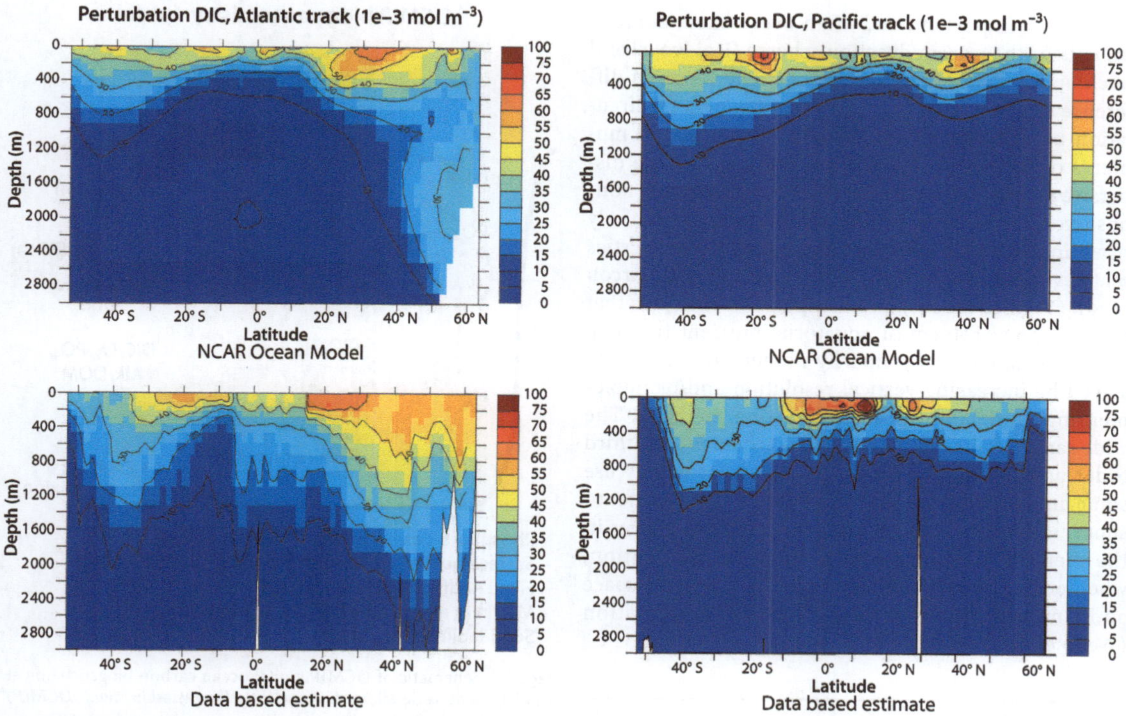

Fig. 9.3. Depth vs. latitude contour plots of anthropogenic CO_2 (mmol C m^{-3}). The panels show the simulated results from the NCAR CSM ocean model and the empirical, observation based estimates (N. Gruber 2000, pers. comm.) each for an Atlantic and Pacific section along the main path of the thermohaline circulation (see Fig. 9.2). Note that depth is limited to 3 000 m

Table 9.1.
Estimated basin inventories of anthropogenic DIC (Pg C)

Ocean	NCAR CSM ocean model	Data-based C* estimates	Data reference
Indian	22.1	20 ±3	Sabine et al. (1999)
Atlantic	39.5	40 ±6	Gruber (1998)
Pacific	46.7	46 ±5?	Feely and Sabine (pers. comm.)
Total	108.4	106 ±8?	Feely and Sabine (pers. comm.)

Toggweiler 1999). The incorporation of active biology tests new facets of the physical solutions, especially the surface air-sea fluxes and boundary layer dynamics (Large et al. 1994; Doney 1996; Doney et al. 1998) and their interaction with the interior mesoscale field (Gent and McWilliams 1990).

The desired horizontal resolution for ocean carbon cycle models is often a contentious issue, involving tradeoffs between model fidelity/realism and computational constraints. Most global climate models used for long integrations (i.e., the multi-decade to centennial and longer timescales often of interest to the ocean carbon community) have relatively coarse horizontal resolution of one to a few degrees and thus do not explicitly represent key processes such as deep-water overflows and mesoscale eddies. Increasing the resolution of this class of models is an important objective but is not a general panacea for a number of reasons. First, computational costs increase dramatically; for every factor of two increase in horizontal resolution, the integration time goes up by roughly a factor of 8. Basin-scale, eddy-resolving biological simulations at such resolution are only now becoming computationally feasible and only for short integrations. Second, very high resolution on the order of 1/10° appears to be required to correctly capture the dynamics (not just presence) of the mesoscale eddies (e.g., eddy kinetic energy; eddy-mean flow interactions) (Smith et al. 2000), and some numerical errors persist even as resolution is decreased (Roberts and Marshall 1998). One solution is to incorporate the effect of the unresolved processes using more sophisticated sub-grid scale parameterizations. For example, the Gent and McWilliams (1990) isopycnal mixing scheme tends to greatly reduce the resolution dependence and improves, in both eddy permitting and non-eddy resolving solutions, the simulated meridional heat transport, an important physical diagnostic likely relevant for nutrients and carbon as well as heat.

Another important, and often overlooked, numerical issue is the tracer advection scheme (Haidvogel and Beckmann 1999; Griffes et al. 2000). The centered difference schemes used in most 3-D ocean general circulation models, while conserving first and second moments of the tracer distribution, tend to produce dispersive errors (e.g., under and overshoots, ripples, non-positive definite tracer fields), which can be particularly troubling for biogeochemical and biological properties that have sharp vertical gradients (Oschlies and Garçon 1999). Oschlies (2000), for example, demonstrates that the common problem of equatorial nutrient trapping (Najjar et al. 1992) is primarily numerical and can be solved by increasing vertical resolution and/or implementing more sophisticated advection methods. The wide range of alternative advection schemes (e.g., third order upwinding, flux corrected transport) mostly use some amount of diffusion (only first order accurate) to suppress the dispersion errors. The main differences in the methods are the magnitude of the dissipation, whether it is applied uniformly or selectively in space and time, and the exact numerical implementation (Webb et al. 1998; Hecht et al. 2000).

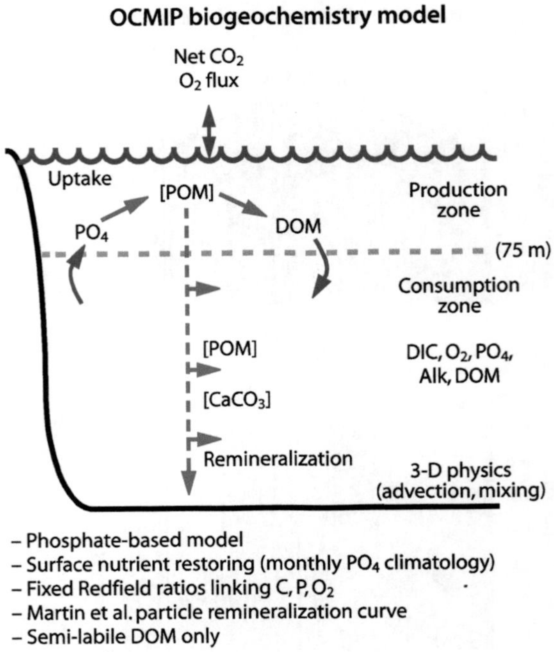

Fig. 9.4. Schematic of OCMIP global ocean carbon biogeochemical model. For more details see text and (*http://www.ipsl.jussieu.fr/OCMIP*)

9.3 Global Biogeochemical Cycles

The net anthropogenic ocean carbon uptake occurs on top of the large background DIC inventory and ocean gradients, air-sea fluxes, biological transformations, and internal transports driven by the natural carbon cycle (Fig. 9.1). Beginning with a series of global biogeochemical simulations in the early 1990s (Bacastow and Maier-Reimer 1990; Najjar et al. 1992; Maier-Reimer 1993), numerical models have played key roles in estimating basin and global-scale patterns and rates of biogeochemical processes (e.g., export production, remineralization). The primary measure for evaluating such models has been the large-scale fields of inorganic nutrients, oxygen, and DIC (Levitus et al. 1993; Conkright et al. 1998; Wallace 2001). As more robust global estimates of biogeochemical rates (e.g., new production, Laws et al. 2000) are developed from the JGOFS field data and satellite remote sensing, they too are being included in model-data comparisons (Gnanadesikan et al. 2001). Numerical biogeochemical models are also valuable tools for exploring specific hypotheses (e.g., iron fertilization; Joos et al. 1991), estimating interannual variability (Le Quéré et al. 2000), and projecting future responses to climate change (Sarmiento et al. 1998).

With a few exceptions (Six and Maier-Reimer 1996), the treatment of biology in these global biogeochemical models to date has been rather rudimentary. This is exhibited in Fig. 9.4 by a schematic of the biotic carbon model from OCMIP Phase 2. The OCMIP model consists of five prognostic variables, a limiting nutrient PO_4, dissolved inorganic carbon DIC, total alkalinity TALK, semi-labile dissolved organic matter DOM, and dissolved oxygen. Upper ocean production (0–75 m) is calculated by restoring excess model PO_4 toward a monthly nutrient climatology (Louanchi and Najjar 2000). The production is split with 1/3 going into rapidly sinking particles and the remainder into the DOM pool. The sinking particles are remineralized in the subsurface consumption zone (>75 m) using an empirical particle flux depth curve similar in form (though with different numerical parameters; Yamanaka and Tajika 1996) to that found by Martin et al. (1987) from sediment trap data. The DOM decays back to phosphate and DIC using first order kinetics with a 6 month time-scale throughout the water column. Most of the DOM is remineralized within the surface production zone but a fraction is mixed or subducted downward prior to decay and thus contributes to overall export production. Surface $CaCO_3$ production is set at a uniform 7% of particulate organic matter production, and all of the $CaCO_3$ is export as sinking particles which are remineralized with a deeper length-scale relative to organic matter. The relative uptake and release rates of PO_4, DIC, and O_2 from the organic pools are set by fixed, so-called Redfield elemental ratios, and CO_2 and O_2 are exchanged with the atmosphere via surface air-sea gas fluxes computed using the quadratic wind-speed gas exchange relationship of Wanninkhof (1992).

Despite its simplicity, the OCMIP model captures to a degree many of the large-scale ocean biogeochemical

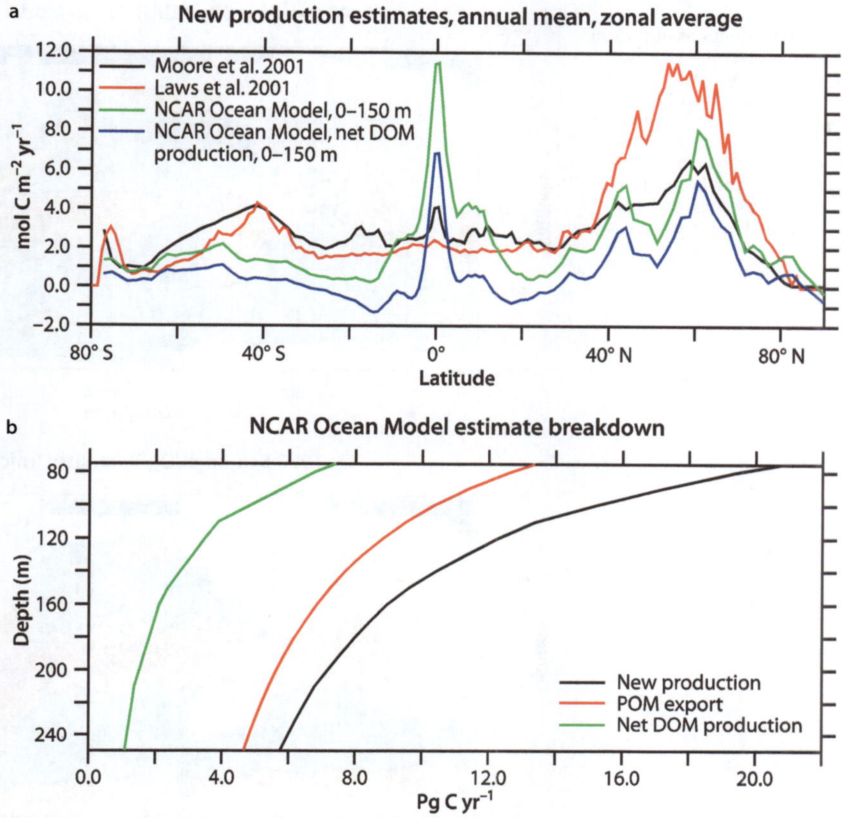

Fig. 9.5. Annual averaged new production estimates. In the *upper panel* (**a**) the NCAR model total production (particle plus net semi-labile DOM creation) and net DOM creation computed to 150 m are compared against recent new/export production estimates from Laws et al. (2000) (satellite primary production and ecosystem model based *f*-ratios) and Moore et al. (2002a) (global ecosystem model including DOM loss from downwelling and seasonal mixed layer shoaling). In the *lower panel* (**b**), the NCAR model global integral total, particle and DOM new production rates are shown as a function of the bottom limit of the depth integration

patterns found in nature. The model, zonally averaged, total new production (particle export plus net DOM production) is compared in Fig. 9.5a with recent new/export production estimates from Laws et al. (2000) (satellite primary production and ecosystem model based *f*-ratios) and Moore et al. (2002a,b) (global ecosystem model; see below for more details). The NCAR model estimate has been recomputed at 150 m rather than 75 m as specified in the OCMIP formulation to be more consistent with data based and the other model estimates. The global integrated new production estimates from the GCM (9.6 Pg C at 150 m), satellite diagnostic calculation (12.6 Pg C), and ecosystem model (11.9 Pg C) are comparable but with significant regional differences. The Moore et al. and Laws et al. curves have similar patterns with high values in the Northern Hemisphere temperate and subpolar latitudes, low levels in the tropics and subtropics and slightly elevated rates in the Southern Ocean around 40° S. The GCM results are considerably larger in the equatorial upwelling band and lower in the subtropics, reflecting in part net production, horizontal export and subsequent remineralization of organic matter. The Laws et al. (2000) estimates are based on two components: satellite derived primary production rates from CZCS ocean color data and the Behrenfeld and Falkowski (1997) algorithm, and a functional relationship of *f*-ratio to temperature and primary production from an ecosystem model. As discussed by Gnanadesikan et al. (2001), the Laws et al. (2000) values in the equatorial region are sensitive to assumptions about the maximum growth rate as a function of temperature (and implicitly nutrients), and alternative formulations can give higher values.

A significant fraction of the GCM export production at mid- to high latitudes is driven by net DOM production followed by downward transport (global integral at 150 m of 2.4 Pg C) (Fig. 9.5a and 9.5b). This has been observed in the field at a number of locations (Carlson et al. 1994; Hansell and Carlson 1998), and is thought to be an important mechanism north of the Antarctic Polar Front supporting a significant fraction of the organic matter remineralization in the upper thermocline (Doval and Hansell 2000). Because the semi-labile DOM in the model is advected by the horizontal currents, the local sum of new production and remineralization do not always balance leading to regional net convergence/divergence of nutrients and DIC. Some ocean inversion transport estimates, for example, suggest that there are net horizontal inputs of organic nutrients into subtropical areas from remote sources (Rintoul and Wunsch 1991). Another factor to consider when looking at the model production estimates and model-data compari-

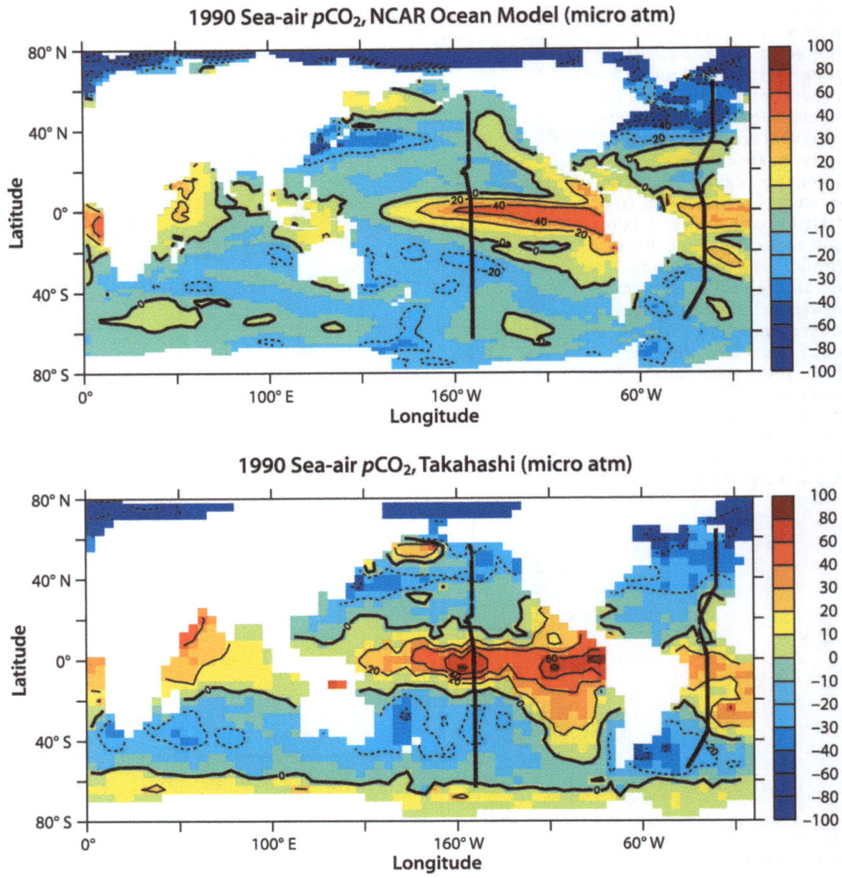

Fig. 9.6. Spatial distributions of present (1990), annual mean surface sea-air pCO_2 difference (μatm) from (*top*) the NCAR CSM Ocean Model and (*bottom*) the Takahashi et al. (1997) climatology. The two *lines* indicate the Atlantic and Pacific transects used for the horizontal sections in Fig. 9.3 and 9.7

sons is the sensitivity of new production to the depth surface chosen for the vertical integral. The cumulative (surface to depth) new production drops off significantly with depth below 75 m in the model because of the assumed rapid decrease in the sinking particle flux and relatively shallow penetration of DOM governed mostly by seasonal convection (Fig. 9.5b). For most field studies, the vertical mixing and advection terms are difficult to quantify, and the new production is computed typical at either the base of the euphotic zone (100 m to 125 m) or the shallowest sediment trap (~150 m).

Another important measure of model skill is the surface water pCO_2 field (Sarmiento et al. 2000), which can be compared to extensive underway pCO_2 observations (Takahashi et al. 1997, 1999) and atmospheric CO_2 data sets (Keeling et al. 1996; Rayner et al. 1999). The model surface water pCO_2 field is the thermodynamic driving force for air-sea gas exchange and is governed by biological DIC drawdown, physical transport, surface temperature (and salinity), and air-sea fluxes. Figure 9.6 shows the annual mean air-sea ΔpCO_2 field from the model for 1990 (pre-industrial equilibrium plus anthropogenic perturbation) and the Takahashi et al. (1997) climatology. The large-scale patterns are similar with CO_2 outgassing from the equatorial regions, where cold DIC rich water is brought to the surface by upwelling, and CO_2 uptake in the western boundary currents, Antarctic Circumpolar Current, and North Atlantic deep water formation zones. The most striking regional model-data difference is the predicted larger (smaller) model uptake in the Southern Ocean (North Atlantic), compared to the Takahashi et al. (1997) climatology, and the indication of net outgassing right along the Antarctic coast in the observations. Interestingly, the model Southern Ocean results are more in line with recent atmospheric inversion results from the IGBP/GAIM atmospheric transport model intercomparison, TRANSCOM (S. Denning, per. comm. 2000). All three approaches (ocean model, pCO_2 data climatology, and atmospheric inversion) have their own unique uncertainties and potential biases, and more effort should be given to resolving these apparent discrepancies using a combination of improved numerical models and enhanced field data collection.

The model subsurface nutrient, DIC and oxygen fields can also be compared with observations, in this case historical hydrographic data sets and the JGOFS/WOCE global CO_2 survey. The preindustrial DIC results are

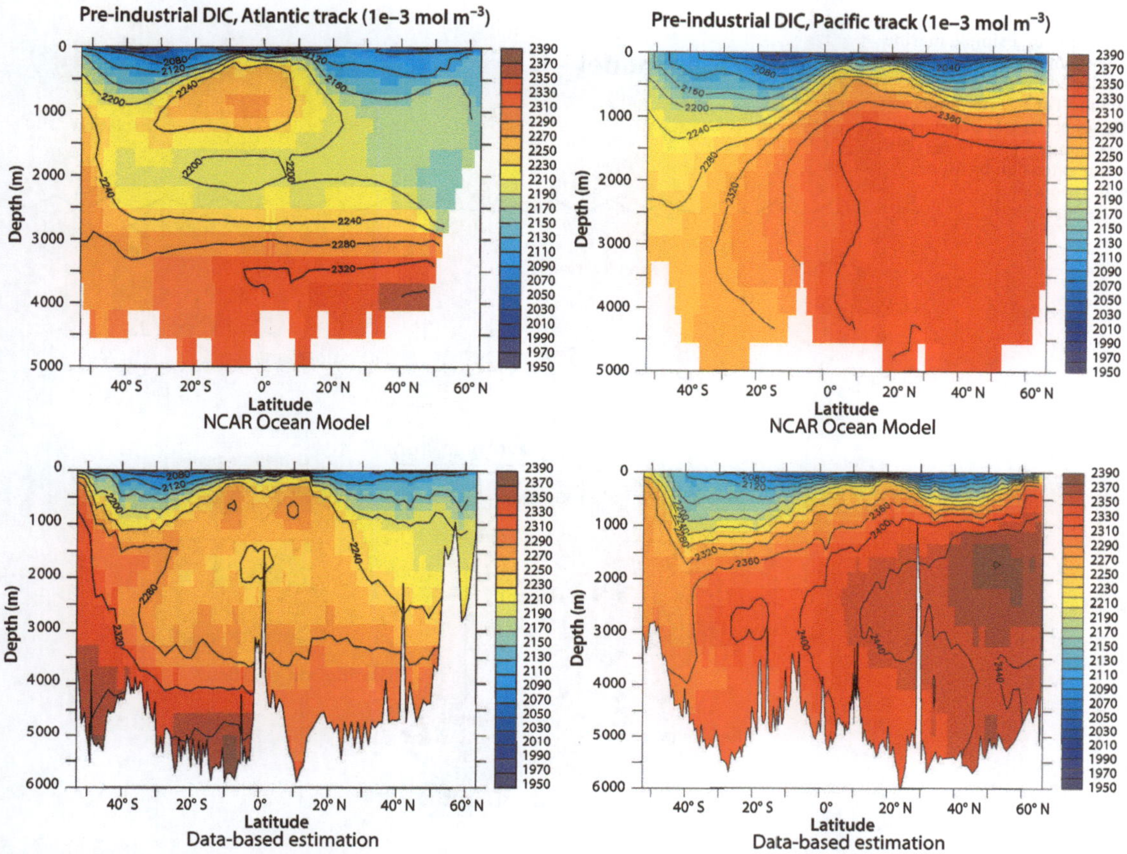

Fig. 9.7. Depth vs. latitude contour plots of pre-industrial DIC (mmol C m^{-3}). The panels show the simulated results from the NCAR CSM ocean model and observed DIC fields with the anthropogenic DIC component removed using the C* technique (N. Gruber 2000, pers. comm.) for an Atlantic and Pacific section along the main path of the thermohaline circulation (see Fig. 9.6)

shown in the same format as for anthropogenic DIC (Fig. 9.3), i.e., meridional sections in the Atlantic and Pacific (Fig. 9.7). The model surface to deep water DIC vertical gradient, which is comparable to the observations, results from contributions of about 2/3 from the biological export ('biological pump') and 1/3 from the physics ('solubility pump'). The horizontal gradients in the deep-water are determined by a mix of the thermohaline circulation and the subsurface particle remineralization rate, and the NCAR-OCMIP model captures most of the broad features. Several of the key model-data differences can be ascribed, at least partly, to problems with the model physics (e.g., too shallow outflow of North Atlantic Deep Water, Large et al. 1997; overly weak production of Antarctic bottom water, Doney and Hecht 2002). The WOCE/JGOFS carbon survey and historical data sets can also be used to estimate the horizontal transport of biogeochemical species within the ocean (e.g., Brewer et al. 1989; Rintoul and Wunsch 1991; Broecker and Peng 1992; Holfort et al. 1998; Wallace 2001), providing another constraint for ocean biogeochemical models (Murnane et al. 1999; Sarmiento et al. 2000; Gruber et al. 2001).

9.4 Ecosystem Dynamics

If the simple OCMIP biogeochemical model captures the zeroth-order state of the ocean carbon cycle then what are the important areas for progress? An obvious deficiency of the OCMIP straw man is the lack of explicit, prognostic biological dynamics to drive surface production, export and remineralization. By linking to a fixed surface nutrient climatology, we have avoided specifying the details of how the surface nutrient field is controlled (e.g., grazing, iron fertilization, mesoscale eddies) or how it might evolve under altered forcing. While useful for the purposes of OCMIP, clearly a more mechanistic approach is desired for many applications. For example, looking toward the next several centuries, future changes in ocean circulation and biogeochemistry may lead to large alterations in the background carbon cycle that could strongly impact projected ocean carbon sequestration (Denman et al. 1996; Sarmiento et al. 1998; Doney and Sarmiento 1999; Boyd and Doney 2003). Realistic projections will require coupled ecosys-

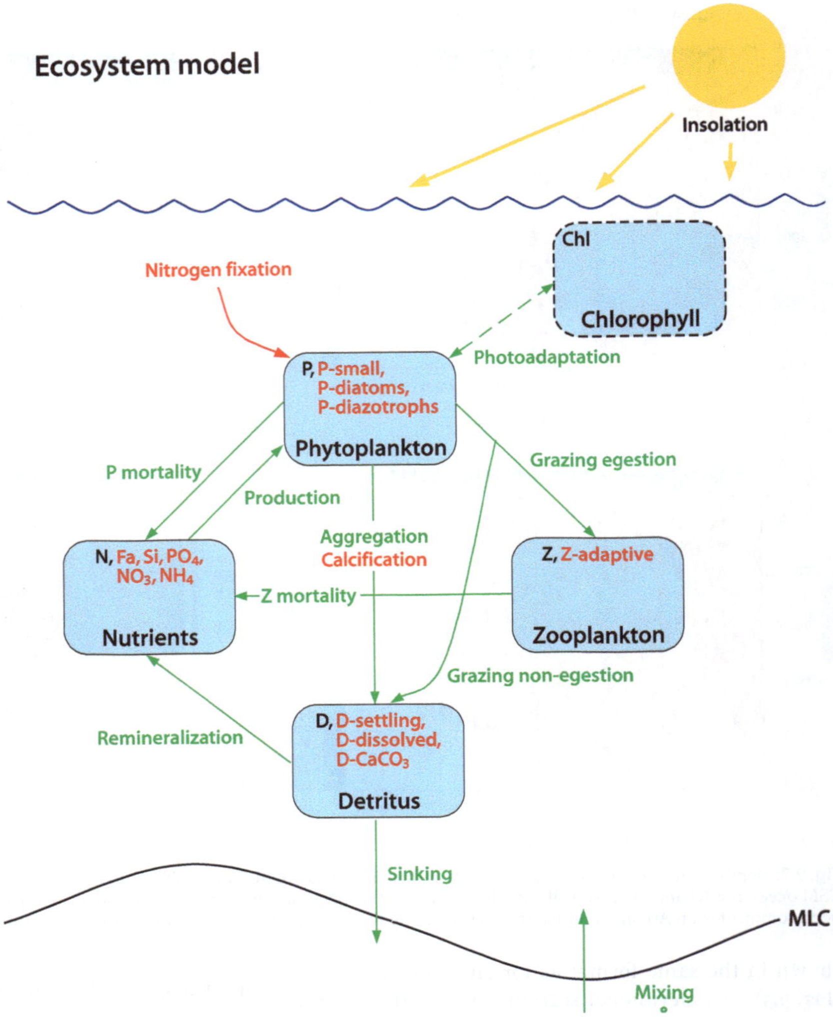

Fig. 9.8.
Schematic of a simple marine ecosystem model originally developed for the Bermuda Atlantic Time-Series Study site (Doney et al. 1996; Doney et al., pers. comm.) and (in *red*) the recent extension by Moore et al. (2001a)

tem-biogeochemical models that include the main processes thought to be sensitive to climate change (e.g., atmospheric dust, nitrogen fixation, community structure).

As an example of a typical marine ecosystem, consider the schematic shown in Fig. 9.8. The model developed for vertical 1-D simulations of the Sargasso Sea by Doney et al. (1996) incorporates five prognostic variables: phytoplankton, zooplankton, nutrient, detritus and chlorophyll (a so-called PZND model). As is common, the model aggregates populations and species of organisms into broadly defined trophic compartments. The equations are based on the flow of a single limiting currency, in this case the concentration of nitrogen (mol N m^{-3}), among compartments rather than individual organisms. The various source/sink terms (e.g., photosynthesis, zooplankton grazing, detrital remineralization) are calculated using standard, though not always well agreed upon, sets of empirical functional forms and parameters derived either from limited field data or laboratory experiments, the latter often with species and conditions of limited relevance to the actual ocean (Fasham 1993; Evans and Fasham 1993; Evans and Garçon 1997). This type of compartment ecosystem model has been used extensively in oceanography (Steele 1974) and theoretical ecology (May 1973; Case 2000) since the early 1970s but has roots much further back in the literature (e.g., Riley 1946; Steele 1958). The area was revitalized about the time of the inception of JGOFS by the seminal work of Evans and Parslow (1985), Frost (1987), Fasham et al. (1990), and Moloney and Field (1991).

Despite its simplifications, the PZND model (Fig. 9.8) does an adequate job capturing the vertical structure and broad seasonal patterns of bulk biogeochemical properties in Bermuda field data (e.g., chlorophyll – Fig. 9.9; nitrate; Doney et al. 1996). Specific features include: a winter phytoplankton bloom following nutrient injection via deep convection; low surface nutrients and chlorophyll during the stratified summer period; and the formation of a sub-

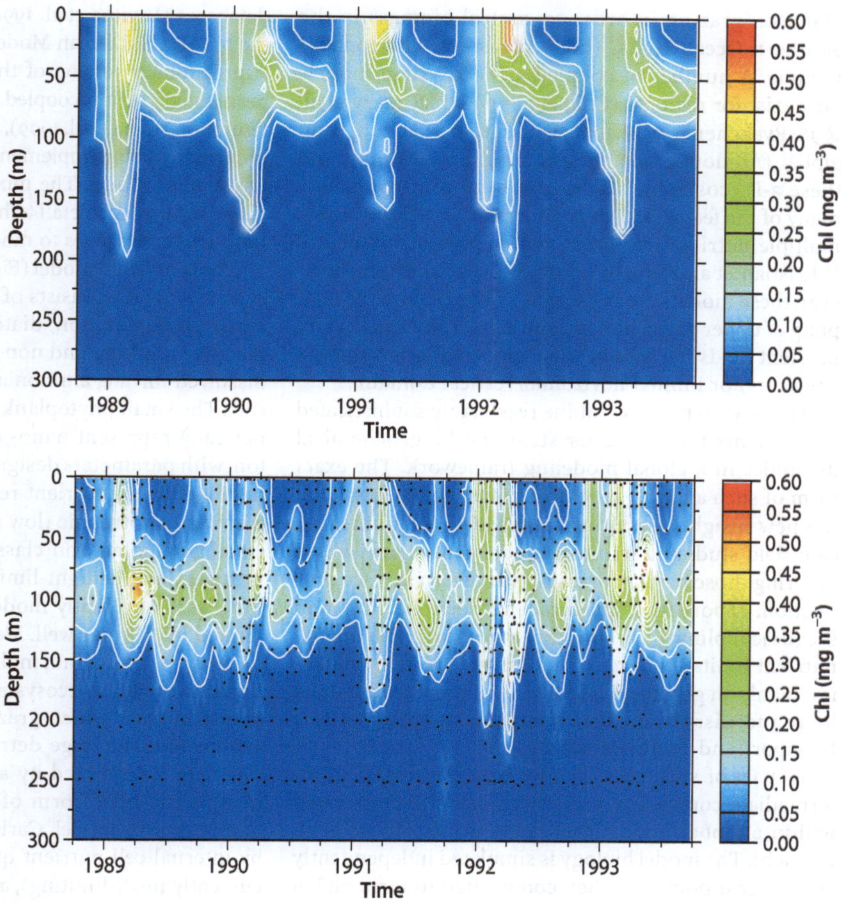

Fig. 9.9. A comparison of the modeled and observed time-depth chlorophyll distribution for the Bermuda Atlantic Time-Series Study site in the western subtropical North Atlantic. The 1-D coupled biological-physical model is based on Doney et al. (1996) and Doney (1996)

surface chlorophyll maximum at the top of the nutricline. The 1-D coupled biological-physical model, based on surface forcing and physics described by Doney (1996), also reproduces aspects of the observed interannual variability driven by the depth of the winter convection.

Variants on the PZND theme have been successfully applied in vertical 1-D form in a diverse range of biogeographical regimes from oligotrophic subtropical gyres (Bissett et al. 1994) to seasonal bloom regimes (Fasham 1995) to subarctic high-nitrate, low chlorophyll regions (McClain et al. 1996; Pondaven et al. 2000). The construction of the 1-D physical framework (vertical mixing, temperature etc.) requires explicit consideration (Archer 1995; Doney 1996; Evans and Garçon 1997), but in general 1-D coupled models have resulted in useful test-beds for exploring ecological processes and implementing biological data assimilation techniques. It has been known for a while that the relatively simple PZND dynamics belie the ecological complexity of the real system, and recent idealized and local 1-D coupled models include increasing levels of ecological sophistication. Models are incorporating a range of factors such as: size and community structure (Armstrong 1994, 1999a; Bissett et al. 1999), iron limitation (Armstrong 1999b; Leonard et al. 1999; Denman and Pena 1999; Pondaven et al. 2000), and nitrogen fixation (Hood et al. 2001; Fennel et al. 2002). One problem, however, is that most 1-D coupled models are developed and evaluated for a single site, and the generality of these models and their derived parameter values for basin and global simulations remains an open question.

Early three-dimensional basin and global scale calculations (Sarmiento et al. 1993; Six and Maier-Reimer 1996) were conducted with single, uniform PZND ecosystem models applied across the entire domain. These experiments demonstrated that large-scale features such as the contrast between the oligotrophic subtropical and eutrophic subpolar gyres could be simulated qualitatively. Some problems arose, however, with the details. For example, the incorporation of the Fasham et al. (1990) model into a North Atlantic circulation model by Sarmiento et al. (1993) showed too low production and biomass in the oligotrophic subtropics and too weak a spring bloom at high latitudes. The Six and Maier-Reimer (1996) result required careful tuning of the phytoplankton growth temperature sensitivity and zooplankton grazing in order to control biomass in the Southern Ocean HNLC (high nitrate-low chlorophyll) regions. A number of coupled 3-D ecosystem models now exist for regional (Chai et al. 1996; McCreary et al. 1996; Ryabchenko et al. 1998; Dutkiewicz et al. 2001) and global (Aumont et al., pers. comm.) applications, and these 3-D ecosystem models are beginning to include many of the features already addressed in 1-D, including multiple nutrient limitation and community structure (Christian et al. 2001a,b; Gregg et al. 2002). Often, however, these models are not used to fully explore the coupling of upper ocean biology and subsurface carbon and nutrient fields because of the short integration time (a few years) or limited horizontal/vertical domain.

The next step is to combine reasonably sophisticated components for both ecosystem and biogeochemical dynamics in a global modeling framework. The exact form of such a model is yet to be determined. Based on the new insights emerging from JGOFS and other recent field studies, a minimal model can be envisioned covering those basic processes that govern surface production, export flux, subsurface remineralization, and the (de)coupling of carbon from macronutrients (multinutrient limitation; size structure and trophic dynamics; plankton geochemical functional groups; microbial loop and dissolved organic matter cycling; particle transport and remineralization).

As part of such a project, we have developed an intermediate complexity, ecosystem model incorporated within a global mixed layer framework (Moore et al. 2002a,b). The model biology is simulated independently at each grid point and then composited to form global fields. The model has a low computational overhead, and thus can be used for extensive model evaluation and exploration. Sub-surface nutrient fields are from climatological databases, and the mixed layer model captures the local processes of turbulent mixing, vertical advection at the base of the mixed layer, seasonal mixed layer entrainment/detrainment, but not horizontal advection. Other forcings include sea surface temperature, percent sea ice cover, surface radiation, and the atmospheric deposition of iron (Fung et al. 2000; Fig. 9.10). The physical forcings are prescribed from climatological databases (Levitus et al. 1994; Conkright et al. 1998) and the NCAR CSM Ocean Model (NCOM) (Large et al. 1997). A preliminary version of the ecosystem model also has been tested in a fully coupled, 3-D North Atlantic Basin configuration (Lima et al. 1999), and the full ecosystem model is currently being implemented in the new global NCAR-Los Alamos model. The mixed layer ecosystem model is discussed in some detail to highlight new modeling directions and approaches to model-data evaluation.

The ecosystem model (Fig. 9.8) is adapted from Doney et al. (1996) and consists of eleven main compartments, small phytoplankton, diatoms, and diazotrophs; zooplankton; sinking and non-sinking detrital classes; and dissolved nitrate, ammonia, phosphorus, iron, and silicate. The small phytoplankton size class is meant to generically represent nano- and pico-sized phytoplankton, with parameters designed to replicate the rapid and highly efficient nutrient recycling found in many subtropical, oligotrophic (low nutrient) environments. The small phytoplankton class may be iron, phosphorus, nitrogen, and/or light-limited. The larger phytoplankton class is explicitly modeled as diatoms and may be limited by silica as well. Many of the biotic and detrital compartments contain multiple elemental pools to track flows through the ecosystem. The model has one zooplankton class which grazes the three phytoplankton groups and the large detritus. Phytoplankton growth rates are determined by available light and nutrients using a modified form of the Geider et al. (1998) dynamic growth model. Carbon fixation rate is governed by internal cell nutrient quotas (whichever nutrient is currently most-limiting), and the cell quotas computed relative to carbon are allowed to vary dynamically as the phytoplankton adapt to changing light levels and nutrient availability. There is good laboratory evidence for a relationship between cell quotas (measured as nutrient/C ratios) and specific growth rates (Sunda and Huntsman 1995; Geider et al. 1998). Photoadaptation is modeled according to Geider et al. (1996, 1998) with a dynamically adaptive Chl/C ratio. The diazotrophs are assumed to fix all required nitrogen from N_2 gas following Fennel et al. (2002) and are limited by iron, phosphorus, light or temperature. Calcification is para-

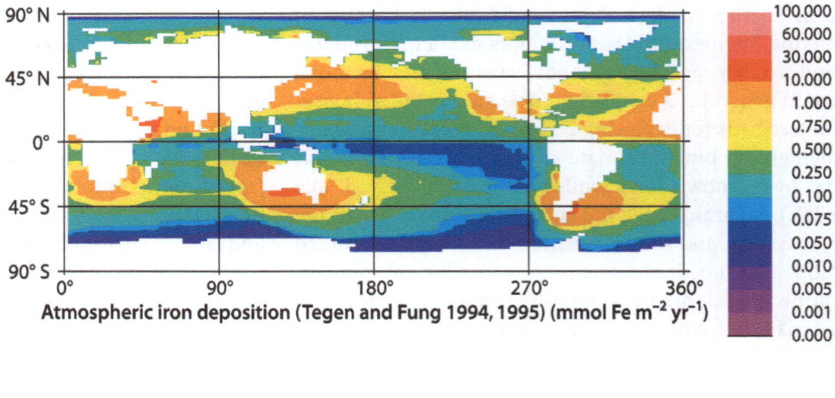

Fig. 9.10. Annual mean map of atmospheric iron deposition to the ocean adapted from Tegen and Fung (1995) model estimates (reprinted from Deep-Sea Res II 49, Moore et al. (2002) Iron cycling and nutrient limitation patterns in surface waters of the world ocean. pp 463–507, © 2002, with permission from Elsevier Science)

meterized as a time-varying fraction of the small (pico/nano) plankton production as a function of ambient temperature and nutrient concentrations. Based on Harris (1994) and Milliman et al. (1999) we assume that grazing processes result in substantial dissolution of $CaCO_3$ in the upper water column.

The model output is in generally good agreement with the bulk ecosystem observations (e.g., total biomass; productivity; nutrients) across diverse ecosystems that include both macro-nutrient and iron-limited regimes and very different physical environments from high latitude sites to the mid-ocean gyres. The detailed, local data sets from JGOFS and historical time-series stations (Kleypas and Doney 2001) have been important for developing parameterizations, testing hypotheses, and evaluating model performance. As an example, a comparison of model simulated and observed mixed layer seasonal cycle for nitrate is shown in Fig. 9.11 for nine locations across the globe. The time-series stations and regional JGOFS process studies (e.g., EqPAC, Arabian Sea) often provide invaluable constraints on biological fluxes (primary productivity profiles, export flux, zooplankton grazing, not shown) as well, parameters that are typically sampled too sparsely to construct global data sets. The variables that are available from observations on a global scale are more limited, including seasonal (now monthly) nutrient fields (Conkright et al. 1998), satellite remotely sensed surface chlorophyll (McClain et al. 1998) (Fig. 9.12) and diagnostic model derived products such as satellite based integrated primary production (Behrenfeld and Falkowski 1997) and *f*-ratio (Laws et al. 2000) estimates. Compared with the satellite estimates, the model produces realistic global patterns of both primary and export production.

The incorporation of iron limitation plays a critical part in the model skill of reproducing the observed high nitrate and low phytoplankton biomass conditions in the Southern Ocean and the subarctic and equatorial Pacific regions. A small number of desert regions (e.g., China, Sahel), mostly in the Northern Hemisphere, provide the main sources of atmospheric dust (and thus iron) to the ocean, and the estimated iron deposition rate to oceanic HNLC environments can be orders of magnitude lower than other locations (Fig. 9.10). At such low deposition rates, upwelling of subsurface iron likely contributes a significant fraction of the total bioavailable

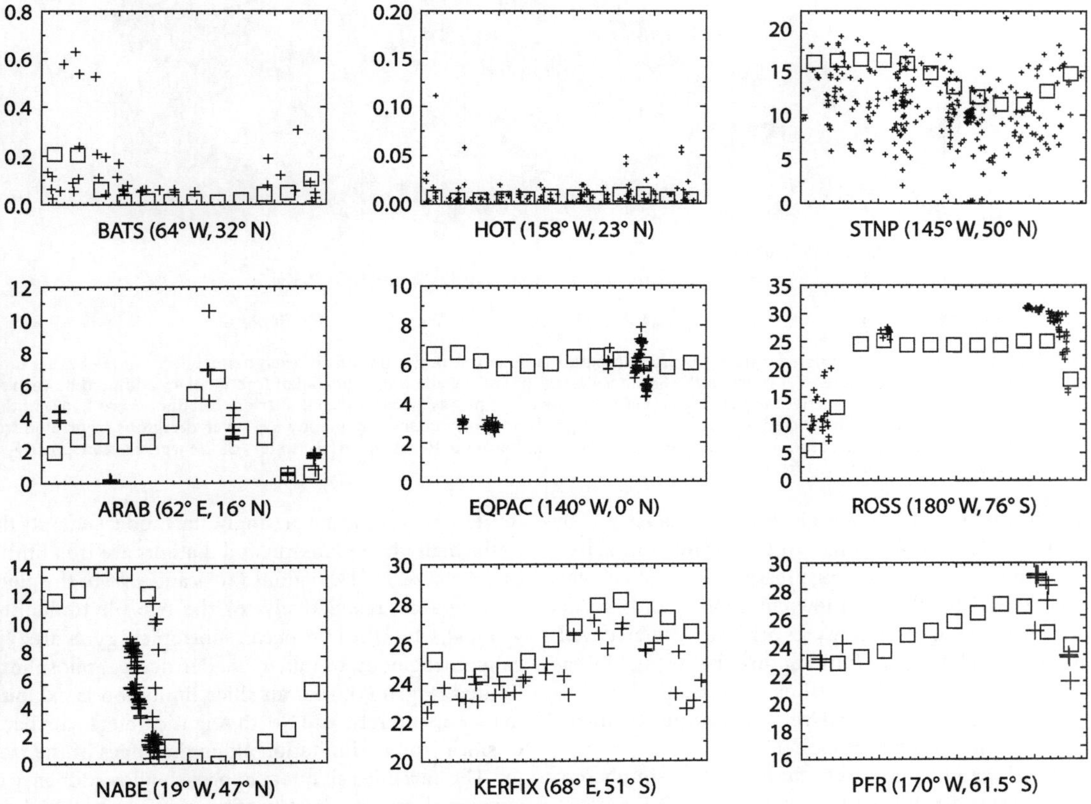

Fig. 9.11. Comparison of simulated and observed seasonal nitrate cycle at nine JGOFS time-series stations across a range of biogeographical regimes (Kleypas and Doney 2001). The model results are from a global mixed layer ecosystem model with uniform biological coefficients (reprinted from Deep-Sea Res II 49, Moore et al. (2002) An intermediate complexity marine ecosystem model for the global domain. pp 403–462, © 2002, with permission from Elsevier Science)

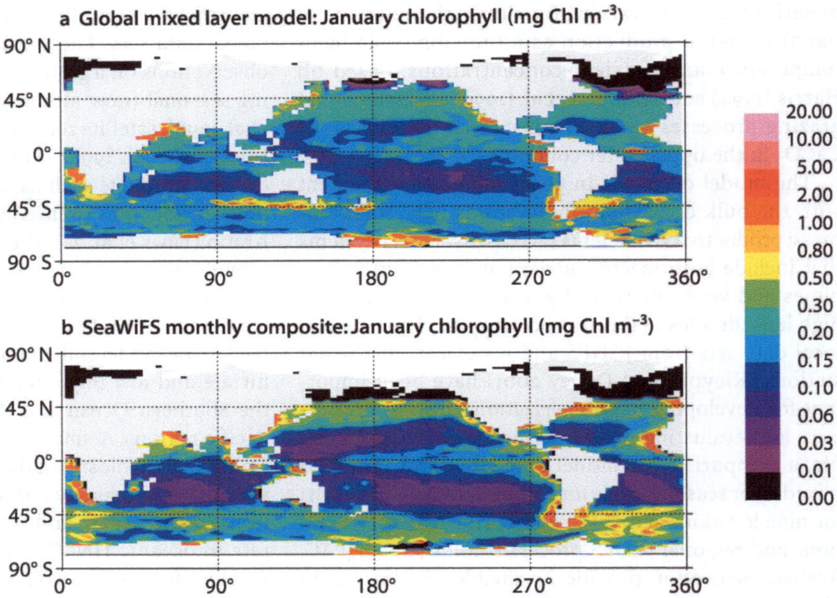

Fig. 9.12. Global field of monthly mean surface chlorophyll concentration for January from SeaWiFS and a global mixed layer ecosystem model (reprinted from Deep-Sea Res II 49, Moore et al. (2002) Iron cycling and nutrient limitation patterns in surface waters of the world ocean. pp 463–507, © 2002, with permission from Elsevier Science)

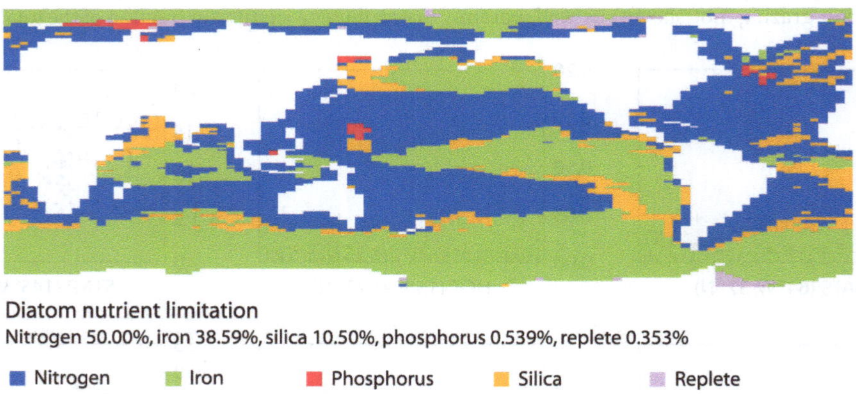

Fig. 9.13. Ecosystem model simulated nutrient limitation patterns during summer months in each hemisphere (June–August in the Northern Hemisphere, December–February in the Southern Hemisphere) for diatoms. The global fractional area limited by each nutrient is listed below the plot. Nutrient replete areas (here arbitrarily defined as areas where all nutrient cell quotas are >90% of their maximum values) are largely restricted to areas of extreme light-limitation under permanently ice-covered regions (reprinted from Deep-Sea Res II 49, Moore et al. (2002) Iron cycling and nutrient limitation patterns in surface waters of the world ocean. pp 463–507, © 2002, with permission from Elsevier Science)

iron. In the model, these regions are characterized by strong iron limitation of diatom growth and modest iron limitation and strong grazing pressure on the small phytoplankton. The observed low chlorophyll and low nitrate levels in oligotrophic gyres are also simulated well, but the model does not fully capture the strong blooms in some of the coastal upwelling regions, most likely a result of the weak vertical velocities input from the coarse resolution physics model.

Models should allow us to do more than simply replicate what is already known, by posing new (and testable) hypotheses of how the ocean functions at the system level. As an example, the global mixed layer model predicts the degree and time/space patterns of nutrient limitation. Not too surprisingly, the model suggests that both small phytoplankton and diatoms are iron limited in the classic HNLC regions (40% and 52% of the global surface area, respectively for the two phytoplankton groups), while the mid-ocean subtropical gyres are typically nitrogen or, to much smaller degree, phosphorus limited (Fig. 9.13). Diatom silica limitation is exhibited in the subantarctic and North Atlantic waters with bands of silica-iron co-limitation along the edges of the tropics. The variable cell quota approach allows for easy diagnosis of varying degrees of nutrient stress, which can be compared in the near future with global nutrient stress fields to be derived from the MODIS natural fluorescence measurements (Letelier and Abbott 1996).

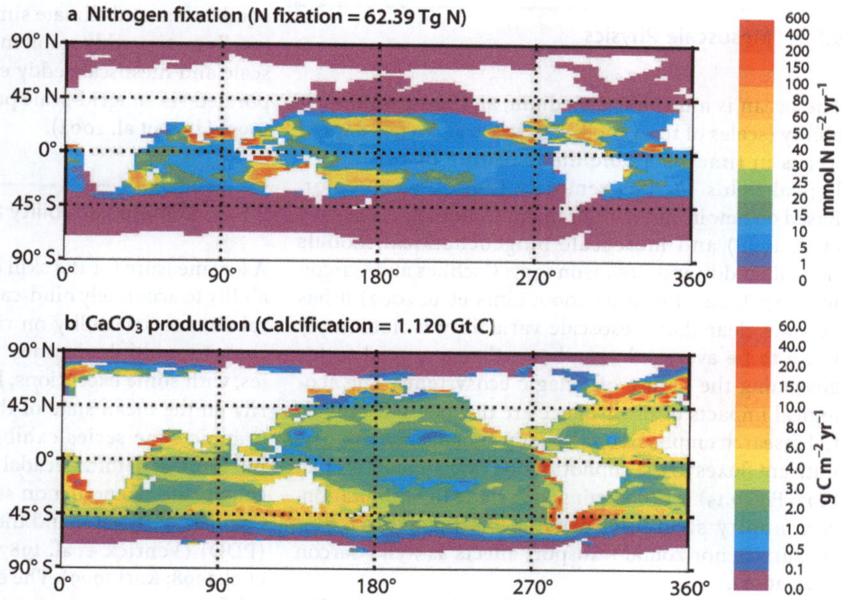

Fig. 9.14. Model simulated annual mean nitrogen fixation and calcification fields (reprinted from Deep-Sea Res II 49, Moore et al. (2002) Iron cycling and nutrient limitation patterns in surface waters of the world ocean. pp 463–507, © 2002, with permission from Elsevier Science)

The other new aspect of the global model is the inclusion of community structure through planktonic geochemical functional groups, namely diatoms (export flux and silica ballast), diazotrophs (nitrogen fixation), and calcifiers (alkalinity and ballast). The model spatial patterns of annual nitrogen fixation (Fig. 9.14) agree well with the limited information known from in situ work (Capone et al. 1997), high trichodesmium biomass and/or nitrogen fixation rates reported in the Caribbean Sea and eastern tropical North Atlantic (Lipschulz and Owens 1996) as well as in the subtropical North Pacific (Letelier and Karl 1996, 1998; Karl et al. 1997). The total model nitrogen fixation of 58 Tg N, which accounts only for the mixed layer production, is somewhat less than, though of comparable magnitude to, the 80 Tg N estimate of Capone et al. (1997) and the Gruber and Sarmiento (1997) geochemical estimates of >100 Tg N.

The parameterization of phytoplankton calcification is an active research topic, but the spatial patterns shown in Fig. 9.14 are generally similar to those estimated by Milliman (1993) and Milliman et al. (1999). $CaCO_3$ production/export is lower in the mid-ocean gyres and higher in the North Atlantic, coastal upwelling zones and mid-latitude Southern Ocean waters. The high latitude North Atlantic in particular is known to be a region with frequent coccolithophore blooms (Holligan et al. 1993). The model production/export is lower in the equatorial Pacific and Indian ocean compared with Milliman et al. (1999), but the global sinking export of 0.46 Gt C is in good agreement with their integrated estimate.

Two main factors limiting progress on ecosystem modeling are the conceptualization of key processes at a mechanistic level and the ability to verify model behavior through robust and thorough model-data comparisons (Abbott 1995). The phytoplankton iron limitation story is an illuminating example. Atmospheric dust/iron deposition flux estimates vary considerably (perhaps as large as a factor of ten or more in some areas) and the bioavailable fraction of the dust iron is not well known. Surface and subsurface ocean iron measurements are limited (particularly from a global modeler's perspective), and there remain serious analytical and standardization issues. Organic ligands may play a role in governing both bioavailability and subsurface iron concentrations. Not enough is known about the effect of iron limitation and variability on species competition at ambient low iron levels. A host of other processes may be relevant, but are currently poorly characterized, including: iron release by photochemistry and zooplankton grazing, release of iron from ocean margin sediments, and iron remineralization from sinking particles.

9.5 Other Topics

In a recent review paper, Doney (1999) described a set of key marine ecological and biogeochemical modeling issues to be addressed in the next generation of numerical models: multi-element limitation and community structure; large-scale physical circulation; mesoscale space and time variability; land, coastal, and sediment exchange with the ocean; and model-data evaluation and data assimilation. In the preceding three sections we have presented in some detail the nature of several of these challenges and specific initial progress made by our group. Below we more briefly outline some of the remaining items.

9.5.1 Mesoscale Physics

The ocean is a turbulent medium, and mesoscale variability (scales of 10 to 200 km in space and a few days to weeks in time) is a ubiquitous feature of ocean biological fields such as remotely sensed ocean color. Based on new in situ measurement technologies (Dickey et al. 1998) and mesoscale biogeochemical models (McGillicuddy and Robinson 1997; Oschlies and Garçon 1998; Spall and Richards 2000; Lima et al. 2002) it has become clear that mesoscale variability is not simply noise to be averaged over, but rather a crucial factor governing the nature of pelagic ecosystems. The ecological impacts of disturbance are diverse, and the initial research emphasis on the eddy enhancement of new nutrient fluxes to the euphotic zone (McGillicuddy et al. 1998; Fig. 9.15) is broadening to include light limitation, community structure, organic matter export, and subsurface horizontal transport effects as well (Garçon et al. 2001).

Quantifying the large-scale effect of such variability will require concerted observational, remote sensing and numerical modeling programs with likely heavy reliance on data assimilation. The computational demands of truly eddy resolving basin to global calculations are significant, however. Recent high resolution physical simulations of the North Atlantic show that dramatic improvement in eddy statistics and western boundary current dynamics is reached only at 1/10 of a degree resolution (Smith et al. 2000), and even higher resolution may be needed for biology if submesoscale (0.5–10 km) processes are as important as suggested by preliminary results (Levy et al. 1999; Mahadevan and Archer 2000; Lima et al. 2002). Over the near term, long time-scale equilibrium and climate simulations will be limited primarily to non-eddy resolving models in which submesoscale and mesoscale eddy effects will have to be incorporated via subgrid-scale parameterizations (Levy et al. 1999; Lima et al. 2002).

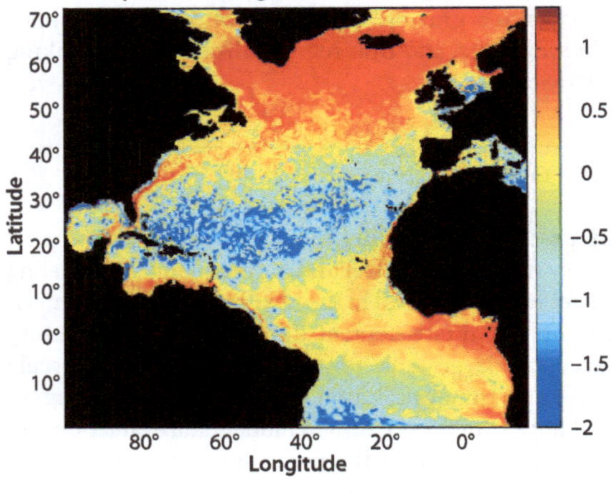

Fig. 9.15. Daily snapshot of new production from a Los Alamos-POP 1/10° mesoscale simulation of the North Atlantic (Dennis McGillicuddy, pers. comm.)

9.5.2 Climate Variability and Secular Change

A key measure for the skill of numerical models is their ability to accurately hind-cast oceanic responses to natural climate variability on timescales from the seasonal cycle to multiple decades. Large-scale modeling studies, with some exceptions, have tended to focus primarily on the mean state of the ocean. Biological oceanographic time series exhibit significant variability on interannual to interdecadal scales associated with physical climate phenomenon such as the El Niño-Southern Oscillation (ENSO) and the Pacific Decadal Oscillation (PDO) (Venrick et al. 1987; Karl et al. 1995; McGowan et al. 1998; Karl 1999). The ecosystem response to physical forcing may be quite nonlinear, manifesting in the North Pacific, for example, as a major biological regime shift in the mid-1970s due to the PDO (Francis and Hare 1994). Comparable climate related biological shifts are also inferred for the North Atlantic (Reid et al. 1998). Retrospective models can help explain the underlying mechanisms of such phenomena (Polovina et al. 1995). Because of an interest in separating terrestrial and oceanic signals in the atmospheric CO_2 network, there is also a growing effort to model the oceanic contribution to atmospheric variability, which appears to be small except for the tropical ENSO signal (Rayner et al. 1999; Le Quéré et al. 2000).

Numerical models are also being used to project the potential marine biogeochemical responses to anthropogenic climate change (Sarmiento et al. 1998; Matear and Hirst 1999). Coupled ocean-atmosphere model simulations differ considerably in their details, but most models suggest general warming of the upper ocean and thermocline, increased vertical stratification in both the low latitude (warming) and high latitude (freshening) surface waters, and weakening of the thermohaline circulation. Combined, the physical effects lead to a 30–40% drop in the cumulative anthropogenic CO_2 uptake over the next century partly compensated by changes in the strength of the natural biological carbon pump. Given the low level of biological sophistication used in these early simulations, such projections must be considered preliminary, demonstrating the potential sensitivity of the system and posing important questions to be addressed through future research.

Preliminary ecosystem simulations (Bopp et al. 2001) show different regional climate change responses to enhanced stratification with decreased subtropical productivity (nutrient limited) and increased subpolar productivity (light limited) reminiscent of the PDO signal (Polovina et al. 1995). Other environmental factors to consider include alterations of aeolian trace metal deposition due to changing land-use and hydrological cycle, variations in cloud cover and solar and UV irradiance, coastal eutrophication, and lower surface water pH and carbonate ion concentrations due to anthropogenic CO_2 uptake (Kleypas et al. 1999). The decadal time-scale biogeochemical and ecological responses to such physical and chemical forcings are not well understood in detail, and prognostic numerical models will be relied on heavily along with historical and paleoceanographic climate variability reconstructions (Doney and Sarmiento 1999; Boyd and Doney 2003).

9.5.3 Land, Coastal Ocean, and Sediment Interactions

The coastal/margins zone interacts strongly and complexly with the land, adjacent atmosphere, continental shelves and slopes, and open-ocean. The specific rates of productivity, biogeochemical cycling, and organic/inorganic matter sequestration are higher than those in the open ocean, with about half of the global integrated new production occurring over the continental shelves and slopes (Walsh 1991; Smith and Hollibaugh 1993). The high organic matter deposition to, and close proximity of the water column to, the sediments raises the importance of sedimentary chemical redox reactions (e.g., denitrification, trace metal reduction and mobilization), with implications for the global carbon, nitrogen, phosphorus and iron cycles. Finally, the direct and indirect human perturbations to the coastal environment (e.g., pollution, nutrient eutrophication, fisheries) are large, with important impacts on marine ecosystems (harmful algal blooms, coral reefs, spawning grounds) and society (e.g., commercial fisheries, tourism, and human health and aesthetics).

Because of the topographic complexity, smaller time/space scales, and specific regional character of coastal environments, basin to global scale models typically do not fully account for biogeochemical fluxes and dynamics on continental margins and in the coastal ocean. Thus coastal/open-ocean exchange and the large-scale influence on the ocean are not well quantified except in a few locations (Falkowski et al. 1994; Liu et al. 2000). Regional coastal ecosystem models have been moderately successful (Robinson et al. 2001; Fig. 9.16), and an obvious next step is to meld open ocean and coastal domains through more adaptable grid geometries such as unstructured (spectral) finite element grids (Haidvogel et al. 1997) or by embedding regional domain, higher-resolution models (Spall and Holland 1991). Dynamic marine sediment geochemistry models (Heinze et al. 1999) are needed both for the coastal problem and for large-scale paleoceanographic applications, an example being the compensation of the sediment $CaCO_3$ to changes in ocean carbon chemistry on millennial time-scales (Archer and Maier-Reimer 1994; Archer et al. 2000).

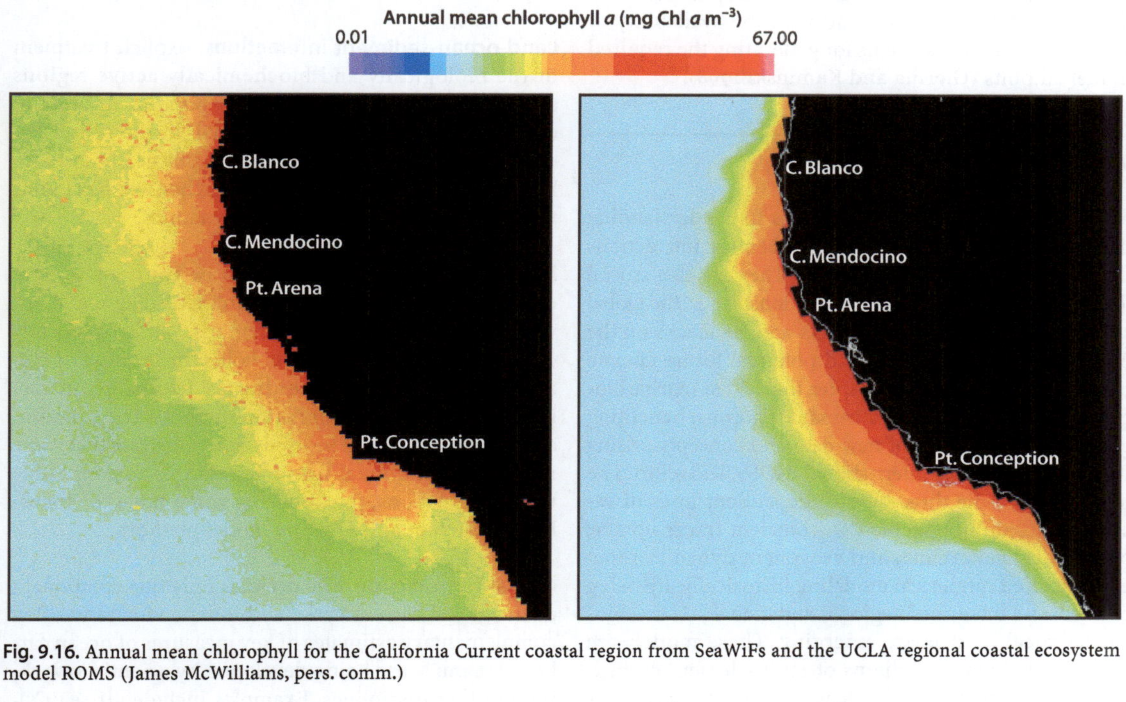

Fig. 9.16. Annual mean chlorophyll for the California Current coastal region from SeaWiFs and the UCLA regional coastal ecosystem model ROMS (James McWilliams, pers. comm.)

9.5.4 Inverse Modeling and Data Assimilation

The emerging techniques of inverse modeling and data assimilation, which more formally compare and meld model results and data, are becoming essential in model development and evaluation (U.S. JGOFS 1992; Kasibhatla et al. 2000). In theory data assimilation provides a solution, if it exists, that is dynamically consistent with both the observations and model equations within the estimated uncertainties. Much of the art of data assimilation lies in assigning relative error weights to different data types and to the model equations themselves, the so-called cost function problem (U.S. JGOFS 1992). A number of recent studies have used this approach to better constrain or optimize parameters for marine biogeochemical box and one-dimensional models, particularly with time series data (Matear 1995; Fasham and Evans 1995; Hurtt and Armstrong 1996; Spitz et al. 1998; Fennel et al. 2001). Applications to three-dimensional models are more limited but include efforts to assimilate satellite ocean color data into ecosystem models (Ishizaka 1990) or to estimate poorly measured fluxes such as dissolved organic phosphorus transport/remineralization (Matear and Holloway 1995), surface export production (Schlitzer 2000), and air-sea oxygen fluxes (Gruber et al. 2001) from the large-scale nutrient distributions and physical circulation flow fields. The utility of data assimilation will continue to grow with the import and refinement of numerical methods from meteorology and physical oceanography to interdisciplinary problems (Robinson 1996) and with the availability of automated software systems for generating the required model adjoints (Giering and Kaminski 1998).

9.6 Summary

Numerical models are essential tools for understanding the complex physical, biological and chemical interactions that govern the ocean carbon cycle. They are also crucial for extrapolating local/regional relationships to the global scale and for projecting the effects and feedbacks on the ocean carbon cycle of past and potential future climate change. As outlined in this chapter, the field of marine biogeochemical modeling is alive and vigorous, benefiting greatly from the surge of new data and concepts arising from the decade long international JGOFS field effort. The boundaries of the three quasi-independent lines of research (i.e., anthropogenic CO_2 transient tracer uptake; biogeochemical cycling; and ecosystem dynamics) that characterized numerical modeling historically are being blurred, and integrated regional and global 3-D eco-biogeochemical models are emerging. These models are based on the new paradigms of multi-element cycling, community structure and geochemical functional groups (e.g., nitrogen fixers, calcifiers), key to addressing hypotheses of how the ocean might alter or drive long term changes in atmospheric CO_2. Growing utilization of retrospective or hindcasting experiments will be used to evaluate model skill relative to historical interannual and paleoclimate variability. Significant progress is also being made in process and regional models on issues such as biological-physical sub- and mesoscale interactions as well as coastal ecosystem and biogeochemical dynamics.

A number of major challenges remain for the next decade(s):

Ecological sophistication. Ocean models must be grounded at a more fundamental level by ecological and evolutionary hypotheses. The current emphasis is often on simulating chemical and biochemical analyses: phytoplankton treated simply as concentrations of pigments and organic carbon, zooplankton as grazers, and physics as a mechanism for providing nutrients. A more mechanistic understanding is needed of how individual organisms and species interact to form pelagic ecosystems, how food webs affect biogeochemical fluxes, and how the structure of food webs and corresponding biogeochemical fluxes will change in the coming decades.

High frequency variability. The importance of high frequency spatial and temporal variability (e.g., fronts, mesoscale eddies) on the large-scale carbon cycle needs to be better characterized. This will require a combination of subgrid-scale parameterizations, nested models, and dedicated very high-resolution computations.

Land-ocean-sediment interactions. Explicit treatment of the biologically and biochemically active regions along continents needs to be incorporated. At present, coastal modeling is often 'parochial' in the sense that each region is treated as unique both physically and ecologically. The computational approaches will be similar to those outlined for mesoscale dynamics.

Model-data fusion. Models must be confronted more directly with data using a hierarchy of diagnostic, inverse, and data assimilation methods. While technically challenging, data assimilation holds the promise of creating evolving, 4-D 'state estimates' for the ocean carbon cycle. Further, assimilation methods (e.g., parameter optimization) can be used to demonstrate that some models or functional forms are simply incompatible with observations, thus offering some hope for focusing the current and growing model plethora.

Global carbon cycle. The ocean is only one component of the global carbon cycle, and independent and often complementary estimates of key measures of ocean carbon dynamics are being developed by scientists working in other disciplines. Examples include air-sea CO_2

fluxes based on atmospheric inversions and seasonal marine net community production based on atmospheric O_2/N_2 ratios. Similar to progress made in ocean-atmosphere modeling, one solution is to emphasize and attempt to reconcile model fluxes that occur between the ocean-atmosphere and land-ocean. Another is to actively pursue adding integrated carbon cycle dynamics into coupled (ocean-atmosphere-land) climate models.

Acknowledgements

We would like to thank our numerous colleagues and collaborators who have contributed to this chapter through formal and informal discussions, comments, and research material. We especially thank P. Falkowski, D. Feely, I. Fung, D. Glover, N. Gruber, J. Kleypas, I. Lima, F. Mackenzie, D. McGillicuddy, J. McWilliams, C. Sabine, and S. Smith. We also thank the editor of this volume, M. Fasham, and an anonymous reviewer for providing guidance and suggestions on the text. This work is supported in part by a NASA US Ocean Carbon Model Intercomparison Project (OCMIP Phase II) grant (NASA W-19,274), a NOAA-OGP Global Carbon Cycle grant (NOAA-NA96GP0360), and the NSF US JGOFS Synthesis and Modeling Project management grant (NSF 97308). The National Center for Atmospheric Research is sponsored by the US National Science Foundation.

References

Abbott MR (1995) Modeling the southern ocean ecosystem. GLOBEC Planning Report, 18, U.S. Global Ocean Ecosystem Dynamics (GLOBEC) Program, Berkeley, California, 63 pp

Archer D (1995) Upper ocean physics as relevant to ecosystem dynamics: a tutorial. Ecol Appl 5:724–739

Archer D, Maier-Raimer E (1994) Effect of deep-sea sedimentary calcite preservation on atmospheric CO_2 concentration. Nature 367:260–263

Archer D, Kheshgi H, Maier-Reimer E (1998) Dynamics of fossil fuel CO_2 neutralization by marine $CaCO_3$. Global Biogeochem Cy 12:259–276

Archer D, Eshel G, Winguth A, Broecker W (2000) Atmospheric CO_2 sensitivity to the biological pump in the ocean. Global Biogeochem Cy 14:1219–1230

Armstrong RA (1994) Grazing limitation and nutrient limitation in marine ecosystems: steady state solutions of an ecosystem model with multiple food chains. Limnol Oceanogr 39:597–608

Armstrong RA (1999a) Stable model structures for representing biogeochemical diversity and size spectra in plankton communities. J Plankton Res 21:445–464

Armstrong RA (1999b) An optimization-based model of iron-light ammonium colimitation of nitrate uptake and phytoplankton growth. Limnol Oceanogr 44:1436–1446

Bacastow R, Maier-Reimer E (1990) Ocean-circulation model of the carbon cycle. Clim Dynam 4:95–125

Behrenfeld MJ, Falkowski PG (1997) Photosynthetic rates derived from satellite-based chlorophyll concentration. Limnol Oceanogr 42:1–20

Bissett WP, Meyers MB, Walsh JJ (1994) The effects of temporal variability of mixed layer depth on primary productivity around Bermuda. J Geophys Res 99:7539–7553

Bissett WP, Walsh JJ, Carder KL (1999) Carbon cycling in the upper waters of the Sargasso Sea: I. Numerical simulation of differential carbon and nitrogen fluxes. Deep-Sea Res Pt I 46:205–269

Bopp L, Monfray P, Aumont O, Dufresne J-L, Le Treut H, Madec G, Terray L, Orr JC (2001) Potential impact of climate change on marine export production. Global Biogeochem Cy 15:81–99

Boyd P, Doney S (2003) The impact of climate change and feedback process on the ocean carbon cycle. Chap. 7. Springer-Verlag, (this volume)

Brewer PG, Goyet C, Dyrssen D (1989) Carbon dioxide transport by ocean currents at 25° N latitude in the Atlantic Ocean. Science 246:477–479

Broecker WS, Peng T-H (1992) Interhemispheric transport of carbon dioxide by ocean circulation. Nature 356:587–589

Broecker W, Lynch-Stieglitz J, Archer D, Hofmann M, Maier-Reimer E, Marchal O, Stocker T, Gruber N (1999) How strong is the Harvardton-Bear constraint? Global Biogeochem Cy 13:817–820

Capone DG, Zehr JP, Paerl HW, Bergman B, Carpenter EJ (1997) Trichodesmium: a globally significant cyanobacterium. Science 276:1221–1229

Carlson CA, Ducklow HW, Michaels AF (1994) Annual flux of dissolved organic carbon from the euphotic zone in the northwest Sargasso Sea. Nature 371:405–408

Case TJ (2000) An illustrated guide to theoretical ecology. Oxford University Press 4, 49 pp

Chai F, Lindley ST, Barber RT (1996) Origin and maintenance of high nutrient condition in the equatorial Pacific. Deep-Sea Res Pt II 42:1031–1064

Christian JR, Verschell MA, Murtugudde R, Busalacchi AJ, McClain CR (2001a) Biogeochemical modelling of the tropical Pacific Ocean. I. Seasonal and interannual variability. Deep-Sea Res Pt II 49:509–543

Christian JR, Verschell MA, Murtugudde R, Busalacchi AJ, McClain CR (2001b) Biogeochemical modelling of the tropical Pacific Ocean. II. Iron biogeochemistry. Deep-Sea Res Pt II 49:545–565

Conkright ME, Levitus S, O'Brien T, Boyer TP, Stephens C, Johnson D, Stathoplos L, Baranova O, Antonov J, Gelfeld R, Burney J, Rochester J, Forgy C (1998) World ocean atlas database 1998; CD-ROM data set documentation. National Oceanographic Data Center, Silver Spring, MD

Danabasoglu G, McWilliams JC, Gent PR (1994) The role of mesoscale tracer transports in the global ocean circulation. Science 264:1123–1126

Denman K, Hofmann E, Marchant H (1996) Marine biotic responses to environmental change and feedbacks to climate. In: Houghton JT, Meira LG Filho, Callander BA, Harris N, Kattenberg A, Maskell K (eds) Climate change 1995. IPCC, Cambridge University Press, pp 487–516

Denman KL, Pena MA (1999) A coupled 1-D biological/physical model of the northeast subarctic Pacific Ocean with iron limitation. Deep-Sea Res Pt II 46:2877–2908

Dickey T, Frye D, Jannasch H, Boyle E, Manov D, Sigurdson D, McNeil J, Stramska M, Michaels A, Nelson N, Siegel D, Chang G, Wu J, Knap A (1998) Initial results from the Bermuda testbed mooring program. Deep-Sea Res Pt I 45:771–794

Doney SC (1996) A synoptic atmospheric surface forcing data set and physical upper ocean model for the U.S. JGOFS Bermuda Atlantic Time-Series Study (BATS) site. J Geophys Res 101: 25615–25634

Doney SC (1999) Major challenges confronting marine biogeochemical modeling. Global Biogeochem Cy 13:705–714

Doney SC, Hecht MW (2002) Antarctic bottom water formation and deep water chlorofluorocarbon distributions in a global ocean climate model. J Phys Oceanogr 32:1642–1666

Doney SC, Sarmiento JL (eds) (1999) Synthesis and modeling project; ocean biogeochemical response to climate change. U.S. JGOFS Planning Report 22, U.S. JGOFS Planning Office, Woods Hole, MA, 105 pp

Doney SC, Glover DM, Najjar RG (1996) A new coupled, one-dimensional biological-physical model for the upper ocean: applications to the JGOFS Bermuda Atlantic Time Series (BATS) site. Deep-Sea Res Pt II 43:591–624

Doney SC, Large WG, Bryan FO (1998) Surface ocean fluxes and water-mass transformation rates in the coupled NCAR Climate System Model. J Climate 11:1422–1443

Doval M, Hansell DA (2000) Organic carbon and apparent oxygen utilization in the western South Pacific and central Indian Oceans. Mar Chem 68:249–264
Dutay J-C, Bullister JL, Doney SC, Orr JC, Najjar R, Caldeira K, Champin J-M, Drange H, Follows M, Gao Y, Gruber N, Hecht MW, Ishida A, Joos F, Lindsay K, Madec G, Maier-Reimer E, Marshall JC, Matear RJ, Monfray P, Plattner G-K, Sarmiento J, Schlitzer R, Slater R, Totterdell IJ, Weirig M-F, Yamanaka Y, Yool A (2001) Evaluation of ocean model ventilation with CFC-11: comparison of 13 global ocean models. Ocean Modelling 4: 89–120
Dutkiewicz S, Follows M, Marshall J, Gregg WW (2001) Interannual variability of phytoplankton abundances in the North Atlantic. Deep-Sea Res Pt II 48:2323–2344
England MH (1995) Using chlorofluorocarbons to assess ocean climate models. Geophys Res Lett 22:3051–3054
England MH, Maier-Reimer E (2001) Using chemical tracers to assess ocean models. Rev Geophys 39:29–70
Evans GT, Fasham MJR (ed) (1993) Towards a model of ocean biogeochemical processes. Springer-Verlag, New York
Evans GT, Garçon VC (ed) (1997) One-dimensional models of water column biogeochemistry. JGOFS Report 23/97, 85 pp., JGOFS, Bergen, Norway
Evans GT, Parslow JS (1985) A model of annual plankton cycles. Biolo Oceanogr 3:327–347
Falkowski PG, Biscaye PE, Sancetta C (1994) The lateral flux of biogenic particles from the eastern North American continental margin to the North Atlantic Ocean. Deep-Sea Res Pt II 41: 583–601
Fasham MJR (1993) Modelling the marine biota. In: Heimann M (ed) The global carbon cycle. Springer-Verlag, Heidelberg, pp 457–504
Fasham MJR (1995) Variations in the seasonal cycle of biological production in subarctic oceans: a model sensitivity analysis. Deep-Sea Res Pt I 42:1111–1149
Fasham MJR, Evans GT (1995) The use of optimisation techniques to model marine ecosystem dynamics at the JGOFS station at 47° N and 20° W. Philos T Roy Soc B 348:206–209
Fasham MJR, Ducklow HW, McKelvie SM (1990) A nitrogen-based model of plankton dynamics in the oceanic mixed layer. J Mar Res 48:591–639
Fennel K, Losch M, Schröter J, Wenzel M (2001) Testing a marine ecosystem model: sensitivity analysis and parameter optimization. J Marine Syst 28:45–63
Fennel K, Spitz YH, Letelier RM, Abbott MR, Karl DM (2002) A deterministic model for N_2-fixation at the HOT site in the subtropical North Pacific. Deep-Sea Res II 49:149–174
Francis RC, Hare SR (1994) Decadal-scale regime shifts in the large marine ecosystems of the North-east Pacific: a case for historical science. Fish Oceanogr 3:279–291
Frost BW (1987) Grazing control of phytoplankton stock in the subarctic Pacific: a model assessing the role of mesozooplankton, particularly the large calanoid copepods, *Neocalanus* spp. Mar Ecol Prog Ser 39:49–68
Fung IY, Meyn SK, Tegen I, Doney SC, John JG, Bishop JKB (2000) Iron supply and demand in the upper ocean. Global Biogeochem Cy 14:281–295
Garçon VC, Oschlies A, Doney SC, McGillicuddy D, Waniek J (2001) The role of mesoscale variability on plankton dynamics. Deep-Sea Res Pt II 48:2199–2226
Geider RJ, MacIntyre HL, Kana TM (1996) A dynamic model of photoadaptation in phytoplankton. Limnol Oceanogr 41:1–15
Geider RJ, MacIntyre HL, Kana TM (1998) A dynamic regulatory model of phytoplankton acclimation to light, nutrients, and temperature. Limnol Oceanogr 43:679–694
Gent PR, McWilliams JC (1990) Isopycnal mixing in ocean circulation models. J Phys Oceanogr 20:150–155
Gent PR, Bryan FO, Danabasoglu G, Doney SC, Holland WR, Large WG, McWilliams JC (1998) The NCAR Climate System Model global ocean component. J Climate 11:1287–1306
Giering R, Kaminski T (1998) Recipes for adjoint code construction. Acm t math software 24:437–474
Gnanadesikan A (1999) A global model of silicon cycling: sensitivity to eddy parameterization and dissolution. Global Biogeochem Cy 13:199–220
Gnanadesikan A, Toggweiler JR (1999) Constraints placed by silicon cycling on vertical exchange in general circulation models. Geophys Res Lett 26:1865–1868
Gnanadesikan A, Slater R, Gruber N, Sarmiento JL (2001) Oceanic vertical exchange and new production: a comparison between models and observations. Deep-Sea Res Pt II 49:363–401
Gregg WW (2002) Tracking the SeaWiFS record with a coupled physical/biogeochemical/radiative model of the global oceans. Deep-Sea Res Pt II 49:81–105
Griffes SM, Böning C, Bryan FO, Chassignet EP, Gerdes R, Hasumi H, Hirst A, Treguer A-M, Webb D (2000) Developments in ocean climate modelling, vol. 2. pp 123–192
Gruber N (1998) Anthropogenic CO_2 in the Atlantic Ocean. Global Biogeochem Cy 12:165–191
Gruber N, Sarmiento JL (1997) Global patterns of marine nitrogen fixation and denitrification. Global Biogeochem Cy 11: 235–266
Gruber N, Sarmiento JL, Stocker TF (1996) An improved method for detecting anthropogenic CO_2 in the oceans. Global Biogeochem Cy 10:809–837
Gruber N, Gloor M, Fan SM, Sarmiento JL (2001) Air-sea flux of oxygen estimated from bulk data: implications for the marine and atmospheric oxygen cycle. Global Biogeochem Cy 15(4): 783–803
Haidvogel DB, Beckmann A (1999) Numerical ocean circulation modeling. Imperial College Press, London, 318 pp
Haidvogel DB, Curchitser E, Iskandarani M, Hughes R, Taylor M (1997) Global modeling of the ocean and atmosphere using the spectral element method. Atmos Ocean 35:505–531
Hansell DA, Carlson CA (1998) Net community production of dissolved organic carbon. Global Biogeochem Cy 12:443–453
Harris RP (1994) Zooplankton grazing on the coccolithophore *Emiliania huxleyi* and its role in inorganic carbon flux. Mar Biol 119:431–439
Harrison DE (1996) Vertical velocity variability in the tropical Pacific: a circulation model perspective for JGOFS. Deep-Sea Res Pt II 43:687–705
Hecht MW, Wingate BA, Kassis P (2000) A better, more discriminating test problem for ocean tracer transport. Ocean Modelling 2:1–15
Heinze C, Maier-Reimer E, Schlosser P (1998) Transient tracers in a global OGCM: source functions and simulated distributions. J Geophys Res 103:15903–15922
Heinze C, Maier-Reimer E, Winguth AME, Archer D (1999) A global oceanic sediment model for long-term climate studies. Global Biogeochem Cy 13:221–250
Holfort J, Johnson KM, Wallace DWR (1998) Meridional transport of dissolved inorganic carbon in the South Atlantic Ocean. Global Biogeochem Cy 12:479–499
Holligan PM, Fernandez E, Aiken J, Balch WM, Boyd P, Burkill PH, Finch M, Groom SB, Malin O, Muller K, Purdie DA, Robinson C, Trees CC, Turner SM, van del Wal P (1993) A biogeochemical study of the coccolithophore *Emiliania huxleyi* in the North Atlantic. Global Biogeochem Cy 7:879–900
Hood RR, Bates NR, Olson DB (2001) Modeling the seasonal to interannual biogeochemical and N_2 fixation cycles at BATS. Deep-Sea Res Pt II 48:1609–1648
Hurtt GC, Armstrong RA (1996) A pelagic ecosystem model calibrated with BATS data. Deep-Sea Res Pt II 43:653–683
Ishizaka J (1990) Coupling of Coastal Zone Color Scanner data to a physical-biological model of the Southeastern U.S. continental shelf ecosystem, 3, nutrient and phytoplankton fluxes and CZCS data assimilation. J Geophys Res 95:20201–20212
Joos F, Siegenthaler U, Sarmiento JL (1991) Possible effects of iron fertilization in the Southern Ocean on atmospheric CO_2 concentration. Global Biogeochem Cy 5:135–150
Joos F, Plattner O-K, Schmittner A (1999) Global warming and marine carbon cycle feedbacks on future atmospheric CO_2. Science 284:464–467
Karl DM (1999) A sea of change: biogeochemical variability in the North Pacific subtropical gyre. Ecosystems 2:181–214
Karl DM, Letelier R, Hebel D, Tupas L, Dore J, Christian J, Winn C (1995) Ecosystem changes in the North Pacific subtropical gyre attributed to the 1991–1992 El Niño. Nature 378:230–234

Karl D, Letelier R, Tupas L, Dore J, Christian J, Hebel D (1997) The role of nitrogen fixation in biogeochemical cycling in the subtropical North Pacific Ocean. Nature 388:533–538

Kasibhatla P, Heimann M, Rayner P, Mahowald N, Prinn RG, Hartley DE (ed) (2000) Inverse methods in global biogeochemical cycles. AGU Geophysical. Monograph Series, American Geophysical Union, Washington D.C., 324 pp

Keeling RF, Piper SC, Heimann M (1996) Global and hemispheric CO_2 sinks deduced from changes in atmospheric O_2 concentration. Nature 381:218–221

Kleypas JA, Doney SC (2001) Nutrients, chlorophyll, primary production and related biogeochemical properties in the ocean mixed layer – a compilation of data collected at nine JGOFS sites. NCAR Technical Report, NCAR/TN-447+STR, 53 pp

Kleypas JA, Buddemeier RW, Archer D, Gattuso J-P, Langdon C, Opdyke BN (1999) Geochemical consequences of increased atmospheric carbon dioxide on coral reefs. Science 284:118–120

Large WG, McWilliams JC, Doney SC (1994) Oceanic vertical mixing: a review and a model with a nonlocal boundary layer parameterization. Rev Geophys 32:363–403

Large WG, Danabasoglu G, Doney SC, McWilliams JC (1997) Sensitivity to surface forcing and boundary layer mixing in a global ocean model: annual-mean climatology. J Phys Oceanogr 27:2418–2447

Laws EA, Falkowski PG, Smith WO Jr., Ducklow H, McCarthy JJ (2000) Temperature effects on export production in the open ocean. Global Biogeochem Cy 14:1231–1246

Le Quéré C, Orr JC, Monfray P, Aumont O, Madec G (2000) Interannual variability of the oceanic sink of CO_2 from 1979 though 1997. Global Biogeochem Cy 14:1247–1265

Leonard CL, McClain CR, Murtugudee R, Hofmann EE, Harding JLW (1999) An iron-based ecosystem model of the central equatorial Pacific. J Geophys Res 104:1325–1341

Letelier RM, Abbott MR (1996) An analysis of chlorophyll fluorescence for the Moderate Resolution Imaging Spectrometer (MODIS). Remote Sens Environ 58:215–223

Letelier R, Karl D (1996) Role of *Trichodesmium* spp. in the productivity of the subtropical North Pacific Ocean. Mar Ecol Prog Ser 133:263–273

Letelier R, Karl D (1998) *Trichodesmium* spp. physiology and nutrient fluxes in the North Pacific subtropical gyre. Aquat Microb Ecol 15:265–276

Levitus S, Conkright ME, Reid JL, Najjar RG, Mantilla A (1993) Distribution of nitrate, phosphate and silicate in the world oceans. Prog Oceanogr 31:245–273

Levitus S, Burgett R, Boyer T (1994) World atlas 1994. NOAA Atlas NESDIS, U.S. Dept. of Commerce, Washington D.C.

Levy M, Memery L, Madec G (1999) Combined effects of mesoscale processes and atmospheric high-frequency variability on the spring bloom in the MEDOC area. Deep-Sea Res Pt I 47:27–53

Lima I, Doney S, Bryan F, McGillicuddy D, Anderson L, Maltrud M (1999) Preliminary results from an eddy-resolving ecosystem model for the North Atlantic. EOS, Transactions AGU, 80(49), Ocean Sciences Meeting Supplement, OS28

Lima ID, Olson DB, Doney SC (2002) Biological response to frontal dynamics and mesoscale variability in oligotrophic environments: a numerical modeling study. J Geophys Res (in press)

Lipschultz F, Owens NJP (1996) An assessment of nitrogen fixation as a source of nitrogen to the North Atlantic. Biogeochemistry 35:261–274

Liu K-K, Atkinson L, Chen CTA, Gao S, Hall J, Macdonald RW, Talaue McManus L, Quiñones R (2000) Exploring continental margin carbon fluxes on a global scale. EOS, Transactions of the American Geophysical Union 81:641–644

Louanchi F, Najjar RG (2000) A global monthly mean climatology of phosphate, nitrate and silicate in the upper ocean: spring-summer production and shallow remineralization. Global Biogeochem Cy 14:957–977

Mahadevan A, Archer D (2000) Modeling the impact of fronts and mesoscale circulation on the nutrient supply and biogeochemistry of the upper ocean. J Geophys Res 105:1209–1225

Maier-Reimer E (1993) Geochemical cycles in an ocean general circulation model. Preindustrial tracer distributions. Global Biogeochem Cy 7:645–677

Maier-Reimer E, Hasselmann K (1987) Transport and storage in the ocean – an inorganic ocean-circulation carbon cycle model. Clim Dynam 2:63–90

Martin JH, Knauer GA, Karl DM, Broenkow WW (1987) VERTEX: carbon cycling in the northeast Pacific. Deep-Sea Res 34:267–285

Matear RJ (1995) Parameter optimization and analysis of ecosystem models using simulated annealing: a case study at Station P. J Mar Res 53:571–607

Matear RJ, Hirst AC (1999) Climate change feedback on the future oceanic CO_2 uptake. Tellus B 51:722–733

Matear RJ, Holloway G (1995) Modeling the inorganic phosphorus cycle of the North Pacific using an adjoint data assimilation model to assess the role of dissolved organic phosphorus. Global Biogeochem Cy 9:101–119

May RM (1973) The stability and complexity of model ecosystems. Princeton University Press, Princeton New Jersey, 265 pp

McClain CR, Arrigo K, Turk D (1996) Observations and simulations of physical and biological processes at ocean weather station P, 1951–1980. J Geophys Res 101:3697–3713

McClain CR, Cleave ML, Feldman GC, Gregg WW, Hooker SB, Kuring N (1998) Science quality SeaWiFS data for global biosphere research. Sea Technol 39:10–14

McCreary JP, Kohler KH, Hood RR, Olson DB (1996) A four compartment ecosystem model of biological activity in the Arabian Sea. Prog Oceanogr 37:193–240

McGillicuddy DJ Jr., Robinson AR (1997) Eddy-induced nutrient supply and new production. Deep-Sea Res Pt I 44:1427–1450

McGillicuddy DJ Jr., Robinson AR, Siegel DA, Jannasch HW, Johnson R, Dickey TD, McNeil J, Michaels AF, Knap AH (1998) Influence of mesoscale eddies on new production in the Sargasso Sea. Nature 394:263–266

McGowan JA, Cayan DR, Dorman LM (1998) Climate-ocean variability and ecosystem response in the Northeast Pacific. Science 281:210–217

McWilliams JC (1996) Modeling the oceanic general circulation. Annu Rev Fluid Mech 28:215–248

Milliman JD (1993) Production and accumulation of calcium carbonate in the ocean: budget of a nonsteady state. Global Biogeochem Cy 7:927–957

Milliman JD, Troy PJ, Balch WM, Adams AK, Li YH, Mackenzie FT (1999) Biologically mediated dissolution of calcium carbonate above the chemical lysocline? Deep-Sea Res Pt I 46:1653–1669

Moloney CL, Field JG (1991) The size-based dynamics of plankton food webs. I. A simulation model of carbon and nitrogen flows. J Plankton Res 13:1003–1038

Moore JK, Doney SC, Kleypas JA, Glover DM, Fung IY (2001a) An intermediate complexity marine ecosystem model for the global domain. Deep-Sea Res Pt II 49:403–462

Moore JK, Doney SC, Kleypas JA, Glover DM, Fung IY (2001b) Iron cycling and nutrient limitation patterns in surfce waters of the world ocean. Deep-Sea Res Pt II 49:463–507

Murnane RJ, Sarmiento JL, Le Quéré C (1999) Spatial distribution of air-sea CO_2 fluxes and the interhemispheric transport of carbon by the oceans. Global Biogeochem Cy 13:287–305

Najjar RG, Sarmiento JL, Toggweiler JR (1992) Downward transport and fate of organic matter in the ocean: simulations with a general circulation model. Global Biogeochem Cy 6:45–76

Oeschger H, Siegenthaler U, Guglemann A (1975) A box-diffusion model to study the carbon dioxide exchange in nature. Tellus 27:168–192

Orr JC, Maier-Reimer E, Mikolajewicz U, Monfray P, Sarmiento JL, Toggweiler JR, Taylor NK, Palmer J, Gruber N, Sabine CL, Le Quéré C, Key RM, Boutin J (2001) Estimates of anthropogenic carbon uptake from four three-dimensional global ocean models. Global Biogeochem Cy 15:43–60

Oschlies A (2000) Equatorial nutrient trapping in biogeochemical ocean models: the role of advection numerics. Global Biogeochem Cy 14:655–667

Oschlies A, Garçon V (1998) Eddy-induced enhancement of primary production in a model of the North Atlantic Ocean. Nature 394:266–269

Oschlies A, Garçon V (1999) An eddy-permitting coupled physical-biological model of the North Atlantic-1. Sensitivity to advection numerics and mixed layer physics. Global Biogeochem Cy 13:135–160

Polovina JJ, Mitchum GT, Evans GT (1995) Decadal and basin-scale variation in mixed layer depth and the impact on biological production in the Central and North Pacific, 1960–88. Deep-Sea Res Pt I 42:1701–1716

Pondaven P, Ruiz-Pino D, Jeandel C (2000) Interannual variability of Si and N cycles at the time-series station KERFIX between 1990 and 1995 – a 1-D modelling study. Deep-Sea Res Pt I 47:223–257

Rayner PJ, Enting IG, Francey RJ, Langenfelds R (1999) Reconstructing the recent carbon cycle from atmospheric CO_2, *d*13C and O2/N2 observations. Tellus B 51:213–232

Reid PC, Edwards M, Hunt HG, Warner AJ (1998) Phytoplankton change in the North Atlantic. Nature 391:546

Riley GA (1946) Factors controlling phytoplankton populations on Georges Bank. J Mar Res 6:54–73

Rintoul SR, Wunsch C (1991) Mass, heat, oxygen and nutrient fluxes and budgets in the North Atlantic Ocean. Deep-Sea Res 38 (suppl.) S355–S377

Roberts M, Marshall D (1998) Do we require adiabatic dissipation schemes in eddy-resolving ocean models? J Phys Oceanogr 28:2050–2063

Robinson AR (1996) Physical processes, field estimation and an approach to interdisciplinary ocean modeling. Earth-Sci Rev 40:3–54

Robinson AR, McCarthy JJ, Rothschild BJ (2001) The sea: biological-physical interactions in the ocean. John Wiley & Sons, New York

Ryabchenko VA, Gorchakov VA, Fasham MJR (1998) Seasonal dynamics and biological productivity in the Arabian Sea euphotic zone as simulated by a three-dimensinoal ecosystem model. Global Biogeochem Cy 12:501–530

Sabine CL, Key RM, Goyet C, Johnson KM, Millero FJ, Poisson A, Sarmiento JL, Wallace DWR, Winn CD (1999) Anthropogenic CO_2 inventory in the Indian Ocean. Global Biogeochem Cy 13:179–198

Sarmiento JL, Sundquist ET (1992) Revised budget for the oceanic uptake of anthropogenic carbon dioxide. Nature 356:589–593

Sarmiento JL, Wofsy SC (1999) A U.S. Carbon Cycle Science Plan. (U.S. CCSP 1999), U.S. Global Change Research Program, Washington DC, 69 pp

Sarmiento JL, Orr JC, Siegenthaler U (1992) A perturbation simulation of CO_2 uptake in an ocean general circulation model. J Geophys Res 97:3621–3646

Sarmiento JL, Slater RD, Fasham MJR (1993) A seasonal three-dimensional ecosystem model of nitrogen cycling in the North Atlantic euphotic zone. Global Biogeochem Cy 7:417–450

Sarmiento JL, Hughes TMC, Stouffer RJ, Manabe S (1998) Simulated response of the ocean carbon cycle to anthropogenic climate warming. Nature 393:245–249

Sarmiento JL, Monfray P, Maier-Reimer E, Aumont O, Murnane RJ, Orr JC (2000) Sea-air CO_2 fluxes and carbon transport: a comparison of three ocean general circulation models. Global Biogeochem Cy 14:1267–1281

Schimel D, Enting IG, Heimann M, Wigley TML, Raynaud D, Alves D, Siegenthaler U (1995) CO_2 and the carbon cycle. In: Houghton JT, Meira Filho LG, Bruce J, Lee H, Callander BA, Haites E, Harris N, Maskell K (eds) Climate change 1994. Intergovernmental Panel on Climate Change. Cambridge University Press, pp 39–71

Schlitzer R (2000) Applying the adjoint method for global biogeochemical modeling. In: Kasibhatla P, et al. (eds) Inverse methods in global biogeochemical cycles. AGU Geophysical Monograph Series, American Geophysical Union, Washington D.C., pp 107–124

Siegenthaler U, Joos F (1992) Use of a simple model for studying oceanic tracer distributions and the global carbon cycle. Tellus B 44:186–207

Siegenthaler U, Oeschger H (1978) Predicting future atmospheric carbon dioxide levels. Science 199:388–395

Siegenthaler U, Sarmiento JL (1993) Atmospheric carbon dioxide and the ocean. Nature 365:119–125

Six KD, Maier-Reimer E (1996) Effects of plankton dynamics on seasonal carbon fluxes in an ocean general circulation model. Global Biogeochem Cy 10:559–583

Smith RD, Maltrud ME, Bryan FO, Hecht MW (2000) Numerical simulation of the North Atlantic at 1/10°. J Phys Oceanogr 30:1532–1561

Smith SV, Hollibaugh JT (1993) Coastal metabolism and the oceanic organic carbon balance. Rev Geophys 31:75–89

Spall MA, Holland WR (1991) A nested primitive equation model for oceanic applications. J Phys Oceanogr 21:205–220

Spall SA, Richards KJ (2000) A numerical model of mesoscale frontal instabilities and plankton dynamics. I. Model formulation and initial experiments. Deep-Sea Res Pt I 47:1261–1301

Spitz YH, Moisan JR, Abbott MR, Richman JG (1998) Data assimilation and a pelagic ecosystem model: parameterization using time series observations. J Marine Syst 16:51–68

Steele JH (1958) Plant production in the northern North Sea. Mar Res 7:1–36

Steele JH (1974) The structure of marine ecosystems. Harvard University Press, Cambridge, MA, 128 pp

Stephens BB, Keeling RF (2000) The influence of Antarctic sea ice on glacial/interglacial CO_2 variations. Nature 404:171–174

Stocker TF, Broecker WS, Wright DG (1994) Carbon uptake experiments with a zonally-averaged global ocean circulation model. Tellus B 46:103–122

Sunda WG, Huntsman SA (1995) Iron uptake and growth limitation in oceanic and coastal phytoplankton. Mar Chem 50:189–206

Takahashi T, Feely RA, Weiss RF, Wanninkhof RH, Chipman DW, Sutherland SC, Takahashi TT (1997) Global air-sea flux of CO_2, An estimate based on measurements of sea-air pCO_2 difference. P Natl Acad Sci USA 94:8929–8299

Takahashi T, Wanninkhof RH, Feely RA, Weiss RF, Chipman DW, Bates N, Olafson J, Sabine C, Sutherland SC (1999) Net air-sea CO_2 flux over the global oceans: an improved estimate based on the sea-air pCO_2 difference. In: Center for Global Environmental Research, National Institute for Environmental Studies (ed) Proceedings of the 2nd International Symposium on CO_2 in the Oceans. Tsukuba, Japan, pp 9–15

Tegen I, Fung I (1995) Contribution to the atmospheric mineral aerosol load from land surface modification. J Geophys Res 100:18,707–18,726

Toggweiler JR (1999) Variation of atmospheric CO_2 by ventilation of the ocean's deepest water. Paleoceanography 14:571–588

Toggweiler JR, Dixon K, Bryan K (1989a) Simulations of radiocarbon in a coarse-resolution world ocean model; 1. Steady state pre-bomb distribution. J Geophys Res 94:8217–8242

Toggweiler JR, Dixon K, Bryan K (1989b) Simulations of radiocarbon in a coarse-resolution world ocean model; 2. Distributions of bomb-produced carbon-14. J Geophys Res 94:8243–8264

US Joint Global Ocean Flux Study (US JGOFS) (1992) Report of the U.S. JGOFS Workshop on Modeling and Data Assimilation. Planning Report Number 14, U.S. JGOFS Planning Office, Woods Hole, MA, 28 pp

Venrick EL, McGowan JA, Cayan DR, Hayward TL (1987) Climate and chlorophyll *a*: longterm trends in the central North Pacific Ocean. Science 238:70–72

Wallace DWR (1995) Monitoring global ocean carbon inventories. Ocean Observing System Development Panel Background, Texas A&M University, College Station, TX, 54 pp

Wallace DWR (2001) Storage and transport of excess CO_2 in the oceans: the JGOFS/WOCE Global CO_2 survey. In: Siedler G, Gould J, Church J (eds) Ocean circulation and climate: observing and modeling the global ocean. Academic Press, New York

Walsh JJ (1991) Importance of continental margins in the marine biogeochemical cycling of carbon and nitrogen. Nature 350:53–55

Wanninkhof R (1992) Relationship between wind speed and gas exchange over the ocean. J Geophys Res 97:7373–7382

Wanninkhof R, Doney SC, Peng T-H, Bullister J, Lee K, Feely RA (1999) Comparison of methods to determine the anthropogenic CO_2 invasion into the Atlantic Ocean. Tellus B, 51:511–530

Watson AJ, Orr JC (2003) Carbon dioxide fluxes in the global ocean. Springer-Verlag, (this volume)

Webb DJ, deCuevas BA, Richmond CS (1998) Improved advection schemes for ocean models. J Atmos Ocean Tech 15:1171–1187

Yamanaka Y, Tajika E (1996) The role of the vertical fluxes of particulate organic matter and calcite in the oceanic carbon cycle: studies using an ocean biogeochemical general circulation model. Global Biogeochem Cy 10:361–382

Chapter 10

Temporal Studies of Biogeochemical Processes Determined from Ocean Time-Series Observations During the JGOFS Era

David M. Karl · Nicholas R. Bates · Steven Emerson · Paul J. Harrison · Catherine Jeandel · Octavio Llinás
Kon-Kee Liu · Jean-Claude Marty · Anthony F. Michaels · Jean C. Miquel · Susanne Neuer · Y. Nojiri · Chi Shing Wong

10.1 Introduction

A comprehensive understanding of the global carbon cycle is required to address contemporary scientific issues related to the atmospheric accumulation of greenhouse gases and their cumulative effects on global environmental change. Consequently, detailed in situ investigations of terrestrial and marine ecosystems are necessary prerequisites for developing a predictive capability of future environmental variability and the effects of human-induced perturbations. These investigations need to address broad questions regarding the distribution, abundance, diversity and control of key plant, animal and microbe populations and their interactions with their habitats. They must be conducted with an explicit recognition of the interdisciplinary connections between physics, chemistry, biology and geology in each ecosystem. Ideally, these field studies should be conducted at strategic sites that are representative of large biomes or in regions that are likely to exhibit substantial interannual variability over large areas. However, it is more important that the unique features of each site elucidate representative processes that underpin the dynamics of the wider ocean. Furthermore, these field investigations should be conducted for at least several decades, in order to distinguish natural variability from that induced by human activities.

In response to a growing awareness of the ocean's role in climate and global change research and the need for comprehensive oceanic time-series measurements, the International Geosphere-Biosphere Programme: A Study of Global Change (IGBP) was established in 1986. One of the essential core components of IGBP, the Joint Global Ocean Flux Study (JGOFS) project, was established in 1987 to improve our understanding of the oceanic carbon cycle. More formally stated, "JGOFS seeks to determine and understand, on a global scale, the processes controlling the time-varying fluxes of carbon and associated biogenic elements in the ocean, to evaluate the related exchanges with the atmosphere, sea floor and continental boundaries, and to develop a capability to predict the response of oceanic biogeochemical processes to anthropogenic perturbations, in particular those related to climate change." To achieve these goals, four separate program elements were defined: (*1*) process studies designed to capture key, regular events; (*2*) time-series observations at strategic sites; (*3*) a global inventory of carbon dioxide (CO_2) concentrations in the surface ocean; and (*4*) vigorous data synthesis, data assimilation and modeling efforts.

Long-term biogeochemical studies in oceanography are predicated on the relatively straightforward notion that certain processes, such as climate-driven changes in community structure and productivity, and natural or anthropogenic changes in nutrient loading and habitat changes, are long-term processes and must be studied as such. If the phenomena of greatest interest are episodic, rare, complex or characterized by thresholds or feedback control mechanisms, then long-term observations are mandatory (Risser 1991). Despite this need and strong justification, there are relatively few long-term biogeochemical studies of the world's ocean. This paucity of robust observations, especially in the remote regions of the open ocean, has led to an incomplete mechanistic understanding of the global carbon cycle, especially for key issues related to carbon sequestration in the ocean's interior. During the past two decades several ocean time-series programs have emerged and the data sets provided by these studies have collectively contributed to our increasing understanding of biogeochemical processes in the sea.

This chapter will focus on the ocean time-series components of JGOFS, especially the relationships between nutrient dynamics and productivity, and the relationships between productivity and carbon export. We will focus on the time-series program similarities, while acknowledging the important differences. Numerous papers have already been written on these topics, including at least four special issues of Deep-Sea Research, Topical Studies in Oceanography (Karl et al. 1996; Boyd and Harrison 1999a; Siegel et al. 2001; Marty et al. 2002). Our goal here is to report significant cross-ecosystem comparisons of the important processes of carbon production, export and remineralization and to investigate and understand the underlying mechanisms controlling carbon sequestration in the sea. This will be achieved by the presentation of sample data sets, and several biogeochemical 'case studies'. The structure, efficiency and controls of the ocean's various biological pumps will be a common theme in this comparative study.

10.2 The Oceanic Carbon Cycle and the Biological Carbon Pump

The large and dynamic oceanic reservoir of carbon, approximately 4×10^{19} g, is distributed unequally among dissolved and particulate constituents with various redox states and plays an important role in global biogeochemical cycles. These pools include dissolved inorganic carbon (DIC = $[H_2CO_3] + [CO_2] + [HCO_3^-] + [CO_3^{2-}]$), and the less oxidized pools of mostly uncharacterized dissolved organic carbon (DOC) and partially characterized particulate organic carbon (POC). The latter pool includes both living organisms and non-living particulate organic detritus; particulate inorganic carbon (PIC), mostly calcium carbonate ($CaCO_3$), is also present as a component of both living and non-living particulate matter. A chemical disequilibrium between oxidized DIC and reduced organic matter is produced and maintained by biological processes. The reversible, usually biologically-mediated, inter-conversions between dissolved and particulate carbon pools in the sea collectively define the oceanic carbon cycle.

The global distributions and movements of carbon in the sea are governed by two major processes with independent controls (Fig. 10.1). The solubility carbon pump, which transports DIC, is mediated by physical processes such as vertical mixing, advection and diffusion which are only partially understood. In particular, DIC pool dynamics in relation to mesoscale and sub-

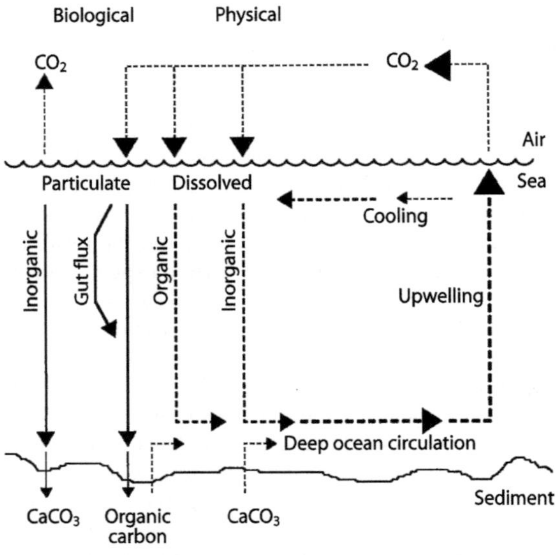

Fig. 10.1. A schematic view of the ocean's carbon pump showing the interactions between the surface ocean and the atmosphere above, and the surface ocean and the deep sea below. There are both physical and biological components of the ocean's carbon pump; the latter involve a complex set of poorly constrained ecological processes, some of which are discussed in more detail in the chapter (redrawn from SCOR (1990))

mesoscale features and the importance of horizontal processes are not well resolved. Likewise, the air-to-sea exchange of CO_2, especially the importance of short-term variability in wind speed, is not well constrained. Similarly, the key processes of the biological carbon pump are only partially understood.

The observed vertical distribution of DIC in the sea cannot be accurately reproduced by existing circulation models without the inclusion of biological processes. Furthermore, it is well known that the rate and efficiency of the oceanic uptake of atmospheric CO_2 can be dramatically affected by the structure and dynamics of the phytoplankton community. Because most, if not all, ecological interactions are non-linear, they are not easily studied or modeled. We consider it axiomatic that an explicit understanding of the biological carbon pump is required to understand and to model carbon dynamics in the sea. To adequately resolve the biological pump, we must know the individual mechanisms, the rates and controls, the regional and temporal variations and the sensitivities of the various biological pump components to habitat perturbations, including global environmental change. These are, we believe, the most important and as yet unresolved components of the ocean's carbon cycle.

Primary conversion of oxidized DIC to reduced organic matter (both DOC and POC) and $CaCO_3$ is generally restricted to the euphotic zone of the world ocean via the process of photosynthesis. The annual rate of global ocean primary production, about 90–92 Pg C yr^{-1} (1 Pg = 10^{15} g), is nearly 50 times the amount of carbon entering the ocean from CO_2 build-up in the atmosphere which (estimated to be ~2 Pg C yr^{-1}). The supply of reduced carbon and energy required to support subeuphotic zone metabolic processes is ultimately derived from the upper ocean and is transported downward by advection and diffusion of dissolved organic matter (Hansell et al. 1997), gravitational settling of particulate inorganic and organic matter (McCave 1975), and by the vertical migrations of pelagic animals (Longhurst and Harrison 1989) and phytoplankton (Villareal et al. 1993). These diverse processes collectively define the biological carbon pump.

In theory, the export flux of carbon from the euphotic zone includes both dissolved and particulate matter. However, as Margalef (1978) has so eloquently stated, "any atom is more likely to travel downwards when in a particle than in solution." This predicts a more significant role for processes that favor export of particulate matter. The ability of larger phytoplankton to aggregate (the diatom branch; Fig. 10.2) and sink out of the euphotic zone is, perhaps, the most efficient means of carbon export. As we shall see later, there are variations on this theme that can further enhance net carbon sequestration. Alternatively, when the phytoplankton assemblage is dominated by small picoplankton (the *Prochlorococcus* branch; Fig. 10.2) export is reduced. Most models of the

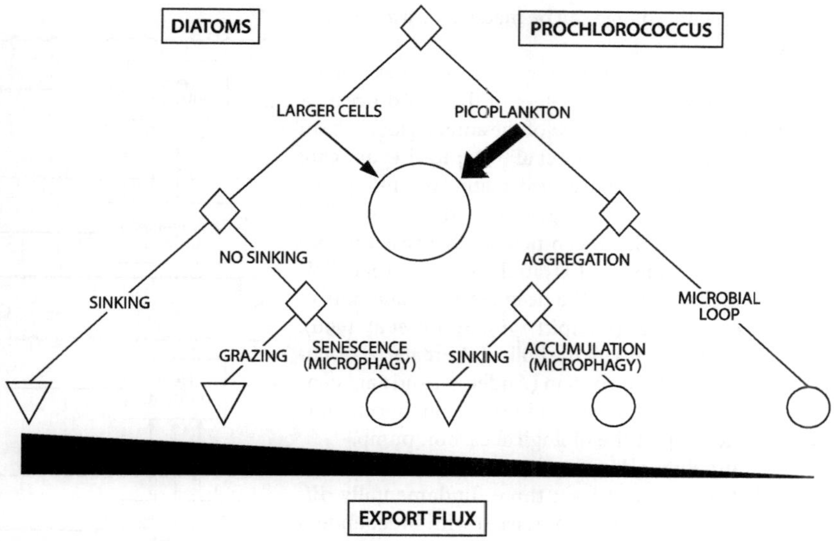

Fig. 10.2. Flows of chemical energy and materials through a hypothetical pelagic food web showing two major branches: diatom-dominated and *Prochlorococcus*-dominated biomes. Branch points, indicated by *diamonds*, were called 'hydrodynamic singularities' by the original authors (Legendre and Le Fèvre 1989). Branches to the *left* indicate a higher probability for export and branches to the *right* favor local regeneration and, hence, low export. The two major fates of carbon are export (*triangles*) or accumulation (*circles*) (redrawn from Cullen et al. (2002))

biological carbon pump assume that short food chains lead to high new production and high export, whereas complex microbial food webs lead to high recycling and low export.

Not all plankton carbon is created equal. The two major classes are: (1) organic carbon ($C_{org.}$), which itself is quite diverse, ranging from small lipid molecules containing only carbon and hydrogen to nucleic acids containing carbon, nitrogen, hydrogen, phosphorus and oxygen in varying proportions, and (2) inorganic carbon ($C_{inorg.}$), which is present mainly as $CaCO_3$ primarily in the skeletons of coccolithophorids, foraminifera and pteropods. The production of organic matter during photosynthesis removes CO_2 in an equimolar carbon stoichiometry ($CO_2 + H_2O \longleftrightarrow CH_2O + O_2$), whereas the production of biogenic $CaCO_3$ in the pelagic realm releases CO_2 and reduces alkalinity ($Ca^{2+} + 2HCO_3^- \longrightarrow CaCO_3 + CO_2 + H_2O$). During massive surface blooms of coccolithophorids, the net carbon balance can swing towards net CO_2 production if the $C_{org.}:C_{inorg.}$ ratio in the exported particulate materials is <1.0 (e.g., Bates et al. 1996a). Theoretically, these two carbon-containing phases have different bulk densities, chemical behaviors including solubility, and subeuphotic zone remineralization length scales. However, in reality, the biogenic $CaCO_3$ is associated with organic carbon which affects its predicted chemical reactivity. Nevertheless, $C_{inorg.}$ is generally exported to greater depths in the water column than $C_{org.}$, so the $C_{org.}:C_{inorg.}$ ratio tends to decrease with depth. Furthermore, the ratio of Si-to-Ca or, more formally, the opal (biogenic Si) to calcium carbonate ratio in exported particulate matter, provides information on the structure of the phytoplankton assemblage in the euphotic zone. The diatom, or silicate mode, vs. the coccolithophorid, or carbonate mode is a key variable of the biological pump. Even in low nutrient, low chlorophyll (LNLC) regions where the phytoplankton assemblage is dominated by prokaryotic picoplankton, these larger eukaryotes determine the tempo and mode of particulate export (Fig. 10.2). Among the various diatom species, maximum growth rate, iron requirements and propensity to form large blooms and subsequently aggregate are all key variables. In the Southern Ocean for example, two congeners, *Chaetocerous dichaeta* and *C. brevis* have adopted different ecological strategies: the former has a rapid growth potential and blooms episodically in response to the pulsed atmospheric deposition of iron (Fe), whereas the latter grows best, but slowly, under the climatological Fe-deficient conditions of these high nutrient, low chlorophyll (HNLC) habitats (Timmermans et al. 2001). Community structure does matter, perhaps even down to the species level, both on the short-term and over much larger time scales.

When particulate carbon export from the euphotic zone is expressed as a proportion of contemporaneous primary production, this value is termed the export ratio or '*e*-ratio' (Baines et al. 1994). Results from broad-scale, cross-ecosystem analyses suggest that the *e*-ratio in oceanic habitats is a positive, non-linear function of total integrated primary production (Suess 1980; Pace et al. 1987; Martin et al. 1987; Wassman 1990), with values ranging from less than 0.10 in oligotrophic waters to greater than 0.50 in productive coastal regions. It should be emphasized, however, that the field data from which the existing export production models were derived are extremely limited and that open ocean habitats, in particular, are under-represented. It is important to understand the mechanisms that control the biological carbon pump in a variety of functionally distinct biomes so that accurate and meaningful predictions of the response of the oceanic carbon cycle to global environmental change can be made. This was a stated JGOFS goal in 1987.

Each year, the biological pump removes an estimated 7–15 Pg C from the surface waters of the world ocean, a value that is equivalent to ~10% of the annual global ocean primary production (Martin et al. 1987; Karl et al. 1996; Laws et al. 2000). Microbial transformation of sinking particles in the thermocline that gives rise to increased C:N and C:P ratios with depth can potentially drive a net atmosphere-to-ocean flux of CO_2 in the surface ocean. Episodic flux 'events' carry to the deep sea large amounts of 'fresh' organic matter (Lampitt 1985; Smith et al. 1996). These events may represent the bulk of the material reaching depths greater than 1 000 m (Anderson and Sarmiento 1994), making processes within the main thermocline also dependent upon the biological carbon pump.

After more than a decade of JGOFS research, we now realize that there are at least three fundamentally different biological carbon pumps, each with independent controls and fundamentally distinct biogeochemical consequences (Fig. 10.3):

1. *'Redfield Ratio – Dissolved/Particulate Carbon Pump'* wherein the C:N:P stoichiometry of the exported material exactly balances the C:N:P stoichiometry of the subeuphotic zone supply of inorganic nutrients: Partitioning of the exported carbon into separate dissolved and particulate matter components is crucial because these two separate pools generally have different bulk stoichiometries, and different remineralization rates and fates. However, only particulate matter fluxes can be directly measured in the field. A characteristic of this pump is the vertical attrition of particulate matter mass, which can be modeled as a normalized power function of the form $F_z = F_{150\,\mathrm{m}}(Z/150)^b$, where Z is water depth and F and $F_{150\,\mathrm{m}}$ are fluxes at depth Z and 150 m, respectively (Martin et al. 1987; Knauer et al. 1990; Karl et al. 1996). In this model, the regeneration length scales of C vs. N and P are critical to the carbon sequestration process (Christian et al. 1997). If there is no spatial or temporal separation between C and the export of the production rate limiting nutrients N and P, there can be no net carbon sequestration. Even though the Redfield pump operates worldwide and is probably a dominant component of the ocean's carbon cycle, it largely sustains a bi-directional reflux of bioelements, rather than net export. In other words, even though there is a measurable downward flux of carbon and associated bioelements (e.g., N and P) from the euphotic zone, there is a stoichiometrically equivalent importation of regenerated nutrients from below which over the long term (months to years) cancels out. As we shall see below, other export mechanisms may be more important for net carbon sequestration in the sea.

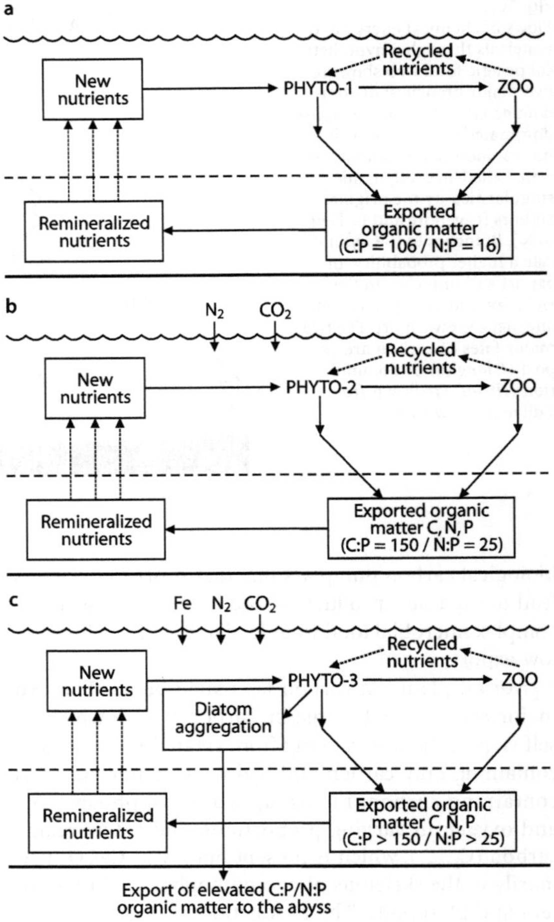

Fig. 10.3. The three major variations of the biological carbon pump in the sea. **a** The Redfield ratio – dissolved/particulate carbon pump with a balanced C:N:P stoichiometry for imported and exported materials. The PHYTO-1 assemblage is a diverse mixture of eukaryotic and prokaryotic algae and bacteria. Under these hypothetical conditions of a Redfield balanced ocean, the flux of C, N and P out of the euphotic zone is balanced by the importation of regenerated C, N and P from below, so there is no net carbon export. **b** The N_2-primed prokaryote carbon pump with a portion of the new N derived from dissolved N_2 gas. The PHYTO-2 assemblage has selected for potentially diverse free-living (unicellular and filamentous) and symbiotic N_2-fixing prokaryotes capable of surviving under conditions of fixed N depletion. Under these conditions the exported particulate matter is characteristically enriched in C and N, relative to P, and there is an imbalance between the import and export nutrient balance in favor of net C removal. **c** The event-driven, Fe-stimulated diatom aggregation carbon pump driven by the stochastic atmospheric delivery of bioavailable Fe. The PHYTO-3 assemblage has selected for large diatom species which use the Fe to assimilate nitrate or, in certain cases, to stimulate the metabolism of intracellular N_2-fixing symbiotic cyanobacteria who then manufacture fixed N for the host. The latter would be expected to occur in habitats that are both Fe- and fixed N-stressed (e.g., central gyres). Following nutrient (Fe, P or Si) exhaustion, the diatoms aggregate and sink carrying carbon to the abyss (see also diatom branch; Fig. 10.2). If N_2-fixation and Fe-stimulation co-occur, then the production stoichiometry also favors C export see **b**, above. This combined Fe-stimulated, N_2-fixing diatom pump is likely to be the most efficient means for sequestering atmospheric CO_2 in the world ocean

2. *'N_2-primed Prokaryote Carbon Pump'* wherein microbiological fixation of N_2 temporarily relieves the ecosystem of fixed-N limitation resulting in P (or Fe) control of new and export production: Under these new habitat conditions, there is a selection for the growth of microorganisms with altered ecological stoichiometry, one which produces dissolved and particulate matter with elevated C:P ratios (>250–300:1) relative to the expected Redfield ratio (C:P = 106). Export of this non-Redfield organic matter provides a mechanism for the net, long-term (centennial to millennial) sequestration of carbon into the mesopelagic zone, or deeper (Fig. 10.3b). The time scales of this sequestration will be determined by the balance between N_2 fixation and denitrification and the time-scale for ventilation of the midwater zones where some of the denitrification occurs. Variability in this process will depend upon the stability of the climate parameters that encourage the growth of the N_2-fixing prokaryotes, and on a continued supply of bioavailable P and Fe or to the physiological limits in the non-Redfield stoichiometry. A re-supply of P and, perhaps, Fe could be satisfied by vertically-migrating, N_2-fixing microorganisms like *Trichodesmium* (Karl et al. 1992). A biogeochemical effect similar to that described here for the N_2-primed Prokaryote Carbon Pump (i.e., net removal of C from the surface water relative to N/P upwelled from depth) can also occur by the selective subeuphotic zone remineralization of N/P relative to C in the absence of N_2 fixation. However, it is unlikely that this differential remineralization process could ever result in effective C:P ratios as high as those produced by the N_2-primed prokaryote carbon pump mechanism.
3. *'Event-driven, Mass Sedimentation Carbon Pump'* wherein a specific physical or biological perturbation to the biogeochemical steady-state results in a rapid pulse of export to the sea floor. This is most often characterized by Fe-stimulated diatom aggregations where atmospheric deposition of Fe-rich dust results in the rapid growth of large chain- or aggregate-forming diatoms followed by mass export. Massive blooms of filter-feeding planktonic animals, like salps, will have the same flux impact. A characteristic feature of these rapid growth-export events is the efficient delivery of fresh organic matter to the deep sea floor (the so called phytodetritus pulse). Even if the stoichiometry of these exported materials conforms to the Redfield ratio, there is a net removal of carbon from the ocean's surface on time scales of a few decades to centuries and, hence, net carbon sequestration. Although aggregation events have been reported to occur in both coastal and open ocean habitats the physical, chemical and biological controls are poorly understood. While the diatom bloom events were once thought to be coupled to the turbulent-driven supply of new nutrients by deep mixing events or eddy-induced upwelling, it is now evident that diatoms can propagate, aggregate and sink even from stable open ocean waters (Scharek et al. 1999b; Cullen et al. 2002). These latter, stratified open ocean blooms are likely to be manifestations of the convergence of the N_2-primed prokaryote and diatom aggregation pump processes (Fig. 10.3c), effected through endo-symbiotic associations between N_2-fixing cyanobacteria and oceanic diatoms (e.g., *Rhizosolenia-Richelia* associations; Martinez et al. 1983). In this regard, the endosymbiont-containing diatoms facilitate the utilization of diatomic N and should be called 'diatomic-diatoms' to distinguish them ecologically from the more common nitrate-utilizing diatom populations.

Of these three separate biological carbon pumps, only the first has been studied or modeled in any detail. Unfortunately, it is the most predictable of the three and, for many reasons, the least important for variations in the ocean's carbon cycle. It now appears that the non-Redfield carbon pumps may be important in many open ocean habitats. If the N_2-primed Prokaryote Carbon Pump and the Event-driven, Mass Sedimentation Carbon Pump are found to be present in other regions of the global ocean, we may need to alter our most basic dogma on nutrient biogeochemistry in the sea.

Carbon export from the upper regions of the euphotic zone can be measured directly using surface-tethered, free-drifting, particle-interceptor traps (PITs; Knauer et al. 1979), or indirectly using oxygen mass balance estimation (Jenkins 1982, and subsequent papers e.g., Emerson et al. 1991; and Michaels et al. 1994), dissolved inorganic carbon isotope mass balance estimation (Zhang and Quay 1997) or ^{238}U-^{234}Th disequilibrium measurements (Coale and Bruland 1987; Buesseler 1998; Benitez-Nelson et al. 2001). Each approach has fundamental advantages and limitations, and each measures a slightly different set of ecosystem processes. For example, PITs cannot assess the rates of downward diffusion of DOC or active vertical migration processes, so this approach will tend to underestimate total downward carbon flux. PITs can also suffer from hydrodynamic collection bias, especially in high turbulent regions. Net oxygen production, carbon isotope mass balance and ^{234}Th scavenging are all indirect surrogates for carbon flux, so model assumptions must be applied. Most of these indirect methods also suffer from a general lack of distinction between particulate and dissolved organic matter export; only PITs provide a return sample for quantitative chemical, microscopic or molecular/biochemical measurement. However, taken together, this suite of carbon export techniques can provide complementary, redundant information and a robust constraint on the carbon export process.

10.3 Global Inventory of JGOFS Time-Series Programs

During the JGOFS era eight ocean time-series programs were initiated (Fig. 10.4): Bermuda Atlantic Time-series Study (BATS), European Station for Time-series in the Ocean, Canary Islands (ESTOC), Dynamique des Flux Atmosphériques Méditerranée (DYFAMED), Hawaii Ocean Time-series (HOT), Kerguelen Point Fixe (KERFIX), Kyodo Northwest Pacific Ocean Time-series (KNOT), Ocean Station Papa (OSP) and the Southeast Asia Time-series (SEATS) programs. Other JGOFS-relevant programs, some pre-dating the official commencement of JGOFS in 1987, also exist, including hydrographic and biological monitoring programs established at marine laboratories, autonomous bottom-moored sediment traps, physical-biogeochemical moorings and underway data collected during repeat transits of merchant or military vessels. This chapter will focus on the eight JGOFS-sponsored research efforts (Fig. 10.5). These ecosystem-based studies include measurements of nutrient and population dynamics, primary and secondary productivity rates and controls, as well as seascape structure and variability, including climate. Each program has unique roots, experimental design, organizational structure and local-to-global connectivity. Despite these fundamental differences, they do share common JGOFS objectives and implementation procedures, including standard core measurements and protocols (Table 10.1). However, it should be pointed out that not all of the 'recommended' JGOFS core measurements are conducted at every site (e.g., ^{14}C primary production, 'new production' by ^{15}N uptake, microzoo-plankton herbivory rates are not conducted at many sites). Likewise, some of the most exciting emergent data sets from selected JGOFS time-series programs (e.g., temporal changes in the N:P stoichiometry of total dissolved and particulate organic matter, rates of N_2 fixation and atmospheric Fe deposition rates) were not even recommended as key

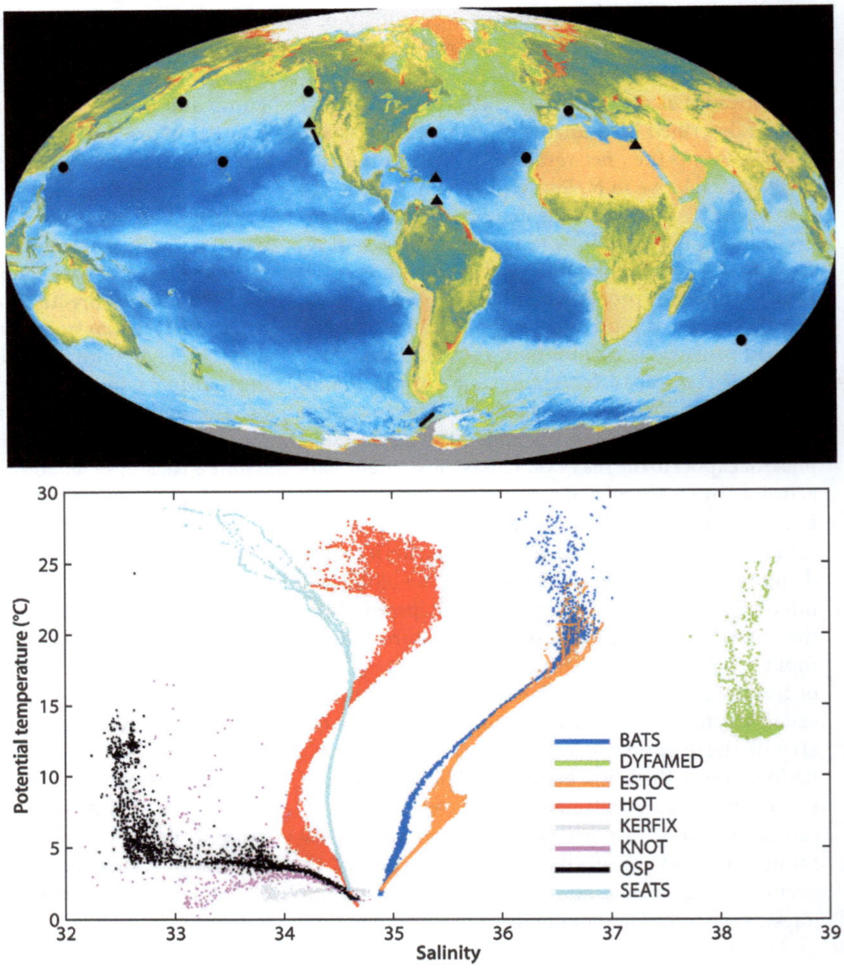

Fig. 10.4. The global distribution of JGOFS ocean time-series stations and related biogeochemical programs. Shown at *top* are the positions of selected study sites in relation to the surface distribution of Chl as determined from the SeaWiFS color satellite (*http://seawifs.gsfc.nasa.gov/SEAWIFS.html*). The JGOFS sites discussed in this chapter are shown as *solid circles*; the *solid triangles* and *solid rectangles* show the locations of non-JGOFS, fixed-point biogeochemical ocean time-series sites or regional surveys, respectively. Shown at the *bottom* are the potential temperature-salinity relationships for each of the JGOFS sites showing the characteristic water mass structures for each site

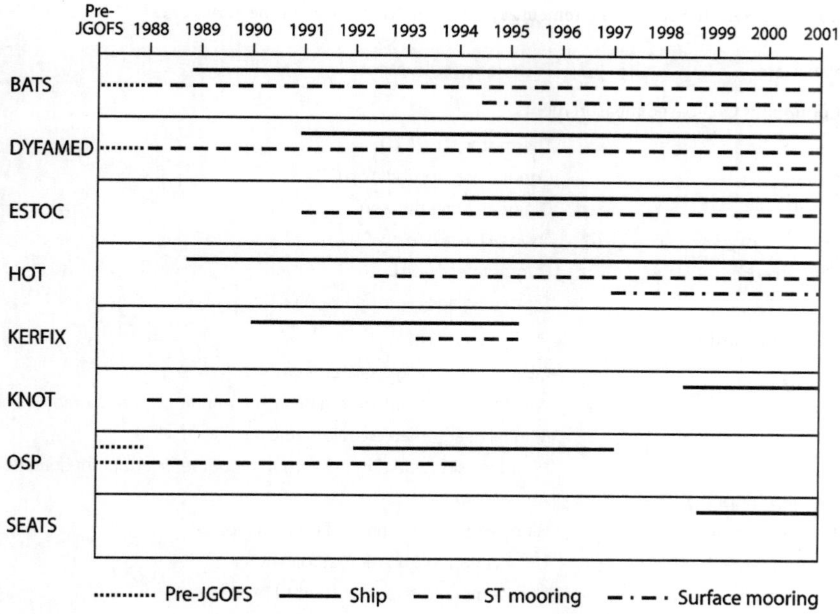

Fig. 10.5. Timelines for the eight JGOFS ocean time-series stations discussed in this chapter. Shown as *coded horizontal lines* are the periods of ship-based biogeochemical sampling and the deployments of bottom-moored sediment traps (*ST*) and surface meteorological-biogeochemical moorings

variables at the start of the JGOFS program. This is clearly 'work in progress.'

10.3.1 Bermuda Atlantic Time-Series Study (BATS)

Location: The primary station is located at 31° 50' N, 64° 10' W approximately 82 km SE of St. David's Light, Bermuda in 4500 m of water. The BATS sampling site is near, but not co-located with, Hydrostation 'S' (32° 10' N, 64° 30' W), one of the longest running hydrographic-chemical ocean time-series stations in existence. The BATS time-series station is intended to be representative of the western North Atlantic subtropical gyre (Sargasso Sea).

Inauguration and present status: BATS field sampling began in October 1988 and has continued to the present (Fig. 10.5). The current JGOFS sampling program builds on the Hydrostation 'S' program that was initiated in June 1954, but current BATS sampling is conducted at a new station approximately three times the distance from shore to eliminate any biogeochemical influences of the coastal zone. A bottom-moored sediment trap program located near the current BATS site began in 1976 (Deuser 1986; Conte et al. 2001), and independent physical-biogeochemical mooring and atmospheric sampling programs presently complement the BATS sampling program (Dickey et al. 2001).

Objectives: (1) to observe and interpret hydrographic, chemical and biological variations that occur on time scales ranging from subseasonal to decadal, (2) to estimate the dynamics of gaseous, dissolved inorganic, dissolved organic and particulate organic carbon pools, (3) to determine the impact of mode water intrusion on biogeochemical processes, and (4) to understand the linkages between the atmospheric delivery of nutrients and organic matter export processes.

Sampling frequency and methods: Approximately bi-weekly to monthly field observations are conducted using the UNOLS vessel R/V *Weatherbird II* which is based in Bermuda and operated by the Bermuda Biological Station for Research, Inc. (BBSR). The interdisciplinary station work includes physical, chemical and biological observations and rate measurements.

Logistical management and funding sources: BATS is maintained by scientists and staff from the BBSR as well as several key guest investigators. Funding is provided primarily from the US National Science Foundation.

Data availability and key contact persons: All BATS core measurement data are publicly available approximately 1 year after collection (*http://www.bbsr.edu*), or by request from the US National Oceanic Data Center. Annual Data Reports are published and distributed through the US JGOFS-Planning Office in Woods Hole, MA. The principal contact persons are A. Knap (*knap@sargasso.bbsr.edu*) and N. Bates (*nick@sargasso.bbsr.edu*).

Key references: Michaels and Knap (1996), Steinberg et al. (2001).

Table 10.1. Selected core measurements made at the JGOFS ocean time-series sites[1]

Parameter	Technique, instrument
Continuous CTD-based measurements	
Temperature	Thermistor(s) on CTD
Salinity	Conductivity sensor(s) on CTD
Depth	Digiquartz pressure sensor on CTD
Dissolved oxygen	Oxygen electrode on CTD
Fluorescence	Fluorometer on CTD
PAR	Scalar irradiance sensor on CTD
Discrete measurements	
Salinity	Conductivity, using a salinometer
Oxygen	Winkler titration with automated photometric or potentiometric endpoint detection
Total CO_2	Manual or automated coulometric analysis
Alkalinity	High precision acid titration with potentiometric endpoint detection
Nitrate, phosphate, silicate	Autoanalyzer
Dissolved organic C	High-temperature combustion-oxidation
Total dissolved N and P	UV or chemical oxidation, autoanalyzer
Particulate C	High-temperature combustion, CHN analyzer
Particulate N	High-temperature combustion, CHN analyzer
Particulate biogenic silica	Chemical digestion, autoanalyzer
Fluorometric chlorophyll a	Acetone extraction, fluorescence detection
Phytoplankton pigments	HPLC, fluorescence detection
Bacteria	Fluorescence microscopy or flow cytometry
Zooplankton	Net tows, wet and dry weights, CHN analyzer
Rate measurements	
Primary production	Trace-metal clean, in situ incubation, ^{14}C uptake
Particle fluxes	Free-drifting cylindrical sediment traps (PITs)
Mass	Gravimetric analysis
Total carbon	CHN analysis
Total nitrogen	CHN analysis

10.3.2 Dynamique des Flux Atmosphérique en Méditerranée (DYFAMED)

Location: The primary station is located at 43° 25' N, 7° 52' E in the northwestern sector of the Mediterranean Sea (Ligurian Sea) approximately 45 km south of Cape Ferrat, France, in 2350 m of water. This region is believed to be free from coastal zone fluxes, but it does receive a significant atmospheric input from the deserts of North Africa and from the industrialized countries bordering the Mediterranean Sea. These atmospheric fluxes are measured at nearby Cape Ferrat.

Inauguration and present status: Field sampling began in 1986, pre-JGOFS, with a sediment trap mooring and atmospheric deposition survey (Miguel et al. 1994; Migon et al. 2002) that were enhanced with a ship-based biogeochemistry measurement program and benthic survey in 1991 (Guidi-Guilvard 2002); all three components have continued to the present.

Objectives: (*1*) to study variations of hydrography and biogeochemistry at the seasonal and interannual scale, (*2*) to investigate the ecosystem response to atmospheric deposition events and to long-term environmental/climate forcing, (*3*) to investigate and understand the ecological effects of meteorological forcing, especially the transition in community structure from spring mesotrophy to summer oligotrophy and (*4*) to estimate the air-to-sea exchange of carbon dioxide.

1 JGOFS has established a suite of biogeochemical measurements that has been recommended as the core parameters necessary for a comprehensive investigation of the ocean's carbon cycle. Nevertheless, the exact list of core measurements and specific analytical methods used varies considerably from site to site, and has even changed within a given site over time. Furthermore, some sites routinely measure parameters that are not part of the 'JGOFS core' (e.g., inherent and apparent optical properties, natural CN isotopes, radionuclides, atmospheric deposition, etc.). The program objectives and the science questions should always be the motivation behind the sampling and measurement programs.

Sampling frequency and methods: Approximately monthly field observations are conducted using the French R/V *Tethys II* which is operated by the Centre National de la Recherche Scientifique (CNRS) Institut National des Sciences de l'Univers (INSU) and based in Marseille, France. The interdisciplinary station work includes physical, chemical and biological observations and rate measurements. In March 1999, a meteorological buoy was deployed with plans to add in-water optical and biogeochemical sensors in the near future.

Logistical management and funding sources: DYFAMED is maintained by scientists from the Laboratoire d'Océanographie de Villefranche and of IAEA Marine Environment Laboratory in Monaco. Funding is provided by INSU/CNRS.

Data availability and key contact persons: Most DYFAMED data sets can be obtained at: *http://www.obs-vlfr.fr/jgofs2/sodyf/home/htm.* The principal contact person is J.-C. Marty (*marty@obs-vlfr.fr*).

Key reference: Marty et al. (2002).

10.3.3 European Station for Time-Series in the Ocean Canary Islands (ESTOC)

Location: The primary station is located at 29° 10' N, 15° 30' W approximately 100 km north of the islands of Gran Canaria and Tenerife in 3600 m of water. This time-series station is intended to be representative of the eastern boundary regime of the Northeast Atlantic Ocean.

Inauguration and present status: ESTOC field sampling began in February 1994 and has continued to the present.

Objectives: (*1*) to investigate the long-term changes of stratification and circulation on seasonal and interannual time scales, and (*2*) to investigate biogeochemical cycles in this region to understand controls on flux of carbon and associated bioelements on seasonal and interannual time scales.

Sampling frequency and methods: Approximately monthly field observations are conducted using the Spanish R/V *Taliarte* which is operated by the Instituto Canario de Ciencias Marinas and based in Telde, Gran Canaria. German research vessels have complemented these baseline observations. These latter cruises also maintain two long-term moorings; one deep-sea mooring supports current meters and thermistors and the other supports sediment traps. The interdisciplinary station work includes physical, chemical and biological observations and rate measurements.

Logistical management and funding sources: ESTOC is maintained by a consortium of four institutions: (*1*) Instituto Canario de Ciencias Marinas, Telde Gran Canaria, Spain (ICCM), (*2*) Instituto Español de Oceanografia, Madrid, Spain (IEO), (*3*) Institut für Meereskunde, Kiel, Germany (IFMK) and (*4*) Fachbereich Geowissenschaften, Universität Bremen, Germany (UBG). ESTOC has received funding in part from the German Ministry for Education and Research (BMBF).

Data availability and key contact persons: Most ESTOC data sets can be obtained at: (*1*) *http://www.iccm.rcanaria.es/estocing.htm*, (*2*) *http://www.ifm.uni-kiel.de/general/estoc.htm* and (*3*) *http://www.pangea.de/Projects/ESTOC/*. Key contact persons include O. Llinás (*ollinas@iccm.rcanaria.es*), G. Siedler (*gsiedler@ifm.uni-kiel.de*) and G. Wefer (*gwefer@marum.de*).

Key references: Neuer et al. (1997), Davenport et al. (1999) and Freudenthal et al. (2001).

10.3.4 Hawaii Ocean Time-Series (HOT)

Location: The primary deep ocean station, Sta. ALOHA (A Long-term Oligotrophic Habitat Assessment), is located at 22° 45' N, 158° W approximately 100 km north of Kahuku Point, Oahu, in 4740 m of water. A coastal station is also maintained at a location approximately 10 km offshore from Kahe Point, Oahu in 1500 m of water. Sta. ALOHA is believed to be representative of the North Pacific Subtropical Gyre biome.

Inauguration and present status: HOT field sampling began in October 1988 and has continued to the present. A bottom-moored sediment trap program was added in June 1992, and between January 1997 and June 2000 a meteorological-physical-biogeochemical mooring was deployed for high frequency atmospheric and in-ocean observations.

Objectives: (*1*) to document seasonal and interannual variability in water mass structure, (*2*) to relate water mass variations to gyre fluctuations, (*3*) to develop a climatology of biogeochemical rates and processes including microbial community structure, primary and export production and nutrient inventories and (*4*) to estimate the annual air-to-sea flux of carbon dioxide.

Sampling frequency and methods: Approximately monthly field observations are conducted primarily using the UNOLS vessel R/V *Moana Wave*, based in Honolulu and operated by the University of Hawaii (1988–2000) and the State of Hawaii owned vessel R/V *Kaimikai-O-Kanaloa* (2000 to present). Over the 13-year lifetime of HOT at least 10 different public and private research vessels have been employed in the field effort. The sediment trap mooring is serviced annually from these same vessels. The interdisciplinary station work includes physical, chemical and biological observations and rate measurements.

Logistical management and funding sources: HOT is maintained by scientists from the University of Hawaii. Funding derives primarily from the US National Science Foundation and the State of Hawaii.

Data availability and key contact persons: All HOT core measurement data are publicly available approximately 1 year after collection (*http://hahana.soest.hawaii.edu*), or by request from the US National Oceanic Data Center. Annual data reports are published and distributed through the US JGOFS Planning Office in Woods Hole, MA, or as downloadable PDF files at the above referenced website. The principal contact persons are D. Karl (*dkarl@soest.hawaii.edu*) and R. Lukas (*rlukas@soest.hawaii.edu*).

Key references: Karl and Winn (1991), Karl and Lukas (1996) and Karl (1999).

10.3.5 Kerguelen Point Fixe (KERFIX)

Location: The primary station is located at 50° 40' S, 68° 25' E, approximately 100 km southwest of Kerguelen Islands in 1700 m of water. This site, south of the polar front, is characteristic of the Permanently Open Ocean Zone (POOZ) of the Southern Ocean. A second site (Bio-Station) located 24 km off Kerguelen was also used for selected biological rate measurements.

Inauguration and present status: KERFIX field sampling began in January 1990 and was terminated as a JGOFS effort in March 1995. Since the end of KERFIX, a new hydrographic (temperature and salinity) sampling program called CLIOKER (CLImat Océanique a KERguelen; Y Park, Principal Investigator) has emerged at the KERFIX site as a component of the international CLIVAR program. The biogeochemical studies, with a special focus on air-to-sea CO_2 gas exchange have continued under the OISO (Océan Indien Service d'Observation; N. Metzl, Principal Investigator), but with only two cruises per year.

Objectives: (1) to parameterize the air-sea flux of CO_2 and O_2, (2) to understand the physical and biological processes that control these exchanges and (3) to observe and interpret the seasonal and interannual variability in production, flux and decomposition of carbon and associated elements.

Sampling frequency and methods: Approximately monthly field observations were conducted using the French R/V *La Curieuse* which is operated by the French Polar Institute (IFRTP) and based in Port-aux-Français (Kerguelen). From April 1993 to March 1995 a mooring was also in place for continuous measurements of the downward flux of particulate matter and current velocity and direction. The mooring deployments and maintenance were supported by the R/V *Marion Dufresne* which is operated by IFRTP and based in Port-aux-Français.

Logistical management and funding sources: KERFIX is supported locally by staff from the Biologie Marine (BIOMAR) on Kerguelen, with financial support from the Scientific Mission of the Terres Australes and Antarctiques Francaise (1990–1992), and subsequently by the Center National de la Recherche Scientifique (INSU/France-JGOFS; 1991–1995) and the IFRTP (1992–1995).

Data availability and key contact persons: Additional information on KERFIX can be obtained at *http://www.obs-vlfr.fr/jgofs/html/bdjgofs.htm*. KERFIX was initiated by A. Poisson (LPCM, Paris) and since 1993 has been led by C. Jeandel (*Catherine.Jeandel@cnes.fr*).

Key references: Jeandel et al. (1999) and Louanchi et al. (1999).

10.3.6 Kyodo Northwest Pacific Ocean Time-Series (KNOT)

Location: The primary station is located at 44° N, 155° E, approximately 400 km northeast of Hokkaido Island, Japan in 4900 m of water. This time-series station is intended to be representative of the southwestern subarctic gyre.

Inauguration and present status: KNOT field sampling began in June 1998 and was terminated in October 2000. Research at this site evolved from a previous (1988–1991) sediment trap experiment at the same location (Noriki 1999). The establishment of KNOT was a joint JGOFS-Japan and JGOFS-NPTT (North Pacific Task Team) effort. Kyodo is the Japanese word for collaboration.

Objectives: (1) to investigate the inorganic carbon system dynamics in response to variations in hydrography and biological processes, (2) to investigate the response of the biological pump to climate forcing and (3) to provide a data set from the western subarctic Pacific gyre for comparison to OSP in the eastern subarctic Pacific gyre.

Sampling frequency and methods: Approximately monthly field observations were conducted using the Japanese vessels T/S *Hokusei Maru* of Hokkaido University, R/V *Bosei Maru* of Tokai University, R/V *Mirai* of JAMSTEC, R/V *Hakuho Maru* of Tokyo University, and the *Hakurei Maru II* of the Metal Mining Agency of Japan.

Key contact person: Y. Nojiri (*nojiri@ees.hokudai.ac.jp*).

Key reference: Tsurushima et al. (1999).

10.3.7 Ocean Station Papa (OSP or Sta. P)

Location: The primary station is located at 50° N, 145° W, approximately 1 500 km due west from the approaches to the Juan de Fuca Strait at 125° W in 4 200 m of water. JGOFS relevant sampling is also periodically conducted along the transect from Vancouver to OSP at several locations along 'line P.' Conditions at OSP are believed to be representative of the northeast subarctic Pacific Ocean (southern edge of the Alaska gyre).

Inauguration and present status: JGOFS relevant research at OSP evolved from the co-located Canadian weathership sampling program (1956–1981), the Subarctic Pacific Ecosystem Research (SUPER) program (1984–1988) and the Canadian JGOFS program commencing in 1992 and terminating in 1997. From 1998 to the present, the Department of Fisheries and Oceans, Canada (DFO) has continued to support two or three cruises to OSP per year to continue the previous decades of seasonal sampling. A bottom-moored sediment trap program at OSP began in 1983 and ran through 1994. A deep-sea mooring was deployed from 1995 to 1997 which recorded T, S, light and solar-induced fluorescence at ~30 m.

Objectives: (1) to document seasonal, interannual and decadal variations in hydrographic and key biogeochemical parameters and determine their relationship to carbon export, (2) to investigate the role of atmospherically-deposited iron on ecosystem dynamics, including carbon dioxide drawdown and (3) to determine the impact of El Niño events on biogeochemical cycling.

Sampling frequency and methods: Ship-based field observations at OSP were obtained approximately three times per annum using the Canadian Coast Guard vessel *John P. Tully* based in Victoria, British Columbia, Canada. These cruises were usually in the periods February, May–June, and August–September to coincide with winter, late spring and late summer periods, respectively. Line 'P', 12 stations from Victoria to OSP, occupied from 1959–1981, was also re-occupied during the JGOFS campaign.

Logistical management and funding resources: The scientific programs at OSP are maintained by scientists and technical support in the Ocean Science and Productivity Division of the DFO. Salary and shiptime are provided by DFO. Research and time-series activities are funded by the Panel for Energy Research and Development (PERD) of Natural Resources Canada under programs of Time-series and Oceanic CO_2 Uptake. Monitoring of water properties is funded by the Ocean Climate Program of DFO.

Data availability and contact persons:
C.S. Wong (*WongCS@pac.dfo-mpo.gc.ca*) and
P.J. Harrison (*pharrisn@unixg.ubc.ca*).

Key references: Whitney et al. (1998), Wong et al. (1995), Wong et al. (1999) and Harrison et al. (1999).

10.3.8 South East Asia Time-Series Station (SEATS)

Location: The primary station is located at 18° N, 116° E about 700 km southwest of Taiwan (Shiah et al. 1999). The SEATS time-series station is in the South China Sea (SCS), the largest ice-free marginal sea in the world (the ice-covered Arctic Sea is the largest). It has a wide continental shelf to the south, significant runoff from several large rivers, including the Mekong River and the Pearl River, and a deep (>3 000 m) basin. The SCS is subject to physical forcing of the alternating southeastern Asian monsoons (Shaw and Chao 1994), typhoons, strong internal waves (Liu et al. 1998) and ENSO (Chao et al. 1996). SEATS is potentially sensitive to climate change because of its locality between the third pole of the world, namely the Tibet Plateau, and the western Pacific warm pool, which are two of the most important heat engines of the earth's climate.

Inauguration and present status: The pilot study of the SEATS project began in August 1998 (Shiah et al. 1999). A suite of stations just west of the Luzon Strait was occupied on bimonthly cruises between August 1998 and June 1999. Because the hydrography of these stations are strongly influenced by the Kuroshio intrusion through the strait as well as the monsoon driven upwelling off northwest Luzon (Shaw et al. 1996), the location of SEATS was changed to the site mentioned above in August 1999. The observational program, including shipboard measurements and moored instruments for physical and biogeochemical measurements, is yet to be fully developed.

Objectives: (1) to understand how monsoonal forcing controls biogeochemical cycles in the SCS and how ENSO modulation of the monsoon strength influences it, (2) to monitor how the episodic events, such as typhoons or mesoscale eddies, affect biogeochemical processes in the upper water column, and (3) to link the present day biogeochemical processes with paleorecords preserved in sediment cores taken by IMAGES and the Ocean Drilling Project for better understanding of the effect of climate change on the ocean biogeochemistry in the marginal sea.

Sampling frequency and methods: Ship-based hydrographic and biogeochemical surveys are conducted every 2–3 months. Moored ADCP and thermister chains have been deployed since 1997. Moored light sensors and fluorometers have been deployed since October 2001.

Logistical management and funding sources: SEATS is maintained by scientists and staff from the National Center for Ocean Research (NCOR), Taipei. Funding is provided by the National Science Council, Republic of China (ROC).

Data availability and key contact persons: The data archive of SEATS is being compiled and will be available from NCOR (*http://www.ncor.ntu.edu.tw*). The principal contact person is K. K. Liu (*kkliu@ccms.ntu.edu.tw*).

10.4 Some Practical Lessons Learned from the JGOFS Time-Series Programs

There is a very broad range of natural ecosystem variability in the sea. Each part of this variance spectrum is derived from specific physical, biogeochemical and biological forcing, and each has unique consequences for the ocean's carbon cycle. For example, diel variations are tied to the daily and seasonal solar cycles and are very predictable for a given latitude. However, for phototrophic microorganisms in the sea, there can be large day-to-day variations in total irradiance at a given reference depth as a result of inertial period oscillations, internal waves and clouds. On subseasonal time scales there are discrete nutrient-enhancing upwelling and mixing events that can occur on mesoscale (100–1 000 km) spatial scales (McGillicuddy et al. 1998, 1999). In the North Pacific Ocean, there are also decade-scale transitions or regime shifts such as the Pacific Decadal Oscillation (PDO; Mantua et al. 1997) that can fundamentally alter community structure and nutrient dynamics (Karl et al. 2001a). In the North Atlantic Ocean, the North Atlantic Oscillation (NAO) similarly influences upper ocean hydrography and biogeochemistry at BATS (Bates 2001). Finally, there are gradual, continuous, unidirectional changes such as the increasing atmospheric and oceanic burdens of CO_2 (Winn et al. 1994; Bates et al. 1996b; Bates 2001).

Ecosystem variance will respond simultaneously, in a non-linear fashion, to the combination of all these independent forces. The JGOFS ocean time-series programs were designed to observe and interpret these complex interactions on time scales from months to years. At the very least, these repeat biogeochemical measurements provide – largely for the first time – the data necessary to define a mean state or climatology from which a quantitative anomaly field can be prepared. These anomalies are then used to assess change in the biogeochemical state variables and carbon fluxes. The longer the time-series record, the more valid the climatology and the more relevant and diagnostic the anomalies. The mission and method of the JGOFS time-series programs are straightforward, but in practice these are easier said than done. The large-scale, low-frequency changes, in particular, will likely be the most difficult to observe and interpret. Another complication is the assimilation of Eulerian, fixed-point time-series data into a four-dimensional context with various mesoscale and submesoscale natural variability (see McGillicuddy et al. 1998, 1999; Siegel et al. 1999). Nevertheless, we submit that ocean time-series measurement programs remain the most effective means for studying seasonal, interannual and decadal scale physical-biogeochemical processes. They also remain the most cost effective and efficient programs for understanding local and regional scale ecosystem dynamics and serve as floating platforms of opportunity for the support of complementary ocean and atmosphere research.

Sampling should, ideally, be continuous in order to resolve all relevant high- and low-frequency scales of habitat variability. Unfortunately, many biogeochemical parameters and most carbon fluxes cannot be measured remotely and a continuous human-operated station would be prohibitively expensive. Consequently, in JGOFS we have compromised with approximately monthly-to-quarterly cruises, supplemented with continuous moored observations to the extent possible. Despite their recognized importance, comprehensive ocean time-series field sampling programs are very difficult to sustain. Sampling during the past 40 years at OSP is a case in point. During the period 1965–1981, seawater was collected twice per week; the temporal resolution was excellent, but the core measurement list was relatively limited. This initial period of sampling was well situated for investigating intraseasonal and interannual variability, but the ecological consequences of the event-scale habitat perturbations that were observed could not be constrained. As complementary programs emerged to enhance the core measurement suite, the weathership program was discontinued. During the JGOFS era, the two-to-three cruises per year measurement frequency at OSP was insufficient to capture episodic events despite the comprehensive core measurement program that was assembled for this purpose.

Another frustrating example of the tradeoffs between time-series sampling design and logistics is the California Cooperative Oceanic Fisheries Investigation (CalCOFI). This sampling program began as a physical-biological observation program in 1949 to study the factors controlling the abundance of major pelagic fish stocks in the Southern California Current System. From 1950 to 1960, monthly cruises were conducted with few interruptions. But in 1961, the sampling frequency was reduced to approximately quarterly and this frequency was maintained through 1968. It was soon realized that this reduced frequency was insufficient for the intended program objectives, but fiscal resources precluded a return to the higher-frequency sampling mode. So in 1969, the CalCOFI sampling reverted back to monthly cruises, but only every third year (1969, 1972, 1975, 1978, 1981). This sampling strategy inadvertently imposed serious limitations on an otherwise robust data set (Chelton et al. 1982). For example, the 1976–1977 phase shift in the PDO occurred during a period of infrequent sampling, and the major 1982–83 El Niño event was 'conveniently scheduled' between sampling years. From 1984, the spatial grid was reduced significantly as a compromise to accommodate quarterly cruises every year. This has been retained as the CalCOFI sampling frequency to the present. These sampling changes were forced primarily by financial and logistical considerations, not by science. Likewise, the North Pacific Ocean Climax time-series sampling program (Hayward et al. 1983), which lasted from 1968 to 1985, also had serious logistical constraints. Of the 22 cruises during the 17-year observation period, four were in 1973, three in 1985, two each in 1971, 1972, 1974, 1976 and 1983, one each in 1968, 1969, 1977, 1982 and none in 1970, 1975, 1978, 1979, 1981 and 1984. In this data set, the impact of the major 1982–1983 El Niño event may have, therefore, been over-represented relative to the longer term climatology.

There are two points regarding sampling frequency that are relevant to the JGOFS era time-series programs. First, of the eight programs discussed in this chapter, only three (BATS, DYFAMED and HOT) have a continuous, approximately monthly frequency, decade-long record of core biogeochemical parameters (Fig. 10.5). One program (SEATS) has just begun, and routine sampling at three of the sites (KERFIX, KNOT and OSP) has already been terminated as high-frequency biogeochemical sampling programs (Fig. 10.5). Even Hydrostation 'S,' which started in 1954, lost funding for a brief period (1977–1979), as it was considered not important science, 'just monitoring.' In light of the past lessons learned, this is unfortunate because it is impossible to re-sample the past. Adding another year to these emergent records is probably more valuable than starting a time-series measurement program at a new location, if funding is the limiting factor. Second, the JGOFS era (1987–present) has coincided with a period of unprecedented El Niño favorable conditions (Karl et al. 2001a), and warm-phase Pacific Decadal Oscillation and North Pacific Mode indices (Barlow et al. 2001). The JGOFS era has also been a period of predominantly positive NAO, which may have significant implications for the North Atlantic time-series programs (BATS and ESTOC). To the extent that these large-scale climate variations impact upper ocean ecosystem dynamics, and there is strong evidence that they do, then the biogeochemical data sets collected during JGOFS era may not be wholly representative of the long-term (century scale) ocean climatology at these respective sites. The recent shift to a cold phase of the PDO may create limitations on how easily these time-series data can be used to understand future ocean dynamics.

One advantage of the JGOFS time-series data sets may be their use for evaluating and predicting higher frequency biogeochemical variability, e.g., seasonality. The systematic repeat core measurements provide unique data for diagnostic, prognostic and predictive syntheses, and for developing and testing models of biological/biogeochemical processes. These model results provide an interpretational context for extrapolating our knowledge of key mechanisms to the biomes and regions of interest.

An important research consideration is the statistical treatment of time-series data sets once in hand. For many ecological studies, central tendency analyses (e.g., regression, ANOVA) are employed to estimate the relationships between dependent or independent variables, or both. If time-series data sets are available, more sophisticated statistical analyses are also possible. For ecological and biogeochemical processes that are characteristically 'noisy,' that means that trend analysis is more robust and, generally, more meaningful. The successful separation of signal from noise depends on the choice and application of any of a number of statistical data filtering or smoothing techniques that are key elements in the analysis of time-series observations (Diggle 1990). An important question still remains. How do we treat data that are unusual (e.g., fluctuations of more than ±2 or 3 standard deviations from the mean); flag them? remove them? focus future studies on them? Single point anomalies, particularly when they could be significant parts of an annual signal, may be important. Some of the most significant advances in the existing time-series studies came as the scientists were confronted with repeats of previous (one-off) patterns from isolated expeditions and forced to explain these patterns.

Finally, it is essential to emphasize that the combination of sufficient funding and credible research goals is a necessary but insufficient condition for the successful operation of an ocean time-series program. These programs also require a staff of dedicated, well-trained support personnel who are not prone to seasickness. These skilled technicians maintain the high level of data quality in spite of unexpected and demanding circumstances and the inevitable doldrums that come from repeated sampling of the same system. The ocean sciences community owes a debt of gratitude to these dedicated staff members. As time-series programs mature, it has become apparent to the principal investigators that the human infrastructure is one of the most challenging components of running and sustaining a time-series program.

10.5 Cross Ecosystem Habitat Comparisons: Nutrient, Chlorophyll and Production-Export Relationships

Cross-habitat comparisons of selected state variables and carbon fluxes have proven invaluable in ecological research. Key parameters relevant to the functioning of the ocean's biological pump include near-surface and

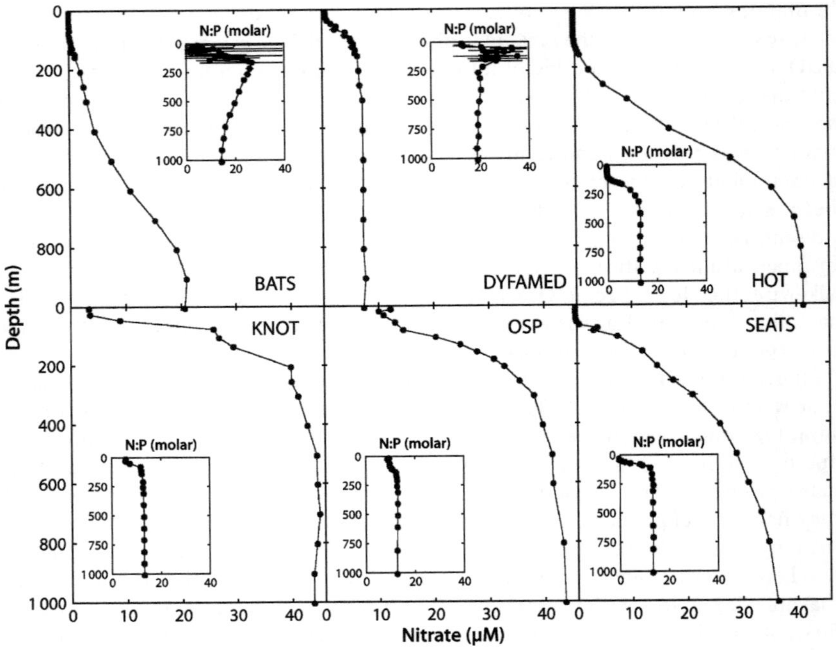

Fig. 10.6.
Nitrate concentration vs. water column depth profiles for BATS, DYFAMED, HOT, KNOT, OSP and SEATS time-series study sites. The data shown are climatological mean values and 95% confidence intervals which are generally less than the size of the symbols used. Inserts show the depth variation in the dissolved molar nitrate-to-phosphate (N:P) ratios as mean values and 95% confidence intervals. Note the high variability in the BATS N:P data set in the 0–200 m portion of the water column and the generally opposing trends in the N:P ratio profiles (i.e., decreasing for BATS vs. increasing for HOT with increasing water depth)

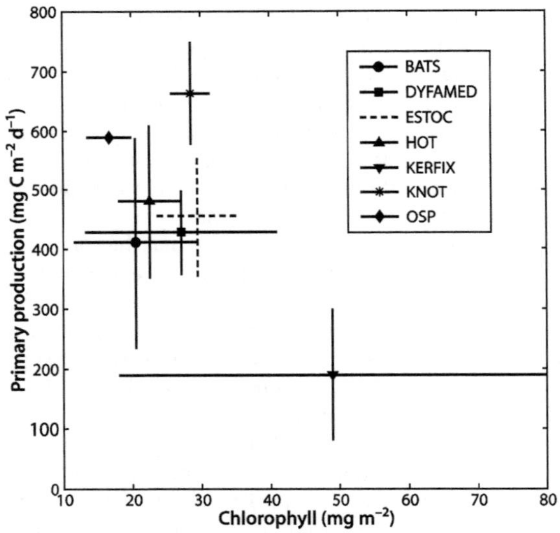

Fig. 10.7. Cross-site ecosystem comparisons during the JGOFS era. Data shown are the euphotic zone depth integrated chlorophyll *a* concentrations vs. contemporaneous rates of primary production presented as mean ±1 standard deviation for most data sets (see Table 10.2 for values)

euphotic zone depth-integrated nitrate and Chl *a* inventories, and rates of primary, new and export production (Fig. 10.6–10.9, Tables 10.2 and 10.3).

Due to space limitations and, in part, to the proprietary nature of selected data sets we have not attempted a comprehensive integration or cross-site ecological interpretation of these biogeochemical data sets; however, a few general comments are warranted. First, despite the broad geographic range and cross-site variations in physical forcing and biological community structure, there is a fairly narrow range in standing stocks of photoautotrophs (as measured by euphotic zone Chl *a*) and in rates of primary production (Table 10.2 and Fig. 10.7). Integrated Chl *a* concentrations varied by approximately a factor of two and rates of primary production by, at most, a factor of three. The rate measurements are harder to compare because of the different methodologies that were employed; for those programs utilizing the standard ^{14}C JGOFS measurement protocol (BATS, DYFAMED, HOT and OSP) the rates of primary production were nearly identical with values ranging from 416 ±178 mg C m^{-2} d^{-1} at BATS to 589 mg C m^{-2} d^{-1} at OSP. These values are much higher than previous (pre-JGOFS) estimates would have implied, which may be a reflection of improved techniques or, possibly, habitat change.

The measured values for particulate carbon export from the euphotic zone were more variable, as would be expected, given the uncertainties and potential bias of free-drifting sediment traps (Table 10.2 and Fig. 10.8). Nevertheless, methodology aside, there appears to be a large variation in the *e*-ratio both between sites, and over time at a given site (Fig. 10.8 and 10.9). For example, for the two most extensive data sets (BATS and HOT; Fig. 10.9) there are substantial aseasonal, multi-year changes in the *e*-ratio of nearly an order of magnitude, suggesting variations in biogeochemical processes including, but not limited to, possible changes in community structure. The means and standard deviations for the *e*-ratios at BATS and HOT are 0.072 ±0.038

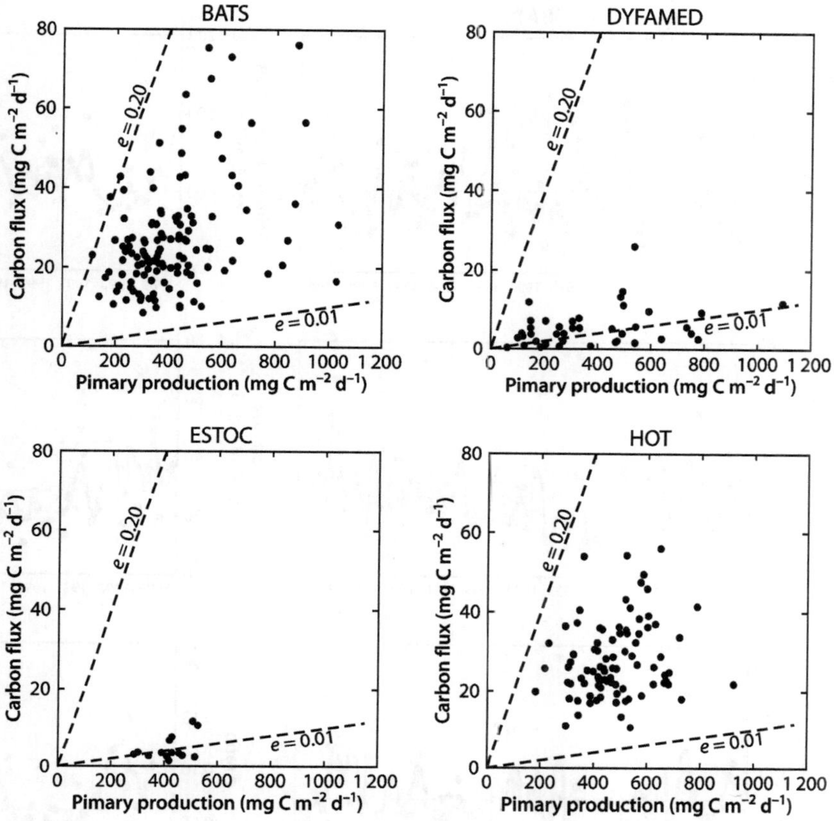

Fig. 10.8. Cross-site ecosystem comparisons during the JGOFS era. Data shown are the individual values of primary production and contemporaneous carbon flux for field experiments conducted at BATS, DYFAMED, ESTOC and HOT sites. Each graph also shows lines that are equivalent to *e*-ratios of 0.20 and 0.01 for comparison. The particle fluxes, measured using drifting sediment traps, are from either 150 m (BATS and HOT) or 200 m (DYFAMED and ESTOC). Values are presented in Table 10.2

Table 10.2. A cross-ecosystem comparison of selected JGOFS core parameters. Data presented as minimum and maximum values for each site. This generally reflects a regular seasonal variation in the parameter noted, e.g., summer-to-winter extremes. However, it is important to emphasize that the values of the individual parameters do not always align with summer vs. winter. For example, mixed-layer depth is maximum in winter but temperature and, generally, chlorophyll values are highest in summer. The data presented are simply the climatological extremes or, in certain data entries, the mean ±1 standard deviation

Location	Mixed layer properties				Euphotic zone properties		
	Depth (m)	T (°C)	[NO_3^- + NO_2^-] (µM)	Chl *a* (µg l^{-1})	Chl *a* (mg m^{-2})	Primary production[a] (mg C m^{-2} d^{-1})	C export (mg C m^{-2} d^{-1})
BATS	10 – 300	19 – 29	<0.05 – 0.5	0.05 – 0.3	20.5 ±8.8	416 ±178	27.2 ±13.9
DYFAMED	10 – 500	12 – 25	0.1 – 2	0.02 – 1.5	27.1 ±13.9	427 ±70	11.5 ±4.5
ESTOC	35 – 150	17 – 24	<0.05	0.05 – 0.4	29.4 ±5.7	456 ±98	4.5 ±2.9
HOT	20 – 100	22 – 30	0.01 – 0.03	0.05 – 0.1	22.5 ±4.5	480 ±129	28.3 ±9.9
KERFIX	<50 – 250	1.5 – 4.5	23 – 28	0.3 – 1.5	18 – 80	189 ±110	1.9 – 2.9
KNOT	15 – 60	3 – 7.5	10 – 23	–	28.6 ±2.9	663 ±86	–
OSP	40 – 120	5.5 – 13	8.5 – 16	0.2 – 0.4	13.5 – 20	589	18.2 ±7.2

[a] Primary production is estimated by: (a) ^{14}C or ^{13}C incubation techniques (BATS, DYFAMED, HOT, OSP), (b) spring-to-summer nutrient drawdown (KERFIX, KNOT), and (c) bio-optical model that considered both chlorophyll and temperature (ESTOC; S. Neuer, pers. comm.).

(range = 0.016–0.215; $n = 125$) and 0.062 ±0.026 (range = 0.020–0.149; $n = 89$), respectively. The *e*-ratio variability at both sites also has a lower frequency signal with higher than average *e*-ratios in 1989–1990 and 1997–1999, and lower than average *e*-ratios in the mid-decade especially 1991–1992 and 1995–1996 (Fig. 10.7 and 10.8). These changes in *e*-ratio are driven mostly by changes in particulate C flux rather than by changes in primary production; the latter displays distinct seasonality at both sites, but no secular trend. An accurate estimate of the annual organic matter export from the zone of net oxygen production of the ocean requires time-series meas-

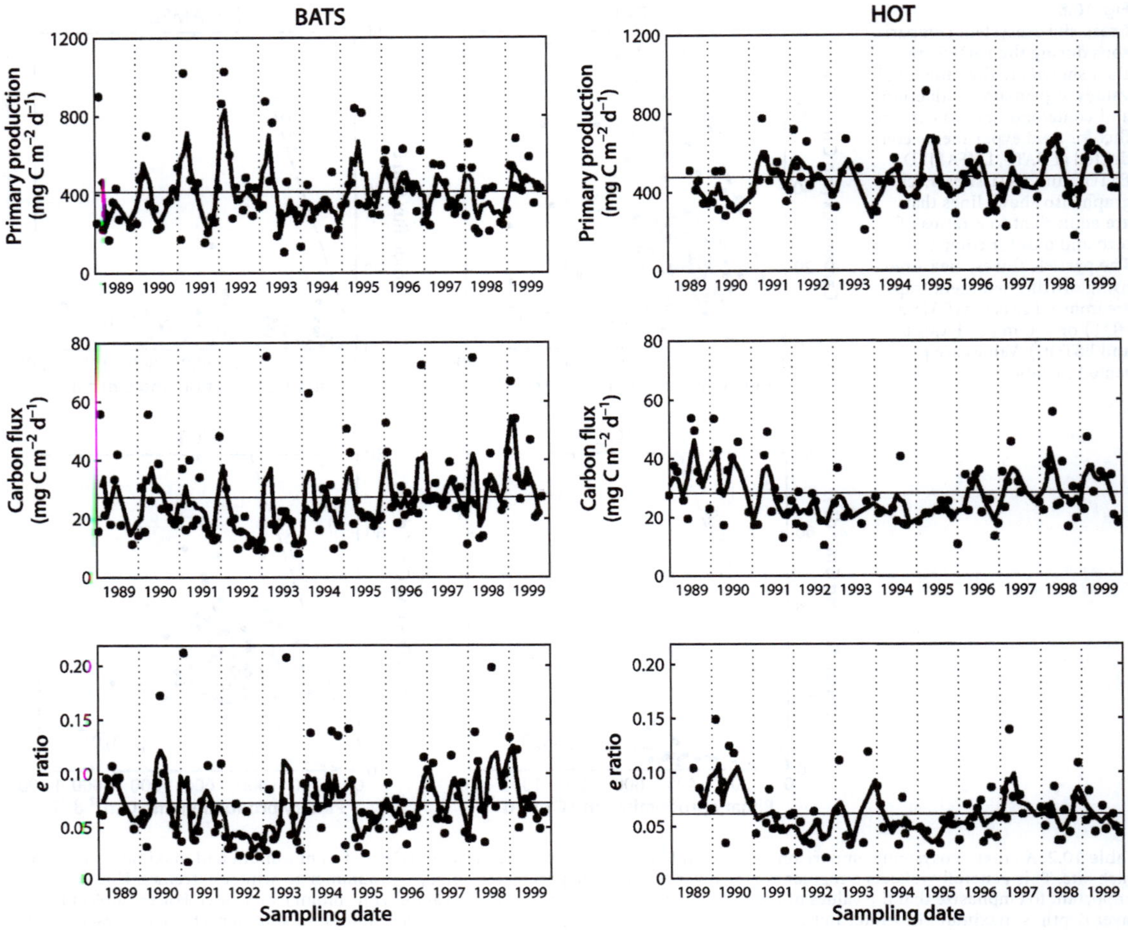

Fig. 10.9. Temporal variations of primary production (measured by the ^{14}C technique) and carbon flux (measured using free-drifting sediment traps positioned at the base of the euphotic zone) for an 11-year period at BATS and HOT time-series sites. Also shown at the *bottom* is the corresponding *e*-ratio (flux ÷ production). The *solid symbols* represent the individual cruise data for each parameter and the *heavy solid line* is the 3-point running mean. *Left:* BATS data sets showing the climatological mean values (*horizontal lines*): primary production = 416 mg C m^{-2} d^{-1}, carbon flux = 27.2 mg C m^{-2} d^{-1}, and *e*-ratio = 0.072. *Right:* HOT data sets showing the climatological mean values (*horizontal lines*): primary production = 480 mg C m^{-2} d^{-1}, carbon flux = 28.3 mg C m^{-2} d^{-1}, and *e*-ratio = 0.062

Table 10.3. Inorganic carbon pool dynamics at selected JGOFS time-series sites (data are presented as the measured range (minima and maxima) or as mean and, in parentheses, 95% confidence intervals, as shown. *n* = number of observations)

Location	Measurement period	Annual N-DIC range[a] (µmol kg^{-1})	Secular N-DIC change (µmolkg^{-1} yr^{-1})	Annual fCO_2 range (µatm)	Secular fCO_2 change (µatm yr^{-1})	Reference
BATS	10/88 to 12/99 (11 yr)	35 – 45	+1.6	60 – 120	+1.4	Bates (2001)
DYFAMED	2/98 to 2/00 (2 yr)	100	–	120	–	Copin-Montégut and Begovic (2002)
ESTOC	10/95 to 12/00 (5 yr)	20 – 25	1.2 ±0.2	60 – 80	+1.1 (±0.2)	Gonzalez-Davila (unpubl.)
HOT	10/88 to 12/99 (11 yr)	15 – 20	+1.18 (0.79 – 1.58), *n* = 94	25 – 60	+2.51 (1.59 – 3.44), *n* = 86	Dore et al. (2002)
KNOT	6/88 to 2/00 (1.5 yr)	107	–	–	–	Nojiri et al. (unpubl.)

[a] N-DIC is dissolved inorganic carbon concentration normalized to a constant salinity of 35.0.

urements, and in this regard, the ocean is still dramatically under-sampled.

Just prior to the start of JGOFS there was a somewhat heated debate regarding the validity of the ^{14}C method, as generally applied, for the determination of primary production. The controversy derived, in part, from estimates of rates of oxygen utilization in the intermediate depths of the ocean (Jenkins 1982; Jenkins and Goldman 1985) and assumptions that were applied to derive estimates of new and export production under steady-state conditions. Platt and Harrison (1985) later reconciled the data sets by concluding that the *f*-ratio (new / total production) was an ecosystem variable that positively scaled on ambient nitrate concentration. Application of their model to the 4-year nitrate data set (1959–1963) from Hydrostation S near Bermuda, predicted that the *f*-ratio could vary from 0.03 to 0.53, with an annual mean of 0.31 (Platt and Harrison 1985). This variable and higher than previously assumed mean value, when considered along with measured rates of primary production, fully accommodated the export production (assuming new production ≈ export production) that was needed to balance the subeuphotic zone oxygen consumption rates. From this analysis, Platt and Harrison (1985) made a very important conclusion regarding the biological carbon pump, namely that it is "untenable to speak about a particular ocean province, and certainly not the Sargasso Sea, as being oligotrophic (also see Lipschultz et al. 2002). Rather, a given oceanic province can be expected, locally in (x, y, z and t), to manifest a range of characteristics from apparent extreme oligotrophy to eutrophy as evidenced by the nearly 20-fold range in *f*-ratios." The data sets from the JGOFS time-series programs (Fig. 10.8), especially BATS and HOT (Fig. 10.9) confirm the Platt and Harrison (1985) prediction of variable *e*-ratios, and the implied *f*-ratios.

Below we present four JGOFS case studies that provide selected vignettes of key oceanic processes that have been explored at the time-series stations during JGOFS.

10.5.1 Case Study 1: Estimates of the Biological Carbon Pump at Ocean Times Series Sites

The mass balance of carbon in the upper ocean includes the fluxes of DIC, DOC and PC, which are influenced by numerous physical and biological processes including gas exchange, upwelling, horizontal advection, eddy diffusion, and the nature of the biological carbon pump. The strength of the carbon pump can be determined by a variety of methods at the ocean time-series sites for comparison to contemporaneous photosynthetic carbon production (Table 10.4). Net oxygen flux estimates are derived from periodic (typically monthly) measurements of dissolved O_2, argon (Ar) and nitrogen (N_2) and a model of processes that affect dissolved gas concentrations (gas exchange, seasonal temperature change, mixing, bubble collapse). A comparison of O_2 and Ar saturation state can be used to estimate the impact of biological processes on O_2 gas concentration (Fig. 10.10). This is because O_2 is produced by net photosynthesis whereas Ar is biologically inert. For the HOT site, about half of the shallow O_2 maximum is the result of biological processes (Emerson et al. 1995). Particle fluxes were determined from monthly floating sediment trap deployments at 100–200 meters. Other methods include carbon isotope mass balance, ^{15}N incubations and mass balances of both particulate and dissolved organic matter.

At Sta. ALOHA the values determined by O_2 mass balance are within the uncertainties (±50%) of those measured by organic carbon fluxes, and carbon isotope mass balances (Emerson et al. 1997). At the OSP site the mean annual organic carbon export rate from O_2 mass balance is about 40% lower than that determined from recent annual measurements of ^{14}C productivity and ^{15}N uptake (Boyd and Harrison 1999b; Varela and Harrison 1999b), which is also likely within the uncertainties of these

Table 10.4. Upper water column carbon fluxes as measured by a variety of techniques: A cross-ecosystem comparison. Total primary production and particulate fluxes are from compilations of the time-series data. Net O_2 production estimates are from BATS (Spitzer and Jenkins 1989); HOT (Emerson et al. 1997); and OSP (Emerson et al. 1991). Other methods are: BATS, particulate + DOC fluxes (Carlson et al. 1994); HOT, $\delta^{13}C$ mass balance and particulate and organic carbon fluxes (Emerson et al. 1997); KNOT, seasonal NO_3 change (Wong et al. 2002); and OSP, calculated from ^{14}C primary production and a ^{15}N *f*-ratio measurements (Boyd and Harrison 1999b; Varela and Harrison 1999a)

Location	Total primary production (mol C m^{-2} yr^{-1})	Particulate C flux (mol C m^{-2} yr^{-1})	Net O_2 production method (mol C m^{-2} yr^{-1})	Other methods
BATS	12.7	0.83	3.3 ±1.1	1.8 ±1.0
DYFAMED	13.0	0.35	–	–
ESTOC	13.9	0.14	–	–
HOT	14.6	0.86	2.7 ±1.7	1.6 ±0.9, 2.0 ±1.0
KERFIX	5.7	0.07	–	–
KNOT	20.2	–	–	6–7
OSP	17.9	0.55	2.0 ±1.1	2.8

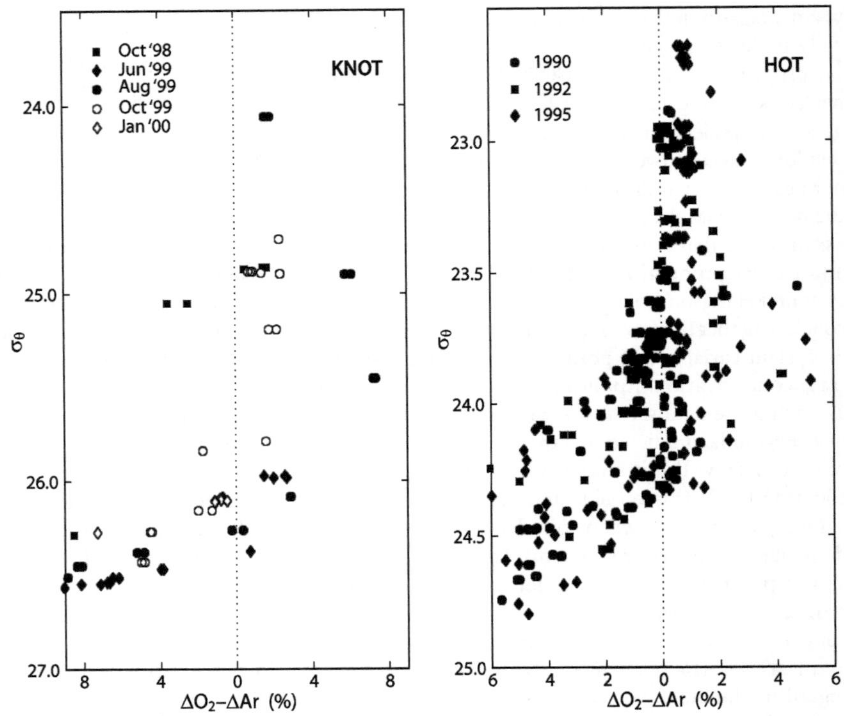

Fig. 10.10. Dissolved oxygen and argon data for the upper water columns at the KNOT (*left*) and HOT (*right*) study sites. Data presented are the difference in the degree of saturation, Δ (in %), between oxygen (O_2) and argon (Ar) as a function of potential density, σ_θ, indicating a maximum in net biological O_2 production beneath the mixed-layer (*note:* for KNOT a σ_θ of 25.0 is approximately 39 m and for HOT, a σ_θ of 24.0 is approximately 100 m) (data from Emerson et al. (1997) and Wong et al. (1999))

determinations (Table 10.4). This type of closure has not been achieved for BATS where the O_2 mass-balance estimates continue to be about a factor of two greater than that determined from organic matter mass balances (Carlson et al. 1994). This discrepancy may have to do with the depths over which each of these techniques integrate the net carbon export, but this problem remains an open question.

The preliminary results from KNOT are based on the summertime decrease in nitrate concentration determined by surface water measurements across the subarctic Pacific on a ship of opportunity 8–9 times per year since 1995 (Goes et al. 1999; Wong et al. 1999, 2002). This estimate agrees with independent determinations of new production in this area based on satellite-determined Chl *a* and sea surface temperature and also an east-west comparison of sediment trap observations (Goes et al. 1999). Other estimates of the net carbon export rate are presently being determined by a time-series of measurements of ^{13}C primary production and new production and O_2, Ar and N_2 measurements (see Fig. 10.10).

Values for the biological pump determined at these time series sites indicate some very interesting contrasts. The fact that the export production in the subtropical and subarctic Pacific are the same to within the error of our measurements was unexpected based on satellite color measurements (Falkowski et al. 1998), but may be partially explained by recent studies of the effect of temperature on carbon export (Laws et al. 2000). The apparent difference in the biological carbon pump between the western and eastern subarctic Pacific begs the question about the processes controlling export at these high latitudes. Are the very high estimates from the west caused by greater mixed-layer depths and thus more efficient nutrient transport from below, or are they due to the enhanced role of large diatoms caused by the supply of Fe from the Asian continent?

An estimate of the global net biological carbon export of 10–15 Gt yr^{-1} has been made, based on a summary of these estimates (Table 10.4), including independent estimates for values in the Equatorial and near shore regions, and the assumption that these measured values are representative of their different ocean provinces. This value is much larger than originally determined by Eppley and Peterson (1979), but is consistent with the model-based estimates of Laws et al. (2000) which fall within this range. A remaining discrepancy in these approaches, however, is the role of subtropical oceans. If the measurements at HOT are typical of subtropical values, up to 40% of the biological pump may be located in these regions (Emerson et al. 1997). As we shall see below (Case Study 3), biogeochemical conditions at the HOT site are time-variable and climate-sensitive (also see Karl et al. 2001a,b). It appears that we are reaching consensus for the value for the global carbon export flux, but there are still great uncertainties in the global distribution of the biological pump that will require estimates from a greater number of time-series stations.

10.5.2 Case Study 2: A 'Bermuda Triangle' Carbon Mystery with Global Implications

When photoautotrophic microorganisms produce new organic matter during the process of photosynthesis, current models predict that they remove dissolved inorganic C, N, P and other essential bioelements in proportions equivalent to the new organic matter that is formed. When careful mass balance studies are conducted, the removal of dissolved nutrients has been shown to equal the sum of newly formed euphotic zone dissolved and particulate organic matter, plus that which has been exported from the local source. During the initiation of the spring bloom in the BATS study area there is a systematic, simultaneous drawdown of all required nutrients, which is fully anticipated. The mystery begins to develop after the nitrate, upwelled to the surface during deep winter mixing events, disappears from the water column as the annual summer oligotrophy returns to Sargasso Sea (Fig. 10.11, top). As the presumptive rate-limiting nutrient for organic matter production, the exhaustion of nitrate should cause an immediate cessation of new primary production and coupled organic matter export.

The continued disappearance of salinity normalized dissolved inorganic carbon (N-DIC) in the absence of nitrate was first reported by Michaels et al. (1994). They reasoned that if nitrate was added by episodic wind mixing or mesoscale eddy motions, the nitrate would be delivered along with DIC, so simple enhancements of nitrate-supported new and export production could not be responsible for the repeatable summertime N-DIC disappearance at BATS. They also evaluated, then rejected, the potential role of CO_2 outgassing during local warming of the near surface ocean, and the accumulation of nascent carbon-enriched dissolved organic matter as mechanisms for a net removal of N-DIC (Michaels et al. 1994; Bates et al. 1996b). More recently the role of short-term wind variability on air-to-sea CO_2 gas exchange (Bates et al. 1998) and the vertical migration of zooplankton (Steinberg et al. 2000) have been suggested as potentially important processes, but neither flux

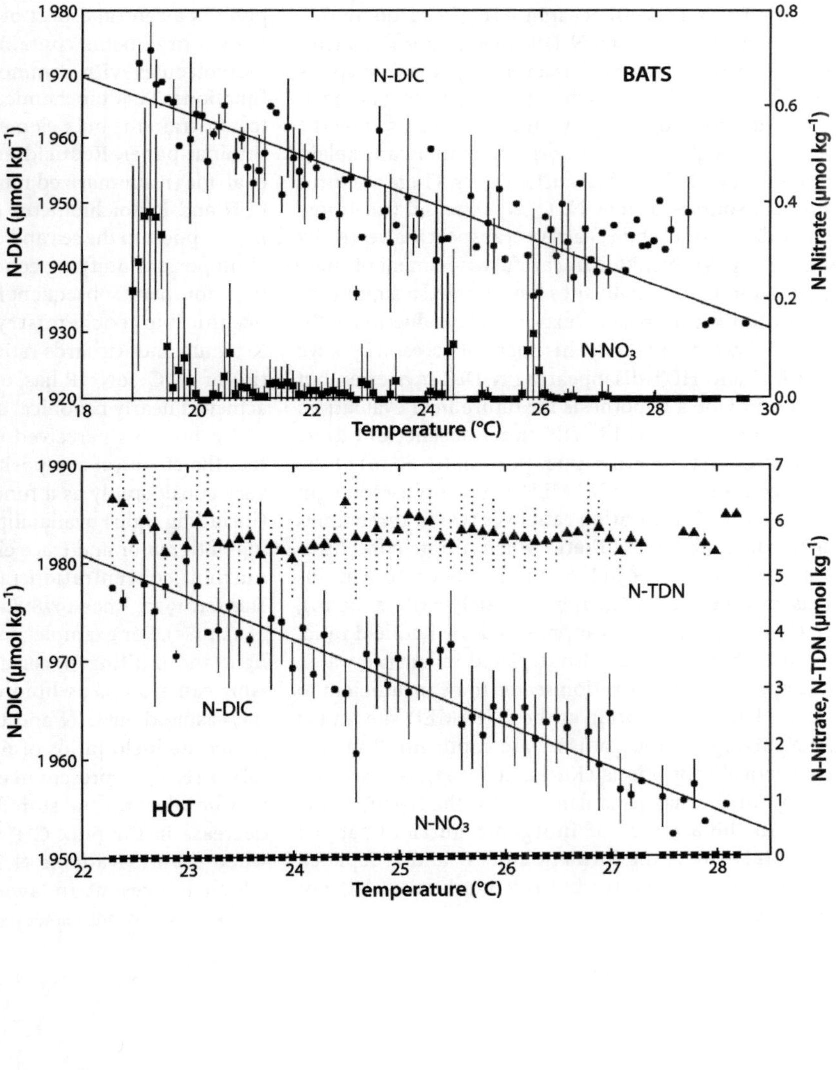

Fig. 10.11. Relationships between carbon removal and fixed nitrogen in the surface waters of the subtropical North Atlantic (*top:* BATS) and subtropical North Pacific (*bottom:* HOT) during the period of summertime warming of the sea surface. At the BATS site, DIC normalized to a salinity of 35 (N-DIC) shows a systematic decrease with increasing temperature even in the absence of nitrate. At the HOT site, a similar summertime N-DIC drawdown in the absence of nitrate (the surface water nitrate concentration at Sta. ALOHA during the summer period is always less than 0.01 μmol kg^{-1}). Analyses of salinity-normalized total dissolved N (N-TDN) also failed to document a simultaneous loss of fixed N from the much larger pool of dissolved organic N

seems large enough to account for the observed upper ocean carbon imbalance. Regional horizontal gradients were weak (Bates et al. 1996c) and, therefore, insufficient for local N-DIC removal (Ono et al. 2001); where is this substantial pool of carbon (~35–40 μmol kg^{-1}) going, and what process(es) are responsible for the recurrent summertime loss of N-DIC in the absence of nitrate?

A summertime drawdown of N-DIC of nearly equivalent magnitude is also observed at the HOT site (Fig. 10.11, bottom). At this oligotrophic Pacific Ocean Station, the nitrate is even lower than it is near Bermuda, with surface concentrations always ≤ 10 nM (Karl et al. 2001b). Because there is a fairly large (5–6 μM) but poorly characterized pool of dissolved organic nitrogen that dominates (>99%) the total dissolved N (TDN) pool at Sta. ALOHA, it is conceivable that its utilization could support the summertime drawdown of N-DIC. For a measured C:N export ratio of 8 at Sta. ALOHA (Karl et al. 1996), the DON pool if fully utilized could support the removal of approximately 45 μmol DIC kg^{-1}, and could fully reconcile the 'mystery.' However, an analysis of the N-TDN data set for Sta. ALOHA indicates that the summertime concentration changes very little, if at all, during the period of N-DIC disappearance (Fig. 10.11, bottom).

The variability in total N-DIC in seawater is controlled by at least three processes: photosynthesis/respiration, air-to-sea CO_2 gas exchange and vertical and horizontal advection/diffusion. Of these processes, only the net biological production of organic matter can explain the Bermuda (and now Hawaii) mystery. There are three potential sources of new N: (*1*) N_2-fixation, (*2*) atmospheric deposition, and (*3*) active transport via vertically migrating phytoplankton. A careful assessment of these potential sources at both sites has revealed a significant role for N_2 fixation as a new export and production pathway (Michaels et al. 2000). This does not necessarily solve the BATS and HOT disappearing N-DIC mysteries, but it does provide a hypothesis for future field evaluation.

The consumption of N-DIC in the absence of nitrate was also observed in the upper portion (0–20 m) of the water column at the DYFAMED site (Copin-Montégut 2000). The C:N utilization ratio was two to three times higher than the Redfield ratio, when nitrate concentration was less than 0.5 μM. As nitrate appeared with increasing water depth (to approximately 6 μM at 40 m), the C:N utilization ratio approached the Redfield ratio. Because phosphate was also depleted in surface water, the hypothesis of N_2 fixation seems to be insufficient to explain DIC consumption at the DYFAMED site unless the N_2-fixing microorganisms are capturing P via active vertical migrations (Karl et al. 1992), or by other means. An alternative explanation for the N-DIC drawdown in the absence of inorganic nutrients at the DYFAMED site is the formation of N- and P-depleted dissolved organic matter, but this mechanism has not been confirmed.

A recent global ocean analysis of the "disappearing DIC in the absence of measurable nitrate" has shown it to be a recurrent feature of all tropical and subtropical marine habitats (Lee 2001). Application of several fairly well-constrained assumptions, enable the author to estimate a global rate of new carbon production that is supported by N_2 fixation of 0.8 ±0.2 Gt C yr^{-1} for subtropical and tropical marine habitats. Apparently the summertime drawdown of N-DIC in the absence of nitrate, first reported near Bermuda from the BATS data sets, is a ubiquitous global phenomenon. This process takes on an added significance because it can decouple C, N and P cycles, and provides a mechanism for the net sequestration of atmospheric carbon as defined by the N_2-primed Prokaryote Carbon Pump (Fig. 10.3b).

10.5.3 Case Study 3: Decade-Scale, Climate-Driven Changes in the N_2-Primed Prokaryote Carbon Pump

Nutrient dynamics and their role in the stoichiometric variability of dissolved and particulate organic matter pools is a central aspect of biogeochemical studies. All known organisms contain a nearly identical suite of biomolecules with common structural and metabolic functions. This biochemical uniformitarianism serves to constrain the bulk elemental composition of life. In a seminal paper, Redfield and his colleagues (Redfield et al. 1963) summarized much of the earlier research on C, N and P stoichiometry of dissolved and particulate matter pools in the sea and combined these data sets into an important unifying concept which has served as the basis for many subsequent field and modeling studies in oceanic biogeochemistry. The so-called 'Redfield, Ketchum and Richards ratio' (or simply, the Redfield ratio) of 106C:16N:1P has, over the intervening decades, achieved nearly canonical status in aquatic sciences.

Despite this perceived uniformity, it is well known that the chemical composition of living organisms can vary considerably as a function of growth rate, energy (including light) availability, ambient nutrient (including both major and trace elements) concentrations and nutrient concentration ratios (Sakshaug and Holm-Hansen 1977; Rhee 1978; Laws and Bannister 1980; Tett et al. 1985). For example, under conditions of saturating light and limiting N, certain photoautotrophic organisms can store C as lipid or as carbohydrate, thereby increasing their C:N and C:P ratios relative to the expected Redfield ratios of 6.6:1 and 106:1, respectively. Likewise, if P is present in excess of cellular demands, it may be taken up and stored as polyphosphate causing a decrease in the bulk C:P and N:P ratios. Conversely, when the bioavailable N:P ratio is greater than that which is present in 'average' organic matter (i.e., >16N:1P by atoms) selected groups of microorganisms can reduce their cell quotas of P, relative to C and N, and effect net biomass production with C:P and N:P ratios significantly greater than the canonical Redfield ratios of 106:1 and 16:1, respectively.

Open ocean ecosystems are characterized by low concentrations of fixed, bioavailable N which would appear to make them a suitable niche for the proliferation of N_2-fixing prokaryotes. In their now classic treatise on nutrient dynamics in the sea, Dugdale and Goering (1967) introduced the unifying concept of 'new' (i.e., nutrients imported to the local environment from surrounding regions) vs. 'regenerated' (i.e., nutrients that are locally remineralized) forms of nitrogen. They were careful to emphasize that there were several potential sources of new N for the euphotic zone, each of equal value but with potentially different ecological consequences. Since there were few data on N_2 fixation rates when their paper was published, importation of nitrate from below the euphotic zone was considered to provide the majority of new N in the sea.

Now, thirty years after the new production concept was introduced, there is increasing evidence that rates of oceanic N_2 fixation may have been systematically underestimated (see Case Study 2, above), or, perhaps, have increased in relative importance over time. This new evidence comes from several independent lines of investigation. One of the most interesting and provocative modern data sets is that derived from the application of novel molecular methods to detect the presence and abundance of N_2-fixing microbes either by hybridization or amplification of nitrogenase (*nif*; the enzyme system used to reduce N_2 to ammonia) genes (Zehr et al. 1998, 2000). Application of these methods to open ocean biomes in the North Atlantic and North Pacific Oceans has revealed a spectrum of previously uncharacterized *nif* gene phylotypes. Furthermore, significant *nif* phylotype diversity is apparent both within and between open ocean ecosystems. These novel data sets, when considered in concert with other recent reports of high rates of oceanic N_2 fixation, support the hypothesis that N_2 fixation is a major source of new N over vast regions of the world ocean (Michaels et al. 1996; Gruber and Sarmiento 1997; Karl et al. 1997, 2002).

Several lines of evidence from Sta. ALOHA suggest that N_2 fixation is an important contemporary source of new nitrogen for the pelagic ecosystem of the North Pacific Ocean. These independent measurements and data syntheses include: (*a*) N-DIC drawdown in absence of nitrate or other forms of fixed N (Case Study 2, above); (*b*) *Trichodesmium* (the putative, dominant open ocean N_2 fixer) population abundances and estimates of their potential rates of biological N_2 fixation; (*c*) assessment of the molar N:P stoichiometries of surface-ocean dissolved and particulate matter pools and development of a one-dimensional model to calculate N and P mass balances; (*d*) seasonal variations in the natural ^{15}N isotopic abundances of particulate matter exported to the deep sea and collected in bottom-moored sediment traps; and (*e*) observations on secular changes in soluble reactive P (SRP), soluble nonreactive P (SNP) and dissolved organic N (DON) pools during the period of increased rates of N_2 fixation (Karl et al. 1997).

The production, export and remineralization of N-enriched (relative to P) dissolved and particulate organic matter (Fig. 10.12) is a diagnostic characteristic of a N_2-supported ecosystem. Following the selection for N_2-fixing microorganisms in response to N starvation, the ecosystem will eventually replenish the fixed N deficit via the remineralization of high N:P from exported organic matter. In theory, the N:P ratio in dissolved nutrients beneath the euphotic zone will increase above the Redfield ratio of 16N:1P, and the subsequent replenishment of surface waters with this high N:P ratio water will serve to repress further N_2 fixation and thereby select against these populations. Because the residence time of nutrients in the top of the thermocline of the North Pacific subtropical gyre is on the order of a few decades, one might predict an episodic selection for, followed by selection against N_2-fixing microbial assemblages. This feedback will be bounded by the, yet unknown, constraints on the elemental stoichiometry of the full oceanic ecosystem.

Apparently, the HOT program began at or near the beginning of one of these 1–2 decade long episodes of enhanced N_2 fixation and may be moving towards the opposite phase. The best evidence for this comes from a time-series of the N:P ratio for suspended/sinking particulate matter and dissolved matter near the top of the thermocline (200–250 m). The continued production and export of particulate matter with a N:P ratio higher than the Redfield ratio is strong evidence for N_2 fixation (Fig. 10.12). However, at the beginning of the HOT program in 1988, the N:P ratio in the thermocline was approximately 16, which suggests that the biogeochemical trends observed during the HOT program cannot be representative of the longer term (100 yr) climatology. During the 1990s, however, the N:P ratio has increased significantly at a sustained increase equivalent to nearly 3–4 Redfield 'units' per decade (i.e., from a N:P of 16 in 1989 to nearly 20 in 2000). Although N budget estimates suggest that N_2 fixation may presently supply up to half of the N required to sustain particulate matter export from the euphotic zone, these processes are clearly not in steady-state. The relatively high percentage of N_2-supported production may represent a transient ecosystem state reflecting either oceanic variability or, perhaps, an unusual state established in response to the well-documented, decade-long shift in North Pacific climate (Karl 1999).

There is now ample evidence to suggest that major changes in the structure of the NPSG can occur over decadal time scales (Emerson et al. 2001); both the po-

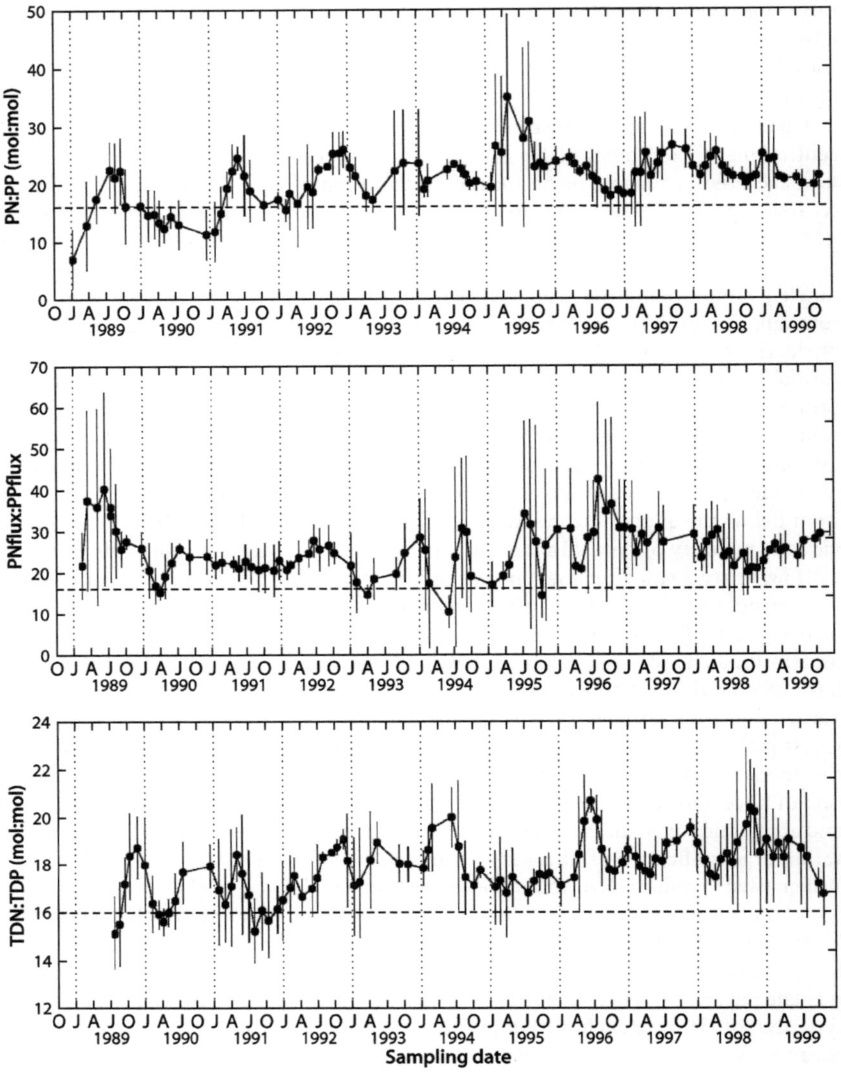

Fig. 10.12.
The average N:P stoichiometry of suspended particulate matter in the upper (0–100 m) portion of the water column (*top*), the average N:P ratio for exported matter (*center*) collected at 150 m reference depth, and the average N:P ratio for total dissolved matter at the top of the nutricline (200–250 m depth interval) for Sta. ALOHA at the HOT site. Data are presented as three point running mean values ±1 standard deviation of the mean. The Redfield ratio (N:P = 16) is shown as a horizontal dashed line in all three panels

sitions and intensities of major ocean currents (e.g., Kuroshio Current and extension) and atmospheric circulation features (e.g., Aleutian Low) can have profound effects on ocean biogeochemical processes. However, the precise linkages between climate and the biological carbon pump are not well documented, in part, due to lack of relevant time-series data sets on community structure and carbon fluxes. This altered view of biogeochemical dynamics in the gyre may have a profound influence on how one models ecosystem processes, including the potential impacts of natural or human-induced environmental change and its relationship to carbon sequestration.

Trichodesmium, a non-heterocystous filamentous N_2-fixing cyanobacterium with buoyancy regulation capabilities, is a major contributor to global N_2 fixation (Capone et al. 1997), especially in subtropical and tropical marine habitats. It is well known that rates of N_2 fixation by *Trichodesmium*, and perhaps other N_2 fixing microorganisms as well, are enhanced under periods of low turbulence. Massive blooms of *Trichodesmium*, easily recognizable from aircraft and satellites, have been reported during extended periods of low wind and calm seas (Carpenter and Price 1976; Karl et al. 1992). For the NPSG, we have used meteorological data from the NOAA-NDBC buoy #51001 (23° 24' N, 162° 18' W) near Sta. ALOHA to hindcast periods that would be conducive for the growth of *Trichodesmium*. We used diel variations in sea surface temperature (ΔT), defined as the daily maximum SST (SST-max) minus the minimum (SST-min), as an indirect measure of the combined influence of wind, wave height and degree of ocean stratification (Karl et al. 1992). During rough periods, ΔT would be small (<0.2 °C) because of an efficient dis-

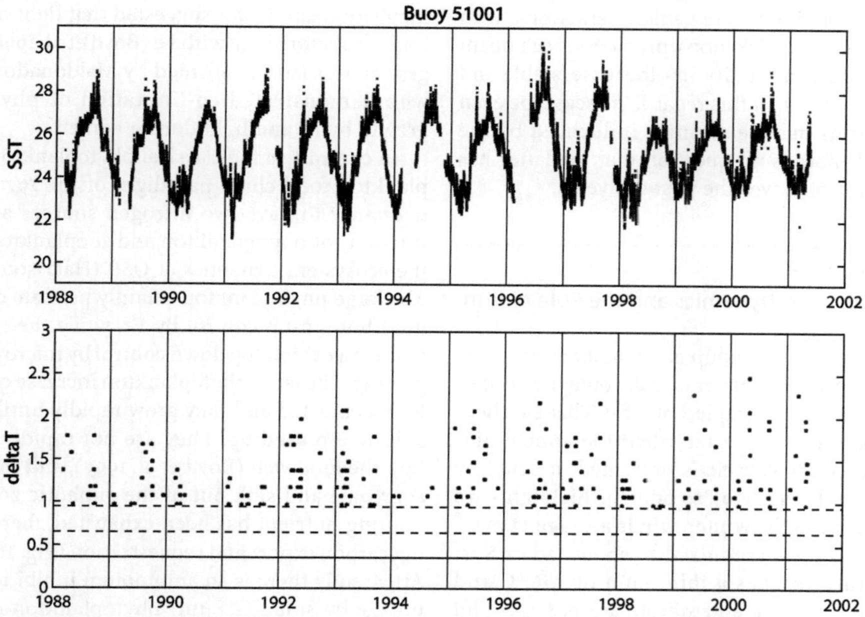

Fig. 10.13. Time-series measurement of sea surface temperature (SST) at the 1 m reference depth in the North Pacific Subtropical Gyre as observed at the NOAA-NDBC 6-m Nomad buoy at 23° 24' N, 162° 16' W. The *panel at the top* shows the full data set during the JGOFS (HOT program) period. In addition to the expected seasonal variability, there are significant interannual differences including the warmer than normal summers in 1996 and 1997 and the colder than normal summers in 1998–2000. The *bottom panel* shows the daily temperature excursions, expressed as ΔT (°C) as determined by daily maximum minus daily minimum SST. The *solid horizontal line* is the mean ΔT for the complete data set (0.36 °C) and the *dashed lines* are the +1 and +2 standard deviations (1 standard deviation = 0.29 °C), respectively for the 13.5 yr climatology. Also shown are those individual data points that are > +2 standard deviations above the mean ΔT. These large temperature excursions reflect well-stratified ocean conditions that are known to be conducive for the growth of N_2-fixing microorganisms at the HOT site

sipation of the daily solar heating. During calm periods, ΔT can reach values of 2 °C, or greater (Fig. 10.13). This nearly continuous record reveals the following trends: (*1*) SST has a regular seasonal, as well as irregular interannual and subdecadal trends that are related to climate variations in the North Pacific and, more important for the discussion here; (*2*) the frequency of large positive ΔT excursions (the frequency of calm seas) has increased during the period of the HOT program investigations (Fig. 10.13). These meteorological data support the biogeochemical data sets and indicate an enhancement of N_2 fixation favorable conditions. While this discussion has focused on *Trichodesmium*, the same mechanism of local habitat stratification, dust deposition and N_2-fixing cyanobacterial bloom formation would also hold for large diatoms with N_2-fixing endosymbionts as well as for other free-living N_2-fixing picocyanobacteria.

With this new general understanding of the meteorological controls on potential rates of N_2 fixation in the NPSG, it may soon be possible to provide forecasts of N_2-based new production and carbon sequestration by the combined N_2-primed prokaryote and Fe-stimulated diatom aggregation carbon pumps. A 'hypothetical' forecast might be: "light trade winds with a diurnal SST excursion of 3 °C and a 50% probability of significant N_2 fixation, increasing to 90% during periods of aperiodic dust (Fe) deposition!"

Dugdale and Goering (1967) were careful to warn that if N_2 fixation was (later) found to be a quantitatively important pathway for nutrient supply, then a revision of the new vs. recycled N conceptual framework would be necessary; it may now be time for a reconsideration of this paradigm (Karl 2000; Lipschultz et al. 2002). First, the net rate of CO_2 sequestration into the interior portion of the ocean is directly controlled by the source(s) of new N (Fig. 10.3). If the nitrate flux dominates and nitrate is completely removed from the surface ocean, then the bidirectional mass fluxes of C and N would be nearly in balance as predicted by the new production – export production model (i.e., the 'Redfield ratio – dissolved/particulate carbon pump'). On the other hand, if N_2 fixation sustains a significant amount of new and export production in open ocean ecosystems, then net CO_2 will be sequestered (Michaels et al. 2000). Furthermore, N_2 fixation in the world's oceans may be controlled by the atmospheric deposition of Fe and the degree of surface ocean stratification, both of which are variable, climate sensitive parameters (see also Case Study 4). Total atmospheric dust transport is also affected by humankind, including population demographics, global economies and land use patterns. These complex natural and anthropogenic interactions, with multiple potential feedback loops, provide a mechanism for biogeochemical variability in otherwise 'stable and homogeneous' biomes. In this regard, the seascape even in remote regions may be strongly influenced by the landscape, and there is no question that the latter has changed significantly over the past 250 years.

10.5.4 Case Study 4: OSP Ecosytem Dynamics and the Role of Iron

During the 25 year weathership era, temperature, salinity, dissolved inorganic nutrients, chlorophyll *a* (Chl *a*) and zooplankton were sampled at OSP. This excellent temporal coverage firmly established the annual biogeochemical cycle. This cycle is presented in detail by Whitney and Freeland (1999), so only the highlights are summarized here. In the winter, winds average 12 m s^{-1} and the surface waters are mixed to about 120 m. Surface temperature reaches a minimum of 5–6 °C and maximum winter nitrate and silicate are 15.8 ±2.3 and 24.0 ±3.4 µM, respectively; winter Chl *a* is approximately 0.2 µg l^{-1}. As radiant energy increases in spring, and the mixed-layer shoals, Chl *a* increases from 0.2 to 0.4 µg l^{-1}, but, in contrast to the North Atlantic Ocean, no significant 'spring bloom' occurs.

OSP is a good example of an HNLC region, poised to fuel the biological carbon pump, but not currently functioning at maximum capacity. Initially it was thought that mesozooplankton grazing was responsible for the lack of a spring bloom, but this hypothesis was tested and rejected during the SUPER research program (Frost 1987). More recently, it has been suggested that dissolved Fe, present at a concentration of approximately 0.1 nM (Martin and Fitzwater 1988), limits the growth of large bloom forming diatoms and, thereby controls the efficiency of the biological carbon pump. Despite daily summertime production rates of 500–600 mg C m^{-2} d^{-1}, the phytoplankton community consists mainly of nanoflagellates that utilize primarily ammonium, not nitrate (Harrison et al. 1999).

At the beginning of the Canadian JGOFS OSP project in 1992, the issue of Fe limitation was not broadly accepted (Banse 1990; Miller et al. 1991; Miller 1993). During the early 1990s, further shipboard Fe enrichment experiments by Boyd and colleagues confirmed that Fe limitation did limit the drawdown of nitrate in May and September experiments (Boyd et al. 1996). When Fe was added in their experiments, mainly large (>18 µm) pennate diatoms (primarily *Pseudonitzschia* sp.) grew up, confirming the Martin and Fitzwater (1988) observations. Iron limitation was also confirmed by an increase in the molecular biomarker for Fe limitation, flavodoxin (LaRoche et al. 1996). In February, when Boyd added Fe, little or no increase in Chl *a* was obtained after a 5 day incubation and it was suggested that light may be a co-limiting factor along with Fe (Boyd et al. 1996). This suggestion was later confirmed by Maldonado et al. (1999) who demonstrated co-limitation of phytoplankton growth by Fe and light during winter.

In contrast to the classical phytoplankton-mesozooplankton food chain paradigm of the 1970s, it is now necessary to have two nitrogen sources and two size fractions of phytoplankton and zooplankton to explain the ecosystem dynamics at OSP (Harrison et al. 1999). The large phytoplankton (mainly pennate diatoms) exhibit bottom-up control by Fe, while the small phytoplankton exhibit top down control by microzooplankton grazing. The large phytoplankton increase quickly when Fe is deposited and they grow rapidly until nitrate and silicate are used up. They are not rapidly assimilated into the food web (Boyd et al. 1999), and therefore they aggregate and sink out of the euphotic zone after the limiting nutrient has been exhausted, thereby enhancing carbon export and sequestration (Fig. 10.2 and 10.3). Apparently there is an ammonium inhibition of nitrate uptake by small (<2 µm) phytoplankton cells (Varela and Harrison 1999a), and an iron limitation of nitrate uptake by larger (>2 µm) phytoplankton cells, including diatoms (Harrison et al. 1999). Fe is not required for ammonium or urea assimilation, but it is needed for growth on nitrate. Large diatoms cannot compete with the smaller cells for the uptake of ammonium and possibly urea also, so Fe availability controls the metabolic activity and growth of large diatoms.

Interannual variation in nitrate and silicate has been well documented in the 1970s due to the weekly sampling by weathership personnel (Whitney et al. 1998). During the summers of 1972, 1976 and 1979, silicate was depleted to <1 µM compared to the longer term climatology of about 10 µM (Wong and Matear 1999). Both 1972 and 1976 were high silicate and nitrate utilization years, while 1976 had low nitrate utilization relative to silicate utilization. It has been suggested that these periods of complete removal of silicate and nitrate may be manifestations of a large, pulsed Fe deposition event, but no direct evidence is available. The dominant source of new Fe is atmospheric deposition (Duce and Tindale 1991); episodic storm events over China lead to dust transport and fallout into the NE Pacific less than one week later. Because the NE subarctic Pacific is a Fe-limited ecosystem, pulsed deposition of bioavailable Fe from atmospheric sources would be expected to result in pulses of primary production, selection for large diatoms and a subsequent export pulse of organic matter, including biogenic silica. This Fe supply – carbon export prediction was tested by Boyd et al. (1998) using field data from OSP. While an explicit cause-and-effect could not be firmly established due to the episodic timing and short duration of the dust deposition events, historical data showing an aperiodic, order of magnitude, increase in mixed-layer chlorophyll and the flux patterns from bottom-moored sediment traps (which had a continuous record of pulsed export events) suggest a frequency of approximately three to five major dust events per decade. With this superannual frequency, the design of the field sampling program would need to be carefully evaluated and reliance on remote, continuous instrumentation imperative. While emphasis has been on atmospheric inputs of Fe from dust, transport of coastal waters offshore cannot be ruled out. Even though drifter studies show that surface water transport at OSP is eastward (i.e., shoreward; Bograd et al. 1999), recent TOPEX/POSEIDON satellite images show that very large eddies (~200 km in diameter) form near the southwest corner of the Queen Charlotte Islands and move offshore in the direction of OSP (Thomson and Gower 1998). These eddies could carry Fe-rich coastal water seaward and they could explain the increases in Chl *a* that are occasionally observed along Line P (Harrison et al. 1999).

A warming of 1.2 °C per century and freshening of 0.2 psu per century in the surface waters has been estimated for OSP (Freeland et al. 1997). From these data, they calculated that the mean mixed-layer depth has also decreased significantly from 130 m in the 1960s to 100 m in the late 1990s. Whitney and Freeland (1999) compared the nitrate and silicate concentrations of the 1970s to the 1990s and observed that the winter nitrate has decreased by 2.5 μM and silicate by 3.6 μM. Their removal rate between February and September has declined from 7.8 to 6.5 μM NO_3 and 8.5 to 6.0 μM Si (Si:NO_3 ratio decreased from 1.08 to 0.92). The larger decrease in silicate uptake (29%) relative to the decrease in nitrate uptake (17%) indicates that there was a marked decrease in diatom growth, suggesting that the supply of Fe to these latitudes may have also declined during these two decades, or a shift in diatom community composition to species having lower Si:N ratios. The decrease in the mixed-layer depth likely explains the decrease in the winter nutrient concentrations, but one would expect an increase in Fe because the atmospherically deposited Fe would be mixed into a smaller volume of surface water.

The critical role of large, eukaryotic phytoplankton (especially large diatoms) in the production of exportable particulate matter cannot be overstated (see Fig. 10.2 and 10.3), and in this context the importance of aperiodic, pulsed events is paramount. However, Legendre and Le Fèvre (1989) have already shown that there is no a priori direct equivalence between the new production and export production concepts. The former concerns particulate matter production by photoautotrophs and the latter is controlled by the creation of large, relatively dense sinking particles by numerous trophic levels and processes. Consequently, the formation of a near surface phytoplankton bloom via natural fertilization by micro- or macronutrient addition is probably a necessary but insufficient condition to result in pulsed carbon export. Documentation of the export phase of the implied production-export process is as important as the documentation of the growth phase. Furthermore, the fate of these export pulses is critical to the geochemical implications of blooms.

At Sta. ALOHA, there also appears to be an episodic, but recurrent, diatom aggregation-sinking event in late summer of every year (Scharek et al. 1999a, 1999b). This could be the manifestation of a N_2-primed 'echo' bloom following disappearance, by autolysis, of the seasonally accumulated, N_2-fixing microbial assemblage, or could be the de novo growth of eukaryotic phytoplankton (including diatoms) following the episodic deposition of bioavailable Fe (e.g., DiTullio and Laws 1991). Regardless of the mechanism, the diatom species selected under these oligotrophic conditions (summertime at Sta. ALOHA) are mostly those with endosymbiotic N_2-fixing cyanobacteria. The aggregates that are formed eventually, upon nutrient exhaustion (Fe, P or Si) sink rapidly (>200 m d^{-1}) and reach the 5 000-m seabed as 'bioavailable' organic matter (e.g., Fig. 10.3c). These large, exportable N_2-supported organic aggregates also remove bioavailable Fe and P that would otherwise be retained in the euphotic zone by efficient grazing and remineralization processes. Because of the rapid sinking rates and high percentage of living biomass, these aggregates also largely escape mesopelagic zone mineralization and carry their exported carbon and associated bioelements to the abyss (Fig. 10.3c). This dramatic diatomic diatom dump emphasizes the importance of episodic export in the NPSG, and the complexities of modeling an ecosystem where high-export cell aggregation and low-export microbial loop processes can occur simultaneously, and probably in direct competition.

Natural or artificial Fe fertilization of both HNLC and LNLC ocean regions could, theoretically, lead to an increase in the ocean's capacity to assimilate atmospheric CO_2 but by different ecological processes. In high latitude HNLC regions like OSP, the pulsed Fe additions would lead to the growth and eventual export of large pennate diatoms, as discussed above. In the LNLC subtropical gyres like HOT, the effect is less direct and is manifest through a stimulation of N_2-fixing microorganisms and a change in the total stock of nitrate in the ocean. This combined N_2-primed prokaryote and Fe-stimulated diatom aggregation carbon pump (see Fig. 10.3c) may be one of the most efficient export processes in the sea.

10.6 Beyond JGOFS: a Prospectus

All biogeochemical processes in the sea reside in a temporal domain and each process has characteristic time scales of variability. A major achievement of the Joint

Global Ocean Flux Study (JGOFS) is an improved understanding of the time-varying fluxes of carbon and associated biogenic elements, both within the ocean and the exchanges of carbon between the ocean and the atmosphere. This legacy derives, in part, from a network of ocean time-series stations located in representative biogeochemical provinces ranging from low-latitude, subtropical ocean gyres to high latitude coastal and oceanic regions. The JGOFS time-series programs were designed to capture low frequency (>1 year) changes, stochastic events and complex processes that may have multiple causes and unconstrained biogeochemical consequences. A basic underpinning of these time-series programs is the ecosystem concept; by investigating selected habitats and their inhabitants, and by conducting relevant cross-ecosystem inter-comparisons, a general ecological understanding will emerge.

Significant biogeochemical features include variations in the mechanisms of nutrient supply, especially the ecological consequences of 'pulsed' nutrient delivery including atmospheric dust (Fe), mesoscale eddy-induced upwelling of nutrients, and the nutrification of low latitude regions in the absence of turbulence (e.g., enhanced N_2 fixation). In this regard, microbial community structure is one of the most important ecosystem variables for predicting carbon export. Additionally, the bioelemental stoichiometry of exported organic matter (i.e., the C:N:P:Si:Ca:Fe ratio), and the selective remineralization of particles beneath the euphotic zone (i.e., C vs. P regeneration length scales) together constrain the efficiency the biological pump as a mechanism for the net sequestration of atmospheric carbon dioxide. The decoupling of primary organic matter production, particulate matter export and remineralization processes in time and space, and the detection of decade-scale, climate-driven ecosystem perturbations and feedbacks combine to reveal a time-varying, biogeochemical complexity that is just now becoming evident in our independent ocean time-series data sets.

A crucial issue as existing time-series programs mature, is data management as well as timely, user-friendly data accessibility. A related issue – that of data ownership – is less straightforward but equally important. Beyond careful archiving and peer review publication, the long-term data sets should be used to generate new ideas that serve as the basis for controlled experimentation or other forms of hypothesis testing (Coull 1985). Generation of hypotheses by systematic analyses of the collected data is as important as the subsequent hypothesis-testing. Science must be the driving force if any time-series program is to succeed beyond a few years. As G. Likens (1983) concluded, "A real danger of long-term research with reliable funding is that it could become static, pedantic, generally uninteresting and unproductive." These JGOFS era ocean observation programs have already yielded invaluable ecological insights even without deliberate manipulation or other forms of direct hypothesis-testing. Nevertheless it is apparent that whole ecosystem experiments would add further to the value of the field programs, and in this regard their introduction to these ongoing time-series research portfolios would be desirable. Furthermore, these sites provide invaluable research opportunities for other scientists, and the ongoing encouragement of extensive ancillary programs helps to keep the science of these time-series stations timely and interdisciplinary.

JGOFS did not begin as a field program focused on Fe control of carbon export, non-Redfield organic matter production stoichiometry, N_2 fixation or the physiological ecology of the recently discovered planktonic *Archaea*, but these are important biogeochemical issues in the twilight of JGOFS era. Quite frankly, without many of the JGOFS time-series data sets that presently exist, we would have never questioned the extant ecological paradigms, or realized that new ecological understanding was even needed. As the JGOFS era of ocean exploration comes to a close, new observational and hypothesis-driven research programs must emerge to assimilate and extend these retired and ongoing field studies. Planning is now underway for the development of a complementary set of integrated measurements including satellite altimetry (Jason), a global array of profiling floats (Argo) and a network of fixed point time-series stations. The eventual establishment of a comprehensive global ocean observation network of key biogeochemical parameters would provide unlimited opportunities for basic and applied research, ocean climate prediction and for marine science education. Human-operated time-series stations will be the intellectual heart and soul of this new age of ocean exploration.

Acknowledgements

Sufficient space does not exist to express the collective debt of gratitude that we owe to the numerous scientists, technical staff, computer specialists/data managers, students and ship officers and crew who have assisted in the collection, analysis and interpretation of our respective time-series data sets. We likewise thank our various public and private sector sponsors for their generous financial support of the time-series programs. The lead author thanks L. Lum and L. Fujieki for their assistance with the preparation of text, tables and graphics, and C. Benitez-Nelson and K. Björkman for their helpful comments on an earlier draft of the chapter. Finally, we thank Professor Michael Fasham for his efforts compiling this volume, and for his kind patience during our long gestation period. This is US JGOFS contribution #768.

References

Anderson LA, Sarmiento JL (1994) Redfield ratios of remineralization determined by nutrient data analysis. Global Biogeochem Cy 8:65–80

Baines SB, Pace ML, Karl DM (1994) Why does the relationship between sinking flux and planktonic primary production differ between lakes and oceans? Limnol Oceanogr 39:213–226

Banse K (1990) Does iron really limit phytoplankton production in the offshore subarctic Pacific? Limnol Oceanogr 35:772–775

Barlow M, Nigam S, Berbery EH (2001) ENSO, Pacific decadal variability, and U.S. summertime precipitation, drought, and stream flow. J Climate 14:2105–2128

Bates NR (2001) Interannual variability of oceanic CO_2 and biogeochemical properties in the Western North Atlantic subtropical gyre. Deep-Sea Res Pt II 48:1507–1528

Bates NR, Michaels AF, Knap AH (1996a) Alkalinity changes in the Sargasso Sea: geochemical evidence of calcification? Mar Chem 51:347–358

Bates NR, Michaels AF, Knap AH (1996b) Seasonal and interannual variability of oceanic carbon dioxide species at the U.S. JGOFS Bermuda Atlantic Time-series Study (BATS) site. Deep-Sea Res Pt II 43:347–383

Bates NR, Michaels AF, Knap AH (1996c) Spatial variability of CO_2 species in the Sargasso Sea. Caribb J Sci 32:303–304

Bates NR, Takahashi T, Chipman DW, Knap AH (1998) Variability of pCO_2 on diel to seasonal timescales in the Sargasso Sea. J Geophys Res 103:15567–15585

Benitez-Nelson C, Buesseler KO, Karl DM, Andrews J (2001) A time-series study of particulate matter export in the North Pacific Subtropical Gyre based on ^{234}Th:^{238}U disequilibrium. Deep-Sea Res Pt I 48:2595–2611

Bograd SJ, Thomson RE, Rabinovich AB, Leblond PH (1999) Near-surface circulation of the northeast Pacific Ocean derived from WOCE-SVP satellite-tracked drifters. Deep-Sea Res Pt II 46:2371–2403

Boyd P, Harrison PJ (eds) (1999a) Canadian JGOFS in the NE Subarctic Pacific. Deep-Sea Res Pt II special issue, vol. 46, issue 11–12, pp 2345–3017, Elsevier Science, Oxford, UK

Boyd P, Harrison PJ (1999b) Phytoplankton dynamics in the NE subarctic Pacific. Deep-Sea Res Pt II 46:2405–2432

Boyd PW, Muggli DL, Varela DE, Goldblatt RH, Chretian R, Orians KJ, Harrison PJ (1996) In vitro iron enrichment experiments in the NE subarctic Pacific. Mar Ecol Prog Ser 136:179–193

Boyd PW, Wong CS, Merrill J, Whitney F, Snow J, Harrison PJ, Gower J (1998) Atmospheric iron supply and enhanced vertical carbon flux in the NE subarctic Pacific: Is there a connection? Global Biogeochem Cy 12:429–441

Boyd PW, Goldblatt RH, Harrison PJ (1999) Mesozooplankton grazing manipulations during in vitro iron enrichment studies in the NE subarctic Pacific. Deep-Sea Res Pt II 46: 2645–2668

Buesseler KO (1998) The decoupling of production and particle export in the surface ocean. Global Biogeochem Cy 12:297–310

Capone DG, Zehr JP, Paerl HW, Bergman B, Carpenter EJ (1997) *Trichodesmium* a globally significant marine cyanobacterium. Science 276:1221–1229

Carlson CA, Ducklow HW, Michaels AF (1994) Annual flux of dissolved organic carbon from the euphotic zone in the northwestern Sargasso Sea. Nature 371:405–408

Carpenter EJ, Price CC IV (1976) Marine *Oscillatoria* (*Trichodesmium*): explanation for aerobic nitrogen fixation without heterocysts. Science 191:1278–1280

Chao SY, Shaw PT, Wu SY (1996) El Niño modulation of the South China Sea circulation. Prog Oceanogr 38:51–93

Chelton DB, Bernal PA, McGowan JA (1982) Large-scale interannual physical and biological interaction in the California Current. J Mar Res 40:1095–1125

Christian JR, Lewis MR, Karl DM (1997) Vertical fluxes of carbon, nitrogen and phosphorus in the North Pacific Subtropical gyre near Hawaii. J Geophys Res 102:15667–15677

Coale KH, Bruland KW (1987) Oceanic stratified euphotic zone as elucidated by ^{234}Th:^{238}U disequilibria. Limnol Oceanogr 32: 189–200

Conte MH, Ralph N, Ross EH (2001) Seasonal and interannual variability in deep ocean particle fluxes at the Oceanic Flux Program (OFP)/Bermuda Atlantic Time Series (BATS) site in the western Sargasso Sea near Bermuda. Deep-Sea Res Pt II 48: 1471–1505

Copin-Montégut C (2000) Consumption and production on scales of a few days of inorganic carbon, nitrate and oxygen by the planktonic community: results at the Dyfamed Station in the northwestern Mediterranean Sea (May 1995). Deep-Sea Res Pt I 47:447–477

Copin-Montégut C, Begovic M (2002) Distributions of carbonate properties and oxygen along the water column (0–2000 m) in the central part of the Northwestern Mediterranean Sea (Dyfamed site). Influence of the winter vertical mixing on air-sea CO_2 and O_2 exchanges. Deep-Sea Res Pt II 49:2049–2066

Coull BC (1985) The use of long-term biological data to generate testable hypotheses. Estuaries 8:84–92

Cullen JJ, Franks PJS, Karl DM, Longhurst A (2002) Physical influences on marine ecosystem dynamics. In: Robinson AR, McCarthy JJ, Rothschild BJ (eds) The sea, vol. 12. John Wiley & Sons, New York, pp 297–336

Davenport R, Neuer S, Hernández-Guerra A, Rueda MJ, Llinás O, Fischer G, Wefer G (1999) Seasonal and interannual pigment concentration in the Canary Islands region from CZCS data and comparison with observations from ESTOC. Int J Remote Sens 20:1419–1433

Deuser WG (1986) Seasonal and interannual variations in deep-water particle fluxes in the Sargasso Sea and their relation to the surface hydrography. Deep-Sea Res Pt I 33:225–246

Dickey T, Zedler S, Yu X, Doney SC, Frye D, Jannasch H, Manov D, Sigurdson D, McNeil JD, Dobeck L, Gilboy T, Bravo C, Siegel DA, Nelson N (2001) Physical and biogeochemical variability from hours to years at the Bermuda Testbed Mooring site: June 1994–March 1998. Deep-Sea Res Pt II 48:2105–2140

Diggle PJ (1990) Time series: a biostatistical introduction. Oxford University Press, New York, 257 pp

DiTullio GR, Laws EA (1991) Impact of an atmospheric-oceanic disturbance on phytoplankton community dynamics in the North Pacific Central gyre. Deep-Sea Res Pt I 38:1305–1329

Dore JE, Carrillo CJ, Hebel DV, Karl DM (2002) Carbon cycle observations at the Hawaii Ocean Time-series Station ALOHA. In: Nojiri Y, Feely R (eds) Proceedings of the PICES North Pacific CO_2 Data Synthesis Symposium, Tsukuba, Japan, October 2000 (in press)

Duce RA, Tindale NW (1991) Atmospheric transport of iron and its deposition in the ocean. Limnol Oceanogr 36:1715–1726

Dugdale RC, Goering JJ (1967) Uptake of new and regenerated forms of nitrogen in primary productivity. Limnol Oceanogr 12:196–206

Emerson S, Quay PD, Stump C, Wilbur D, Knox M (1991) O_2, Ar, N_2 and ^{222}Rn in surface waters of the subarctic ocean: net biological O_2 production. Global Biogeochem Cy 5:49–69

Emerson S, Quay PD, Stump C, Wilbur D, Schudlich R (1995) Chemical tracers of productivity and respiration in the subtropical Pacific Ocean. J Geophys Res 100:15,873–15,887

Emerson S, Quay P, Karl D, Winn C, Tupas L, Landry M (1997) Experimental determination of the organic carbon flux from open-ocean surface waters. Nature 389:951–954

Emerson S, Mecking S, Abell J (2001) The biological pump in the subtropical North Pacific Ocean: nutrient sources, Redfield ratios, and recent changes. Global Biogeochem Cy 15:535–554

Eppley RW, Peterson BJ (1979) Particulate organic matter flux and planktonic new production in the deep ocean. Nature 282:677–680

Falkowski PG, Barber RT, Smetacek U (1998) Biogeochemical controls and feedbacks on ocean primary production. Science 281:200–206

Freeland HJ, Denman K, Whitney F, Jacques R (1997) Evidence of change in the mixed layer in the northeast Pacific Ocean. Deep-Sea Res Pt I 44:2117–2129

Freudenthal T, Neuer S, Meggers H, Davenport R, Wefer G (2001) Influence of lateral particle advection and organic matter degradation on sediment accumulation and stable nitrogen isotope ratios along a productivity gradient in the Canary Islands region. Mar Geol 177:93–109

Frost BW (1987) Gazing control of phytoplankton stock in the open subarctic Pacific Ocean: a model assessing the role of mesozooplankton, particularly the large calanoid copepods *Neocalanus* spp. Mar Ecol Prog Ser 39:49–68

Goes JI, Saino T, Gomes HR, Ishizaka J, Nojiri Y, Wong CS (1999) A comparison of new production in the north Pacific ocean for 1997 and 1998 estimated from remotely sensed data. In: Nojiri Y (ed) Proceedings of the 2nd International symposium on CO_2 in the Oceans. NIES, Tsukuba, Japan, pp 141–146

Gruber N, Sarmiento JL (1997) Global patterns of marine nitrogen fixation and denitrification. Global Biogeochem Cy 11:235–266

Guidi-Guilvard LD (2002) DYFAMED-BENTHOS, a long time-series benthic survey at 2347 m depth in the NW Mediterranean: general introduction. Deep-Sea Res Pt II 49:2183–2193

Hansell DA, Carlson CA, Bates NR, Poisson A (1997) Horizontal and vertical removal of organic carbon in the equatorial Pacific Ocean: a mass balance assessment. Deep-Sea Res Pt II 44:2115–2130

Harrison PJ, Boyd PW, Varela DE, Takeda S, Shiomoto A, Odate T (1999) Comparison of factors controlling phytoplankton productivity in the NE and NW subarctic Pacific gyres. Prog Oceanogr 43:205–234

Hayward TL, Venrick EL, McGowan JA (1983) Environmental heterogeneity and plankton community structure in the central North Pacific. J Mar Res 41:711–729

Jeandel C, Ruiz-Pino D, Gjata E, Poisson A, Brunet C, Charriaud E, Dehairs F, Delille D, Fiala M, Fravalo C, Miquel JC, Park Y-H, Pondaven P, Quéginer B, Razouls S, Shauer B, Tréguer P (1999) KERFIX, a time-series station in the Southern Ocean: a presentation. J Marine Syst 17:555–569

Jenkins WJ (1982) Oxygen utilization rates in North Atlantic subtropical gyre and primary production in oligotrophic systems. Nature 300:246–248

Jenkins WJ, Goldman J (1985) Seasonal oxygen cycling and primary production in the Sargasso Sea. J Mar Res 43:465–491

Karl DM (1999) A sea of change: Biogeochemical variability in the North Pacific subtropical gyre. Ecosystems 2:181–214

Karl DM (2000) A new source of 'new' nitrogen in the sea. Trends Microbiol 8:301

Karl DM, Lukas R (1996) The Hawaii Ocean Time-series (HOT) program: background, rationale and field implementation. Deep-Sea Res Pt II 43:129–156

Karl DM, Winn CD (1991) A sea of change: monitoring the ocean's carbon cycle. Environ Sci Technol 25:1976–1981

Karl DM, Letelier R, Hebel DV, Bird DF, Winn CD (1992) *Trichodesmium* blooms and new nitrogen in the North Pacific gyre. In: Carpenter EJ, et al. (eds) Marine pelagic cyanobacteria: *Trichodesmium* and other Diazotrophs. Kluwer Academic, Dordrecht, pp 219–237

Karl DM, Christian JR, Dore JE, Hebel DV, Letelier RM, Tupas LM, Winn CD (1996) Seasonal and interannual variability in primary production and particle flux at Station ALOHA. Deep-Sea Res Pt II 43:539–568

Karl D, Letelier R, Tupas L, Dore J, Christian J, Hebel D (1997) The role of nitrogen fixation in biogeochemical cycling in the subtropical North Pacific Ocean. Nature 388:533–538

Karl DM, Bidigare RR, Letelier RM (2001a) Long-term changes in plankton community structure and productivity in the North Pacific Subtropical Gyre: the domain shift hypothesis. Deep-Sea Res Pt II 48:1449–1470

Karl DM, Björkman KM, Dore JE, Fujieki L, Hebel DV, Houlihan T, Letelier RM, Tupas LM (2001b) Ecological nitrogen-to-phosphorus stoichiometry at station ALOHA. Deep-Sea Res Pt II 48:1529–1566

Karl DM, Michaels A, Bergman B, Capone D, Carpenter E, Letelier R, Lipschultz F, Paerl H, Sigman D, Stal L (2002) Dinitrogen fixation in the world's oceans. Biogeochemistry 57/58:47–98

Knauer GA, Martin JH, Bruland KW (1979) Fluxes of particulate carbon, nitrogen and phosphorus in the upper water column of the northeast Pacific. Deep-Sea Res Pt I 26:97–108

Knauer GA, Redalje DG, Harrison WG, Karl DM (1990) New production at the VERTEX time-series site. Deep-Sea Res Pt I 37: 1121–1134

Lampitt RS (1985) Evidence for the seasonal deposition of detritus to the deep-sea floor and its subsequent resuspension. Deep-Sea Res Pt I 32:885–897

LaRoche J, Boyd PW, Mckay RML, Gelder RJ (1996) Flavodoxin as in situ marker for iron stress in phytoplankton. Nature 382:802–805

Laws E, Falkowski P, Smith WO, Ducklow H, McCarthy JJ (2000) Temperature effects on export production in the open ocean. Global Biogeochem Cy 14:1231–1246

Laws EA, Bannister TT (1980) Nutrient- and light-limited growth of *Thalassiosira fluviatilis* in continuous culture, with implications for phytoplankton growth in the ocean. Limnol Oceanogr 25:457–473

Lee K (2001) Global net community production estimated from the annual cycle of surface water total dissolved inorganic carbon. Limnol Oceanogr 46:1287–1297

Legendre L, Le Fèvre J (1989) Hydrodynamical singularities as controls of recycled versus export production in oceans. In: Berger WH, Smetacek VS, Wefer G (eds) Productivity of the ocean: present and past. John Wiley & Sons, New York, pp 49–63

Likens GE (1983) A priority for ecological research. Bull Ecol Soc Am 64(4):234–243

Lipschultz FJ, Bates NR, Carlson CA, Hansell DA (2002) New production in the Sargasso Sea: history and current status. Global Biogeochem Cy (in press)

Liu AK, Chang YS, Hsu MK, Liang NK (1998) Evolution of nonlinear internal waves in the East and South China Seas. J Geophys Res 103:7995–8008

Longhurst AR, Harrison WG (1989) The biological pump: profiles of plankton production and consumption in the upper ocean. Prog Oceanogr 22:47–123

Louanchi F, Ruiz-Pino DP, Poisson A (1999) Temporal variations of mixed-layer oceanic CO_2 at JGOFS-KERFIX time-series station: physical versus biogeochemical processes. J Mar Res 57: 165–187

Maldonado MT, Boyd PW, Harrison PJ, Price NM (1999) Co-limitation of phytoplankton growth by light and Fe during winter in the subarctic Pacific Ocean. Deep-Sea Res Pt II 46:2475–2485

Mantua NJ, Hare SR, Zhang Y, Wallace JM, Francis RC (1997) A Pacific interdecadal climate oscillation with impacts on salmon production. B Am Meteorol Soc 78:1069–1079

Margalef R (1978) Life forms of phytoplankton as survival alternatives in an unstable environment. Oceanol Acta 1:493–509

Martin JH, Fitzwater SE (1988) Iron deficiency limits phytoplankton growth in the north-east Pacific subarctic. Nature 331:341–343

Martin JH, Knauer GA, Karl DM, Broenkow WW (1987) VERTEX: carbon cycling in the northeast Pacific. Deep-Sea Res Pt I 34: 267–285

Martinez L, Silver MW, King JM, Alldredge AL (1983) Nitrogen fixation by floating diatom mats: a source of new nitrogen to oligotrophic ocean waters. Science 221:152–154

Marty JC, Chiavérini J, Pizay MD, Avril B (2002) Seasonal and interannual dynamics of nutrients and phytoplankton pigments in the Western Mediterranean Sea at the DYFAMED time-series station (1991–1999). Deep-Sea Res Pt II 49:1965–1985

McCave IN (1975) Vertical flux of particles in the ocean. Deep-Sea Res Pt I 22:491–502

McGillicuddy DJ, Robinson AR, Siegel DA, Jannasch HW, Johnson R, Dickey TD, McNeil JD, Michaels AF, Knap AH (1998) New evidence for the impact of mesoscale eddies on biogeochemical cycling in the Sargasso Sea. Nature 394:263–265

McGillicuddy DJ Jr., Johnson RJ, Siegel DA, Michaels AF, Bates N, Knap AH (1999) Mesoscale variability of ocean biogeochemistryin the Sargasso Sea. J Geophys Res 104:13381–13394

Michaels A, Knap A (1996) Overview of the U.S.-JGOFS Bermuda Atlantic Time-series Study and Hydrostation S program. Deep-Sea Res Pt II 43:157–198

Michaels AF, Bates NR, Bueseler KO, Carlson CA, Knap AH (1994) Carbon-cycle imbalances in the Sargasso Sea. Nature 372:537–540

Michaels AF, Olson D, Sarmiento JL, Ammerman JW, Fanning K, Jahnke R, Knap AH, Lipschultz F, Prospero JM (1996) Inputs, losses and transformations of nitrogen and phosphorus in the pelagic North Atlantic Ocean. Biogeochemistry 35:181–226

Michaels AF, Karl DM, Knap AH (2000) Temporal studies of biogeochemical dynamics in oligotrophic oceans. In: Hanson RB, Ducklow HW, Field JG (eds) The changing ocean carbon cycle: a midterm synthesis of the Joint Global Ocean Flux Study. Cambridge University Press, Cambridge, UK, pp 392–413

Migon C, Sandroni V, Marty JC, Gasser B, Miquel JC (2002) Transfer of atmospheric matter through the euphotic layer in the northwestern Mediterranean: seasonal pattern and driving forces. Deep-Sea Res Pt II 49:2125–2141

Miquel JC, Fowler SW, LaRosa J, Buat-Menard P (1994) Dynamics of the downward flux of particles and carbon in the open northwestern Mediterranean Sea. Deep-Sea Res I 41:243–261

Miller CB (1993) Pelagic production processes in the subarctic Pacific. Prog Oceanogr 32:1–15

Miller CB, Frost BW, Wheeler PA, Landry ML, Welschmeyer N, Powell TM (1991) Ecological dynamics in the subarctic Pacific, a possibly iron-limited system. Limnol Oceanogr 33:1600–1615

Neuer S, Ratmeyer V, Davenport R, Fischer G, Wefer G (1997) Deep water particle flux in the Canary Island region: seasonal trends in relation to long-term satellite derived pigment data and lateral sources. Deep-Sea Res Pt I 44:1451–1466

Noriki S (1999) Particulate fluxes at Stn. KNOT in the western North Pacific during 1988–1991. In: Nojiri Y (ed) Proceedings of the 2nd International Symposium on CO_2 in the Oceans. National Institute for Environmental Studies, Tsukuba, Japan, pp 331–337

Ono S, Najjar R, Ennyu A, Bates NR (2001) Shallow remineralization in the Sargasso Sea estimated from seasonal variations in oxygen and dissolved inorganic carbon. Deep-Sea Res Pt II 48:1567–1582

Pace ML, Knauer GA, Karl DM, Martin JH (1987) Primary production, new production and vertical flux in the eastern Pacific Ocean. Nature 325:803–804

Platt T, Harrison WG (1985) Biogenic fluxes of carbon and oxygen in the ocean. Nature 318:55–58

Redfield AC, Ketchum BH, Richards FA (1963) The influence of organisms on the composition of seawater. In: Hill MN (ed) The sea, ideas and observations on progress in the study of the seas, vol. 2. Interscience, New York, pp 26–77

Rhee G-Y (1978) Effects of N:P atomic ratios and nitrate limitation on algal growth, cell composition, and nitrate uptake. Limnol Oceanogr 23:10–25

Risser PG (ed) (1991) Long-term ecological research: an international perspective. John Wiley & Sons, New York, 294 pp

Sakshaug E, Holm-Hansen O (1977) Chemical composition of *Skeletonema costatum* (Grev.) Cleve and *Pavlova (Monochrysis) lutheri* (Droop) Green as a function of nitrate-, phosphate-, and iron-limited growth. J Exp Mar Biol Ecol 29:1–34

Scharek R, Tupas LM, Karl DM (1999a) Diatom fluxes to the deep sea in the oligotrophic North Pacific gyre at Station ALOHA. Mar Ecol Prog Ser 182:55–67

Scharek R, Latasa M, Karl DM, Bidigare RR (1999b) Temporal variations in diatom abundance and downward vertical flux in the oligotrophic North Pacific gyre. Deep-Sea Res Pt I 46:1051–1075

SCOR (1990) Oceans, carbon and climate change: an introduction to the Joint Global Ocean Flux Study. Scientific Committee on Oceanic Research, Halifax, Canada, 61 pp

Shaw P-T, Chao S-Y (1994) Surface circulation in the South China Sea. Deep-Sea Res Pt I 41:1663–1683

Shaw P-T, Chao S-Y, Liu K-K, Pai S-C, Liu C-T (1996) Winter upwelling off Luzon in the north-eastern South China Sea. J Geophys Res 101:16435–16448

Shiah F-K, Liu K-K, Tang TY (1999) South East Asia Time-series Station established in South China Sea. U.S. JGOFS Newsletter 10(1):8–9

Siegel DA, Fields E, McGillicuddy DJ Jr. (1999) Mesoscale motions, satellite altimetry and new production in the Sargasso Sea. J Geophys Res 104:13,359–13,379

Siegel DA, Karl DM, Michaels AF (2001) Interpretations of biogeochemical processes from the US JGOFS Bermuda and Hawaii time-series sites. Deep-Sea Res Pt II 48:1403–1404

Smith CR, Hoover DJ, Doan SE, Pope RH, DeMaster DJ, Dobbs FC, Altabet MA (1996) Phytodetritus at the abyssal seafloor across 10° of latitude in the central equatorial Pacific. Deep-Sea Res Pt II 43:1309–1338

Spitzer WS, Jenkins WJ (1989) Rates of vertical mixing, gas exchange and new production: estimates from seasonal gas cycles in the upper ocean near Bermuda. J Mar Res 47:169–196

Steinberg DK, Carlson CA, Bates NR, Goldthwait SA, Madin LP, Michaels AF (2000) Zooplankton vertical migration and the active transport of dissolved organic and inorganic carbon in the Sargasso Sea. Deep-Sea Res Pt I 47:137–158

Steinberg DK, Carlson CA, Bates NR, Johnson RJ, Michaels AF, Knap AH (2001) Overview of the US JGOFS Bermuda Atlantic Time-series Study (BATS): a decade-scale look at ocean biology and biogeochemistry. Deep-Sea Res Pt II 48:1405–1447

Suess E (1980) Particulate organic carbon flux in the oceans – surface productivity and oxygen utilization. Nature 288:260–263

Tett P, Heaney SI, Droop MR (1985) The Redfield ratio and phytoplankton growth rate. J Mar Biol Assoc Uk 65:487–504

Thomson RE, Gower JFR (1998) A basin-scale instability event in the Gulf of Alaska. J Geophys Res 103:3033–3040

Timmermans KR, Gerringa LJA, de Baar HJW, van der Wagt B, Veldhuis MJW, de Jong JTM, Croot PL (2001) Growth rates of large and small Southern Ocean diatoms in relation to availability of iron in natural seawater. Limnol Oceanogr 46:260–266

Tsurushima N, Nojiri Y, Watanabe S (1999) New ocean time series KNOT in the western sub arctic Pacific from June 1998. In: Nojiri Y (ed) Proceedings of the 2nd International Symposium on CO_2 in the Oceans. National Institute for Environmental Studies, Tsukuba, Japan, pp 605–609,

Varela DE, Harrison PJ (1999a) Effect of ammonium on nitrate utilization by *Emiliania huxleyi* a coccolithophore from the northeastern Pacific. Mar Ecol Prog Ser 186:67–74

Varela DE, Harrison PJ (1999b) Seasonal variability in nitrogenous nutrition of phytoplankton assemblages in the northeastern subarctic Pacific Ocean. Deep-Sea Res Pt II 46:2505–2538

Villareal TA, Altabet MA, Culver-Rymsza K (1993) Nitrogen transport by migrating diatom mats in the North Pacific Ocean. Nature 363:709–712

Wassman P (1990) Relationship between primary and export production in the boreal coastal zone of the North Atlantic. Limnol Oceanogr 35:464–471

Whitney FA, Freeland HJ (1999) Variability in upper-ocean water properties in the NE Pacific Ocean. Deep-Sea Res Pt II 46: 2351–2370

Whitney FA, Wong CS, Boyd PW (1998) Interannual variability in nitrate supply to surface waters of the Northeast Pacific Ocean. Mar Ecol Prog Ser 170:15–23

Winn CD, Mackenzie FT, Carrillo CJ, Sabine CL, Karl DM (1994) Air-sea carbon dioxide exchange in the North Pacific subtropical gyre: implications for the global carbon budget. Global Biogeochem Cy 8:157–163

Wong CS, Matear RJ (1999) Silicate limitation of phytoplankton productivity in the northeast subarctic Pacific. Deep-Sea Res Pt II 46:2539–2555

Wong CS, Whitney FA, Iseki K, Page JS, Zeng J (1995) Analysis of trends in primary production and chlorophyll-*a* over two decades at Ocean Station P (50° N, 145° W) in the subarctic northeast Pacific Ocean. In: Beamish RJ (ed) Climate change and northern fish populations. Canadian special publication of Journal of Fisheries and Aquatic Sciences 121:107–117

Wong CS, Whitney FA, Crawford DW, Iseki K, Matear RJ, Johnson WK, Page JS, Timothy D (1999) Seasonal and interannual variability in particle fluxes of carbon, nitrogen and silicon from time series of sediment traps at Ocean Station P, 1982–1993: relationship to changes in subarctic primary productivity. Deep-Sea Res Pt II 46:2735–2760

Wong CS, Waser NAD, Nojiri Y, Whitney FA, Page JS, Zeng J (2002) Seasonal cycles of nutrients and dissolved inorganic carbon at high and mid latitudes in the North Pacific Ocean: determination of new production and nutrient uptake ratios. Deep-Sea Res Pt II 49:5317–5338

Zehr JP, Mellon MT, Zani S (1998) New nitrogen-fixing microorganisms detected in oligotrophic oceans by amplification of nitrogenase (nifH) genes. Appl Environ Microb 64:3444–3450

Zehr JP, Carpenter EJ, Villareal TA (2000) New perspectives on nitrogen-fixing microorganisms in tropical and subtropical oceans. Trends Microbiol 8:68–73

Zhang J, Quay PD (1997) The total organic carbon export rate based on DIC and DIC13 budgets in the equatorial Pacific Ocean. Deep-Sea Res Pt II 44:2163–2190

Chapter 11

JGOFS: a Retrospective View

Michael J. R. Fasham

The international JGOFS project was initiated in 1987 and carried out its first major process study in the North Atlantic in 1989. Since then major process studies have been made in the Equatorial Pacific, the Arabian Sea, the Southern Ocean, and the North Pacific, major time series have been initiated in the Atlantic, Pacific, and Southern Oceans, and a world-wide survey of ocean CO_2 has been made in conjunction with WOCE. During the last couple of years the vast datasets from these activities have begun to be synthesised into a coherent picture of the carbon, and other element, cycles in the ocean. Early results of this synthesis are summarised in recent reviews (Field et al. 2000; Fasham et al. 2001) and are the subject of the chapters in this book. I do not intend to repeat here the scientific conclusions of these reviews but will look back over the last 15 years to give an overall survey of what work has been carried out by JGOFS, and discuss how well the goals and expectations of the original plans have been realised.

11.1 The JGOFS Science Plan

All properly brought-up international science projects are now required to produce a detailed science plan before any funding agency will provide any money. JGOFS was exceedingly ill behaved in that it organised the North Atlantic Bloom Experiment (NABE) a year before it published its Science Plan (SCOR 1990). This was possible because of the previous planning that had already been done by the US JGOFS community and because of the tremendous enthusiasm on the part of the international community for the general concept of JGOFS. The Science Plan was further developed in the JGOFS Implementation Plan (SCOR 1992). I will now discuss the various components of the Science Plan and review how the national and international JGOFS programmes have contributed to these components.

11.2 The Process Studies

The central aim of the process studies was to provide the basic data for conceptual models of the biogeochemical processes needed to explain and predict global-scale patterns of element cycling between the atmosphere, ocean, and sediments. In the Science Plan it was envisaged that four major process studies would be carried out in (*1*) the North Atlantic, (*2*) the Equatorial Pacific, (*3*) the Northwest Indian Ocean, and (*4*) the Southern Ocean.

As mentioned earlier, the first JGOFS process study, the North Atlantic Bloom Experiment (NABE), took place in 1989. It was based around a series of stations along 20° W from 18° N to 72° N originally planned by German oceanographers to celebrate the centenary of Victor Hensen's 1889 Plankton Expedition (Ducklow 1989). The main international effort (Germany, Netherlands, UK and US) was focused on 47° N and 60° N covering the period from late April to late August (Ducklow and Harris 1993). In the spring of the same year the Canadian component took place further west (40° W–50° W) between latitudes 32° N and 47° N (Harrison et al. 1993). During 1990–1992 there were further UK cruises in the NE Atlantic (Savidge et al. 1992, 1995; Balch et al. 1996a, 1996b), while the French EUMELI project investigated a 1 000 km transect westwards from the eutrophic North African upwelling area to an oligotrophic open ocean site (Morel 1996). Finally in 1992–1993 three trans-Atlantic transects from Nova Scotia to Canary Islands were surveyed by Canadian JGOFS (Harrison et al. 2001). A Canadian JGOFS process study was also carried out in the shelf waters of the Gulf of St. Lawrence (Roy et al. 2000) during 1992–1994.

During these cruises, much was learnt about the role of biology in the draw-down of CO_2, the factors controlling the spring bloom, the dominant role of microzooplankton in the grazing of phytoplankton, and the links between bacterial production and dissolved organic matter (DOM) distribution. Some syntheses of these data have been carried out (Bender et al. 1992; Koeve and Ducklow 2001, and papers therein), and some of the datasets have been used in local model studies (Fasham and Evans 2000, and references therein; Dadou et al. 2001). However, much remains to be done to meld all these datasets into a consistent picture of North Atlantic biogeochemistry.

The next JGOFS process study was designed to investigate the physical and biogeochemical mechanisms regulating primary production, CO_2 out-gassing, and export fluxes in the Equatorial Pacific Ocean. The bulk of the cruise activity took place between the years 1991–1994 (Murray et al. 1997). The main US JGOFS work was conducted on a N-S transect along 140° W from 12° N to 12° S (Murray et al. 1995), while most of the Australian, French, and Japanese cruises were made in the western Equatorial Pacific warm pool region (Mackey et al. 1995; Le Borgne and Rodier 1997; Ishizaka et al. 1997). The period of study covered part of the 1991–1994 ENSO event, which reached its maximum intensity in the spring of 1992. This temporal coverage demonstrated the strong modulating effect of the ENSO cycle on air-sea CO_2 exchange in the Pacific (e.g., Feely et al. 1997). Another highlight was the detailed shipboard observations on a convergent front that produced elevated chlorophyll concentrations observable from space (Yoder et al. 1994). To date four special issues of Deep-Sea Research II have been devoted to the results from these cruises (vols. 42(2–3), 43(4–6), 44(9–10), and 49(13–14)).

The NW Indian Ocean process study addressed the variations in primary and new production and export fluxes that arise from the seasonally reversing monsoonal circulation. The first cruises were carried out by Netherlands JGOFS in 1992–1993 and concentrated on the Somalia upwelling region and the Gulf of Aden (Veldhuis et al. 1997). The main multi-national focus was in the Arabian Sea in 1994–1997 with cruises by the US (Smith et al. 1998; Smith 2001), UK (Burkill 1999), India (Prasanna Kumar et al. 2000), Pakistan (Ahmed et al. 1995), and Germany (Pfannkuche and Lochte 2000). To date there have been seven special issues of Deep-Sea Research II devoted to these studies (vols. 45(10–11), 46(3–4), 46(8–9), 47(7–8), 47(14), 48(6–7), and 49(12) while additional Indian JGOFS results have appeared in Current Science (vol. 71(11)) and Earth and Planetary Sciences (vol. 109(4)). The JGOFS studies in the Arabian Sea represent by far the largest concentration of resources in this area since the International Indian Ocean Expedition in 1962–1965 (Zeitzschel 1973), and the datasets being assembled, covering both monsoon and inter-monsoon periods, will provide a rich resource for oceanographers and modellers studying this region for years to come.

The fourth large JGOFS process study took place in the Southern Ocean and had the aim of investigating the role of the Southern Ocean in the global carbon balance, the reasons for spatial variability in production, the persistence of high nitrate levels, and the possible role of iron as a limiting nutrient. In this study the various countries involved worked in different sectors. The UK JGOFS programme took place in the Bellinghausen Sea to the west of the Antarctic Peninsula during austral spring 1992 (Turner and Owens 1995). German and Dutch JGOFS scientists took part on a cruise of R.V. *Polarstern* in the South Atlantic and Weddell Sea also during austral spring 1992, with most of the sampling conducted on a N-S transect along 6° W between 48° S and 57° S (Smetacek et al. 1997). The French ANTARES-I cruise took place in the South Indian and Southern Oceans in austral autumn 1993 (Gaillard 1997). Further cruises in the ANTARES programme took place in 1994, 1995 (Strass et al. 2002), and 1999 (Le Févre and Tréguer 1998; Tréguer and Pondaven 2002). The US JGOFS programme focused on two areas, the Ross Sea continental shelf and the Polar Front along 170° W (Smith Jr. et al. 2000; Anderson and Smith Jr 2001) during 1996–1998. The South African JGOFS programme ran between 1991–1995 and consisted of additional JGOFS observations along a WOCE transect between Cape Town and the South African Antarctic base, plus some detailed surveys in the Subtropical Convergence and the Polar Front (Lucas et al. 1993; Bathmann et al. 2000). There were two Australian contributions to Southern Ocean JGOFS: the Sub-Antarctic Zone (SAZ) project consisting of a study of biogeochemical processes along 140° W from Tasmania to 55° S during austral summers 1997–1998 (Trull et al. 2001) and the BROKE survey off the Coast of East Antarctica in January–March 1996 (Nicol et al. 2000). There was also a Japanese cruise off Adélie Land that ran concurrently with the BROKE programme (Chiba et al. 2000).

If the Southern Ocean iron fertilisation experiments (see below) are also included then it is clear that more effort has been devoted to the Southern Ocean during JGOFS than to the other three major process studies combined. Although some initial syntheses of these datasets have been made (Bathmann et al. 2000, and papers in Deep-Sea Research II 49(9–10) and 49(16)), much still remains to be done in bringing together results from different national programmes.

In the Science Plan seven other areas were suggested as possible sites for process studies. One of these was the Antarctic ice margin, which was in fact studied extensively during a number of the Southern Ocean cruises discussed above. Three other suggested areas, the Gulf Stream, the central gyres, and the eastern boundary upwelling areas did not have large-scale process studies dedicated to them, mainly due to the large time and resource commitment by the community to the four main process studies.

The northern North Atlantic Ocean (Greenland and Norwegian Seas) was seen as an important area as it plays a pivotal role in the drawdown of atmospheric CO_2 due to the formation of North Atlantic Deep Water. There is also high productivity and export flux during the summer. Although there has been no multi-national work in this area, Norwegian JGOFS have been studying the CO_2 system for some time (Anderson et al. 2000), while German oceanographers maintained a sediment trap time-series in the Norwegian Sea for several years (Bodungen et al. 1995).

In the original Science Plan the subarctic North Pacific gyre was not planned to be a major study area because there had recently been a major project in this area (the SUPER project, Miller 1993). However, there was considerable interest in carrying out further JGOFS studies in the North Pacific by Canadian and Japanese JGOFS communities, and so a North Pacific Task Team was set up by JGOFS to coordinate a science programme (Saino et al. 2002). Most of the Canadian work took place during 1992–1997 with observations along a transect from the coast of British Columbia to Station Papa (50° N, 145° W) (Boyd et al. 1999b). The Japanese programme focused on the formation and dispersion of the North Pacific Intermediate Water (NPIW) and its effects on primary production and CO_2 drawdown (SAGE, Hiroe et al. 2002), while a second project used ship-of-opportunity transects across the North Pacific to study hydrography and air-sea CO_2 exchange (Zeng et al. 2002). Finally in the eastern South Pacific, Chilean JGOFS have been conducting process studies and sediment trap time series in the Humboldt Current region off Chile since 1991 (R. Quiñones, pers. comm.).

There is no doubt that the largest fraction of the national resources allocated to JGOFS was used to fund process studies. This allowed the international community to plan large multinational programmes in the chosen priority areas that far outstripped anything previously undertaken. The numerous papers published in the many JGOFS special issues are a measure of the scientific resources devoted to the process studies and the multi-faceted science that has resulted from them. However, to date, most of the published work has concentrated on a particular national cruise or multi-cruise programme. In the coming years, it is extremely important that a broader synthesis of these vast datasets be carried out to develop new conceptual models of ocean biogeochemical cycles and to provide the data to validate the emerging global simulation models.

11.3 Iron Fertilisation Experiments

Over the last few years, studying the role of iron in marine ecosystems has been one of the expanding areas of oceanography. It is perhaps surprising therefore that iron was only briefly mentioned in the Science Plan. However, Martin and Fitzwater's seminal paper had only recently been published (Martin and Fitzwater 1988) and it was not until the special ASLO symposium in 1991 that iron moved to centre-stage. At that meeting, Watson et al. suggested that an open ocean iron fertilisation experiment was feasible using an SF_6 tracer to label a patch of water (Watson et al. 1991). The US community moved quickly to plan such an experiment and the first IronEx experiment was carried out in the eastern Equatorial Pacific in 1993 (Martin et al. 1994). A subsequent experiment in 1995 (Coale et al. 1996) unequivocally demonstrated that iron was limiting the growth rate of at least the larger phytoplankton and that fertilisation produced a large bloom that decreased surface CO_2 fugacity (Behrenfeld et al. 1996; Cooper et al. 1996). Further shipboard iron fertilisation experiments were made during the Canadian JGOFS work at Station Papa (Boyd et al. 1999a) and during a number of the Southern Ocean JGOFS cruises (de Baar and Boyd 2000; Olson et al. 2000). In February 1999 the first iron fertilisation experiment in the Southern Ocean was carried out south of the Polar Front at 61° S, 140° E (Boyd and Law 2001), and in December 2000 a second German-led experiment (EISENEX) took place 1500 km to the south of the Cape of Good Hope.

As a result of all these studies it is now firmly established that, in many areas, iron is a limiting nutrient for phytoplankton growth. However, all the experiments to date have only demonstrated the transient response of the ecosystem to iron fertilisation, and none of the experiments has been run long enough to determine the longer-term changes that would result once the zooplankton has adapted to the new iron-rich phytoplankton community. It is this information that is needed if we are to understand the climate change consequences of iron fertilisation and it may be that this question can only be answered by modeling studies (LeFévre and Watson 1999; Popova et al. 2000; Hannon et al. 2001).

11.4 The Time Series Stations

The JGOFS Science Plan recognised the need to have long-term time series observations to complement the short time-scale process studies. The US JGOFS programme initiated two such time-series stations in 1988, station ALOHA in the North Pacific Ocean gyre at 30° N, 140° W (Karl and Lukas 1996; Karl et al. 2001) and the Bermuda Atlantic Time-Series Study at 31° 45' N, 64° 10' W (Michaels and Knap 1996; Siegel et al. 2001). These stations have produced exciting observations on the temporal increase in DIC, the role of nitrogen fixation, and the role of DOM in export (see Chap. 10).

Other time-series stations have been initiated as part of national JGOFS programmes. The French JGOFS KERFIX station was occupied from 1990 to 1995 at 50° 40' S, 68° 25' E, 60 miles southwest of the Kerguelen Islands (Jeandel et al. 1998). This has provided valuable information on seasonal biogeochemical cycles in an iron-limited area and it is planned to restart this time-series in the near future. There has also been a long-running French JGOFS time-series in the Northwestern Mediterranean (DYFAMED station, Marty et al. 2001). In the North Atlantic, Spanish and German oceanographers have collaborated to set up the ESTOC station, about 100 km north of Gran Canaria (29° 10' N, 15° 30' W), that has been in operation since 1994 (Davenport et al. 1999). As part of the Canadian JGOFS programme, the long-running time-series at Station Papa (50° N, 145° W) has been extended into the 1990s (Wong et al. 1999). In 1998, Japanese JGOFS initiated the KNOT time-series station in the southwest margin of the North Pacific subarctic gyre at 44° N, 155° E (Tsurushima et al. 2002). Finally, the South-East Asia Time Series (SEATS), located in the South China Sea at 18° N, 116° E, was initiated in 1998 by the China Taipei JGOFS programme. Further details of all these stations and some of the key scientific findings are given in Chap. 10.

Some of these time-series stations have now been in operation continuously for 12 years and have already yielded much exciting new information about decadal scale regime shifts in biogeochemical cycles (Karl et al. 2001). These time-series also provide key datasets for model validations (see Chap. 9). In view of the importance for climate studies of long time-series that can span decadal time-scales, it is extremely important that these stations remain in operation into the future. Unfortunately at present there are only firm plans to do this for the HOT and BATS sites. It is often difficult to persuade national funding agencies of the need for long-term monitoring and so perhaps international plans for monitoring the ocean being developed under the GOOS (Global Ocean Observing System) programme (*http://ioc.unesco.org/goos*) hold out the best hope for the future.

11.5 The Global Survey

The JGOFS Science Plan considered three main contributions to the JGOFS Global Survey. These were (*a*) a global survey of oceanic dissolved inorganic carbon (DIC) distribution in cooperation with WOCE, (*b*) an extensive set of surface pCO_2 measurements capable of resolving the seasonal cycle, and (*c*) a global array of sediment traps and benthic measurements to provide lower boundary conditions for the models.

In May 1990, JGOFS and WOCE agreed that two berths should be made available on all WOCE global survey cruises to JGOFS scientists for the purpose of measuring two or more of the components (TCO_2, pCO_2, pH, alkalinity) that fully specify ocean DIC. By the end of the WOCE survey, DIC had been measured on 80% of all cruises. In addition, because high-quality nutrient and oxygen measurements were also made on the WOCE cruises, it is also possible to calculate the excess CO_2 (i.e., the anthropogenic CO_2) that has been added to the ocean since the industrial revolution (Wallace 2001). These datasets will provide strong validity checks on global ocean carbon models (see Chap. 9).

Measurements of pCO_2 in the surface ocean pre-dated JGOFS but the temporal and spatial coverage of such measurements increased dramatically during the JGOFS era. The most recent compilation of all these results (Takahashi et al. 1999) used approximately 2.5 million measurements and gave an estimate of the global CO_2 air-sea exchange of 2.2 Pg C yr^{-1}, providing another strong restraint on model results. The first two components of the JGOFS Global Survey have thus been carried out as planned and can be considered one of the solid achievements of JGOFS.

The planned geographic network of sediment traps and benthic observations was not fully achieved during JGOFS. However, many time-series sediment traps were deployed and benthic studies carried out in conjunction with some of the process studies, providing the ability to study the continuity of carbon fluxes from the surface layers to the ocean floor (Chap. 6 and 8).

11.6 Remote Sensing

If the Process Studies, the Time-Series stations, and the Global Survey are counted as the big successes of JGOFS, then the lack of remote sensing of ocean colour during the major part of JGOFS was a serious disappointment. The need for ocean colour data was recognised at the original 1983 GOFS workshop. In 1987, the Goddard Space Flight Center published the first global composite of the Coastal Zone Color Scanner (CZCS) observations and the power of this new tool was very apparent. The CZCS sensor had ceased operation in 1986 and NASA originally planned to launch a new SeaWiFS sensor in 1992/1993. Unfortunately the project suffered long delays from locating a commercial sponsor and overcoming development problems with the Pegasus launcher, and so SeaWiFS was not launched until September 1997, by which time most of the JGOFS field programme had been completed. There is no disguising the fact that the JGOFS project was the poorer for the absence of ocean colour data. Firstly, it was more difficult to set the results from each process study or time-series in a larger, regional or basin-scale picture. Secondly, during 1986–1997, we lacked global-scale information on chlorophyll distributions and primary production that would have been invaluable in investigating the effects on global biogeochemistry of ENSO events or the eruption of Mt Pinatubo in 1991.

It is hoped that the remote sensing of ocean colour will not suffer from such temporal gaps in the future. One of the aims of the International Ocean Colour Coordinating Group (IOCCG), established by the International Oceanographic Commission in 1996, is to provide the ocean community with a strong lobby to the various national space organisations. There are presently seven ocean colour sensors in operation with six more scheduled for launch over the next 5 years (see the IOCCG website: *www.ioccg.org*). Truly a case of rags to riches!

11.7 Benthic Studies

The objectives of the JGOFS benthic processes programme were stated in the Science Plan as (*a*) to determine and understand the controls on and rates of transfer of carbon and associated biogenic elements between the water column and bottom sediments and the relation to the sedimentary record, and (*b*) to examine the Quaternary palaeo-oceanographic record to determine the relationship of ocean circulation, palaeoproductivity and CO_2 content of the atmosphere, and to aid in the prediction of CO_2-related climate change. Lochte et al. have stated in Chap. 8 that the study of benthic processes was not always a major priority of national JGOFS programmes. Despite this, benthic sampling programmes were carried out during many of the process studies and there were also some cruises dedicated solely to benthic sampling (e.g., Shimmield et al. 1995; Gage et al. 2000; Pfannkuche and Lochte 2000). The contribution that these studies made to JGOFS objectives is discussed fully in Chap. 8.

11.8 Continental Margins

One of the objectives of the Science Plan was to estimate the exchange of carbon and nutrients at continental margins. It is true to say, though, that planning for projects to investigate such exchanges took some time to initiate. As this was also an area of interest to the IGBP LOICZ (Land-Ocean Interactions in the Coastal Zone) project, a joint JGOFS-LOICZ Continental Margins Task Team (CMTT) was established. In 1996, the CMTT organised an international meeting (Hall et al. 1996) with the aim of using existing datasets to produce C, N, and P budgets for four well-studied areas, namely the East China Sea, North Sea, Peru-Chile coast, and Gulf of Guinea Shelf.

A number of national studies of continental margins have been carried out under the auspices of JGOFS. These include the Taiwanese KEEP (Kuroshio Edge Exchange Processes) study (Wong et al. 2000) and the French ECOMARGE (ECOsystemes de MARGEs continentales) study in the southern Bay of Biscay (Monaco et al. 1999). Other studies of ocean margins, such as the European OMEX study, were not formally part of JGOFS and so lie outside the scope of this review.

Synthesis of the results from these and other studies is just beginning (Liu et al. 2000b; Chap. 3). A key question is whether there is a significant exchange of atmospheric CO_2 to the deep ocean via the continental shelves (Tsunogai et al. 1999; Liu et al. 2000a) because the magnitude of this flux may effect how measurements of global air-sea exchange are reconciled with model calculations of the uptake of anthropogenic CO_2 by the oceans (Sarmiento and Sundquist 1992).

11.9 Data Archiving

The JGOFS Science Plan strongly emphasised that the success of JGOFS depended critically on the establishment of international data exchange and management. Indeed the Science Plan stated that read access to the JGOFS database should be without restriction to any interested user as long as credit is fully given to the originator of that data. The JGOFS SSC set up a Data Management Task Team (DMTT) to coordinate international data archiving. Have we lived up to those high ideals? In practice there is a wide diversity in the way national JGOFS programmes have managed their data and made it available to the wider community. Some countries (USA and France) have developed systems that allow web access to most of their data. Others (Australia, Canada, India, Japan, South Africa, and UK) have deposited their data in National Data Centres and data is available on request. In addition, many of the National Data Centres have produced CD-ROMs to provide wider access. Yet other countries (e.g., Germany, Netherlands, and Norway) have set up data centres in marine institutes. The present status of national archiving is given in the most recent reports of the DMTT (Lowry and Baliño 1999; Conkwright and Baliño 2000).

In addition to this, the JGOFS International Project Office (IPO) has produced a metadata catalogue that is accessible on the web that lists all the JGOFS or JGOFS-related cruises since 1986. The IPO has also begun to develop meta-datasets that summarise the data available for different geographical areas and provide access links to these data.

The present state of JGOFS archiving might be described as a job only half-done. Much of the data from the process studies and time-series stations has been, or will soon be, safely archived at a national level. However, two important tasks remain to be done. The first is to produce an international JGOFS master dataset that could be made available either on the web or on CD-ROM. The amount of work involved in this should not be under-estimated, but it is an essential requirement if the JGOFS data are to be used in basin-scale or global synthesis. It is interesting to note that even US researchers have found that their otherwise excellent database system does not make it easy to assemble data for synthesis purposes. There is a clear need to make a transition from database systems that are cruise orientated to ones that are synthesis orientated. The second task is to provide for the long-term stewardship of the data in a way that will be immune from changes in software and hardware technologies and from changes in funding priorities. This can probably only be done by one or more of the major national and world data centres. Both of these problems are being actively addressed by the DMTT.

11.10 Models and Synthesis

A browse through the JGOFS special issues referred to in this chapter gives some idea of the breadth of ocean science carried out during the process studies, the time series stations, and the global survey. In 1998, the JGOFS SSC initiated the synthesis phase of the JGOFS project with the aim of using this vast array of data to develop a new unified picture of global biogeochemical cycles. The Regional Planning Groups were asked to transform themselves into Synthesis Groups with the aim of developing synthesis projects in their given area and encouraging the publication of synthesis papers. The first product of this exercise was the Deep-Sea Research special issue on North Atlantic synthesis and modeling (Koeve and Ducklow 2001) and other issues are shortly to be published or are planned. The chapters in this book are also a contribution to the JGOFS synthesis.

A useful first step in any synthesis exercise is to use data from a process study or time series to calculate a carbon or nitrogen budget for some area of the ocean for which an extensive dataset has been obtained. However, so far only a few of such studies have been made (Chipman et al. 1993; Zhang and Quay 1997; Berelson et al. 1997) and an attempt to determine an observational carbon budget for the Equatorial Pacific (Quay 1997) showed that uncertainties of ±30–50% for individual carbon fluxes reduces the likelihood that a balanced carbon budget can be obtained during a JGOFS process-type study. If this is generally the case, then such budgets may best be determined by assimilating observations into a suitable model (e.g., Fasham et al. 1999).

The field of biogeochemical modeling has seen great progress since the Science Plan was written. In 1990, the first basin-scale models were still in the development stage and were based on simple ecosystem models using a single currency, nitrogen (Sarmiento et al. 1993). Since then, global carbon-nitrogen ocean models have been developed by many groups and more complex ecosystem models are being developed that allow for iron limitation of phytoplankton growth and different size-classes of organisms (see Chap. 9 and papers in Deep-Sea Research II 49(1–3)). A key requirement of any marine ecosystem model is to have a good physical model within which to embed the biogeochemical equations. It is fortunate therefore that the whole field of physical modeling has shown great progress in the last few years and that many physical oceanographers were involved in the JGOFS field programmes. One of the key interactions between the physics and the biogeochemistry is in determining the supply of nutrients to the euphotic zone. It is becoming clear that a number of different physical processes contribute to this supply on many different space-scales (Chap. 2). At the present time the resolution of the global biogeochemical models is too coarse to resolve all these processes and this may put a limit on our ability to predict future changes in the ocean carbon cycle.

The priority for the JGOFS modeling community must be to use the JGOFS datasets to test and validate an ever-developing and improving set of ocean coupled physical and biogeochemical models that will form the basis of the earth system models needed by the international scientific community to guide the debate on the earth's future.

11.11 Overall Conclusions

Having given this summary of the activities carried out under the auspices of JGOFS we should now try and come to a view on whether the founding goals of JGOFS have been achieved. The first goal of JGOFS was: "To determine and understand on a global scale the processes controlling the time-varying fluxes of carbon and associated biogenic elements in the ocean, and to evaluate the related exchanges with the atmosphere, sea floor, and continental boundaries."

I believe that the unprecedented international effort of the last 12 years will enable us to achieve this goal. However, the process of synthesising the data from these activities will need to be pursued with vigour for some years before the full fruits of this effort are realised. To date most synthesis studies have only been applied to a single cruise or time series station, or as part of the JGOFS/WOCE Global Survey. We need to develop the ability to work on a broader canvas and this will require state-of-the-art data archiving and retrieval software, methods of assimilating observations into regional and global models and, perhaps most importantly, the willingness of funding agencies to be aware of the need to continue funding JGOFS synthesis activities after the JGOFS field programme has finished. To date only a few national funding agencies have had the wisdom to do this.

The second JGOFS goal was "To develop the capability to predict on a global scale the response of oceanic biogeochemical processes to anthropogenic perturbations, in particular those related to climate change."

This goal can be seen as the societal pay-off for the JGOFS funding. Global ocean carbon models are already being used in complex models of the earth's carbon cycle to predict the evolution of the earth's climate for the next 100–150 years (Sarmiento and LeQuéré 1996; Cox et al. 2000). Because of the computational demands of such models it is only possible to use extremely simplified models of the ocean biological pump that may miss some of the important feedbacks between marine ecosystems and the carbon cycle (see Chap. 7). One of our challenges over the coming years is to continue the dialogue between modellers and observationalists to ensure that, as computer power increases, ocean models include sufficient complexity to increase our confidence, and that of society, in our predictions of the planet's future.

References

Ahmed SI, Khan N, Saleem M, Ali A (1995) Changes in biomass and size-fractionated primary productivity in the northern Arabian Sea in response to winter and summer monsoons. In: Thompson M-F, Tirmizi NM (eds) The Arabian Sea: living marine resources and the environment. Balkema Rotterdam, pp 479–495

Anderson LG, Drange H, Chierici M, Fransson M, Johannessen T, Skjelvan I, Rey F (2000) Annual carbon fluxes in the upper Greenland Sea based on measurements and a box model approach. Tellus B 52:1013–1024

Anderson RF, Smith WO Jr. (2001) The US Southern Ocean Joint Global Ocean Flux Study: vol. 2. Deep-Sea Res Pt II 48:3883–3890

Balch WM, Kilpatrick KA, Holligan P, Harbour D, Fernandez E (1996a) The 1991 coccolithophore bloom in the central North Atlantic. 2. Relating optics to coccolith concentration. Limnol Oceanogr 41:1684–1696

Balch WM, Kilpatrick KA, Trees CC (1996b) The 1991 coccolithophore bloom in the central North Atlantic. 1. Optical properties and factors affecting their distribution. Limnol Oceanogr 41:1669–1683

Bathmann U, Priddle J, Tréguer P, Lucas M, Hall J, Parslow J (2000) Plankton ecology and biogeochemistry in the Southern Ocean: a review of Southern Ocean JGOFS. In: Hanson RB, Ducklow HW, Field JG (eds) The changing ocean carbon cycle: a midterm synthesis of the Joint Global Ocean Flux Study. Cambridge University Press, Cambridge, pp 300–339

Behrenfeld MJ, Bale AJ, Kolber ZS, Aiken J, Falkowski PG (1996) Confirmation of iron limitation of phytoplankton photosynthesis in the equatorial Pacific Ocean. Nature 383:508–511

Bender ML, Ducklow HW, Kiddon J, Marra J, Martin JH (1992) The carbon balance during the 1989 spring bloom in the North Atlantic Ocean at 47°N 20°W. Deep-Sea Res 39:1707–1725

Berelson WM, Anderson RF, Dymond J, Demaster D, Hammond DE, Collier R, Honjo S, Leinen M, McManus J, Pope R, Smith C, Stephens M (1997) Biogenic budgets of particle rain, benthic remineralization and sediment accumulation in the equatorial Pacific. Deep-Sea Res Pt II 44:2251–2283

Bodungen Bv, Antia A, Bauerfeind E, Haupt OJ, Koeve W, Machado E, Peeken I, Peinert R, Reitmeier S, Thomsen C, Voss M, Wunsch M, Zeller U, Zeitzschel B (1995) Pelagic processes and vertical flux of particles: an overview of a long-term comparative study in the Norwegian Sea and Greenland Sea. Geol Rundsch 84:11–27

Boyd PW, Law CS (2001) The Southern Ocean Iron RElease Experiment (SOIREE) – introduction and summary. Deep-Sea Res Pt II 48:2425–2438

Boyd PW, Goldblatt RH, Harrison PJ (1999a) Mesozooplankton grazing manipulations during in vitro iron enrichment studies in the NE subarctic Pacific. Deep-Sea Res Pt II 46:2645–2668

Boyd PW, Harrison PJ, Johnson BD (1999b) The Joint Global Ocean Flux Study (Canada) in the NE subarctic Pacific. Deep-Sea Res Pt II 46:2345–2350

Burkill PH (1999) ARABESQUE: an overview. Deep-Sea Res Pt II 46:529–547

Chiba ST, Hirawake T, Ushio S, Horimota N, Satoh R, Nakajima Y, Ishimaru T, Yamaguchi Y (2000) An overview of the biological/oceanographic survey by the RTV Umitaka-Maru III off Adélie Land, Antarctica, between January–February 1996. Deep-Sea Res Pt II 47:2589–2613

Chipman DW, Marra J, Takahashi T (1993) Primary production at 47° N 20° W in the North Atlantic Ocean: a comparison between the ^{14}C incubation method and the mixed layer carbon budget. Deep-Sea Res Pt II 40:151–170

Coale KH, Johnson KS, Fitzwater SE, Gordon RM, Tanner S, Chavez FP, Ferioli L, Sakamoto C, Rogers P, Millero F, Steinberg P, Nightingale P, Cooper D, Cochlan WP, Landry MR, Constantinou J, Rollwagen G, Trasvina A, Kudela R (1996) A massive phytoplankton bloom induced by an ecosystem-scale iron fertilization experiment in the equatorial Pacific Ocean. Nature 383: 495–501

Conkwright M, Baliño BM (2000) Meeting of the JGOFS Data Management Task Team, 5–6 June 2000. JGOFS International Project Office, 40 pp

Cooper DJ, Watson AJ, Nightingale PD (1996) Large decrease in ocean-surface CO_2 fugacity in response to in situ iron fertilization. Nature 383:511–513

Cox PM, Betts RA, Jones CD, Spall SA, Tottedell IJ (2000) Acceleration of global warming due to carbon-cycle feedbacks in a coupled climate model. Nature 408:184–187

Dadou I, Lamy F, Rabouille C, Ruiz-Pino D, Andeesen V, Bianchi M, Garçon V (2001) An integrated biological pump model from the euphotic zone to the sediment: a 1-D application in the Northeast tropical Atlantic. Deep-Sea Res Pt II 48:2345–2381

Davenport R, Neuer S, Hernadez-Guerra A, Rueda MJ, Llinas O, Fischer G, Wefer G (1999) Seasonal and interannual pigment concentration in the Canary Island region from CZCS data and comparison with observations from the ESTOC. Int J Remote Sens 20:1419–1433

de Baar HJW, Boyd PW (2000) The role of iron in plankton ecology and carbon dioxide transfer in the global oceans. In: Hanson RB, Ducklow HW, Field JG (eds) The changing ocean carbon cycle: a midterm synthesis of the Joint Global Ocean Flux Study. Cambridge University Press, Cambridge, pp 61–140

Ducklow HW (1989) Joint Global Ocean Flux Study: the 1989 North Atlantic bloom Experiment. Oceanography 2:4–8

Ducklow HW, Harris RP (1993) Introduction to the JGOFS North Atlantic Bloom Experiment. Deep-Sea Res Pt II 40:1–8

Fasham MJR, Evans GT (2000) Advances in ecosystem modelling within JGOFS. In: Hanson RG, Ducklow HW, Field JG (eds) The changing ocean carbon cycle: a midterm synthesis of the Joint Global Ocean Flux Study. Cambridge University Press Cambridge, pp 417–446

Fasham MJR, Boyd PW, Savidge G (1999) Modeling the relative contributions of autotrophs and heterotrophs to carbon flow at a Lagrangian JGOFS station in the northeast Atlantic: the importance of DOC. Limnol Oceanogr 44:80–94

Fasham MJR, Baliño BM, Bowles MC (2001) A new vision of ocean biogeochemistry after a decade of the Joint Global Ocean Flux Study (JGOFS). Ambio Special Report, 3–31

Feely RA, Wanninkhof R, Goyet C, Archer DE, Takahashi T (1997) Variability of CO_2 distributions and sea-air fluxes in the central and eastern equatorial Pacific during the 1991–1994 El Niño. Deep-Sea Res Pt II 44:1851–1867

Field JG, Ducklow HW, Hanson RB (2000) Some conclusion and highlights of JGOFS mid-project achievements. In: Hanson RB, Ducklow HW, Field JG (eds) The changing ocean carbon cycle: a midterm synthesis of the Joint Global Ocean Flux Study. Cambridge University Press, Cambridge, pp 493–499

Gage JD, Levin LA, Wolff GA (2000) Benthic processes in the deep Arabian Sea: introduction and overview. Deep-Sea Res Pt II 47:1–8

Gaillard J-F (1997) ANTARES-I: a biogeochemical study of the Indian sector of the Southern Ocean. Deep-Sea Res Pt II 44:951–961

Hall J, Smith SV, Boudreau PR (1996) Report on the International Workshop on Continental Shelf Fluxes of carbon, nitrogen and phosphorus. LOICZ, Texel, The Netherlands, 50 pp

Hannon E, Boyd PW, Silvoso M, Lancelot C (2001) Modeling the bloom evolution and carbon flows during SOIREE: implications for future in situ iron-enrichments in the Southern Oceans. Deep-Sea Res Pt II 48:2745–2773

Harrison WG, Head EJH, Horne EPW, Irwin B, Li WKW, Longhurst AR, Paranjape MA, Platt T (1993) The western North Atlantic Bloom Experiment. Deep-Sea Res Pt II 40:278–305

Harrison WG, Arístegui J, Head EJH, Li W, Longhurst AR, Sameoto DD (2001) Basin-scale variability in plankton biomass and community metabolism in the sub-tropical North Atlantic Ocean. Deep-Sea Res Pt II 48:2241–2293

Hiroe Y, Yasuda I, Komatsu K, Kawasaki I, Joyce TM, Bahr F (2002) Transport of North Pacific intermediate water in the Kuroshio-Oyashio interfrontal zone. Deep-Sea Res II 49:5353–5364

Ishizaka J, Harada K, Ishikawa K, Kiyosawa H, Furusawa H, Watanabe Y, Ishida H, Susuki K, Handa N, Takahashi M (1997) Size and taxonomic plankton community structure and carbon flow at the equator, 175° E during 1990–1994. Deep-Sea Res Pt II 44:1927–1950

Jeandel C, Ruiz-Pino D, Gjata E, Poisson A, Brunet C, Charriaud E, Dehairs F, Delille D, Fiala M, Fravalo C, Miquel JC, Park Y-H, Pondaven P, Quéguiner B, Razouls S, Shauer B, Tréguer P (1998) KERFIX, a time-series station in the Southern Ocean: a presentation. J Marine Syst 17:555–569
Karl DM, Lukas R (1996) The Hawaii Ocean Time-series (HOT) program: background, rationale and field implementation. Deep-Sea Res Pt II 43:129–156
Karl DM, Bidigare RR, Letelier RM (2001) Long-term changes in plankton community structure and productivity in the North Pacific Subtropical gyre: the domain shift hypothesis. Deep-Sea Res Pt II 48:1449–1470
Koeve W, Ducklow HW (2001) JGOFS synthesis and modeling: the North Atlantic Ocean. Deep-Sea Res Pt II 48:2141–2154
Le Borgne R, Rodier M (1997) Net zooplankton and the biological pump: a comparison between the oligotrophic and mesotrophic Pacific. Deep-Sea Res Pt II 44:2003–2024
Le Févre J, Tréguer P (1998) Carbon fluxes and dynamic processes in the Southern Ocean: present and past. J Marine Syst 17:1–4
Le Févre N, Watson AJ (1999) Modeling the geochemical cycle of iron in the oceans and its impact on atmospheric CO_2 concentrations. Global Biogeochem Cy 13:727–736
Liu KK, Atkinson L, Chen CTA, Gao S, Hall J, MacDonald RW, McManus LT, Quiñones R (2000a) Exploring continental margin carbon fluxes on a global scale. Eos 81:641–644
Liu KK, Iseki K, Chao S-Y (2000b) Continental margin carbon fluxes. In: Hanson RB, Ducklow HW, Field JG (eds) The changing ocean carbon cycle: a midterm synthesis of the Joint Global Ocean Flux Study. Cambridge University Press, Cambridge, pp 187–239
Lowry RK, Baliño BM (1999) JGOFS Data Management and Synthesis Workshop, 25–27 September 1998. JGOFS Report No. 29, 45 pp
Lucas MI, Lutjeharms JRE, Field JG, McQuaid CD (1993) A new South African research programme in the Southern Ocean. S Afr J Sci 83:61–67
Mackey DJ, Parslow J, Higgins HW, Griffiths FB, O'Sullivan JE (1995) Plankton productivity and biomass in the western equatorial Pacific: biological and physical controls. Deep-Sea Res Pt II 42:499–533
Martin JH, Fitzwater SE (1988) Iron deficiency limits phytoplankton growth in the north-east Pacific subarctic. Nature 331: 341–343
Martin JH, Coale KH, Johnson KS, Fitzwater SE, Gordon RM, Tanner SJ, Hunter CN, Elrod VA, Nowicki JL, Coley TL, Barber RT, Lindley S, Watson AJ, Van Scoy K (1994) Testing the iron hypothesis in ecosystems of the equatorial Pacific Ocean. Nature 371:123–129
Marty JC, Chiavérini MD, Pizay MD, Avril B (2001) Seasonal and interannual dynamics of nutrients and phytoplankton pigments in the Western Mediterranean Sea at the DYFAMED time-series station (1991–1999). Deep-Sea Res Pt II 49/11:1965–1985
Michaels AF, Knap AH (1996) Overview of the US JGOFS Bermuda Atlantic Time-Series Study and the Hydrostation 'S' program. Deep-Sea Res Pt II 43:157–198
Miller CB (1993) Pelagic processes in the Subarctic Pacific. Prog Oceanogr 32:1–15
Monaco A, Biscaye PE, Laborde P (1999) The ECOFER (ECOsystème du canyon du cap-FERret) experiment in the Bay of Biscay: introduction, objectives and major results. Deep-Sea Res Pt II 46:1967–1978
Morel A (1996) An ocean flux study in eutrophic, mesotrophic and oligotrophic situations: the EUMELI program. Deep-Sea Res Pt I 43:1185–1190
Murray JW, Johnson E, Garside C (1995) A U.S. JGOFS Process Study in the equatorial Pacific (EqPac): introduction. Deep-Sea Res Pt II 42:275–294
Murray JW, Borgne R Le, Dandoneau Y (1997) JGOFS studies in the equatorial Pacific. Deep-Sea Res Pt II 44:1759–1764
Nicol S, Pauly T, Bindoff NL, Strutton PG (2000) 'BROKE' a biological/oceanographic survey off the coast of East Antarctica (80–150° E) carried out in January–March 1996. Deep-Sea Res Pt II 47:2281–2298
Olson RJ, Sosik HM, Chekalyuk AM, Shalpyonok A (2000) Effects of iron enrichment on phytoplankton in the Southern Ocean during late summer: active fluorescence and flow cytometric analyses. Deep-Sea Res Pt II 47:3179–3200
Pfannkuche O, Lochte K (2000) The biogeochemistry of the deep Arabian Sea: overview. Deep-Sea Res Pt II 47:2615–2628
Popova E, Ryabchenko VA, Fasham MJR (2000) Biological pump and vertical mixing in the Southern Ocean: their impact on atmospheric CO_2. Global Biogeochem Cy 14:477–498
Prasanna Kumar SP, Madhupratap M, Dileep Kumar M, Gauns M, Muraleedharan PM, Sarma VVSS, De Souza SN (2000) Physical control of primary productivity on a seasonal scale in the central and eastern Arabian Sea. P Indian As-Earth 109: 433–442
Quay P (1997) Was a carbon balance measured in the equatorial Pacific during JGOFS? Deep-Sea Res Pt II 44:1765–1783
Roy S, Sundby B, Vézina AF, Legendre L (2000) A Canadian JGOFS process study in the Gulf of St. Lawrence. Introduction. Deep-Sea Res Pt II 47:377–384
Saino T, Kusakabe M (2002) JGOFS Studies on carbon cycle in the North Pacific. Institute for Hydrospheric-Atmospheric Sciences, Nagoya University, (in press)
Saino T, Bychkov A, Chen C-TA, Harrison PJ (2002) The Joint Global Flux Study in the North Pacific. Deep-Sea Research II 49:5297–5301
Sarmiento JL, Le Quéré C (1996) Oceanic carbon dioxide in a model of century-scale global warming. Science 274:1346–1349
Sarmiento JL, Sundquist ET (1992) Revised budget for the oceanic uptake of anthropogenic carbon. Nature 356:589–593
Sarmiento JL, Slater RD, Fasham MJR, Ducklow HW, Toggweiler JR, Evans GT (1993) A seasonal three-dimensional ecosystem model of nitrogen cycling in the North Atlantic euphotic zone. Global Biogeochem Cy 7:417–450
Savidge G, Turner DR, Burkill PH, Watson AJ, Angel MV, Pingree RD, Leach H, Richards KJ (1992) The BOFS 1990 spring bloom experiment: temporal evolution and spatial variability of the hydrographic field. Prog Oceanogr 29:235–281
Savidge G, Boyd P, Pomroy A, Harbour D, Joint I (1995) Phytoplankton production and biomass estimates in the northeast Atlantic Ocean, May–June 1990. Deep-Sea Res Pt I 42: 599–617
SCOR (1990) The Joint Global Ocean Flux Study science plan. Scientific Committee on Oceanic Research. JGOFS Report No. 5, 61 pp
SCOR (1992) Joint Global Ocean Flux Study implementation plan. International Geosphere-Biosphere Programme. IGBP Report No. 23 and JGOFS Report No. 9, 38 pp
Shimmield GB, Brand TD, Ritchie GD (1995) The benthic geochemical record of the late Holocene carbon flux in the northeast Atlantic. Philos T Roy Soc B 348:221–227
Siegel DA, Karl DM, Michaels AF (2001) Interpretations of biogeochemical processes from the US JGOFS Bermuda and Hawaii time-series sites. Deep-Sea Res Pt II 48:1403–1404
Smetacek V, De Baar HJW, Bathmann UV, Lochte K, Van Der Loeff MMR (1997) Ecology and biogeochemistry of the Antarctic Circumpolar Current during austral spring: a summary of Southern Ocean JGOFS cruise ANT X/6 of R.V. Polarstern. Deep-Sea Res Pt II 44:1–21
Smith SL (2001) Understanding the Arabian Sea: reflections on the 1994–1996 Arabian Sea expedition. Deep-Sea Res Pt II 48: 1385–1402
Smith SL, Codispoti LA, Morrison JM, Barber RT (1998) The 1994–1996 Arabian Sea Expedition: an integrated, interdisciplinary investigation of the response of the Northwestern Indian Ocean to monsoonal forcing. Deep-Sea Res Pt II 45:1905–1915
Smith WO Jr., Anderson RF, Moore JK, Codispoti LA, Morrison JM (2000) The US Southern Ocean Joint Global Ocean Flux Study: an introduction to AESOPS. Deep-Sea Res Pt II 47: 3073–3093
Strass VH, Bathmann UV, Rutgers van der Loeff MM, Smetacek V (2002) Mesoscale physics, biogeochemistry and ecology of the Antarctic Polar Front, Atlantic sector: an introduction to and summary of cruise ANT XIII/2 of R.V. Polarstern. Deep-Sea Res II 49:3707–3711

Takahashi T, Wanninkhof RH, Feely RA, Weis RF, Chipman DW, Bates NR, Olafsson J, Sabine C, Sutherland SC (1999) Net sea-air CO_2 flux over the global oceans: an improved estimate based on the air-sea pCO_2 difference. In: Nojiri Y (ed) Second International Symposium on CO_2 in the Oceans. National Institute for Environmental Studies, Tsukuba Tsukuba, Japan, pp 9–15

Tréguer P, Pondaven P (2002) Climatic changes and the Southern Ocean: a step forward. Deep-Sea Res Pt II 49:1597–1610

Trull T, Sedwick PN, Griffiths FB, Rintoul SR (2001) Introduction to special section: SAZ project. J Geophys Res 106(C12):31425–31429

Tsunogai S, Watanabe S, Sato T (1999) Is there a 'continental shelf pump' for the absorption of atmospheric CO_2? Tellus B 51:701–712

Tsurushima N, Nojiri Y, Imai K, Watanabe S (2002) Seasonal variations of carbon dioxide system and nutrients in the surface mixed layer at station KNOT (44°N, 155°E) in the subarctic western North Pacific. Deep-Sea Res II 49:5377–5394

Turner DR, Owens NJP (1995) A biogeochemical study in the Bellingshausen Sea: overview of the STERNA 1992 expedition. Deep-Sea Res Pt II 42:907–932

Veldhuis MJW, Kraay GW, Van Bleijswijk JDL, Baars MA (1997) Seasonal and spatial variability in phytoplankton biomass, productivity and growth in the northwestern Indian Ocean: the southwest and northeast monsoon, 1992–1993. Deep-Sea Res Pt I 44:425–449

Wallace DWR (2001) Storage and transport of excess CO_2 in the oceans: the JGOFS/WOCE Global CO_2 Survey. In: Siedler G, Church J, Gould J (eds) Ocean circulation and climate. Academic Press, pp 489–521

Watson AJ, Liss P, Duce R (1991) Design of a small-scale in situ iron fertilisation experiment. Limnol Oceanogr 36:1960–1965

Wong CS, Whitney FA, Crawford DW, Iseki F, Matear RJ, Johnson WK, Page JS, Timothy D (1999) Seasonal and interannual variability in particle fluxes of carbon, nitrogen and silicon from time series of sediment traps at Ocean Station P, 1982–1993: relationship to changes in subarctic primary production. Deep-Sea Res Pt II 46:2735–2760

Wong GTF, Chao S-Y, Li Y-H, Shiah F-K (2000) The Kuroshio edge exchange processes (KEEP) study – an introduction to hypotheses and highlights. Cont Shelf Res 20:335–347

Yoder JA, Ackleson SG, Barber RT, Flament P, Balch WM (1994) A line in the sea. Nature 371:689–692

Zeitzschel B (1973) The biology of the Indian Ocean. Springer-Verlag, Berlin, 549 pp

Zeng J, Nojiri Y, Murphy PP, Wong CS, Fujinuma Y (2002) A comparison of pCO_2 distributions in the northern North Pacific using results from a commercial vessel in 1995–1999. Deep-Sea Res II 49:5303–5315

Zhang J, Quay PD (1997) The total organic carbon export based on ^{13}C and ^{12}C of DIC budgets in the equatorial Pacific region. Deep-Sea Res Pt II 44:2163–2190

Index

C

D

E

F

I

J

K

L

N

O

P

Q

R

S

T

U

V

W

X, Y, Z

MIX
Papier aus verantwortungsvollen Quellen
Paper from responsible sources
FSC® C105338

If you have any concerns about our products,
you can contact us on
ProductSafety@springernature.com

In case Publisher is established outside the EU,
the EU authorized representative is:
Springer Nature Customer Service Center GmbH
Europaplatz 3, 69115 Heidelberg, Germany

Printed by Libri Plureos GmbH
in Hamburg, Germany

Springer Proceedings in Mathematics & Statistics

Volume 123

Springer Proceedings in Mathematics & Statistics

This book series features volumes composed of selected contributions from workshops and conferences in all areas of current research in mathematics and statistics, including operation research and optimization. In addition to an overall evaluation of the interest, scientific quality, and timeliness of each proposal at the hands of the publisher, individual contributions are all refereed to the high quality standards of leading journals in the field. Thus, this series provides the research community with well-edited, authoritative reports on developments in the most exciting areas of mathematical and statistical research today.

More information about this series at http://www.springer.com/series/10533

Ingrid Bauer · Shelly Garion
Alina Vdovina
Editors

Beauville Surfaces and Groups

Springer

Editors
Ingrid Bauer
Lehrstuhl Mathematik VIII
University of Bayreuth
Bayreuth
Germany

Alina Vdovina
School of Mathematics and Statistics
Newcastle University
Newcastle-upon-Tyne
UK

Shelly Garion
Institut für Mathematische Logik und
Grundlagenforschung
University of Münster
Münster
Germany

ISSN 2194-1009 ISSN 2194-1017 (electronic)
Springer Proceedings in Mathematics & Statistics
ISBN 978-3-319-34425-6 ISBN 978-3-319-13862-6 (eBook)
DOI 10.1007/978-3-319-13862-6

Mathematics Subject Classification: 11 Number theory (11E57, 11G32), 14 Algebraic Geometry (14J10, 14J29, 14J50, 14H30, 14L30), 20 Group Theory and Generalizations (20B25, 20D06, 20D15, 20E42, 20F29, 20G40)

Springer Cham Heidelberg New York Dordrecht London

Softcover reprint of the hardcover 1st edition 2015

Printed on acid-free paper

Springer International Publishing AG Switzerland is part of Springer Science+Business Media (www.springer.com)

Contents

Introduction

Ingrid Bauer, Shelly Garion and Alina Vdovina

"Beauville Surfaces" are certain rigid regular surfaces of general type, which can be described in a purely algebraic combinatorial way. Thus they play a very important role in different fields of mathematics such as algebraic geometry, group theory and number theory. The notion of Beauville surface was introduced by Fabrizio Catanese in 2000 and after the first systematic study of these surfaces by Ingrid Bauer, Fabrizio Catanese and Fritz Grunewald, there has been an increasing interest in the subject.

The conference "Beauville Surfaces and Groups 2012" was held in the University of Newcastle, UK, from 7 to 9 June 2012. This conference brought together, for the first time, experts from different fields of mathematics interested in Beauville surfaces, as well as young researchers and Ph.D. students, from the UK, USA, Germany, Italy and Spain, in order to share the status of the art and to discuss further developments in the study of these fascinating surfaces.

These conference proceedings include 11 chapters discussing Beauville surfaces (and generalizations of them), from different points of view, algebro-geometric and group-theoretic. Some of the chapters are expository survey papers and the others are research chapters on recent developments in this area. The chapters reflect the topics of the lectures presented during the workshop, and discuss various open problems and conjectures related to Beauville surfaces.

We briefly describe the content of the chapters.

The first chapter, by Ingrid Bauer, Fabrizio Catanese and Davide Frapporti, gives a comprehensive overview of what is currently known about the fundamental groups and, in particular, about the first homology groups of Beauville surfaces, and more generally, surfaces isogenous to a higher product of curves. The authors provide a computer script which calculates the first homology groups of regular surfaces of general type, hence in particular for Beauville surfaces.

The chapter by Nathan Barker, Nigel Boston, Norbert Peyemirnhoff and Alina Vdovina presents new interesting examples of finite groups admitting unmixed ramification structures giving rise to regular algebraic surfaces isogenous to a higher product of curves, generalizing Beauville surfaces. These groups are constructed as certain p-quotients of particular groups with special presentation related to finite projective planes and to expander graphs.

Nigel Boston overviews the important role played by a very interesting class of groups, namely p-groups, in the study of Beauville surfaces. The chapter describes recent results as to which p-groups admit a Beauville structure, with emphasis on ones of small order (due to Barker, Boston and Fairbairn) and ones that form inverse systems (due to Barker, Boston, Peyerimhoff and Vdovina).

Ben Fairbairn gives new interesting examples of finite groups which define real Beauville surfaces, and in particular, new examples of infinite families of groups admitting strongly real Beauville structures. The chapter discusses certain finite simple groups as well as abelian and nilpotent groups. It moreover deals with characteristically simple groups and almost simple groups.

Shelly Garion surveys the probabilistic group-theoretical approach towards proving well-known conjectures of Bauer, Catanese and Grunewald regarding Beauville surfaces arising from finite simple groups, by describing the following three works. The first is the work of Garion, Larsen and Lubotzky, showing that almost all finite simple groups of Lie type admit a Beauville structure. The second is the work of Garion and Penegini on Beauville structures of alternating groups, based on results of Liebeck and Shalev, and the third is the case of the group $\mathrm{PSL}_2(\mathrm{q})$, in which the author gives bounds on the probability of generating a Beauville structure.

The contribution of Christian Gleissner gives a complete classification of the connected components of the moduli space of surfaces of general type corresponding to regular unmixed surfaces S isogenous to a product of curves with $\chi(S) = 2$. These surfaces are natural generalizations of Beauville surfaces and the results require similar techniques from computational group theory. The surfaces studied in this research paper are no longer rigid, whence they come up in families, which are completely described.

Gareth Jones presents new interesting infinite families of Beauville surfaces arising from characteristically simple groups. More precisely, the author proves that if G is a cartesian power of a finite simple group of Lie type of low Lie rank or a sporadic simple group, then G is a Beauville group if and only if it has two generators and is not isomorphic to A_5. The author moreover conjectures that any cartesian power of a finite simple group of Lie type, which is 2-generated and not isomorphic to A_5, admits a Beauville structure.

The chapter by Kay Magaard and Chris Parker discusses the natural question of which extensions of the finite simple groups admit Beauville structures. This follows the previous exciting results due to Guralncik and Malle and to Fairbairn, Maggard and Parker, that all finite simple (respectively, quasi-simple) groups (except A_5 and $\mathrm{SL}_2(5)$) admit Beauville structures, thus proving a well-known conjecture of Bauer, Catanese and Grunewald. The chapter gives various interesting new examples of Frattini covers and semidirect products of finite simple groups admitting Beauville structures. As an application the authors deduce that every finite quotient of the profinite group $\mathrm{SL}_{\mathrm{d}}(\mathbb{Z}_{\mathrm{p}})$, where $d \geq 9$, admits a Beauville structure.

Matteo Penegini gives asymptotic estimates on the number of connected components of the moduli space of surfaces of general type, focusing on interesting families of surfaces isogenous to a higher product of curves, and Beauville surfaces in particular, and describes recent results due to Garion and Penegini. The proof is based on group theoretical methods, using the group theoretical terminology of the so-called "ramification structures", and Beauville structures in particular.

The paper by Roberto Pignatelli surveys the state of the art on quasi-étale quotients of a product of two curves, i.e. surfaces which are the quotient of a product of two curves by a finite group, acting freely outside a finite set of points. This class of surfaces has been successfully used in the last years to construct many interesting examples of surfaces of general type in a systematic computer aided way. The chapter reviews the principal results and gives a complete list of minimal quasi-étale surfaces of general type with geometric genus equal to the irregularity ≤ 2.

Francesco Polizzi gives a survey on recent work on isotrivially fibred surfaces and their numerical invariants. One of the main goals is, if $\alpha : X \to C$ is a relatively minimal isotrivial fibration of the surface X over a curve C, to relate the invariants K_X^2 and $\chi(\mathcal{O}_X)$ by some inequalities. Since surfaces X admitting an isotrivial fibration are of the form $X = C_1 \times C_2/G$, where $C_i, i = 1, 2$ are curves of respective genera at least two, and G is a finite group acting faithfully (but not necessarily freely) on $C_1 \times C_2$, these surfaces are natural generalizations of Beauville surfaces.

The editors would like to thank all the authors contributing to this volume and the referees for their assistance. The editors are grateful to Springer Proceedings in Mathematics and Statistics (PROMS) for publication of this collection of chapters.

The editors are grateful to all the speakers and participants in the workshop "Beauville Surfaces and Groups 2012", to the University of Newcastle for hosting the workshop, and to Nathan Barker for his assistance in organizing the workshop. The editors want to thank the London Mathematical Society (LMS) and the Deutsche Forschungsgemeinschaft (DFG) for their financial support.

These conference proceedings are dedicated to the memory of Fritz Grunewald, *who was the first to realize how fascinating Beauville surfaces are and inspired many mathematicians from different areas to investigate these surfaces. He is deeply missed.*

The Fundamental Group and Torsion Group of Beauville Surfaces

Ingrid Bauer, Fabrizio Catanese and Davide Frapporti

Abstract We give a survey on the fundamental group of surfaces isogenous to a higher product. If the surfaces are regular, e.g. if they are Beauville surfaces, the first homology group is a finite group. We present a MAGMA script which calculates the first homology groups of regular surfaces isogenous to a product.

1 Introduction

One aim of this note is to give an overview of what is known about the fundamental groups and more specifically about the first homology groups of Beauville surfaces and of their generalizations. In particular, we wrote a MAGMA script which calculates the torsion group of Beauville surfaces, or more generally of regular surfaces isogenous to a higher product.

Fundamental groups of algebraic varieties are a very interesting area of research and they are still quite mysterious objects, since usually it may be very hard to determine them. On the other hand, if one can calculate them in some specific cases (e.g., for surfaces of general type) they are a powerful tool to conclude that two surfaces (of general type) are not contained in the same connected component of their moduli space.

Here we will restrict ourselves to the following situation: let C_1 and C_2 be projective algebraic curves of respective genera g_1, g_2 at least 2 and let G be a subgroup of the automorphism group $\mathrm{Aut}(C_1 \times C_2)$. We denote by X the quotient $(C_1 \times C_2)/G$; when X is singular, we denote by S the minimal resolution of singularities of X.

We define $G^0 := G \cap (\mathrm{Aut}(C_1) \times \mathrm{Aut}(C_2))$. Then (compare [10] for this and the following assertions) G^0 is a subgroup of index at most 2 inside $G < \mathrm{Aut}(C_1 \times C_2)$ and acts on each factor and "diagonally" on the product $C_1 \times C_2$ (i.e., for each $\gamma \in G^0$, we can write $\gamma = (\gamma_1, \gamma_2)$, and $\gamma(x_1, x_2) = (\gamma_1(x_1), \gamma_2(x_2))$).

I. Bauer (✉) · F. Catanese · D. Frapporti
Lehrstuhl Mathematik VIII, Mathematisches Institut der Universität Bayreuth, NW II, Universitätsstr. 30, 95447 Bayreuth, Germany
e-mail: Ingrid.bauer@uni-bayreuth.de

I. Bauer et al. (eds.), *Beauville Surfaces and Groups*, Springer Proceedings in Mathematics & Statistics 123, DOI 10.1007/978-3-319-13862-6_1

By [10], it is always possible to assume that G^0 acts faithfully on both factors: in this case we say that $(C_1 \times C_2)/G$ is a *minimal realization*, and this minimal realization is moreover unique. From now on we shall assume that we have indeed a minimal realization. Then (see [10]) there are two cases.

The *mixed case* is the case where the action of G exchanges the two factors: in this case $C_1 \cong C_2$ and $G^0 \lhd_2 G$.

The *unmixed case* is the case where $G = G^0$, and its two projections into $\mathrm{Aut}(C_j)$, $j = 1, 2$, are injective. Identifying then G to a subgroup $G \subset \mathrm{Aut}(C_j)$, for $j = 1, 2$, we say that G acts diagonally: this means that we view G as the diagonal subgroup inside $G \times G \subset (\mathrm{Aut}(C_1) \times \mathrm{Aut}(C_2))$.

In the unmixed case, S is called a *product-quotient surface* and X is its *quotient model*; while in the mixed case, S is a *mixed surface* and X is a *mixed quotient*.

In this note we want to focus on the unmixed case, so from now on we implicitly assume that $G = G^0$. Some of the results can be extended/generalized to the mixed case, but the approach is slightly different. We refer to [13, 14] for further details.

In the first section we describe product-quotient surfaces and the algebraic data which determine them. In the second section we comment on the results on the fundamental group of product-quotient surfaces and their higher-dimensional analogues. The third section is dedicated to the description of a MAGMA [8] script which, given as input a finite group G and two ordered tuples

$$T_1 = (n_1, \ldots, n_r), \quad T_2 = (m_1, \ldots, m_s),$$

has as output

- one representative surface $S = (C_1 \times C_2)/G$ for each irreducible connected component of the moduli space, where G is acting freely with signatures (T_1, T_2),
- for each such representative, the first homology group $H_1(S, \mathbb{Z})$ of S.

Finally, in the last section, we show several concrete and explicit calculations, obtained through a direct application of our program.

2 Product-Quotient Surfaces

Let G be a finite group and let S be a *product-quotient surface* with *quotient model* $X = (C_1 \times C_2)/G$. If G acts freely, then X is smooth, and we have the following:

Definition 2.1 A surface S is said to be *isogenous to a (higher) product* if S is the quotient $(C_1 \times C_2)/G$, where $g_i := genus(C_i) \geq 2$, and G is a finite group acting freely on $C_1 \times C_2$.

The word "higher" means that the respective genera of C_1, C_2 are ≥ 2, in particular this implies that S is of general type (and with ample canonical divisor). However, for commodity, from now on we shall drop the word "higher".

In the last years, a huge amount of new surfaces of general type with $p_g = q$ have been constructed as the quotient of a product of two curves by the action of a finite group; see [1, 3, 4, 6, 7, 13, 14] for $p_g = 0$, [11, 13, 17, 19, 20] for $p_g = 1$, [18, 23] for $p_g = 2$.

In particular, we have a complete classification of surfaces isogenous to a product with $p_g = q$. The case where G does not act freely is still open (and particularly difficult in the regular case $q = 0$), in spite of several results. A difficulty which is peculiar of the regular case is that the following can happen: the minimal resolution S of singularities X may not be minimal surface. Up to now there are almost no techniques to decide whether there are exceptional curves of the first kind (i.e., smooth rational curves with self intersection (-1)) on S and how to find them explicitly (cf. [7]).

The description of product-quotient surfaces is accomplished through the theory of Galois coverings between projective curves (also named "Riemann surfaces").

Definition 2.2 Let $g \geq 0$ and $m_1, \ldots, m_r > 1$ be integers. The *orbifold surface group of signature* $(g; m_1, \ldots, m_r)$ is defined as:

$$\mathbb{T}(g; m_1, \ldots, m_r) := \langle a_1, b_1, \ldots, a_g, b_g, c_1, \ldots, c_r \mid c_1^{m_1}, \ldots, c_r^{m_r}, \textstyle\prod_{i=1}^{g} [a_i, b_i] \cdot c_1 \cdots c_r \rangle .$$

Given a finite group H, a *generating vector* for H of signature $(g; m_1, \ldots, m_r)$ is a $(2g + r)$-tuple of elements of H:

$$V := (d_1, e_1, \ldots, d_g, e_g; h_1, \ldots, h_r)$$

such that

- V generates H,
- $\prod_{i=1}^{g} [d_i, e_i] \cdot h_1 \cdot h_2 \cdots h_r = 1$ and
- there exists a permutation $\sigma \in \mathfrak{S}_r$ such that $\mathrm{ord}(h_i) = m_{\sigma(i)}$ for $i = 1, \ldots, r$.

If $g = 0$, then $V := (h_1, \ldots, h_r)$ is called a *spherical system of generators* of H of signature $(m_1, \ldots, m_r)$.

To give a generating vector of signature $(g; m_1, \ldots, m_r)$ for a finite group H is equivalent to give an *appropriate orbifold homomorphism*

$$\psi \colon \mathbb{T}(g; m_1, \ldots, m_r) \longrightarrow H,$$

i.e., a surjective homomorphism ψ such that $\psi(c_i)$ has order m_i.

Remark 2.3 By Riemann's existence theorem (see [5]), any curve C of genus g together with an action of a finite group H on it, such that C/H is a curve C' of genus g', is determined (modulo automorphisms) by the following data:

(1) the branch point set $\{p_1, \ldots, p_r\} \subset C'$;

(2) generators $\alpha_1, \dots, \alpha_{g'}, \beta_1, \dots, \beta_{g'}, \gamma_1, \dots, \gamma_r$ of $\pi_1(C' \setminus \{p_1, \dots, p_r\})$, where each γ_i is a simple geometric loop around p_i and

$$\prod_{i=1}^{g'} [\alpha_i, \beta_i] \cdot \gamma_1 \cdot \dots \cdot \gamma_r = 1;$$

(3) a generating vector V for H of signature $(g'; m_1, \dots, m_r)$ with the property that *Hurwitz's formula* holds:

$$2g - 2 = |H| \left(2g' - 2 + \sum_{i=1}^{r} \frac{m_i - 1}{m_i} \right). \tag{2.1}$$

Remark 2.4 Let C be a curve with an action of a finite group H on it and consider the Galois covering $c: C \to C/H$. The appropriate orbifold homomorphism $\psi: \mathbb{T}(g'; m_1, \dots, m_r) \to H$ associated to V is induced by the monodromy of the Galois étale H-covering $c^0: C^0 \to C'^0$ induced by c, where C'^0 is the curve obtained from C' by removing the branch points $p_1, \dots, p_r$ of c, and $C^0 := c^{-1}(C'^0)$.

Let $h_i := \psi(c_i) \in H$: h_i is called the *local monodromy element* and generates the stabilizer of a point in $c^{-1}(p_i)$. We also define

$$\Sigma(V) := \bigcup_{g \in G} \bigcup_{j \in \mathbb{Z}} \bigcup_{i=1}^{r} \{g \cdot h_i^j \cdot g^{-1}\},$$

as the union of the stabilizers for the action of H on C.

Remark 2.5 Let $S \to X := (C_1 \times C_2)/G$ be a product-quotient surface. The two Galois coverings $C_i \to C_i/G =: C_i'$ determine a pair (V_1, V_2) of generating vectors of G. The action of G on $C_1 \times C_2$ is free if and only if (V_1, V_2) is *disjoint*, that is

$$\Sigma(V_1) \cap \Sigma(V_2) = \{1\}.$$

Conversely, a pair (V_1, V_2) of generating vectors of G of respective signatures $T_1 = (g_1'; m_1, \dots, m_r)$ and $T_2 = (g_2'; n_1, \dots, n_s)$, determines a family of product-quotient surfaces of dimension $M_1 + r + M_2 + s$, where $M_i = 3g_i' - 3$.

In particular, the product-quotient surface can only be rigid (i.e., has no non-trivial deformations), if $r = s = 3$ and $g_1' = g_2' = 0$. If the action is non free, then it can happen that the above family has dimension 0, still the surface is not rigid, i.e., it has non trivial deformations which are no longer product-quotient surfaces.

Instead, if S is isogenous to a product, it follows by the result of the second author below that S is rigid if and only if $r = s = 3$ and $g_1' = g_2' = 0$.

Theorem 2.6 (see [10]) *(a) A projective smooth surface S is isogenous to a product of two curves of respective genera $g_1, g_2 \geq 2$, if and only if the following two conditions are satisfied:*

(1) there is an exact sequence

$$1 \to \Pi_{g_1} \times \Pi_{g_2} \to \pi = \pi_1(S) \to G \to 1,$$

where G is a finite group and where Π_{g_i} denotes the fundamental group of a projective curve of genus $g_i \geq 2$;

(2) $e(S)(= c_2(S)) = \frac{4}{|G|}(g_1 - 1)(g_2 - 1)$.

(b) Write $S = (C_1 \times C_2)/G$. Any surface X with the same topological Euler number and the same fundamental group as S is diffeomorphic to S and is also isogenous to a product. There is a smooth proper family with connected smooth base manifold T, $p\colon \mathcal{X} \to T$ having two fibres respectively isomorphic to X, and Y, where Y is one of the four surfaces $S = (C_1 \times C_2)/G$, $S_{+-} := (\overline{C_1} \times C_2)/G$, $\overline{S} = (\overline{C_1} \times \overline{C_2})/G$, $S_{-+} := (C_1 \times \overline{C_2})/G = \overline{S_{+-}}$.

(c) The corresponding subset of the moduli space of surfaces of general type $\mathfrak{M}_S^{top} = \mathfrak{M}_S^{diff}$, corresponding to surfaces orientedly homeomorphic, resp. orientedly diffeomorphic to S, is either irreducible and connected or it contains two connected components which are exchanged by complex conjugation.

In particular, if S′ is orientedly diffeomorphic to S, then S′ is deformation equivalent to S or to $\overline{S}$.

Definition 2.7 Let $S = (C_1 \times C_2)/G$ be a surface isogenous to a product. S is called a *Beauville surface* if $r = s = 3$ and $g(C_1/G) = g(C_2/G) = 0$.

A corollary of the above Theorem 2.6 (see [10]) is that a Beauville surface is rigid, it has no nontrivial deformations.

Remark 2.8 Observe that, by [9] (cf. also [21]), if $S \to X := (C_1 \times C_2)/G$ is a product-quotient surface, then $q(S) = g(C_1/G) + g(C_2/G)$. Therefore, any Beauville surface is regular.

The concept of Beauville surfaces was introduced by the second author in 1997, and their global rigidity was shown in [10]. They were first systematically studied in [2], where the connection between the algebro-geometric background and the group theoretic description was explained in more detail, many new examples were constructed, and a lot of conjectures were presented. Quite a number of these conjectures have been solved in the meantime and there exists a substantial literature on Beauville surfaces nowadays, which is reflected by the several contributions to the present volume.

3 The Fundamental Group

Let $S := (C_1 \times C_2)/G$ be the minimal realization of a surface isogenous to a product: then the surface $C_1 \times C_2$ is a Galois étale covering space of S with group G and we have already observed (in Theorem 2.6) that the fundamental group of S sits into an exact sequence

$$1 \longrightarrow \pi_1(C_1) \times \pi_1(C_2) \longrightarrow \pi_1(S) \longrightarrow G \longrightarrow 1.$$

If one drops the assumption about the freeness of the action of G on the product $C_1 \times C_2$, there is no reason that the behaviour of the fundamental group of the quotient should be similar to the above situation. Nevertheless, in the unmixed case, it turns out that the fundamental group $\pi_1(X)$ admits a very similar description.

Definition 3.1 We call the fundamental group $\Pi_g := \pi_1(C)$ of a projective curve of genus g a *(genus g) surface group.*

Theorem 3.2 (see [4]) *Let C_1 and C_2 be projective curves of genus at least 2 and let G be a finite group acting on each C_i as a group of automorphisms and diagonally on the product $C_1 \times C_2$.*

Let S be the minimal desingularization of $X := (C_1 \times C_2)/G$. Then the fundamental group $\pi_1(X) \cong \pi_1(S)$ has a normal subgroup $\mathcal{N}$ of finite index which is isomorphic to the product of surface groups, i.e., there are integers $h_1, h_2 \geq 0$ such that $\mathcal{N} \cong \Pi_{h_1} \times \Pi_{h_2}$.

Remark 3.3 (1) The previous theorem holds also in dimension $n \geq 3$, see [4, 12].
(2) In the case of surfaces isogenous to a product we have that h_i equals the genus of C_i, for $i = 1, 2$.

Especially the case of infinite fundamental groups is the one where the above structure theorem turns out to be extremely helpful in order to give an explicit description of these groups. As we will explain in the sequel, it is easy to get a presentation of the fundamental group of a product-quotient surface. But, in general, a presentation of a group does not say much about the group. Even the problem whether it is trivial or not is in general an undecidable problem. Here, since we know that π_1 is a "surface times surface by finite" group, we go through its normal subgroups of finite index until such a subgroup $\mathcal{N}$ appears.

We recall now how to compute the fundamental group of a product-quotient surface $S \to X = (C_1 \times C_2)/G$ starting from the associated algebraic data:

$$\psi_i : \mathbb{T}_i \longrightarrow G \qquad i = 1, 2.$$

Remark 3.4 The kernel of ψ_i is isomorphic to the fundamental group $\pi_1(C_i)$, and the action of $\pi_1(C_i)$ on the universal cover $u: \hat{C}_i \to C_i$ extends to a properly discontinuous action of $\mathbb{T}_i$. Moreover, u is ψ_i-equivariant and $\hat{C}_i/\mathbb{T}_i \cong C_i/G_i$.

There are two short exact sequences:

$$1 \longrightarrow \pi_1(C_i) \longrightarrow \mathbb{T}_i \xrightarrow{\psi} G \longrightarrow 1, \quad i = 1, 2. \tag{3.1}$$

We define the fibre product

$$\mathbb{H} := \mathbb{H}(G; \psi_1, \psi_2) := \{(x, y) \in \mathbb{T}_1 \times \mathbb{T}_2 \mid \psi_1(x) = \psi_2(y)\}.$$

The exact sequences in (3.1) induce then an exact sequence:

$$1 \to \pi_1(C_1) \times \pi_1(C_2) \to \mathbb{H}(G; \psi_1, \psi_2) \to \Delta_G \cong G \to 1,$$

where $\Delta_G \subset G \times G$ denotes the diagonal subgroup.

Definition 3.5 Let H be a group. Then $\mathrm{Tors}(H)$ is the normal subgroup generated by the torsion elements of H (i.e., the elements of finite order in H).

Proposition 3.6 ([4, Proposition 3.4]) *Let $S \to X = (C_1 \times C_2)/G$ be a product-quotient surface and let*

$$\psi_i : \mathbb{T}_i \longrightarrow G \quad i = 1, 2$$

be the associated appropriate orbifold homomorphisms. Then

$$\pi_1(S) = \pi_1(X) = \mathbb{H}/\mathrm{Tors}(\mathbb{H}).$$

A very special and theoretically easy case is:

Corollary 3.7 *Let $S = (C_1 \times C_2)/G$ be a surface isogenous to a product. Then $\pi_1(S) = \mathbb{H}$ and $H_1(S, \mathbb{Z}) = \mathbb{H}^{ab}$.*

In the literature there are several articles on the construction and classification of surfaces birational to the quotient of a product of curves; we have already cited most of them. In some cases the authors provide a computer-script to compute the fundamental group (or the first homology group) of the surfaces they study.

Nevertheless, in this note we take the opportunity to give a simple algorithm in the special case, which includes the case Beauville surfaces, where $S = (C_1 \times C_2)/G$ is a *regular* surface isogenous to a product (equivalently, $g'_1 = g'_2 = 0$).

We now recall how to determine whether two regular surfaces isogenous to a product of unmixed type are deformation equivalent (cf. [1, 3]).

By Remark 2.8, and as observed above, S regular means that $C_i/G \cong \mathbb{P}^1, i = 1, 2$. We denote by $\mathcal{B}(G; T_1, T_2)$ the set of disjoint pairs (V_1, V_2) of spherical systems of generators of G of respective signatures T_1 and T_2.

Let $V := [h_1, \ldots, h_r]$ be a r-tuple of elements of G and $1 \leq i \leq r$. We consider the usual *Hurwitz move* $\sigma_i(V)$ defined by

$$\sigma_i(V) := [h_1, \ldots, h_{i-1}, h_i h_{i+1} h_i^{-1}, h_i, h_{i+2}, \ldots, h_r].$$

It is well known that $\sigma_1, \ldots, \sigma_r$ generate the braid group on r letters $\mathbf{B}_r$, and that $\mathbf{B}_r$ maps spherical system of generators to spherical system of generators, and preserving the signature. Also the automorphism group $\mathrm{Aut}(G)$ of G acts on the set of spherical systems of generators of a fixed signature by simultaneous application of an automorphism φ to the coordinates of a r-tuple: $\varphi(V) := [\varphi(h_1), \ldots, \varphi(h_r)]$. We get the following action of $\mathbf{B}_r \times \mathbf{B}_s \times \mathrm{Aut}(G)$ on $\mathcal{B}(G; T_1, T_2)$: let $(\gamma_1, \gamma_2, \varphi) \in \mathbf{B}_r \times \mathbf{B}_s \times \mathrm{Aut}(G)$ and $(V_1, V_2) \in \mathcal{B}(G; T_1, T_2)$, where T_1 has length r and T_2 has length s, we set

$$(\gamma_1, \gamma_2, \varphi) \cdot (V_1, V_2) := (\varphi(\gamma_1(V_1)), \varphi(\gamma_2(V_2))).$$

We denote this action by $\mathcal{H}$.

Theorem 3.8 (cf. [3, Theorem 5.2]) *Let S and S' be two regular surfaces, both isogenous to a product of unmixed type, and with associate pairs of spherical system of generators (V_1, V_2), respectively (V_1', V_2').*

Then S and S' are deformation-equivalent if and only if the respective groups are isomorphic, $G(S) \cong G(S')$, and either (V_1, V_2) and (V_1', V_2') belong to the same $\mathcal{H}$-orbit or (V_1, V_2) and (V_2', V_1') do.

Remark 3.9 Given a finite group G and two signatures T_1 and T_2 of length r and s, it is then work for a computer to determine $\mathcal{B}(G; T_1, T_2)$, its orbits for the $\mathcal{H}$-action and the first homolgy group $H_1(S, \mathbb{Z})$ of the associate surfaces S.

We have written a MAGMA [8] script (see Appendix section "Appendix: The Script") which takes as input G, T_1 and T_2, and returns as output a representative $(V_1, V_2) \in \mathcal{B}(G; T_1, T_2)$ for each $\mathcal{H}$-orbit, together with the first homology group $H_1(S, \mathbb{Z})$ of the associate surface S. The outline of the script is the following:

- Step 1: we compute all the spherical systems of generators of G of respective signatures T_1 and T_2 and we collect them in orbits for the action of the braid group $\mathbf{B}_r$ (resp. $\mathbf{B}_s$).
- Step 2: we discard the pairs of orbits of non-disjoint spherical systems of generators.
- Step 3: the remaining pairs yield surfaces isogenous to a product. We consider the action of *Aut*(G) on them, hence we get the $\mathcal{H}$-orbits in $\mathcal{B}(G; T_1, T_2)$. Observe that indeed it suffices to consider only the action of *Out*(G), since the *Inn*(G)-action was already taken care of in the first step.
- Step 4: we run over the outputs of Step 3 and we compute their first homology group $H_1(S, \mathbb{Z})$.

4 Some Applications

Surfaces isogenous to a product of unmixed type with $p_g = 0$ have been classified in [3]. We run our script for these surfaces and we get:

Table 1 Surfaces isogenous to a product of unmixed type with $p_g = q = 0$

G	$Id(G)$	T_1	T_2	N	D	$H_1(S, \mathbb{Z})$
$\mathcal{A}_5$	$\langle 60, 5\rangle$	[2, 5, 5]	[3, 3, 3, 3]	1	1	$(\mathbb{Z}_3)^2 \times (\mathbb{Z}_{15})$
$\mathcal{A}_5$	$\langle 60, 5\rangle$	[5, 5, 5]	[2, 2, 2, 3]	1	1	$(\mathbb{Z}_{10})^2$
$\mathcal{A}_5$	$\langle 60, 5\rangle$	[3, 3, 5]	[2, 2, 2, 2, 2]	1	2	$(\mathbb{Z}_2)^3 \times \mathbb{Z}_6$
$\mathcal{S}_4 \times \mathbb{Z}_2$	$\langle 48, 48\rangle$	[2, 4, 6]	[2, 2, 2, 2, 2, 2]	1	3	$(\mathbb{Z}_2)^4 \times \mathbb{Z}_4$
G(32)	$\langle 32, 27\rangle$	[2, 2, 4, 4]	[2, 2, 2, 4]	1	2	$(\mathbb{Z}_2)^2 \times \mathbb{Z}_4 \times \mathbb{Z}_8$
$(\mathbb{Z}_5)^2$	$\langle 25, 2\rangle$	[5, 5, 5]	[5, 5, 5]	1	0	$(\mathbb{Z}_5)^3$
$\mathcal{S}_4$	$\langle 24, 12\rangle$	[3, 4, 4]	[2, 2, 2, 2, 2, 2]	1	3	$(\mathbb{Z}_2)^4 \times \mathbb{Z}_8$
G(16)	$\langle 16, 3\rangle$	[2, 2, 4, 4]	[2, 2, 4, 4]	1	2	$(\mathbb{Z}_2)^2 \times \mathbb{Z}_4 \times \mathbb{Z}_8$
$D_4 \times \mathbb{Z}_2$	$\langle 16, 11\rangle$	[2, 2, 2, 4]	[2, 2, 2, 2, 2, 2]	1	4	$(\mathbb{Z}_2)^3 \times (\mathbb{Z}_4)^2$
$(\mathbb{Z}_2)^4$	$\langle 16, 14\rangle$	[2, 2, 2, 2, 2]	[2, 2, 2, 2, 2]	1	4	$(\mathbb{Z}_4)^4$
$(\mathbb{Z}_3)^2$	$\langle 9, 2\rangle$	[3, 3, 3, 3]	[3, 3, 3, 3]	1	2	$(\mathbb{Z}_3)^5$
$(\mathbb{Z}_2)^3$	$\langle 8, 5\rangle$	[2, 2, 2, 2, 2]	[2, 2, 2, 2, 2, 2]	1	5	$(\mathbb{Z}_2)^4 \times (\mathbb{Z}_4)^2$

Theorem 4.1 *Let $S = (C_1 \times C_2)/G$ be a surface isogenous to a product of unmixed type, with $p_g(S) = 0$, then G is one of the groups in the Table 1 and the signatures are listed in the table. The number N of components in the moduli space, their dimension D and the first homology group of S are given in the last three columns.*

Remark 4.2 (1) In a previous paper [1] (containing the classification of surfaces isogenous to a product of unmixed type with $p_g = q = 0$ and G abelian) there is a mistake, an erroneous statement about commutators in a product. The error resulted into finding only a proper quotient of the actual first homology group.
(2) The correct calculation of the first homology group was done more than one year ago by the first author, who used a MAGMA script to find the correct answer.
(3) Another correction to do to [1] is that the authors forgot the possibility of swapping factors, hence there is only one isomorphism class of surfaces in the case $G = (\mathbb{Z}_5)^2$.
(4) Our computer calculations confirm the result of the paper [22], dedicated to the calculation of the first homology groups of surfaces isogenous to a product of unmixed type with $p_g = q = 0$ and G abelian.

In [16], the author classifies the regular surfaces isogenuos to a product of unmixed type with $\chi(\mathcal{O}) = 2$; in particular he classifies the unmixed Beauville surfaces with $p_g = 1$. Applying our program we get the following:

Theorem 4.3 (cf. [16]) *Let $S = (C_1 \times C_2)/G$ be a Beauville surface of unmixed type, with $p_g(S) = 1$. Then G is one of the groups in Table 2 and the signatures are listed in the table. The number N of components in the moduli space and the first homology group of S are given in the last two columns.*

Table 2 Unmixed Beauville surfaces with $p_g = 1$

G	$Id(G)$	T_1	T_2	N	$H_1(S, \mathbb{Z})$
$PSL(2,7) \times \mathbb{Z}_2$	$\langle 336, 209\rangle$	$[2, 3, 14]$	$[4, 4, 4]$	2	$(\mathbb{Z}_4)^2$
$PSL(2,7)$	$\langle 168, 42\rangle$	$[7, 7, 7]$	$[3, 3, 4]$	2	$\mathbb{Z}_7 \times \mathbb{Z}_{21}$
$PSL(2,7)$	$\langle 168, 42\rangle$	$[3, 3, 7]$	$[4, 4, 4]$	2	$\mathbb{Z}_4 \times \mathbb{Z}_{12}$
$G(128, 36)$	$\langle 128, 36\rangle$	$[4, 4, 4]$	$[4, 4, 4]$	2	$(\mathbb{Z}_2)^3 \times (\mathbb{Z}_4)^2$

Table 3 Some unmixed Beauville surfaces with group $G = PSL(2, q)$

G	$Id(G)$	T_1	T_2	N	$H_1(S, \mathbb{Z})$	$\chi(\mathcal{O}_S)$
$PSL(2,7)$	$\langle 168, 42\rangle$	$[3, 3, 4]$	$[7, 7, 7]$	2	$\mathbb{Z}_7 \times \mathbb{Z}_{21}$	2
		$[3, 4, 4]$	$[7, 7, 7]$	1	$\mathbb{Z}_7 \times \mathbb{Z}_{28}$	4
		$[4, 4, 4]$	$[7, 7, 7]$	2	$(\mathbb{Z}_{28})^2$	6
		$[4, 4, 4]$	$[3, 7, 7]$	4	$\mathbb{Z}_4 \times \mathbb{Z}_{28}$	4
		$[4, 4, 4]$	$[3, 3, 7]$	2	$\mathbb{Z}_4 \times \mathbb{Z}_{12}$	2
$PSL(2,8)$	$\langle 504, 156\rangle$	$[2, 7, 7]$	$[3, 3, 9]$	3	$\mathbb{Z}_3 \times \mathbb{Z}_{21}$	6
		$[2, 7, 7]$	$[3, 9, 9]$	3	$\mathbb{Z}_3 \times \mathbb{Z}_{63}$	12
		$[2, 7, 7]$	$[9, 9, 9]$	7	$\mathbb{Z}_9 \times \mathbb{Z}_{63}$	18
		$[7, 7, 7]$	$[9, 9, 9]$	14	$(\mathbb{Z}_{63})^2$	48
		$[7, 7, 7]$	$[2, 9, 9]$	6	$\mathbb{Z}_7 \times \mathbb{Z}_{63}$	20
		$[7, 7, 7]$	$[2, 3, 9]$	6	$\mathbb{Z}_7 \times \mathbb{Z}_{21}$	4
		$[7, 7, 7]$	$[3, 9, 9]$	6	$\mathbb{Z}_{21} \times \mathbb{Z}_{63}$	32
		$[7, 7, 7]$	$[3, 3, 9]$	6	$(\mathbb{Z}_{21})^2$	16
$PSL(2,9) \cong \mathcal{A}_6$	$\langle 360, 118\rangle$	$[3, 3, 5]$	$[4, 4, 4]$	1	$(\mathbb{Z}_{12})^2$	3
		$[3, 5, 5]$	$[4, 4, 4]$	1	$\mathbb{Z}_4 \times \mathbb{Z}_{60}$	6
		$[5, 5, 5]$	$[4, 4, 4]$	2	$\mathbb{Z}_{20} \times \mathbb{Z}_{60}$	9
				2	$(\mathbb{Z}_{20})^2$	
		$[5, 5, 5]$	$[3, 3, 5]$	2	$\mathbb{Z}_5 \times \mathbb{Z}_{15}$	3
				2	$(\mathbb{Z}_{15})^2$	

Let $G = \mathrm{PSL}(2, q)$, where q and is a prime power and $q \leq 9$. For $q \leq 5$, no group has a disjoint pair of spherical generators (see [2]), while in the other cases the "admissible" pairs of signatures are classified (see [15]). The outputs of our script are collected in Table 3. We use the same notation of the previous tables and we add a column reporting the Euler characteristic $\chi(\mathcal{O}_S)$.

In Table 4, we collect the outputs in some other easy cases: $G \in \{\mathcal{S}_5, \mathcal{S}_6, (\mathbb{Z}_7)^2\}$. One can prove (using a simple MAGMA script), that if $X = (C_1 \times C_2)/G$ is a Beauville surface with $G \in \{\mathcal{S}_5, \mathcal{S}_6, (\mathbb{Z}_7)^2\}$, then its signature (T_1, T_2) is in Table 4.

Table 4 Other unmixed Beauville surfaces with a given group

G	$Id(G)$	T_1	T_2	N	$H_1(S,\mathbb{Z})$	$\chi(\mathcal{O}_S)$
$\mathcal{S}_5$	$\langle 120, 34\rangle$	[4, 4, 5]	[3, 6, 6]	1	$\mathbb{Z}_3 \times \mathbb{Z}_{24}$	3
$\mathcal{S}_6$	$\langle 720, 763\rangle$	[5, 6, 6],	[4, 6, 6]	8	$(\mathbb{Z}_6)^2$	35
		[5, 6, 6],	[4, 4, 6]	16	$\mathbb{Z}_2 \times \mathbb{Z}_{24}$	28
		[5, 6, 6],	[4, 4, 4]	8	$\mathbb{Z}_4 \times \mathbb{Z}_{12}$	21
		[3, 6, 6],	[4, 4, 4]	5	$(\mathbb{Z}_{12})^2$	15
		[2, 5, 6],	[4, 4, 4]	1	$(\mathbb{Z}_4)^2$	6
$(\mathbb{Z}_7)^2$	$\langle 49, 2\rangle$	[7, 7, 7]	[7, 7, 7]	7	$(\mathbb{Z}_7)^3$	4

Appendix: The Script

```
// TuplesOfGivenOrder creates a sequence of the same length as the input
// sequence type, whose entries are subsets of the group in the input,
// and precisely the subsets of elements of order the corresponding
// entry of type

TuplesOfGivenOrders:=function(G,type)
SEQ:=[];
for i in [1..#type-1] do
  EL:={g: g in G| Order(g) eq type[i]};
  if IsEmpty(EL) then return [{}];
   else Append(~SEQ,EL);
end if; end for;
return SEQ;
end function;

// This function transforms a tuple into a sequence.

TupleToSeq:=func<tuple|[x: x in tuple]>;

// Now we create all sets of spherical generators of a group of the
// prescribed signature.

VectGens:=function(G, type)
Vect:={}; SetCands:=TuplesOfGivenOrders(G,type);
for cand in CartesianProduct(SetCands) do
  if Order(&*cand) eq type[#type] then
    if #sub<G|TupleToSeq(cand)> eq #G then
      Include(~Vect, Append(TupleToSeq(cand),(&*cand)^-1));
end if; end if; end for;
return Vect;
end function;

// HurwitzOrbit, starting from a sequence seq of elements of a group,
// creates all sequences of elements which are equivalent to the given one
// for the equivalence relation generated by the Hurwitz moves
// and returns (to spare memory) only the ones whose entries have
// orders disposed as the ones in seq.
```

```
HurwitzMove:= func<seq,j|Insert(Remove(seq,j),j+1,seq[j]^seq[j+1])>;

HurwitzOrbit:=function(seq)
orb:={ }; shortorb:={  }; Trash:={ seq};
repeat ExtractRep(~Trash,~gens); Include(~orb, gens);
  for k in [1..#seq-1] do
    newgens:=HurwitzMove(gens,k);
    if newgens notin orb then Include(~Trash, newgens);
end if; end for; until IsEmpty(Trash);
for gens in orb do test:=true;
  for k in [1..#seq] do
    if not Order(gens[k]) eq Order(seq[k]) then
      test:=false; break k;
  end if; end for;
  if test then Include(~shortorb, gens); end if;
end for;
return shortorb;
end function;

// OrbitsVectGens creates a sequence
// whose entries are subsets of the group in the input,
// and precisely the subsets  are the orbits under the
// Hurwitz moves.

OrbitsVectGens:=function(G, type)
Orbits:=[];   Vects:=VectGens(G, type);
while not IsEmpty(Vects) do
  v:=Rep(Vects);  orb:=HurwitzOrbit(v);
  Append(~Orbits, orb); Vects:=Vects diff orb;
end while;
return Orbits;
end function;

// Disjoint checks if two spherical systems of generators are disjoint.

Stab:= function(seq, G)
M:= {} ;
for i in [1..#seq] do g:=seq[i];
  for n in [1 .. (Order(g)-1) ] do
    M := M join Conjugates(G,g^n) ;
end for; end for;
return M;
end function ;

Disjoint:=func<G,seq1, seq2 | IsEmpty( Stab(seq1,G) meet Stab(seq2,G))>;

// Homology computes the first homology group
// of the surface associated to the disjoint pair
// of spherical systems of generators (seq1, seq2) of G.

Poly:=function(seq, gr)
F:=FreeGroup(#seq);  Rel:={}; Q:=Id(F);
for i in {1..#seq} do
  Q:=Q*F.(i); Include(~Rel,F.(i)^(Order(seq[i])));
end for;
Include(~Rel,Q); P:=quo<F|Rel>;
return P, hom<P->gr|seq>;
end function;
```

```
Homology:=function(G, seq1,  seq2)
T1,f1:=Poly(seq1,G);       T2,f2:=Poly(seq2,G);
T1xT2:=DirectProduct(T1,T2);
 GxG,inG:=DirectProduct(G,G);
if Category(G) eq GrpPC then
  n:=NPCgens(G);
  else n:=NumberOfGenerators(G);
end if;
Diag:= hom<G->GxG| [inG[1](G.j)*inG[2](G.j): j in [1..n]]>(G);
f:=hom<T1xT2->GxG| inG[1](seq1) cat inG[2](seq2)>;
Pi1:=Rewrite(T1xT2,Diag@@f);
return AbelianQuotient(Pi1);
end function;

// Surfaces is the main function.
// It takes as input a group G and two signatures.
// It calls the previous function and moreover identifies
// distinct pairs of Hurwitz's orbits of spherical systems of
// generators under the action of Aut(G).
// Note that we consider only the action of Out(G), since
// the Inn(G)-action was already taken care of in HurwitzOrbit.

Surfaces:=function(G, type1,type2)
R:={}; Aut:=AutomorphismGroup(G);
F,q:=FPGroup(Aut);      O1,p:=OuterFPGroup(Aut);
Out,k:=PermutationGroup(O1);
V1:=OrbitsVectGens(G, type1);
if type1 eq type2 then V2:=V1;
  else V2:=OrbitsVectGens(G,type2);
end if;
W:=Set(CartesianProduct({1..#V1},{1..#V2}));
for pair  in W do
  v1:= Rep(V1[pair[1]]); v2:= Rep(V2[pair[2]]);
  if not Disjoint(G, v1,v2) then Exclude(~W, pair);
 end if;
end for;
while not IsEmpty(W) do
  pair:=Rep(W);
  v1:= Rep(V1[pair[1]]); v2:= Rep(V2[pair[2]]);Include(~R,[v1,v2]);
  for phi in Out do
    if exists(x){y: y in W | (q(phi@@(p*k)))(v1) in V1[y[1]] and
     (q(phi@@(p*k)))(v2) in V2[y[2]]} then Exclude(~W, x);
      if type1 eq type2 then Exclude(~W, <x[2],x[1]>);
    end if; end if;
    if IsEmpty(W) then break phi; end if;
end for; end while;
printf "Number of components: %o\n", #R;
for r in R do
  printf "Spherical generators:\n%o\n%o\n", r[1],r[2];
  printf "Homology:\n%o\n", Homology(G,r[1],r[2]);
end for;
return {};
end function;
```

References

1. I. Bauer, Some new surfaces with $p_g = q = 0$, in *The Fano Conference*, Turin University, Torino, pp. 123–142 (2004)
2. I. Bauer, F. Catanese, F. Grunewald, Beauville surfaces without real structures, *Geometric Methods in Algebra and Number Theory*, volume 235 of Progress in Mathematics (Birkhäuser, Boston, 2005), pp. 1–42
3. I. Bauer, F. Catanese, F. Grunewald, The classification of surfaces with $p_g = q = 0$ isogenous to a product of curves. Pure Appl. Math. Q. **4**(2), 547–586 (2008)
4. I. Bauer, F. Catanese, F. Grunewald, R. Pignatelli, Quotients of products of curves, new surfaces with $p_g = 0$ and their fundamental groups. Am. J. Math. **134**(4), 993–1049 (2012)
5. I. Bauer, F. Catanese, R. Pignatelli, Surfaces of general type with geometric genus zero: a survey, in *Complex and Differential Geometry*. Springer Proceedings in Mathematics, vol. 8 (Springer, Berlin, 2011), pp. 1–48
6. I. Bauer, R. Pignatelli, The classification of minimal product-quotient surfaces with $p_g = 0$. Math. Comput. **81**(280), 2389–2418 (2012)
7. I. Bauer, R. Pignatelli, Product-quotient surfaces: new invariants and algorithms (2013). arXiv:1308.5508
8. W. Bosma, J. Cannon, C. Playoust, The Magma algebra system I. The user language. J. Symbolic Comput. **24**(3–4), 235–265 (1997)
9. F. Catanese, Everywhere nonreduced moduli spaces. Invent. Math. **98**(2), 293–310 (1989)
10. F. Catanese, Fibred surfaces, varieties isogenous to a product and related moduli spaces. Am. J. Math. **122**(1), 1–44 (2000)
11. G. Carnovale, F. Polizzi, The classification of surfaces with $p_g = q = 1$ isogenous to a product of curves. Adv. Geom. **9**(2), 233–256 (2009)
12. T. Dedieu, F. Perroni, The fundamental group of a quotient of a product of curves. J. Group Theory **15**(3), 439–453 (2012)
13. D. Frapporti, R. Pignatelli, Mixed quasi-étale quotients with arbitrary singularities. Glasg. Math. J. **57**(1), 143–165 (2015)
14. D. Frapporti, Mixed quasi-étale surfaces, new surfaces of general type with $p_g = 0$ and their fundamental group. Collect. Math. **64**(3), 293–311 (2013)
15. S. Garion, On Beauville structures for PSL (2,q) (2010). arXiv:1003.2792
16. C. Gleißner, in *The classification of regular surfaces isogenous to a product of curves with* $\chi(\mathcal{O}_S) = 2$. ed. by I. Bauer, S. Garion, A. Vdovina. Beauville Surfaces and Groups (Springer, Cham, 2015)
17. E. Mistretta, F. Polizzi, Standard isotrivial fibrations with $p_g = q = 1$ II. J. Pure Appl. Algebra **214**(4), 344–369 (2010)
18. M. Penegini, The classification of isotrivially fibred surfaces with $p_g = q = 2$. Collect. Math. **62**(3), 239–274 (2011) (With an appendix by Sönke Rollenske)
19. F. Polizzi, On surfaces of general type with $p_g = q = 1$ isogenous to a product of curves. Commun. Algebra **36**(6), 2023–2053 (2008)
20. F. Polizzi, Standard isotrivial fibrations with $p_g = q = 1$. J. Algebra **321**(6), 1600–1631 (2009)
21. F. Serrano, Isotrivial fibred surfaces. Ann. Mat. Pura Appl. **171**(4), 63–81 (1996)
22. T.I. Shabalin, Homology of some surfaces with $p_g = q = 0$ isogenous to a product. Izv. RAN. Ser. Mat. **78**(6), 211–221 (2014)
23. F. Zucconi, Surfaces with $p_g = q = 2$ and an irrational pencil. Can. J. Math. **55**(3), 649–672 (2003)

Regular Algebraic Surfaces, Ramification Structures and Projective Planes

N. Barker, N. Boston, N. Peyerimhoff and A. Vdovina

Abstract Regular algebraic surfaces isogenous to a higher product of curves can be obtained from finite groups with ramification structures. We find unmixed ramification structures for finite groups constructed as p-quotients of particular infinite groups with special presentation related to finite projective planes.

2000 Mathematics Subject Classification: 14L30 · 20F32 · 51E24

1 Introduction

An algebraic surface is *isogenous to a higher product* (of curves) if it admits a finite unramified covering which is isomorphic to a product of curves $C_1 \times C_2$ of genera $g(C_i) \geq 2$. It was shown in [10] that every such surface S has a unique *minimal* realisation $S \cong (C_1 \times C_2)/G$, where G is a finite group acting freely on $C_1 \times C_2$ and C_1 and C_2 have the smallest possible genera. Moreover, G respects the product structure by either acting diagonally on each factor (unmixed case) or there are elements in G interchanging the factors (mixed case). Surfaces isogenous to a higher product are always minimal and of general type. Particularly interesting examples

N. Barker (✉) · A. Vdovina
Department for Pure Mathematics and Mathematical Statistics,
University of Cambridge, CB3 OWB, Cambridge, UK
e-mail: nb443@cam.ac.uk

A. Vdovina
e-mail: alina.vdovina@ncl.ac.uk

N. Boston
Department of Mathematics,
University of Wisconsin, 303 Van Vleck Hall, 480 Lincoln Drive,
Madison, WI 53706, USA
e-mail: boston@math.wisc.edu

N. Peyerimhoff
Department of Mathematical Sciences, Science Laboratories,
Durham University, South Road, DH1 3LE, Durham, UK
e-mail: norbert.peyerimhoff@durham.ac.uk

I. Bauer et al. (eds.), *Beauville Surfaces and Groups*, Springer Proceedings in Mathematics & Statistics 123, DOI 10.1007/978-3-319-13862-6_2

are *Beauville surfaces*, i.e., algebraic surfaces isogeneous to a higher product which are rigid (i.e., do not admit nontrivial deformations).

The *irregularity* $q(S)$ of a surface S is the difference between its geometric and its algebraic genus, and agrees with the Hodge number $h^{1,0}(S)$. Surfaces with vanishing irregularity are called *regular*. Since $q(S) = g(C_1/G) + g(C_2/G)$ (see [24, Proposition 2.2]), we have $C_i/G \cong \mathbb{P}^1$ for both curves in the minimal realisation of a regular surface.

Every surface S isogenous to a higher product gives rise to a finite group G via its minimal realisation. This process can be reversed. Starting with a finite group G, the existence of a so called *ramification structure* can be used to construct a regular surface of the form $(C_1 \times C_2)/G$. We will discuss ramification structures and the construction of the associated surfaces in Sect. 2. Bauer, Catanese and Grunewald [5] used this group theoretical description to classify all regular surfaces S isogenous to a product of curves with vanishing geometric genus $p_g(S) = h^{2,0}(S)$. The process in [5] was aided by the reduction of the search of ramification structures to groups of order less than 2000, for which the MAGMA library of small groups could then be used. They saw this classification as the solution in a very special case to the open problem posed by Mumford: "Can a computer classify all surfaces of general type with $p_g = 0$?"

The infinite group in [15, Example 6.3] given by the presentation

$$G_0 := \langle x_0, \ldots, x_6 \mid x_i x_{i+1} x_{i+3} \ (i \in \mathbb{Z}_7) \rangle \tag{1}$$

was used in [1, 2] to construct finite 2-groups with special unmixed and mixed ramification structures, giving rise to unmixed and mixed Beauville surfaces. These finite 2-groups were the maximal 2-quotients of 2-class k of both the group G_0 and its index two subgroup H_0 generated by x_0 and x_1. We like to mention that very little is known about *mixed* Beauville structures and it is generally assumed that they are very rare. Results in [13, 14] show that the symmetric groups S_n and all almost simple groups with sporadic derived groups cannot have mixed structures. The examples in [2] are the first known *infinite* family of 2-groups with mixed Beauville structures. It is immediate from the definition that a p-group can only admit a mixed structure if $p = 2$. The only other known construction of groups admitting mixed Beauville structures was given in [4]. But this general construction is very different in nature and does not provide examples of 2-*groups* admitting mixed Beauville structures. In this paper we restrict our considerations, however, to the *unmixed case*.

The above group G_0 belongs to a family called *groups with special presentation*. These groups were introduced by Howie [15] and are related to projective planes over finite fields (see Sect. 3 for more details). It was proved in [12] that all groups with special presentation are just infinite (i.e., they are infinite groups all of whose non-trivial normal subgroups have finite index). A natural question arose: *Do any other groups with special presentations give rise to finite groups with particular ramification structures?*

In this article we consider finite index subgroups of the groups listed in [11, Example 3.3], an index 3 subgroup of the following group with special presentation from [15, Example 6.4], the group

$$G := \langle x_0, \dots, x_{12} \mid x_i^3, x_i x_{i+1} x_{i+4} \ (i \in \mathbb{Z}_{13}) \rangle \tag{2}$$

and the group given in [18, Example 2] constructed from a polyhedral presentation (a generalization of the triangle presentations defined in [8]). We use the computer program MAGMA to search for unmixed ramification structures in maximal p-quotients of p-class k of the above mentioned groups for various primes p. These ramification structures give then rise to particular regular surfaces isogenous to a higher product. Moreover, these results lead to natural conjectures about infinite families of p-groups admitting unmixed ramification structures.

Let us present an example of our results. The subgroup H of G defined in (2), and generated by x_0, x_1, x_2 has index 3. Let $H_{3,k}$ be the maximal 3-quotient of 3-class k. For simplicity, let us denote the elements in $H_{3,k}$ corresponding to x_0, x_1, x_2, again, by x_0, x_1, x_2. Let $y_0 = x_0 x_1^2 x_2^2$, $y_1 = x_0^2 x_1 x_2^2$ and $y_2 = x_1 x_2^{-1} x_2^{x_0}$. Then we have the following result.

Theorem 1.1 *For $k = 2, \dots, 60$, the groups $H_{3,k}$ are of order 3^{a_k} and admit unmixed ramification structures (T_1, T_2) of type $([3,3,3,3^{d_k}], [3^{b_k},3^{b_k},3^{b_k},3^{b_k}])$, where $T_1 = (x_0, x_1, x_2, (x_0x_1x_2)^{-1})$, $T_2 = (y_0, y_1, y_2, (y_0y_1y_2)^{-1})$, $b_k = 1 + [\log_3 \frac{3k}{4}]$, $d_k = 1 + [\log_3 k]$, and*

$$a_k = \begin{cases} 8j & \text{if } k = 3j, \\ 8j+3 & \text{if } k = 3j+1, \\ 8j+6 & \text{if } k = 3j+2. \end{cases}$$

Here $[x]$ denotes the largest integer $\leq x$.

This result motivates the following conjecture.

Conjecture 1.2 *Let H be the index 3 subgroup of the group (2), generated by x_0, x_1, x_2. Then, for all $k \geq 2$, the maximal 3-quotients $H_{3,k}$ of 3-class k are of order 3^{a_k} and admit unmixed ramification structures (T_1, T_2) of type $([3,3,3,3^{d_k}], [3^{b_k},3^{b_k},3^{b_k},3^{b_k}])$, where a_k, b_k, d_k and T_1 and T_2 are explicitly given in Theorem 1.1.*

A promising approach to prove this conjecture is to employ the matrix representation of (2) in Appendix 2: "Representation for the Group G". (Such a matrix representation was key in [2] to prove that infinitely many 2-quotients of the group (1) admit mixed Beauville structures.)

The article is organised as follows. Section 2 presents fundamental facts about ramification structures and algebraic surfaces. Our results on ramification structures are presented in Sect. 3 below. Appendix 1: "Expanders Associated to the Group G_0" is concerned with the derivation of an explicit matrix representation of the group (2). Finally, Appendix 1: "Expanders Associated to the Group G_0" provides a brief survey about explicit recent expander constructions related to the group (1).

2 Ramification Structures and Associated Surfaces

2.1 Group Theoretical Structures

Following [5] closely, we give the definition of an (unmixed) ramification structure of a finite group G.

An r-tuple $T = [g_1, \dots, g_r]$ of non-trivial elements of G is called a *spherical system of generators*, if $g_1, \dots, g_r$ generate G and $g_1 g_2 \cdots g_r = 1$. The r-tuple $[m_1, \dots, m_r]$ of non-decreasing orders of the elements g_i is called the *type* of the spherical system T of generators, i.e., $2 \leq m_1 \leq m_2 \leq \cdots \leq m_r$ and there is a permutation $\tau \in Sym(r)$ such that $m_i = \text{ord}(g_{\tau(i)})$. Let

$$\Sigma(T) := \bigcup_{g \in G} \bigcup_{j=0}^{\infty} \bigcup_{i=1}^{r} \{g g_i^j g^{-1}\}$$

be the union of all conjugates of the cyclic subgroups generated by the elements g_i of the spherical system. Two spherical systems of generators $T_1 = [g_1, \dots, g_r]$ and $T_2 = [g'_1, \dots, g'_s]$ are called *disjoint* if $\Sigma(T_1) \cap \Sigma(T_2) = \{1\}$. An unmixed ramification structure is defined as follows.

Definition 2.1 *(Unmixed ramification structures, see* [5, Definition 1.1]) Let $A_1 = [m_1, \dots, m_r]$ and $A_2 = [n_1, \dots, n_s]$ be tuples of natural numbers with $2 \leq m_1 \leq \cdots \leq m_r$ and $2 \leq n_1 \leq \cdots \leq n_s$. An *unmixed ramification structure of type* (A_1, A_2) for a finite group G is a pair (T_1, T_2) of disjoint spherical systems of generators such that T_1 has type A_1 and T_2 has type A_2.

The disjointness of the pair (T_1, T_2) of an unmixed ramification structure guarantees that G acts freely on the product $C_1 \times C_2$ of associated algebraic curves (see Sect. 2.2 and the references therein). In this article we will only consider unmixed ramification structures and their associated surfaces. For examples of the mixed case see, e.g., [1, 4, 5].

Recall that *unmixed Beauville structures* are unmixed ramification structures with two spherical systems (T_1, T_2) of length 3, i.e., $r = s = 3$. They are of particular interest, since they give rise to the Beauville surfaces mentioned in the Introduction.

2.2 From Ramification Structures to Algebraic Surfaces

In this section we explain how to construct an algebraic surface $S = (C_{T_1} \times C_{T_2})/G$ from a given finite group G with an unmixed ramification structure (T_1, T_2).

Let G be a finite group and $T = [g_1, \dots, g_r]$ be a spherical system of generators with $m_i = \text{ord}(g_{\tau(i)})$. For $1 \leq i \leq r$, let $P_1, \dots, P_l \in \mathbb{P}^1$ be a sequence of points ordered counterclockwise around a base point P_0 and $\gamma_i \in \pi(\mathbb{P}^1 - \{P_1, \dots, P_r\}, P_0)$ be represented by a simple counterclockwise loop around P_i, such that $\gamma_1\gamma_2\dots\gamma_r = 1$. By Riemann's existence theorem, we obtain a surjective homomorphism

$$\Phi : \pi(\mathbb{P}^1 - \{P_1, \dots, P_r\}, P_0) \to G$$

with $\Phi(\gamma_i) = g_i$ and a Galois covering $\lambda : C_T \to \mathbb{P}^1$ with ramification indices equal to the orders of the elements $g_1, \dots, g_r$. These data induce a well defined action of G on the curve C_T, and by the Riemann-Hurwitz formula, we have

$$g(C_T) = 1 + \frac{|G|}{2}\left(r - 2 - \sum_{l=1}^{r} \frac{1}{m_l}\right). \tag{3}$$

Now, we assume that G admits an unmixed ramification structure (T_1, T_2). This leads to a diagonal action of G on the product $C_{T_1} \times C_{T_2}$, and the disjointness of the two spherical systems of generators ensures that G acts freely on the product of curves. The associated algebraic surface S is the quotient $(C_{T_1} \times C_{T_2})/G$. By the Theorem of Zeuthen-Segre, we have for the topological Euler number

$$e(S) = 4\frac{(g(C_{T_1}) - 1)(g(C_{T_2}) - 1)}{|G|},$$

as well as the relations (see [10, Theorem 3.4]),

$$\chi(S) = \frac{e(S)}{4} = \frac{K_S^2}{8},$$

where K_S^2 is the self intersection number of the canonical divisor and $\chi(S) = 1 + p_g(S) - q(S)$ is the holomorphic Euler-Poincaré characteristic of S. Assume that (T_1, T_2) is of the type (A_1, A_2) with $A_1 = [m_1, \dots, m_r]$ and $A_2 = [n_1, \dots, n_s]$. Then the above relations imply for the associated surface S that

$$\chi(S) = \frac{|G|}{4}\left(r - 2 - \sum_{l=1}^{r} \frac{1}{m_l}\right)\left(s - 2 - \sum_{l=1}^{s} \frac{1}{n_l}\right).$$

3 Groups with Special Presentations

As mentioned in [11], small cancellation groups are generalizations of surface groups and satisfy many of the nice properties of those groups. It was proved in [11] (with a small list of exceptions) that almost all groups with a presentation satisfying the small cancellation conditions $C(3)$ and $T(6)$ contain a free subgroup of rank 2.

Further, [15] proved that most $C(3)$, $T(6)$ groups G (namely, the ones which do *not* have special presentations) are SQ-universal. (A group G is called SQ-universal if every countable group can be embedded in a quotient group of G.) In that article, a group presentation was called *special* if every relator has length 3 and the star graph is isomorphic to the incidence graph of a finite projective plane. (See [21, p. 61] for a textbook reference on the *star graph* of a presentation. In short, the star graph of a presentation $\langle \mathbf{x} \mid \mathbf{r} \rangle$ is defined as follows (see [15]): its vertex set are the elements in $\mathbf{x} \cup \mathbf{x}^{-1}$ and there is an edge from x to y if there is a cyclically reduced word beginning in x and ending in y, which is a cyclic permutation of a relator or its inverse.) Moreover, it was asked (see [15, Question 6.11]) whether any or all of the groups with special presentations are SQ-universal. It was proved in [12] that groups with special presentation are just infinite (i.e., all non-trivial normal subgroups have finite index) and, therefore, cannot be SQ-universal. (Note that special presentations in the sense of [15] are (3, 3)-special in the sense of [12].)

Howie [15] also set up an example machine (see Theorem 3.1 below) to create infinitely many groups with special presentations. More precisely, he constructed a special presentation with star graph isomorphic to the incidence graph of the projective plane over every finite field $\mathbb{F}_q$, where q is a prime power. Until then, only seven examples of special presentations were known (see [11, Example 3.3]), and each of them has a star graph isomorphic to the Heawood graph (i.e., the incidence graph of the 7-point projective plane over $\mathbb{F}_2$).

Given a finite field $K = \mathbb{F}_q$ (for q a prime power), a *positive* presentation with star graph isomorphic to this incidence graph of the Desarguesian projective plane over K is formed.

The construction takes a cubic extension of K, namely $F = \mathbb{F}_{q^3}$, and identifies the cyclic group $C_m = F^\times / K^\times$ with the points of the projective pane $\mathcal{P}$ over K, where $m = q^2 + q + 1$.

The group C_m acts on $\mathcal{P}$ via multiplication in F, and this action is regular on both the points and lines of $\mathcal{P}$ i.e. C_m is a Singer group, see [16]. The lines of $\mathcal{P}$ can be identified with the subset σL of C_m, where σ ranges over C_m and L is a fixed line or perfect difference set i.e. a set of residues $a_1, \dots, a_{q+1} \mod m$ such that every non-zero residue modulo $m = q^2 + q + 1$ can be expressed uniquely in the form $a_i - a_j$.

Theorem 3.1 ([15, Theorem 6.2]) *Let q be a prime power and $m = q^2 + q + 1$. Then there exists a subset l of $q + 1$ elements of $\mathbb{Z}_m$ such that*

$$\langle x_0, \dots, x_{m-1} \mid x_i x_{i+\lambda} x_{i+\lambda+q\lambda} \ (i \in \mathbb{Z}_m, \lambda \in l) \rangle$$

is a special presentation whose star graph is isomorphic to the incidence graph of the projective plane over $GF(q)$.

Let us now present results on ramification structures of finite groups obtained from particular groups G with special presentations. These finite groups are generated via the lower, exponent p-central series, i.e.,

$$G = P_0(G) \geq \cdots \geq P_{i-1}(G) \geq P_i(G) \geq \dots,$$

where $P_i(G) = [P_{i-1}(G), G]P_{i-1}(G)^p$ for $i \geq 1$. The finite groups $G_{p,k}$ under considerations are then the maximal p-quotients of p-class k, denoted by $G_{p,k}$ and given by $G_{p,k} = G/P_k(G)$.

The results discussed below are obtained via the computer program MAGMA (see [6]). Note that the algorithm `pQuotient` constructs, for a given group G, a consistent power-conjugate presentation for $G_{p,k}$.

3.1 Ramification Structures for the Group in [15, Example 6.3]

The group G_0 in (1) with seven generators $x_0, \ldots, x_6$ appeared as Example 6.3 in [15]. (G_0 is constructed using Theorem 3.1 with $q = 2$, $m = 7$ and $l = \{1, 2, 4\}$.) The subgroup H_0 generated by x_0, x_1 has index two. In [1, Theorems 4.1 and 4.2], we presented unmixed ramification structures for the 2-groups $(H_0)_{2,k}$ for $3 \leq k \leq 64$ (for k not a power of 2), as well as mixed ramification structures for the 2-groups $(G_0)_{2,k}$ for $3 \leq k \leq 10$ (again, for k not a power of 2). Since the involved spherical systems of generators consist of three elements, these ramification structures are actually Beauville structures and lead to new examples of Beauville surfaces.

Moreover, [2] presents a rigorous proof that an *infinite family* of 2-quotients of G_0 admits mixed Beauville structures. [2] uses a faithful matrix representation of G_0 by infinite upper triangular matrices (created in an analogous way as described in Appendix 1: "Expanders Associated to the Group G_0"), and the quotients are obtained via truncations at upper diagonals. MAGMA calculations show that the first 100 of these quotients are isomorphic to the quotients $(G_0)_{2,k}$. It is natural to conjecture that all of these quotients of G_0 constructed in these two different ways are pairwise isomorphic (see [22, Conjecture 1]).

The group G_0 appears also in [9, Sect. 4] as the group A.1. (The articles [8, 9] are concerned with simply transitive group actions on the vertices of $\widetilde{A}_2$-buildings.) The index two subgroup H_0 was also used in [17, 22] to construct families of expander graphs of vertex degrees 4 and 3. These expander graph constructions are briefly described in Appendix 1: "Expanders Associated to the Group G_0".

3.2 Ramification Structures for the Groups in [11, Example 3.3]

There, a list of seven special group presentations $G_i = \langle \mathbf{x} \mid \mathbf{r}_i \rangle$, $1 \leq i \leq 7$, with

$$r_1 = \{ab^{-1}d, a^{-1}dc, d^{-1}ea, b^2 f, ceg, cgf, efg\},$$
$$r_2 = \{a^{-1}df, b^{-1}ed, c^{-1}fe, a^2 g, bdg, bec, cgf\},$$
$$r_3 = \{abc, ade, afg, cge, bef, bdg, dfc\},$$
$$r_4 = \{a^2 b, acd, bde, bfc, ceg, dgf, efg\},$$

$$r_5 = \{a^2b, acd, bde, bfc, ceg, dg^2, ef^2\},$$
$$r_6 = \{a^2b, acb, bef, bge, d^2f, c^2g, egf\},$$
$$r_7 = \{abc, adb, acd, bef, cge, dfg, egf\},$$

were given ($\mathbf{x} = \{a, b, c, d, e, f, g\}$). The star graphs of all seven presentations are isomorphic to the incidence graph of the 7-point projective plane. See Fig. 1 for the star graph of the group G_1.

Our group G_0 in (1) coincides with their group G_3, which was discussed in Sect. 3.1. It is stated in [11] that the only isomorphism between abelianised groups G_i^{ab} is between G_4^{ab} and G_6^{ab}. However, if one looks at the commutator subgroup C_4 and C_6 of the groups G_4 and G_6, then $C_4^{ab} \cong \mathbb{Z}/4\mathbb{Z}$ and $C_6^{ab} \cong \mathbb{Z}/2\mathbb{Z}$. Therefore, G_4 can not be isomorphic to G_6.

We use the computer program MAGMA to search in the maximal p-quotients of maximal class k for $1 \leq k \leq 10$ of certain finite index subgroups of the groups G_i for unmixed ramification structures.

Special Presentation G_1

There is a subgroup H_1 of index 4 in G_1 generated by b. Thus, as $H_1 \cong \mathbb{Z}$ all maximal p-quotients of p-class k of H_1 are cyclic groups of order p^k. Therefore, there will be no unmixed ramification structures coming from the groups $(H_1)_{p,k}$.

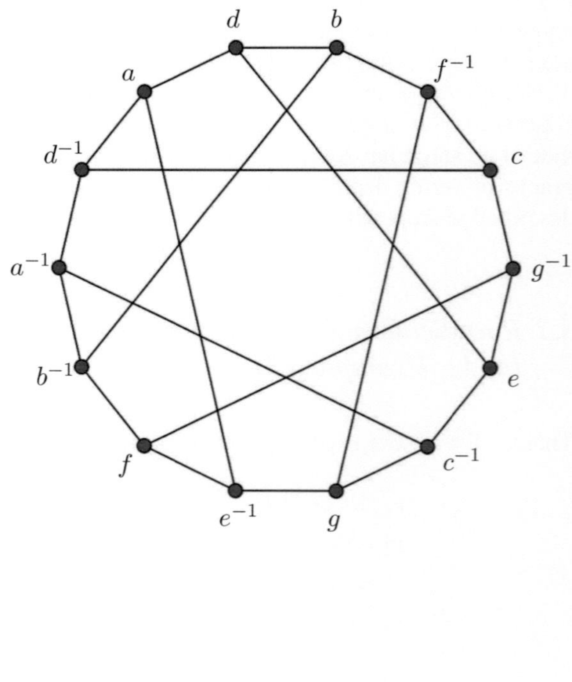

Fig. 1 Star graph of $G_1 = \langle \mathbf{x} \mid \mathbf{r}_1 \rangle$

Special Presentation G_2 and G_4

There is a subgroup H_2 of index 16 in G_2 (the commutator subgroup of G_2) generated by $h_{2,0} = bd^{-1}a^{-1}bc^{-1}$, $h_{2,1} = abd^{-1}abd^{-1}bc^{-1}$, $h_{2,2} = (bc^{-1})^2$, $h_{2,3} = adc^{-1}$ and $h_{2,4} = db^{-1}a^{-1}\, db^{-1}a^{-1}bc^{-1}$. The maximal 7-quotients of 7-class k, written as $(H_2)_{7,k}$, gives rise to a disjoint pair of spherical generators of length 3 given by the tuples,

$$T_{2,1} = [h_{2,0}, h_{2,1}, h_{2,1}^{-1}h_{2,0}^{-1}] \text{ and } T_{2,2} = [h_{2,0}h_{2,1}^{2}, h_{2,0}h_{2,1}^{3}, h_{2,1}^{-3}h_{2,0}^{-1}h_{2,2}^{-2}h_{2,0}^{-1}],$$

for $1 \leq k \leq 10$. To simplify notation, we denoted the images of $h_{2,0}$ and $h_{2,1}$ in $(H_2)_{7,k}$, again, by $h_{2,0}$ and $h_{2,1}$. Thus, the groups $(H_2)_{7,k}$ have unmixed ramification structures.

The group G_4 of [11, Example 3.3] coincides with the group C.1. in [9, Sect. 5] (via the identification $a_0 = a, a_1 = f, a_2 = c, a_3 = d, a_4 = e, a_5 = g, a_6 = b$). We find a subgroup H_4 of index 48 in G_4 generated by $h_{4,0} = da^{-1}bc^{-1}$, $h_{4,1} = bc^{-1}bc^{-1}ea^{-1}bf^{-1}$, $h_{4,2} = cb^{-1}ae^{-1}cf^{-1}$, $h_{4,3} = (ea^{-1}bf^{-1})^2$, $h_{4,4} = cb^{-1}cb^{-1}ea^{-1}bf^{-1}$ and $h_{4,5} = ea^{-1}bc^{-1}ea^{-1}bc^{-1}ea^{-1}bf^{-1}$. For $1 \leq k \leq 10$, the maximal 7-quotients of 7-class k, $(H_4)_{7,k}$, gives rise to a disjoint pair of spherical generators of length 3 given by the tuples,

$$T_{4,1} = [h_{4,0}, h_{4,1}, h_{4,1}^{-1}h_{4,0}^{-1}] \text{ and } T_{4,2} = [h_{4,0}h_{4,1}^{2}, h_{4,0}h_{4,1}^{3}, h_{4,1}^{-3}h_{4,0}^{-1}h_{4,2}^{-2}h_{4,0}^{-1}].$$

The groups H_2 and H_4 have the same maximal 7-quotients of maximal 7-class k for $1 \leq k \leq 10$. However, the abelianizations of the infinite groups are $H_2^{ab} \cong \mathbb{Z}_7 \times \mathbb{Z}_{21}$ and $H_4^{ab} \cong \mathbb{Z}_7 \times \mathbb{Z}_7$. The following theorem summarizes the above observations.

Theorem 3.2 *For $r = 2,4$, $k = 1,\ldots,10$, the groups $(H_r)_{7,k}$ are of order 7^a and admit unmixed ramification structures $(T_{r,1}, T_{r,2})$ of type $([7^b,7^b,7^b], [7^b,7^b,7^b])$ for*

$$a = \begin{cases} 2k & \text{if } k = 1, 4, 5, 8, 9, \\ 2k-1 & \text{if } k = 2, 3, 6, 7, 10. \end{cases} \quad \text{and } b = \begin{cases} 2 & \text{if } 1 \leq k \leq 4, \\ 3 & \text{if } 5 \leq k \leq 8, \\ 5 & \text{if } 9 \leq k \leq 10. \end{cases}$$

This result is strong evidence that the following conjecture is true.

Conjecture 3.3 *For $r = 2,4$ and all $k \in \mathbb{N}$, the maximal 7-quotients $(H_r)_{7,k}$ of 7-class k admit unmixed ramification structures with disjoint spherical systems $(T_{r,1}, T_{r,2})$ introduced above.*

The unmixed ramification structures given for the groups $(H_2)_{7,k}$, $(H_4)_{7,k}$ above give rise to unmixed Beauville surfaces $S = (C_{T_1} \times C_{T_2})/(H_n)_{7,k}$ for $n = 2, 4$. For example, the order of the group $(H_2)_{7,1}$ and $(H_4)_{7,1}$ is 7^2. Therefore, the genera of the curves C_{T_i} is (see (3))

$$g(C_{T_1}) = g(C_{T_2}) = 1 + 2 \times 7 = 15,$$

and the holomorphic Euler-Poincaré characteristic of S is

$$\chi(S) = \frac{(g(C_{T_1}) - 1)(g(C_{T_2}) - 1)}{|G|} = 4.$$

Remark 3.4 (see [10, Beauville's Example 3.22]) The groups $(H_2)_{7,1}$, $(H_4)_{7,1}$ are isomorphic to the group $(\mathbb{Z}/7\mathbb{Z})^2$ and the two curves $C_{T_1} = C_{T_2}$ are given by the Fermat curve $x^7 + y^7 + z^7 = 0$ of degree 7. The group $(\mathbb{Z}/7\mathbb{Z})^2$ acts on $C_{T_1} \times C_{T_2}$ by the following rule

$$(\alpha, \beta) \cdot ([x : y : z], [u : v : w]) = ([\xi^{\alpha} x : \xi^{\beta} y : z], [\xi^{\alpha+2\beta} u : \xi^{\alpha+3\beta} v : w]),$$

where $\xi = e^{\frac{2\pi i}{7}}$ and $\alpha, \beta \in \mathbb{Z}/7\mathbb{Z}$. We identify $h_{n,0} \mapsto \alpha$ and $h_{n,1} \mapsto \beta$ for $n = 2, 4$.

Special Presentation G_5

The group G_5 coincides with the group A.2 in [9, Sect. 5]. We find a subgroup H_5 of index 3, generated by $h_{5,0} = ba^{-1}$, $h_{5,1} = ca^{-1}$, $h_{5,2} = da^{-1}$, $h_{5,3} = ea^{-1}$, $h_{5,4} = fa^{-1}$ and $h_{5,5} = ga^{-1}$ which have the same maximal 2-quotients of 2-class k as the group G_0 in (1) for $1 \le k \le 10$. However, the abelianization of this group is $H_5^{ab} \cong \mathbb{Z}_2 \times \mathbb{Z}_2 \times \mathbb{Z}_{14}$ which is not isomorphic $G_0^{ab} \cong \mathbb{Z}_2 \times \mathbb{Z}_2 \times \mathbb{Z}_6$.

In addition, we have a subgroup F_5 in H_5 of index 2, which appears to have the same maximal 2-quotients of 2-class k as H_0 (the subgroup of G_0 generated by x_0, x_1) for $1 \le k \le 10$. The abelianization of this group is $F_5^{ab} \cong \mathbb{Z}_4 \times \mathbb{Z}_{28}$ which is not isomorphic $G_0^{ab} \cong \mathbb{Z}_4 \times \mathbb{Z}_{14}$.

These results lead naturally to the following conjecture.

Conjecture 3.5 *Let H_5 be the index 3 subgroup of G_5 introduced above and let G_0 be the group given in (1). Let F_5 be the above mentioned index 2 subgroup of H_5, and H_0 the index 2 subgroup of G_0 generated by x_0, x_1. Even though H_5 and G_0 are not isomorphic, all corresponding maximal 2-quotients of H_5 and G_0 agree. Moreover, the same curious fact holds true for their subgroups F_5 and H_0.*

Special Presentation G_6 and G_7

The group G_6 coincides with the group B.2 in [9, Sect. 5]. We have the group specified by relations $\mathbf{r}_6 = \{a^2b, acb, bef, bge, d^2f, c^2g, egf\}$ on 7 generators but can be rewritten to a group generated by $\{x = e, y = f\}$ with relations

$$\mathbf{r}_6' = \{y^{-1}x^{-1}y^2x^{-2}y^{-3}x^{-1}, x^3yxyx^{-2}y^2\}$$

(see [3, Sect. 2.7]). The group G_7 coincides with the group B.1 in [9, Sect. 5].

We see that both groups have a subgroup H_6, H_7 of index 24 in G_6, G_7, respectively, which gives rise to maximal 3-quotients of 3-class k for $1 \leq k \leq 10$. However, the 3-groups are too large to successfully search for unmixed ramification structures. The abelianization of both groups is $H_6^{ab} \cong H_7^{ab} \cong \mathbb{Z}_3 \times \mathbb{Z}_3 \times \mathbb{Z}_3 \times \mathbb{Z}_3$.

3.3 Ramification Structures for the Group in [15, Example 6.4]

The group G in (2) with 13 generators $x_0, \ldots, x_{12}$ appeared as Example 6.4 in [15]. (G is constructed using Theorem 3.1 with $q = 3$, $m = 13$ and $l = \{0, 1, 3, 9\}$.) The subgroup H generated by x_0, x_1, x_2 has index 3. Again, the group G can also be found in [9, Sect. 4] as the group 1.1 (via the identification $a_i = x_{2i}$, where the indices are taken modulo 13).

The existence of ramification structures for $k = 2, \ldots, 60$ for the finite groups $H_{3,k}$ was already formulated in the Introduction (see Theorem 1.1) and was obtained by using MAGMA. See Conjecture 1.2 in the Introduction for the associated conjecture.

The ramification structures of $H_{3,k}$ in Theorem 1.1 give rise to algebraic surfaces $S = (C_{T_1} \times C_{T_2})/H_{3,k}$. For example, the order of the group $H_{3,2}$ is $a_2 = 3^6$. Therefore, the genera of the curves C_{T_i} is (see (3))

$$g(C_{T_1}) = g(C_{T_2}) = 1 + 3^5 = 244,$$

and the holomorphic Euler-Poincaré characteristic of S is

$$\chi(S) = \frac{(g(C_{T_1}) - 1)(g(C_{T_2}) - 1)}{|G|} = 81.$$

3.4 Ramification Structures for the Groups of Theorem 3.1 with $q \geq 4$

The construction given by Theorem 3.1 is for any q a prime power. For $q = 4$ the group below is given.

Example 3.6 ([15, Example 6.5]) We have that $q^2 + q + 1 = 21$ and so $\mathbb{F}_{q^3}^{\times}/\mathbb{F}_q^{\times}$ is identified with $\mathbb{Z}_{21}$. The group $\widehat{G}$ is given by the presentation,

$$\widehat{G} := \langle x_0, \ldots, x_{20} | x_i x_{i+7} x_{i+14}, x_i x_{i+14} x_{i+7}, x_i x_{i+3} x_{i+15} \text{ for } i \in \mathbb{Z}_{21} \rangle. \quad (4)$$

The abelianization of this group is $\widehat{G}^{ab} \cong \mathbb{Z}_2 \times \mathbb{Z}_2 \times \mathbb{Z}_2 \times \mathbb{Z}_2 \times \mathbb{Z}_6 \times \mathbb{Z}_6$. The group has maximal 2-quotients of 2-class k for $1 \leq k \leq 10$. However, it is extremely difficult to search for ramification structures of the maximal p-quotients of p-class k for $q \geq 4$. The finite groups are too large and have too many conjugacy classes, which leads to a computational expensive search.

3.5 Ramification Structures for the Group in [18, Example 2]

In [18], a new construction of groups presentations based on finite projective planes was introduced, generalizing the triangle presentations of [8, 9]. For the reader's convenience, we explain this briefly. The construction is based on the following general definition.

Definition 3.7 (*see* [18]) Let $\mathcal{P}_1, \ldots, \mathcal{P}_n$ be n disjoint finite projective planes of order q. Let P_i and L_i be the sets of points and lines respectively in $\mathcal{P}_i$. Let $P = \cup P_i$, $L = \cup L_i$, $P_i \cap P_j = \emptyset$ for $i \neq j$ and let λ be a bijection $\lambda : P \to L$.

A set $\mathcal{K}$ of k-tuples $(x_1, \ldots, x_k)$ will be called a *polyhedral presentation* over P compatible with λ if

(1) given $x_1, x_2 \in P$ then $(x_1, \ldots, x_k) \in \mathcal{K}$ for some $x_3, \ldots, x_k$ if and only if x_2 and $\lambda(x_1)$ are incident;
(2) $(x_1, \ldots, x_k) \in \mathcal{K}$ implies that $(x_2, \ldots, x_k, x_1) \in \mathcal{K}$;
(3) given $x_1, x_2 \in P$, then $(x_1, \ldots, x_k) \in \mathcal{K}$ for at most one $x_3 \in P$.

We call λ a *basic bijection*.

A polyhedral presentation $\mathcal{K}$ gives rise to a group presentation $G_{\mathcal{K}}$ in the following way: the generators of $G_{\mathcal{K}}$ are given by $\cup P_i$ and the relations are the k-tuples of $\mathcal{K}$, each written as a product.

Example 3.8 The *triangle presentations* listed in [9] can be seen as special cases of polyhedral presentations for $n = 1$, $k = 3$ and $q = 2, 3$.

We now discuss the case $n = 1, q = 2$. We enumerate the points of the projective plane by $1, 2, \ldots, 6$. The following array illustrates a basic projection λ:

$$\begin{array}{l} 0:1\ 4\ 2 \\ 1:3\ 2\ 5 \\ 2:4\ 3\ 6 \\ 3:0\ 4\ 5 \\ 4:1\ 5\ 6 \\ 5:0\ 2\ 6 \\ 6:0\ 1\ 3. \end{array}$$

Here, every point k represents a row and is followed by the points contained in the associated line $\lambda(k)$. For example, the line $\lambda(3)$ consists of the points $0, 4, 5$.

A triangle presentation $\mathcal{T}$ for the group A.1 in [9] is given by

$$(0, 1, 3), (1, 2, 4), (2, 3, 5), (3, 4, 6), (4, 5, 0), (5, 6, 1), (6, 0, 2),$$

and all the cyclic permutations, i.e. for $(0, 1, 3) \in \mathcal{T}$ we also have $(1, 3, 0), (3, 0, 1) \in \mathcal{T}$. The associated group presentation $G_{\mathcal{T}}$ agrees with the presentation of G_0 in (1).

Example 3.9 [18, Example 2] The projective plane $\mathcal{P}$ of order 4 can be partitioned by three projective planes of order two (see [7]). We denote points of the subplane $\mathcal{P}_i$ for $i = 1, 2, 3$ by numbers from $7i - 6$ to $7i$. Note that lines in $\mathcal{P}$ consist of five points, while the lines in $\mathcal{P}_i$ consist of three points. A basic projection λ for $\mathcal{P}$ is given below. Note that each subplane $\mathcal{P}_i$ has its own basic projection, denoted by λ_i, satisfying $\lambda_i(k) \subset \lambda(k)$. In the array below, the row associated to the point k lists first the three points in the associated line via the basic bijection in the subplane, followed up by the two remaining points in $\lambda(k)$.

4 :	5 6 7	12 18		9 :	12 13 14	1 15
7 :	1 2 5	8 21		11 :	8 9 12	3 17
2 :	3 4 5	14 16		14 :	10 11 12	2 16
5 :	1 3 6	10 19		12 :	8 10 13	4 18
1 :	2 4 6	9 15		10 :	9 11 13	5 19
3 :	1 4 7	11 17		13 :	8 11 13	6 20
6 :	2 3 7	13 20		8 :	9 10 14	7 21

18 :	19 20 21	4 12
21 :	15 16 19	7 8
16 :	17 18 19	2 14
19 :	15 17 20	5 10
15 :	16 18 20	1 9
17 :	15 18 21	3 11
20 :	16 17 21	6 13

The above basic projections give rise to the following polyhedral presentation $\mathcal{K}$ for a projective plane of order 4, induced by polyhedral presentations of projective planes of order 2

$$(1, 9, 15), (1, 15, 9), (2, 14, 16), (2, 16, 14), (3, 11, 17), (3, 17, 11), (4, 12, 18),$$
$$(4, 18, 12), (5, 10, 19), (5, 19, 10), (6, 13, 20), (6, 20, 13), (7, 8, 21), (7, 21, 8),$$
$$(1, 2, 3), (1, 4, 5), (1, 6, 7), (3, 4, 6), (3, 7, 5), (2, 5, 6), (2, 4, 7), (8, 9, 12), (8, 10, 13),$$
$$(8, 14, 11), (9, 14, 10), (9, 13, 11), (12, 13, 14), (10, 11, 12), (15, 16, 17), (15, 18, 19),$$
$$(17, 18, 20), (17, 21, 19), (16, 19, 20), (16, 18, 21),$$

and all their cyclic permutations.

All relators in the group presentation $G_{\mathcal{K}}$ given by $\mathcal{K}$ are of length 3 and the star graph is isomorphic to the incidence graph of a finite projective plane of order 4. This means, by [15], that the group given by this presentation $G_{\mathcal{K}}$ is a *special presentation*.

It also can be seen that this group acts on a Euclidean building where the vertex links are incidence graphs of projective planes of order 4, see [18].

There are remarkable connections between the group $G_{\mathcal{K}}$ and the group G_0 discussed in Sect. 3.1. Firstly, we can present the group $G_{\mathcal{K}}$ in an alternative way with different generators:

$$G_{\mathcal{K}} = \langle w_0, \ldots, w_6, y_0, \ldots, y_6, z_0, \ldots, z_6 \mid w_i w_{i+1} w_{i+3}, y_i y_{i+1} y_{i+3}, z_i z_{i+1} z_{i+3}, \\ w_i^{-1} y_{6(1+i)} z_i^{-1}, w_i^{-1} z_i^{-1} y_{6(1+i)} (i \in \mathbb{Z}_7) \rangle, \quad (5)$$

where each of the three subsets of generators has very similar relators like those appearing for the group G_0 in (1), with only two more series of relators added representing connections between the generators of different subsets.

Secondly, the maximal 2-quotients of 2-class k of the group $G_{\mathcal{K}}$ are isomorphic to the maximal 2-quotients of 2-class k for the group G_0 (given by the presentation in (1)) for $1 \leq k \leq 20$. However, the groups $G_{\mathcal{K}}$ and G_0 are not isomorphic, as they have different abelianized groups $G_{\mathcal{K}}^{ab} \cong \mathbb{Z}_2 \times \mathbb{Z}_6 \times \mathbb{Z}_6$, while $G_0^{ab} \cong \mathbb{Z}_2 \times \mathbb{Z}_2 \times \mathbb{Z}_6$. This gives rise to the following conjecture.

Conjecture 3.10 *Let $G_{\mathcal{K}}$ be the group introduced above (e.g., by the presentation (5)), and G_0 be the group given in (1). Even though $G_{\mathcal{K}}$ and G_0 are not isomorphic, all corresponding maximal 2-quotients of these two groups agree.*

Remark 3.11 If we replace the relators $w_i^{-1} y_{6(1+i)} z_i^{-1}, w_i^{-1} z_i^{-1} y_{6(1+i)}$ in (5) by the relators $x_i y_i z_i, x_i z_i y_i$ we obtain a group G' with the following presentation

$$G' = \langle w_0, \ldots, w_6, y_0, \ldots, y_6, z_0, \ldots, z_6 \mid w_i w_{i+1} w_{i+3}, y_i y_{i+1} y_{i+3}, z_i z_{i+1} z_{i+3}, \\ w_i y_i z_i, w_i z_i y_i (i \in \mathbb{Z}_7) \rangle.$$

This group G' is isomorphic to the group $\widehat{G}$ given by the presentation (4) under the identifications

$$\begin{array}{lll} x_0 \mapsto z_0^{-1}, & x_7 \mapsto w_0^{-1}, & x_{14} \mapsto y_0^{-1}, \\ x_1 \mapsto w_1^{-1}, & x_8 \mapsto y_1^{-1}, & x_{15} \mapsto z_1^{-1}, \\ x_2 \mapsto y_2^{-1}, & x_9 \mapsto z_2^{-1}, & x_{16} \mapsto w_2^{-1}, \\ x_3 \mapsto z_3^{-1}, & x_{10} \mapsto w_3^{-1}, & x_{17} \mapsto y_3^{-1}, \\ x_4 \mapsto w_4^{-1}, & x_{11} \mapsto y_4^{-1}, & x_{18} \mapsto z_4^{-1}, \\ x_5 \mapsto y_5^{-1}, & x_{12} \mapsto z_5^{-1}, & x_{19} \mapsto w_5^{-1}, \\ x_6 \mapsto z_6^{-1}, & x_{13} \mapsto w_6^{-1}, & x_{20} \mapsto y_6^{-1}. \end{array}$$

Acknowledgments We thank Donald Cartwright for the representations and method given in Appendix 2 and helpful correspondences. The first author also wishes to thank Uzi Vishne for useful correspondences.

Appendix 1: Expanders Associated to the Group G_0

Expander graphs are defined with the help of the edge expansion rate.

Definition A. 1 Let $\mathcal{G} = (V, E)$ be a combinatorial graph with vertex set V and edge set E. Then the *edge expansion rate* $h(\mathcal{G})$ is defined as

$$h(\mathcal{G}) = \inf_{\text{finite} A \subset V} \frac{|\partial A|}{\min(|A|, |V \backslash A|)},$$

where $\partial A \subset E$ is the set of all edges connecting a vertex of A with a vertex of $V \backslash A$.

Expanders are infinite families of finite graphs which are both sparse and highly connected. They are not only theoretically important but have also applications in computer science for, e.g., robust network designs.

Definition A. 2 A sequence $\mathcal{G}_n = (V_n, E_n)$ of connected finite graphs with $|V_n| \to \infty$ is called a *family of expanders* if there exists $k \geq 2$ and $\epsilon > 0$ such that

(a) all graphs $\mathcal{G}_n$ are k-regular,
(b) $h(\mathcal{G}_n) \geq \epsilon$ for all n.

It was observed in [22] that the subgroup H_0 of G_0 generated by x_0, x_1 has index 2, and that both groups H_0 and G_0 are just infinite and have Kazhdan property (T). Property (T) implies that, for a fixed choice of generators, the Cayley graphs of all quotients by finite index normal subgroups have a uniform positive lower bound for their edge expansion rate (see [19, Proposition 3.3.1]). A presentation of the subgroup H_0 is given by

$$H_0 = \langle x_0, x_1 \mid r_1, r_2, r_3 \rangle,$$

where

$$\begin{aligned} r_1 &= (x_1 x_0)^3 x_1^{-3} x_0^{-3}, \\ r_2 &= x_1 x_0^{-1} x_1^{-1} x_0^{-3} x_1^2 x_0^{-1} x_1 x_0 x_1, \\ r_3 &= x_1^3 x_0^{-1} x_1 x_0 x_1 x_0^2 x_1^2 x_0 x_1 x_0. \end{aligned}$$

We have the following Cayley graph expanders obtained from finite groups with just two generators and four relations.

Theorem A. 3 (cf. [22, Theorem 1]) *The groups*

$$H_k = \langle x_0, x_1 \mid r_1, r_2, r_3, [x_1, \underbrace{x_0, \ldots, x_0}_{k}] \rangle$$

are finite with $|H_k| \to \infty$, and the associated Cayley graphs with respect to the generators x_0, x_1 define an infinite family of expanders of vertex degree 4.

Using the faithful matrix representation of H_0 by infinite upper triangular matrices and their truncations at the kth upper diagonal as mentioned in Sect. 3.1, we obtain another family $\widetilde{H}_k$ of finite *nilpotent* groups whose associated Cayley graphs $\mathcal{G}_k$ with respect to the generators x_0, x_1 are another family of expander graphs which form a tower of coverings

$$\cdots \mathcal{G}_k \to \mathcal{G}_{k-1} \to \cdots \to \mathcal{G}_1 \to \mathcal{G}_0,$$

whose covering indices are powers of 2 (for more details, see [22]). It was conjectured in [22, Conjecture 2] that the covering indices follow the pattern 4, 8, 4, 8, 8, 4, 8, 8, 4, 8, 8, ... See Fig. 2 for the graph $\mathcal{G}_2$. We use the notation $z_1 = [x_0, x_1]$, $z_2 = x_0^2$, $z_3 = x_1^2$ and $z_{ij} = z_i z_j$ and $z_{ijk} = z_i z_j z_k$. The elements expressed by z_i lie in the centre of $\widetilde{H}_2$. The same graph was illustrated in [22, Fig. 4], but the illustration given here is more symmetric. Solid edges from vertices with label i to vertices with label $i+1 \pmod 4$ represent right multiplication by x_0, while dashed edges from vertices with label i to vertices with label $i+1 \pmod 4$ represent right multiplication by x_1. Note that the solid 4-cycles as well as the dashed 4-cycles in $\mathcal{G}_3$ are consequences of $x_0^4 = x_1^4 = 1$ in $\widetilde{H}_2$.

Another construction of 3-regular expanders was given in [17]. Starting from the same groups $\widetilde{H}_k$, we now consider the associated Cayley graphs X_k with respect to

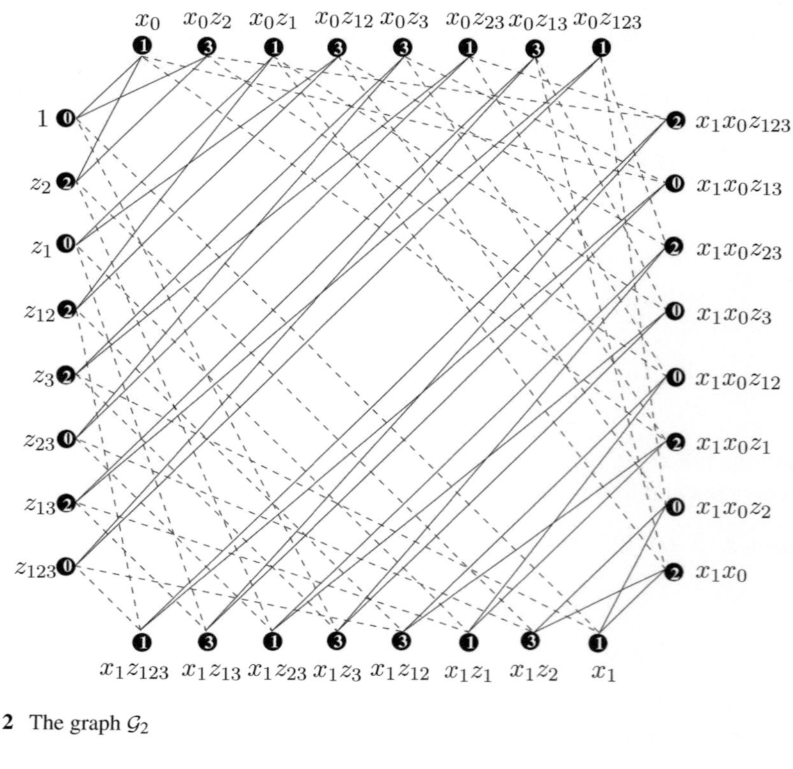

Fig. 2 The graph $\mathcal{G}_2$

the generators x_0, x_1, x_3 where $x_3 = x_1^{-1} x_0^{-1}$. The graphs X_k are 6-regular and $\mathcal{G}_k$ is a subgraph of X_k with the same number of vertices. Property (T) guarantees that the graphs X_k are also a family of expanders. One can check that X_k forms a tessellation of a closed Riemann surface by triangles and by 2^l-gons (with l only depending on k) and that every edge of X_k belongs to precisely one triangle of X_k. Now we apply a $\Delta - Y$ transformation to the graphs X_k to obtain new graphs T_k. The $\Delta - Y$ transformation removes the edges of every triangle in the original graph X_k, adds a new vertex in its centre, and connects this new central vertex with new edges to the 3 original vertices of the triangle. It turns out that the vertex set of the new graph T_k is twice as large as the vertex set of old graph X_k, and that T_k is 3-regular. Moreover, there is an explicit connection between the eigenfunctions of the adjacency matrix of X_k and the eigenfunctions of the adjacency matrix of T_k (see [17, Theorem 2.1]). The spectral characterisation of expander graphs then implies that the new family T_k of 3-regular graphs is, again, a family of expanders.

For yet another expander graphs construction from the group G_0 see [20, 23].

Appendix 2: Representation for the Group G

We include a representation for the group G (given by (2)) in $\mathrm{GL}(9, \mathbb{F}_3[1/Y])$, which may be useful in the future (as the matrix representations for the group G_0 with presentation (1) were useful for several works [8, 20, 22]). The representation is due to Donald Cartwright and the algebra program REDUCE. Recall that the group G coincides with the group 1.1 in [9], where we relate the generators by $a_i = x_{2i}$ for $i = 0, \dots, 12$, with indices taken modulo 13. We set

$$x_0 : \begin{pmatrix} 1&1&1&0&2&2&0&1&1 \\ 0&1&2&0&0&1&0&0&2 \\ 0&0&1&0&0&0&0&0&0 \\ 0&0&0&1&1&1&0&2&2 \\ 0&0&0&0&1&2&0&0&1 \\ 0&0&0&0&0&1&0&0&0 \\ 0&0&0&0&0&0&1&1&1 \\ 0&0&0&0&0&0&0&1&2 \\ 0&0&0&0&0&0&0&0&1 \end{pmatrix} + \frac{1}{Y} \begin{pmatrix} 0&2&2&0&1&1&0&2&2 \\ 0&0&1&0&0&2&0&0&1 \\ 0&0&0&0&0&0&0&0&0 \\ 0&1&1&0&2&2&0&1&1 \\ 0&0&2&0&0&1&0&0&2 \\ 0&0&0&0&0&0&0&0&0 \\ 0&2&2&0&1&1&0&2&2 \\ 0&0&1&0&0&2&0&0&1 \\ 0&0&0&0&0&0&0&0&0 \end{pmatrix}$$

and

$$\tau : \begin{pmatrix} 1&0&0&0&0&0&0&0&0\\ 0&1&0&0&0&0&0&0&0\\ 0&0&1&0&0&0&0&0&0\\ 0&0&0&0&1&0&0&0&0\\ 0&0&0&0&1&1&0&0&0\\ 0&0&0&1&0&1&0&0&0\\ 0&0&0&0&0&0&2&2&0\\ 0&0&0&0&0&0&0&1&2\\ 0&0&0&0&0&0&2&0&1 \end{pmatrix},$$

where the other generators $x_1, \ldots, x_{12}$ are formed via conjugation of x_0 by τ, i.e. $x_i = \tau^i x_0 \tau^{-i}$ for $i = 1, \ldots, 12$.

The idea in creating this representation is to write $\mathbb{F}_{27} = \mathbb{F}_3(\theta)$, where θ is a primitive element on $\mathbb{F}_{27}$ satisfying $\theta^3 = \theta + 1$, and to use the basis $\{\theta^i \sigma^j | i, j = 0, 1, 2\}$ for the divison algebra $\mathcal{A}$ over $\mathbb{F}_{27}(Y)$ for an indeterminate Y (in the order $1, \theta, \theta^2, \sigma, \theta\sigma, \ldots, \theta^2\sigma^2$). Here σ is assumed to satisfy $\sigma^3 = Y - 1$ (which implies $(1+\sigma)^{-1} = (1/Y)(1 - \sigma + \sigma^2)$) and $\sigma\theta\sigma^{-1} = \theta^3$. The generators of $\mathcal{T}_{\mathcal{K}}$, where $\mathcal{K}$ is a triangle presentation from [8, 9], are the $a_u = u^{-1}(1+\sigma)u$, where $u \in \mathbb{F}_{27}^\times / \mathbb{F}_3^\times$. Since $\mathbb{F}_{27}^\times = \mathbb{F}_3^\times \cdot \{1 = \theta^{13}, \theta, \ldots, \theta^{12}\}$, we choose $\alpha_k = \theta^{-k}(1+\sigma)\theta^k$ as in [9, p. 178]. The α_k's act on $\mathcal{A}$ by conjugation. A straightforward calculation yields

$$\begin{aligned} \alpha_k \theta^i \sigma^j \alpha_k^{-1} &= \theta^i \sigma^j \frac{1}{Y} + \left(\theta^{3i+2k} - \theta^{i+2\cdot 3^j k}\right) \sigma^{j+1} \frac{1}{Y} \\ &+ \left(\theta^{i+8\cdot 3^j k} - \theta^{3i+2k+2\cdot 3^{j+1}k}\right) \sigma^{j+2} \frac{1}{Y} + \theta^{3i+2k+8\cdot 3^{j+1}k} \sigma^j \frac{Y-1}{Y}. \end{aligned}$$

Expressing the conjugation by α_k with respect to the above basis of $\mathcal{A}$ then gives rise to a representation as a 9×9 matrix over the field $\mathbb{F}_3(1/Y)$. We conclude from [9] that the matrices associated to the α_k satisfy the relations of our generators x_k. Note, finally, that the above matrix for τ represents the conjugation by θ in $\mathcal{A}$, i.e., $z \mapsto \theta^{-1} z \theta$.

References

1. N. Barker, N. Boston, N. Peyerimhoff, A. Vdovina, New examples of Beauville surfaces. Monatsh. Math. **166**(3–4), 319–327 (2012)
2. N. Barker, N. Boston, N. Peyerimhoff A. Vdovina, An infinite family of 2-groups with mixed Beauville structures. Int. Math. Res. Not. (2014). doi:10.1093/imrn/rnu045
3. S. Barre, Polyédres de rang deux, Thesis ENS Lyon, December (1996). http://web.univ-ubs.fr/lmam/barre/these1.pdf
4. I.C. Bauer, F. Catanese F. Grunewald, Beauville surfaces without real structures, in *Geometric Methods in Algebra and Number Theory*. Progress in Mathematics, vol. 235 (Birkhäuser, Boston, 2005)
5. I.C. Bauer, F. Catanese, F. Grunewald, The classification of surfaces with $p_g = q = 0$ isogenous to a product of curves. Pure Appl. Math. Q. **4**(2), 547–586 (2008)

6. W. Bosma, J. Cannon, C. Playoust, The Magma algebra system. I. The user language. J. Symb. Comput. **24**(3–4), 235–265 (1997)
7. R.H. Bruck, Quadratic extensions of cyclic planes, in *Proceedings Symposium in Applied Mathematics*, pp. 15–44 (1960)
8. D.I. Cartwright, A.M. Mantero, T. Steger, A. Zappa, Groups acting simply transitively on the vertices of a building of type $\widetilde{A}_2$, I. Geom. Dedicata **47**(2), 143–166 (1993)
9. D.I. Cartwright, A.M. Mantero, T. Steger, A. Zappa, Groups acting simply transitively on the vertices of a building of type $\widetilde{A}_2$, II. Geom. Dedicata **47**(2), 167–223 (1993)
10. F. Catanese, Fibred surfaces, varieties isogenous to a product and related moduli spaces. Am. J. Math. **122**(1), 1–44 (2000)
11. M. Edjvet, J. Howie, Star graphs, projective planes and free subgroups in small cancellation groups. Proc. Lond. Math. Soc. **57**(2), 301–328 (1988)
12. M. Edjvet, A. Vdovina, On the SQ-universality of groups with special presentations. J. Group Theory **13**(6), 923–931 (2010)
13. B. Fairbairn, Some exceptional Beauville structures. J. Group Theory **15**, 631–639 (2012)
14. Y. Fuertes, G. González-Diez, On Beauville structures on the groups S_n and A_n. Math. Z. **264**, 959–968 (2010)
15. J. Howie, On the SQ-universality of $T(6)$-groups. Forum Math. **1**(3), 251–272 (1989)
16. D.R. Hughes, F.C. Piper, *Projective Planes* (Springer, New York, 1973)
17. I. Ivrissimtzis, N. Peyerimhoff, A. Vdovina, Trivalent expandes and hyperbolic surfaces (2012) arXiv:1202.2304
18. S. Immervol, A. Vdovina, Partitions of projective planes and construction of polyhedra, Max-Planck-Institut fur Mathematik, Bonn, Preprint Series 23 (2001)
19. A. Lubotzky, *Discrete Groups, Expanding Graphs and Invariant Measures* (Birkhäuser, Basel, 2010)
20. A. Lubotzky, B. Samuels, U. Vishne, Explicit construction of Ramanujan complexes of type $\widetilde{A}_d$. Eur. J. Comb. **26**(6), 965–993 (2005)
21. R.C. Lyndon, P.E. Schupp, *Combinatorial Group Theory, Classics in Mathematics*. Reprint of the 1977 edition (Springer, Berlin, 2001)
22. N. Peyerimhoff, A. Vdovina, Cayley graph expanders and groups of finite width. J. Pure Appl. Algebra **215**(11), 2780–2788 (2011)
23. A. Sarveniazi, Explicit construction of a Ramanujan $(n_1, n_2, \ldots, n_{d-1})$-regular hypergraph. Duke Math. J. **139**(1), 141–171 (2007)
24. F. Serrano, Isotrivial fibred surfaces. Ann. Mat. Pura Appl. **171**(4), 63–81 (1996)

A Survey of Beauville p-Groups

Nigel Boston

Abstract This paper describes recent results as to which p-groups are Beauville, with emphasis on ones of small order (joint with N. Barker and B. Fairbairn) and ones that form inverse systems (joint with N. Barker, N. Peyerimhoff, and A. Vdovina).

2000 Mathematics Subject Classification. 20D15 · 14J29 · 14L30 · 20F05

1 Introduction

Much research has gone into the fundamental question of which finite groups are Beauville. This has been carried out mostly for simple (and related) groups. In a sense, however, most finite groups are p-groups (for example, about 99.2 % of the groups of order ≤ 2000 have order 2^{10}). Moreover, as is described below, p-groups form a class of groups with a great deal of structure. It is, therefore, natural to consider which finite p-groups are Beauville.

Until very recently, this would have made for a very short survey. The seminal work of Bauer, Catanese, and Grunewald [5, 6] included the classification of all abelian Beauville groups. Among these, the p-groups arising are precisely $\mathbf{Z}/p^k \times \mathbf{Z}/p^k$ for $p \geq 5$ and any $k \geq 1$. There have, however, been only a handful of nonabelian Beauville p-groups documented, in [6], where two examples of order 2^8 were presented, and by Fuertes, González-Diez, and Jaikin-Zapirain [8], who gave examples of order 2^{12} and 3^{12}. It turns out, however, that there is much more to be said.

After recalling the definition of a Beauville group, this paper begins with a description of the now standard way to organize the class of finite 2-generator p-groups into a rooted tree. This then facilitates (computational and theoretical) investigations into the location of Beauville groups within this tree. We focus on three particular aspects, namely whether being Beauville or not is inherited by neighbours in the tree, the

N. Boston (✉)
Department of Mathematics, University of Wisconsin, 303 Van Vleck Hall, 480 Lincoln Drive, Madison, WI 53706, USA
e-mail: boston@math.wisc.edu

I. Bauer et al. (eds.), *Beauville Surfaces and Groups*, Springer Proceedings in Mathematics & Statistics 123, DOI 10.1007/978-3-319-13862-6_3

existence of infinite paths through the tree consisting entirely or mostly of Beauville or non-Beauville groups, and the density of Beauville groups among groups of order p^n for fixed n as $p \to \infty$.

2 Definitions

2.1 Beauville Structures

Definition 2.1 (*Unmixed Beauville structure*) Let G be a finite group. Given $x, y \in G$, let $\Sigma(x, y)$ denote the union of all conjugacy classes of all powers of x, y, xy. In the unmixed case, we call G *Beauville* if it has two pairs of generators, $\{x, y\}$ and $\{u, v\}$, such that the following holds:

$$(*) \quad \Sigma(x, y) \cap \Sigma(u, v) = \{1\}$$

In terms of (unmixed) Beauville surfaces minimally represented as $(C_1 \times C_2)/G$ where the map $C_i \to C_i/G = \mathbf{P}^1(\mathbf{C})$ is ramified over $0, 1, \infty$ ($i = 1, 2$), the fact that G is 2-generated is equivalent to it being a quotient of $\pi_1(\mathbf{P}^1(\mathbf{C}) - \{0, 1, \infty\})$ whilst $(*)$ is equivalent to G acting freely on $C_1 \times C_2$.

In the mixed case, the group G can interchange the curves C_i and then the subgroup H of transformations stabilizing each curve has index 2 in G.

Definition 2.2 (*Mixed Beauville structure*) We call G (mixed) *Beauville* if it has a subgroup H of index 2 which has a pair of generators $T = \{x, y\}$ such that, for every $g \in G - H$, $\Sigma(T) \cap \Sigma(gTg^{-1}) = \{1\}$ and $g^2 \notin \Sigma(T)$.

Note that if G is a p-group with mixed Beauville structure, then necessarily $p = 2$.

2.2 p-Groups

As for the basics on p-groups, we focus on a certain filtration by normal subgroups.

Definition 2.3 (*Lower p-central series*) If G is a finite p-group, then it has a normal series

$$\{1\} = P_c(G) \subseteq \ldots \subseteq P_1(G) \subseteq P_0(G) = G,$$

called the *lower p-central series* of G, given by $P_{i+1}(G) = [G, P_i(G)]P_i(G)^p$. The smallest c such that $P_c(G) = \{1\}$ is called the *p-class* of G.

The first term $P_1(G)$ is sometimes called the *Frattini subgroup* of G. It is best to think in terms of the corresponding sequence of quotients of G,

$$G = G/P_c(G) \to G/P_{c-1}(G) \to \dots \to G/P_1(G) = (\mathbf{Z}/p)^{d(G)}$$

where $d(G)$ denotes the *minimum number of generators* of G. In this paper we will be exclusively interested in the case $d(G) = 2$. For each p we will create a rooted tree in which the arrows above correspond to edges leading to the root $(\mathbf{Z}/p)^2$. The group $G/P_1(G)$ is elementary p-abelian by the definition of $P_1(G)$—its rank is $d(G)$ by the following classical result.

Lemma 2.4 (Burnside's Basis Theorem) *$x, y \in G$ generate G if and only if their images in $G/P_1(G)$ generate $G/P_1(G)$.*

Suppose we fix p and d. Then there is a rooted tree, due to O'Brien [11], whose vertices are the isomorphism classes of finite p-groups G with $d(G) = d$. A group G of p-class c is joined to all groups H of p-class $c + 1$ with $H/P_c(H) \cong G$ (these are called the *children* of G). If G has minimal presentation F/R, then these are constructed as quotients of a covering group $G^* = F/R^*$. To each finite p-group is attached a nonnegative integer called its *nuclear rank*, which is defined to be the rank of the elementary abelian group $P_c(G^*)$. This is 0 if and only if the group has no children. See [11] for the technical details.

3 Beauville Groups in O'Brien Trees

3.1 An Example

To illustrate the general theory, consider the case of $p = 5, d = 2$. The root of the O'Brien tree is $\mathbf{Z}/5 \times \mathbf{Z}/5$, which is Beauville as noted in the introduction (indeed it is the first example, found by Beauville himself). It has seven children, of which three have order 5^3, three order 5^4, and one order 5^5 (namely the free 2-generator 5-group of 5-class 2). Of these, we compute using the computer algebra package MAGMA [7] that there are four Beauville groups, namely one of order 5^3, two of order 5^4 (including $\mathbf{Z}/25 \times \mathbf{Z}/25$), and the one of order 5^5.

Consider the three groups of order 5^3. One is $\mathbf{Z}/5 \times \mathbf{Z}/25$, which is not Beauville. It has two children, one childless, the other having two children, one childless, and so on. None of these groups is Beauville. The other group of order 5^3 that is not Beauville is childless. Let us focus on the remaining group of order 5^3, which is Beauville. It has 16 children, four of order 5^4 (one Beauville, three not) and twelve of order 5^5 (nine Beauville, three not).

Note that a child of a Beauville group may or may not be Beauville. As seen with the first group of order 5^3 above, a child of a group that is not Beauville can also not be Beauville. Since $\mathbf{Z}/3 \times \mathbf{Z}/3$ is not Beauville, yet Beauville 3-groups exist, there must exist a group that is not Beauville but which has a Beauville child. There does, however, appear to be some sort of correlation between nuclear rank and being Beauville, as I now explain.

Of the 16 children in the last but one paragraph, one has nuclear rank 3. It is Beauville of order 5^5. It has 116 children. Of these, 8 have order 5^6—the 5 with nuclear rank 0 are not Beauville, the rest are. Of the 116 groups, 61 have order 5^7—the 17 with nuclear rank 2 are not Beauville, the rest are. Of the 116 groups, 47 have order 5^8—the 5 with nuclear rank 4 are not Beauville, the rest are.

There is no exact rule—for instance, of the 16 children in the last but second paragraph, there are five of order 5^5 with nuclear rank 1, two Beauville, three not—but clearly something is going on here, possibly for large enough p-class.

We gather our conclusions here:

Theorem 3.1 *A child of a Beauville p-group may or may not be Beauville. A child of a non-Beauville p-group may or may not be Beauville. There exists a p-group all of whose infinitely many descendants are not Beauville.*

3.2 Infinite Paths Through the O'Brien Trees

The infinite ends of the O'Brien tree correspond to infinite 2-generator pro-p groups, which are obtained as the inverse limit of the groups in the path. Alternatively one might start with a discrete 2-generator group and investigate its pro-p completion.

In this way it was found that, along an infinite ray of the tree, the property of being Beauville or not can alternate infinitely often, as noted in my joint work with Barker, Peyerimhoff, and Vdovina [2, 3]. We considered the groups

$$H = \langle x_0, x_1 \rangle \subseteq \Gamma = \langle x_0, \ldots, x_6 \mid x_i x_{i+1} x_{i+3} = 1, i \in \mathbf{Z}/7 \rangle$$

and conjectured that if k is not a power of 2, then $H/P_k(H)$ is unmixed Beauville and $H/P_k(H) \subseteq G/P_k(G)$ is mixed Beauville. This was checked computationally for several k [2] and proven in general in [3]. Note that the case $k = 3$ gives one of the two mixed Beauville groups of order 256 originally given in [6], SmallGroup (256, 3679), the other being SmallGroup (256, 3678).

In [4], we also found similar results for other groups with special presentation (defined by Howie [10] to be groups for which every relator has length 3 and the star graph is isomorphic to the incidence graph of a finite projective plane). There are also simpler examples to be had, for instance by exploiting the lemma below of Fuertes and Jones.

Lemma 3.2 ([9, Lemma 4.2]) *If $x, y, u, v \in G$ have images $\overline{x}, \overline{y}, \overline{u}, \overline{v} \in G/N$ yielding a Beauville structure on G/N, then if*

$$\langle x \rangle \cap N = \langle y \rangle \cap N = \langle xy \rangle \cap N = \{1\},$$

then x, y, u, v yield a Beauville structure on G.

An easy induction immediately gives:

Corollary 3.3 *If $p \geq 5$ and Γ is the triangle group*

$$\langle x, y, z \mid x^p = y^p = z^p = xyz = 1 \rangle,$$

then its p-central quotients $\Gamma / P_k(\Gamma)$ are all Beauville.

I would conjecture that likewise:

Conjecture 3.4 *If $p \geq 5$ and Γ is either*

(i) the free product $\langle x, y \mid x^p, y^p \rangle$ or
(ii) the free group $\langle x, y \rangle$,
then its p-central quotients $\Gamma / P_k(\Gamma)$ are all Beauville.

4 Beauville Groups of Small Order

A basic question often asked is: are most 2-generated groups Beauville? Since the evidence overwhelmingly indicates that most groups (2-generated or otherwise) are p-groups, we address this question here. In [8], the authors said it was very plausibly true that most 2-generator p-groups are Beauville. One way to make sense of the question is to see whether this holds for groups of order p^n for fixed n as p varies. This is joint work with Barker and Fairbairn [1].

To begin with, we sought the smallest (nonabelian) Beauville p-group for each p.

Theorem 4.1 ([1, Corollary1.9])

(1) The smallest Beauville 2-group is SmallGroup(128, 36) (also discovered in [6]);
(2) the smallest Beauville 3-group is SmallGroup(243, 3);
(3) for $p \geq 5$ the smallest nonabelian Beauville p-group is SmallGroup(p^3, 3).

Definition 4.2 Let $g_n(p)$ denote the number of groups G of order p^n with $d(G) = 2$. Let $h_n(p)$ denotes the number of Beauville groups of order p^n. Let $\delta_n(p) = h_n(p)/g_n(p)$, which denotes the density of Beauville groups among all 2-generated groups of order p^n.

We found using the computer algebra package MAGMA [7] that:

Theorem 4.3

$$g_2(p) = 1; \quad h_2(p) = 1 \textit{ if } p \geq 5$$

$$g_3(p) = 3; \quad h_3(p) = 1 \textit{ if } p \geq 5$$

$$g_4(p) = 9 \textit{ if } p \geq 3; \quad h_4(p) = 3 \textit{ if } p \geq 5$$

$$g_5(p) = p + 26 + 2gcd(p-1,3) + gcd(p-1,4); \quad h_5(p) \geq p + 8 \textit{ if } p \geq 5$$

$$g_6(p) = 10p + 62 + 14gcd(p-1,3) + 7gcd(p-1,4) + 2gcd(p-1,5);$$

$$h_6(p) \leq g_6(p) - (p-1) \textit{ if } p \geq 5$$

Corollary 4.4 *The limit as* $p \to \infty$ *of* $\delta_n(p)$ *is* 1 *if* $n = 2$ *or* 5*, but strictly less than* 1 *if* $n = 3, 4,$ *or* 6.

Conjecture 4.5 $h_5(p) = p + 10$ *if* $p \geq 5$.

There is clearly plenty of room for further investigations into the questions above. Preliminary calculations suggest that the limit is less than 1 if $n = 7$.

Acknowledgments I thank my co-authors Nathan Barker, Ben Fairbairn, Norbert Peyerimhoff, and Alina Vdovina for their encouragement and hard work in helping develop this field.

References

1. N. Barker, N. Boston, B. Fairbairn, A note on Beauville p-groups. Exp. Math. **21**(3), 298–306 (2012)
2. N. Barker, N. Boston, N. Peyerimhoff, A. Vdovina, New examples of Beauville surfaces. Monatsh. Math. **166**(3–4), 319–327 (2012)
3. N. Barker, N. Boston, N. Peyerimhoff, A. Vdovina, An infinite family of 2-groups with mixed Beauville structures (2013). arXiv:1304.4480
4. N. Barker, N. Boston, N. Peyerimhoff, A. Vdovina, Regular algebraic surfaces, ramification structures, and projective planes, submitted to this volume
5. I.C. Bauer, F. Catanese, F. Grunewald, Beauville surfaces without real structures, in *Geometric Methods in Algebra and Number Theory*. Progress in Mathematics, vol. 235 (Birkhäuser Boston, 2005)
6. I.C. Bauer, F. Catanese, F. Grunewald, The classification of surfaces with $p_g = q = 0$ isogenous to a product of curves. Pure Appl. Math. Q. **4**(2), 547–586 (2008)
7. W. Bosma, J. Cannon, C. Playoust, The Magma algebra system, I. The user language. J. Symb. Comput. **24**(3–4), 235–265 (1997)
8. Y. Fuertes, G. González-Diez, A. Jaikin-Zapirain, On Beauville surfaces. Groups Geom. Dyn. **5**(1), 107–119 (2011)
9. Y. Fuertes, G.A. Jones, Beauville surfaces and finite groups. J. Algebra **340**(1), 13–27 (2011)
10. J. Howie, On the SQ-universality of $T(6)$-groups. Forum Math. **1**(3), 251–272 (1989)
11. E.A. O'Brien, The p-group generation algorithm. J. Symb. Comput. **9**, 677–698 (1990)

Strongly Real Beauville Groups

Ben Fairbairn

Abstract A strongly real Beauville group is a Beauville group that defines a real Beauville surface. Here we discuss efforts to find examples of these groups, emphasising on the one extreme finite simple groups and on the other abelian and nilpotent groups. We will also discuss the case of characteristically simple groups and almost simple groups. *En route* we shall discuss several questions, open problems and conjectures as well as giving several new examples of infinite families of strongly real Beauville groups.

1 Introduction

We first issue an apology/assurance. It is the nature of Beauville constructions that this article is likely to be of interest to both geometers and group theorists. The author is painfully aware of this. As a consequence there will be times when we make statements that may seem obvious or elementary to the group theorist but may seem quite surprising to the geometer.

We begin with the usual definitions to establish notation and terminology.

Definition 1 A surface $\mathcal{S}$ is a **Beauville surface of unmixed type** if

- the surface $\mathcal{S}$ is isogenous to a higher product, that is, $\mathcal{S} \cong (\mathcal{C}_1 \times \mathcal{C}_2)/G$ where $\mathcal{C}_1$ and $\mathcal{C}_2$ are algebraic curves of genus at least 2 and G is a finite group acting faithfully on $\mathcal{C}_1$ and $\mathcal{C}_2$ by holomorphic transformations in such a way that it acts freely on the product $\mathcal{C}_1 \times \mathcal{C}_2$ and
- each $\mathcal{C}_i/G$ is isomorphic to the projective line $\mathbb{P}_1(\mathbb{C})$, and the covering map $\mathcal{C}_i \to \mathcal{C}_i/G$ is ramified over three points.

What makes these surfaces so easy to work with is the fact that the definition above can be translated into purely group theoretic terms—the following definition imposes equivalent conditions on the group G.

B. Fairbairn (✉)
Department of Economics, Mathematics and Statistics, Birkbeck,
University of London, Malet Street, London WC1E 7HX, UK
e-mail: b.fairbairn@bbk.ac.uk

I. Bauer et al. (eds.), *Beauville Surfaces and Groups*, Springer Proceedings
in Mathematics & Statistics 123, DOI 10.1007/978-3-319-13862-6_4

Definition 2 Let G be a finite group. Let $x, y \in G$ and let

$$\Sigma(x, y) := \bigcup_{i=1}^{|G|} \bigcup_{g \in G} \{(x^i)^g, (y^i)^g, ((xy)^i)^g\}.$$

An **unmixed Beauville structure** for G is a pair of generating sets of elements $\{\{x_1, y_1\}, \{x_2, y_2\}\} \subset G \times G$ such that $\langle x_1, y_1 \rangle = \langle x_2, y_2 \rangle = G$ and

$$\Sigma(x_1, y_1) \cap \Sigma(x_2, y_2) = \{e\} \qquad (\dagger)$$

where e denotes the identity element of G. If G has a Beauville structure, then we say that G is a **Beauville group**. Furthermore we say that the structure has **type** $((o(x_1), o(y_1), o(x_1y_1)), (o(x_2), o(y_2), o(x_2y_2)))$.

In the author's experience, upon seeing the above definition, group theorists often retort "why record just the orders of the elements and not precisely which classes the elements belong to?" Determining precisely which class an element belongs to is much harder than determining its order. Furthermore, in practice, when ensuring that a set of elements satisfies condition (†) the easiest way often is to show that $o(x_1)o(y_1)o(x_1y_1)$ is coprime to $o(x_2)o(y_2)o(x_2y_2)$. This simple observation has been used to great effect by several authors—see [18, 20, 22, 25] among others. Furthermore, the type alone encodes substantial amounts of geometric information: the Riemann-Hurwitz formula

$$g(\mathcal{C}_i) = 1 + \frac{|G|}{2}\left(1 - \frac{1}{o(x_i)} - \frac{1}{o(y_i)} - \frac{1}{o(x_iy_i)}\right)$$

tells us the genus of each of the curves used to define the surface $\mathcal{S}$. Indeed, whilst some groups have generating pairs that by the above formula define surfaces with the property that $g(\mathcal{C}) \leq 1$, condition (†) ensures that for each i we have that $g(\mathcal{C}_i) \geq 2$. Furthermore, a theorem of Zeuthen-Segre also gives us the Euler number of the surface $\mathcal{S}$ since

$$e(\mathcal{S}) = 4\frac{(g(\mathcal{C}_1) - 1)(g(\mathcal{C}_2) - 1)}{|G|},$$

which in turn gives us the holomorphic Euler-Poincaré characteristic of $\mathcal{S}$ from the relation $4\chi(\mathcal{S}) = e(\mathcal{S})$—see [14, Theorem 3.4].

In light of the above, we make the following non-standard definition which will be of use in what follows.

Definition 3 We say that a Beauville structure $\{\{x_1, y_1\}, \{x_2, y_2\}\}$ is **coprime** if $o(x_1)o(y_1)o(x_1y_1)$ and $o(x_2)o(y_2)o(x_2y_2)$ are coprime.

Given any complex surface $\mathcal{S}$ it is natural to consider the complex conjugate surface $\overline{\mathcal{S}}$. In particular, it is natural to ask whether the surfaces are biholomorphic.

Definition 4 Let $\mathcal{S}$ be a complex surface. We say that $\mathcal{S}$ is **real** if there exists a biholomorphism $\sigma : \mathcal{S} \to \overline{\mathcal{S}}$ such that σ^2 is the identity map.

As is often the case with Beauville surfaces, the above geometric condition can be translated into purely group theoretic terms.

Definition 5 Let G be a Beauville group. We say that G is **strongly real** if there exists a Beauville structure $X = \{\{x_1, y_1\}, \{x_2, y_2\}\}$ such that there exists an automorphism $\phi \in \mathrm{Aut}(G)$ and elements $g_i \in G$ for $i = 1, 2$ such that

$$g_i\phi(x_i)g_i^{-1} = x_i^{-1} \text{ and } g_i\phi(y_i)g_i^{-1} = y_i^{-1}$$

for $i = 1, 2$. In this case we also say that the Beauville structure X is a strongly real Beauville structure.

In practice we can always replace one generating pair by some conjugate of it and so we can take $g_1 = g_2 = e$ and often this is what is done in practice.

In [6] Bauer, Catanese and Grunewald show that a Beauville surface is real if, and only if, the corresponding Beauville group and structure are strongly real.

Example 6 In [15] Catanese classified the abelian Beauville groups by proving the following.

Theorem 7 *If G is an abelain group, then G is a Beauville group if, and only if, $G \cong \mathbb{Z}_n \times \mathbb{Z}_n$ where gcd(n,6)=1 and $\mathbb{Z}_n$ denotes the cyclic group of order $n > 1$.*

This theorem immediately gives us the following.

Corollary 8 *Every abelian Beauville group is a strongly real Beauville group making any Beauville structure for these groups strongly real.*

Proof If H is an abelian group, then the map $H \to H, x \mapsto -x$ is an automorphism. □

More recent (and group theoretic) motivation comes from the following. The absolute Galois group $\mathrm{Gal}(\overline{\mathbb{Q}}/\mathbb{Q})$ is very poorly understood. Indeed, The Inverse Galois Problem—arguably the hardest open problem in algebra today—forms just one small part of efforts to understand $\mathrm{Gal}(\overline{\mathbb{Q}}/\mathbb{Q})$ (it amounts to showing that every finite group arises as the quotient of $\mathrm{Gal}(\overline{\mathbb{Q}}/\mathbb{Q})$ by a topologically closed normal subgroup). When confronted with the task of understanding a group it is natural to consider an action of the group on some set. The group $\mathrm{Gal}(\overline{\mathbb{Q}}/\mathbb{Q})$ acts on the set of Beauville surfaces thanks to Grothendieck's theory of Dessins d'enfants ("children's drawings"). See [27, Sect. 11] for a more detailed discussion of this and related matters.

Henceforth we shall use the standard ATLAS notation for group theoretic concepts (aside from occasional deviations to minimise confusion with geometric concepts) as described in some detail in the introductory sections of [16]. In particular, given two groups A and B we use the following notation.

- We write $A \times B$ for the direct product of A and B, that is, the group whose members are ordered pairs (a, b) with $a \in A$ and $b \in B$ such that for $(a, b), (a', b') \in A \times B$ we have the multiplication $(a, b)(a', b') = (aa', bb')$. Given a positive integer k we write A^k for the direct product of k copies of A.
- We write $A.B$ for the extension of A by B, that is, a group with a normal subgroup isomorphic to A whose quotient is B (such groups are not necessarily direct products—for instance $\mathrm{SL}_2(5) = 2.\mathrm{PSL}_2(5)$).
- We write $A : B$ for a semi-direct product of A and B, also known as a split extension A and B, that is, there is a homomorphism $\phi\colon B \to Aut(A)$ with elements of this group being ordered pairs (b, a) with $a \in A$ and $b \in B$ such that for $(b, a), (b', a') \in A : B$ we have the multiplication $(a, b)(a', b') = (bb', a^{\phi(b')}a')$.
- We write $A \wr B$ for the wreath product of A and B, that is, if B is a permutation group on n points then we have the split extension $A^n : B$ with B acting in a way that permutes the n copies of A.

In several places we shall refer to 'straightforward computations' or calculations that readers can easily reproduce for themselves. On these occasions either of Magma [10] or GAP [32] can easily be used to do this.

In Sect. 2 we will discuss the finite simple groups and in particular a conjecture of Bauer, Catanese and Grunewald concerning which of these groups are strongly real Beauville groups. In Sect. 3 our attention turns to the characteristically simple groups and in particular the recent work of Jones which we push further in the cases of the symmetric and alternating groups in Sect. 4. We go on in Sect. 5 to discuss which of the almost simple groups are strongly real Beauville groups. Finally, in Sect. 6 we briefly discuss nilpotent groups and p-groups.

2 The Finite Simple Groups

Naturally, a necessary condition for being a strongly real Beauville group is being a Beauville group. Furthermore, a necessary condition for being a Beauville group is being 2-generated: we say that a group G is 2-generated if there exist two elements $x, y \in G$ such that $\langle x, y\rangle = G$. It is an easy exercise for the reader to show that the alternating groups A_n for $n \geq 3$ are 2-generated. In [31] Steinberg proved that the simple groups of Lie type are 2-generated and in [1] Aschbacher and Guralnick showed that the sporadic simple groups are 2-generated. We thus have that all of the non-abelian finite simple groups are 2-generated making them natural candidates for Beauville groups. This lead Bauer, Catanese and Grunewald to conjecture that aside from A_5, which is easily seen to not be a Beauville group, every non-abelian finite simple group is a Beauville group—see [6, Conjecture1] and [7, Conjecture 7.17]. This suspicion was later proved correct [19, 20, 24, 25], indeed the full theorem proved by the author, Magaard and Parker in [20] is actually a more general statement about quasisimple groups (recall that a group G is quasisimple if it is generated by its commutators and the quotient by its center $G/Z(G)$ is a simple group).

Having found that almost all of the non-abelian finite simple groups are Beauville groups, it is natural to ask which of the non-abelian finite simple groups are strongly real Beauville groups. In [6, Sect. 5.4] Bauer et al. wrote

> There are 18 finite simple nonabelian groups of order $\leq 15{,}000$. By computer calculations we have found strongly [real] Beauville structures on all of them with the exceptions of A_5, $PSL_2(7)$, A_6, A_7, $PSL_3(3)$, $U_3(3)$ and the Mathieu group M_{11}.

On the basis of these computations they conjectured that all but finitely many non-abelian finite simple groups are strongly real Beauville groups. Several authors have worked on this and many special cases are now known to be true.

- In [21] Fuertes and González-Diez showed that the alternating groups A_n ($n \geq 7$) and the symmetric groups S_n ($n \geq 5$) are strongly real Beauville groups by explicitly writing down permutations for their generators and the automorphisms and applying some of the classical theory of permutation groups to show that their elements had the properties they claimed. Subsequently the alternating group A_6 was also shown to be a strongly real Beauville group.
- In [22] Fuertes and Jones prove that the simple groups $PSL_2(q)$ for prime powers $q > 5$ and the quasisimple groups $SL_2(q)$ for prime powers $q > 5$ are strongly real Beauville groups. As with the alternating and symmetric groups, these results are proved by writing down explicit generators, this time combined with a celebrated theorem usually (but historically inaccurately) attributed to Dickson for the maximal subgroups of $PSL_2(q)$. General lemmas for lifting Beauville structures from a group to its covering groups are also used.
- Settling the case of the sporadic simple groups makes no impact on the above conjecture, there being only 26 of them. Nonetheless, for reasons we shall return to below, in [18] the author determined which of the sporadic simple groups are strongly real Beauville groups, including the '27th sporadic simple group', the Tits group ${}^2F_4(2)'$. Of all the sporadic simple groups only the Mathieu groups M_{11} and M_{23} are not strongly real. For all of the other sporadic groups smaller than the Baby Monster group $\mathbb{B}$ explicit words in the 'standard generators' [34] for a strongly real Beauville structure are given. (For those unfamiliar with standard generators, we will describe these in Sect. 5.) For the Baby Monster group $\mathbb{B}$ and Monster group $\mathbb{M}$ character theoretic methods are used.

As we can see from the above bullet points, several of the groups that Bauer, Catanese and Grunewald could not find strongly real Beauville structures for do indeed have strongly real Beauville structures. In particular, we note that the group $PSL_2(9) \cong A_6$ is in fact strongly real.

Using the results mentioned above, combined with unpublished calculations, the author has pushed Bauer, Catanese and Grunewald's original computations to every non-abelian finite simple group of order at most 100,000,000 and, as we noted above, several much larger ones in [18]. Many of the smaller groups seemed to require the use of outer automorphisms to make their Beauville structures strongly real, which explains much of the above difficulty in finding strongly real Beauville structures in certain groups. Slightly larger groups had enough conjugacy classes for inner automorphisms to be used instead. Consequently, it seems that 'small' non-abelian finite simple groups fail to be strongly real if they have too few conjugacy classes (as is the case with A_5 and as we would intuitively expect) or if they have no outer automorphisms—a phenomenon that is extremely rare. We are thus lead to the following somewhat stronger conjecture.

Conjecture 1 *All non-abelian finite simple groups apart from A_5, M_{11} and M_{23} are strongly real Beauville groups.*

To add further weight to this conjecture we verify this conjecture for the Suzuki groups ${}^2B_2(2^{2n+1})$. Let $q = 2^{2n+1}$.

Theorem 9 *Each of the groups ${}^2B_2(q)$ has a strongly real Beauville structure of type $((q-1, q-1, q-1), (d_1, d_2, 2))$ where d_1 and d_2 are odd and coprime to $q-1$.*

Throughout the following we shall be using the natural 4-dimensional representation of ${}^2B_2(q)$ over the field of order q as described in some detail in [35, Sect. 4.2]. To prove Thoerem 9 we will use knowledge of the maximal subgroups of the Suzuki groups. The following lemma was proved by Suzuki—see [35, Theorem 4.1]. Here we write E_q for the elementary abelian group of order q. Furthermore, by 'subfield subgroup' we mean either the subgroup ${}^2B_2(q_0)$ consisting of matrices whose entries come from a subfield of the field $\mathbb{F}_q$ of order $\mathbb{F}_{q_0}$ where $q_0 > 1$ divides q or one of its conjugates, those appearing in Lemma 10(v) being precisely the maximal subfield subgroups.

Lemma 10 *If $n > 1$, then the maximal subgroups of ${}^2B_2(q)$ are (up to conjugacy).*

(i) $E_q.E_q : \mathbb{Z}_{q-1}$, *the subgroup of lower triangular matrices*
(ii) $D_{2(q-1)}$
(iii) $\mathbb{Z}_{q+\sqrt{2q}+1} : 4$
(iv) $\mathbb{Z}_{q-\sqrt{2q}+1} : 4$
(v) ${}^2B_2(q_0)$ *where $q = q_0^r$, r is prime and $q_0 > 2$.*

From the above the following can easily be deduced.

Lemma 11 *(a) If $x, y \in {}^2B_2(q)$ are two elements with the property that*

$$o(x) = o(y) = o(xy) = q - 1,$$

then $\langle x, y\rangle = \mathbb{Z}_{q-1}$, $E_q.E_q : \mathbb{Z}_{q-1}$ or ${}^2B_2(q)$.
(b) If $x, y \in {}^2B_2(q)$ are two elements such that $o(x)$ and $o(y)$ are have orders dividing $q \pm \sqrt{2q} + 1$ and $o(xy) = 2$, then $\langle x, y\rangle = {}^2B_2(q)$ or a subfield subgroup.

Proof of Theorem 9 For our first generating pair we consider the following elements of ${}^2B_2(q)$ each of which are easily checked to have order 2 by direct calculation.

$$t_1 := \begin{pmatrix} 0 & 0 & 0 & 1 \\ 0 & 0 & 1 & 0 \\ 0 & 1 & 0 & 0 \\ 1 & 0 & 0 & 0 \end{pmatrix} \qquad t_2 := \begin{pmatrix} 0 & 0 & 0 & \beta^{-1} \\ 0 & 0 & \beta^{-2^{n+1}+1} & 0 \\ 0 & \beta^{2^{n+1}-1} & 0 & 0 \\ \beta & 0 & 0 & 0 \end{pmatrix}$$

$$t_3 := \begin{pmatrix} 1 & 0 & 0 & 0 \\ 0 & 1 & 0 & 0 \\ \alpha^{2^{n+1}} & 0 & 1 & 0 \\ \alpha^2 & \alpha^{2^{n+1}} & 0 & 1 \end{pmatrix}$$

where α and β are generators of the multiplicative group $\mathbb{F}_q^{\times}$. The element $x_1 = t_1t_2$ has order $q-1$. The characteristic polynomial of $y_1 = t_1t_3$ is

$$p_1(\lambda) = \lambda^4 + \alpha^2\lambda^3 + \alpha^{2^{n+2}}\lambda^2 + \alpha^2\lambda + 1$$

and if we set $\gamma := \beta + \beta^{-1} + \beta^{2^{n+1}-1} + \beta^{1-2^{n+1}}$ then the characteristic polynomial of t_1t_2 is

$$p_2(\lambda) = \lambda^4 + \gamma\lambda^3 + (\beta^{2^{n+1}} + \beta^{-2^{n+1}+2} + \beta^{2^{n+1}-2} + \beta^{-2^{n+1}})\lambda^2 + \gamma\lambda + 1$$

Comparing p_1 with p_2 we see that the two polynomials are equal if we have

$$\gamma = \alpha^2 \text{ and}$$

$$\beta^{2^{n+1}} + \beta^{-2^{n+1}+2} + \beta^{2^{n+1}-2} + \beta^{-2^{n+1}} = \alpha^{2^{n+2}} = (\alpha^2)^{2^{n+1}} = \gamma^{2^{n+1}}.$$

Since $a \mapsto a^2$ is an automorphism of our underlying field we see that the first of these equalities immediately implies the second if

$$(\beta + \beta^{-1} + \beta^{2^{n+1}-1} + \beta^{1-2^{n+1}})^{2^{n+1}} = \beta^{2^{n+1}} + \beta^{-2^{n+1}+2} + \beta^{2^{n+1}-2} + \beta^{-2^{n+1}}.$$

Since $\beta^{(2^{n+1}-1)2^{n+1}} = (\beta^{2^{2n+1}})^2\beta^{-2^{n+1}} = \beta^{2-2^{n+1}}$ we can choose α and β to satisfy the above condition, so in particular we have that t_1t_2 and t_1t_3 have the same characteristic polynomial and thus both have order $q-1$. Furthermore, these are both inverted by conjugation by t_1 since t_1, t_2 and t_3 all have order 2. Similarly we find that $t_1t_2t_1t_3$ has characteristic polynomial of the correct form to have order $q-1$. From Lemma 11(a) we see that these elements generate the group since x_1 and y_1 are not both contained in a cyclic subgroup (one of them is diagonal) and by direct calculation no one-dimensional subspace in the natural module is preserved by them so there is no proper subgroup containing each of these elements.

For the second triple we consider the matrices

$$x_2 := \begin{pmatrix} 0 & 0 & 0 & 1 \\ 0 & 0 & 1 & 0 \\ 0 & 1 & 0 & \delta^4 \\ 1 & 0 & \delta^4 & \delta^2 \end{pmatrix} \qquad y_2 := \begin{pmatrix} \epsilon^2 & \epsilon^4 & 0 & 1 \\ \epsilon^4 & 0 & 1 & 0 \\ 0 & 1 & 0 & 0 \\ 1 & 0 & 0 & 0 \end{pmatrix}$$

where $\delta, \epsilon \in \mathbb{F}_q$ are chosen so that $\delta \neq \epsilon$ and these do not have the correct form for these elements to have order $q-1$. Direct calculation shows that these elements do not have orders 2 or 4 and that $o(x_2 y_2) = 2$. These elements must, therefore, have orders that divide $q \pm \sqrt{2q}+1$. Furthermore their traces are ϵ^2 and δ^2 which can be chosen to be in no proper subfield since $x \mapsto x^2$ is an automorphism of the field $\mathbb{F}_q$. These elements must therefore generate the group by Lemma 11(b). Further direct calculation shows that $x_2^{t_1} = x_2^{-1}$ and $y_2^{t_1} = y_2^{-1}$. □

3 Characteristically Simple Groups

Another class of finite groups that has recently been studied from the viewpoint of Beauville constructions, and seems like fertile ground for providing further examples of strongly real Beauville groups, are the characteristically simple groups that we define as follows (the definition commonly given is somewhat different from that below but in the case finite groups it is equivalent to this).

Definition 12 A finite group G is said to be **characteristically simple** if G is isomorphic to some direct product H^k where H is a finite simple groups.

For example, as we saw in Theorem 7, if $p > 3$ is prime then the abelian Beauville groups isomorphic to $\mathbb{Z}_p \times \mathbb{Z}_p$ are characteristically simple.

Characteristically simple Beauville groups have recently been investigated by Jones in [28, 29] where the following conjecture is discussed.

Conjecture 2 *Let G be a finite non-abelian characteristically simple group. Then G is a Beauville group if and only if it is a* 2*-generator group not isomorphic to* A_5.

In particular, the main results of [28, 29] verify this conjecture in the cases where H is any of the alternating groups; the linear groups $PSL_2(q)$ and $PSL_3(q)$; the unitary groups $PSU_3(q)$; the Suzuki groups ${}^2B_2(2^{2n+1})$; the small Ree groups ${}^2G_2(3^{2n+1})$ and the sporadic simple groups.

For large values of k, the group H^k will not be 2-generated despite the fact that H will be as discussed in Sect. 2. The values of k for which H^k is 2-generated can be surprisingly large. For example, a special case of the results alluded to in the previous paragraph is the somewhat amusing fact that

$$A_5 \times A_5 \times A_5 \times A_5 \times A_5 \times A_5 \times A_5 \times A_5 \times A_5 \times A_5 \times$$

$$A_5 \times A_5 \times A_5 \times A_5 \times A_5 \times A_5 \times A_5 \times A_5 \times A_5$$

is a Beauville group, despite the fact that A_5 itself is not a Beauville group.

In general, the full automorphism group of H^k will be the wreath product $\mathrm{Aut}(H) \wr S_k$ where S_k is the kth symmetric group acting on the product by permuting the groups H. This bounteous supply of automorphisms makes it likely that characteristically simple Beaville groups are in general strongly real.

Question 1 *Which characteristically simple Beauville groups are strongly real?*

As a more specific conjecture on these matters we assert the following.

Conjecture 3 *If H is a finite simple group of order greater than 3, then the group $H \times H$ is a strongly real Beauville group.*

Note that Corollary 8 tells us that this is true for all abelian characteristically simple Beauville groups. For the nonabelian characteristically simple Beauville groups this conjecture seems rather distant given that, at the time of writing, we have neither a solution to Conjecture 1 nor do we know if $H \times H$ for a simple group H is even a Beauville group, let alone a strongly real one. Anyone tempted to extend the above conjecture to the products of a larger number of copies of simple groups should see the remarks following Lemma 15, although some hope is provided by the results proven in Sect. 4.

Theorem 13 *Let G be a strongly real Beauville group with coprime strongly real Beauville structure $\{\{x_1, y_1\}, \{x_2, y_2\}\}$. Furthermore, suppose that there exists an automorphism $\phi \in \mathrm{Aut}(G)$ such that*

$$\phi(x_1) = x_1^{-1},\ \phi(y_1) = y_1^{-1},\ \phi(x_2) = x_2^{-1} \text{ and } \phi(y_2) = y_2^{-1}.$$

Then the group $G \times G$ is a strongly real Beauville group.

Proof Consider the following elements of $G \times G$

$$g_1 = (x_1, x_2),\ h_1 = (y_1, y_2),\ g_2 = (x_2, x_1) \text{ and } h_2 = (y_2, y_1).$$

The pair $\{g_1, h_1\}$ generate the whole of $G \times G$ since the elements $g_1^{o(x_2)}$ and $h_1^{o(y_2)}$ generate the first factor whilst the elements $g_1^{o(x_1)}$ and $h_1^{o(y_1)}$ generate the second factor thanks to our hypothesis that $o(x_1)o(x_2)o(x_1y_1)$ is coprime to $o(x_2)o(y_2)o(x_2y_2)$. Similarly $\langle g_2, h_2\rangle = G \times G$.

We define an automorphism $\psi \in \mathrm{Aut}(G \times G)$ such that for every $(g, h) \in G \times G$

$$\psi(g, h) = (\phi(g), \phi(h)).$$

This automorphism clearly makes the above Beauville structure for $G \times G$ a strongly real Beauville structure. □

Corollary 14 *Conjecture 3 is true for each of the following groups.*

(a) The alternating groups A_n for $n \geq 6$;
(b) The linear groups $PSL_2(q)$ for prime powers $q > 5$;
(c) The Suzuki groups ${}^2B_2(2^{2n+1})$;
(d) All simple groups of order at most 100,000,000;
(e) The sporadic simple groups.

Proof For part (a) the results proved in Sect. 4 provide a strongly real Beauville structure for $A_n \times A_n$ for sufficiently large n, the smaller cases being straightforward calculations that are easily performed separately.

For part (b) we note that the strongly real Beauville structures constructed by Fuertes and Jones in [22] for the groups $PSL_2(q)$ satisfy the hypotheses of Theorem 13.

For part (c) we note that the strongly real Beauville structures for ${}^2B_2(2^{2n+1})$ we constructed in Theorem 9 are coprime and satisfy the hypotheses of Theorem 13.

For part (d) we note that the author's computations alluded to in Sect. 2 were performed in such a way that the hypotheses of Theorem 13 are satisfied.

Finally for part (e) we observe that for all the sporadic groups, apart from the Mathieu groups M_{11} and M_{23}, the structures given by the author in [18] satisfy the hypotheses of Theorem 13. The groups M_{11} and M_{23} are dealt with separately in Lemma 15. □

We remark that the strongly real Beauville structures for the quasisimple groups $SL_2(q)$ where $q > 5$ constructed by Fuertes and Jones in [22] also satisfy the hypotheses of Theorem 13 and so the groups $SL_2(q) \times SL_2(q)$ are also strongly real.

Unfortunately, Theorem 13 cannot be applied to the strongly real Beauville structures constructed by Fuertes and González-Diez in [21] for the symmetric and alternating groups. This is because the types of the Beauville structures in [21] fail to satisfy the coprime hypothesis since their structures use several elements of order 2. We return to this point in Sect. 4.

Comparing the statement of Conjecture 1 with the statement of Conjecture 3 the reader should immediately be asking "what about the alternating group A_5 and the Matheiu groups M_{11} and M_{23}?" This concern is immediately addressed by the following further piece of evidence for Conjecture 3.

Lemma 15 *The groups $A_5 \times A_5$, $M_{11} \times M_{11}$ and $M_{23} \times M_{23}$ are strongly real Beauville groups.*

Proof This is a straightforward computational calculation. Consider the following permutations.

$$x_1 := (1,2,3,4,5)(6,7,8,9,10) \qquad y_1 := (2,3,4)(7,10)(6,9),$$
$$x_2 := (1,4,3,2,5)(7,8,9) \qquad y_2 := (1,2)(4,5)(6,9,8,7,10) \text{ and}$$
$$a := (1,5)(2,4)(6,10)(7,9)$$

The set $\{\{x_1, y_1\}, \{x_2, y_2\}\}$ gives a Beauville structure for the group $A_5 \times A_5$ of type ((5,6,5),(15,10,15)) acting intransitively on $5+5$ points as a subgroup of the symmetric group S_{10}. The automorphism α defined by conjugation by a has the property that

$$\alpha(x_1) = x_1^{-1}, \alpha(y_1) = y_1^{-1}, \alpha(x_2) = x_2^{-1} \text{ and } \alpha(y_2) = y_2^{-1}$$

from which we have that this Beauville structure is strongly real.

Next, the group $M_{11} \times M_{11}$. Consider the following permutations.

$$x_1 := (1,2,3,4,5,6,7,8,9,10,11)(12,13,14,15,16,17,18,19,20,21,22)$$
$$y_1 := (1,6,10,5,2,7,4,9,11,8,3)(12,14,19,16,21,18,13,17,22,20,15)$$
$$x_2 := (1,3,9,11,10,7,2,4)(5,8)(12,14,20,22,19,21,16,13)(15,18)$$
$$y_2 := (2,6,9,4,8,3,7,5)(10,11)(12,13)(14,17,21,18,16,20,15,19)$$
$$a := (1,22)(2,21)(3,20)(4,19)(5,18)(6,17)(7,16)(8,15)(9,14)(10,13)(11,12)$$

The set $\{\{x_1, y_1\}, \{x_2, y_2\}\}$ gives a Beauville structure for the group $M_{11} \times M_{11}$ of type ((11,11,11),(8,8,8)) acting intransitively on 11+11 points as a subgroup of the symmetric group S_{22}. The automorphism α defined by conjugation by a has the property that

$$\alpha(x_1) = x_1^{-1}, \alpha(y_1) = y_1^{-1}, \alpha(x_2) = x_2^{-1} \text{ and } \alpha(y_2) = y_2^{-1}$$

from which we have that this Beauville structure is strongly real.

Finally for $M_{23} \times M_{23}$ we similarly have that the permutations

$$x_1 := (1,22,5,17,6,10,18,16,19,8,9,15,13,14,21,4,3,7,23,20,2,12,11)$$
$$(24,40,44,43,26,33,34,32,38,39,28,31,29,37,41,30,42,25,46,36,35,45,27)$$
$$y_1 := (1,16,3,14,7,15,18,22,21,8,20,10,4,17,19,13,5,6,23,9,2,12,11)$$
$$(24,41,42,34,28,30,43,37,27,39,26,25,29,32,40,33,44,31,46,36,35,45,38)$$
$$x_2 := (1,3,19,7,18,4,11,21,16,14,6)(2,23,9,17,15,20,22,10,13,12,8)$$
$$(24,45,39,35,34,37,25,27,32,30,38)(26,36,43,29,40,28,44,46,41,33,31)$$

$$\begin{aligned} y_2 := &(1,6,22,9,16,17,5,19,11,18,2)(3,14,13,23,4,12,15,10,7,21,8)\\ &(24,34,33,44,39,26,40,37,32,35,43)(25,41,46,45,29,36,28,42,30,\\ &\ 31,38)\\ a\ := &(1,46)(2,45)(3,44)(4,43)(5,42)(6,41)(7,40)(8,39)(9,38)(10,37)\\ &(11,36)(12,35)(13,34)(14,33)(15,32)(16,31)(17,30)(18,29)(19,28)\\ &(20,27)(21,26)(22,25)(23,24) \end{aligned}$$

define a strongly real Beauville structure of type ((23,23,23),(11,11,11)) for the group $M_{23} \times M_{23}$ acting intransitively on $23 + 23$ points as a subgroup of the symmetric group S_{46}. □

We remark that in the examples of the above lemma, the automorphisms used are outer automorphisms that interchange the two factors. The lack of automorphisms that stop both M_{11} and M_{23} being strongly real will therefore also stop the groups $M_{11} \times M_{11} \times M_{11}$ and $M_{23} \times M_{23} \times M_{23}$ being strongly real Beauville groups. Furthermore it is easy to see that the permutations given in the proof of Lemma 15 can be adapted to construct a strongly real Beauville structure of type ((88,88,88),(88,88,88)) for the group $M_{11} \times M_{11} \times M_{11} \times M_{11}$ and to construct a strongly real Beauville structure of type ((253,253,253),(253,253,253)) for the group $M_{23} \times M_{23} \times M_{23} \times M_{23}$. Similarly $A_5 \times A_5 \times A_5$ is not a strongly real Beauville group. It follows that any extension of Conjecture 3 to products of a larger number of copies of simple groups will necessarily have a much more complicated statement. It is likely that similar remarks apply to M_{11}^{2k+1}, M_{11}^{2k}, M_{23}^{2k+1}, M_{23}^{2k}, A_5^{2k+1} and A_5^{2k} for small values of k.

In light of the above it is natural to ask the following.

Question 2 *Let H be a finite simple group, $n \in \mathbb{Z}^+$ and $G = H^n$. When are inner automorphism sufficient to make G strongly real and when are outer automorphisms interchanging the factors required? Moreover does this have any geometric significance for the corresponding surfaces?*

4 The Symmetric and Alternating Groups

In the last section we discussed characteristically simple groups of the form $H \times H$ for some simple group H. In this section we prove slightly stronger results in the case of the alternating groups and a related result for the symmetric groups. In each of the below results conjugacy of elements is taken care of by the well known fact that two elements of the symmetric group are conjugate if, and only if, they have the same cycle type. We will use the following recent results of Jones.

Lemma 16 *Let $H \leq S_n$.*

(*a*) *If H is primitive and contains a cycle that fixes at least three points then $H \geq A_n$.*
(*b*) *If H is transitive and contains an m-cycle where $m > n/2$ and m is coprime to n then H is primitive.*

(c) *If H is primitive and contains a cycle fixing two points then either $H \geq A_n$ or $PGL_2(q) \leq H \leq P\Gamma L_2(q)$ with $n = q + 1$ for some prime power q.*

Proof See [30] and [28, Sect. 6]. □

Before proving our main results we recall some facts about generating pairs in simple and characteristically simple groups. Let H be a finite simple group. In [26] Philip Hall showed that the largest k such that the characteristically simple group H^k is 2-generated is equal to the number of orbits of $Aut(H)$ on generating pairs of H. (He proved similar results for more general n-tuples but we will not be needing these results here.) To show that H^k for some k is generated by a pair of elements it is sufficient to show that each of the 'coordinates' of these elements (that is, the parts of each permutation that correspond to each of the factors) are inequivalent under the action of $Aut(H)$. For $n \neq 1, 2, 6$ we have that $Aut(A_n) = S_n$ and in the case $n = 6$ the symmetric group S_6 is an index 2 subgroup of $Aut(A_6) \cong P\Gamma L_2(9)$.

Lemma 17 *Let $n \geq 11$ be odd and let $k \leq (n-6)/2$ be positive integers. Then A_n^k is a strongly real Beauville group.*

Proof Since A_n is simple for every $n > 5$ we have that, by the remarks of the previous paragraph, it is sufficient to find k pairs of generating pairs $T_{ij} = \{x_{ij}, y_{ij}\}$ $i = 1, 2$, $j = 1, \ldots, k$ such that for a fixed i no T_{ij} is an image of $T_{ij'}$ under the actions of automorphisms of A_n for distinct j and j'.

For our first pairs we set

$$x_{1j} = (1, \ldots, 2j+3) \text{ and } y_{1j} = (2j+3, \ldots, n)$$

for $1 \leq j < (n-6)/4$. These are cycles of odd length and are thus even permutations. Their product is an n-cycle. It is easy to check that $\langle x_{1j}, y_{1j} \rangle$ is primitive and thus equal to A_n since the group contains a cycle with at least three fixed points by Lemma 16(a). These elements are both inverted by the automorphism defined by conjugation by

$$t = (1, 2j)(2, 2j-1) \cdots (j, j+1)(2j+2, n)(2j+3, n-1) \cdots ((2j+n+1)/2, (2j+n+3)/2)$$

which has only one fixed point, namely $2j + 1$.

For our second pairs we consider the permutations

$$x_{2j} = (1, \ldots, n-2) \text{ and } y_{2j} = (j+1, j+2)(n-j-1, n-j-2) \times ((n-1)/2, n-1)((n+1)/2, n)$$

for $1 \leq j < (n-5)/2$. These are again both even permutations, their product this time being an $(n-2)$-cycle. To confirm that these elements generate the group we note that $\langle x_{2j}, y_{2j} \rangle$ is clearly transitive and so by Lemma 16(b) must be primitive. It now follows from Lemma 16(c) that $\langle x_{2j}, y_{2j} \rangle = A_n$ since $n > 9$ and the only elements of order 2 in $P\Gamma L_2(q)$ do not have the same cycle type as y_{2j}. (In the case

$n = 9$ these permutations generate $PSL_2(8)$.) These elements are both inverted by the automorphism defined by conjugation by

$$(2, n-2)(3, n-3)\cdots((n-1)/2, (n+1)/2)(n-1, n)$$

which has only one fixed point, namely 1, and thus differs from t solely by an inner automorphism. □

Lemma 18 *Let $n \geq 12$ be an even integer and let $k \leq (n-8)/4$. Then A_n^k is a strongly real Beauville group.*

Proof Again, we seek a collection of generating pairs that are not mapped to one another by automorphisms of A_n.

For our first pairs we consider the following elements.

$$x_{1j} = (1, \ldots, 2j+5) \text{ and } y_{1j} = (n, \ldots, 2j+5, 2j+4)$$

for $1 \leq j \leq (n-8)/4$. These are cycles of odd length and are thus even permutations. Their product is an $(n-1)$-cycle. It is easy to check that $\langle x_{1j}, y_{1j}\rangle$ is primitive and thus equal to A_n since the group contains a cycle with at least three fixed points by Lemma 16(a). These elements are both inverted by the automorphism defined by conjugation by

$$t = (2j+4, 2j+5)(1, 2j+3)\cdots(j+1, j+3)(2j+6, n)\cdots \\ ((n+2j+4)/2-1, (n+2j+4)/2+1)$$

which has precisely two fixed points namely $j+2$ and $(n+2j+4)/2$.

For our second pair of generators we consider the permutations

$$x_{2j} = (1, \ldots, n-2)(n-1, n) \text{ and}$$

$$y_{2j} = (n/2, (n-2)/2, n-1)(j+1, j, n, n-j-1, n-j-2)$$

for $1 \leq j < (n-2)/2$. These are both even permutations and their product is an $(n-3)$-cycle that fixes the points $j, n-j-2$ and $(n-2)/2$. The subgroup $\langle x_{2j}y_{2j}, y_{2j}^3\rangle$ fixes the point $(n-2)/2$ and is transitive on the remaining $n-1$ points. It follows that the group $\langle x_{2j}, y_{2j}\rangle$ is 2-transitive and is therefore primitive. Since the group $\langle x_{2j}, y_{2j}\rangle$ also contains the 3-cycle y_{2j}^5 (and the 5-cycle y_{2j}^3) it is equal to A_n by Lemma 16(a). These elements are both inverted by the automorphism defined by conjugation by

$$(1, n-2)(2, n-3)\cdots((n-2)/2, n/2)$$

which has precisely two fixed points, namely $n-1$ and n and thus differs from t solely in an inner automorphism. □

When considering the alternating groups it is natural to seek similar results for the symmetric groups. Unfortunately, here we are somewhat limited: if $k > 2$ then S_n^k is not 2-generated since its abelianisation is $\mathbb{Z}_2^k$ and this is not 2-generated. It follows that we can only find analogous results for $k \leq 2$ and since $k = 1$ comes straight from the work of Fuertes and González-Diez we are left only to consider the case $k = 2$. Note that in this case the outer automorphism of $S_n \times S_n$ that interchanges the two factors is useless since the only permutations that are inverted by this automorphism are even.

Lemma 19 *For $n \geq 5$ the group $S_n \times S_n$ is strongly real.*

Proof We will explicitly construct our Beauville structure in the case of n even and then describe the differences in the case of n odd.

For our first pair we consider the following elements.

$$x_1 = (1, \ldots, n-1)(2n-1, 2n) \text{ and } y_1 = (n, n-1)(n+1, \ldots, 2n-1)$$

The product of these permutations is a pair of n-cycles. It is easy to check that $\{x_1^2, y_1^{n-1}\}$ generates the first of the two factors whilst $\{x_1^{n-1}, y_1^2\}$ generates the second and so $\langle x_1, y_1 \rangle$ is the whole group. These elements are both inverted by the automorphism defined by conjugation by

$$t = (1, n-2)(2, n-3) \cdots (n/2, n/2-1)(n+1, 2n-2) \cdots (3n/2, 3n/2-1)$$

which has precisely four fixed points, namely $n-1$, n, $2n-1$ and $2n$.

For our second pair of generators we consider the permutations

$$x_2 = (1, 2, 3)(4, \ldots, n)(n+4, n+3, n+2, n+1)$$

and

$$y_2 = (4, 3, 2, 1)(n+1, n+2, n+3)(n+4, \ldots, 2n)$$

The product of these permutations is a pair of $n-2$ cycles. It is easy to check that $\{x_1^4, y_1^{3(n-3)}\}$ generates the first factor (this group is easily seen to be 2-transitive, and thus primitive, by considering conjugates of $y_1^{3(n-1)}$ under powers of x_1^4 and since this group contains a 4-cycle it contains the whole of S_n by [30, Corollary1.3]). Similarly $\{x_1^{3(n-1)}, y_1^4\}$ generates the second factor. These elements are both inverted by the automorphism defined by conjugation by

$$(1, 3)(5, n) \cdots (n/2+2, n/2+3)(n+1, n+3)(n+5, 2n) \cdots (3n/2+2, n/2+3)$$

which has precisely four fixed points, namely 2, 4, $n+2$ and $n+4$ and thus differs from t solely in an inner automorphism. (The case $n = 6$ requires a little care—using

$$x_2 = (1, 2, 3, 4)(10, 11, 12), y_2 = (4, 5, 6)(7, 8, 9, 10),$$

and the automorphism defined by $(1,3)(5,6)(7,9)(11,12)$ avoids being trapped inside copies of S_5 acting transitively on six points.)

If n is odd then for the first pair we need only replace the $(n-1)$-cycles with n-cycles to ensure that the elements have odd parity and for the automorphism instead use

$$t = (n, n-1)(n-1, 1)\cdots((n-1)/2-1, (n-1)/2+1)\cdots$$

$$(2n, 2n-1)(2n-2, n+1)\cdots((3n-1)/2-1, (3n-1)/2+1).$$

For the second generating pair we must now replace the 4-cycles with some longer p-cycle whose length is coprime to $3(n-3)$ (if 5 fails then $n \geq 2\times 5+3$ and we can try 7; if both 5 and 7 fail then $n \geq 2\times 5\times 7+3$ and we can try 11 etc.). The permutation of order 2 for the automorphism needs to be adjusted in the obvious manner (i.e. $(1,3)(4,p)(5,p-1)\cdots(p+1,n)(p+2,n-1)\cdots$). Again, the smallest case needs separate attention but it is easily checked that if

$$x_1 = (1,4)(2,5)(6,10)(7,8,9) \text{ and } y_1 = (1,5)(2,3,4)(6,9)(7,10)$$

and

$$x_2 = (1,2,3,4,5)(6,7,9,10) \text{ and } y_2 = (5,4,2,1)(10,9,8,7,6)$$

then $\{\{x_1, y_1\}, \{x_2, y_2\}\}$ is a strongly real Beauville structure whose elements are inverted by conjugation by the element $(1,5)(2,4)(6,10)(7,9)$. □

5 Almost Simple Groups

Let G be a group. Recall that we say G is almost simple if there exists a simple group S such that $S \leq G \leq \mathrm{Aut}(S)$. For example, any simple group is almost simple, as are the symmetric groups. Given our earlier remarks on the finite simple groups it is natural to ask the following.

Question 3 *Which of the almost simple groups are strongly real?*

This is particularly pertinent in light of Fuertes and González-Diez proof that the symmetric groups S_n for $n > 5$ are strongly real. Unfortunately, the general picture here is much more complicated with many almost simple groups not even being Beauville groups, let alone strongly real Beauville groups. Worse, infinitely many of the almost simple groups are not even 2-generated: the smallest example is $\mathrm{PSL}_4(9)$ whose outer automorphism group is $\mathbb{Z}_2 \times \mathrm{D}_8$ (and more generally, if p is an odd prime and r is an even positive integer then $Aut(PSL_4(p^r))$ is not 2-generated). We can at least add the following to the list.

Theorem 20 *The non-simple almost simple sporadic groups are strongly real Beauville groups.*

Before proceeding to the proof of Theorem 20 we make the following remarks for those who are unfamiliar with standard generators of finite groups (those who are familiar with them may skip to the proof in the next paragraph). Any given group will have many generating sets and in particular if $x, y \in G$ are such that $\langle x, y\rangle = G$ then $\langle x^g, y^g\rangle = G$ for any $g \in G$. To provide some standardisation to computational group theory, Wilson [34] introduced the notion of 'standard generators'. These are generators for a group that are unusually easy to find and are specified in terms of which conjugacy classes that they belong to (which can often be determined solely from their orders) and which classes some word(s) in these elements belong to. Representatives for many of the finite simple groups and various other groups closely related to them may be found in explicit permutations and/or matrices for many of their most useful representations on the web-based Atlas of Group Representations [33].

To construct our Beauville structures that prove Theorem 20 we proceed as follows. We first recall some well-known facts about the sporadic simple groups. If G is one of the 27 sporadic simple groups (including the Tits group ${}^2\mathrm{F}_4(2)'$) then the outer automorphism group of G has order at most 2 and that in all cases in which there exists a non-trivial outer automorphism $Aut(G)$ is a non-split extension, apart from the Tits group ${}^2\mathrm{F}_4(2)'$ and may thus be written $G:2$ in ATLAS notation (see Sect. 1). Let G be a simple group such that $\mathrm{Aut}(G) = G:2$. Let $t, t' \in G:2$ have order 2 such that one of the elements lies in G and the other lies in $G:2\setminus G$. For $i=1,2$ we define the elements $x_i = tt'^{g_i}$ for some $g_i \in G:2$. If for $i=1,2$ $u(i) \in C_G(t)$, then we can further define the elements $y_i = (x_i^{j(i)})^{u(i)}$ for some positive integers $j(i)$. Note that since $u(i)$ commutes with t the automorphism defined by conjugation by t inverts both x_i and y_i. Using knowledge of the subgroup structure of $G:2$ it is often possible to choose the elements $g_1, g_2, u(1)$ and $u(2)$ in such a way that $\langle x_1, y_1\rangle = \langle x_2, y_2\rangle = G:2$. Unfortunately, in the case of almost simple groups we must have that the orders of x_i and y_i all have even order and so verifying the conjugacy condition (†) of Definition 2 is more difficult than simply showing that $o(x_1)o(y_1)o(x_1y_1)$ is coprime to $o(x_2)o(y_2)o(x_2y_2)$. For some of the larger groups verifying that they generate the whole group can also be difficult. In these cases, generation is verified by finding words in our elements with the property that no proper subgroup can contain them (in many cases the maximal subgroups for these groups may be found in [16]). The words defining our Beauville structures are given in Table 1 and their types are given in Table 2.

We remark that this construction will not work in cases that are non-split extensions. This includes the almost simple 'sporadic' Tits group ${}^2\mathrm{F}_4(2)$. Straightforward computations verify that this group is not a strongly real Beauville group.

In light of the above we make the following tentative conjecture.

Conjecture 4 *A split extension of a simple group is a Beauville group if, and only if, it is a strongly real Beauville group.*

Table 1 Words in the standard generators providing strongly real Beauville structures for each of the non-simple almost simple sporadic groups

G	t_1	t_2	x_1	x_2
$M_{12}:2$	c	$(cd)^6$	t_1t_2	$t_1t_2^d$
$M_{22}:2$	$((cd)^2d)^5$	d^2	t_1t_2	$t_1t_2^c$
$J_2:2$	c	$(cd^2(cd)^2)^6$	t_1t_2	$t_1t_2^{d^4}$
HS: 2	c	$((cd)^3cd^2)^5$	t_1t_2	$t_1t_2^d$
$J_3:2$	c	$(cd)^{12}$	t_1t_2	$t_1t_2^d$
McL: 2	c	$((cd)^2(cd^2)^2(cd)^2d)^2$	t_1t_2	$t_1t_2^{(dcd)^2}$
He: 2	c	d^3	t_1t_2	$t_1t_2^{cd(cd^2)^2c}$
Suz: 2	c	$(cd)^{14}$	t_1t_2	$t_1t_2^{(dc)^2(d^2c)^2d^2}$
O'N: 2	c	d^2	t_1t_2	$t_1t_2^{cd}$
$Fi_{22}:2$	$(cd^4)^{10}$	$(cd^3)^{15}$	$t_1t_2^{dcd^6}$	$t_1t_2^{dcd}$
HN: 2	c	$(cd^3(cd)^2)^{12}$	$t_1t_2^{(dcd)^2d^2}$	$t_1t_2^{dcd^4cd^2}$
Fi_{24}	d^4	$((cd)^2d^3)^33$	$t_1t_2^{d^4c}$	$t_1t_2^{dcd^2c}$
G	u_1	u_2	$j(1)$	$j(2)$
$M_{12}:2$	$[c,(dc)^2d^2]^3$	$(dc)^2d[c,(dc)^2d]^2$	1	1
$M_{22}:2$	$cd^2cd[t_1,cd^2cd]^5$	$[t_1,c]^2$	5	9
$J_2:2$	$d[c,d]^3$	$d[c,d]^3$	1	9
HS: 2	$d[c,d]$	$d[c,d]$	1	1
$J_3:2$	$d[c,d]^4$	$d[c,d]^4$	21	1
McL: 2	$d^2[c,d^2]^7$	$dcd[c,dcd]^7$	1	7
He: 2	$d[c,d]^7$	$d[c,d]^7$	15	19
Suz: 2	$d[c,d]^3$	$d[c,d]^3$	9	3
O'N: 2	$[c,d]^5$	$[c,d]^5$	7	1
$Fi_{22}:2$	$[t_1,d^3]^3$	$[t_1,d]^3$	3	1
HN: 2	$d^2[cd^2]^10$	$d[c,d]^4$	1	39
Fi_{24}	$c[t_1,c]$	$c[t_1,c]$	7	25

We remark that some (unpublished) progress on this conjecture has been made by the author's PhD student, Emilio Pierro, whilst the question of which of the groups $PGL_2(q)$ are Beauville is discussed by Garion in [23].

Table 2 The types of the Beauville structures specified by the words in Table 1

G	Type	G	Type
$M_{12}:2$	((4,4,5),(6,6,3))	He : 2	((16,16,7),(30,30,5))
$M_{22}:2$	((12,12,4),(10,10,5))	Suz : 2	((10,10,3),(8,8,13))
$J_2:2$	((24,24,15),(14,14,7))	O'N : 2	((38,38,19),(56,56,28))
HS : 2	((8,8,8),(6,6,15))	$Fi_{22}:2$	((10,10,11),(12,12,4))
$J_3:2$	((34,34,17),(24,24,4))	HN : 2	((18,18,25),(44,44,22))
McL : 2	((8,8,3),(10,10,5))	Fi_{24}	((66,66,33),(84,84,26))

6 Nilpotent Groups

It is immediate that the direct product of two Beauville groups of coprime order is again a Beauville group (though slightly more is true—see [4, Lemma 1.3]). Recall that a finite group is nilpotent if and only if it is the direct product of its Sylow subgroups. Since Sylow subgroups for different primes will have coprime orders this observation reduces the study of nilpotent Beauville groups to that of Beauville p-groups.

There is another motivation for wanting to study Beauville p-groups and that is to study how finite groups in general behave from the point of view of Beauville constructions. We saw in Sect. 2 that among the non-abelian finite simple groups only one fails to be a Beauville group and of the rest only two fail to be strongly real. This immediately raises the following question.

Question 4 *Are most Beauville groups strongly real Beauville groups?*

Finite simple groups are rare gems in the rough—for every positive integer n there are at most two finite simple groups of order n and for most values of n there are none at all. Taking our lead from their behaviour is therefore somewhat dangerous.

Few mathematicians outside finite group theory seem to realise that in some sense most finite groups are p-groups, indeed most finite groups are 2-groups. There are 49,910,529,484 groups of order at most 2,000. Of these 49,487,365,422 have order precisely 1024—that's more than $99 \cdot 1\,\%$ of the total! When we throw in the other 2-groups of order at most 1 024 and the other p-groups of order at most 2,000 we have almost all of them. Determining which of the Beauville p-groups are strongly real Beauville p-groups thus goes a long way to answering the above question for groups in general. (For details of these extraordinary computational feats and a historical discussion of the problem of enumerating groups of small order, which has been worked on for almost a century and a half, see the work of Besche et al. in [8, 9].)

Theorem 7 and Corollary 8 tell us that if $p \geq 5$ is prime then there are infinitely many strongly real Beauville p-groups—just let n be any power of p. These results are, however, useless for the primes 2 and 3. As far as the author is aware there are no known examples.

Problem 1 *Find strongly real Beauville* 2*-groups and* 3*-groups.*

The only known infinite family of Beauville 2-groups are those recently constructed by Barker, Boston, Peyerimhoff and Vdovina in [2]. One of the main results of [2] is that the groups constructed there are not strongly real. Furthermore there remain only finitely many known examples of Beauville 3-groups.

In general, p-groups have large outer automorphism groups [12, 13], so it seems likely that most Beauville p-groups are in fact strongly real. Again, as far as the author is aware, this matter remains largely uninvestigated.

Problem 2 *Find non-abelian strongly real Beauville p-groups.*

The best general discussion of work on Beauville p-groups is Boston's contribution to these proceedings [11]. The work of the Barker, Boston and the author in [4] and the work of Barker et al. in [2, 3, 5] are also worth consulting.

Acknowledgments The author wishes to express his deepest gratitude to the organisers of the Beauville Surfaces and Groups 2012 conference held in the University of Newcastle without which this volume, and thus the opportunity to present these results here, would not have been possible. The author also wishes to thank Professor Gareth Jones for many invaluable comments on earlier drafts of this article, particularly regarding the results concerned with products of symmetric and alternating groups. Finally, the author wishes to thank the anonymous referee whose comments and suggestions have substantially improved the readability of this paper, particularly bearing in mind the wide breadth of the audience for this work.

References

1. M. Aschbacher, R. Guralnick, Some applications of the first cohomology group. J. Algebra **90**(2), 446–460 (1984)
2. N. Barker, N. Boston, N. Peyerimhoff, A. Vdovina, An infinite family of 2-groups with mixed Beauville structures, to appear in Int. Math. Res. Not. arXiv:1304.4480
3. N. Barker, N. Boston, N. Peyerimhoff, A. Vdovina, Regular algebraic surfaces isogenous to a higher product constructed from group representations using projective planes, preprint (2011). arXiv:1109.6053
4. N.W. Barker, N. Boston, B.T. Fairbairn, A note on Beauville p-groups. Exp. Math. **21**(3), 298–306 (2012)
5. N. Barker, N. Boston, N. Peyerimhoff, A. Vdovina, New examples of Beauville surfaces. Monatsh. Math. **166**(3–4), 319–327 (2012). doi:10.1007/s00605-011-0284-6
6. I. Bauer, F. Catanese, F. Grunewald, Beauville surfaces without real structures, *Geometric Methods in Algebra and Number Theory*, vol. 235, Progress in Mathematics (Birkhuser, Boston, 2005)
7. I. Bauer, F. Catanese, F. Grunewald, Chebycheff and Belyi polynomials, Dessins d'Enfants, Beauville surfaces and group theory. Mediterr. J. Math. **3**, 121–146 (2006)
8. H.U. Besche, B. Eick, E.A. O'Brien, The groups of order at most 2 000. Electron. Res. Announc. Am. Math. Soc. **7**, 1–4 (2001)
9. H.U. Besche, B. Eick, E.A. O'Brien, A millennium project: constructing small groups. Int. J. Algebra Comput. **12**(5), 623–644 (2002)

10. W. Bosma, J. Cannon, C. Playoust, The Magma algebra system. I. The user language. J. Symb. Comput. **24**, 235–265 (1997)
11. N. Boston, A survey of Beauville p-groups, in *Proceedings of Conference on Beauville Surfaces and Group*, ed. by I. Bauer, S. Garion, A. Vdovina (Newcastle, 2012)
12. N. Boston, M.R. Bush, F. Hajir, Heuristics for p-class towers of imaginary quadratic fields, preprint (2011). arXiv:1111.4679v1
13. N. Boston, Embedding 2-groups in groups generated by involutions. J. Algebra **300**(1), 73–76 (2006)
14. F. Catanese, Fibered surfaces, varieties isogenous to a product and related moduli spaces. Am. J. Math. **122**(1), 1–44 (2000)
15. F. Catanese, Moduli spaces of surfaces and real structures. Ann. Math. **158**(2), 577–592 (2003)
16. J.H. Conway, R.T. Curtis, S.P. Norton, R.A. Parker, R.A. Wilson, *An ATLAS of Finite Groups* (Clarendon Press, Oxford, 1985)
17. J.D. Dixon, B. Mortimer, *Permutation Groups* (Springer, New York, 1996)
18. B.T. Fairbairn, Some exceptional Beauville structures. J. Group Theory **15**(5), 631–639 (2012). arXiv:1007.5050
19. B.T. Fairbairn, K. Magaard, C.W. Parker, Corrigendum to Generation of finite simple groups with an application to groups acting on Beauville surfaces, to appear in the Proc. Lond. Math. Soc
20. B.T. Fairbairn, K. Magaard, C.W. Parker, Generation of finite simple groups with an application to groups acting on Beauville surfaces. Proc. Lond. Math. Soc. **107**(5), 1220 (2013). doi:10.1112/plms/pdt037
21. Y. Fuertes, G. González-Diez, On Beauville structures on the groups S_n and A_n. Math. Z. **264**(4), 959–968 (2010)
22. Y. Fuertes, G.A. Jones, Beauville surfaces and finite groups. J. Algebra **340**, 13–27 (2011)
23. S. Garion, On Beauville structures for $PSL(2, q)$, preprint (2010). arXiv:1003.2792
24. S. Garion, M. Larsen, A. Lubotzky, Beauville surfaces and finite simple groups. J. Reine Angew. Math. **666**, 225–243 (2012)
25. R. Guralnick, G. Malle, Simple groups admit Beauville structures. J. Lond. Math. Soc. (2) **85**(3), 694–721 (2012)
26. P. Hall, The Eulerian functions of a group. Q. J. Math. **7**, 134–151 (1936)
27. G.A. Jones, Beauville surfaces and groups: a survey, to appear Fields Inst. Commun
28. G.A. Jones, Characteristically simple Beauville groups I: Cartesian powers of alternating groups. preprint (2013). arXiv:1304.5444
29. G.A. Jones, Characteristically simple Beauville groups II: low rank and sporadic groups. preprint (2013). arXiv:1304.5450
30. G.A. Jones, Primitive permutation groups containing a cycle, preprint (2012). arXiv:1209.5169v1
31. R. Steinberg, Generators for simple groups. Can. J. Math. **14**, 277–283 (1962)
32. The GAP Group, GAP—Groups, Algorithms, and Programming, Version 4.5.7; 2012. http://www.gap-system.org
33. R.A. Wilson et al., Atlas of finite group representations, version 3. http://brauer.maths.qmul.ac.uk/Atlas/ (2005) onwards
34. R.A. Wilson, Standard generators for sporadic simple groups. J. Algebra **184**(2), 505–515 (1996)
35. R.A. Wilson, *The Finite Simple Groups*, vol. 251, Graduate Texts in Mathematics (Springer, London, 2009)

Beauville Surfaces and Probabilistic Group Theory

Shelly Garion

Abstract A Beauville surface is a complex algebraic surface that can be presented as a quotient of a product of two curves by a suitable action of a finite group. Bauer, Catanese and Grunewald have been able to intrinsically characterize the groups appearing in minimal presentations of Beauville surfaces in terms of the existence of a so-called "Beauville structure". They conjectured that all finite simple groups, except A_5, admit such a structure. This conjecture has recently been proved by Guralnick-Malle and Fairbairn-Magaard-Parker. In this survey we demonstrate another approach towards the proof of this conjecture, based on probabilistic group-theoretical methods, by describing the following three works. The first is the work of Garion, Larsen and Lubotzky, showing that the above conjecture holds for almost all finite simple groups of Lie type. The second is the work of Garion and Penegini on Beauville structures of alternating groups, based on results of Liebeck and Shalev, and the third is the case of the group $\mathrm{PSL}_2(p^e)$, in which we give bounds on the probability of generating a Beauville structure. We also discuss other related problems regarding finite simple quotients of hyperbolic triangle groups and present some open questions and conjectures.

2000 Mathematics Subject Classification: 20D06 · 20H10 · 14J10 · 14J29 · 30F99

1 Beauville Surfaces and Beauville Structures

A *Beauville surface* S (over $\mathbb{C}$) is a particular kind of surface isogenous to a *higher product of curves*, i.e., $S = (C_1 \times C_2)/G$ is a quotient of a product of two smooth curves C_1 and C_2 of genus at least two, modulo a free action of a finite group G

The author was supported by the SFB 878 "Groups, Geometry and Actions".

S. Garion (✉)
Fachbereich Mathematik und Informatik, Universität Münster, Einsteinstrasse 62, 48149 Münster, Germany
e-mail: shelly.garion@uni-muenster.de

I. Bauer et al. (eds.), *Beauville Surfaces and Groups*, Springer Proceedings in Mathematics & Statistics 123, DOI 10.1007/978-3-319-13862-6_5

which acts faithfully on each curve. For Beauville surfaces the quotients C_i/G are isomorphic to $\mathbb{P}^1$ and both projections $C_i \to C_i/G \cong \mathbb{P}^1$ are coverings branched over three points. A Beauville surface is in particular a minimal surface of general type.

Beauville [4] constructed a minimal surface of general type S with $K_S^2 = 8$ and $p_g = q = 0$ in the following way: take two curves $C_1 = C_2$ given by the Fermat equation $x^5 + y^5 + z^5 = 0$ and G the group $(\mathbb{Z}/5\mathbb{Z})^2$ acting on $C_1 \times C_2$ by

$$(a, b) \cdot ([x : y : z], [u : v : w]) = ([\xi^a x : \xi^b y : z], [\xi^{a+3b} u : \xi^{2a+4b} v : w]),$$

where $\xi = e^{\frac{2\pi i}{5}}$ and $a, b \in \mathbb{Z}/5\mathbb{Z}$. Then define S by the quotient $(C_1 \times C_2)/G$. Moreover $C_i \to C_i/G \cong \mathbb{P}^1_{\mathbb{C}}$ and both covers are branched in exactly three points. Curves with such properties are said to be *triangle curves*.

Inspired by this construction Catanese [5] observed that in general if C_1 and C_2 are two triangle curves with group G, if the action of G on the product $C_1 \times C_2$ is free, then $S = (C_1 \times C_2)/G$ is a *strongly rigid surface*, i.e., if S' is another surface homotopically equivalent to S then S' is either biholomorphic or antibiholomorphic to S. He proposed to name these surfaces *Beauville surfaces*.

A Beauville surface S is either of *mixed* or *unmixed* type according respectively as the action of G exchanges the two factors (and then C_1 and C_2 are isomorphic) or G acts diagonally on the product $C_1 \times C_2$. The subgroup G_0 (of index ≤ 2) of G which preserves the ordered pair (C_1, C_2) is then respectively of index 2 or 1 in G. Any Beauville surface S can be presented in such a way that the subgroup G_0 of G acts effectively on each of the factors C_1 and C_2. Catanese called such a presentation *minimal* and proved its uniqueness in [5]. In this survey we shall consider only Beauville surfaces of unmixed type so that $G_0 = G$.

An extensive research on Beauville surfaces was initiated by the collaboration of Bauer, Catanese and Grunewald [2, 3]. They have been able to intrinsically characterize the groups appearing in minimal presentations of unmixed Beauville surfaces in terms of the existence of the so-called unmixed *Beauville structure*.

Definition 1 An *unmixed Beauville structure* for a finite group G is a quadruple $(x_1, y_1; x_2, y_2)$ of elements of G, which determines two triples (x_1, y_1, z_1) and (x_2, y_2, z_2) satisfying:

(i) $x_1 y_1 z_1 = 1$ and $x_2 y_2 z_2 = 1$,
(ii) $\langle x_1, y_1 \rangle = G$ and $\langle x_2, y_2 \rangle = G$,
(iii) $\Sigma(x_1, y_1, z_1) \cap \Sigma(x_2, y_2, z_2) = \{1\}$,
where $\Sigma(x, y, z)$ is the union of the conjugacy classes of all powers of x, all powers of y, and all powers of z.

Moreover denoting the order of an element g in G by $|g|$, we define the *type* τ of (x, y, z) to be the triple $(|x|, |y|, |z|)$. In this situation, we say that G admits an *unmixed Beauville structure of type* (τ_1, τ_2).

The question whether a finite group admits an unmixed Beauville structure of a given type is closely related to the question whether it is a quotient of certain triangle groups. More precisely, a necessary condition for a finite group G to admit an unmixed Beauville structure of type $(\tau_1, \tau_2) = \big((r_1, s_1, t_1), (r_2, s_2, t_2)\big)$ is that G is a quotient with torsion free-kernel of the triangle groups T_{r_1,s_1,t_1} and T_{r_2,s_2,t_2}, where

$$T_{r,s,t} = \langle x, y, z : x^r = y^s = z^t = xyz = 1 \rangle.$$

Indeed, conditions (i) and (ii) of Definition 1 are equivalent to the condition that G is a quotient of each of the triangle groups $T_{|x_i|,|y_i|,|z_i|}$, for $i \in \{1, 2\}$, with torsion-free kernel.

When investigating the existence of an unmixed Beauville structure for a finite group, one can consider only types (τ_1, τ_2), where for $i \in \{1, 2\}$, $\tau_i = (r_i, s_i, t_i)$ satisfies $1/r_i + 1/s_i + 1/t_i < 1$. Then T_{r_i,s_i,t_i} is a (infinite non-soluble) hyperbolic triangle group and we say that τ_i is *hyperbolic*. Indeed, if $1/r_i + 1/s_i + 1/t_i > 1$ then T_{r_i,s_i,t_i} is a finite group, and moreover, it is either dihedral or isomorphic to one of A_4, A_5 or S_4. By [2, Proposition 3.6 and Lemma 3.7], in these cases G cannot admit an unmixed Beauville structure. If $1/r_i + 1/s_i + 1/t_i = 1$ then T_{r_i,s_i,t_i} is one of the (soluble infinite) "wall-paper" groups, and by [2, Sect. 6], none of its finite quotients can admit an unmixed Beauville structure.

Observe that condition (iii) of Definition 1 is clearly satisfied under the assumption that $r_1 s_1 t_1$ is coprime to $r_2 s_2 t_2$. However this assumption is not always necessary, as demonstrated by many examples, such as Beauville's original construction, abelian groups [2, Theorem 3.4], alternating groups [20, Theorem 1.2] and the group $\mathrm{PSL}_2(p^e)$ [18].

2 Beauville Surfaces and Finite Simple Groups

A considerable effort has been made to classify the finite simple groups which admit an unmixed Beauville structure. We recall that by *the classification theorem of finite simple groups*, any finite simple group belongs to one of the following families: the cyclic groups Z_p of prime order; the alternating permutation groups $A_n (n \geq 5)$; the finite simple groups of Lie type, defined over finite fields (e.g. $\mathrm{PSL}_n(q)$); and finally the 26 so-called sporadic groups.

A finite abelian simple group clearly does not admit an unmixed Beauville structure as given a prime p, any pair (a, b) of elements of the cyclic group Z_p of prime order p generating it satisfies $\Sigma(a, b, c) = Z_p$. In fact Bauer, Catanese and Grunewald showed in [2, Theorem 3.4] that the only finite abelian groups admitting an unmixed Beauville structure are the abelian groups of the form $Z_n \times Z_n$ where n is a positive integer coprime to 6. (Here Z_n denotes a cyclic group of order n.)

In [2] the authors also provide the first results on finite non-abelian simple groups admitting an unmixed Beauville structure. More precisely they show that the alternating groups of sufficiently large order admit an unmixed Beauville structure, as well as the projective special linear groups $\mathrm{PSL}_2(p)$ where $p > 5$ is a prime. Moreover using computational methods, they checked that every finite non-abelian simple group of order less than 50,000 admits an unmixed Beauville structure with the exception of the alternating group A_5. Based on these results and the latter observation, they conjectured that all finite non-abelian simple groups admit an unmixed Beauville structure with the exception of A_5.

This conjecture has received much attention and has recently been proved to hold. Concerning the simple alternating groups, it was established in [16] that A_5 is indeed the only one not admitting an unmixed Beauville structure. In [17, 20], the conjecture is shown to hold for the projective special linear groups $\mathrm{PSL}_2(q)$ (where $q > 5$), the Suzuki groups ${}^2B_2(q)$ and the Ree groups ${}^2G_2(q)$ as well as other families of finite simple groups of Lie type of small rank. More precisely, the projective special and unitary groups $\mathrm{PSL}_3(q)$, $\mathrm{PSU}_3(q)$, the simple groups $G_2(q)$ and the Steinberg triality groups ${}^3D_4(q)$ are shown to admit an unmixed Beauville structure if q is large (and the characteristic p is greater than 3 for the simple exceptional groups of type G_2 or ${}^3D_4$). The next major result concerning the investigation of the conjecture with respect to the finite simple groups of Lie type was pursued by Garion, Larsen and Lubotzky who showed in [19] that the conjecture holds for finite non-abelian simple groups of sufficiently large order. The final step regarding the investigation of the conjecture was carried out by Guralnick and Malle [25] and Fairbairn et al. [14] who established its veracity in general, namely,

Theorem 2 ([25]) *Any finite non-abelian simple group, except A_5, admits an unmixed Beauville structure.*

There has also been an effort to classify the finite quasisimple groups and almost simple groups which admit an unmixed Beauville structure. Recall that a finite group G is *quasisimple* provided $G/Z(G)$ is a non-abelian simple group and $G = [G, G]$. In [17] it was shown that $\mathrm{SL}_2(q)$ (for $q > 5$) admits an unmixed Beauville structure. Fairbairn, Magaard and Parker [14] proved the following general result.

Theorem 3 ([14]) *With the exceptions of* $\mathrm{SL}_2(5)$ *and* $\mathrm{PSL}_2(5) \cong \mathrm{SL}_2(4) \cong A_5$, *every finite quasisimple group admits an unmixed Beauville structure.*

Recall that a group G is called *almost simple* if there is a non-abelian simple group G_0 such that $G_0 \leq G \leq \mathrm{Aut}(G_0)$. By [3, 16] the symmetric groups S_n (where $n \geq 5$) admit an unmixed Beauville structure, and by [18] the group $\mathrm{PGL}_2(p^e)$ admits such a structure.

Moreover for the alternating and symmetric groups Garion and Penegini [20] proved another conjecture that Bauer, Catanese and Grunewald proposed in [2], that almost all of these groups admit a Beauville structure with fixed *type*, namely,

Theorem 4 ([20, Theorem 1.2]) *If $\tau_1 = (r_1, s_1, t_1)$ and $\tau_2 = (r_2, s_2, t_2)$ are two hyperbolic types, then almost all alternating groups A_n admit an unmixed Beauville structure of type (τ_1, τ_2).*

A similar theorem also applies for symmetric groups, see [20], and a similar conjecture was raised in [20], replacing A_n by a finite simple classical group of Lie type of sufficiently large Lie rank, namely,

Conjecture 5 ([20, Conjecture 1.7]) *Let $\tau_1 = (r_1, s_1, t_1)$ and $\tau_2 = (r_2, s_2, t_2)$ be two hyperbolic types. If G is a finite simple classical group of Lie type of Lie rank large enough, then it admits an unmixed Beauville structure of type (τ_1, τ_2).*

In contrast, when the Lie rank is very small, as in the case of $\mathrm{PSL}_2(q)$, such a conjecture does not hold, as demonstrated in [18], where there is a characterization of the types of Beauville structures for these groups.

It is well known that almost all pairs of elements in a finite simple (non-abelian) group are generating pairs [12, 26, 31], hence the following question was raised in the workshop "Beauville surfaces and groups 2012".

Question 6 Let G be a finite (non-abelian) simple group. What is the probability $P(G)$ that for four random elements $x_1, y_1, x_2, y_2 \in G$ the quadruple $(x_1, y_1; x_2, z_2)$ is an unmixed Beauville structure for G?

In particular, is it true that if $G = A_n$ or $G = G_n(q)$, a finite simple group of Lie type of Lie rank n, then $P(G) \to 1$ as $n \to \infty$?

Two interesting comments were made during the workshop regarding this question. The first comment, due to Malle, is that for finite simple groups of Lie type of bounded Lie rank, $P(G)$ does not go to 1, and it is bounded above by a function of the rank. The second comment, due to Magaard, is that the techniques in [14] demonstrate that one can generate many unmixed Beauville structures for the finite simple groups of Lie type, allowing to obtain a constant lower bound on $P(G)$, when G is a finite simple classical group. In the specific case where $G = \mathrm{PSL}_2(q)$ we give the following bounds on the probability of $P(G)$ (see Sect. 4.2 for the proof).

Theorem 7 *Let $G = \mathrm{PSL}_2(q)$.*

- *If q is odd then $\frac{1}{32} - \epsilon_q \leq P(G) \leq \frac{15}{16} + \epsilon_q$,*
- *If q is even then $\frac{1}{32} - \epsilon_q \leq P(G) \leq \frac{35}{36} + \epsilon_q$,*

where $\epsilon_q \to 0$ as $q \to \infty$.

3 Hyperbolic Triangle Groups and Their Finite Quotients

Since for a finite group G which admits an unmixed Beauville structure there exists an epimorphism from a hyperbolic triangle group to G, we recall in this section some results on finite quotients of hyperbolic triangle groups.

A *hyperbolic triangle group* T is a group with presentation

$$T = T_{r,s,t} = \langle x, y, z : x^r = y^s = z^t = xyz = 1 \rangle,$$

where (r, s, t) is a triple of positive integers satisfying the condition $1/r + 1/s + 1/t < 1$. Geometrically, let Δ be a hyperbolic triangle group having angles of sizes

π/r, π/s, π/t, then T can be viewed as the group generated by rotations of angles π/r, π/s, π/t around the corresponding vertices of Δ in the hyperbolic plane $\mathbb{H}^2$. Moreover, a hyperbolic triangle group $T_{r,s,t}$ has positive measure $\mu(T_{r,s,t})$ where $\mu(T_{r,s,t}) = 1 - (1/r + 1/s + 1/t)$. As hyperbolic triangle groups are infinite and non-soluble it is interesting to study their finite quotients, particularly the simple ones.

A hyperbolic triangle group $T_{r,s,t}$ has minimal measure when $(r, s, t) = (2, 3, 7)$. The group $T_{2,3,7}$ is also called the (2, 3, 7)-triangle group and its finite quotients are also known as *Hurwitz groups*. These are named after Hurwitz who showed in the late nineteenth century that if S is a compact Riemann surface of genus $h \geq 2$ then $|\mathrm{Aut}\, S| \leq 84(h-1)$ and this bound is attained if and only if $\mathrm{Aut}\, S$ is a quotient of the triangle group $T_{2,3,7}$. Following this result, much effort has been given to classify Hurwitz groups, especially the simple ones, see for example [8] for a historical survey, and [9, 46] for the current state of the art.

Most alternating groups are Hurwitz as shown by Conder (following Higman) who proved in [7] that if $n > 167$ then the alternating group A_n is a quotient of $T_{2,3,7}$. Concerning the finite simple groups of Lie type, there is a dichotomy with respect to their occurrence as quotients of $T_{2,3,7}$ depending on whether the Lie rank is large or not. Indeed as shown in [36] many classical groups of large rank are Hurwitz (and there is no known example of classical groups of large rank which are not Hurwitz). As an illustration by [37] if $n \geq 267$ then the projective special linear group $\mathrm{PSL}_n(q)$ is Hurwitz for any prime power q. The behavior of finite simple groups of Lie type of relatively low rank with respect to the Hurwitz generation problem is rather sporadic. As an illustration by respective results of [6, 38, 39, 46], $\mathrm{PSL}_3(q)$ is Hurwitz if and only if $q = 2$, $\mathrm{PSL}_4(q)$ is never Hurwitz, $G_2(q)$ is Hurwitz for $q \geq 5$, and $\mathrm{PSL}_2(p^e)$ is Hurwitz if and only if $e = 1$ and $p \equiv 0, \pm 1 \bmod 7$, or $e = 3$ and $p \equiv \pm 2, \pm 3 \bmod 7$. Therefore, unlike the alternating groups, there are finite simple groups of Lie type of large order which are not quotients of $T_{2,3,7}$. As for the 26 sporadic finite simple groups, 12 of them are Hurwitz (including the Monster [49]) while the other 14 groups are not.

Turning to general hyperbolic triples (r, s, t) of integers, Higman had already conjectured in the late 1960s that every hyperbolic triangle group has all but finitely many alternating groups as quotients. This was eventually proved by Everitt [13], namely,

Theorem 8 ([13]) *For any hyperbolic triangle group $T = T_{r,s,t}$, if $n \geq n_0(r, s, t)$ then the alternating group A_n is a quotient of T.*

Later, Liebeck and Shalev [32] gave an alternative proof to Higman's Conjecture based on probabilistic group theory, and moreover they have conjectured the following.

Conjecture 9 ([33]) *For any hyperbolic triangle group $T = T_{r,s,t}$, if $G = G_n(q)$ is a finite simple classical group of Lie rank $n \geq n_0(r, s, t)$, then the probability that a randomly chosen homomorphism from T to G is an epimorphism tends to* 1 *as $|G| \to \infty$.*

This conjecture has been proved by Marion [42, 43] for certain families of groups of small Lie rank and certain triples (r, s, t). For example, take (r, s, t) to be a hyperbolic triple of odd primes and $G = \mathrm{PSL}_3(q)$ or $\mathrm{PSU}_3(q)$ containing elements of orders r, s and t, then the conjecture holds. As another example, if (r, s, t) is a hyperbolic triple of primes and $G = {}^2B_2(q)$ or ${}^2G_2(q)$ contains elements of orders r, s and t, then the conjecture also holds.

However, for finite simple groups of small Lie rank such a conjecture does not hold *in general*, and it fails to hold in the case of $\mathrm{PSL}_2(q)$. Indeed, Langer and Rosenberger [28] and Levin and Rosenberger [30] had generalized the aforementioned result of Macbeath, and determined, for a given prime power $q = p^e$, all the triples (r, s, t) such that $\mathrm{PSL}_2(q)$ is a quotient of $T_{r,s,t}$, with torsion-free kernel. It follows that if (r, s, t) is hyperbolic, then for almost all primes p, there is precisely one group of the form $\mathrm{PSL}_2(p^e)$ or $\mathrm{PGL}_2(p^e)$ which is a homomorphic image of $T_{r,s,t}$ with torsion-free kernel. We note that this result can also be obtained by using other techniques. Firstly, Marion [40] has recently provided a proof for the case where r, s, t are primes relying on probabilistic group theoretical methods. Secondly, it also follows from the representation theoretic arguments of Vincent and Zalesski [48, Theorems 2.9 and 2.11]. Such methods can be used for dealing with other families of finite simple groups of Lie type, see for example [10, 41, 45, 47, 48].

Recently, a new approach was presented by Larsen et al. [29], based on the theory of representation varieties (via deformation theory). They prove a conjecture of Marion [41] showing that various finite simple groups are *not* quotients of $T_{r,s,t}$, as well as positive results showing that many finite simple groups are quotients of $T_{r,s,t}$.

4 Beauville Structures for the Group $\mathrm{PSL}_2(q)$

In this section we discuss the specific case of $\mathrm{PSL}_2(q)$, and briefly sketch the proof of Garion and Penegini [20] for the following theorem, which is based on results of Macbeath [38].

Theorem 10 ([17, 20]). *Let p be a prime number, and assume that $q = p^e$ is at least* 7. *Then the group* $\mathrm{PSL}_2(q)$ *admits an unmixed Beauville structure.*

In addition, we bound the probability that four random elements in $\mathrm{PSL}_2(q)$ generate an unmixed Beauville structure and prove Theorem 7.

4.1 Sketch of the Proof of Theorem 10

In order to construct an unmixed Beauville structure for $\mathrm{PSL}_2(q)$ one needs to find a quadruple $(A_1, B_1; A_2, B_2)$ of elements of $\mathrm{PSL}_2(q)$ satisfying the three conditions given in Definition 1. This can be done directly by finding specific elements in the group satisfying these conditions (see [17]) or indirectly by using the following results of Macbeath [38] (see [18, 20]).

Theorem 11 ([38, Theorem 1]) *For every* $\alpha, \beta, \gamma \in \mathbb{F}_q$ *there exist three matrices* $A, B, C \in \mathrm{SL}_2(q)$ *satisfying* $\mathrm{tr}(A) = \alpha$, $\mathrm{tr}(B) = \beta$, $\mathrm{tr}(C) = \gamma$ *and* $ABC = I$.

This theorem immediately implies Condition (i).

Moreover, Macbeath [38] classified the pairs of elements in $\mathrm{PSL}_2(q)$ in a way which makes it easy to decide what kind of subgroup they generate. By [38, Theorem 2], a triple $(\alpha, \beta, \gamma) \in \mathbb{F}_q^3$ is *singular*, namely $\alpha^2+\beta^2+\gamma^2-\alpha\beta\gamma-4 = 0$, if and only if for the corresponding triple of matrices (A, B, C), the group generated by the images of A and B is a *structural subgroup* of $\mathrm{PSL}_2(q)$, that is a subgroup of the Borel or a cyclic subgroup.

Hence, in order to verify Condition (ii), one needs to show that the subgroup generated by $A, B \in \mathrm{PSL}_2(q)$ is neither a structural subgroup (using the aforementioned result of Macbeath), not a dihedral subgroup, not one of the small subgroups A_4, S_4 or A_5, and not a subfield subgroup (namely, isomorphic to $\mathrm{PSL}_2(q_1)$ or to $\mathrm{PGL}_2(q_1)$, where $q = q_1^m$) hence it must be $\mathrm{PSL}_2(q)$ itself, as the subgroup structure of $\mathrm{PSL}_2(q)$ is well-known (see e.g. [11, 44]). For example, Condition (ii) is always satisfied when $q \geq 13$ and the orders $|A| = |B| = |C| = (q-1)/d$ or $(q+1)/d$, where $d = \gcd(2, q-1)$.

Condition (iii) is clearly satisfied under the assumption that the product of the orders $|A_1| \cdot |B_1| \cdot |C_1|$ is coprime to $|A_2| \cdot |B_2| \cdot |C_2|$. For example, for any $q > 7$ the group $\mathrm{PSL}_2(q)$ admits unmixed Beauville structures of types

$$\left(\left(\frac{q-1}{d}, \frac{q-1}{d}, \frac{q-1}{d}\right), \left(\frac{q+1}{d}, \frac{q+1}{d}, \frac{q+1}{d}\right) \right),$$

and

$$\left(\left(\frac{q-1}{d}, \frac{q-1}{d}, \frac{q-1}{d}\right), \left(\frac{q+1}{d}, \frac{q+1}{d}, p\right) \right),$$

appearing in [17, 20] respectively.

However this assumption is not always necessary. Indeed, by [18], $\mathrm{PSL}_2(q)$ (where $q = p^{2e}$, p an odd prime) always admits unmixed Beauville structures of types $((p, p, t_1), (p, p, t_2))$ for certain t_1, t_2 dividing $(q-1)/2$, $(q+1)/2$ respectively.

This approach can be effectively used to construct many unmixed Beauville structures for $\mathrm{PSL}_2(q)$, and in [18] there is a characterization of the types of unmixed Beauville structures for this group.

4.2 Proof of Theorem 7

The proof relies on considering the various types of elements in $G = \mathrm{PSL}_2(q)$. Recall that an element in G is called *split* if its order divides $(q-1)/d$ (where $d = \gcd(2, q-1)$), *non-split* if its order divides $(q+1)/d$, and *unipotent* if its order is p (and then the trace of its pre-image in $\mathrm{SL}_2(q)$ equals ± 2).

It is well-known that there are roughly q^3 matrices in $\mathrm{SL}_2(q)$, and moreover, for any $\alpha \in \mathbb{F}_q$ the number of matrices $A \in \mathrm{SL}_2(q)$ with $\mathrm{tr}(A) = \alpha$ is roughly q^2 (see for example [1, Table 1]). In addition, the probability that a random element in $\mathbb{F}_q$ is a trace of a split (respectively, non-split) matrix in $\mathrm{SL}_2(q)$ goes to $1/2$ as $q \to \infty$ (see [38, Lemma 2]). Therefore, probabilistically, the number of unipotents in G is negligible, and moreover, if we denote by P_q^s (respectively P_q^n) the probability that a random element in G is split (respectively, non-split) then $P_q^s \to \frac{1}{2}$ and $P_q^n \to \frac{1}{2}$ as $q \to \infty$.

By [1, Proposition 7.2] it follows that for any non-singular triple $(\alpha, \beta, \gamma) \in \mathbb{F}_q^3$, the number of triples $(A, B, C) \in \mathrm{SL}_2(q)^3$ satisfying $\mathrm{tr}(A) = \alpha, \mathrm{tr}(B) = \beta, \mathrm{tr}(C) = \gamma$ and $ABC = I$ is roughly q^3. By [38, Lemma 3] almost all triples in $\mathbb{F}_q^3$ are non-singular. Since the probability that a random triple of element in $\mathbb{F}_q^3$ contains only traces of split (respectively, non-split) matrices goes to $\frac{1}{8}$ as $q \to \infty$, it follows that the probability that two random elements $A, B \in G$ satisfy that A, B and AB are all split (respectively, non-split), goes to $\frac{1}{8}$ as $q \to \infty$.

In order to obtain a lower bound, observe that the probability that two random elements $A, B \in G$ do **not** generate G goes to 0 as $q \to \infty$ (this can be deduced from the aforementioned results of Macbeath [38], or alternatively, as a specific case of [26]). Namely,

$$\frac{\#\{(A, B) \in G^2 : \langle A, B\rangle \neq G\}}{|G|^2} \leq \epsilon'_q,$$

where $\epsilon'_q \to 0$ as $q \to \infty$. Hence,

$$\frac{\#\{(A, B) \in G^2 : A, B, AB \text{ are split and } \langle A, B\rangle = G\}}{|G|^2} \geq \frac{1}{8} - \epsilon'_q,$$

and

$$\frac{\#\{(A, B) \in G^2 : A, B, AB \text{ are non-split and } \langle A, B\rangle = G\}}{|G|^2} \geq \frac{1}{8} - \epsilon'_q.$$

Since $(q-1)/d$ and $(q+1)/d$ are relatively prime then Condition (iii) is immediately satisfied when A_1, B_1, A_1B_1 are all split and A_2, B_2, A_2B_2 are all non-split (and vice-versa). Therefore,

$$\frac{\#\{(A_1, B_1, A_2, B_2) \in G^4 : (A_1, B_1; A_2, B_2) \text{ is a Beauville structure}\}}{|G|^4}$$

$$\geq \left(\frac{1}{8} - \epsilon'_q\right)^2 + \left(\frac{1}{8} - \epsilon'_q\right)^2 \geq \frac{1}{32} - \epsilon_q,$$

where $\epsilon_q \to 0$ as $q \to \infty$.

Assume that q is odd. In order to obtain an upper bound, observe that if the orders of A_1 and A_2 are both even then Condition (iii) is not satisfied, see [18, Lemma 4.2]. Denote by P_q^e the probability that a random element in G has even order, then $P_q^e \geq \frac{1}{4} - \epsilon'_q$, where $\epsilon'_q \to 0$ as $q \to \infty$. Indeed, if $(q-1)/2$ (respectively, $(q+1)/2$) is even, then at least half of the split (respectively, non-split) elements are of even order, since any split (respectively, non-split) element belongs to a cyclic subgroup of order $(q-1)/2$ (respectively, $(q+1)/2$), and at least half the elements in a cyclic group of even order are of even order. Therefore,

$$\frac{\#\{(A_1, B_1, A_2, B_2) \in G^4 : (A_1, B_1; A_2, B_2) \text{is } \textbf{not} \text{ a Beauville structure}\}}{|G|^4}$$

$$\geq \left(\frac{1}{4} - \epsilon'_q\right)^2 \geq \frac{1}{16} - \epsilon_q,$$

where $\epsilon_q \to 0$ as $q \to \infty$.

When q is even the proof is similar, replacing P_q^e by the probability that a random element in G has order divisible by 3, which is at least $\frac{1}{6} - \epsilon'_q$, where $\epsilon'_q \to 0$ as $q \to \infty$.

5 Beauville Structures for Finite Simple Groups

In this section we briefly describe the probabilistic group-theoretical methods in proving Theorems 2 and 4. We shall mainly sketch the proof of Garion, Larsen and Lubotzky [19] that the conjecture of Bauer, Catanese and Grunewald (Theorem 2) holds for *almost* all finite simple groups of Lie type, as well as present the proof of Garion and Penegini [20] regarding Beauville structures of alternating groups (Theorem 4), which is based on the probabilistic results of Liebeck and Shalev [32].

As in the probabilistic approach we can ignore finitely many simple groups, we will not deal here with the sporadic groups, whose unmixed Beauville structures can be found in [14, 25]. Thus we shall consider only the alternating groups and the finite simple groups of Lie type.

Recall that in order to construct an unmixed Beauville structure for a finite simple (non-abelian) group G one needs to find a quadruple $(x_1, y_1; x_2, y_2)$ of elements of G satisfying the three conditions given in Definition 1.

5.1 Choosing Disjoint Conjugacy Classes

One usually starts by looking for proper conjugacy classes $X_1, Y_1, Z_1, X_2, Y_2, Z_2$ such that

$$\Sigma(x_1, y_1, z_1) \cap \Sigma(x_2, y_2, z_2) = \{1\}$$

for any $x_i \in X_i,\ y_i \in Y_i,\ z_i \in Z_i$ $(i = 1, 2)$, so that Condition (iii) is satisfied.

For finite simple groups of Lie type, one can choose two maximal tori T_1 and T_2, such that if C_i denotes the set of all conjugates of elements of T_i, then $C_1 \cap C_2 = \{1\}$.

For example, let $G = \mathrm{SL}_{r+1}(q)$ $(r > 1)$, and let t_1 and t_2 denote elements of G whose characteristic polynomials are respectively irreducible (of degree $r + 1$) and the product of irreducible polynomials of degree 1 and r. If T_1 and T_2 denote the centralizers of t_1 and t_2 respectively, then, by [19, Proposition 7], for all $g \in G$, $T_1 \cap g^{-1}T_2 g = Z(G)$, thus Condition (iii) is satisfied for $\mathrm{PSL}_{r+1}(q)$.

In fact, for the finite simple groups of Lie type, one can choose several maximal tori T_1 and T_2 such that Condition (iii) is satisfied, see the various choices in [14, 19, 25]. However, the number of conjugacy classes of maximal tori is bounded above by a function of the Lie rank r, and any maximal torus is isomorphic to a product of at most r cyclic groups, so a similar argument to the one presented in Sect. 4.2 implies that the probability that four random elements generate an unmixed Beauville structure is bounded above by a function of r.

One can also choose proper conjugacy classes for the alternating groups (see [14, 17, 20, 25]). In [20], Garion and Penegini used the so-called *almost homogeneous* conjugacy classes introduced by Liebeck and Shalev [32].

Definition 12 ([32]) Conjugacy classes in S_n of cycle-shape (m^k), where $n = mk$, namely, containing k cycles of length m each, are called *homogeneous*. A conjugacy class having cycle-shape $(m^k, 1^f)$, namely, containing k cycles of length m each and f fixed points, with f bounded, is called *almost homogeneous*.

By [20, Algorithm 3.5] one can construct for any six integers $k_1, l_1, m_1, k_2, l_2, m_2 \geq 2$, six distinct almost homogeneous conjugacy classes $X_1, Y_1, Z_1, X_2, Y_2, Z_2$ in A_n whose elements are of orders $k_1, l_1, m_1, k_2, l_2, m_2$ respectively, such that all of them have different numbers of fixed points, thus satisfying Condition (iii).

5.2 Frobenius Formula and Witten's Zeta Function

Condition (i) follows from a classical formula of Frobenius: If X, Y and Z are conjugacy classes in a finite group G, then the number $N_{X,Y,Z}$ of solutions of $xyz = 1$ with $x \in X$, $y \in Y$ and $z \in Z$ is given by

$$N_{X,Y,Z} = \frac{|X| \cdot |Y| \cdot |Z|}{|G|} \sum_{\chi \in \mathrm{Irr}(G)} \frac{\chi(x)\chi(y)\chi(z)}{\chi(1)}, \tag{1}$$

where $\mathrm{Irr}(G)$ denotes the set of complex irreducible characters of G.

Usually, the main contribution to this character sum comes from the trivial character, and the absolute sum on all other characters is negligible. In order to show this, one needs to bound the absolute value $|\chi(g)|$ for an irreducible character χ and an element g, from any of the above conjugacy classes (see Sect. 5.3 for details). If this value can be effectively bounded then one deduces that the value of the sum of the contribution of all *non-trivial* characters to (1) is bounded above by some global constant (depending only on the sizes of G and the conjugacy classes) multiplied by the sum $\sum_{\chi\neq 1}\chi(1)^{-1}$. So it remains to prove that the letter sum converges to 0 as $|G|\to\infty$.

Therefore, a key role in the probabilistic approach is played by the so called *Witten zeta function*, which is defined by

$$\zeta^G(s)=\sum_{\chi\in\mathrm{Irr}(G)}\chi(1)^{-s}.$$

It was originally defined and studied by Witten [50] for Lie groups. For finite simple groups it was studied and applied in detail by Liebeck and Shalev [32–34], who proved the following desired results.

Theorem 13 ([33, Theorem 1.1], [34, Corollary 1.3] and [32, Corollary 2.7]) *Let G be a finite simple group.*

- *If $s>1$ then $\zeta^G(s)\to 1$ as $|G|\to\infty$.*
- *If $s>2/3$ and $G\neq \mathrm{PSL}_2(q)$ then $\zeta^G(s)\to 1$ as $|G|\to\infty$.*
- *If $s>1/2$ and $G\neq \mathrm{PSL}_2(q),\mathrm{PSL}_3(q),\mathrm{PSU}_3(q)$ then $\zeta^G(s)\to 1$ as $|G|\to\infty$.*
- *If $s>0$ and $G=A_n$ then $\zeta^G(s)\to 1$ as $|G|\to\infty$. Moreover, $\zeta^G(s)=1+O(n^{-s})$.*

An alternative approach to prove Condition (i) for the finite simple groups of Lie type, which was successfully applied in [14, 25] and [19, Sect. 4], is based on the following result of Gow [22]: if X and Y are conjugacy classes of regular semisimple elements in a finite Lie type group G, then the set XY contains every non-central semisimple element of G.

5.3 Character Estimates in Finite Simple Groups

A crucial part in the proof is to estimate character values in finite simple groups. More precisely, one needs to bound the absolute value $|\chi(g)|$ for an irreducible character χ and an element g, from any of the conjugacy classes chosen in Sect. 5.1. We therefore recall some useful results for the finite simple groups of Lie type and the alternating groups.

Lemma 14 ([19], Corollary 4) *Let χ be an irreducible character of $G=\mathrm{SL}_{r+1}(q)$.*

- *If $t_1\in G$ has an irreducible characteristic polynomial, then $|\chi(t_1)|\leq\frac{2(r+1)^2}{\sqrt{3}}$.*
- *If $r\geq 2$ and the characteristic polynomial of $t_2\in G$ has an irreducible factor of degree r, then $|\chi(t_2)|\leq\frac{2r^2}{\sqrt{3}}$.*

More generally, by [19, Proposition 7], there exist an absolute constant c such that for every sufficiently large group of Lie type G (of Lie rank r), there exist maximal tori T_1 and T_2 as in Sect. 5.1, and for every regular $t \in T_1 \cup T_2$ and every irreducible character χ of G,

$$|\chi(t)| \leq cr^3.$$

Lemma 15 ([32], Proposition 2.12) *Let $\pi \in S_n$ have cycle-shape $(m^k, 1^f)$. Then for any $\chi \in \mathrm{Irr}(S_n)$ we have*

$$|\chi(\pi)| \leq c \cdot (2n)^{\frac{f+1}{2}} \chi(1)^{\frac{1}{m}},$$

where c depends only on m.

It is also interesting to provide an upper bound on the *character ratio* $|\chi(g)/\chi(1)|$, where G is a finite simple group, $g \in G$ and χ is an irreducible character. Gluck and Magaard [21] computed these bounds for the finite classical groups. Such bounds play a crucial role in the proof of a conjecture of Guralnick and Thompson [23], which is related to the inverse Galois problem, namely, which finite groups occur as Galois groups of algebraic number fields (for details see [15]).

5.4 Finding Generating Pairs

In order to prove Condition (ii) one should show that the set of solutions of $xyz = 1$ with $x \in X$, $y \in Y$ and $z \in Z$ contains a generating pair of G, namely, one should avoid pairs (x, y) contained in maximal subgroups of G. Probabilistically, one expects such a result to hold since almost all pairs of elements in a finite simple (non-abelian) group are generating pairs (see [12, 26, 31]).

Namely, one should estimate the sum

$$\sum_{M \in \max G} |\{(x, y, z) \mid x \in X \cap M, y \in Y \cap M, z \in Z \cap M, xyz = 1\}|,$$

where $\max G$ denotes the set of maximal proper subgroups of G. This quantity is bounded above by

$$\sum_{M \in \max G} |M|^2 = |G|^2 \sum_{M \in \max G} \frac{1}{[G:M]^2} \leq \frac{|G|^2}{m(G)^{1/2}} \sum_{M \in \max G} [G:M]^{-3/2},$$

where $m(G)$ is the minimal index of a proper subgroup of G or, equivalently, the minimal degree of a non-trivial permutation representation of G. By estimates of Landazuri and Seitz [27], there exists an absolute constant c such that if G is a finite simple group of Lie type G of Lie rank r then $m(G) \geq cq^r$. Hence, again one should estimate a 'zeta function' encoding the indices of maximal subgroups of finite simple groups of Lie type, which was investigated by Liebeck et al. [35], who proved the following desired result.

Theorem 16 [35] *If G is a finite simple group, and $s > 1$, then*

$$\lim_{|G|\to\infty} \sum_{M\in\max G} [G : M]^{-s} \to 0.$$

Alternatively, in [14, 25] they relied on more delicate results about maximal subgroups in finite simple groups of Lie type containing special elements, called *primitive prime divisors*, of Guralnick et al. [24].

For sufficiently large alternating groups, Conditions (i) and (ii) follow from the following result of Liebeck and Shalev [32]. If (k, l, m) is hyperbolic then the probability that three random elements $x, y, z \in A_n$, with product 1, from almost homogeneous classes X, Y, Z, of orders k, l, m will generate A_n, tends to 1 as $n \to \infty$. Moreover, this probability is $1 - O(n^{-\mu})$, where $\mu = 1 - (1/k + 1/l + 1/m)$.

Acknowledgments The author would like to thank her co-organizers, Ingrid Bauer and Alina Vdovina, and all the participants in the workshop "Beauville surfaces and groups 2012" for their assistance and useful conversations, as well as the University of Newcastle for hosting the workshop. The author is grateful to the referee for pointing out further relevant references.

References

1. T. Bandman, S. Garion, Surjectivity and Equidistribution of the word $x^a y^b$ on PSL $(2, q)$ and SL $(2, q)$. Int. J. Algebra Comput. **22**(2) (2012)
2. I. Bauer, F. Catanese, F. Grunewald, Beauville surfaces without real structures, *Geometric Methods in Algebra and Number Theory*, vol. 235, Progress in Mathematics (Birkhäuser, Boston, 2005), pp. 1–42
3. I. Bauer, F. Catanese, F. Grunewald, Chebycheff and Belyi polynomials, dessins d'enfants, Beauville surfaces and group theory. Mediterr. J. Math. **3**(2), 121–146 (2006)
4. A. Beauville, Surfaces algébriques complexes, Astérisque, **54** Paris, (1978)
5. F. Catanese, Fibred surfaces, varieties isogenous to a product and related moduli spaces. Am. J. Math. **122**, 1–44 (2000)
6. J. Cohen, On non-Hurwitz groups and non-congruence subgroups of the modular group. Glasg. Math. J. **22**, 1–7 (1981)
7. M.D.E. Conder, Generators for alternating and symmetric groups. J. Lond. Math. Soc. **22**, 75–86 (1980)
8. M.D.E. Conder, Hurwitz groups: a brief survey. Bull. Am. Math. Soc. **23**, 359–370 (1990)
9. M.D.E. Conder, An update on Hurwitz groups. Groups, Complex. Cryptol. **2**, 35–40 (2010)
10. L. Di Martino, M.C. Tamburini, A.E. Zalesski, On Hurwitz groups of low rank. Commun. Algebra **28**(11), 5383–5404 (2000)
11. L.E. Dickson, *Linear Groups with an Exposition of the Galois Field Theory* (Teubner, Leipzig, 1901)
12. J.D. Dixon, The probability of generating the symmetric group. Math. Z. **110**, 199–205 (1969)
13. B. Everitt, Alternating quotients of Fuchsian groups. J. Algebra **223**, 457–476 (2000)
14. B. Fairbairn, K. Magaard, C. Parker, Generation of finite simple groups with an application to groups acting on Beauville surfaces, to appear in Proc. Lond. Math. Soc

15. D. Frohardt, K. Magaard, About a conjecture of Guralnick and Thompson, in *Groups, Difference Sets, and the Monster, Proceedings of the Ohio State Conference on Groups and Geometrie*, ed. by K.T. Arasu, et al. (Walter de Gruyter, Berlin, 1996), pp. 43–54
16. Y. Fuertes, G. González-Diez, On Beauville structures on the groups S_n and A_n. Math. Z. **264**, 959–968 (2010)
17. Y. Fuertes, G. Jones, Beauville surfaces and finite groups. J. Algebra **340**, 13–27 (2011)
18. S. Garion, On Beauville structures for PSL $(2, q)$. arXiv:1003.2792
19. S. Garion, M. Larsen, A. Lubotzky, Beauville surfaces and finite simple groups. J. Reine Angew. Math. **666**, 225–243 (2012)
20. S. Garion, M. Penegini, New Beauville surfaces and finite simple groups. Manuscripta Math. **142**, 391–408 (2013)
21. D. Gluck, K. Magaard, Character and fixed point ratios in finite classical groups. Proc. Lond. Math. Soc. **71**, 547–584 (1995)
22. R. Gow, Commutators in finite simple groups of Lie type. Bull. Lond. Math. Soc. **32**(3), 311–315 (2000)
23. R. Guralnick, J. Thompson, Finite groups of genus zero. J. Algebra **131**, 303–341 (1990)
24. R. Guralnick, C. Praeger, T. Penttila, J. Saxl, Linear groups with orders having certain large prime divisors. Proc. Lond. Math. Soc. **78**, 167–214 (1999)
25. R. Guralnick, G. Malle, Simple groups admit Beauville structures. J. Lond. Math. Soc. **85**(3), 694–721 (2012)
26. W.M. Kantor, A. Lubotzky, The probability of generating a finite classical group. Geom. Dedicata **36**, 67–87 (1990)
27. V. Landazuri, G.M. Seitz, On the minimal degrees of projective representations of the finite Chevalley groups. J. Algebra **32**, 418–443 (1974)
28. U. Langer, G. Rosenberger, Erzeugende endlicher projektiver linearer Gruppen. Results Math. **15**(1–2), 119–148 (1989)
29. M. Larsen, A. Lubotzky, C. Marion, Deformation theory and finite simple quotients of triangle groups I. arXiv:1301.2949
30. F. Levin, G. Rosenberger, Generators of finite projective linear groups, II. Results Math. **17**(1–2), 120–127 (1990)
31. M.W. Liebeck, A. Shalev, The probability of generating a finite simple group. Geom. Dedicata **56**, 103–113 (1995)
32. M.W. Liebeck, A. Shalev, Fuchsian groups, coverings of Riemann surfaces, subgroup growth, random quotients and random walks. J. Algebra **276**, 552–601 (2004)
33. M.W. Liebeck, A. Shalev, Fuchsian groups, finite simple groups and representation varieties. Invent. Math. **159**(2), 317–367 (2005)
34. M.W. Liebeck, A. Shalev, Character degrees and random walks in finite groups of Lie type. Proc. Lond. Math. Soc. (3) **90**(1), 61–86 (2005)
35. M.W. Liebeck, B.M.S. Martin, A. Shalev, On conjugacy classes of maximal subgroups of finite simple groups, and a related zeta function. Duke Math. J. **128**(3), 541–557 (2005)
36. A. Lucchini, M.C. Tamburini, Classical groups of large rank as Hurwitz groups. J. Algebra **219**, 531–546 (1999)
37. A. Lucchini, M.C. Tamburini, J.S. Wilson, Hurwitz groups of large rank. J. Lond. Math. Soc. **61**, 81–92 (2000)
38. A.M. Macbeath, Generators of the linear fractional groups, Number Theory. Proc. Sympos. Pure Math. **XII**, Houston, Tex. (1967), Am. Math. Soc., Providence, R.I. (1969), 14–32
39. G. Malle, Hurwitz groups and $G_2(q)$. Can. Math. Bull. **33**, 349–357 (1990)
40. C. Marion, Triangle groups and PSL $_2(q)$. J. Group Theory **12**, 689–708 (2009)
41. C. Marion, On triangle generation of finite groups of Lie type. J. Group Theory **13**, 619–648 (2010)
42. C. Marion, Triangle generation of finite exceptional groups of low rank. J. Algebra **332**, 35–61 (2011)
43. C. Marion, Random and deterministic triangle generation of three-dimensional classical groups I. Commun. Algebra **41**(3), 797–852 (2013)

44. M. Suzuki, *Group Theory I* (Springer, Berlin, 1982)
45. M.C. Tamburini, A.E. Zalesski, Classical groups in dimension 5 which are Hurwitz, Finite groups 2003 (Walter de Gruyter, Berlin, 2004)
46. M.C. Tamburini, M. Vsemirnov, Hurwitz groups and Hurwitz generation, in *Handbook of Algebra*, vol. 4, ed. by M. Hazewinkel (Elsevier, Amsterdam, 2006), pp. 385–426
47. M.C. Tamburini, M. Vsemirnov, Irreducible (2, 3, 7)-subgroups of PGL $_n(F)$ for $n \leq 7$. J. Algebra **300**(1), 339–362 (2006)
48. R. Vincent, A.E. Zalesski, Non-Hurwitz classical groups. LMS J. Comput. Math. **10**, 21–82 (2007)
49. R.A. Wilson, The Monster is a Hurwitz group. J. Group Theory **4**(4), 367–374 (2001)
50. E. Witten, On quantum gauge theories in two dimensions. Commun. Math. Phys. **141**, 153–209 (1991)

The Classification of Regular Surfaces Isogenous to a Product of Curves with $\chi(\mathcal{O}_S) = 2$

Christian Gleißner

Abstract A complex surface S is said to be isogenous to a product if S is a quotient $S = (C_1 \times C_2)/G$ where the C_i's are curves of genus at least two and G is a finite group acting freely on $C_1 \times C_2$. In this article we classify all regular surfaces isogenous to a product with $\chi(\mathcal{O}_S) = 2$ under the assumption that the action of G is unmixed i.e. no element of G exchange the factors of the product $C_1 \times C_2$.

1 Introduction

A complex surface S is said to be *isogenous to a product* if S is a quotient

$$S = (C_1 \times C_2)/G,$$

where the C_i's are curves of genus at least two, and G is a finite group acting freely on $C_1 \times C_2$. Due to Catanese [8] there are two possibilities how G can act on the product $C_1 \times C_2$:

- For all $g \in G$ and $(z, w) \in C_1 \times C_2$ we have $g(z, w) = (g(z), g(w))$. In this case the action is called *diagonal*.
- There exists $g \in G$, such that $g(z, w) = (g(w), g(z))$ for all $(z, w) \in C_1 \times C_2$. In this case the curves C_1 and C_2 are isomorphic.

If the action of G is diagonal, we say that S is of *unmixed type*, else we say that S is of *mixed type*.

Since these surfaces were introduced by Catanese [8] there has been produced a considerable amount of literature. In particular the surfaces isogenous to a product with $\chi(\mathcal{O}_S) = 1$ are completely classified: The Bogomolov-Miyaoka-Yau inequality $K_S^2 \leq 9\chi(\mathcal{O}_S)$ together with Debarre's inequality $K_S^2 \geq 2p_g(S)$ gives $0 \leq q(S) = p_g(S) \leq 4$. By Beauville [3] all minimal surfaces S of general type with $p_g(S) =$

C. Gleißner (✉)
Lehrstuhl Mathematik VIII, Universität Bayreuth, Universitätsstraße 30,
95447 Bayreuth, Germany
e-mail: christian.gleissner@uni-bayreuth.de

I. Bauer et al. (eds.), *Beauville Surfaces and Groups*, Springer Proceedings in Mathematics & Statistics 123, DOI 10.1007/978-3-319-13862-6_6

$q(S) = 4$ are a product of two genus two curves. A minimal surface S of general type with $p_g(S) = q(S) = 3$ is either the symmetric square of a genus three curve or $S = (F_2 \times F_3)/\tau$, where F_g is a curve of genus g and τ is of order two acting on F_2 as an elliptic involution and on F_3 as a fixed point free involution [9, 14, 19]. The classifications of surfaces isogenous to a product in the remaining cases are: the case $p_g = 0$, $q = 0$, due to Bauer et al. [1], $p_g = 1$, $q = 1$, due to Carnovale and Polizzi [7] and $p_g = 2$, $q = 2$, due to Penegini [17].

Our aim is to give a classification in the case $\chi(\mathcal{O}_S) = 2$ under the assumption that S regular and of unmixed type. We want to mention that these surfaces have the invariants $q(S) = 0$ and $p_g(S) = 1$ like $K3$ surfaces and there are recent constructions of $K3$ surfaces with non-symplectic automorphisms as *product quotient surfaces* by Garbagnati and Penegini [10]. Our main result is the following (see also the Table in 5.9):

Theorem 1.1 *There are exactly* 49 *families of regular surfaces isogenous to a product of unmixed type with* $\chi(\mathcal{O}_S) = 2$.

We will now explain how the paper is organized. In Sect. 2 we explain the basics about surfaces isogenous to a product of curves. In Sect. 3 we recall *Riemann's existence theorem* and introduce the necessary tools from group theory and combinatorics. These facts are used to show that there is an entirely group theoretic description of surfaces isogenous to a product. In Sect. 4 we recall the description of the moduli space of surfaces isogenous to a product, due to Catanese. In Sect. 5 we give an algorithm, which we use to classify all regular surfaces isogenous to a product of unmixed type with $\chi(\mathcal{O}_S) = 2$. The computations are performed with the computer algebra system MAGMA [15].[1] In particular the *Database of Small Groups* and the *Database of Perfect Groups* is used. Afterwards we discuss the output of the computation. Finally we show the classification result.

2 Surfaces Isogenous to a Product

In this section we explain some basic facts about surfaces isogenous to a product. We work over the field of complex numbers $\mathbb{C}$ and use the standard notations from the theory of complex surfaces, see for example [4]. The *self-intersection of the canonical class* is denoted by K_S^2, the *topological Euler number* by $e(S) = \sum_{i=1}^4 (-1)^i h^i(S, \mathbb{C})$, $\kappa(S)$ is the Kodaira dimension. The *holomorphic-Euler-Poincaré-characteristic* is defined as $\chi(\mathcal{O}_S) := 1 - q(S) + p_g(S)$, where $q(S) := h^1(S, \mathcal{O}_S)$ and $p_g(S) := h^2(S, \mathcal{O}_S)$. The basic objects we consider are the following:

Definition 2.1 A surface S is said to be *isogenous to a product* if S is a quotient

$$S = (C_1 \times C_2)/G,$$

[1] The source code is available at www.staff.uni-bayreuth.de/~bt300503/.

where the C_i's are smooth projective curves of genus at least two, and $G \leq Aut(C_1 \times C_2)$ is a finite group of automorphisms acting *freely* on $C_1 \times C_2$.

Immediate consequences of this definition are: The surface S is smooth, projective and of general type, i.e. $\kappa(S) = 2$. The canonical class K_S is ample. In particular S is minimal.

The *self-intersection of the canonical class* K_S^2, the *topological Euler number* and the *holomorphic-Euler-Poincaré-characteristic* can be expressed in terms of the genera $g(C_i)$:

Proposition 2.2 ([8, Theorem 3.4]) *Let $S = (C_1 \times C_2)/G$ be a surface isogenous to a product, then*

$$K_S^2 = \frac{8(g(C_1)-1)(g(C_2)-1)}{|G|}, \quad e(S) = \frac{1}{2}K_S^2 \ \ \textit{and} \ \ \chi(\mathcal{O}_S) = \frac{1}{8}K_S^2.$$

In our case $p_g(S) = 1$ and $q(S) = 0$ we get $K_S^2 = 16$ and $e(S) = 8$, moreover we have the useful relation

$$2|G| = (g(C_1)-1)(g(C_2)-1). \tag{1}$$

Remark 2.3 For the rest of the paper we consider the *unmixed* case, where the action of G on the product $C_1 \times C_2$ is diagonal, i.e.

$$G = G \cap (Aut(C_1) \times Aut(C_2)).$$

Since we consider unmixed actions only, we obtain two Galois coverings

$$f_1 : C_1 \to C_1/G, \quad f_2 : C_2 \to C_2/G.$$

By [8, Proposition 3.13] one can assume without loss of generality, that G acts faithfully on C_1 and on C_2. We want to relate the invariants $p_g(S)$ and $q(S)$ with the genera of the curves. To do this, we need the following theorem.

Theorem 2.4 *Let X be a smooth projective variety and G be a finite group acting faithfully on X. If $Y = X/G$ is smooth, then*

$$H^0(Y, \Omega_Y^p) \simeq H^0(X, \Omega_X^p)^G.$$

A proof of this result can be found in [12].

By Künneth's formula [13, pp. 103–104]:

$$H^0(C_1 \times C_2, \Omega^1_{C_1 \times C_2})^G = H^0(C_1, \Omega^1_{C_1})^G \oplus H^0(C_2, \Omega^1_{C_2})^G.$$

According to the previous theorem $q(S) = h^0(C_1 \times C_2, \Omega^1_{C_1 \times C_2})^G$. Since $q(S) = 0$ by assumption, we conclude that $g(C_i/G) = 0$ for both $i = 1, 2$. Thus the holomorphic maps f_i from above are Galois coverings of $\mathbb{P}^1_{\mathbb{C}}$.

3 Group Theory, Riemann Surfaces and Combinatorics

To give a purely group theoretic description of surfaces isogenous to a product, we introduce the required notation from group theory and combinatorics and recall *Riemann's existence theorem*.

Definition 3.1 Let $T = [m_1, \ldots, m_r] \in \mathbb{N}^r$ be an r-tuple. We define

$$\Theta(T) := -2 + \sum_{i=1}^{r} \left(1 - \frac{1}{m_i}\right)$$

and in case $\Theta(T) \neq 0$

$$\alpha(T) := \frac{4}{\Theta(T)}.$$

For $r \geq 3$ we denote by $\mathcal{N}_r$ the set of all r-tuples $[m_1, \ldots, m_r]$ with the following properties:

- $2 \leq m_1 \leq \ldots \leq m_r$
- $\Theta([m_1, \ldots, m_r]) > 0$
- $\alpha([m_1, \ldots, m_r]) \in \mathbb{N}$
- $m_i \big| \alpha([m_1, \ldots, m_r])$ for all $1 \leq i \leq r$.

The union of all $\mathcal{N}_r$ is defined as $\mathcal{N} := \bigcup_{r \geq 3} \mathcal{N}_r$. An element in $\mathcal{N}$ is called a type. A type T contained in $\mathcal{N}_r$ is said to be of length r, we write $l(T) = r$. Moreover for a type $T = [m_1, \ldots, m_r]$ we use the notation $T = [m_1, \ldots, m_r]_{\alpha(T)}$.

For simplicity we write

$$[a_1^{k_1}, \ldots, a_r^{k_r}] := [\underbrace{a_1, \ldots, a_1}_{k_1\text{-times}}, \ldots, \underbrace{a_r, \ldots, a_r}_{k_r\text{-times}}].$$

In the following lemma we give a classification of all types which satisfy the conditions from Definition 3.1 above. This is the starting point of the classification of regular surfaces isogenous to a product with $\chi(\mathcal{O}_S) = 2$ (see also 3.5).

Lemma 3.2 *There are no types of length r if $r = 7$ or $r \geq 9$. The set $\mathcal{N}$ is finite and given by:*

$$\mathcal{N} = \left\{ \begin{array}{lllll} [2,3,7]_{168}, & [2,3,8]_{96}, & [2,4,5]_{80}, & [2,3,9]_{72}, & [2,3,10]_{60}, \\ [2,3,12]_{48}, & [2,4,6]_{48}, & [3^2,4]_{48}, & [2,3,14]_{42}, & [2,5^2]_{40}, \\ [2,3,18]_{36}, & [2,4,8]_{32}, & [2,3,30]_{30}, & [2,5,6]_{30}, & [3^2,5]_{30}, \\ [2,4,12]_{24}, & [2,6^2]_{24}, & [3,4^2]_{24}, & [3^2,6]_{24}, & [2^3,3]_{24}, \\ [3^2,7]_{21}, & [2,4,20]_{20}, & [2,5,10]_{20}, & [2,6,9]_{18}, & [3^2,9]_{18}, \\ [2,8^2]_{16}, & [4^3]_{16}, & [2^3,4]_{16}, & [3,5^2]_{15}, & [3^2,15]_{15}, \\ [2,7,14]_{14}, & [2,12^2]_{12}, & [3,4,12]_{12}, & [3,6^2]_{12}, & [4^2,6]_{12}, \\ [2^3,6]_{12}, & [2^2,3^2]_{12}, & [5^3]_{10}, & [2^3,10]_{10}, & [3,9^2]_{9}, \\ [4,8^2]_{8}, & [2^2,4^2]_{8}, & [2^5]_{8}, & [7^3]_{7}, & [2^2,6^2]_{6}, \\ [2,3^2,6]_{6}, & [3^4]_{6}, & [2^4,3]_{6}, & [4^4]_{4}, & [2^3,4^2]_{4}, \\ [2^6]_{4}, & [3^5]_{3}, & [2^8]_{2} & & \end{array} \right\}$$

Proof We use the fourth property in the case $i = r$:

$$\text{From } m_r \Big| \frac{4}{-2 + \sum_{i=1}^{r}\left(1 - \frac{1}{m_i}\right)} \text{ it follows } \sum_{i=1}^{r}\left(1 - \frac{1}{m_i}\right) \leq 2 + \frac{4}{m_r} \quad (*).$$

Since $m_i \geq 2$ for all $1 \leq i \leq r$, we get

$$r - 2 \leq \sum_{i=1}^{r} \frac{1}{m_i} + \frac{4}{m_r} \leq \frac{r}{2} + 2,$$

and therefore $r \leq 8$. We now investigate two cases: $r = 3$ and $r \geq 4$.

- If $r = 3$, we claim that $m_2 \geq 3$. Suppose $m_2 = 2$, then also $m_1 = 2$ and

  $$0 < \Theta([m_1, m_2, m_3]) = 1 - \frac{1}{m_1} - \frac{1}{m_2} - \frac{1}{m_3} = -\frac{1}{m_3},$$

  a contradiction. The inequality $(*)$ in the case $r = 3$ reads

  $$\frac{5}{m_3} \geq 1 - \frac{1}{m_1} - \frac{1}{m_2} \geq 1 - \frac{1}{2} - \frac{1}{3} = \frac{1}{6}.$$

  From this we conclude $m_3 \leq 30$.
- In the second case $r \geq 4$, we use the formula $(*)$ again:

$$r - \sum_{i=1}^{r} \frac{1}{m_i} \leq 2 + \frac{4}{m_r}$$

and get

$$\frac{5}{m_r} \geq r - 2 - \sum_{i=1}^{r-1} \frac{1}{m_i} \geq \frac{r-3}{2}.$$

Because of this, we always have $10 \geq \frac{10}{r-3} \geq m_r$ in that case.

Now, it suffices to check only a finite number of types. This can be easily done with a computer.

Definition 3.3 Let G be a finite group, $2 \leq m_1 \leq \ldots \leq m_r$ integers. A *spherical system* of generators of G of type $[m_1, \ldots, m_r]$ is an r-tuple $A = (g_1, \ldots, g_r)$ of elements of G, such that:

- $G = \langle g_1, \ldots, g_r \rangle, \quad g_1 \cdot \ldots \cdot g_r = 1_G$.
- There exists a permutation $\tau \in \mathfrak{S}_r$, such that $ord(g_i) = m_{\tau(i)}$.

The *stabilizer set* of A is defined as

$$\Sigma(A) := \bigcup_{h \in G} \bigcup_{j \in \mathbb{Z}} \bigcup_{i=1}^{r} \{h g_i^j h^{-1}\}.$$

A pair (A_1, A_2) of spherical systems of generators of G is called *disjoint*, if and only if

$$\Sigma(A_1) \cap \Sigma(A_2) = \{1_G\}.$$

The geometry behind this definition is known as *Riemann's existence theorem*. A detailed explanation can be found in [16, chapter 3, Sects. 3 and 4]. We will use the following version of this theorem:

Theorem 3.4 (Riemann's existence theorem) *A finite group G acts as a group of automorphisms on a compact Riemann surface C of genus $g(C) \geq 2$, such that $C/G \simeq \mathbb{P}^1_{\mathbb{C}}$ if and only if there exists a spherical system of generators A of G of type $T = [m_1, \ldots, m_r]$, such that the following Riemann-Hurwitz formula holds:*

$$2g(C) - 2 = |G|\Theta(T).$$

By Riemann's existence theorem we have a *group theoretical description* of surfaces isogenous to a product: Given $S = (C_1 \times C_2)/G$, isogenous to a product, we can attach a disjoint pair of spherical systems of generators

$$(A_1(S), A_2(S)) \text{ of type } (T_1(S), T_2(S)).$$

Geometrically, disjoint means that G acts without fixed points on $C_1 \times C_2$. Conversely, the data above determine a surface isogenous to a product.

Next, we want to show that the types $(T_1(S), T_2(S))$, attached to a regular surface S isogenous to a product with $p_g(S) = 1$ of unmixed type, satisfy the conditions of 3.1. The proof of this fact is similar to the proof given in [1]. For convenience of the reader we will present the proof.

Theorem 3.5 *Let S be a surface isogenous to a product of curves of unmixed type with $p_g(S) = 1$ and $q(S) = 0$. Let $T_1(S) = [m_1, \dots, m_r]$ and $T_2(S) = [n_1, \dots, n_s]$ be the corresponding types, then*

- $\Theta(T_i(S)) > 0$ *for* $i = 1, 2$.
- $\alpha(T_i(S)) \in \mathbb{N}$ *for* $i = 1, 2$.
- $m_i \big| \alpha(T_1(S))$ *for all* $1 \leq i \leq r$ *and* $n_i \big| \alpha(T_2(S))$ *for all* $1 \leq i \leq s$.

Proof We consider the holomorphic maps $f_i : C_i \to C_i/G$ and apply the Riemann-Hurwitz formula

$$2g(C_i) - 2 = |G|\Theta(T_i(S)), \quad i = 1, 2. \tag{2}$$

This already shows the first claim, since $g(C_1) \geq 2$ and $g(C_2) \geq 2$. From (2) and $2|G| = (g(C_1) - 1)(g(C_2) - 1)$ (1), we deduce

$$\alpha(T_1(S)) = \frac{4}{\Theta(T_1(S))} = g(C_2) - 1 \text{ and } \alpha(T_2(S)) = \frac{4}{\Theta(T_2(S))} = g(C_1) - 1$$

so the second claim follows. It remains to prove the third claim. Let $A_1(S) = (g_1, \dots, g_r)$ be a corresponding ordered spherical system of generators of G of type $T_1(S)$. The cyclic group $\langle g_i \rangle$ of order m_i acts on C_1 with at least one fixed point, but the action on the product $C_1 \times C_2$ is free. Therefore $\langle g_i \rangle$ acts on C_2 freely. The map $C_2 \to C$ of degree m_i, where $C := C_2/\langle g_i \rangle$ is unramified. In this case we have $2g(C) - 2 = \frac{2g(C_2)-2}{m_i} > 0$, due to Riemann-Hurwitz. Hence

$$g(C_2) - 1 = \alpha(T_1(S)) = m_i(g(C) - 1),$$

for all $i = 1, \dots, r$. With the same argument we can show that $n_i \big| \alpha(T_2(S))$ for all $i = 1, \dots, s$.

4 Moduli Spaces

In this section we want to describe the moduli space of surfaces isogenous to a product. We follow the papers [1, S31, S8–9] and [18, Appendix]. Due to the work of Gieseker [11] there exists a quasi-projective moduli space of minimal smooth projective surfaces of general type with fixed invariants K_S^2 and $\chi(\mathcal{O}_S)$, which is denoted by $\mathfrak{M}_{(\chi(\mathcal{O}_S), K_S^2)}$. For a fixed finite group G and a fixed pair of types (T_1, T_2) we denote the subset of $\mathfrak{M}_{(2,16)}$ of isomorphism classes of surfaces isogenous to a product, which admit a disjoint pair of spherical systems of generators (A_1, A_2) of type (T_1, T_2), by $\mathfrak{M}_{(G,T_1,T_2)}$.

Theorem 4.1 ([1, Remark 5.1])

- *The subset $\mathfrak{M}_{(G,T_1,T_2)} \subset \mathfrak{M}_{(2,16)}$ consists of a finite number of connected components of the same dimension, which are irreducible in the Zariski topology.*
- *The dimension $d(G, T_1, T_2)$ of any component in $\mathfrak{M}_{(G,T_1,T_2)}$ is*

$$d(G, T_1, T_2) = l(T_1) - 3 + l(T_2) - 3.$$

The problem to determine the number n of the connected components of $\mathfrak{M}_{(G,T_1,T_2)}$ can be translated in a group theoretical problem. We recall the following definition:

Definition 4.2 Let $r \in \mathbb{N}$ be a positive integer. We define the *Artin-Braid group* $\boldsymbol{B}_r$ as

$$\boldsymbol{B}_r := \left\langle \sigma_1, \ldots, \sigma_{r-1} \;\middle|\; \begin{matrix} \sigma_i\sigma_j = \sigma_j\sigma_i \quad if \quad |i-j| > 1 \\ \sigma_i\sigma_{i+1}\sigma_i = \sigma_{i+1}\sigma_i\sigma_{i+1} \quad for \quad i = 1, \ldots, r-2 \end{matrix} \right\rangle.$$

Let G be a finite group and T be a type of length $l(T) = r$. We denote the set of spherical systems of generators of G of type T by $\mathcal{B}(G, T)$. The Artin-Braid group $\boldsymbol{B}_r$ acts on $\mathcal{B}(G, T)$ as follows:

$$\sigma_i(A) := (g_1, \ldots, g_{i-1}, g_i \cdot g_{i+1} \cdot g_i^{-1}, g_i, g_{i+2}, \ldots, g_r),$$

for all $A = (g_1, \ldots, g_r) \in \mathcal{B}(G, T)$ and $1 \leq i \leq r-1$. This determines a well-defined action, which is called the *Hurwitz action*. There is also a natural action of $Aut(G)$ on $\mathcal{B}(G, T)$:

$$\varphi(A) := (\varphi(g_1), \ldots, \varphi(g_r)),$$

for all $\varphi \in Aut(G)$. Let $(\gamma_1, \gamma_2, \varphi) \in \boldsymbol{B}_r \times \boldsymbol{B}_s \times Aut(G)$ be a triple and $(A_1, A_2) \in \mathcal{B}(G, T_1) \times \mathcal{B}(G, T_2)$, we define

$$(\gamma_1, \gamma_2, \varphi) \cdot (A_1, A_2) := (\varphi(\gamma_1(A_1)), \varphi(\gamma_2(A_2))).$$

It is easy to verify that this defines an action of $\boldsymbol{B}_r \times \boldsymbol{B}_s \times Aut(G)$ on $\mathcal{B}(G, T_1) \times \mathcal{B}(G, T_2)$. We denote this action by $\mathfrak{T}$, and its restriction to the first factor $\mathcal{B}(G, T_1)$ by $\mathfrak{T}_1$ and to the second factor $\mathcal{B}(G, T_2)$ by $\mathfrak{T}_2$.[2]

Theorem 4.3 ([1, Proposition 5.2]) *Let S and S' be surfaces isogenous to a product of unmixed type with $q(S) = q(S') = 0$. The surfaces S and S' are in the same irreducible component if and only if*

- *$G(S) \simeq G(S')$,*
- *$(T_1(S), T_2(S)) = (T_1(S'), T_2(S'))$,*
- *$(A_1(S), A_2(S))$ and $(A_1(S'), A_2(S'))$ are in the same $\mathfrak{T}$-orbit, or $(A_1(S), A_2(S))$ and $(A_2(S'), A_1(S'))$ are in the same $\mathfrak{T}$-orbit.*

[2]We want to stress that the action of $Aut(G)$ and the Hurwitz action commute.

In theory we now have a method to compute the number of connected components of $\mathfrak{M}_{(G;T_1,T_2)}$: first we compute a representative (A_1, A_2) for each orbit of the action $\mathfrak{T}$. Then we determine the pairs, where the intersection $\Sigma(A_1) \cap \Sigma(A_2)$ is trivial. If $T_1(S) = T_2(S)$ and there is more than one pair, we have to consider the $\mathbb{Z}_2$ action corresponding to the exchange of the curves.[3] The number of the remaining pairs is the number of connected components of $\mathfrak{M}_{(G;T_1,T_2)}$. However, the set $\mathcal{B}(G, T_1) \times \mathcal{B}(G, T_2)$ can be very large. Even with a computer it is not possible, or at least very time consuming, to perform this calculation. To improve the speed of the calculation we use an idea of Penegini and Rollenske, which is based on the following lemma:

Lemma 4.4 ([18, Appendix 6.1]) *Let* $(A_1, A_2), (B_1, B_2) \in \mathcal{B}(G, T_1) \times \mathcal{B}(G, T_2)$.

- *If* A_i *and* B_i *are in the same orbit of the Hurwitz action, then also the pairs* (A_1, A_2) *and* (B_1, B_2) *are in the same* $\mathfrak{T}$*-orbit.*
- *If* A_1 *and* B_1 *are in different* $\mathfrak{T}_1$*-orbits, then also* (A_1, A_2) *and* (B_1, B_2) *are in different* $\mathfrak{T}$ *orbits.*

Now we have an effective algorithm:

- Compute a set $\mathfrak{R}_1$ of representatives of the action $\mathfrak{T}_1$ on $\mathcal{B}(G, T_1)$.
- Compute a set $\mathfrak{R}_2$ of representatives of the Hurwitz action on $\mathcal{B}(G, T_2)$.
- Determine the set of tuples $(A_1, A_2) \in \mathfrak{R}_1 \times \mathfrak{R}_2$ which satisfy:

$$\Sigma(A_1) \cap \Sigma(A_2) = \{1_G\}.$$

This set is denoted by $\mathfrak{R}$.

We achieve the following: Every orbit of $\mathfrak{T}$ has at least one representative in $\mathfrak{R}$ by 4.4. Hence we have an upper bound for the number n of $\mathfrak{T}$-orbits. We also have a lower bound for n. The pairs (A_1, A_2) and (B_1, B_2) are in different orbits of $\mathfrak{T}$, if

1. $A_1 \neq B_1$ or if
2. A_2 and B_2 are in different orbits of $\mathfrak{T}_2$ (cf. 4.4).

- In most cases it is possible to determine the number n of orbits of $\mathfrak{T}$ using this method. If this is not possible, then we exchange the pairs T_1 and T_2 and compute the set $\mathfrak{R}$ again. If it is still not possible to determine n we have to identify the pairs of the smaller set using the action $\mathfrak{T}$.

5 The Algorithm and the Classification Result

In this section we explain our algorithm, which allows us to classify all regular surfaces isogenous to a product of curves of unmixed type with $\chi(\mathcal{O}(S)) = 2$. For the implementation of the algorithm the computer algebra system MAGMA [15]

[3] This happens in only one of our examples. Thus we consider only the $\mathfrak{T}$ action in our program and treat the exceptional example separately 5.8.

is used. The program is based on the program in the appendix of [2]. After the explanation of the algorithm, we discuss the output of the computations. Finally we give our classification result.

Let $S = (C_1 \times C_2)/G$ be a regular surface isogenous to a product of unmixed type with $\chi(\mathcal{O}_S) = 2$. From $2|G| = (g(C_1)-1)(g(C_2)-1)$, $\alpha(T_1(S)) = g(C_2)-1$ and $\alpha(T_2(S)) = g(C_1) - 1$ for the attached types $T_i(S)$, it follows

$$|G| = \frac{1}{2}\alpha(T_1)\alpha(T_2).$$

According to the list in Lemma 3.2, $\alpha(T_i) \leq 168$ and therefore $|G| \leq 14112$. The group order is also bounded in terms of the genera $g(C_i)$, in fact $|Aut(C_i)| \leq 84(g(C_i) - 1)$ due to Hurwitz' famous theorem. For small $g(C_i)$ there are better bounds. In Breuer's book [5, p. 91] there is a table which gives the maximum order of $|Aut(C_i)|$ in case $2 \leq g(C_i) \leq 48$.

Definition 5.1 Let $A = [m_1, \ldots, m_r]$ be an r-tuple of integers $m_i \geq 2$. The *polygonal group* $\mathbb{T}(m_1, \ldots, m_r)$ is defined as

$$\mathbb{T}(m_1, \ldots, m_r) = < t_1, \ldots, t_r \mid t_1 \cdot \ldots \cdot t_r = t_1^{m_1} = \ldots = t_r^{m_r} = 1 > .$$

A group G admitting a spherical system of generators of type $[m_1, \ldots, m_r]$ is a quotient of $\mathbb{T}(m_1, \ldots, m_r)$. The following lemma will be used in the sequel. The proof of it is elementary and will be omitted.

Lemma 5.2 *Let G be a group and H a quotient of G, then:*

- H^{ab} *is a quotient of* G^{ab}.
- *The commutator subgroup* $[H, H]$ *is a quotient of* $[G, G]$.
- *If G is a quotient of* $\mathbb{T}(2, 3, 7)$, *then G is perfect.*

We can now describe the algorithm briefly. We perform the following steps:

Step 1: The program computes the set $\mathcal{N}$ of types given in Lemma 3.2. For every integer $m \leq 14112$ we compute the set of all triples of the form (m, T_1, T_2) up to permutation of T_1 and T_2, where $T_1, T_2 \in \mathcal{N}$ and $m = \frac{1}{2}\alpha(T_1)\alpha(T_2)$.
Step 2: For every triple (m, T_1, T_2) the script computes $g(C_2) = \alpha(T_1(S)) + 1$ and $g(C_1) = \alpha(T_2(S)) + 1$. If $2 \leq g(C_i) \leq 48$ for at least one i, we check if m is less or equal to the maximum group order of the automorphism group $Aut(C_i)$ allowed by Breuer's table.
Step 3: For every triple (m, T_1, T_2) passing this test, the script searches the list of groups of order m for a group admitting a spherical system of generators of type T_1 and one of type T_2.
Step 4: For each triple (m, T_1, T_2) and each group G of order m, admitting a spherical system of generators of type T_1 and of type T_2, the script computes the number of orbits of the action $\mathfrak{T}$ on $\mathcal{B}(G, T_1) \times \mathcal{B}(G, T_2)$, using the method explained after Lemma 4.4.

In **Step 3** we face two computational difficulties:

- In MAGMA's *Database of Small Groups* all groups of order $|G| \leq 2000$ are contained, except the groups of order $|G| = 1024$ (which can not occur). In the case $2001 \leq |G| \leq 14112$, there is no MAGMA *Database* containing all groups of these orders. Only the perfect groups with $|G| \leq 50000$ are contained in MAGMA's *Database of Perfect Groups*.
- Groups of order
$$|G| \in \{1920, 1152, 768, 512, 384, 256\}$$
can occur. Despite the fact, that we have access to all groups of these orders, it is not efficient to search through all of them for spherical systems of generators, because the number of these groups is too high.[4]

Due to these difficulties we split the program into two *main routines* namely $Mainloop1$ and $Mainloop2$.

- The function $Mainloop1$ treats the cases where
$$|G| \geq 2001 \quad \text{or} \quad |G| \in \{1920, 1152, 768, 512, 384, 256\}.$$

(i) The case $m = |G| \geq 2001$. For all triples (m, T_1, T_2), we search the *Database of Perfect Groups* for a perfect group of order m, admitting a spherical system of generators of type T_1 and T_2. If neither $T_1 = [2, 3, 7]$ nor $T_2 = [2, 3, 7]$ we can not yet decide if there is a non-perfect group of order m, admitting a spherical system of generators of type T_i (cf. 5.2). The script then saves the triple in the file $exceptional.txt$. We have to investigate these cases with theoretical arguments (see Sect. 5.1), and we will show these do not occur.

(ii) The case $|G| \in \{1920, 1152, 768, 512, 384, 256\}$. Each group G, admitting a spherical system of generators of type $T_1 = [n_1, \ldots, n_r]$ and of type $T_2 = [m_1, \ldots, m_s]$ is a quotient of
$$\mathbb{T}(T_1) := \mathbb{T}(n_1, \ldots, n_r) \quad \text{and} \quad \mathbb{T}(T_2) := \mathbb{T}(m_1, \ldots, m_s).$$
According to Lemma 5.2 the group G^{ab} is a quotient of
$$\mathbb{T}(n_1, \ldots, n_r)^{ab} \quad \text{and} \quad \mathbb{T}(m_1, \ldots, m_s)^{ab}.$$
The script first computes all possible abelianizations of G from the types. With this step we can exclude the groups which don't have the right abelianization. Then we search through the remaining groups if there is a group, admitting spherical systems of generators of type T_i.

[4]Indeed there are 12.059.590 groups G, such that $|G| \in \{1920, 1152, 768, 512, 384, 256\}$.

- The function *Mainloop*2 treats the case where

$$|G| \leq 2000 \text{ and } |G| \notin \{1920, 1152, 768, 512, 384, 256\}.$$

The output is written in the file *loop*2*.txt*. Due to the high use of memory, if one of the types is $[2^8]$, we split the computation into two parts:

(i) With the command $Mainloop2(n_1, n_2, 1)$ the script classifies, in the sense of above, all surfaces where $n_1 \leq |G| \leq n_2$ and $T_i \neq [2^8]$.
(ii) With the command $Mainloop2(n_1, n_2, 0)$ the script classifies all surfaces where $n_1 \leq |G| \leq n_2$ and $T_1 = [2^8]$.

Also in this main routine, if $T_1 = [2^8]$ we have to treat some cases separately. Our workstation has not enough memory to compute the set $\mathcal{B}(G, [2^8])$, if

$$|G| \in \{168, 96, 48\}.$$

The occurring triples (n, T_1, T_2), where n is one of the group orders above are marked by the script as exceptional. We treat these cases in Sect. 5.2.

Before we discuss the exceptional cases of the output, we explain some notation from group theory, that will be used in this and in the next section.

- We use the following MAGMA notation: $\langle a, b\rangle$ denotes the group of order a having number b in the database of Small Groups [15].
- The group $U(4, 2) \leq Gl(4, \mathbb{F}_2)$ is defined to be the subgroup of upper triangle matrices $A = (a_{ij})$ with $a_{ii} = 1$, for all $1 \leq i \leq 4$.
- The group $G(128, 36) := \langle 128, 36\rangle$ is given in a polycyclic presentation:

$$\langle 128, 36\rangle = \left\langle g_1, \ldots, g_7 \;\middle|\; \begin{array}{l} g_1^2 = g_4, \quad g_2^2 = g_5, \quad g_2^{g_1} = g_2g_3 \\ g_3^{g_1} = g_3g_6, \; g_3^{g_2} = g_3g_7, \; g_4^{g_2} = g_4g_6 \\ g_5^{g_1} = g_5g_7 \end{array} \right\rangle$$

Here $g_i^{g_j}$ means $g_j^{-1}g_ig_j$. The squares of the generators $g_1, \ldots, g_7$, which are not mentioned in the presentation are equal to 1. If $g_i^{g_j} = g_i$, this relation is omitted in the presentation.

5.1 Exceptional Cases for Mainloop1:

The following cases are skipped by the program, because the group order is greater than 2000, and saved in the file *exceptional.txt*:

$\|G\|$	T_1	T_2
4608	[2, 3, 8]	[2, 3, 8]
3840	[2, 3, 8]	[2, 4, 5]
3456	[2, 3, 8]	[2, 3, 9]
3200	[2, 4, 5]	[2, 4, 5]
2880	[2, 3, 9]	[2, 4, 5]
2880	[2, 3, 8]	[2, 3, 10]
2592	[2, 3, 9]	[2, 3, 9]
2400	[2, 3, 10]	[2, 4, 5]
2304	[2, 3, 8]	[2, 4, 6]
2304	[2, 3, 8]	[3, 3, 4]
2304	[2, 3, 8]	[2, 3, 12]
2160	[2, 3, 9]	[2, 3, 10]

Our aim is to show, that the cases above can not occur.

Proposition 5.3 *There are no finite groups G admitting a disjoint pair of spherical systems of generators (A_1, A_2) of type (T_1, T_2), where $|G|$, T_1 and T_2 are in the table above.*

Proof Our MAGMA code has already excluded the cases where G is perfect. We only treat the case $|G| = 4608$, $T_1 = [2, 3, 8]$ and $T_2 = [2, 3, 8]$. The other cases can be excluded using similar methods. The ideas we use are from [1]. The group G is a quotient of $\mathbb{T}(2, 3, 8)$ and there is a surjective homomorphism $\phi^{ab} : \mathbb{T}(2, 3, 8)^{ab} \to G^{ab}$. Since G is not perfect, $G^{ab} \neq \{1_G\}$. Similarly, the commutator subgroup $G' = [G, G]$ is also a quotient of $\mathbb{T}(2, 3, 8)' = [\mathbb{T}(2, 3, 8), \mathbb{T}(2, 3, 8)]$. We have

$$\mathbb{T}(2, 3, 8)^{ab} \simeq \mathbb{Z}_2, \quad \text{and} \quad \mathbb{T}(2, 3, 8)' \simeq \mathbb{T}(3, 3, 4).$$

This implies $G^{ab} \simeq \mathbb{Z}_2$, thus $|G'| = 2304$ and the group G' is a quotient of $\mathbb{T}(3, 3, 4)$.

$$\mathbb{T}(3, 3, 4)^{ab} \simeq \mathbb{Z}_3, \quad \text{and} \quad \mathbb{T}(3, 3, 4)' \simeq \mathbb{T}(4, 4, 4).$$

Since $|G'| = 2^8 \cdot 3^2$ the group G' is solvable, due to Burnsides theorem [6]. We have $G'^{ab} \simeq \mathbb{Z}_3$, thus: $|G''| = 768$ and G'' is a quotient of $\mathbb{T}(4, 4, 4)$. This is impossible according to the following computational fact, which can be verified with MAGMA.

Lemma 5.4 ([1, Lemma 4.11]) *There are* 1090235 *groups of order* 768. *None of them is a quotient of* $\mathbb{T}(4, 4, 4)$.

5.2 Exceptional Cases for Mainloop2:

Here $T_1 = [2^8]$ and $T_2 \in \mathcal{N}$. We have $|G| = \frac{1}{2}\alpha(T_1)\alpha(T_2) = \alpha(T_2) \leq 168$. Since our workstation has not enough memory to compute the set $\mathcal{B}(G, [2^8])$ if $|G| \in \{168, 96, 48\}$, these cases are marked as exceptional. We receive the following output:

T_2	G	No. of $\mathfrak{T}$-orbits	
[2, 3, 7]	⟨168, 42⟩	0	Exceptional case
[2, 3, 8]	⟨96, 64⟩	0	Exceptional case
[2, 4, 6]	⟨48, 48⟩	2	Exceptional case
$[3, 4^2]$	⟨24, 12⟩	1	
$[2^3, 4]$	⟨16, 11⟩	2	
$[2^5]$	⟨8, 5⟩	1	

Next, we will investigate the exceptional cases.

Proposition 5.5 *The group* ⟨168, 42⟩ *has no disjoint pair of spherical systems of generators of type* $([2^8], [2, 3, 7])$.

Proof A MAGMA computation shows, that this group has only one conjugacy class of elements of order 2. Hence, for every pair (A_1, A_2) of generators of type $([2^8], [2, 3, 7])$ the intersection $\Sigma(A_1) \cap \Sigma(A_2)$ is nontrivial.

Proposition 5.6 *The group* ⟨96, 64⟩ *has no disjoint pair of spherical systems of generators of type* $([2^8], [2, 3, 8])$.

Proof A MAGMA calculation shows, that this group has two conjugacy classes of elements of order 2. We denote them by K_1 and K_2. We have $|K_2| = 3$ and $\langle K_2 \rangle \simeq \mathbb{Z}_2 \times \mathbb{Z}_2$. Since $|\langle K_2 \rangle| = 4$, there is no spherical system of generators $A_1 = (h_1, \ldots, h_8)$ of G of type $[2^8]$, with $h_i \in K_2$ for all $1 \leq i \leq 8$. Every spherical system of generators of G contains elements of K_1. A further MAGMA calculation shows, that there is no spherical system of generators $A_2 = (g_1, g_2, g_3)$ of G of type [2, 3, 8], with $g_1 \in K_2$. □

Proposition 5.7 *The group* ⟨48, 48⟩ *admits disjoint pairs of spherical systems of generators of type* $([2^8], [2, 4, 6])$. *The number of* $\mathfrak{T}$*-orbits is two.*

Proof The group G has 19 elements of order 2, they are contained in 5 conjugacy classes. We denote them by $K_1, \ldots, K_5$ and the set of elements of order two by M.

Class	Rep	Length
K_1	$G.2$	1
K_2	$G.4$	3
K_3	$G.2 * G.4$	3
K_4	$G.1$	6
K_5	$G.1 * G.2$	6

A MAGMA calculation shows, that $g_1 \in K_4 \cup K_5$ for each spherical system of generators $(g_1, g_2, g_3) \in \mathcal{B}(G, [2, 4, 6])$. Since we are interested in disjoint pairs of spherical systems of generators, it is not necessary to compute the whole set $\mathcal{B}(G, [2^8])$. All elements $(h_1, \ldots, h_8) \in \mathcal{B}(G, [2^8])$, which contain some $h_i \in K_4$, as well as some $h_j \in K_5$ are irrelevant. For each $1 \leq l \leq 5$ we define subsets

$$\mathcal{B}_l := \{[h_1, \ldots, h_8] \in \mathcal{B}(G, [2^8]) \mid h_i \in M \setminus K_l\}$$

of $\mathcal{B}(G, [2^8])$ and compute them in the cases $l = 4$ and $l = 5$. We have $|\mathcal{B}_4| = |\mathcal{B}_5| = 9.213.120$. Since the Hurwitz action acts via conjugation and permutation of elements, we can restrict it to $\mathcal{B}_4$ and to $\mathcal{B}_5$. Next we compute for each orbit of the restricted actions a representative. The sets of representatives are denoted by $\mathcal{R}_4$ and $\mathcal{R}_5$. We have $|\mathcal{R}_4| = |\mathcal{R}_5| = 10$. All elements in $\mathcal{B}(G, [2, 4, 6])$ are contained in two orbits of the Hurwitz action. We denote by A_1 and A_2 two representatives of these orbits. There are two disjoint pairs in $\{A_1, A_2\} \times \mathcal{R}_4$, and two disjoint pairs in $\{A_1, A_2\} \times \mathcal{R}_5$. According to 4.4, we have at least one representative for each $\mathfrak{T}$-orbit. Using the function"Orbi" we can identify the pairs above, which are $\mathfrak{T}$-equivalent. We find two equivalence classes. To verify the above computations a source code can be found at www.staff.uni-bayreuth.de/~bt300503/ in the file *script.txt*.

Remark 5.8 From the output files we can see that there is exactly one occurrence with $T_1(S) = T_2(S)$ and $n \geq 2$. In this case $G = G(128, 36)$. The types are $T_i = [4, 4, 4]$ and $n = 2$. We denote by (A_1, A_2) and (B_1, B_2) the representatives for the $\mathfrak{T}$-orbits from the output file *loop2.txt*. It remains to check if (A_2, A_1) and (B_1, B_2) are in the same $\mathfrak{T}$-orbit. A MAGMA computation shows that this is not the case. The source code for this computation is available at www.staff.uni-bayreuth.de/~bt300503/ in the file *script.txt*.

Now we can give our main theorem, which implies in particular Theorem 1.1.

Theorem 5.9 *Let $S = (C_1 \times C_2)/G$ be a regular surface isogenous to a product of unmixed type with $\chi(\mathcal{O}_S) = 2$. Then $g(C_1)$, $g(C_2)$, the group G and the corresponding types $T_1(S)$, $T_2(S)$ are:*

$g(C_1)$	$g(C_2)$	G	Id	$T_1(S)$	$T_2(S)$	n
17	43	$PSL(2, \mathbb{F}_7) \times \mathbb{Z}_2$	$\langle 336, 209\rangle$	$[2, 3, 14]$	$[4^3]$	2
49	9	$(\mathbb{Z}_2)^3 \rtimes_\varphi \mathfrak{S}_4$	$\langle 192, 955\rangle$	$[2^2, 4^2]$	$[2, 4, 6]$	2
49	8	$PSL(2, \mathbb{F}_7)$	$\langle 168, 42\rangle$	$[7^3]$	$[3^2, 4]$	2
17	22	$PSL(2, \mathbb{F}_7)$	$\langle 168, 42\rangle$	$[3^2, 7]$	$[4^3]$	2
5	81	$(\mathbb{Z}_2)^4 \rtimes_\varphi \mathcal{D}_5$	$\langle 160, 234\rangle$	$[2, 4, 5]$	$[4^4]$	5
17	17	$G(128, 36)$	$\langle 128, 36\rangle$	$[4^3]$	$[4^3]$	2
9	31	$\mathfrak{S}_5$	$\langle 120, 34\rangle$	$[2, 5, 6]$	$[2^2, 4^2]$	1
5	49	$(\mathbb{Z}_2)^4 \rtimes_\varphi \mathcal{D}_3$	$\langle 96, 195\rangle$	$[2, 4, 6]$	$[4^4]$	1
25	9	$(\mathbb{Z}_2)^4 \rtimes_\varphi \mathcal{D}_3$	$\langle 96, 227\rangle$	$[2^5]$	$[3, 4^2]$	1
9	17	$(\mathbb{Z}_2)^3 \rtimes_\varphi \mathcal{D}_4$	$\langle 64, 73\rangle$	$[2^3, 4]$	$[2^2, 4^2]$	1
9	17	$U(4, 2)$	$\langle 64, 138\rangle$	$[2^3, 4]$	$[2^2, 4^2]$	1
13	11	$\mathcal{A}_5$	$\langle 60, 5\rangle$	$[5^3]$	$[2^2, 3^2]$	1
41	4	$\mathcal{A}_5$	$\langle 60, 5\rangle$	$[3^5]$	$[2, 5^2]$	2
9	16	$\mathcal{A}_5$	$\langle 60, 5\rangle$	$[3, 5^2]$	$[2^5]$	2
5	31	$\mathcal{A}_5$	$\langle 60, 5\rangle$	$[3^2, 5]$	$[2^6]$	1
5	25	$\mathfrak{S}_4 \times \mathbb{Z}_2$	$\langle 48, 48\rangle$	$[2^3, 3]$	$[4^4]$	1
9	13	$\mathfrak{S}_4 \times \mathbb{Z}_2$	$\langle 48, 48\rangle$	$[2^3, 6]$	$[2^2, 4^2]$	1
13	9	$\mathfrak{S}_4 \times \mathbb{Z}_2$	$\langle 48, 48\rangle$	$[2^5]$	$[4^2, 6]$	1
3	49	$\mathfrak{S}_4 \times \mathbb{Z}_2$	$\langle 48, 48\rangle$	$[2, 4, 6]$	$[2^8]$	2
9	9	$(\mathbb{Z}_2)^3 \rtimes_\varphi \mathbb{Z}_4$	$\langle 32, 22\rangle$	$[2^2, 4^2]$	$[2^2, 4^2]$	1
9	9	$\mathcal{D}_4 \times (\mathbb{Z}_2)^2$	$\langle 32, 46\rangle$	$[2^5]$	$[2^5]$	1
17	5	$(\mathbb{Z}_2)^4 \rtimes_\varphi \mathbb{Z}_2$	$\langle 32, 27\rangle$	$[2^3, 4^2]$	$[2^3, 4]$	3
9	9	$(\mathbb{Z}_2)^4 \rtimes_\varphi \mathbb{Z}_2$	$\langle 32, 27\rangle$	$[2^2, 4^2]$	$[2^5]$	1
5	13	$\mathfrak{S}_4$	$\langle 24, 12\rangle$	$[2^2, 3^2]$	$[4^4]$	1
3	25	$\mathfrak{S}_4$	$\langle 24, 12\rangle$	$[3, 4^2]$	$[2^8]$	1
5	9	$\mathcal{D}_4 \times \mathbb{Z}_2$	$\langle 16, 11\rangle$	$[2^5]$	$[2^3, 4^2]$	1
9	5	$(\mathbb{Z}_2)^2 \rtimes_\varphi \mathbb{Z}_4$	$\langle 16, 3\rangle$	$[2^3, 4^2]$	$[2^2, 4^2]$	2
9	5	$(\mathbb{Z}_2)^4$	$\langle 16, 14\rangle$	$[2^6]$	$[2^5]$	2
3	17	$\mathcal{D}_4 \times \mathbb{Z}_2$	$\langle 16, 11\rangle$	$[2^3, 4]$	$[2^8]$	2
7	4	$(\mathbb{Z}_3)^2$	$\langle 9, 2\rangle$	$[3^5]$	$[3^4]$	1
5	5	$(\mathbb{Z}_2)^3$	$\langle 8, 5\rangle$	$[2^6]$	$[2^6]$	1
3	9	$(\mathbb{Z}_2)^3$	$\langle 8, 5\rangle$	$[2^5]$	$[2^8]$	1

Each row in the table corresponds to a union of connected components of the Gieseker moduli space *of surfaces of general type with* $K_S^2 = 16$ *and* $= 2$. *The number* n *of these components is given in the last column.*

Acknowledgments The author thanks Ingrid Bauer and Sascha Weigl for several suggestions, useful discussions and very careful reading of the paper.

References

1. I. Bauer, F. Catanese, F. Grunewald, The classification of surfaces with $p_g = q = 0$ isogenus to a product. Pure Appl. Math. Q. part 1 **4**(2), part 1, 547–586 (2008)
2. I. Bauer, F. Catanese, F. Grunewald, R. Pignatelli, Quotients of product of curves, new surfaces with $p_g = 0$ and their fundamental groups. Am. J. Math. **134**(4), 993–1049 (2012)
3. A. Beauville, L' Inégalité $p_g \geq 2q - 4$ pour les surfaces de type général. Bull. Soc. Math. **110**, 343–346 (1982)
4. A. Beauville, *Complex Algebraic Surfaces*. vol. 68, London Mathematical Society Lecture Note Series (Cambridge University Press, Cambridge, 1983)
5. T. Breuer, *Characters and Automorphism Groups of Compact Riemann Surfaces*. London Mathematical Society Lecture Note Series, vol. 280 (Cambridge University Press, Cambridge, 2000)
6. W. Burnside, On groups of order $p^a q^b$. Proc. Lond. Math. Soc. 388 (1904)
7. G. Carnovale, F. Polizzi, The classification of surfaces with $p_g = q = 1$ isogenus to a product of curves. Adv. Geom. **9**(2), 233–256 (2009)
8. F. Catanese, Fibred surfaces, varieties isogenus to a product and related moduli spaces. Am. J. Math. **122**, 1–44 (2000)
9. F. Catanese, C. Ciliberto, M. Mendes Lopes, On the classification of irregular surfaces of general type with nonbirational bicanonical map. Trans. Am. Math. Soc. **350**, 275–308 (1998)
10. A. Garbagnati, M. Penegini, K3 surfaces with a non-symplectic automorphism and product-quotient surfaces with cyclic groups (2013). arXiv:1303.1653v1
11. D. Gieseker, Global moduli for surfaces of general type. Invent. Math. **43**(3), 233–282 (1977)
12. P.A. Griffiths, Variations on a theorem of Abel. Invent. Math. **35**, 321–390 (1976)
13. P. Griffiths, J. Harris, *Principles of Algebraic Geometry*. A Wiley-interscience publication (Wiley, New York, 1978)
14. C.D. Hacon, R. Pardini, Surfaces with $p_g = q = 3$. Trans. Amer. Math. Soc. **354**(7), 2631–2638 (2002)
15. MAGMA Databases of Groups; http://magma.maths.usyd.edu.au/magma/handbook/text/668
16. R. Miranda, *Algebraic Curves and Riemann Surfaces*. Graduate Studies in Mathematics, vol. 5 (American Mathematical Society, Providence, 1995)
17. M. Penegini, The classification of isotrivially fibred surfaces with $p_g = q = 2$, and topics on Beauville surfaces. PhD thesis, Universität Bayreuth, 2010
18. M. Penegini, The classification of isotrivially fibred surfaces with $p_g = q = 2$. Collect. Math. **62**(3), 239–274 (2011)
19. G.P. Pirola, Surfaces with $p_g = q = 3$. Manuscripta Math. **108**(2), 163–170 (2002)

Characteristically Simple Beauville Groups, II: Low Rank and Sporadic Groups

Gareth A. Jones

Abstract A Beauville surface is a rigid complex surface of general type, isogenous to a higher product by the free action of a finite group, called a Beauville group. Here we consider which characteristically simple groups can be Beauville groups. We show that if G is a cartesian power of a simple group $L_2(q)$, $L_3(q)$, $U_3(q)$, $Sz(2^e)$, $R(3^e)$, or of a sporadic simple group, then G is a Beauville group if and only if it has two generators and is not isomorphic to A_5.

MSC classification: 20B25 (primary) · 14J50 · 20G40 · 20H10 (secondary)

1 Introduction

A Beauville surface $\mathcal{S}$ of unmixed type is a complex algebraic surface which is isogenous to a higher product (that is, $\mathcal{S} = (\mathcal{C}_1 \times \mathcal{C}_2)/G$, where $\mathcal{C}_1$ and $\mathcal{C}_2$ are complex algebraic curves of genus $g_i \geq 2$, and G is a finite group acting freely on their product), and is rigid in the sense that G preserves the factors $\mathcal{C}_i$, with $\mathcal{C}_i/G \cong \mathbb{P}^1(\mathbb{C})$ $(i = 1, 2)$ and the induced covering $\beta_i : \mathcal{C}_i \to \mathbb{P}^1(\mathbb{C})$ ramified over three points. (In this paper we will not consider Beauville surfaces of mixed type, where elements of G transpose the factors $\mathcal{C}_i$.) The first examples, with $\mathcal{C}_1 = \mathcal{C}_2$ Fermat curves, were introduced by Beauville [3, p. 159] in 1978. Subsequently, these surfaces have been intensively studied by geometers such as Bauer et al. and Catanese (see [1, 2, 6], for instance).

Recently, there has been considerable interest in determining which groups G, known as Beauville groups, can arise in this way. It is easily shown that the alternating group A_5 is not a Beauville group. In 2005 Bauer et al. [1] conjectured that every other non-abelian finite simple group is a Beauville group. In 2010, more or less simultaneously, Garion et al. [13] proved that this is true with finitely many possible

G.A. Jones (✉)
School of Mathematics, University of Southampton, Southampton SO17 1BJ, UK
e-mail: G.A.Jones@maths.soton.ac.uk

I. Bauer et al. (eds.), *Beauville Surfaces and Groups*, Springer Proceedings in Mathematics & Statistics 123, DOI 10.1007/978-3-319-13862-6_7

exceptions, while Guralnick and Malle [14] and Fairbairn et al. [11] proved that it is completely true. It is natural to consider other classes of finite groups, and this paper is part of a project to extend this result to characteristically simple groups, those with no characteristic subgroups other than 1 and G.

A finite group is characteristically simple if and only if it is isomorphic to a cartesian power H^k of a simple group H. In [6], Catanese showed that the abelian Beauville groups are those isomorphic to C_n^2 for some n coprime to 6; these are characteristically simple if and only if n is prime, so we may assume that H is a non-abelian finite simple group.

The ramification condition on the coverings β_i implies that a Beauville group is a quotient of a triangle group (in two ways), so it is a 2-generator group. Clearly we require that $G \not\cong A_5$, and in [20] it was conjectured that these two conditions are also sufficient:

Conjecture *Let G be a non-abelian finite characteristically simple group. Then G is a Beauville group if and only if it is a 2-generator group not isomorphic to A_5.*

This was proved in [20] in the case where $G = H^k$ with H an alternating group A_n. (Further details on the background to this problem are given there, while [19] gives a more general survey of Beauville groups.) By the classification of finite simple groups, this leaves the simple groups of Lie type, together with the 26 sporadic simple groups. Our aim here is to verify the conjecture for five families of groups of small Lie rank, specifically the projective special linear groups $L_2(q)$ and $L_3(q)$, the unitary groups $U_3(q)$, the Suzuki groups $Sz(2^e)$ and the 'small' Ree groups $R(3^e)$, together with the sporadic simple groups (see [8, 33] for notation and properties of these and other finite simple groups).

The 2-generator condition on H^k can be restated as follows. For any 2-generator finite group H, there is an integer $c_2(H)$ such that H^k is a 2-generator group if and only if $k \le c_2(H)$. If H is a non-abelian finite simple group (and hence a 2-generator group, by the classification), then $c_2(H)$ is equal to the number $d_2(H)$ of orbits of $\mathrm{Aut}\, H$ on ordered pairs of generators of H. Thus in this case, H^k is a 2-generator group if and only if $k \le d_2(H)$.

For the groups $L_2(q) = SL_2(q)/\{\pm I\}$, which are simple for all prime powers $q \ge 4$, with $L_2(4) \cong L_2(5) \cong A_5$, we will prove:

Theorem 1.1 *Let $G = H^k$, where $k \ge 1$ and $H = L_2(q)$ for a prime power $q \ge 4$. Then the following are equivalent:*

(a) G is a Beauville group;
(b) G is a 2-generator group not isomorphic to A_5;
(c) $k \le d_2(H)$, and if $q \le 5$ then $k > 1$.

The groups $L_3(q)$ and $U_3(q)$ are simple for all prime powers $q \ge 2$ and 3 respectively, while the Suzuki groups $Sz(2^e)$ and the Ree groups $R(3^e)$ are simple for all odd $e \ge 3$. For these and for the sporadic simple groups, we have a slightly simpler formulation:

Theorem 1.2 *Let $G = H^k$, where $k \geq 1$ and H is a simple group $L_3(q)$, $U_3(q)$, $Sz(2^e)$ or $R(3^e)$, or a sporadic simple group. Then the following are equivalent:*

(a) G is a Beauville group;
(b) G is a 2-generator group;
(c) $k \leq d_2(H)$.

In Theorems 1.1 and 1.2, the implications (a) ⇒ (b) ⇒ (c) are straightforward (see Sects. 2 and 3), and the main task is to prove that (c) implies (a). The case where $k = 1$ having been dealt with by others (see above), it is thus sufficient to prove the following:

Theorem 1.3 *If H is a simple group $L_2(q)$, $L_3(q)$, $U_3(q)$, $Sz(2^e)$ or $R(3^e)$, or a sporadic simple group, then H^k is a Beauville group for each $k = 2, \ldots, d_2(H)$.*

It is hoped that the methods developed here and in [20] may allow this result to be extended to all finite simple groups of Lie type, thus proving the above conjecture.

The results proved here and in [20] yield further examples of Beauville groups, as follows. Let us say that two groups are *orthogonal* if they have no non-identity epimorphic images in common. It is straightforward to show that a cartesian product of finitely many mutually orthogonal Beauville groups is a Beauville group: orthogonality ensures that the obvious triples generate the group. It follows that if $G_1, \ldots, G_r$ are characteristically simple Beauville groups, which are cartesian powers of mutually non-isomorphic simple groups, then $G_1 \times \cdots \times G_r$ is also a Beauville group.

2 Background and Method of Proof

The proof of Theorem 1.3 is based on the characterisation by Bauer et al. [1, 2] of Beauville groups as those finite groups G such that

- G is a smooth quotient of two hyperbolic triangle groups

$$\Delta_i = \Delta_i(l_i, m_i, n_i) = \langle A_i, B_i, C_i \mid A_i^{l_i} = B_i^{m_i} = C_i^{n_i} = A_i B_i C_i = 1 \rangle$$

  for $i = 1, 2$,
- $\Sigma_1 \cap \Sigma_2 = 1$, where Σ_i is the set of conjugates of powers of the elements a_i, b_i, c_i of G which are images of the canonical generators A_i, B_i, C_i of Δ_i.

Here 'smooth' means that a_i, b_i and c_i have orders l_i, m_i and n_i, so that the kernel K_i of the epimorphism $\Delta_i \to G$ is torsion-free, and thus a surface group. 'Hyperbolic' means that $l_i^{-1} + m_i^{-1} + n_i^{-1} < 1$, so that Δ_i acts on the hyperbolic plane $\mathbb{H}$. The isomorphism $G \cong \Delta_i/K_i$ induces an action of G as a group of automorphisms of the compact Riemann surface (or complex algebraic curve) $\mathcal{C}_i = \mathbb{H}/K_i$ of genus $g_i > 1$, with the natural projection $\mathcal{C}_i \to \mathcal{C}_i/G \cong \mathbb{H}/\Delta_i \cong \mathbb{P}^1(\mathbb{C})$ branched over three points (corresponding to the fixed points of A_i, B_i and C_i in $\mathbb{H}$). Since Σ_i is the set of elements of G with fixed points in $\mathcal{C}_i$, the condition $\Sigma_1 \cap \Sigma_2 = 1$ is necessary and sufficient for G to act freely on $\mathcal{C}_1 \times \mathcal{C}_2$, so that $\mathcal{S} := (\mathcal{C}_1 \times \mathcal{C}_2)/G$ is a complex surface. The generating triples (a_i, b_i, c_i) for G (always satisfying $a_i b_i c_i = 1$) are said to have *type* (l_i, m_i, n_i), and the pair of them form a *Beauville structure* of type $(l_1, m_1, n_1; l_2, m_2, n_2)$ in G.

Since Σ_i is closed under taking powers, the condition $\Sigma_1 \cap \Sigma_2 = 1$ is equivalent to the condition that

$$\Sigma_1^{(p)} \cap \Sigma_2^{(p)} = \emptyset$$

for every prime p dividing $l_1 m_1 n_1$ and $l_2 m_2 n_2$, where $\Sigma_i^{(p)}$ denotes the set of elements of order p in Σ_i. In many cases, this condition is simpler to verify.

If $G = H^k$ for some H then the members of any generating triple for G are k-tuples $a_i = (a_{ij})$, $b_i = (b_{ij})$, $c_i = (c_{ij})$ such that (a_{ij}, b_{ij}, c_{ij}) is a generating triple for H for each $j = 1, \ldots, k$, and these k triples are mutually inequivalent under the action of $\mathrm{Aut}\, H$. If H is a non-abelian finite simple group then conversely any k mutually inequivalent generating triples for H yield a generating triple for G. Moreover, if $G \not\cong A_5$ this triple is hyperbolic. To prove that G is a Beauville group it is therefore sufficient to find two k-tuples of mutually inequivalent generating triples for H, and then to show that the resulting pair of generating triples for G satisfy $\Sigma_1^{(p)} \cap \Sigma_2^{(p)} = \emptyset$ for all primes p dividing $|H|$.

There are k-tuples of mutually inequivalent generating triples for H if and only if $k \leq d_2(H)$. In [20], suitable generating triples for the simple groups $H = A_n$ were exhibited by explicitly defining their members a_{ij}, b_{ij} and c_{ij}, and then applying results on groups containing such permutations to show that they generate H. Here we adopt a similar approach for the groups $H = L_2(q)$, using Dickson's description of their maximal subgroups [9, Ch. XII] to show that various specific triples generate H. However, for the remaining simple groups it is more convenient to use a character-theoretic formula of Frobenius (see Sect. 6) to prove the existence of triples of various types, and then again to use knowledge of the maximal subgroups of H to show that they generate H.

In each case, the condition that $\Sigma_1^{(p)} \cap \Sigma_2^{(p)} = \emptyset$ for all p is ensured by carefully choosing the two k-tuples of generating triples for H so that powers of a_1, b_1 or c_1 of order p have different supports in $\{1, 2, \ldots, k\}$ from those of a_2, b_2 or c_2. Several results which guarantee this were proved in [20], and for completeness their proofs are outlined in Sect. 4. These results are then applied to the groups $L_2(q)$ in Sect. 5, and to the other families of simple groups in Sects. 7–10. However, before doing this we will briefly discuss in Sect. 3 the values of k for which H^k is a 2-generator group.

3 Generating Cartesian Powers

If H is a finite group, let

$$c_2(H) = \max\{k \in \mathbb{N} \mid H^k \text{ is a 2-generator group}\}.$$

If H^k is a Beauville group then it is a 2-generator group, and thus a quotient of the free group F_2 of rank 2, so

$$k \le c_2(H) \le d_2(H)$$

where $d_2(H)$ denotes the number of normal subgroups N of F_2 with $F_2/N \cong H$. Following Hall [15], let $\phi_2(H)$ be the number of 2-bases (ordered generating pairs) for H. Any 2-base for H determines an epimorphism $\theta : F_2 \to H$, and hence a normal subgroup $N = \ker\theta$ of F_2 with $F_2/N \cong H$; conversely, every such normal subgroup arises in this way, with two 2-bases corresponding to the same normal subgroup N if and only if they are equivalent under an automorphism of H. Thus $d_2(H)$ is equal to the number of orbits of Aut H on 2-bases for H. Since only the identity automorphism can fix a 2-base, this action is semiregular, so we have the following:

Lemma 3.1 *If H is a finite group then*

$$d_2(H) = \frac{\phi_2(H)}{|\text{Aut } H|}.$$

If H is a non-abelian finite simple group, and F_2 has k normal subgroups with quotient H, then their intersection is a normal subgroup with quotient H^k. It follows then that $c_2(H) = d_2(H)$ for such a group H, so we have the following:

Corollary 3.2 *If H is a non-abelian finite simple group and H^k is a Beauville group, then*

$$k \le c_2(H) = d_2(H) = \frac{\phi_2(H)}{|\text{Aut } H|}.$$

Since generating triples (a, b, c) for a group (with $abc = 1$) correspond bijectively to its 2-bases (a, b), we have the following useful characterisation of generating triples for H^k:

Corollary 3.3 *Let H be a non-abelian finite simple group. Then k-tuples $a = (a_j)$, $b = (b_j)$ and $c = (c_j)$ form a generating triple for H^k if and only if their components (a_j, b_j, c_j) for $j = 1, \dots, k$ form k mutually inequivalent generating triples for H.* □

By results of Dixon [10], Kantor and Lubotzky [22], and Liebeck and Shalev [25], a randomly-chosen pair of elements of a non-abelian finite simple group H generate the whole group with probability approaching 1 as $|H| \to \infty$, so

$$d_2(H) \sim \frac{|H|^2}{|\mathrm{Inn}\, H|.|\mathrm{Out}\, H|} = \frac{|H|}{|\mathrm{Out}\, H|} \quad \text{as} \quad |H| \to \infty.$$

For each infinite family of such groups, $|\mathrm{Out}\, H|$ is constant or grows much more slowly than $|H|$, so $d_2(H)$ grows almost as fast as $|H|$. In particular, as $q = p^e \to \infty$ with p prime, the infinite families considered here satisfy

- $d_2(L_2(q)) \sim q^3/d^2e$ while $|L_2(q)| \sim q^3/d$, where $d = \gcd(2, q-1)$,
- $d_2(L_3(q)) \sim q^8/2d^2e$ while $|L_3(q)| \sim q^8/d$, where $d = \gcd(3, q-1)$,
- $d_2(U_3(q)) \sim q^7/2d^2e$ while $|U_3(q)| \sim q^7/d$, where $d = \gcd(3, q+1)$,
- $d_2(Sz(q)) \sim q^5/e$ while $|Sz(q)| \sim q^5$,
- $d_2(R(q)) \sim q^7/e$ while $|R(q)| \sim q^7$.

One can illustrate how close the asymptotic estimate $|H|/|\mathrm{Out}\, H|$ is to $d_2(H)$ as follows. For any finite group H we have

$$\phi_2(H) = |H^2 \setminus \bigcup_M M^2| \geq |H|^2 - \sum_M |M|^2,$$

where M ranges over the maximal subgroups of H. If H is perfect then each such M has $|H : M|$ conjugates, so

$$\phi_2(H) \geq |H|^2 \left(1 - \sum_{i=1}^{r} \frac{1}{|H : M_i|}\right),$$

where M_i ranges over a set of representatives of the r conjugacy classes of maximal subgroups of H. It follows that if H also has trivial centre (and in particular if H is a non-abelian finite simple group) then

$$1 \geq \frac{d_2(H)}{|H|/|\mathrm{Out}\, H|} \geq 1 - \sum_{i=1}^{r} \frac{1}{|H : M_i|}.$$

When H is large, the sum of the right is typically very small. Thus, for the simple groups $H = L_2(q), U_3(q), Sz(q)$ and $R(q)$ we have

$$\sum_{i=1}^{r} \frac{1}{|H : M_i|} \sim \frac{1}{q^m} \quad \text{as} \quad q \to \infty,$$

where $m = 1, 3, 2$ and 3 respectively, with the sum dominated by the term corresponding to the doubly transitive permutation representation of H of degree $q^m + 1$; for $H = L_n(q)$ with $n \geq 3$ the sum grows like $2/q^{n-1}$, corresponding to the two doubly transitive representations of degree $(q^n - 1)/(q - 1)$, on the points and hyperplanes of $\mathbb{P}^{n-1}(q)$.

The situation is similar for the sporadic simple groups: for instance, if H is O'Nan's group $O'N$, with $r = 13$ conjugacy classes of maximal subgroups, we have

$$\frac{|H|}{|\mathrm{Out}\, H|} = \frac{460815505920}{2} = 230407752960$$

and

$$\sum_{i=1}^{r} \frac{1}{|H : M_i|} = 0.00001726863378\ldots,$$

so

$$230407752960 \geq d_2(H) \geq 230403774132.$$

4 Beauville Structures in Cartesian Powers

Corollary 3.3 describes the generating triples in a cartesian power $G = H^k$ of a non-abelian finite simple H, in terms of those for H. To avoid confusion, we will use notations such as $\mathcal{T} = (a, b, c)$ and $T = (x, y, z)$ for generating triples in G and H respectively. We now consider sufficient conditions for two triples $\mathcal{T}_i = (a_i, b_i, c_i)$ in G to satisfy $\Sigma_1^{(p)} \cap \Sigma_2^{(p)} = \emptyset$ for all primes p, so that they form a Beauville structure for G.

If $g = (g_j) \in \Sigma_i^{(p)}$ for some prime p, then g has order p and is conjugate in G to a power of some $d_i = a_i, b_i$ or c_i, so for each $j \in \mathbb{N}_k := \{1, \ldots, k\}$ its jth coordinate g_j has order 1 or p. The support

$$\mathrm{supp}(g) = \{j \in \mathbb{N}_k \mid g_j \neq 1\}$$

of g is then the *p-summit* $S_p(d_i)$ of d_i, defined to be the set of $j \in \mathbb{N}_k$ for which the power of p dividing the order of its jth coordinate d_{ij} is greatest. The p-summit $S_p(\mathcal{T})$ of a triple $\mathcal{T} = (a, b, c)$ in G is the set $\{S_p(a), S_p(b), S_p(c)\}$ of subsets of $\mathbb{N}_k$. The following is obvious:

Lemma 4.1 *Let H be any finite group. If two triples $\mathcal{T}_i = (a_i, b_i, c_i)$ $(i = 1, 2)$ in $G = H^k$ have disjoint p-summits $S_p(\mathcal{T}_i)$ for some prime p, then $\Sigma_1^{(p)} \cap \Sigma_2^{(p)} = \emptyset$.* □

An element of H is *p-full* if its order is divisible by the highest power of p dividing the exponent $\exp(H)$ of H. If d_i has at least one p-full coordinate d_{ij}, then

$$S_p(d_i) = F_p(d_i) := \{j \in \mathbb{N}_k \mid d_{ij} \text{ is } p\text{-full}\}.$$

If each element d_i of a triple $\mathcal{T}_i$ in G has a p-full coordinate, it is easier to determine $S_p(\mathcal{T}_i)$, and hence to ensure that pairs of such triples $\mathcal{T}_i$ satisfy the assumptions of Lemma 4.1.

Given a triple $T = (x, y, z)$ in H, let $\nu_p(T)$ be the number of p-full elements among x, y and z. Two triples T_i $(i = 1, 2)$ in H are *p-distinguishing* if $\nu_p(T_1) \neq \nu_p(T_2)$, and *strongly p-distinguishing* if, in addition, whenever $\nu_p(T_i) = 0$ then either p^2 does not divide $\exp(H)$ or p does not divide any of the three periods of T_i.

Lemma 4.2 *Suppose that a non-abelian finite simple group H has a set $\{(T_{1,s}, T_{2,s}) \mid s = 1, \dots, t\}$ of ordered pairs $(T_{1,s}, T_{2,s})$ of generating triples such that*

1. *for each prime p dividing $|H|$ there is some $s = s(p) \in \{1, \dots, t\}$ such that $T_{1,s}$ and $T_{2,s}$ are a strongly p-distinguishing pair;*
2. *for each $i = 1, 2$ the $3t$ triples consisting of $T_{i,1}, \dots, T_{i,t}$ and their cyclic permutations are mutually inequivalent.*

Then $G := H^k$ is a Beauville group for each $k = 3t, \dots, d_2(H)$.

Proof Let $T_{i,s} = (x_{i,s}, y_{i,s}, z_{i,s})$ for each $i = 1, 2$ and $s = 1, \dots, t$. Since $3t \leq k \leq d_2(H)$, for each i one can form a generating triple $\mathcal{T}_i = (a_i, b_i, c_i)$ for G by using k inequivalent generating triples for H in the different coordinate positions, with $T_{i,s}$ and its two cyclic permutations in the jth positions where $j = 3s - 2, 3s - 1$ or $3s$ for $s = 1, \dots, t$.

Suppose that $g \in \Sigma_1^{(p)} \cap \Sigma_2^{(p)}$ for some prime p. There is a strongly p-distinguishing pair $(T_{1,s}, T_{2,s})$, with $\nu_p(T_{1,s}) \neq \nu_p(T_{2,s})$. First suppose that $\nu_p(T_{i,s}) > 0$ for $i = 1, 2$, so that each element $d_i = a_i, b_i$ or c_i of $\mathcal{T}_i$ has at least one p-full coordinate, and hence $S_p(d_i) = F_p(d_i)$. It follows that

$$|\mathrm{supp}(g) \cap \{3s - 2, 3s - 1, 3s\}| = \nu_p(T_{i,s}) \tag{1}$$

for $i = 1$ and 2, which is impossible since $\nu_p(T_{1,s}) \neq \nu_p(T_{2,s})$. Otherwise, we may suppose without loss of generality that $\nu_p(T_{1,s}) > 0 = \nu_p(T_{2,s})$, with (1) satisfied for $i = 1$ but not for $i = 2$. Some $d_2 = a_2, b_2$ or c_2 must then have a non-p-full coordinate of order divisible by p in position $3s - 2, 3s - 1$ or $3s$, which is impossible if the periods of $T_{2,s}$ are coprime to p or if p^2 does not divide $\exp(H)$, as we are assuming. Thus $\Sigma_1^{(p)} \cap \Sigma_2^{(p)} = \emptyset$, so the triples $\mathcal{T}_i$ form a Beauville structure for G. □

If we simply assume that the pairs $T_{1,s}$ and $T_{2,s}$ are p-distinguishing, we have the following rather weaker conclusion:

Lemma 4.3 *Suppose that a non-abelian finite simple group H has a set $\mathcal{T} = \{(T_{1,s}, T_{2,s}) \mid s = 1, \dots, t\}$ of ordered pairs $(T_{1,s}, T_{2,s})$ of generating triples such that*

1. *for each prime p dividing $|H|$ there is some $s = s(p) \in \{1, \dots, t\}$ such that $T_{1,s}$ and $T_{2,s}$ are a p-distinguishing pair;*
2. *the $6t$ triples consisting of the $2t$ triples $T_{i,s}$ and their cyclic permutations are mutually inequivalent.*

Then $G := H^k$ is a Beauville group for each $k = 6t, \dots, d_2(H)$.

Proof The proof is similar. The first $3t$ coordinates of a_i, b_i and c_i are defined as before, but now those in positions $j = 3t+1, \ldots, 6t$ are the coordinates of a_{3-i}, b_{3-i} or c_{3-i} in positions $j - 3t$. Each of the six generators a_i, b_i, c_i of G now has at least one p-full coordinate for each prime p dividing $|H|$, so any $g \in \Sigma_1^{(p)} \cap \Sigma_2^{(p)}$ satisfies (1) for $i = 1$ and 2, leading to a contradiction as before. □

For the small values of k omitted by these lemmas one can often use the following:

Lemma 4.4 *Let H be a non-abelian finite simple group with a set of $r \geq 2$ mutually inequivalent generating triples of type (l, m, n), where l, m and n are mutually coprime. Then $G := H^k$ is a Beauville group for each $k = 2, \ldots, 6r$.*

Proof If the specified generating triples for H are (x_j, y_j, z_j) for $j = 1, \ldots, r$, then one can form $6r$ mutually inequivalent generating triples for H by cyclically permuting each (x_j, y_j, z_j) and each $(z_j^{-1}, y_j^{-1}, x_j^{-1})$. If $k = 2, \ldots, 6r$ one can form two generating triples

$$a_1 = (x_1, x_2, \ldots), \quad b_1 = (y_1, y_2, \ldots), \quad c_1 = (z_1, z_2, \ldots)$$

and

$$a_2 = (x_1, y_2, \ldots), \quad b_2 = (y_1, z_2, \ldots), \quad c_2 = (z_1, x_2, \ldots),$$

for G, using k of these triples in the different coordinate positions. Since l, m and n are mutually coprime, any prime p dividing l, m or n divides exactly one of them, so if $g \in \Sigma_1^{(p)}$ then $|\mathrm{supp}(g) \cap \{1, 2\}| = 2$ whereas if $g \in \Sigma_2^{(p)}$ then $|\mathrm{supp}(g) \cap \{1, 2\}| = 1$. Thus $\Sigma_1^{(p)} \cap \Sigma_2^{(p)} = \emptyset$, so these two triples form a Beauville structure for G. □

5 The Groups $L_2(q)$

We will now apply the results in the preceding section to show that the simple groups $L_2(q)$ satisfy Theorem 1.3.

Let $G = H^k$ where $H = L_2(q)$ with $q = p_0^e \geq 4$ for some prime p_0, and with $k = 2, \ldots, d_2(H)$. Since $L_2(4) \cong L_2(5) \cong A_5$ and $L_2(9) \cong A_6$, the main theorem of [20], which proves Theorem 1.3 for $H = A_n$ $(n \geq 5)$, allows us to assume that $q = 7$ or 8 or $q \geq 11$. We will use Lemma 4.2 with $t = 1$ to show that G is a Beauville group. However, this lemma does not apply when $k = 2$, so we will deal with this case first by a separate argument.

Let $k = 2$. Results of Macbeath [26] show that there exist generating triples (x_i, y_i, z_i) $(i = 1, 2)$ for H of types (q_1, q_1, p_0) and (q_2, q_2, p_0), where $q_1, q_2 = (q \pm 1)/d$ and $d = \gcd(2, q - 1)$. For instance, in the equation

$$\begin{pmatrix} s & 1 \\ -1 & 0 \end{pmatrix} \begin{pmatrix} 0 & 1 \\ -1 & t \end{pmatrix} = \begin{pmatrix} -1 & s+t \\ 0 & -1 \end{pmatrix}$$

in $SL_2(q)$ one can choose s and t to be the traces of elements x_i and y_i of order $o(x_i) = o(y_i) = q_i$ in H for $i = 1, 2$, with the matrix on the right representing an element $z_i^{-1} = x_i y_i$ of order p_0 provided $s + t \neq 0$. Dickson's description of the subgroups of $L_2(q)$ (see [9, Ch. XII]) shows that the only maximal subgroups containing elements of orders q_i and p_0 are dihedral groups of order q_i when $p_0 = 2$, impossible here since x_i and y_i do not commute, and stabilisers of points in $\mathbb{P}^1(q)$, also impossible since z_i fixes only ∞, which is not fixed by x_i. Thus each triple (x_i, y_i, z_i) generates H. Having different types, these two generating triples are mutually inequivalent, so we obtain a generating triple

$$a_1 = (x_1, x_2), \quad b_1 = (y_1, y_2), \quad c_1 = (z_1, z_2)$$

for G. Similarly G has a second generating triple

$$a_2 = (x_2, z_1), \quad b_2 = (y_2, x_1), \quad c_2 = (z_2, y_1).$$

The primes p dividing $|H| = q(q^2-1)/d$ are p_0 and those dividing q_1 or q_2. Since p_0, q_1 and q_2 are mutually coprime, if $p = p_0$ then any $g \in \Sigma_i^{(p)}$ has $|\mathrm{supp}(g)| = 2$ or 1 for $i = 1$ or 2 respectively, so $\Sigma_1^{(p)} \cap \Sigma_2^{(p)} = \emptyset$. If $g \in \Sigma_i^{(p)}$ with p dividing q_1, then $\mathrm{supp}(g) = \{1\}$ or $\{2\}$ for $i = 1$ or 2, while for p dividing q_2 it is the other way round. Thus $\Sigma_1^{(p)} \cap \Sigma_2^{(p)} = \emptyset$ for all p, so H^2 is a Beauville group.

Now let $k = 3, \ldots, d_2(H)$. Let (x_i, y_i, z_i) $(i = 1, 3)$ be generating triples for H of types (q_1, q_1, p_0) and (q_1, q_2, q_2) for H. For instance, the first could be as above in the case $k = 2$; the second could consist of the images in H of matrices in $SL_2(q)$ of the form

$$\begin{pmatrix} u & 1 \\ -1 & 0 \end{pmatrix}, \quad \begin{pmatrix} v & w \\ 0 & v^{-1} \end{pmatrix} \quad \text{and} \quad \begin{pmatrix} -w & -uw - v^{-1} \\ v & uv \end{pmatrix},$$

where u is the trace of an element of H of order q_1, v generates the multiplicative group of the field $\mathbb{F}_q$, and w is chosen so that $uv - w$ is the trace of an element of H of order q_2. Again, it follows from [9, Ch. XII] that this triple generates H.

We now apply Lemma 4.2, with $t = 1$, to this pair of triples. Again, we must consider the primes p dividing $|H|$. The generator z_1 is p_0-full, since the Sylow p_0-subgroups of H are elementary abelian, while x_1, y_1 and x_3 are p-full for all primes p dividing q_1, and y_3 and z_3 are p-full for primes dividing q_2. The hypotheses of Lemma 4.2 are satisfied, with $t = 1$, so H^k is a Beauville group for each $k = 3, \ldots, d_2(H)$.

6 Counting Triples

For the other families of finite simple groups we shall consider, it is easier to prove the existence of generating triples of various types by means of the following enumerative formula of Frobenius [12] than to exhibit them directly as we did for $H = L_2(q)$.

Proposition 6.1 *If X, Y and Z are conjugacy classes in a finite group H, then the number $\nu_H(X, Y, Z)$ of triples $(x, y, z) \in X \times Y \times Z$ such that $xyz = 1$ is given by*

$$\nu_H(X, Y, Z) = \frac{|X| \cdot |Y| \cdot |Z|}{|H|} \sum_{\chi} \frac{\chi(x)\chi(y)\chi(z)}{\chi(1)},$$

where $x \in X$, $y \in Y$, $z \in Z$, and χ ranges over the irreducible complex characters of H.

Proofs of this and related results can be found in [18, Appendix] or [30, Theorem 7.2.1]. By summing over all choices of conjugacy classes X, Y and Z of elements of given orders l, m and n, one can find the number $\nu_H(l, m, n)$ of triples of type (l, m, n) in H.

An equivalent, and sometimes more convenient, version of Proposition 6.1 is that

$$\nu_H(X, Y, Z) = \frac{|H|^2}{|C_H(x)| \cdot |C_H(y)| \cdot |C_H(z)|} S(X, Y, Z),$$

where $S(X, Y, Z)$ denotes the character sum

$$S(X, Y, Z) = \sum_{\chi} \frac{\chi(x)\chi(y)\chi(z)}{\chi(1)}.$$

In most of the cases we shall consider, one finds that many characters χ take the value 0 on at least one of the three classes X, Y and Z, so they do not contribute to this sum. Even when a non-principal character χ contributes, the absolute values $|\chi|$ on these classes are often so much smaller than the degree $\chi(1)$ that the sum is dominated by the contribution, equal to 1, from the principal character, so that $S(X, Y, Z) \approx 1$. These estimates often show that $S(X, Y, Z) > 0$, so that $\nu_H(X, Y, Z) > 0$ and hence H has such triples (x, y, z).

We are particularly interested in generating triples, those not contained in any maximal subgroup M of H. The number $\phi_H(l, m, n)$ of these of type (l, m, n) clearly satisfies

$$\phi_H(l, m, n) \geq \nu_H(l, m, n) - \sum_{M} \nu_M(l, m, n),$$

where M ranges over the maximal subgroups of M. In particular, if $\nu_M(l, m, n) = 0$ for all such M then

$$\phi_H(l, m, n) = \nu_H(l, m, n).$$

7 The Suzuki Groups

We will apply this method first to the Suzuki groups $Sz(q)$, using the description of their conjugacy classes, subgroups and characters given by Suzuki in [31] (see also [33, Sect. 4.2]).

If $q = 2^e$ for some odd $e \geq 3$, then the group $H = Sz(q) = {}^2B_2(q)$ is a simple group of order $q^2(q^2+1)(q-1)$. It has four conjugacy classes of maximal cyclic subgroups, of mutually coprime orders 4, $q_1 = q + \sqrt{2q} + 1$, $q_2 = q - \sqrt{2q} + 1$ and $q_3 = q - 1$. (Note that $q_1 q_2 = q^2 + 1$.) It follows that any prime p dividing $|H|$ divides precisely one of these four integers, and any non-identity element of G is conjugate to an element of precisely one of these four subgroups. In particular, a pair of generating triples of types $(4, q_2, q_3)$ and (q_1, q_3, q_3), if they exist, will satisfy the conditions of Lemma 4.2 with $t = 1$, and hence prove that H^k is a Beauville group for $k = 3, \ldots, d_2(H)$.

To prove that such triples exist, we use Frobenius's formula (Proposition 6.1), together with the character table of H in [31, Sect. 17]. Conjugacy classes X, Y and Z of elements of orders 4, q_2 and q_3 satisfy $|X| = 4(q^2+1)(q-1)$, $|Y| = q^2 q_1(q-1)$ and $|Z| = q^2(q^2+1)$, so each choice of such classes gives rise to $4q_1|H| > 0$ triples $(x, y, z) \in X \times Y \times Z$ with $xyz = 1$. (Only the principal character χ makes a non-zero contribution to the sum in Proposition 6.1, all other characters vanishing on at least one of the chosen classes.) The subgroups of H are described in [31, Sect. 15]; no proper subgroup contains elements of orders q_2 and q_3, so each of these triples generates H. A similar argument applies to triples of type (q_1, q_3, q_3): each appropriate choice of conjugacy classes yields $q_2(q+1)|H|$ generating triples, with only the irreducible characters of degrees 1 and q^2 contributing to the sum.

The case $k = 2$ is covered by Lemma 4.4: since $|\mathrm{Aut}\, H| = e|H|$ and $4q_1 > 4q > e$ there are at least two inequivalent generating triples of type $(4, q_2, q_3)$, and their periods are mutually coprime. Thus the Suzuki groups satisfy Theorem 1.3.

8 The Small Ree Groups

Similar methods can be applied to the small Ree groups $H = R(q) = {}^2G_2(q)$, where $q = 3^e$ for some odd $e \geq 3$. (These groups, described by Ree in [28], are called 'small' to distinguish them from the 'large' Ree groups ${}^2F_4(2^e)$.) Each of these groups H is simple, of order $q^3(q^3+1)(q-1)$. The Sylow 2-subgroups of H are elementary abelian, of order 8; there is a single conjugacy class of involutions, with centralisers isomorphic to $C_2 \times L_2(q)$, of order $q(q^2-1)$. The Sylow 3-subgroups are non-abelian, of order q^3 and exponent 9; there are three conjugacy classes of elements of order 9, all with centralisers of order $3q$. There are cyclic Hall subgroups A_i of mutually coprime odd orders $q_i := (q-1)/2$, $(q+1)/4$, $q - \sqrt{3q} + 1$ and $q + \sqrt{3q} + 1$ for $i = 0, \ldots, 3$. (Note that $q \equiv 3 \bmod (8)$ and $q^3 + 1 = (q+1)(q^2 - q + 1) = 4q_1 q_2 q_3$.)

We will again use Lemmas 4.2 and 4.4, now applied to triples of types $(2, q_1, q_3)$ and $(9, q_0, q_2)$, with six mutually coprime periods. For each prime p dividing $|H|$, one of these triples contains a single p-full element, while the elements of the other triple all have orders coprime to p. Provided they exist, such triples therefore satisfy the hypotheses of these two lemmas.

One can count triples by applying Frobenius's formula to the character table for H given by Ward in [32]. In the case of triples of type $(9, q_0, q_2)$, there are respectively three, $\varphi(q_0)/2$ and $\varphi(q_2)/6$ conjugacy classes of elements of orders 9, q_0 and q_2, denoted by YT^i $(i = 0, \pm 1)$, R^a and V in [32], with centralisers of orders $3q$, $q-1$ and q_2. Only the principal character makes a non-zero contribution to the character sum in the formula, so the number of triples of this type in H is

$$3 \cdot \frac{\varphi(q_0)}{2} \cdot \frac{\varphi(q_2)}{6} \cdot \frac{|H|^2}{3q(q-1)q_2} = \frac{1}{12}\varphi(q_0)\varphi(q_2)q^5(q^3+1)(q-1)^2 q_3 > 0.$$

A similar calculation applies to triples of type $(2, q_1, q_3)$. The elements of orders $2, q_1$ and q_3, denoted by J, S^a and W in [32], form $1, \varphi(q_1)/6$ and $\varphi(q_3)/6$ conjugacy classes, with centralisers of orders $q(q^2-1)$, $q+1$ and q_3 respectively. The character sum takes the value

$$1 + \frac{q \cdot (-1)^2}{q^3} + 2 \cdot \frac{q_0 \cdot (-1) \cdot 1}{q_0 q_2 \sqrt{q/3}} = 1 + \frac{1}{q^2} - \frac{2}{q_2\sqrt{q/3}},$$

with ξ_3 of degree q^3, and ξ_6 and ξ_8 of degree $q_0 q_2 \sqrt{q/3}$ the only non-principal characters contributing to it. It follows that the number of triples of this type is

$$\frac{\varphi(q_1)}{6} \cdot \frac{\varphi(q_3)}{6} \cdot \frac{|H|^2}{q(q^2-1)(q+1)q_3}\left(1 + \frac{1}{q^2} - \frac{2}{q_2\sqrt{q/3}}\right) > 0.$$

The maximal subgroups of H have been determined by Levchuk and Nuzhin [24] and Kleidman [23] (see also [33, Sect. 4.5.3]). They are as follows:

- the point stabilisers in the doubly transitive permutation representation of degree q^3+1, or equivalently, the normalisers of the Sylow 3-subgroups, of order $q^3(q-1)$;
- the centralisers of involutions, of order $q(q^2-1)$;
- the normalisers of Hall subgroups A_i $(i = 1, 2, 3)$, of order $6q_i$;
- subgroups isomorphic to $R(3^f)$ where e/f is prime.

None of these subgroups contains elements of orders q_0 and q_2, or of orders q_1 and q_3, so each triple of type $(9, q_0, q_2)$ or $(2, q_1, q_3)$ generates H. Dividing the above numbers of triples by $|\mathrm{Aut}\, H| = e|H|$ gives the numbers of equivalence classes of triples of these two types. As in the case of the Suzuki groups, Lemmas 4.2 and 4.4 now show that these Ree groups satisfy Theorem 1.3.

9 The Sporadic Simple Groups

In this section we will prove that the 26 sporadic simple groups satisfy Theorem 1.3. Most of the information we need about these groups can be found in [8] or [33]; in particular, we have adopted the notation of [8] for conjugacy classes, maximal subgroups, etc.

9.1 The Mathieu Groups

It is convenient to deal with the Mathieu groups together, in view of the inclusion relations between them. Each of the Mathieu groups $H = M_n$ ($n = 11, 12, 22, 23$ or 24) has order divisible by at most six primes. This allows us to use Lemma 4.2 with $t = 1$ to deal with the values $k = 3, \ldots, d_2(H)$, while Lemma 4.4 deals with $k = 2$. In each case we can use Frobenius's formula (Proposition 6.1), applied to the character tables in [8], to count triples of a given type, and the lists of maximal subgroups in [8] to eliminate those which do not generate H. Here, as with some of the other sporadic simple groups, we will compute the exact numbers of equivalence classes of generating triples of the required types, even though it is usually sufficient to show that at least one or two exist: this is because it may be useful in other contexts (such as regular maps and hypermaps) to know the number of normal subgroups of a given triangle group with quotient isomorphic to H. For instance, by the calculation in Sect. 9.2.1 there are, up to isomorphism, 18 orientably regular maps of type $\{5, 19\}$ with orientation-preserving automorphism group isomorphic to Janko's group J_1; the Riemann-Hurwitz formula shows that they all have genus 21,715.

M_{11}

Let $H = M_{11}$, a simple group of order $2^4.3^3.5.11$ and exponent $2^3.3.5.11$ with $|\mathrm{Out}\, H| = 1$. Frobenius's formula and the character table in [8] show that there are $2^6.3^3.5^2.11 = 20|H| = 20|\mathrm{Aut}\, H|$ triples of type (3, 8, 11) in H; they all generate H since no maximal subgroup contains elements of orders 8 and 11, so they form 20 equivalence classes.

Similarly, there are $54|\mathrm{Aut}\, H|$ triples of type (5, 11, 11). The only maximal subgroups containing elements of order 5 and 11 are the twelve isomorphic to $L_2(11)$. Applying Frobenius's formula to $L_2(11)$ shows that $8|\mathrm{Aut}\, H|$ of these triples are contained in such maximal subgroups, so the remainder form 46 equivalence classes of generating triples.

A pair of generating triples of types (3, 8, 11) and (5, 11, 11) satisfy the hypotheses of Lemma 4.2, so H^k is a Beauville group for each $k = 3, \ldots, d_2(H)$. Since there are at least two inequivalent generating triples of type (3, 8, 11), Lemma 4.4 proves this for $k = 2$.

M_{12}

The group $H = M_{12}$, of order $2^6.3^3.5.11$ with $|\mathrm{Out}\, H| = 2$, has the same exponent as M_{11}, so we can use the same method, with triples of the same types $(3, 8, 11)$ and $(5, 11, 11)$, provided they generate H. Frobenius's formula shows that H has $96|\mathrm{Aut}\, H|$ triples of type $(3, 8, 11)$. The only maximal subgroups containing elements of orders 8 and 11 are the 24 isomorphic to M_{11}; we have already counted such triples in M_{11}, so we find that there are $20|\mathrm{Aut}\, H|$ non-generating triples, and hence 76 equivalence classes of generating triples of this type.

Similarly there are $158|\mathrm{Aut}\, H|$ triples of type $(5, 11, 11)$ in H. The only maximal subgroups containing elements of orders 5 and 11 are the 24 isomorphic to M_{11} and the 144 isomorphic to $L_2(11)$. Having already counted triples of this type in these two groups, we find that there are $54|\mathrm{Aut}\, H|$ and $4|\mathrm{Aut}\, H|$ triples contained in maximal subgroups respectively isomorphic to M_{11} and to $L_2(11)$. These two sets of triples are disjoint, since any triple of type $(5, 11, 11)$ contained in $L_2(11)$ generates that group, so there are $58|\mathrm{Aut}\, H|$ non-generating triples, and hence 100 equivalence classes of generating triples of this type. From this point, as with the remaining Mathieu groups, the argument follows that for M_{11}.

M_{22}

The group $H = M_{22}$ has order $2^7.3^2.5.7.11$ and exponent $2^3.3.5.7.11$ with $|\mathrm{Out}\, H| = 2$. There are $2^{17}.3^3.5.7.11 = 1536|\mathrm{Aut}\, H|$ triples of type $(5, 11, 11)$ in H. The only maximal subgroups containing elements of order 11 are the $2^5.3.7 = 672$ isomorphic to $L_2(11)$. We have seen that $L_2(11)$ contains $2^5.3.5.11 = 5280$ such triples, and is generated by each of them, so there are $2^{10}.3^2.5.7.11 = 4|\mathrm{Aut}\, H|$ non-generating triples and hence 1532 equivalence classes of generating triples of this type.

Similarly, there are $216|\mathrm{Aut}\, H|$ triples of type $(3, 7, 8)$ in H, each generating H or contained in a maximal subgroup isomorphic to $AGL_3(2)$. There are 330 such maximal subgroups, each containing 224 elements of order 3 (eight in each of the 28 two-point stabilisers, isomorphic to S_4, in the natural affine action of this group), and 384 elements of order 7 (48 in each of the eight point stabilisers, isomorphic to $L_3(2)$). Thus there are at most $330 \times 224 \times 384 = 32|\mathrm{Aut}\, H|$ non-generating triples and hence at least 184 equivalence classes of generating triples of type $(3, 7, 8)$ in H.

M_{23}

The group $H = M_{23}$ has order $2^7.3^2.5.7.11.23$ and exponent $2^3.3.5.7.11.23$ with $|\mathrm{Out}\, H| = 1$. There are $2^{10}.3^2.5.7^2.11^2.23 = 616|\mathrm{Aut}\, H|$ triples of type $(3, 8, 23)$. The only maximal subgroups with elements of order 23 are the normalisers of Sylow 23-subgroups, of order 11.23, so each of these triples generates H, giving 616 equivalence classes.

There are $2^{15}.3^3.5.7.11.23^2 = 17{,}664|\mathrm{Aut}\, H|$ triples of type $(5, 7, 11)$ in H. The only maximal subgroups with elements of orders 5, 7 and 11 are the 23 subgroups isomorphic to M_{22}. Now there are $2^{16}.3^4.5.7.11$ triples of type $(5, 7, 11)$ in M_{22}, each generating M_{22}, so H has $2^{16}.3^4.5.7.11.23 = 4608|\mathrm{Aut}\, H|$ non-generating triples and hence 13,056 equivalence classes of generating triples of this type.

M_{24}

The group $H = M_{24}$ has order $2^{10}.3^3.5.7.11.23$ and $|\mathrm{Out}\, H| = 1$, with the same exponent $2^3.3.5.7.11.23$ as M_{23}, so we can again use triples of types $(3, 8, 23)$ and $(5, 7, 11)$. There are $2^{11}.3^2.5.7.11.23.769 = 1538|\mathrm{Aut}\, H|$ triples of type $(3, 8, 23)$ in H with the generator of order 3 in the class $3A$, and $2^{14}.3^3.5.7.11^2.17.23 = 1992|\mathrm{Aut}\, H|$ with it in class $3B$, giving a total of $3530|\mathrm{Aut}\, H|$ triples. The maximal subgroups of H have been determined by Choi [7]. The only maximal subgroups with elements of orders 8 and 23 are the 24 isomorphic to M_{23}. As we have seen, M_{23} has $2^{10}.3^2.5.7^2.11^2.23$ triples of type $(3, 8, 23)$, each generating M_{23}, so there are $2^{13}.3^3.5.7^2.11^2.23 = 616|\mathrm{Aut}\, H|$ non-generating triples of this type in H, and hence 2914 equivalence classes of generating triples.

There are $22{,}256|\mathrm{Aut}\, H|$ triples of type $(5, 7, 11)$ in H. The only maximal subgroups with elements of orders 5, 7 and 11 are the 24 isomorphic to M_{23} and the $\binom{24}{2} = 276$ isomorphic to $M_{22} : 2 = \mathrm{Aut}\, M_{22}$. These are the stabilisers of points and of unordered pairs of points in the natural representation of H. Any triple of type $(5, 7, 11)$ in $\mathrm{Aut}\, M_{22}$ must be contained in the subgroup M_{22} of index 2 fixing two points, so it is contained in two of the point stabilisers. However, no such triple lies in three point stabilisers, since elements of order 11 fix just two points. We have seen that there are $2^{15}.3^3.5.7.11.23^2 = 17{,}664|\mathrm{Aut}\, M_{23}| = 17{,}664|\mathrm{Aut}\, H|/24$ triples of type $(5, 7, 11)$ in M_{23}, and $2^{16}.3^4.5.7.11 = 192|\mathrm{Aut}\, H|/23$ in M_{22}, so the number of generating triples of this type is

$$22{,}256|\mathrm{Aut}\, H| - 17{,}664|\mathrm{Aut}\, H| + 276 \cdot \frac{192|\mathrm{Aut}\, H|}{23} = 6896|\mathrm{Aut}\, H|,$$

and they form 6896 equivalence classes.

9.2 Other Small Sporadic Simple Groups

The method applied to the Mathieu groups can also be applied to some of the other sporadic simple groups H, provided they are 'small' in the sense that not too many primes divide $|H|$. In these cases, it is sufficient to show that there are inequivalent generating triples satisfying the hypotheses of Lemmas 4.2 and 4.4. We will deal with these groups in ascending order of magnitude.

The Janko Group J_1

The first Janko group $H = J_1$ has order $2^3.3.5.7.11.19$ and exponent $2.3.5.7.11.19$ with $|\mathrm{Out}\, H| = 1$. We will use triples of types $(2, 5, 19)$ and $(3, 7, 11)$. There are $2^4.3^2.5.7.11.19 = 18|\mathrm{Aut}\, H|$ triples of type $(2, 5, 19)$ in H. They are all generating triples, since no maximal subgroup (classified by Janko [17]) contains elements of orders 5 and 19, so they form 18 equivalence classes. Similarly there are $2^5.3.5.7.11.19^2 = 76|\mathrm{Aut}\, H|$ triples of type $(3, 7, 11)$ in H. No maximal subgroup contains elements of orders 7 and 11, so these form 76 equivalence classes of generating triples.

The Janko Group J_2

The second Janko group $H = J_2$ has order $2^7.3^3.5^2.7$ and exponent $2^3.3.5.7$ with $|\mathrm{Out}\, H| = 2$. We will use triples of types $(3, 5, 7)$ and $(7, 7, 8)$. There are $2^{16}.3^4.5^2.7 = 768|\mathrm{Aut}\, H|$ triples of type $(7, 7, 8)$ in H, with only the principal character and that of degree 225 contributing to the character sum. The only triples of this type contained in proper subgroups are those which generate one of the 100 maximal subgroups isomorphic to $U_3(3)$. This group contains $2^{12}.3^3.7$ such triples, so there are $2^{14}.3^3.5^2.7 = 64|\mathrm{Aut}\, H|$ non-generating triples in H, and hence there are 704 equivalence classes of generating triples of this type.

Any triple of type $(3, 5, 7)$ generates H since no maximal subgroup has elements of orders 5 and 7. For simplicity we will consider only those triples for which the element of order 3 is in the $\mathrm{Aut}\, H$-invariant conjugacy class $3B$, consisting of those with a centraliser of order 36, and the element of order 5 is in class $5C$ or $5D$, transposed by $\mathrm{Aut}\, H$, with a centraliser of order 50. We find that there are 49 equivalence classes of such triples, with only the principal character and that of degree 288 appearing in the character sum.

The Higman-Sims Group *HS*

The Higman-Sims group $H = HS$ has order $2^9.3^2.5^3.7.11$ and exponent $2^3.3.5.7.11$ with $|\mathrm{Out}\, H| = 2$. We will use triples of types $(7, 11, 11)$ and $(8, 11, 15)$. The only maximal subgroups containing elements of order 11 are isomorphic to M_{11} or M_{22}; neither of these groups has elements of order 15, and only M_{22} has elements of order 7. There are $1{,}04{,}760|\mathrm{Aut}\, H|$ triples of type $(7, 11, 11)$ in H, with only the principal character and that of degree 3200 contributing to the character sum. There are 100 subgroups isomorphic to M_{22}, in total containing (and generated by) $2048|\mathrm{Aut}\, H|$ such triples, so there are 1,02,712 equivalence classes of triples of this type. There are $50{,}496|\mathrm{Aut}\, H|$ triples of type $(8, 11, 15)$, with only the principal character and that of degree 175 contributing to the character sum. Each of these triples generates H, so they form 50,496 equivalence classes.

The McLaughlin Group M^cL

The McLaughlin group $H = M^cL$ has order $2^7.3^6.5^3.7.11$ and exponent $2^3.3^2.5.7.11$ with $|\mathrm{Out}\, H| = 2$. We will use triples of types $(5, 7, 11)$ and $(8, 9, 9)$. In counting triples of type $(5, 7, 11)$ we will consider only those for which the element of order 5 is in the conjugacy class $5A$, with centralisers of order 750. The only maximal subgroups with elements of orders 7 and 11 are those isomorphic to M_{22}; the permutation character of H on the cosets of such a subgroup, as given in [8], takes the value 0 on elements of $5A$, so these elements do not lie in such subgroups. Thus each of these triples generates H. There are $2^{12}.3^{11}.5^3.7.11 = 3888|\mathrm{Aut}\, H|$ of them, with only the principal character contributing to the character sum, so they form 3888 equivalence classes.

There are $2^{13}.3^6.5^6.7^2.11^2 = 2^5.5^3.7.11|\mathrm{Aut}\, H| = 3{,}08{,}000|\mathrm{Aut}\, H|$ triples of type $(8, 9, 9)$ in H, again with only the principal character contributing to the character sum. The only maximal subgroups containing elements of orders 8 and 9 are those isomorphic to $U_4(3)$, $3_+^{1+4}{:}2S_5$ or $3^4{:}M_{10}$; in the last two cases, any triple of this type must lie in the subgroup $3_+^{1+4}{:}2A_5$ or $3^4{:}A_6$ of index 2, whereas this has no elements of order 8. Thus any non-generating triple of this type must lie in a maximal subgroup isomorphic to $U_4(3)$. There are 275 of these, any two of them intersecting in a subgroup isomorphic to $3^4{:}A_8$ or $L_3(4)$; these have no elements of orders 8 or 9 respectively, so each such triple lies in a unique maximal subgroup. Since $U_4(3)$ has $2^{14}.3^7.5.7.23$ triples of type $(8, 9, 9)$ (with only the principal character and the two of degree 35 contributing to the character sum), there are $2^{14}.3^7.5^3.7.11.23 = 2^7.5.7|\mathrm{Aut}\, H| = 4480|\mathrm{Aut}\, H|$ non-generating triples in H, and hence $3{,}03{,}584|\mathrm{Aut}\, H|$ generating triples, forming 3,03,584 equivalence classes.

One can apply this method to a few other sporadic simple groups, such as J_3, *He*, *Ru*, *O'N* and Co_3, but it is more efficient to deal with them by using a separate method.

9.3 The Larger Sporadic Simple Groups

In the case of the larger sporadic simple groups H, and in particular those of order divisible by more than six primes, Lemma 4.2 may not be applicable, since it may be impossible to find a pair of generating triples satisfying its hypotheses for each of the primes dividing $|H|$. Instead, one can use Lemma 4.3 with $t > 1$ to deal with values $k \geq 6t$, and Lemma 4.4 to deal with small values of k. In each case, we find two mutually coprime integers q_1 and q_2 dividing $|H|$ (usually the largest two prime divisors) such that no maximal subgroup of H contains elements of orders q_1 and q_2. (The maximal subgroups have been completely classified in all cases except that of the Monster simple group M, and in this case sufficient is known about the maximal subgroups to justify this claim.) It follows that for any l, a triple of type (l, q_1, q_2)

in H must generate the whole group. Provided they exist, we can therefore use pairs of triples of types (l, q_1, q_2) for various l to distinguish the primes dividing $|H|$. By Lemma 4.3 this will deal with large values of k, and one of these types, with l coprime to q_1 and q_2, will also deal with small k by Lemma 4.4, provided there are sufficiently many inequivalent triples of that type.

As before, we can use Frobenius's formula (Proposition 6.1) to show that suitable triples exist. In all the relevant cases, inspection of the character table of H shows that the character sum $S(X, Y, Z)$ is very close to 1, with the modulus $|\chi(g)|$ of each non-principal character χ of H for g in each of the chosen conjugacy classes X, Y and Z significantly smaller than the degree $\chi(1)$. Thus $\nu_H(X, Y, Z) > 0$, so suitable generating triples exist, and moreover in sufficient numbers for Lemma 4.4 to deal with the small values of k not covered by Lemma 4.3.

We will give a detailed explanation of this method as it applies to the largest and the smallest of the remaining sporadic simple groups, namely the Monster group and the third Janko group. For the other groups, in order to omit tedious and repetitive numerical details we will simply refer to Table 1, which lists the groups in ascending order of magnitude, together with their exponents and the choices of q_1 and q_2.

Table 1 The larger sporadic simple groups

H	Name	Exponent	q_1, q_2
J_3	Janko	$2^3.3^2.5.17.19$	17, 19
He	Held	$2^5.3.5.7.17$	7, 17
Ru	Rudvalis	$2^3.3.5.7.13.29$	13, 29
Suz	Suzuki	$2^3.3^2.5.7.11.13$	11, 13
$O'N$	O'Nan	$2^4.3.5.7.11.19.31$	19, 31
Co_3	Conway	$2^4.3^2.5.7.11.23$	9, 23
Co_2	Conway	$2^4.3^2.5.7.11.23$	9, 23
Fi_{22}	Fischer	$2^4.3^2.5.7.11.13$	11, 13
HN	Harada-Norton	$2^3.3^2.5^2.7.11.19$	11, 19
Ly	Lyons	$2^3.3^2.5^2.7.11.31.37.67$	37, 67
Th	Thompson	$2^3.3^3.5.7.13.19.31$	19, 31
Fi_{23}	Fischer	$2^4.3^3.5.7.11.13.17.23$	17, 23
Co_1	Conway	$2^4.3^2.5.7.11.13.23$	13, 23
J_4	Janko	$2^4.3.5.7.11.23.29.37.43$	37, 43
Fi'_{24}	Fischer	$2^4.3^3.5.7.11.13.17.23.29$	23, 29
B	Baby monster	$2^5.3^3.5^2.7.11.13.17.19.23.31.47$	31, 47
M	Monster	$2^5.3^3.5^2.7.11.13.17.19.23.29.31.41.47.59.71$	59, 71

The Monster Group M

The Monster group $H = M$ has order $2^{46}.3^{20}.5^9.7^6.11^2.13^3.17.19.23.29.31.41.47.59.71$ and exponent $2^5.3^3.5^2.7.11.13.17.19.23.29.31.41.47.59.71$, with $|\mathrm{Out}\, H| = 1$. We will apply the method described above, with $q_1 = 59$ and $q_2 = 71$. The maximal subgroups of M are not yet completely known, but it is known that the only maximal subgroups of order divisible by 59 or 71 are those isomorphic to $L_2(59)$ or $L_2(71)$ (see [33, Table 5.6]); thus no proper subgroup has order divisible by both 59 and 71, so a triple of type $(l, 59, 71)$ for any l must generate H. Provided they exist, we can therefore use a pair of triples of types $(59, 59, 71)$ and $(59, 71, 71)$ to distinguish the primes 59 and 71, and six pairs of triples of types $(l, 59, 71)$, where $l = 11, 13, 19, 23, 25, 27, 29, 31, 32, 41, 47, 119$ (=7.17), for the remaining thirteen primes $p < 59$ dividing $|H|$. By Lemma 4.3 this will deal with values $k \geq 42$, and one of the latter types, with mutually coprime periods, will cover smaller k by Lemma 4.4. Inspection of the character table of H in [8] shows that for each of the above types one can choose appropriate conjugacy classes X, Y and Z so that the character sum $S(X, Y, Z)$ is very close to 1. This shows that such generating triples exist, and in sufficient numbers for Lemma 4.4 to deal with the cases $k = 2, \dots, 41$.

In fact, all but 14 of the 193 non-principal irreducible characters χ vanish on elements of order 59 or 71, so they do not contribute to any of the character sums, while the non-vanishing characters, of degrees $\chi(1) \geq 8980616927734375$, all take values $\chi(g) = \pm 1$ or $(-1 \pm i\sqrt{59})/2$ on those classes. The character values on the other classes needed are also much smaller than these degrees: for instance, if g has order $l = 11$ then $|\chi(g)| \leq 190$ for all χ. Since there are two conjugacy classes each of elements of orders 59 and 71, and one of order 11, with centralisers of orders 59, 71 and 1045440 respectively, it follows that the number of equivalence classes of generating triples of type $(11, 59, 71)$ is approximately

$$4 \times \frac{|H|}{59.71.1045440} \approx 7 \cdot 38 \times 10^{44}.$$

The estimates for the other types required are even larger, since there are more elements of order l in the cases where $l > 11$.

The Janko Group J_3

The third Janko group $H = J_3$ has order $2^6.3^5.5.17.19$ and exponent $2^3.3^2.5.17.19$ with $|\mathrm{Out}\, H| = 2$. Although there are only five primes dividing $|H|$, it is more convenient to apply the method used for M than that used for the Mathieu groups. We will take $q_1 = 17$ and $q_2 = 19$. The only maximal subgroups of H with elements of order 19 are those isomorphic to $L_2(19)$; this group has no elements of order 17, so for any l, a triple of type $(l, 17, 19)$ must generate H. Provided they exist, one can therefore use a pair of triples of types $(17, 17, 19)$ and $(17, 19, 19)$ to distinguish the primes 17 and 19, a pair of types $(8, 17, 19)$ and $(9, 17, 19)$ for the primes $p = 2$

and 3, and a pair of types (5, 17, 19) and (for instance) (4, 17, 19) for $p = 5$. As in the case of M, the character values guarantee the existence of all the required triples: the only non-principal irreducible characters χ not vanishing on elements of orders 17 or 19, and hence contributing to any of the relevant character sums $S(X, Y, Z)$, are one of degree 324, two of degree 1215 and three of degree 1920; these satisfy $|\chi(g)| \leq 4$ for all elements g of order at least 4, so each $S(X, Y, Z)$ is very close to 1. By Lemma 4.3 this deals with the values $k = 18, \ldots, d_2(H)$. Frobenius's formula shows that there are, for instance, 21,312 equivalence classes of triples of type (5, 17, 19), more than enough for Lemma 4.4 to deal with the values $k = 2, \ldots, 17$.

Table 1 shows how the same method can be applied to the remaining 15 sporadic simple groups. Thus each of the 26 sporadic simple groups satisfies Theorem 1.3, and the proof is complete.

10 The Groups $L_3(q)$ and $U_3(q)$

The method used for the larger sporadic simple groups can also be applied to other families of simple groups, such as the groups $L_3(q)$ and $U_3(q)$.

10.1 $L_3(q)$

The 3-dimensional projective special linear group $H = L_3(q)$ is simple for every prime power q. Since $L_3(2) \cong L_2(7)$ we may assume that $q > 2$. The group H contains elements h_1 and h_2 of orders

$$q_1 = \frac{q^2 + q + 1}{d} \quad \text{and} \quad q_2 = \frac{q^2 - 1}{d}$$

where $d = \gcd(q^2 + q + 1, 3) = \gcd(q - 1, 3)$. The maximal subgroups of H were determined for even q by Hartley [16], and for odd q first by Mitchell [27] and then (using more modern methods) by Bloom [5]: there are none containing elements of orders q_1 and q_2 (see also Kantor's classification [21] of linear groups containing Singer cycles). It follows that for any l, a triple of type (l, q_1, q_2) must generate H. The generic character table for H given by Simpson and Frame in [29] shows that $\chi(h_1)\chi(h_2) = 0$ for every non-principal irreducible character χ of H except the Steinberg character, of degree q^3; for this character we have $\chi(h_1) = 1 = -\chi(h_2)$, and $|\chi(h)| \leq q$ for all $h \neq 1$. Thus $S(X, Y, Z) > 0$ where X is any non-identity conjugacy class, and Y and Z are classes of elements of orders q_1 and q_2, so generating triples of type (l, q_1, q_2) exist whenever H contains elements of order $l \neq 1$.

Now

$$|H| = \frac{1}{d}q^3(q^2+q+1)(q+1)(q-1)^2,$$

so each prime divisor p of $|H|$ divides at least one of $q, q^2+q+1, q+1$ and $q-1$. These four integers are mutually coprime, except that $\gcd(q^2+q+1, q-1) = 3$ if $q \equiv 1 \bmod (3)$, and $\gcd(q+1, q-1) = 2$ if q is odd. It follows from this that q_1 and q_2 are always mutually coprime. A pair of generating triples of types (q_1, q_1, q_2) and (q_1, q_2, q_2) can therefore be used to strongly distinguish all the primes dividing q_1 or q_2. This deals with every prime p dividing $|H|$, except the prime $p = p_0$ dividing q, and the prime $p = 3$ if $q \equiv 4$ or $7 \bmod (9)$ (so that q_1 and q_2 are both coprime to 3, whereas $|H|$ is not). The exponent q_0 of a Sylow p_0-subgroup of H is p_0 if $p_0 > 2$, and 4 if $p_0 = 2$, so a pair of triples of types (q_0, q_1, q_2) and $(3, q_1, q_2)$ will strongly distinguish p_0 if $p_0 \neq 3$, and will strongly distinguish 3 if $q \equiv 4$ or $7 \bmod (9)$; if $p_0 = 3$, so that $q \not\equiv 4$ or $7 \bmod (9)$, we can instead replace the second triple with one of type $(2, q_1, q_2)$ to strongly distinguish p_0. In either case these two pairs of triples satisfy the hypotheses of Lemma 4.2 with $t = 2$, so they deal with the values $k = 6, \ldots, d_2(H)$.

There are at least two inequivalent generating triples of type (p_0, q_1, q_2): for instance, one can choose the first element to be central or non-central in the unique Sylow p_0-subgroup containing it. These triples have mutually coprime periods, so Lemma 4.4(1) deals with $k = 2, \ldots, 12$, and hence the groups $H = L_3(q)$ satisfy Theorem 1.2.

10.2 $U_3(q)$

The proof for the unitary groups $H = U_3(q)$ is essentially the same. These groups have order

$$|H| = \frac{1}{d}q^3\left(q^3+1\right)(q-1) = \frac{1}{d}q^3\left(q^2-q+1\right)(q+1)(q-1)$$

where $d = \gcd(q+1, 3)$, and they are simple for all prime powers $q > 2$. They contain elements h_1 and h_2 of mutually coprime orders

$$q_1 = \frac{q^2-q+1}{d} \quad \text{and} \quad q_2 = \frac{q^2-1}{d},$$

all satisfying $H = \langle h_1, h_2\rangle$ (see [33, Sect. 3.10.9] for the maximal subgroups of H, or [4] for subgroups containing a Singer cycle h_1). As with $L_3(q)$, the generic character table for H in [29] shows that generating triples of type (l, q_1, q_2) exist whenever H contains elements of order $l \neq 1$. A pair of generating triples of types (q_1, q_1, q_2) and (q_1, q_2, q_2) strongly distinguish each prime dividing $|H|$, apart from the prime p_0 dividing q, and the prime 3 if $q \equiv 2$ or $5 \bmod (9)$. The exponent

q_0 of a Sylow p_0-subgroup is as before, so for these primes one can use a pair of triples of types (q_0, q_1, q_2) and either $(3, q_1, q_2)$ or $(2, q_1, q_2)$ as $q \equiv 2, 5 \bmod (9)$ or not. Lemma 4.2, with $t = 2$, then deals with the values $k = 6, \ldots, d_2(H)$, and again Lemma 4.4(1), applied to triples of type (p_0, q_1, q_2), deals with small k. This completes the proof of Theorem 1.2.

References

1. I. Bauer, F. Catanese, F. Grunewald, Beauville surfaces without real structures I, in *Geometric Methods in Algebra and Number Theory*. Progress in Mathematics, vol. 235 (Birkhäuser, Boston, 2005), pp. 1–42
2. I. Bauer, F. Catanese, F. Grunewald, Chebycheff and Belyi polynomials, dessins d'enfants, Beauville surfaces and group theory. Mediterr. J. Math. **3**, 121–146 (2006)
3. A. Beauville, Surfaces algébriques complexes, Astérisque 54, *Soc. Math.* France, (1978)
4. A. Bereczky, Maximal overgroups of Singer elements in classical groups. J. Algebra **234**, 187–206 (2000)
5. D.M. Bloom, The subgroups of $PSL(3, q)$ for odd q. Trans. Am. Math. Soc. **127**, 150–178 (1967)
6. F. Catanese, Fibred surfaces, varieties isogenous to a product and related moduli spaces. Am. J. Math. **122**, 1–44 (2000)
7. C. Choi, On subgroups of M_{24}. II. The maximal subgroups of M_{24}. Trans. Am. Math. Soc. **167**, 29–47 (1972)
8. J.H. Conway, R.T. Curtis, S.P. Norton, R.A. Parker, R.A. Wilson, *ATLAS of Finite Groups* (Clarendon Press, Oxford, 1985)
9. L.E. Dickson, *Linear Groups* (Dover, New York, 1958)
10. J.D. Dixon, The probability of generating the symmetric group. Math. Z. **110**, 199–205 (1969)
11. B.T. Fairbairn, K. Magaard, C.W. Parker, Generation of finite simple groups with an application to groups acting on Beauville surfaces. Proc. London Math. Soc. **107**(3), 744–798 (2013) doi:10.1112/plms/pds097
12. F.G. Frobenius, *Über Gruppencharaktere*, Sitzber. Königlich Preuss. Akad. Wiss. Berlin, pp. 985–1021 (1896)
13. S. Garion, M. Larsen, A. Lubotzky, Beauville surfaces and finite simple groups. J. Reine Angew. Math. **666**, 225–243 (2012)
14. R. Guralnick, G. Malle, Simple groups admit Beauville structures. J. London Math. Soc. **85**(2), 694–721 (2012)
15. P. Hall, The Eulerian functions of a group. Q. J. Math. **7**, 134–151 (1936)
16. R.W. Hartley, Determination of the ternary collineation groups whose coefficients lie in the $GF(2^n)$. Ann. Math. **27**(2), 140–158 (1925)
17. Z. Janko, A new finite simple group with abelian Sylow subgroups and its characterization. J. Algebra **3**, 147–186 (1966)
18. G.A. Jones, Enumeration of homomorphisms and surface-coverings. Q. J. Math. **46**, 485–507 (1995)
19. G.A. Jones, Beauville surfaces and groups: a survey, in *Rigidity and Symmetry*, ed. by R. Connelly, A. Weiss and W. Whiteley, *Fields Inst. Commun.* **70**, pp. 205–225 (2014)
20. G.A. Jones, Characteristically simple Beauville groups, I: cartesian powers of alternating groups. in *Geometry, Groups and Dynamics*, ed. by C.S. Aravinda, W.M. Goldman, et al., Contemp. Math. **639**, to appear (2015)
21. W.M. Kantor, Linear groups containing a Singer cycle. J. Algebra **62**, 232–234 (1980)
22. W.M. Kantor, A. Lubotzky, The probability of generating a finite classical group. Geom. Dedicata **36**, 67–87 (1990)

23. P.B. Kleidman, The maximal subgroups of the Chevalley groups $G_2(q)$ with q odd, of the Ree groups ${}^2G_2(q)$, and of their automorphism groups. J. Algebra **117**, 30–71 (1988)
24. V.M. Levchuk, Ya. N. Nuzhin, Structure of Ree groups (Russian). Algebra Log. **24**, 26–41 (1985)
25. M.W. Liebeck, A. Shalev, The probability of generating a finite simple group. Geom. Dedicata **56**, 103–113 (1995)
26. A.M. Macbeath, Generators of the linear fractional groups, in *Number Theory (Houston 1967)*, ed. by W.J. Leveque, E.G. Straus (Am. Math. Soc., Providence, 1969), pp. 14–32. (Proc. Sympos. Pure Math. 12)
27. H.H. Mitchell, Determination of the ordinary and modular ternary linear groups. Trans. Am. Math. Soc. **12**, 207–242 (1911)
28. R. Ree, A family of simple groups associated with the simple Lie algebra of type (G_2). Am. J. Math. **83**, 432–462 (1961)
29. W.A. Simpson, J.S. Frame, The character tables for $SL(3,q)$, $SU(3,q^2)$, $PSL(3,q)$, $PSU(3,q^2)$. Can. J. Math. **25**, 486–494 (1973)
30. J.-P. Serre, *Topics in Galois Theory*, 2nd edn., A.K. Peters (Wellesley, 2008)
31. M. Suzuki, On a class of doubly transitive groups. Ann. Math. **75**(2), 105–145 (1962)
32. H.N. Ward, On Ree's series of simple groups. Trans. Am. Math. Soc. **121**, 62–89 (1966)
33. R.A. Wilson, *The Finite Simple Groups*. Graduate Texts in Mathematics, vol. 251 (Springer, London, 2009)

Remarks on Lifting Beauville Structures of Quasisimple Groups

Kay Magaard and Christopher Parker

Abstract In previous work, we developed theorems which produce a multitude of hyperbolic triples for finite classical groups. We apply these theorems to prove a conjecture of Bauer, Catanese and Grunewald, which asserts that all non-abelian finite quasisimple groups except for the alternating group of degree five are Beauville groups. Here we show that our results can be used to show that certain split- and Frattini extensions of quasisimple groups are also Beauville groups. We also discuss some open problems for future investigations.

2000 Mathematics Subject Classification. 20E34 · 20F05 · 14J29 · 30F10

1 Introduction

Suppose that G is a group. A *hyperbolic triple* in G is a triple $(x, y, z) \in G \times G \times G$ such that

$G = \langle x, y, z \rangle$ and $xyz = 1$; and
$1/o(x) + 1/o(y) + 1/o(z) < 1$.

A *Beauville structure* $\mathcal{B}$ of G is a pair of hyperbolic triples $((x_1, y_1, z_1), (x_2, y_2, z_2))$ in G such that no non-identity power of x_1, y_1 or z_1 is conjugate in G to a power of x_2, y_2 or z_2. A hyperbolic triple (x, y, z) has *type* $(o(x), o(y), o(z))$ and a Beauville structure $((x_1, y_1, z_1), (x_2, y_2, z_2))$ has *type*

$$((o(x_1), o(y_1), o(z_1)), (o(x_2), o(y_2), o(z_2))).$$

K. Magaard (✉) · C. Parker
School of Mathematics, University of Birmingham, Birmingham, B15 2TT Edgbaston, UK
e-mail: k.magaard@bham.ac.uk

C. Parker
e-mail: c.w.parker@bham.ac.uk

I. Bauer et al. (eds.), *Beauville Surfaces and Groups*, Springer Proceedings in Mathematics & Statistics 123, DOI 10.1007/978-3-319-13862-6_8

In joint work with Fairbairn [2] the authors determined those quasisimple groups which have Beauville structures. The simple groups with Beauville structures have also been determined by Guralnick and Malle [4]. The main result of [2] asserts that, with the exception of $PSL_2(5)$ and $SL_2(5)$, all quasisimple groups do have such structures. Following on from these discoveries, a natural question that arises is

which extensions of the finite simple groups admit Beauville structures?

The purpose of this contribution is to point out some ways to begin to answer this question. One immediate restriction is that if G admits a Beauville structure, then G must be 2 generated. The recent conjecture of Gareth Jones [6, 7] suggests that in many interesting cases this might be sufficient.

One type of extension that generalizes perfect central extensions is called a *Frattini cover*. Particular amongst the Frattini covers are the non-split extensions of irreducible $GF(p)G$-modules. In this note we draw attention to a corollary of a theorem of Guralnick and Tiep [5] which asserts that under certain mild restrictions Beauville structures for a group G lift to Frattini extensions of G. The details of this are in Sect. 2.

In Sect. 3, we consider split extensions $H = GV$ where V is a non-central elementary abelian minimal normal subgroup of H and G is a complement to V in H. We present an elementary criterion which guarantees that if G has a Beauville structure, then this structure lifts to H. Notice that a Beauville structure of GV may not project to a Beauville structure of G as the conjugacy condition may fail.

We choose examples of extensions of $SL_d(p^a)$ to illustrate the type of results we find interesting and present only the most general statements. For instance, we present Beauville structures for p-Frattini covers $PSL_d(p^a)$ for $d \geq 9$. The results we give are not the best results that can easily be obtained, but they are elementary and suffice to make our point. We also explain that if $G \cong SL_d(p^a)$ and V is the natural $GF(p^a)G$-module, then GV has a Beauville structure (again assuming that d is sufficiently large). Our final section poses a short series of questions which we hope will stimulate further research.

2 Frattini Covers of Beauville Groups

Suppose that G and X are groups such that $X/\Phi(X) \cong G$. Then X is called a *Frattini cover* of G. If $\Phi(X)$ is a p-group for some prime p, then X is a *p-Frattini cover.* By the Schur-Zassenhaus Theorem, if X is a p-Frattini cover of G and $\Phi(X) \neq 1$, then p divides $|G|$. The recent investigations in [5, 9] on p-Frattini covers are motivated by their connection with modular towers which were introduced by Michael Fried in connection with the inverse problem of Galois Theory.

Theorem 2.1 (Guralnick and Tiep [5]) *Let p be an odd prime and X be a p-Frattini cover of G. Assume that p does not divide the order of the Schur multiplier of G. Let $g_1, \ldots, g_r \in G$ satisfy*

(i) $G = \langle g_1, ..., g_r \rangle$,
(ii) $g_1 \dots g_r = 1$, *and*
(iii) *the order of each* g_i *is coprime to* p.

Then, for any $f \in \Phi(X)$, *there exist* $x_i \in X$, *with* $x_i\Phi(X) = g_i$ *and* $o(x_i) = o(g_i)$ *such that* $x_1 \dots x_r = f$.

As an immediate corollary to the Guralnick-Tiep Theorem we obtain

Corollary 2.2 *Suppose that G has a Beauville structure* $\mathcal{B}$ *of type* $((r, s, t), (u, v, w))$ *and X is a p-Frattini cover of G. If p is coprime to* $2rstuvw$ *and to the order of the Schur multiplier of G, then X has a Beauville structure of type* $((r, s, t), (u, v, w))$.

Proof Take $f = 1$ in the Guralnick-Tiep Theorem. Then the hyperbolic triples in $\mathcal{B}$ lift to hyperbolic triples of X and provide a Beauville structure of X. □

By Zsigmondy's Theorem, for a natural number $n > 2$ and a prime p such that $(n, p) \neq (6, 2)$, there is a prime which divides $p^n - 1$ which does not divide $p^k - 1$ for $k < n$. Such prime numbers are called *Zsigmondy primes*. Let $\zeta_{n,p}$ be a Zsigmondy prime dividing $p^n - 1$ but not $p^k - 1$ for $k < n$ and choose it maximally from among all such prime numbers. Define $\lambda_{n,p}$ to be the largest power of $\zeta_{n,p}$ which divides $p^n - 1$. We present the following special case of [2, Lemma 4.1].

Lemma 2.3 *Assume that* $d \geq 9$, $G = \mathrm{SL}_d(p^a)$ *and* $Z \leq Z(G)$. *Then*

(i) G/Z *has a hyperbolic triple of type* $(\lambda_{ad,p}, \lambda_{ad,p}, \lambda_{a(d-1),p})$; *and*
(ii) G/Z *has a hyperbolic triple of type*

$$(\lambda_{a(d-3),p}\lambda_{3a,p}, \lambda_{a(d-3),p}\lambda_{3a,p}, \lambda_{a(d-2),p})$$

where $\lambda_{6,2} = \lambda_{3.2,2}$ *is* 7.

In particular, G/Z *has a Beauville structure.*

Combining this with the Guralnick-Tiep Theorem we easily obtain the following theorem.

Theorem 2.4 *Suppose that* p *is an odd prime,* $G \cong \mathrm{PSL}_d(p^a)$ *and that* X *is a* p*-Frattini cover of* G. *If* $d \geq 9$, *then* X *has a Beauville structure.*

Proof Since p is odd and the types of the triples listed in Lemma 2.3 (i) and (ii) are coprime to $p(p^a - 1)$, Corollary 2.2 applies to give the result. □

Suppose that X is the profinite group $\mathrm{SL}_d(\mathbb{Z}_p)$ where $\mathbb{Z}_p$ denotes the ring of p-adic integers. A calculation of Serre [10, Lemma 3 and Exercise 1a), pages IV-23 and IV-27] shows that the non-trivial finite quotients of X are p-Frattini covers whenever $p \geq 5$. Combining this with Theorem 2.4 yields the following statement.

Example 2.5 If $d \geq 9$ and $p \geq 5$, then every finite quotient of $X = \mathrm{SL}_d(\mathbb{Z}_p)$ has a Beauville structure.

3 Semidirect Products with Beauville Structures

In this section, we present some observations about Beauville structures in semidirect products $H = GV$. Lifting hyperbolic triples from G to H is not always possible. However there is an elementary criterium which at least allows us to guarantee that there are lifts of generators of G to generators of H in the case when V is a minimal normal subgroup of H which is abelian.

Lemma 3.1 *Suppose that V is an irreducible, faithful* $\mathrm{GF}(p)G$*-module and let $H = GV$ be the semidirect product of G and V. Assume $\mathcal{H} = (a, b, c)$ is a hyperbolic triple in G. Then*

(i) *For all $v, w \in V$, we have $avbwc(v^{bc}w^c)^{-1} = 1$.*
(ii) *Suppose that $X \subseteq V$ and $Y \subseteq V$ with $|X||Y| > |V||\mathrm{H}^1(G, V)|$. Then there exist $u \in X$ and $v \in Y$ such that $H = \langle au, bv\rangle$.*
(iii) *There exist $u, v \in V$ such that $H = \langle au, bv\rangle$.*

In particular, there are $u, v \in V$ such that $(av, bw, c(v^{bc}w^c)^{-1})$ is a hyperbolic triple in H which lifts $\mathcal{H}$.

Proof We have

$$\begin{aligned} avbwc(v^{bc}w^c)^{-1} &= abcv^{bc}w^c(v^{bc}w^c)^{-1} \\ &= 1 \end{aligned}$$

and so (i) is true.

We next prove (ii). Set $\mathcal{C} = \{\langle au, bv\rangle \mid u \in X, v \in Y\}$ and let $J = \langle au, bv\rangle \in \mathcal{C}$. Since V is an irreducible $\mathrm{GF}(p)G$-module and, since

$$JV = \langle au, bv\rangle V = \langle a, b\rangle V = GV = H,$$

we infer that either J is a complement to V or $J = H$. It follows that either (ii) holds or all members of $\mathcal{C}$ are complements to V in G. Assume that the latter is the case and let $K = \langle au_1, bv_1\rangle \in \mathcal{C}$. If $J = K$, then $(au_1)^{-1}au = u_1^{-1}u \in J \cap V$ and $v_1^{-1}v \in J \cap V$. As J is a complement to V, we deduce that $uu_1^{-1} = vv_1^{-1} = 1$. This yields $|\mathcal{C}| = |X||Y| > |V||\mathrm{H}^1(G, V)|$. However $|\mathcal{C}| \leq |V||\mathrm{H}^1(G, V)|$ which is the total number of complements to V in X. This contradiction proves (ii).

Taking $V = X = Y$, we have $|X||Y| = |V|^2 > |V||\mathrm{H}^1(G, V)|$ by the Aschbacher-Guralnick Theorem [1, Theorem 1]. Thus (iii) follows from (ii).

Lemma 3.2 *Assume that $\mathcal{B} = ((a_1, b_1, c_1), (a_2, b_2, c_2))$ is a Beauville structure of the group G. Suppose that V is an irreducible, faithful* $\mathrm{GF}(p)G$*-module and set $H = GV$ the semidirect product of G and V. If*

$$\dim[V, a_1] + \dim[V, b_1] > \dim V + \dim \mathrm{H}^1(G, V),$$

$$\dim[V, a_2] + \dim[V, b_2] > \dim V + \dim \mathrm{H}^1(G, V)$$

and either

(i) $C_V(c_1) = 0$ *or* $C_V(c_2) = 0$; *or*
(ii) $\langle c_1, c_2^g \rangle$ *does not centralize a vector of* V *for all* $g \in G$,

then H *has a Beauville structure which is a lift of* $\mathcal{B}$.

Proof For $i = 1, 2$, set $X_i = [V, a_i]$ and $Y_i = [V, b_i]$. Then, as $|V||\mathrm{H}^1(G, V)| < |X_i||Y_i|$, Lemma 3.1 (ii) implies that there exists $v_i \in X_i$ and $w_i \in Y_i$ such that $H = \langle a_i v_i, b_i w_i \rangle$. The choice of $v_i \in X_i$ and $w_i \in Y_i$ means that a_i and $a_i v_i$ are conjugate in H and so are b_i and $b_i w_i$. Now to form a hyperbolic triple completing $a_i v_i, b_i w_i$ we need to take a third element $d_i = c_i(v_i^{b_i c_i} w_i^{c_i})^{-1}$. Let r_i be the order of c_i, then d_i has order r_i or pr_i. The only way that $\mathcal{B}$ can fail to lift to a Beauville structure of H is if d_1 has order $r_1 p$ and d_2 has order $r_2 p$ and $e_1 = d_1^{r_1}$ and $e_2 = d_2^{r_2}$ are conjugate in H (and hence by elements of G). In particular, if (i) holds then $\mathcal{B}$ lifts to a Beauville structure of H. So consider the alternative case. Then there exists $g \in G$ such that $e_1 = e_2^g$ is centralized by $\langle d_1, d_2^g \rangle$. But $\langle c_1, c_2^g \rangle$ does not centralize any vector in V. This shows that $\mathcal{B}$ lifts to a Beauville structure of H.

We state the famous theorem of Scott as it shows that the initial restriction on the sizes of the commutators in Lemma 3.2 is not a severe limitation. We present the theorem just for hyperbolic triples.

Theorem 3.3 (L. Scott) *Suppose that* (x_1, x_2, x_3) *is a hyperbolic triple in* G *and let* V *be an irreducible* G*-module of dimension* n. *Assume that, for* $i = 1, 2, 3$, $\dim[V, x_i] = d_i$. *Then the following hold:*

(i) $d_1 + d_2 + d_3 \geq \dim \mathrm{H}^1(G, V) + 2n$; *and*
(ii) *if* x_i *each have order coprime to* p *and the Schur multiplier of* G *has order coprime to* p, *then* $d_1 + d_2 + d_3 \geq \dim \mathrm{H}^1(G, V) + \dim \mathrm{H}^1(G, V^*) + 2n$.

Proof This is a combination of Scott's results [11] Theorem 1 and Proposition 1 (a) for (i) and (b) for (ii) stated for the special case of hyperbolic triples. □

Now we give an example which shows that hyperbolic triples do not always lift so that the elements in the triple have the same order.

Example 3.4 Suppose that $G \cong \mathrm{SL}_3(2)$ and V is the natural 3-dimension G-module. Then G has a hyperbolic triple (a, b, c) with $o(a) = 2$, $o(b) = 3$ and $o(c) = 7$. We know $|[V, a]| = 2$, $|[V, b]| = 2^2$ and $V = [V, c]$. It is easy to check that GV has no hyperbolic triple (a', b', c') such that $o(a') = 2$, $o(b') = 3$, $o(c') = 7$.

Theorem 3.5 *Suppose that* $G \cong \mathrm{SL}_d(p^a)$ *with* $d \geq 9$ *and* V *is the natural* $\mathrm{GF}(p^a)G$*-module. Then* $H = GV$ *has a Beauville structure.*

Proof By Lemma 2.3, G has a hyperbolic triple of type $(\lambda_{ad,p}, \lambda_{ad,p}, \lambda_{a(d-1),p})$ and a hyperbolic triple of type $(\lambda_{a(d-3),p}\lambda_{3a,p}, \lambda_{a(d-3),p}\lambda_{3a,p}, \lambda_{a(d-2),p})$ where $\lambda_{6,2} = \lambda_{3.2,2}$ is 7. The elements of order $\lambda_{ad,p}$ and $\lambda_{a(d-3),p}\lambda_{3a,p}$ act fixed point freely on V and the elements of order $\lambda_{a(d-1),p}$ and $\lambda_{a(d-2),p}$ fix a 1 and a 2-space in V respectively. Furthermore, by [8, Table B], $H^1(G, V) = 0$ and so $H = GV$ has a Beauville structure by Lemma 3.2.

4 Questions

In this section we pose some questions which we hope suggest fruitful research directions for the study of hyperbolic triples and Beauville surfaces. For Frattini covers we ask

Question 4.1 *For some of the small rank quasisimple classical groups G in characteristic p, one of the hyperbolic triples describing the Beauville structure given in* [2] *involves a p-element. This means that we may not apply Theorem 2.1 to lift the hyperbolic triple to a p-Frattini cover of G. Under what hypotheses can a hyperbolic triple be lifted to a hyperbolic triple for G?*

Since the Guralnick-Tiep Theorem does not apply when $p = 2$, it would be nice to know the answer to the following question.

Question 4.2 *Under which additional hypotheses can the Guralnick-Tiep Theorem be extended to 2-Frattini covers?*

The next question, which we assume typically has an affirmative answer, is a special case of the last question.

Question 4.3 *The groups $2_-^{1+2n}.\Omega_{2n}^-(2)$, $2_-^{1+2n}.\Omega_{2n}^-(2)$ and $\mathbb{Z}/4\mathbb{Z} \circ 2_+^{1+2n}.\mathrm{Sp}_{2n}(2)$ are derived subgroups of groups in the Aschbacher class $\mathcal{C}_6$ of maximal subgroups of classical groups. For sufficiently large n, they are 2-Frattini covers of their simple quotients. Do they have Beauville structures?*

The natural questions for semidirect products seem to be

Question 4.4 *For a quasisimple group G and prime p, for which irreducible $\mathrm{GF}(p)G$-modules is the semidirect product GV a Beauville group?*

In [6, 7] Gareth Jones begins the investigation of Beauville structures in direct products H^k of k non-abelian simple groups isomorphic to H (characteristically simple groups). He conjectures that such groups have Beauville structures whenever they can be generated by two elements (other than $H = \mathrm{Alt}(5)$ and $k = 1$). He proves his conjecture for $\mathrm{Alt}(n)$, $n \geq 5$, $\mathrm{PSL}_2(q)$, $q \geq 4$, $\mathrm{PSL}_3(q)$, $\mathrm{PSU}_3(q)$, $q > 2$, ${}^2\mathrm{B}_2(2^e)$, $e > 1$ odd, ${}^2\mathrm{G}_2(3^e)$, $e > 1$ odd, and for the sporadic simple groups. We can pose a similar question for semidirect products which extends Question 4.4.

Question 4.5 *Suppose that G is a quasisimple group, p is a prime and V is an irreducible* $\mathrm{GF}(p)G$*-module. Let nV denote a direct sum of n copies of V. For which n is $G(nV)$ a Beauville group?*

The next example, though not using a hyperbolic triple, shows that we may expect that an answer to Question 4.5 which is similar to that conjectured in [6, 7] for characteristically simple non-abelian groups.

Example 4.6 Let $G = \mathbb{Z}/3\mathbb{Z}$ and V be elementary abelian of order 4 admitting G non-trivially. Then $GV \cong \mathrm{Alt}(4)$ and G has a triple of type $(3, 3, 3)$ (which is not hyperbolic). The universal $(3, 3, 3)$-group is isomorphic to $\mathbb{Z}^2$ extended by a cyclic group of order 3. Now, in $G(nV)$ we see that every element has order either 2 or 3. If a $(3, 3, 3)$ triple for GV could be extended to a $(3, 3, 3)$-triple for $G(nV)$, then that triple would also be a $(3, 3, 3)$-triple. Hence it must generate a subgroup of $G(nV)$ isomorphic to $\mathrm{Alt}(4)$. We conclude that $n = 1$.

Based on our results in the previous section, our most general question involves one-headed groups. Recall that a group X is *one-headed* with *head* G provided it has a unique maximal normal subgroup Y and $X/Y \cong G$ is a non-abelian simple group. Clearly the quasisimple groups, Frattini covers and the semidirect products in Lemma 3.2 are one-headed.

Question 4.7 *Let X be a one-headed group with head G. Under which hypotheses do Beauville structures of G lift to Beauville structures of X?*

Magaard's talk at the meeting in Newcastle raised questions about the number of Beauville structures that a given quasisimple group might admit. We close this article by stating this question explicitly and then setting it in a more specific context.

Question 4.8 (Magaard, 2012) *Given a quasisimple group G, how many Beauville structures does G admit?*

Question 4.9 *Given a non-abelian simple group G, a Beauville structure $\mathcal{B}$ of G and an extension X of G, how many Beauville structures of X project onto $\mathcal{B}$?*

Question 4.10 *Given a group extension X with $X/Y \cong G$ a simple group and a hyperbolic triple $\mathcal{H}$ of G, how many hyperbolic triples project on to $\mathcal{H}$?*

We expect the last three questions may have a probabilistic answer rather than an explicit integer value.

Of course, ultimately these results are closely connected to the 1- and 2-cohomology of G. The 3-cohomology of groups may be needed to investigate Beauville structures of one-headed non-abelian extensions.

References

1. M. Aschbacher, R.M. Guralnick, Some applications of the first cohomology group. J. Algebra **90**(2), 446–460 (1984)
2. B. Fairbairn, K. Magaard, C. Parker, Generation of finite quasisimple groups with an application to groups acting on Beauville surfaces. Proc. LMS appeared online 18 February, (2013)
3. W. Feit, On large Zsigmondy primes. Proc. Am. Math. Soc. **102**(1), 29–36 (1988)
4. R. Guralnick, G. Malle, Simple groups admit Beauville structures. J. Lond. Math. Soc. **85**(3), 694–721 (2012)
5. R.M. Guralnick, P.H. Tiep, Lifting in Frattini covers and a characterization of finite solvable groups. arXiv:1112.4559
6. G.A. Jones, Characteristically simple Beauville groups, I: Cartesian powers of alternating groups. arXiv:1304.5444
7. G.A. Jones, Characteristically simple Beauville groups, II: low rank and sporadic groups. arXiv:1304.5450
8. W. Jones, B. Parshall, On the 1-cohomology of finite groups of Lie type, in *Proceedings of the Conference on Finite Groups* (Univ. Utah, Park City, Utah, 1975)
9. Darren Semmen, The group theory behind modular towers. Séminaires Congrès **13**, 343–366 (2006)
10. J.-P. Serre, Abelian l-adic representations and elliptic curves. McGill University lecture notes written with the collaboration of Willem Kuyk and John Labute, W. A. Benjamin, Inc., New York-Amsterdam (1968)
11. L.L. Scott, Matrices and cohomology, Ann. Math. (2) 105 (1977), no. 3, 473–492

Surfaces Isogenous to a Product of Curves, Braid Groups and Mapping Class Groups

Matteo Penegini

Abstract We present some group theoretical methods to give bounds on the number of connected components of the moduli space of surfaces of general type, focusing on some families of regular surfaces isogenous to a product of curves.

1 Introduction

This article is a revised version of the talk I gave at the conference "*Beauville Surfaces and groups*" held in Newcastle in June 2012. It presents some group theoretical methods to give bounds on the number of connected components of the moduli space of surfaces of general type, focusing on some families of regular surfaces isogenous to a product of curves. Some of the results appearing in this work have been proven in collaboration with Shelly Garion.

We will use the standard notation from the theory of complex algebraic surfaces. Let S be a smooth, complex, projective, *minimal* surface S of *general type*; this means that the canonical divisor K_S of S is *big* and *nef*.

The principal numerical invariants for the study of minimal surfaces of general type are

- the *geometric genus* $p_g(S) := h^0(S, \Omega_S^2) = h^0(S, \mathcal{O}_S(K_S))$,
- the *irregularity* $q(S) := h^0(S, \Omega_S^1)$, and
- the *self intersection of the canonical divisor* K_S^2.

In fact, these determine all the other classical invariants, as

- the *Euler-Poincaré characteristic* $\chi(\mathcal{O}_S) = 1 - q(S) + p_g(S)$,
- the *topological Euler number* $e(S) = 12\chi(S) - K_S^2$, and
- the *plurigenera* $P_n(S) = \chi(S) + \binom{n}{2}K_S^2$.

M. Penegini (✉)
Matteo Penegini, Dipartimento di Matematica "Federigo Enriques", Università degli Studi di Milano, Via Saldini 50, 20133 Milano, Italy
e-mail: matteo.penegini@unimi.it

I. Bauer et al. (eds.), *Beauville Surfaces and Groups*, Springer Proceedings in Mathematics & Statistics 123, DOI 10.1007/978-3-319-13862-6_9

Moreover, we call a surface *regular* if its irregularity vanishes, i.e., $q(S) = 0$.

By a theorem of Bombieri, a minimal surface of general type S with fixed invariants is birationally mapped to a normal surface X in a fixed projective space of dimension $P_5(S)-1$. Moreover, X is uniquely determined and is called the *canonical model* of S. Let us recall Gieseker's Theorem.

Theorem 1.1 *There exists a quasi-projective coarse moduli space $\mathcal{M}_{y,x}$ for canonical models of surfaces of general type S with fixed invariants $y := K_S^2$ and $x := \chi$.*

The union $\mathcal{M}$ over all admissible pairs of invariants (y, x) of these spaces is called the *moduli space of surfaces of general type*. If S is a smooth minimal surface of general type, we denote by $\mathcal{M}(S)$ the subvariety of $\mathcal{M}_{y,x}$, corresponding to surfaces (orientedly) homeomorphic to S. Moreover, we denote by $\mathcal{M}^0_{y,x}$ the subspace of the moduli space corresponding to regular surfaces.

It is known that the number of connected components $\delta(y, x)$ of $\mathcal{M}^0_{y,x}$ is bounded from above by a function of y; more precisely by [7] we have $\delta(y, x) \leq cy^{77y^2}$, where c is a positive constant. Hence we have that the number of components has an exponential upper bound in K^2.

There are also some results regarding the lower bound. In [21], for example, Manetti constructed a sequence S_n of simply connected surfaces of general type, such that the lower bound for the number of the connected components $\delta(S_n)$ of $\mathcal{M}(S_n)$ is given by

$$\delta(S_n) \geq y_n^{\frac{1}{5}log y_n}.$$

Using group theoretical methods we are able to describe the asymptotic growth of the number of connected components of the moduli space of surfaces of general type relative to certain sequences of surfaces. More precisely, we apply the definition and some properties of regular surfaces isogenous to a product of curves and of some special cases of them, Beauville surfaces, to reduce the geometric problem of finding connected components into the algebraic one of counting orbits of some group action, which can be effectively computed.

The paper is organized as follows.

In the first Section we recall the definition and some properties of the mapping class group. Moreover, we describe the Hurwitz moves in the most general setting.

In the second Section we briefly recall the definition of surfaces isogenous to a product of curves. We will recall some of the properties of these surfaces focusing on their moduli space.

In the third part we present some results obtained with Shelly Garion about the number of connected components of the moduli space of surfaces isogenous to a product.

Finally, we also point out some possible future developments.

2 Braid Group and Mapping Class Group

In this section, we first recall the definition of mapping class group. Next we give a presentation of it for $\mathbb{P}^1 - \{p_1, \dots, p_r\}$, and more generally for a curve of genus g' with r marked points. After that, we calculate the Hurwitz moves induced by those groups. We mainly follow the definitions and notation of [9].

Definition 2.1 Let M be a differentiable manifold, then the *mapping class group* of M is the group:

$$\mathrm{Map}(M) := \pi_0(\mathrm{Diff}^+(M)) := \mathrm{Diff}^+(M)/\mathrm{Diff}^0(M),$$

where $\mathrm{Diff}^+(M)$ is the group of orientation preserving diffeomorphisms of M and $\mathrm{Diff}^0(M)$ is the subgroup of diffeomorphisms of M isotopic to the identity.

If M is a compact complex curve of genus g' we use the following notation:

(1) We denote the mapping class group of M without marked points by $\mathrm{Map}_{g'}$.
(2) If we consider r unordered marked points $p_1, \dots, p_r$ on M we define:

$$\mathrm{Map}_{g',[r]} = \pi_0(\mathrm{Diff}^+(M - \{p_1, \dots, p_r\})),$$

and this is known as the *full mapping class group*.

There is a way to present the full mapping class group of a curve using three different types of twists.

Theorem 2.2 *The mapping class group* $\mathrm{Map}_{0,[r]} = \pi_0(\mathrm{Diff}^+(\mathbb{P}^1 - \{p_1, \dots, p_r\}))$ *is isomorphic to the* braid group $\mathbf{B}_r$ *on r strands, which can be presented as*

$$\mathbf{B}_r = \langle \sigma_1, \dots, \sigma_{r-1} | \sigma_i\sigma_{i+1}\sigma_i = \sigma_{i+1}\sigma_i\sigma_{i+1}, \sigma_i\sigma_j = \sigma_j\sigma_i \text{ if } |i-j| \geq 2 \rangle.$$

For a proof of the above Theorem see, for example, [6, Theorem 1.11].

In this way Artin's standard generators σ_i ($i = 1, \dots, r-1$) of $\mathbf{B}_r$ can be represented by the so-called half-twists.

Definition 2.3 The *half-twist* σ_j is a diffeomorphism of $\mathbb{P}^1 - \{p_1, \dots, p_r\}$ isotopic to the homeomorphism given by (see Fig. 1):

- A rotation of 180° on the disk with center $j + \frac{1}{2}$ and radius $\frac{1}{2}$;
- on a circle with the same center and radius $\frac{2+t}{4}$ the map σ_j is the identity if $t \geq 1$ and a rotation of $180(1-t)°$, if $t \leq 1$.

We want to give a similar presentation for a group $\mathrm{Map}_{g'}$ with $g' \geq 1$, so we have to introduce the Dehn twists.

Definition 2.4 Let C be an oriented Riemann surface. Then a *positive Dehn twist* t_α with respect to a simple closed curve α on C is an isotopy class of a diffeomorphism

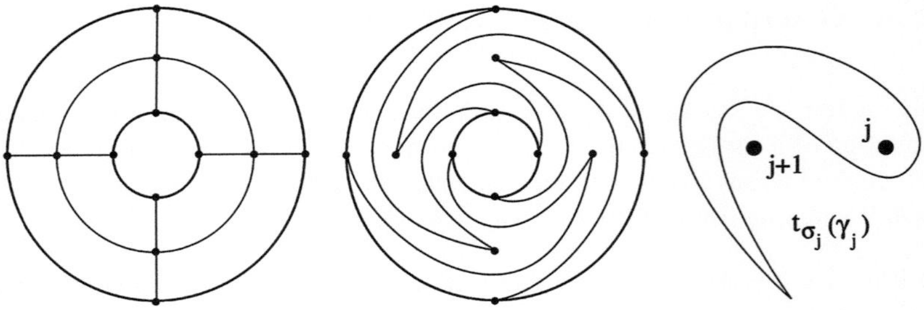

Fig. 1 The half–twist

h of C which is equal to the identity outside a neighborhood of α orientedly homeomorphic to an annulus in the plane, while inside the annulus h rotates the inner boundary of the annulus by 360° to the right and damps the rotation down to the identity at the outer boundary (see Fig. 2).

We have then the following classical results of Dehn [13].

Theorem 2.5 *The mapping class group* $\mathrm{Map}_{g'}$ *is generated by Dehn twists.*

We give the generators of the group $\mathrm{Map}_{g'}$.

Theorem 2.6 *The group* $\mathrm{Map}_{g'}$ *is generated by the Dehn twists with respect to the curves in the Fig. 3.*

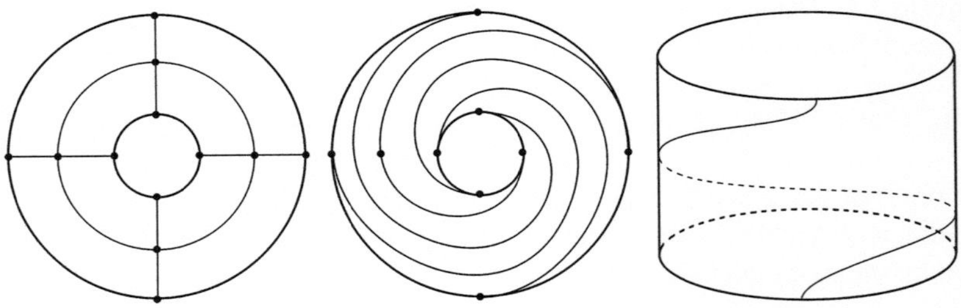

Fig. 2 The positive Dehn twist

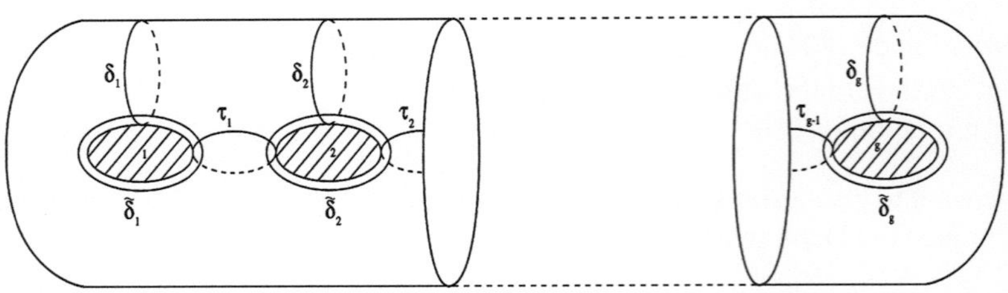

Fig. 3 Generators for $\mathrm{Map}_{g'}$

A proof of the above Theorem can be found in [6, Theorem 4.8].

For a more general situation where the surface has $g' > 0$ and r marked points we need to introduce a third type of twist: The ξ-twists, which link the holes with the marked points. Let us recall the Birman short exact sequence for a Riemann surface C' with $g(C') > 1$ or $r > 1$, setting $\mathcal{B} := \{p_1, \ldots, p_r\}$:

$$1 \longrightarrow \pi_1(C' \setminus \mathcal{B}, p) \xrightarrow{\Xi} \pi_0(\mathrm{Diff}^+(C' \setminus (\mathcal{B} \cup \{p\})) \longrightarrow \pi_0(\mathrm{Diff}^+(C' \setminus \mathcal{B}) \longrightarrow 1. \tag{1}$$

The map Ξ can be described as follows (cf. [5]). Let $[\gamma] \in \pi_1(C' \setminus \mathcal{B}, p)$ and γ be a simple, smooth loop based at p representing $[\gamma]$. Then $\Xi([\gamma])$ is the isotopy class of a ξ-twist with respect to the closed curve γ. In addition, this new twist is isotopic to the identity in $\pi_0(\mathrm{Diff}^+(C \setminus (\mathcal{B}))$. Let us now describe a ξ-twist. We shall consider the annulus $A := \{z = \rho e^{i\theta} \in \mathbb{C} | 1 \leq \rho \leq 2\}$, and we define $h : A \to A$ as follows

$$h(\rho, \theta) : \begin{cases} \big(\rho, \theta - 4\pi(\rho - 1)\big) \ 1 \leq \rho \leq \frac{3}{2} \\ \big(\rho, \theta - 4\pi(2 - \rho)\big) \ \frac{3}{2} \leq \rho \leq 2. \end{cases} \tag{2}$$

Definition 2.7 Let C be a Riemann surface, and α a simple closed curve on C. Let ι be a diffeomorphism between A and a tubular neighborhood of α. Then the *ξ-twist* $\mathbf{t}_\alpha$ with respect to α is defined as $\iota \circ h \circ \iota^{-1} |_{\iota(A)}$ extended to the whole C as the identity on $C \setminus \iota(A)$ (see Fig. 4).

Therefore, for the more general situation we have the following.

Theorem 2.8 ([5, Theorem 3]) *Let $g(C') \neq 0$ and $g(C') > 1$ or $r > 1$ then the group* $\mathrm{Map}_{g',[r]}$ *is generated by the* $3g' - 1$ *Dehn twists with respect to the curves* δ_j, $\tilde{\delta}_j$ *and* τ_j, *by the* $2rg'$ ξ*-twists with respect to the curves* $\xi^l_{j,d}$ *and the* $r-1$ *half-twists about the points* $p_1, \ldots, p_r$ *in Fig. 5.*

Let C' be a Riemann surfaces of genus g', and let $\mathcal{B} := \{p_1, \ldots, p_r\}$ a set of points on C'. A *geometric basis* of $\pi_1(C' \setminus \mathcal{B}, p_0)$ consists of simple non-intersecting (away

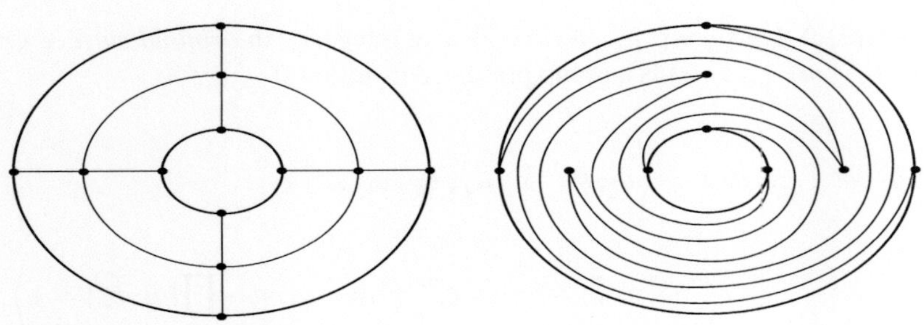

Fig. 4 The ξ–twist

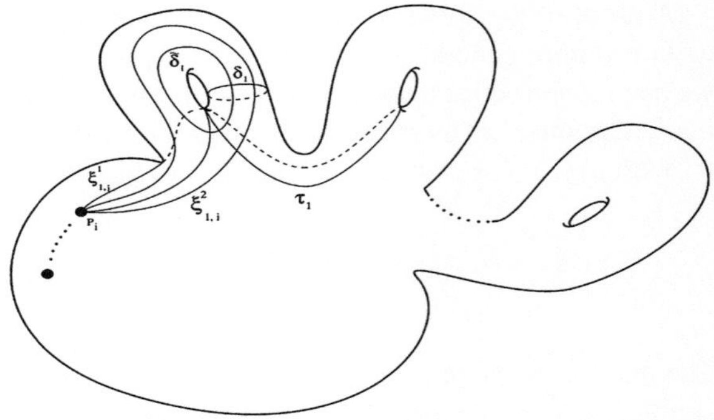

Fig. 5 Generators of $\mathrm{Map}_{g',[r]}$

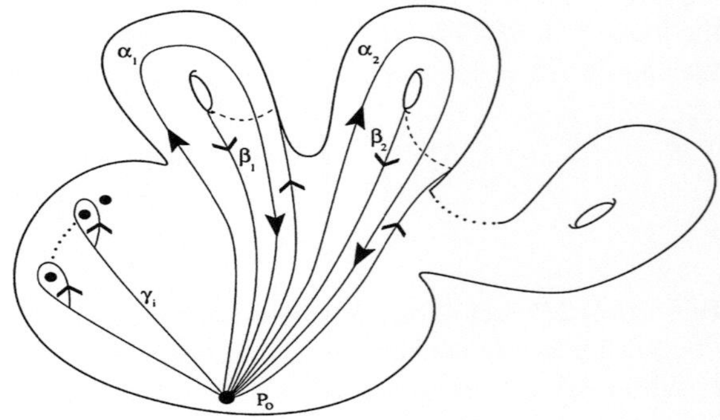

Fig. 6 Geometric basis of $\pi_1(C' \setminus \mathcal{B}, p_0)$

from the base point) loops (see Fig. 6)

$$\gamma_1, \ldots, \gamma_d, \alpha_1, \beta_1, \ldots, \alpha_{g'}, \beta_{g'}$$

such that we get the presentation

$$\pi_1(C \setminus \mathcal{B}, p_0) := \left\langle \alpha_1, \beta_1, \ldots, \alpha_{g'}, \beta_{g'}, \gamma_1, \ldots, \gamma_r | \gamma_1 \cdot \ldots \cdot \gamma_r \cdot \prod_{k=1}^{g'} [\alpha_k, \beta_k] \right\rangle.$$

Definition 2.9 Let $g', m_1, \ldots, m_r$ be positive integers. An *orbifold surface group* of type $(g' \mid m_1, \ldots, m_r)$ is a group presented as follows:

$$\Gamma(g' \mid m_1, \ldots, m_r) := \left\langle a_1, b_1, \ldots, a_{g'}, b_{g'}, c_1, \ldots, c_r | \right.$$

$$\left. c_1^{m_1} = \cdots = c_r^{m_r} = c_1 \cdot \ldots \cdot c_r \cdot \prod_{k=1}^{g'} [a_k, b_k] = 1 \right\rangle.$$

We notice that the choice of a geometric basis yields an obvious epimorphism $\pi_1(C \setminus \mathcal{B}, p_0) \to \Gamma(g' \mid m_1, \ldots, m_r)$.

The following is a reformulation of *Riemann's existence theorem*:

Theorem 2.10 *A finite group G acts as a group of automorphisms on some compact Riemann surface C of genus g if and only if there are natural numbers $g', m_1, \ldots, m_r$, and an orbifold homomorphism*

$$\theta \colon \Gamma(g' \mid m_1, \ldots, m_r) \to G \tag{3}$$

such that $ord(\theta(c_i)) = m_i$ for all i and such that the Riemann-Hurwitz relation holds:

$$2g - 2 = |G| \left(2g' - 2 + \sum_{i=1}^{r} \left(1 - \frac{1}{m_i} \right) \right). \tag{4}$$

If this is the case, then g' is the genus of $C' := C/G$. The G-cover $C \to C'$ is branched at r points $p_1, \ldots, p_r$ with branching indices $m_1, \ldots, m_r$, respectively. Let $\Gamma = \Gamma(g' \mid m_1, ..., m_r)$ be an orbifold surface group with a presentation as in Definition 2.9. If G is a finite group quotient of Γ as in (3), then we say that G is $(g' \mid m_1, \ldots, m_r)-$generated, the image of the generators of Γ in G is called a *system of generators* for G. Finally, θ is called an *admissible epimorphism*.

Definition 2.11 An automorphism $\eta \in \mathrm{Aut}(\Gamma)$ is said to be *orientation preserving* if the action induced on $\langle \alpha_1, \beta_1, \ldots, \alpha_{g'}, \beta_{g'} \rangle^{ab}$ has determinant $+1$, and for all $i \in \{1, \ldots, r\}$ there exists j such that $\eta(\gamma_i)$ is conjugate to γ_j, which implies $ord(\gamma_i) = ord(\gamma_j)$.

The subgroup of orientation preserving automorphisms of Γ is denoted by $\mathrm{Aut}^+(\Gamma)$ and the quotient $\mathrm{Out}^+(\Gamma) := \mathrm{Aut}^+(\Gamma)/\mathrm{Inn}(\Gamma)$ is called the *mapping class group* of Γ.

Theorem 2.12 *Let $\Gamma = \Gamma(g' \mid m_1, \ldots, m_r)$ be an orbifold surface group. Then there is an isomorphism of groups:*

$$\mathrm{Out}^+(\Gamma) \cong \mathrm{Map}_{g',[r]}.$$

This is a classical result cf. e.g., [20, Sect. 4] .

Moreover let G be a finite group $(g' \mid m_1, \ldots, m_r)$—generated. There is a section $s : \mathrm{Out}^+(\Gamma) \to \mathrm{Aut}^+(\Gamma)$, which induces an action of the $\mathrm{Map}_{g',[r]}$ on the generators of Γ. Such action does not depend on s up to simultaneous conjugation, meaning that the action is defined up to inner automorphisms. This action induces an action on the systems of generators of G via composition with admissible epimorphisms.

Definition 2.13 Let G be a finite group $(g' \mid m_1, \ldots, m_r)$—generated. If two systems of generators $\mathcal{V}_1$ and $\mathcal{V}_2$ are in the same $\mathrm{Map}_{g',[r]}$-orbit, we say that they are related by a *Hurwitz move* (or are *Hurwitz equivalent*).

Proposition 2.14 *Let C' be a curve of genus g', $\mathcal{B} := \{p_1, \dots, p_r\}$, and with $g' \neq 0$ and $g' > 1$ or $r > 1$. Up to inner automorphisms, the action of* $\mathrm{Map}_{g',[r]}$ *on $\Gamma(g' \mid m_1 \dots m_r)$ is induced by the following action on a geometric basis of $\pi_1(C' \setminus \mathcal{B}, p_0)$*

$$\mathbf{t}_{\mathrm{j}} : \begin{cases} \alpha_j \mapsto \alpha_j \beta_j^{-1} & \\ \alpha_i \mapsto \alpha_i & \forall i \neq j \\ \beta_i \mapsto \beta_i & \forall i \\ \gamma_i \mapsto \gamma_i & \forall i \end{cases} \qquad \mathbf{t}_{\tilde{\delta}_{\mathrm{j}}} : \begin{cases} \alpha_i \mapsto \alpha_i & \forall i \\ \beta_j \mapsto \beta_j \alpha_j & \\ \beta_i \mapsto \beta_i & \forall i \neq j \\ \gamma_i \mapsto \gamma_i & \forall i \end{cases}$$

$$\mathbf{t}_{\sigma_{\mathrm{h}}} : \begin{cases} \alpha_i \mapsto \alpha_i & \forall i \\ \beta_i \mapsto \beta_i & \forall i \\ \gamma_h \mapsto \gamma_{h+1} & \\ \gamma_{h+1} \mapsto \gamma_{h+1}^{-1} \gamma_h \gamma_{h+1} & \\ \gamma_i \mapsto \gamma_i & \forall i \neq h, h+1 \end{cases} \qquad \mathbf{t}_{\tau_{\mathrm{k}}} : \begin{cases} \alpha_k \mapsto \alpha_k \eta_k^{-1} & \\ \beta_k \mapsto \beta_k^{\eta_k} & \\ \alpha_{k+1} \mapsto \eta_k \alpha_k & \\ \alpha_i \mapsto \alpha_i & \forall i \neq k, k+1 \\ \beta_i \mapsto \beta_i & \forall i \neq k \\ \gamma_i \mapsto \gamma_i & \forall i \end{cases}$$

$$\mathbf{t}_{\xi^1_{j,d}} : \begin{cases} \alpha_j \mapsto \chi_{j,d} \alpha_j & \\ \alpha_i \mapsto \alpha_i & \forall i \neq j \\ \beta_i \mapsto \beta_i & \forall i \\ \gamma_d \mapsto \gamma_d^{\epsilon_{j,d}} & \\ \gamma_i \mapsto \gamma_i & \forall j \neq d \end{cases} \qquad \mathbf{t}_{\xi^2_{j,d}} : \begin{cases} \alpha_i \mapsto \alpha_i & \forall i \\ \beta_j \mapsto \alpha_j^{-1} \chi_{j,d} \alpha_j \beta_j & \\ \beta_i \mapsto \beta_i & \forall i \neq j \\ \gamma_d \mapsto \gamma_d^{\epsilon'_{j,d}} & \\ \gamma_i \mapsto \gamma_i & \forall j \neq d \end{cases}$$

for $1 \leq j \leq g'$, $1 \leq k \leq (g' - 1)$, $1 \leq h \leq (r - 1)$, and $1 \leq d \leq r$. Moreover we set $\eta_k := \beta_k^{-1} \alpha_{k+1} \beta_{k+1} \alpha_{k+1}^{-1}$, $\chi_{j,d} := (\Pi_{k=1}^{j-1} [\alpha_k, \beta_k])^{-1} \gamma_d \Pi_{k=1}^{j-1} [\alpha_k, \beta_k]$, $\epsilon_{j,d} := \gamma_d (\Pi_{k=1}^{j} [\alpha_k, \beta_k]) \alpha_j \beta_j \alpha_j^{-1} (\Pi_{k=1}^{j} [\alpha_k, \beta_k])^{-1}$, and $\epsilon'_{j,d} := \gamma_d (\Pi_{k=1}^{j} [\alpha_k, \beta_k]) \alpha_j^{-1}$ $(\Pi_{k=1}^{j} [\alpha_k, \beta_k])^{-1}$.

In the above proposition the twists $\mathbf{t}_{\tilde{\delta}_j}$, $\mathbf{t}_{\delta_j}$ and $\mathbf{t}_{\tau_j}$ correspond to Dehn twists, $\mathbf{t}_{\xi_{r_{j,d}}}$ and $\mathbf{t}_{\xi_{r_{j,d}}}$ to ξ-twists, and finally $\mathbf{t}_{\sigma_h}$ to half-twists.

Proof One notices that a Riemann surface of genus g' is a connected sum of g' tori. Then one can use the Fig. 7 to calculate the Dehn twists about the curves δ_j, and similarly for the Dehn twists about the curves $\tilde{\delta}_j$. One can use the results given in [22] to calculate the Dehn twists about the curves τ_j. In the Appendix of [12] are described the actions of the ξ-twists. Finally the half-twists action is clear by Fig. 1. □

For the case $g(C') = 1$ and $r = 1$ see e.g., [24] for a proof of the following proposition.

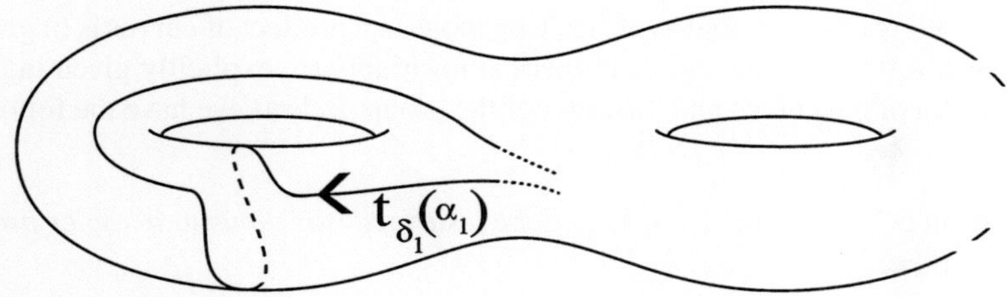

Fig. 7 Example $t_{\delta_1}(\alpha_1)$

Proposition 2.15 ([24, Proposition 1.10]) *Up to inner automorphisms, the action of* $\mathrm{Map}_{1,1}$ *on* $\Gamma(1 \mid m_1)$ *is induced by the following action on a geometric basis of* $\pi_1(C' \setminus \mathcal{B}, p_0)$

$$\mathbf{t}_{\tilde{\delta}} : \begin{cases} \alpha_1 \to \alpha_1 \\ \beta_1 \to \beta_1\alpha_1 \\ \gamma_1 \to \gamma_1 \end{cases} \quad \mathbf{t}_{\delta} : \begin{cases} \alpha_1 \to \alpha_1\beta_1^{-1} \\ \beta_1 \to \beta_1 \\ \gamma_1 \to \gamma_1. \end{cases}$$

Finally we have.

Proposition 2.16 *Up to inner automorphism, the action of* $\mathrm{Map}_{0,[r]}$ *on* $\Gamma(0 \mid m_1, \dots, m_r)$ *is induced by the following action on a geometric basis of* $\pi_1(\mathbb{P}^1 \setminus \mathcal{B}, p_0)$

$$\sigma_i : \begin{cases} \gamma_i \to \gamma_{i+1} \\ \gamma_{i+1} \to \gamma_{i+1}^{-1}\gamma_i\gamma_{i+1} \\ \gamma_j \to \gamma_j \text{ if } j \neq i, i+1, \end{cases}$$

for $i = 1, \dots, r-1$.

3 Surfaces Isogenous to a Product of Curves

A surface S is said to be *isogenous to a (higher) product of curves* if and only if S is a quotient $S := (C_1 \times C_2)/G$, where C_1 and C_2 are curves of genus at least two, and G is a finite group acting freely on $C_1 \times C_2$.

Let S be a surface isogenous to a higher product, and $G^\circ := G \cap (Aut(C_1) \times Aut(C_2))$. Then G° acts on the two factors C_1 and C_2 and diagonally on the product $C_1 \times C_2$. If G° acts faithfully on both curves, we say that $S = (C_1 \times C_2)/G$ is a *minimal realization* of S. In [8], the author proves that any surface isogenous to a higher product admits a unique minimal realization. From now on, we work only with minimal realizations.

There are two cases: the *mixed* case where the action of G exchanges the two factors (in this case C_1 and C_2 are isomorphic and $G^\circ \neq G$); the *unmixed* case (where $G = G^\circ$, and therefore it acts diagonally).

Moreover, we observe that a surface isogenous to a product of curves is of general type. It is always minimal and its numerical invariants are explicitly given in terms of the genera of the curves and the order of the group. Indeed, we have the following proposition.

Proposition 3.1 *Let $S = (C_1 \times C_2)/G$ be a surface isogenous to a higher product of curves, then:*

$$\chi(S) = \frac{(g(C_1) - 1)(g(C_2) - 1)}{|G|}, \quad e(S) = 4\chi(S), \quad K_S^2 = 8\chi(S). \tag{5}$$

The irregularity of these surfaces is easily computed by

$$q(S) = g(C_1/G) + g(C_2/G). \tag{6}$$

By the above formula a surface S isogenous to a product of curves has $q(S) = 0$ if and only if the two quotients C_i/G are isomorphic to $\mathbb{P}^1$. Moreover, if both coverings $C_i \to C_i/G \cong \mathbb{P}^1$ are ramified in exactly 3 points, S is a *Beauville surface*. This last condition is equivalent to saying that Beauville surfaces are rigid, i.e., have no nontrivial deformations.

In the unmixed case G acts separately on C_1 and C_2, and the two projections $\pi_i : C_1 \times C_2 \longrightarrow C_i$ for $i = 1, 2$ induce two isotrivial fibrations $\alpha_i : S \longrightarrow C_i/G$ for $i = 1, 2$, whose smooth fibres are isomorphic to C_2 and C_1, respectively. *We work only with surfaces of unmixed type.*

Working out the definition of surfaces isogenous to a product, one sees that there is a pure group theoretical condition which characterizes the groups of such surfaces: the existence of a "*ramification structure*".

Definition 3.2 Let G be a finite group and $\theta_1 : \Gamma(g_1' \mid m_{1,1}, \ldots, m_{1,r_1}) \twoheadrightarrow G$ an admissible epimorphism. Let $\mathcal{V}_1$ be the system of generators of G induced by θ_1, i.e., the elements of G which are images of the generators of Γ. We say that $\mathcal{V}_1$ is of *type* $\tau_1 := (g_1' \mid m_{1,1}, \ldots, m_{1,r_1})$.

Moreover, let $\theta_2 : \Gamma(g_2' \mid m_{2,1}, \ldots, m_{2,r_2}) \twoheadrightarrow G$ be another admissible epimorphism and $\mathcal{V}_2$ be the system of generators of G induced by θ_2. Then $\mathcal{V}_1$ and $\mathcal{V}_2$ are said to be *disjoint*, if:

$$\Sigma(\mathcal{V}_1) \bigcap \Sigma(\mathcal{V}_2) = \{1\}, \tag{7}$$

where

$$\Sigma(\mathcal{V}_i) := \bigcup_{g \in G} \bigcup_{j=0}^{\infty} \bigcup_{k=1}^{r_i} g \cdot \theta_i(\gamma_k)^j \cdot g^{-1}.$$

Definition 3.3 Let $\tau_i := (g_i' \mid m_{1,i}, \ldots, m_{r_i,i})$ for $i = 1, 2$ be two types. An *unmixed ramification structure* of type (τ_1, τ_2) for a finite group G, is a pair $(\mathcal{V}_1, \mathcal{V}_2)$ of disjoint systems of generators of G, whose types are τ_i, and they satisfy:

$$\mathbb{Z} \ni \frac{|G|\left(2g_i' - 2 + \sum_{l=1}^{r_i}\left(1 - \frac{1}{m_{i,l}}\right)\right)}{2} + 1 \geq 2, \tag{8}$$

for $i = 1, 2$.

We shall denote by $\mathcal{U}(G; \tau_1, \tau_2)$ the set of all pairs $(\mathcal{V}_1, \mathcal{V}_2)$ of disjoint systems of generators of unordered type (τ_1, τ_2). Here *unordered type* τ means that there is a permutation $\sigma \in \mathfrak{S}_r$ such that: $\mathrm{ord}(c_1) = m_{\sigma(1)}, \ldots, \mathrm{ord}(c_r) = m_{\sigma(r)}$. We obtain that the datum of a surface isogenous to a higher product of unmixed type $S = (C_1 \times C_2)/G$ is determined, looking at the monodromy of each covering of C_i/G, by the datum of a finite group G together with an unmixed ramification structure. The condition (7) ensures that the action of G on the product of the two curves $C_1 \times C_2$ is free. We remark here that this can be specialized to $C_i/G \cong \mathbb{P}^1$ in order to obtain regular surfaces isogenous to a product. In this case condition (8) is automaticaly satisfied, see [16, Lemma 2.4]. Moreover, we can also ask $r_i = 3$, and therefore we obtain Beauville surfaces, in this case the ramification structure of G is called a *Beauville ramification structure*.

Remark 3.4 Note that a group G and an unmixed ramification structure (or equivalently a Beauville structure) determine the main invariants of the surface S. Indeed, by (5) and (4) we obtain:

$$4\chi(S) = |G| \cdot \left(2g_1' - 2 + \sum_{k=1}^{r_1}\left(1 - \frac{1}{m_{1,k}}\right)\right) \cdot \left(2g_2' - 2 + \sum_{k=1}^{r_2}\left(1 - \frac{1}{m_{2,k}}\right)\right), \tag{9}$$

and so, in the Beauville case,

$$4\chi(S) = 4(1 + p_g) = |G|(1 - \mu_1)(1 - \mu_2),$$

where

$$\mu_i := \frac{1}{m_{1,i}} + \frac{1}{m_{2,i}} + \frac{1}{m_{3,i}}, \quad (i = 1, 2). \tag{10}$$

The most important property of surfaces isogenous to a product is their weak rigidity property.

Theorem 3.5 ([10, Theorem 3.3, Weak Rigidity Theorem]) *Let $S = (C_1 \times C_2)/G$ be a surface isogenous to a higher product of curves. Then every surface with the same*

- *topological Euler number and*
- *fundamental group*

is diffeomorphic to S. The corresponding moduli space $\mathcal{M}^{top}(S) = \mathcal{M}^{diff}(S)$ of surfaces (orientedly) homeomorphic (resp. diffeomorphic) to S is either irreducible and connected or consists of two irreducible connected components exchanged by complex conjugation.

Thanks to the Weak Rigidity Theorem, we have that the moduli space of surfaces isogenous to a product of curves with fixed invariants—a finite group G and a type (τ_1, τ_2) in the unmixed case—consists of a finite number of irreducible connected components of $\mathcal{M}$. More precisely, let S be a surface isogenous to a product of curves of unmixed type with group G and a pair of disjoint systems of generators of type (τ_1, τ_2). By (9) we have $\chi(S) = \chi(G, (\tau_1, \tau_2))$, and consequently, by (5) $K_S^2 = K^2(G, (\tau_1, \tau_2)) = 8\chi(S)$, and $e(S) = e(G, (\tau_1, \tau_2)) = 4\chi(S)$. Moreover the fundamental group of S fits in the following exact sequence (cf. [8]):

$$1 \longrightarrow \pi_1(C_1) \times \pi_2(C_2) \longrightarrow \pi_1(S) \longrightarrow G \longrightarrow 1.$$

Let us fix a group G and a type (τ_1, τ_2) of an unmixed ramification structure, and denote by $\mathcal{M}_{(G,(\tau_1,\tau_2))}$ the moduli space of isomorphism classes of surfaces isogenous to a product of curves of unmixed type admitting these data, then it is obviously a subset of the moduli space $\mathcal{M}_{K^2(G,(\tau_1,\tau_2)),\chi(G,(\tau_1,\tau_2))}$. By the Weak Rigidity Theorem, the space $\mathcal{M}_{(G,(\tau_1,\tau_2))}$ consists of a finite number of irreducible connected components.

A group theoretical method to count the number of these components is given in [1, Theorem 1.3] in case of surfaces isogenous to a product of curves of unmixed type with $q = 0$ and G abelian. The following theorem is a natural generalization.

Theorem 3.6 ([22, Theorem 5.7]) *Let S be a surface isogenous to a product of unmixed type. Then we attach to S its finite group G (up to isomorphism) and the equivalence class ramification structures $(\mathcal{V}_1, \mathcal{V}_2)$ of type (τ_1, τ_2) of G, under the equivalence relation generated by:*

(1) Hurwitz moves and $\mathrm{Inn}(G)$ *on $\mathcal{V}_1$,*
(2) Hurwitz moves and $\mathrm{Inn}(G)$ *on $\mathcal{V}_2$,*
(3) simultaneous conjugation of $\mathcal{V}_1$ and $\mathcal{V}_2$ by an element $\phi \in \mathrm{Aut}(G)$, i.e., we let $(\mathcal{V}_1, \mathcal{V}_2)$ be equivalent to $(\phi(\mathcal{V}_1), \phi(\mathcal{V}_2))$.

Then two surfaces S and S' are deformation equivalent if and only if the corresponding pairs of systems of generators are in the same equivalence class.

If we fix a finite group G and a pair of types (τ_1, τ_2) of an unmixed ramification structure for G, counting the number of connected components of $\mathcal{M}_{(G,(\tau_1,\tau_2))}$ is then equivalent to the group theoretical problem of counting the number of classes of pairs of systems of generators of G of type (τ_1, τ_2) under the equivalence relation defined in Theorem 3.6. This leads also to the following definition.

Definition 3.7 Denote by $h(G; \tau_1, \tau_2)$ the number of *Hurwitz components*, namely the number of orbits of $\mathcal{U}(G; \tau_1, \tau_2)$ under the action of the group prescribed in Theorem 3.6.

4 Connected Components of the Moduli Space of Surfaces of General Type

We can simplify a lot the discussion of the previous section if we consider only *regular surfaces*, which will be assumed in the whole section. In this case, the Hurwitz moves are given only by the $\mathbf{t}_{\sigma_i}$ of Proposition 2.14, which corresponds to the ones described in Proposition 2.16. By Theorem 2.2 they are given by the braid group of the sphere on r strands $\mathbf{B}_r$ acting on the generators of the orbifold fundamental group as in Fig. 8.

In addition, we recall

Lemma 4.1 ([25, Lemma 9.4]) *The inner automorphism group,* $\mathrm{Inn}(G)$*, leaves each braid orbit invariant.*

This lemma allows us to use the above Theorem 3.6 in the simplified version without the $\mathrm{Inn}(G)$ action on the system of generators $\mathcal{V}_i$. Since the two actions of $\mathbf{B}_r$ and $\mathrm{Aut}(G)$ commute, one gets a double action of $\mathbf{B}_r \times \mathrm{Aut}(G)$ on the set of r-systems of generators for G.

Therefore, for regular surfaces isogenous to a product we have that fixing a finite group G and a pair of types $(\tau_1, \tau_2) = (m_{1,1}, \ldots, m_{1,r_1}, m_{2,1}, \ldots, m_{2,r_2})$ of an unmixed ramification structure for G counting the number of connected components of $\mathcal{M}_{(G,(\tau_1,\tau_2))}$ is then equivalent to the group theoretical problem of counting the number of classes of pairs of systems of generators of G of type (τ_1, τ_2) under the equivalence relation given by the action of $\mathbf{B}_{r_1} \times \mathbf{B}_{r_2} \times \mathrm{Aut}(G)$. In this case the number of Hurwitz components $h(G; \tau_1, \tau_2)$ is given by the number of orbits of $\mathcal{U}(G; \tau_1, \tau_2)$ under the following actions:

if $\tau_1 \neq \tau_2$ the action of $(\mathbf{B}_{r_1} \times \mathbf{B}_{r_2}) \times \mathrm{Aut}(G)$, given by:

$$((\gamma_1, \gamma_2), \phi) \cdot (T_1, T_2) := \big(\phi(\gamma_1(T_1)), \phi(\gamma_2(T_2))\big),$$

where $\gamma_1 \in \mathbf{B}_{r_1}$, $\gamma_2 \in \mathbf{B}_{r_2}$, $\phi \in \mathrm{Aut}(G)$ and $(T_1, T_2) \in \mathcal{U}(G; \tau_1, \tau_2)$.

if $\tau_1 = \tau_2$ the action of $(\mathbf{B}_r \wr \mathbb{Z}/2\mathbb{Z}) \times \mathrm{Aut}(G)$, where $\mathbb{Z}/2\mathbb{Z}$ acts on (T_1, T_2) by exchanging the two factors.

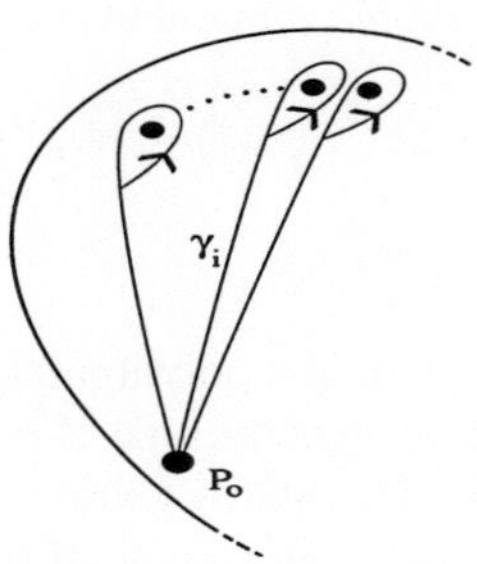

Fig. 8 Generators of of $\pi_1(\mathbb{P}^1 \setminus \mathcal{B}, p_0)$

In case of Beauville surfaces we define h as above substituting r_1 and r_2 with 3.

Proposition 4.2 *Fix r_1 and r_2 in $\mathbb{N}$. Let $\{G_n\}_{n=1}^{\infty}$ be a family of finite groups, which admit an unmixed ramification structure of size (r_1, r_2). Let $\tau_{n,1} = (m_{n,1,1}, \ldots, m_{n,1,r_1})$ and $\tau_{n,2} = (m_{n,2,1}, \ldots, m_{n,2,r_2})$ be sequences of types $(\tau_{n,1}, \tau_{n,2})$ of unmixed ramification structures for G_n, and $\{S_n\}_{n=1}^{\infty}$ be the family of surfaces isogenous to product with $q = 0$ admitting the given data, then as $|G_n| \overset{n\to\infty}{\longrightarrow} \infty$:*

(1) $\chi(S_n) = \Theta(|G_n|)$.
(2) $h(G_n; \tau_{n,1}, \tau_{n,2}) = O(\chi(S_n)^{r_1+r_2-2})$.

Proof (1) Note that, for $i = 1, 2,$

$$\frac{1}{42} \leq -2 + \sum_{j=1}^{r_i} \left(1 - \frac{1}{m_{n,i,j}}\right) \leq r_i - 2.$$

Indeed, for $r_i = 3$, the minimal value for $(1 - \mu_i)$ is $1/42$. For $r_i = 4$, the minimal value for $\left(-2 + \sum_{j=1}^{r_i}\left(1 - \frac{1}{m_{n,i,j}}\right)\right)$ is $1/6$, and when $r_i \geq 5$, this value is at least $1/2$.
Now, by Eq. (9),

$$4\chi(S_n) = |G_n| \cdot \left(-2 + \sum_{j=1}^{r_1}\left(1 - \frac{1}{m_{n,1,j}}\right)\right) \cdot \left(-2 + \sum_{j=1}^{r_2}\left(1 - \frac{1}{m_{n,2,j}}\right)\right),$$

hence

$$\frac{|G_n|}{4 \cdot 42^2} \leq \chi(S_n) \leq \frac{(r_1 - 2)(r_2 - 2)|G_n|}{4}.$$

(2) For $i = 1, 2$, any r_i-system of generators $\mathcal{V}_{n,i}$ contains at most $r_i - 1$ independent elements of G_n. Thus, the size of the set of all unordered pairs of type $(\tau_{n,1}, \tau_{n,2})$ is bounded from above, by

$$|\mathcal{U}(G_n; \tau_{n,1}, \tau_{n,2})| \leq |G_n|^{r_1+r_2-2},$$

and so, the number of connected components is bounded from above by

$$h(G_n; \tau_{n,1}, \tau_{n,2}) \leq |G_n|^{r_1+r_2-2}.$$

Now, the result follows from (1).

Therefore we cannot expect more than a polynomial growth in the number of connected components of surfaces of general type if we count only regular surfaces isogenous to a product. Together with Shelly Garion we investigated the asymptotic growth of the number of connected components of the moduli space of surfaces of general type by using the group theoretical methods described above for regular surfaces isogenous to a product. Our first results already appeared on the ArXiv in 2009 (see arXiv:0910.5402v1). The following are some of the results contained in [16].

Notation 4.3 *Denote:*

- $h(n) = \Omega(g(n))$, *if* $h(n) \geq cg(n)$ *for some positive constant* c, *as* $n \to \infty$.
- $h(n) = \Theta(g(n))$, *if* $c_1 g(n) \leq h(n) \leq c_2 g(n)$ *for some positive constants* c_1, c_2, *as* $n \to \infty$.

Theorem 4.4 *Let* $\tau_1 = (m_{1,1}, \dots, m_{1,r_1})$ *and* $\tau_2 = (m_{2,1}, \dots, m_{2,r_2})$ *be two sequences of natural numbers such that* $m_{k,i} \geq 2$ *and* $\sum_{i=1}^{r_k}(1 - 1/m_{k,i}) > 2$ *for* $k = 1, 2$. *Let* $h(\mathfrak{A}_n, \tau_1, \tau_2)$ *be the number of connected components of the moduli space of surfaces isogenous to a product with* $q = 0$, *with group the alternating group* $\mathfrak{A}_n$, *and with type* (τ_1, τ_2). *Then*

$$(a) \quad h(\mathfrak{A}_n, \tau_1, \tau_2) = \Omega(n^{r_1 + r_2}),$$

and moreover,

$$(b) \quad h(\mathfrak{A}_n, \tau_1, \tau_2) = \Omega\big((\log(\chi))^{r_1 + r_2 - \epsilon}\big).$$

where $0 < \epsilon \in \mathbb{R}$.

Theorem 4.5 *Let* $\tau_1 = (m_{1,1}, \dots, m_{1,r_1})$ *and* $\tau_2 = (m_{2,1}, \dots, m_{2,r_2})$ *be two sequences of natural numbers such that* $m_{k,i} \geq 2$, *at least two of* $(m_{k,1}, \dots, m_{k,r_k})$ *are even and* $\sum_{i=1}^{r_k}(1 - 1/m_{k,i}) > 2$, *for* $k = 1, 2$. *Let* $h(\mathfrak{S}_n, \tau_1, \tau_2)$ *be the number of connected components of the moduli space of surfaces isogenous to a product with* $q = 0$, *with group the symmetric group* $\mathfrak{S}_n$, *and with type* (τ_1, τ_2). *Then*

$$(a) \quad h(\mathfrak{S}_n, \tau_1, \tau_2) = \Omega(n^{r_1 + r_2}),$$

and moreover,

$$(b) \quad h(\mathfrak{S}_n, \tau_1, \tau_2) = \Omega\big((\log(\chi))^{r_1 + r_2 - \epsilon}\big).$$

where $0 < \epsilon \in \mathbb{R}$.

The proofs of part (a) of both Theorems are presented in [16, Sect. 3.2], and are based on results of Liebeck and Shalev [18]. The proofs of part (b) of both theorems appear in [16, Sect. 2].

We can specialize the results above to Beauville surfaces. Recall that a triple $(r, s, t) \in \mathbb{N}^3$ is said to be hyperbolic if

$$\frac{1}{r} + \frac{1}{s} + \frac{1}{t} < 1.$$

Corollary 4.6 *Let* $\tau_1 = (r_1, s_1, t_1)$ *and* $\tau_2 = (r_2, s_2, t_2)$ *be two hyperbolic types and let* $h(\mathfrak{A}_n, \tau_1, \tau_2)$ *be the number of Beauville surfaces with group* $\mathfrak{A}_n$ *and with types* (τ_1, τ_2). *Then*

$$(a) \quad h(\mathfrak{A}_n, \tau_1, \tau_2) = \Omega(n^6),$$

and moreover,

$$(b) \quad h(\mathfrak{A}_n, \tau_1, \tau_2) = \Omega\big((\log(\chi))^{6-\epsilon}\big).$$

where $0 < \epsilon \in \mathbb{R}$.

Corollary 4.7 *Let* $\tau_1 = (r_1, s_1, t_1)$ *and* $\tau_2 = (r_2, s_2, t_2)$ *be two hyperbolic types, assume that at least two of* (r_1, s_1, t_1) *are even and at least two of* (r_2, s_2, t_2) *are even, and let* $h(\mathfrak{S}_n, \tau_1, \tau_2)$ *be the number of Beauville surfaces with group* $\mathfrak{S}_n$ *and with types* (τ_1, τ_2). *Then*

$$(a) \quad h(\mathfrak{S}_n, \tau_1, \tau_2) = \Omega(n^6),$$

and moreover,

$$(b) \quad h(\mathfrak{S}_n, \tau_1, \tau_2) = \Omega\big((\log(\chi))^{6-\epsilon}\big).$$

where $0 < \epsilon \in \mathbb{R}$.

The situation is more interesting for abelian groups. We have the following results which assure the existence of ramification structure for abelian group.

Theorem 4.8 *Let* G *be an abelian group, given as*

$$G \cong \mathbb{Z}/n_1\mathbb{Z} \times \cdots \times \mathbb{Z}/n_t\mathbb{Z},$$

where $n_1 \mid \cdots \mid n_t$. *For a prime* p, *denote by* $l_i(p)$ *the largest power of* p *which divides* n_i *(for* $1 \le i \le t$*).*

Let $r_1, r_2 \ge 3$, *then* G *admits an unmixed ramification structure of size* (r_1, r_2) *if and only if the following conditions hold:*

- $r_1, r_2 \ge t + 1$;
- $n_t = n_{t-1}$;
- *If* $l_{t-1}(3) > l_{t-2}(3)$ *then* $r_1, r_2 \ge 4$;
- $l_{t-1}(2) = l_{t-2}(2)$;
- *If* $l_{t-2}(2) > l_{t-3}(2)$ *then* $r_1, r_2 \ge 5$ *and* r_1, r_2 *are not both odd.*

This theorem is proved in [16, Sect. 3.4]. Once we established the existence of ramification structures, we can put the surfaces in sequences and compute the number of connected components of the moduli space. More precisely, the following holds.

Theorem 4.9 *Let* $\{S_p\}$ *be the family of surfaces isogenous to a product with* $q = 0$ *with group* $G_p := (\mathbb{Z}/p\mathbb{Z})^r$ *admitting ramification structure of type* $\tau_p = (p, \ldots, p)$ *(*p *appears* $(r + 1)-$*times) where* p *is prime. If we denote by* $h(G_p; \tau_p, \tau_p)$ *the number of connected components of the moduli space of isomorphism classes of surfaces isogenous to a product with* $q = 0$ *admitting these data, then*

$$h(G_p; \tau_p, \tau_p) = \Theta(\chi^r(S_p)).$$

Therefore, there exist families of surfaces such that the degree of the polynomial h in χ (and so in K^2) can be arbitrarily large. The proof of this theorem appears in [16, Sect. 2]. Notice that not only the number of connected components increases, but also their sizes. Indeed, we see that $2r - 6$ is the dimension of these connected components.

Again we can specialize the results for Beauville surfaces.

Corollary 4.10 *Let $\{S_p\}$ be the family of Beauville surfaces $G_p := (\mathbb{Z}/p\mathbb{Z})^2$ admitting ramification structure of type $\tau_p = (p, p, p)$ where $p \geq 5$ is prime. If we denote by $h(G_p; \tau_p, \tau_p)$ the number of Beauville surfaces admitting these data. Then*

$$h(G_p; \tau_p, \tau_p) = \Theta(\chi^2(S_p)).$$

Proof Let $(x_1, x_2; y_1, y_2)$ be an unmixed Beauville structure for G. Since x_1, x_2 are generators of G, they are a basis, and without loss of generality x_1, x_2 are the standard basis $x_1 = (1, 0)$, $x_2 = (0, 1)$. Now, let $y_1 = (a, b)$, $y_2 = (c, d)$, then the condition (7) means that any pair of the six vectors yield a basis of G, implying that a, b, c, d must satisfy the following conditions

$$a - b, a + c, c - d, b + d, a + c - b - d, ad - bc \in U \tag{11}$$

Moreover, the number N_p of quadruples (a, b, c, d) satisfy (11) is $N_p = (p - 1)(p-2)(p-3)(p-4)$. The pairs $\big((1, 0), (0, 1); (a, b), (c, d)\big)$, where a, b, c, d satisfy (11), are exactly the representatives for the $\mathrm{Aut}(G)$−orbits in the set $\mathcal{U}(G; \tau, \tau)$.

Now, one should consider the action of $\mathbf{B}_3 \times \mathbf{B}_3$ on $\mathcal{U}(G; \tau, \tau)$, which is equivalent to the action of $\mathfrak{S}_3 \times \mathfrak{S}_3$, since G is abelian. The action of $\mathfrak{S}_3$ on the second component is obvious (there are 6 permutations), and the action of $\mathfrak{S}_3$ on the first component can be translated to an equivalent $\mathrm{Aut}(G)$−action, given by multiplication in one of the six matrices:

$$\begin{pmatrix} 1 & 0 \\ 0 & 1 \end{pmatrix}, \begin{pmatrix} 0 & 1 \\ 1 & 0 \end{pmatrix}, \begin{pmatrix} -1 & 0 \\ -1 & 1 \end{pmatrix}, \begin{pmatrix} 1 & -1 \\ 0 & -1 \end{pmatrix}, \begin{pmatrix} -1 & 1 \\ -1 & 0 \end{pmatrix}, \begin{pmatrix} 0 & -1 \\ 1 & -1 \end{pmatrix},$$

yielding an equivalent representative.

Therefore, the action of $\mathfrak{S}_3$ on the second component yields orbits of length 6, and the action of $\mathfrak{S}_3$ on the first component connects them together, and gives orbits of sizes from 6 to 36. Moreover since one can exchange the vector (x_1, x_2) with the vector (y_1, y_2) we get

$$N_p/72 \leq h \leq N_p/6.$$

By Proposition 4.2, we have as $p \to \infty$:

$$\chi(S_p) = \Theta(p^2),$$

while by the above computation we have

$$h(G_p; \tau_p, \tau_p) = \Theta(p^4).$$

Therefore

$$h(G_p; \tau_p, \tau_p) = \Theta(\chi^2(S_p)).$$

Corollary 4.11 *Let n be an integer such that $(n, 6) = 1$. The number $h = h(G; \tau, \tau)$, where $\tau = (n, n, n)$, of Hurwitz components for $G = (\mathbb{Z}/n\mathbb{Z})^2$, where $n = p_1^{k_1} \cdot \ldots \cdot p_t^{k_t}$, satisfies*

$$N_n/72 \le h \le N_n/6,$$

where $N_n = \prod_{i=1}^t p_i^{4k_i-4}(p_i - 1)(p_i - 2)(p_i - 3)(p_i - 4)$.

Remark 4.12 Notice that if n is divisible by the first l primes $p_i \ge 5$ then since:

$$\lim_{l\to\infty} \prod_i (1 - \frac{1}{p_i}) = 0$$

we have $N_n/n^4 \to 0$ as $l \to \infty$.

In [17] the authors give an explicit formula for the number of isomorphism classes of Beauville surfaces $\Theta(n)$, which we now explain. We shall keep the notation of [17]. Define the following functions for $n, e \in \mathbb{N}$ and p prime.

$$\Theta_1(n) := n^4 \prod_{p|n} \left(1 - \frac{1}{p}\right)\left(1 - \frac{2}{p}\right)\left(1 - \frac{3}{p}\right)\left(1 - \frac{4}{p}\right);$$

$$\Theta_2(p^e) := \begin{cases} p^{2e}\left(1 - \frac{1}{p}\right)\left(1 - \frac{2}{p}\right) & if \quad p \equiv 1 \mod 4, \\ p^{2e}\left(1 - \frac{1}{p}\right)\left(1 - \frac{4}{p}\right) & if \quad p \equiv 3 \mod 4; \end{cases}$$

$$\Theta_3(p^e) := p^{2e}\left(1 - \frac{3}{p}\right)\left(1 - \frac{5}{p}\right);$$

$$\Theta_4(p^e) := \begin{cases} 0 & \text{if } \ p \equiv -1 \mod 3, \\ 2 & \text{if } \ p \equiv 1 \mod 3. \end{cases}$$

Theorem 4.13 ([17, Theorem 2]) *Let $n = p_1^{e_1} \cdot \ldots \cdot p_t^{e_t}$ be an integer such that $(n, 6) = 1$. Then the number of isomorphism classes of Beauville surfaces with group $G = (\mathbb{Z}/n\mathbb{Z})^2$ is*

$$\Theta(n) := \frac{1}{72}\left(\Theta_1(n) + 4\prod_{i=1}^t \Theta_2(p_i^{e_i}) + 6\prod_{i=1}^t \Theta_3(p_i^{e_i}) + 12\prod_{i=1}^t \Theta_4(p_i^{e_i})\right).$$

The case of *irregular* surfaces has not been so intensively investigated. Indeed, counting the number of connected components of the moduli space of surfaces isogenous to a product with fixed data can be more complicated. We have to consider Theorem 3.6 together with the Hurwitz moves described in Proposition 2.14. Indeed, this procedure was applied in two specific cases: For surfaces isogenous to a product with $p_g = q = 1$ or 2 in [22, 24]. In both cases it was used a GAP4 script which can be found in [23]. It would be interesting to consider all the surfaces isogenous to a product with fixed invariants K^2 and χ and count the number of connected components of the moduli space of surfaces of general type they give. Indeed for $K^2 = 8$ and $\chi = 1$ we can count already 94 connected componens. Respectively: 1 component with $p_g = q = 4$, 1 with $p_g = q = 3$, 27 with $p_g = q = 2$, 52 with $p_g = q = 1$ and 13 with $p_g = q = 0$.

In this paper we consider only *unmixed* surfaces isogenous to a product, nevertheless similar methods could be applied also for the *mixed* case. Indeed, there are already works in these direction, see [14].

Acknowledgments The author is grateful to G. Bini for reading and commenting the paper. Moreover the author thanks the organizers of the conference *Beauville surfaces and Groups* N. Barker, I. Bauer, S. Garion and A. Vdovina for the invitation and the kind hospitality.

References

1. I. Bauer, F. Catanese, Some new surfaces with $p_g = q = 0$, in *Proceeding of the Fano Conference*, (Torino, 2002), pp. 123–142
2. I. Bauer, F. Catanese, F. Grunewald, Beauville surfaces without real structures, in *Geometric Methods in Algebra and Number Theory*, Progress in Mathematics, vol. 235 (Birkhäuser, Boston, 2005), pp. 1–42
3. I. Bauer, F. Catanese, F. Grunewald, Chebycheff and Belyi polynomials, dessins d'enfants, Beauville surfaces and group theory. Mediterr. J. Math. **3**, 121–146 (2006)
4. I. Bauer, F. Catanese, F. Grunewald, The classification of surfaces with $p_g = q = 0$ isogenous to a product. Pure Appl. Math. Q. **4**, 547–586 (2008)
5. J.S. Birman, Mapping class groups and their relationship to braid groups. Commun. Pure Appl. Math. **XXII**, 213–238 (1969)
6. J.S. Birman, *Braids, Links, and Mapping Class Groups*, vol. 82, Annals of Mathematics Studies (Princeton University Press, Princeton, 1974)
7. F. Catanese, Chow varieties, Hilbert schemes and moduli spaces of surfaces of general type. J. Algebraics Geom. **1**, 561–595 (1992)
8. F. Catanese, Fibred surfaces, varieties isogenous to a product and related moduli spaces. Am. J. Math. **122**, 1–44 (2000)
9. F. Catanese, Differentiable and deformation type of algebraic surfaces, real and symplectic structures. Symplectic 4-Manifolds and Algebraic Surfaces. Lectures Given at the C.I.M.E. Summer School, Cetraro, Italy, 2–10 September 2003, Lecture Notes in Mathematics, vol. 1938 (Springer, Berlin, 2008), pp. 55–167
10. F. Catanese, Moduli spaces of surfaces and real structures. Ann. Math. **158**, 577–592 (2003)
11. F. Catanese, Trecce, mapping class groups, fibrazioni di Lefschetz e applicazioni al diffeomorfismo di superfici algebriche. Luigi Cremona (1830–1903) (Incontri Studio 36., Milano, 2005), pp. 207–235

12. F. Catanese, M. Lönne, F. Perroni, Irreducibility of the space of dihedral cover of the projective line of a given numerical type. arXiv:1206.5498
13. M. Dehn, Die Gruppe der Abbildungsklassen. (Das arithmetische Feld auf Flächen.). Acta Math. **69**, 135–206 (1938)
14. D. Frapporti, R. Pignatelli, Mixed quasiale quotients with arbitrary singularities. Glasg. Math. J. **57** 143–165 (2015)
15. S. Garion, M. Penegini, New Beauville surfaces and finite simple groups. Manuscripta Math. **142** 391–408 (2013)
16. S. Garion, M. Penegini, Beauville surfaces, moduli spaces and finite groups. Comm. Algebra. **42**, 2126–2155 (2014)
17. G. Gonzales-Diez, G. Jones, D. Torres-Teigell, Beauville surfaces with abelian Beauville group. Preprint. arXiv:1102.4552v3
18. M.W. Liebeck, A. Shalev, Fuchsian groups, coverings of Riemann surfaces, subgroup growth, random quotients and random walks. J. Algebra **276**, 552–601 (2004)
19. M.W. Liebeck, A. Shalev, Fuchsian groups, finite simple groups and representation varieties. Invent. Math. **159**(2), 317–367 (2005)
20. C. Maclachlan, Modular groups and fiber spaces over Teichmüller spaces, in *Proceedings of the 1973 Conference on Discontinuous Groups Riemann Surf.*, University of Maryland, pp. 297–314 (1974)
21. M. Manetti, Iterated double covers and connected components of moduli spaces. Topology **36**, 745–764 (1997)
22. M. Penegini, The classification of isotrivially fibred surfaces with $p_g = q = 2$. With an Appendix of S. Rollenske. Collect. Math. **62**, 239–274 (2011)
23. M. Penegini, The classification of isotrivially fibred surfaces with $p_g = q = 2$, and topics on Beauville surfaces. Ph.D. Thesis, Universität Bayretuh (2010)
24. F. Polizzi, On surfaces of general type with $p_g = q = 1$ isogenous to a product of curves. Commun. Algebra **36**, 2023–2053 (2008)
25. H. Völklein, *Groups as Galois Groups—An Introduction*, vol. 53, Cambridge Studies in Advanced Mathematics (Cambridge University Press, Cambridge, 1996)

On Quasi-Étale Quotients of a Product of Two Curves

Roberto Pignatelli

Abstract A quasi-étale quotient of a product of two curves is the quotient of a product of two curves by the action of a finite group which acts freely out of a finite set of points. A quasi-étale surface is the minimal resolution of the singularities of a quasi-étale quotient. They have been successfully used in the last years by several authors to produce several interesting new examples of surfaces. In this paper we describe the principal results on this class of surfaces, and report the full list of the minimal quasi-étale surfaces of general type with geometric genus equal to the irregularity ≤ 2.

1 Introduction

Throughout this paper a "surface" (resp. curve) is a smooth complex algebraic surface (resp. curve); these are compact complex manifolds of dimension 2 (resp. 1) with an algebraic structure.

We are interested in the birational geometry of surfaces; in other words we look at surfaces modulo the equivalence relation generated by the blow-up in a point.

For sake of simplicity, we will restrict to projective surfaces, so we assume that the surface can be algebraically embedded in a projective space. Most of the statements in this section are classical results in complex algebraic geometry; a good reference is the classical book [3].

For each surface S, we will denote

- by Ω^q_S the sheaf of algebraic q-forms
- by $\mathcal{O}_S$ the structure sheaf Ω^0_S

The author was partially supported by the projects PRIN2010-2011 Geometria delle varietà algebriche and Futuro in Ricerca 2012 Moduli Spaces and Applications.

R. Pignatelli (✉)
Dipartimento di Matematica dell'Università di Trento,
Via Sommarive 14, 38123 Trento, TN, Italy
e-mail: Roberto.Pignatelli@unitn.it

I. Bauer et al. (eds.), *Beauville Surfaces and Groups*, Springer Proceedings in Mathematics & Statistics 123, DOI 10.1007/978-3-319-13862-6_10

- by K_S a canonical divisor, that is the divisor of the zeroes and of the poles of a meromorphic algebraic 2-form
- for each divisor D in S, by $\mathcal{O}_S(D)$ the invertible sheaf associated to it; so $\mathcal{O}_S(K_S) \cong \Omega^2_S$
- for each sheaf $\mathcal{F}$, by $h^q(\mathcal{F})$ the dimension of the qth Cech cohomology group $H^q(\mathcal{F})$.

The "classical" birational invariants are

- the topological fundamental group $\pi_1(S)$
- the geometric genus

$$p_g(S) := h^{2,0}(S) := h^0(\Omega^2_S) = h^0(\mathcal{O}_S(K_S)) = h^2(\mathcal{O}_S)$$

- the irregularity

$$q(S) := h^{1,0}(S) := h^0(\Omega^1_S) = h^1(\mathcal{O}_S)$$

- the Euler characteristic

$$\chi(\mathcal{O}_S) := 1 - q + p_g$$

- the plurigenera

$$P_n(S) := h^0(\mathcal{O}_S(nK_S))$$

- the Kodaira dimension $\kappa(S)$, which is the smallest number κ such that $\frac{P_n}{n^\kappa}$ is bounded from above, known to be at most 2.

Definition 1.1 A surface S is of *general type* if $\kappa(S) = 2$.

The Enriques-Kodaira classification provides a relatively good understanding of the surfaces of *special* type, which are those with Kodaira dimension $\kappa(S) \neq 2$ (this includes rational surfaces, K3 surfaces, Enriques surfaces, Abelian surfaces, elliptic surfaces...). The class of the surfaces of general type may be considered as the class of the surfaces... we do not understand, and therefore the most interesting, at least in the opinion of the author. It is worth mentioning that the method we are going to discuss has been very recently applied also to a different (very interesting) class of surfaces, the K3 surfaces, see [17].

Definition 1.2 A surface S is minimal if K_S is nef, that is if the intersection of K_S with every curve is nonnegative.

In the birational class of a surface of general type S there is exactly one minimal surface, say $\bar{S}$, which is *its minimal model*. The self intersection of the canonical divisor K^2_S, which is not a birational invariant, measures in some sense the distance among S and $\bar{S}$; more precisely S is obtained by $\bar{S}$ by a sequence of exactly $K^2_{\bar{S}} - K^2_S$ blow ups.

If S is of general type, then the Riemann-Roch formula computes all plurigenera $P_n(S)$ from $\chi(\mathcal{O}_S)$ and $K^2_{\bar{S}}$; so knowing p_g, q, $K^2_{\bar{S}}$ and the topological fundamental group is enough to have all the birational invariants mentioned above.

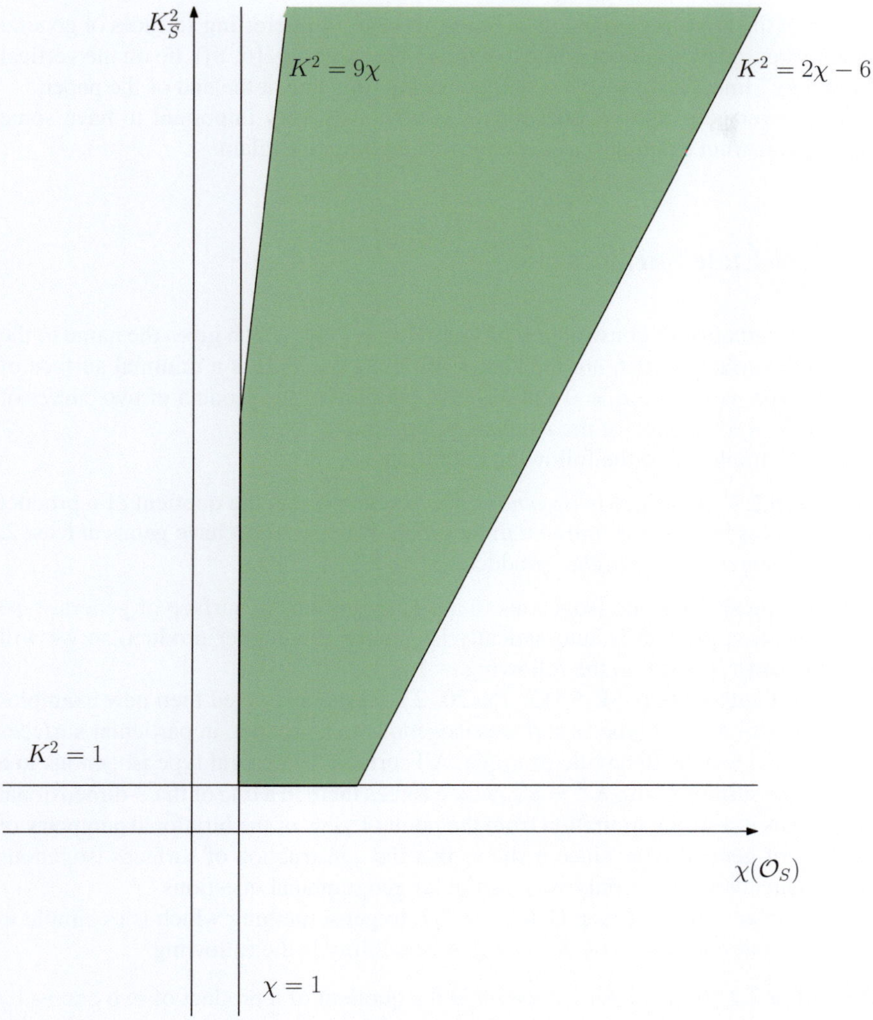

Fig. 1 The geography of the surfaces of general type

This leads us to a famous picture, known as *the geography* of the surfaces of general type. By some famous inequalities (Noether, Bogomolov-Miyaoka-Yau,...) the possible values of the pair (χ, K_S^2) are in the green region of the Fig. 1.

It is still unknown if all integral points in the green region of the Fig. 1 can be obtained by surfaces of general type, although we know we can fill "most" of it: this is an example of *geographical question*. More generally, one would like to know all possible values of the 4-tuple of topological invariants $(p_g(S), q(S), K_S^2, \pi_1(S))$; we are very far from that, but one can hope to answer this question at least for some portion of the green region in Fig. 1. Some of the most interesting surfaces of general type, for reasons we are not going to explain here (see e.g. [6, 8]), lie on the vertical "boundary" line $\chi(\mathcal{O}_S) = 1$; we will be back to this line at the end of the paper.

To answer geographical questions it is obviously very important to have some tools to construct examples, as the one we are going to explain.

2 Quasi-Étale Surfaces

The idea starts from a construction of Beauville in [11], which gives the name to the Beauville surfaces which are the center of this book. This is a minimal surface of general type with $p_g = q = 0$ and $K^2 = 8$, quotient of the product of two curves of genus 6 by a free action of the Abelian group $\mathbb{Z}^2_{/5\mathbb{Z}}$.

This example led to the following definition

Definition 2.1 A surface is *isogenous to a product* if it is the quotient of a product of two curves by a free action of a finite group. If both curves have genus at least 2, then it is isogenous to a *higher* product.

Beauville surfaces are isogenous to a higher product. A surface of general type isogenous to a product is automatically isogenous to a higher product, so we will drop the word 'higher' in the following.

Several authors (e.g. [4, 5, 13, 18, 20, 23, 24]) constructed then new examples of surfaces of general type as surfaces isogenous to a product, in particular surfaces with $p_g = q = 0$ as Beauville example. All surfaces of general type isogenous to a product are minimal with $K^2 = 8\chi$, which forces them in a line of the 2-dimensional Fig. 1. This is a strong limitation from the point of view of the birational geometry of surfaces of general type, since it shows that the construction of surfaces isogenous to a product can answer only very particular geographical questions.

This suggested to weaken Definition 2.1, to get something which is as simple to construct, but not limited by $K^2 = 8\chi$. A possibility is the following:

Definition 2.2 A *quasi-étale quotient* is the quotient of a product of two curves by the action of a finite group G acting freely out of a finite set of points. A *quasi-étale surface* is the minimal resolution of the singularities of a quasi-étale quotient.

Indeed, a quasi-étale quotient is smooth if and only if the action is free, so the surface is isogenous to a product. Each point in the product of the two curves stabilized by a non trivial subgroup of G maps onto a singular point of the quotient. We will always denote by X the singular quotient, and by S the smooth resolution of its singularities.

We have in this case the additional problem to study the singularities of X and their minimal resolution. The advantage is that these surfaces get out of the line $K^2 = 8\chi$. Indeed, it is easy to prove that $K^2 \leq 8\chi$, but apparently no other constraint applies [26] proves $K^2_S \neq 8\chi - 1$ for a quasi-étale unmixed surface, but this does not extend to the mixed case) and we may hope to fill the whole yellow region in Fig. 2.

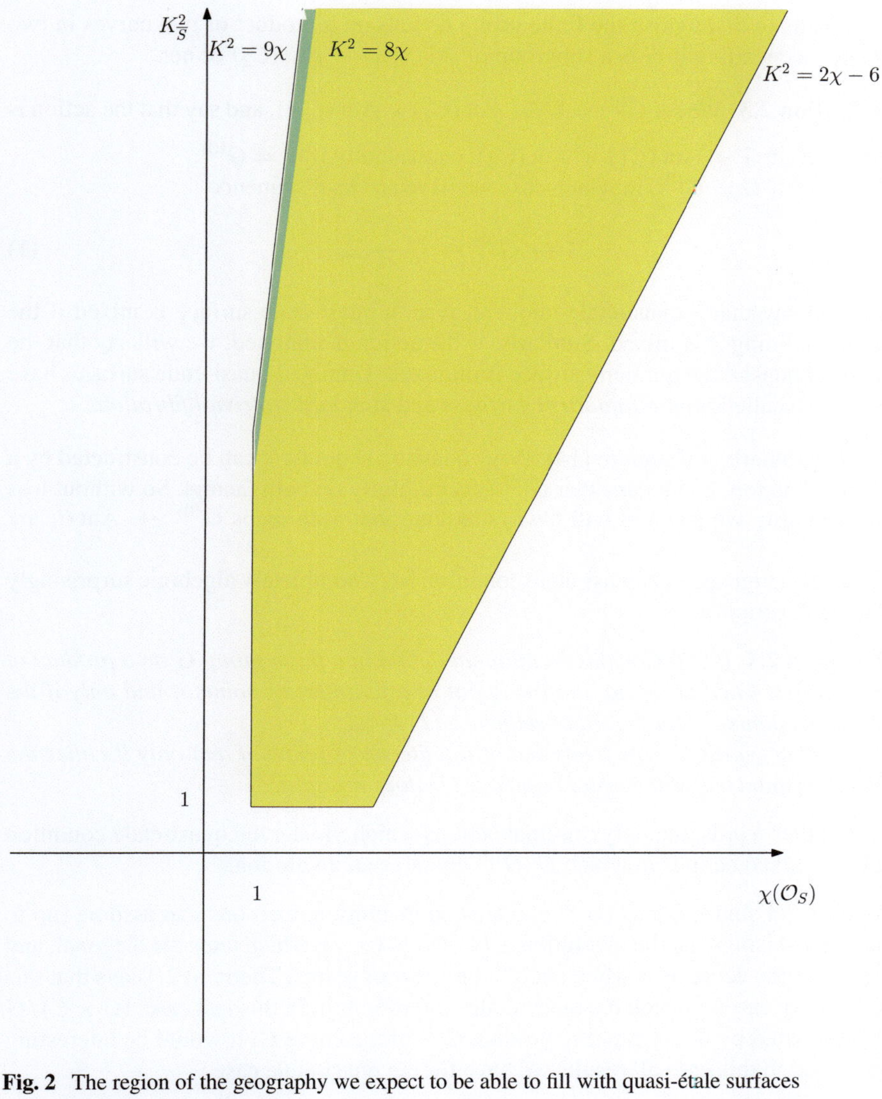

Fig. 2 The region of the geography we expect to be able to fill with quasi-étale surfaces

Quasi-étale surfaces splits naturally in two classes. Indeed (see [14]), if C_1 and C_2 are two algebraic curves

- either C_1 and C_2 are not isomorphic, in which case $\mathrm{Aut}\,(C_1 \times C_2) \cong \mathrm{Aut}\,(C_1) \times \mathrm{Aut}\,(C_2)$
- or $C_1 \cong C_2 \cong C$: in this case $\mathrm{Aut}\,(C \times C) = (\mathrm{Aut}\,C)^2 \rtimes \mathbb{Z}_2$, where the involution generating the (non normal) subgroup on the right is the exchange of the factors.

This leads to distinguish the finite group actions on a product of two curves in two classes, depending if G is a subgroup of $\mathrm{Aut}(C_1) \times \mathrm{Aut}(C_2)$ or not.

Definition 2.3 We set $G^{(0)} = G \cap (\mathrm{Aut}(C_1) \times \mathrm{Aut}(C_2))$, and say that the action is

unmixed if $G < \mathrm{Aut}(C_1) \times \mathrm{Aut}(C_2)$, equivalently if $G = G^{(0)}$.
mixed if $G \neq G^{(0)}$, in which case we have an exact sequence

$$1 \to G^{(0)} \to G \to \mathbb{Z}_2 \to 1 \tag{1}$$

We will say that a quasi-étale quotient, resp. a quasi-étale surface is mixed if the action defining it is mixed. Similarly, if the action is unmixed, we will say that the induced quasi-étale quotient/surface is unmixed. Unmixed quasi-étale surfaces have been also called *product-quotient surfaces* and *standard isotrivial fibrations*.

By a remark of Catanese [14], every quasi-étale quotient can be constructed by a minimal action, that means that $G^{(0)}$ acts faithfully on both factors. So without loss of generality we can and will always assume that both maps $G^{(0)} \to \mathrm{Aut}\, C_i$ are injective.

In this language, the quasi-étale condition has a completely algebraic surprisingly simple description.

Theorem 2.4 ([15]) *Consider a minimal action of a finite group G on a product of two curves. Then G do not act freely out of a finite set of points if and only if the action is unmixed and the exact-sequence (1) splits.*

In other words G acts freely out of a finite set of points if and only if either the action is unmixed or the exact sequence (1) does not split.

In other words, the only minimal actions which violate the quasi-étale condition are the mixed actions in which $G \setminus G^{(0)}$ contains an involution.

Remark 2.5 Indeed, if $G \setminus G^{(0)}$ contains an involution, then one can assume (up to automorphisms) that the involution is $(x, y) \mapsto (y, x)$, which fixes the diagonal, and therefore the action is not free out of a finite set of points. Theorem 2.4 says that this is the only case in which the quasi-étale condition fail. In this last case, $(C \times C)/G$ is dominated by the symmetric product $C^{(2)}$ of the curve C. It would be interesting to extend to this case all results we have for the quasi-étale case.

3 Constructing Curves with a Finite Group Action

In order to construct a quasi-étale surface, we need curves with several automorphisms; indeed the general curve of genus $g > 2$ has no nontrivial automorphisms at all. It is worth mentioning here that if $g \geq 2$, which is the case we are interested in, $\mathrm{Aut}\, C$ is finite by the classical Schwarz Theorem. Anyway the method we are going to describe works also for $g \leq 1$, assuming the order of the group to be finite.

The method for constructing it, based on the classical Riemann Existence Theorem, is the following.

Let C be a smooth algebraic curve, and let $G^{(0)}$ be a finite subgroup of $\operatorname{Aut} C$. Then $C' := C/G$ is a smooth algebraic curve of smaller genus g' and the projection $C \to C'$ has finitely many critical values, say $p_1, \ldots, p_r$. Let us fix one of these p_i; each of its preimages is stabilized by a cyclic subgroup of $G^{(0)}$, and all these subgroups lie in the same conjugacy class; in particular they all have the same cardinality, say m_i.

Removing the p_i from C' and their preimages from C, we remain with a regular topological cover $C^0 \to C' \backslash \{p_i\}$; these are determined by their monodromy map $\pi_1(C' \backslash \{p_i\}) \to G^{(0)}$. The Riemann Existence Theorem shows that each such topological cover may be uniquely extended to a map among compact complex curves $C \to C'$.

This gives a way to construct all pairs $(C, G^{(0)})$ where C is a smooth compact complex curve and $G^{(0)}$ is a finite subgroup of $\operatorname{Aut} C$.

Theorem 3.1 (Consequence of the Riemann Existence Theorem) *Given*

(a) *a compact complex curve* C'
(b) *a finite set* $\{p_i\} \subset C'$
(c) *a surjective homomorphism* $\pi_1(C' \backslash \{p_i\}) \to G^{(0)}$

there is, up to automorphisms, a unique curve C, and a unique inclusion of $G^{(0)}$ in $\operatorname{Aut} C$ such that $C' = C/G^{(0)}$, the critical values of the quotient map $C \to C'$ belong to the set $\{p_i\}$, and such that, removing $\{p_i\}$ from C' and its preimage from C, we get the topological cover whose monodromy map is the map in (c).

The key point here is the homomorphism (c). The more effective way is to construct a *generating vector* (see, e.g., [26]) of $G^{(0)}$. One chooses in a "standard" way a set of generators of $\pi_1(C' \backslash \{p_i\})$, and then gives the map (c) by giving their images in $G^{(0)}$; we need then a set of generators (to ensure the surjectivity) of $G^{(0)}$ respecting some relations reflecting the relations among the chosen generators of $\pi_1(C' \backslash \{p_i\})$.

These relations are summarized in the definition of *generating vector* of *signature* $(g'; m_1, \ldots, m_r)$. We do not repeat here this definition, referring the interested reader, e.g., to [16]; just repeat here that g' is the genus of C', r the number of critical values of the map $C \to C'$, m_i the order of the stabilizer of each preimage of the critical value p_i.

By Hurwitz formula, the genus of C can be computed by the signature of the generating formula, namely

$$2g(C) - 2 = |G| \left(2g' - 2 + \sum_i \left(1 - \frac{1}{m_i} \right) \right) \tag{2}$$

4 Constructing Quasi-Étale Quotients

In the case of unmixed actions (where, as seen in Theorem 2.4, the quasi-étale condition is empty), we need to give two curves C_1, C_2 and inclusions of $G^{(0)}$ in both $\operatorname{Aut}(C_i)$. Following the strategy in Sect. 3 we need to give two generating vectors of $G^{(0)}$, look at their signatures, say $(g_1'; m_1, \dots, m_r)$ and $(g_2'; n_1, \dots, n_s)$, and then choose two curves C_1', C_2' of respective genera g_1', g_2', and two finite subsets $\{p_1, \dots, p_r\} \subset C_1'$, $\{q_1, \dots, q_s\} \subset C_2'$.

In the mixed case, the construction is even simpler; one needs only one generating vector of $G^{(0)}$, look at its signature, say $(g'; m_1, \dots, m_r)$, then choose a curve C' of genus g' and a finite subset $\{p_1, \dots, p_r\} \subset C'$, and finally choose a degree 2 extension G of $G^{(0)}$ as in (1).

These data determine (see [14, 15]) a mixed action on the product of two curves as follows. First of all they give, as in Sect. 3, a curve C and an inclusion of $G^{(0)}$ in $\operatorname{Aut} C$. We choose an element $\tau' \in G \backslash G^{(0)}$, and notice that for each element $g \in G \backslash G^{(0)}$ there is a unique $g_0 \in G^{(0)}$ such that $g = \tau' g_0$. Then we give the mixed action on $C \times C$ as follows: we note that $(\tau')^2 \in G^{(0)}$ and set, $\forall g_0 \in G^{(0)}, \forall x, y \in C$

$$\begin{cases} g_0(x, y) = & (g_0 x, \tau' g_0 \tau'^{-1} y) \\ \tau' g_0(x, y) = & (\tau' g_0 \tau'^{-1} y, \tau'^2 g_0 x) \end{cases}$$

In the mixed case, since we are only interested in the quasi-étale surfaces, we will assume from now on that the extension (1) is unsplit.

4.1 Singularities

We have a recipe to construct quasi-étale quotient; to obtain the quasi-étale surfaces, we need to understand their singularities.

Let then $X := (C_1 \times C_2)/G$ be a quasi-étale quotient. Then [9, 15, 16, 19] the singularities of X are the images of all points in $C_1 \times C_2$ which are stabilized by some nontrivial subgroup of G; these can be computed by the generating vectors, see, e.g., [9] for the unmixed case and [16] for the mixed case. We recall here the analytic type of singularities one can find.

Definition 4.1 For each rational number $0 < \frac{q}{n} < 1$ ($\gcd(q, n) = 1$) a singularity of type $C_{n,q}$ (also called of type $\frac{1}{n}(1, q)$, or of type $\frac{q}{n}$) is an isolated singularity locally isomorphic to the singularity obtained by quotienting $\mathbb{C}^2$ by the action of the diagonal matrix with eigenvalues $e^{\frac{2\pi i}{n}}$ and $e^{\frac{2q\pi i}{n}}$.

The exceptional divisor of the minimal resolution of a singularity of type $C_{n,q}$ is a chain of rational curves $A_1, \dots, A_k$, each intersecting only the previous and the next one transversally in a single point, with respective self intersections $-b_1, \dots, -b_k$ where the b_i are given by the continued fraction

$$\frac{n}{q} = [b_1, \ldots, b_k] = b_1 - \cfrac{1}{b_2 - \cfrac{1}{b_3 - \cdots}}.$$

In other words, the dual graph of the exceptional divisor is

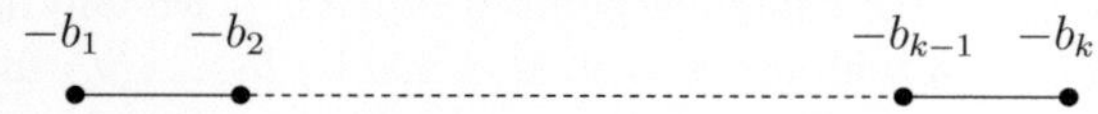

In the mixed case, we have an intermediate unmixed quotient $Y := C^2/G^{(0)}$ which has then only cyclic quotient singularities, and a double cover $Y \to X$, so we can see X as the quotient of Y by an involution.

Since the square of each element of $G \setminus G^{(0)}$ is a nontrivial element of $G^{(0)}$, the points of $C \times C$ stabilized by a nontrivial subgroup of $G^{(0)}$ are the same stabilized by a nontrivial subgroup of G; therefore the singular locus of X is the image of the singular locus of Y and, once we have computed the cyclic quotient singularities of Y, we only have to describe how the involution acts on them.

There are two possibilities. If the involution exchange two singular points, then they are isomorphic, and we get one singularity of the same type $C_{n,q}$ on the quotient. In the other case the following holds

Theorem 4.2 ([16]) *Let P be a singular point of Y which is fixed by the involution. Then P is a singular point of type $C_{n,q}$ with $q^2 \equiv 1$ mod n; in other words the dual graph of the minimal resolution of the singularity of P is a symmetric string*

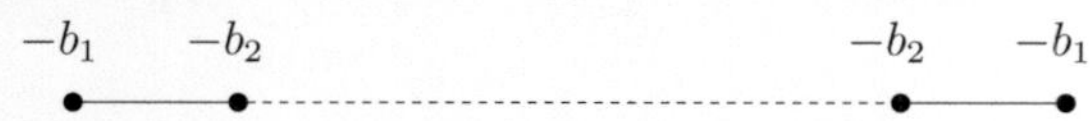

Moreover the number of vertices of the graph (which is the number of components of the exceptional divisor) is odd, say $2h + 1$, and the lift of the involution to the resolution exchanges the extremal curves.

The exceptional divisor of the minimal resolution of the corresponding singular point of X has $h + 3$ component, all rational, and its dual graph is the following

Definition 4.3 We will say that such a singular point of X is of type $D_{n,q}$.

4.2 Invariants

Once we have computed all singular points of X, which means that we know exactly how many singular points of each type has our X, we can compute some of the characteristic numbers of the constructed quasi-étale surface S.

We give here the formulas [9, 16, 19] in terms of the construction data described at the beginning of this section, and of the singularities of X. Recall that the signatures of the generating vectors determine the genera g_i of the induced covers by the formula (2).

In the following, for each singular point of type $C_{n,q}$, we will denote by q' the only integer $0 < q' < n$ with $qq' \equiv 1 \mod n$, and by $b_1, \dots, b_k$ the coefficient of the continued fraction of $\frac{n}{q}$. Then $\frac{n}{q'} = [b_k, \dots, b_1]$ and therefore a singularity of type $C_{n,q}$ is also of type $C_{n,q'}$.

For each singular point of type $C_{n,q}$ (e.g. [9])

$$k_x := -2 + \frac{2+q+q'}{n} + \sum_1^k (b_i - 2) \quad B_x := \frac{q+q'}{n} + \sum_1^k b_i$$

whereas, for singular points of type $D_{n,q}$ [16]

$$k_x := -1 + \frac{2+q+q'}{2n} + \sum_1^k \frac{b_i - 2}{2} \quad B_x := 6 + \frac{q+q'}{2n} + \sum_1^k \frac{b_i}{2}$$

All k_x, B_x are nonnegative. Then

$$K_S^2 = \frac{8(g_1 - 1)(g_2 - 1)}{|G|} - \sum_{x \in \operatorname{Sing} X} k_x$$

$$\chi(\mathcal{O}_S) = \frac{K_S^2}{8} + \frac{\sum_{x \in \operatorname{Sing} X} B_x}{24}$$

Finally, for the irregularity [27], in the unmixed case

$$q(S) = g_1' + g_2'$$

whereas in the mixed case

$$q(S) = g_1' = g_2'$$

The topological fundamental group, finally, can be computed directly by a method due to Armstrong [1, 2]. The computation is rather complicated, but can easily performed by a computer. For example, in [10] there is an implementation for that in the MAGMA [12] language.

So, if the surface is minimal ($S = \overline{S}$), we know all invariants of the constructed surfaces. As we will see in the next sections, this happens often, but not always, and determining $K_{\overline{S}}^2$ is sometimes a challenging task.

5 The Minimality Problem

In the previous sections we have given a method to construct quasi-étale surfaces and to compute, from the construction of S, $\pi_1(S)$, K_S^2, $p_g(S)$ and $q(S)$. If S is minimal, then $K_S^2 = K_{\tilde{S}}^2$ and we can compute by Riemann-Roch Theorem all the plurigenera of S, and locate its position in the "geography".

This is often the case. Indeed K_X (which is the Weil divisor defined as closure of the canonical divisor of the open subset of the smooth points of X) pulls back to $K_{C_1 \times C_2}$ and therefore, if C_1 and C_2 have genus at least 2, it is big and nef. Therefore, if the action of G is free, then $X = S$ is minimal. On the other hand, if the action is not free, the singularities of X induce a discrepancy among K_S and the pull-back of K_X which is big and nef; if this discrepancy is big enough, the surface S may become non minimal or even not of general type. The first examples of non minimal unmixed quasi-étale surfaces of general type have been produced in [19], two surfaces with $p_g = q = 1$. The first example of an unmixed regular non minimal quasi-étale surfaces of general type has been produced in [9], a surface with $p_g = q = 0$; some more examples have been constructed in [7].

It is unclear if there exists a non minimal mixed quasi-étale surface of general type. Indeed we have the following

Theorem 5.1 ([16]) *Every irregular mixed quasi-étale surface of general type is minimal.*

The argument of the proof can't be extended to the regular case. Indeed [16] shows that, if S is a mixed quasi-étale surface of general type containing a smooth rational curve E with self intersection (-1), then the image of E on X pulls back to a curve of $C \times C$ whose image in $C' \times C'$ is a rational curve (here as before $C' := C/G^{(0)}$). Then, since the genus of C' equals the irregularity of the surface, we get a contradiction in the irregular case ($C' \times C'$ does not contain any rational curve), but nothing can be deduced in the regular case.

The unmixed irregular case, as (previously) shown in [19], is not much more complicated. Indeed if both curves C_1' and C_2' are irregular, then $C_1' \times C_2'$ does not contain any rational curve and one concludes as above (indeed this remark in [19] inspired the proof of Theorem 5.1). Else, up to exchanging C_1 and C_2, C_1' is rational, C_2' has genus $q > 0$ and all rational curves on S are contracted by the natural fibration $f_2 \colon S \to C_2'$. On the other hand, the only fibers of f_2 which are not isomorphic to C_1 are those whose image on X contains some singular points: it is not difficult then to compute explicitly their decomposition in irreducible components, and therefore describe all rational curves on S: in this way [19] could determine the minimal model of all the surfaces with $p_g = q = 1$ they constructed.

For example: [19] constructed one surface with $K_S^2 = 1$ whose quotient model had three singular points of respective type $\frac{1}{7}, \frac{2}{7}, \frac{2}{7}$ all in the image of the same fiber of f_2. That fiber depicted in Fig. 3, on the left, contains then all rational curves of S. It has six irreducible components, all smooth rational curves, 5 forming the exceptional divisor of the resolution of the singularities, and the sixth having self-intersection -1.

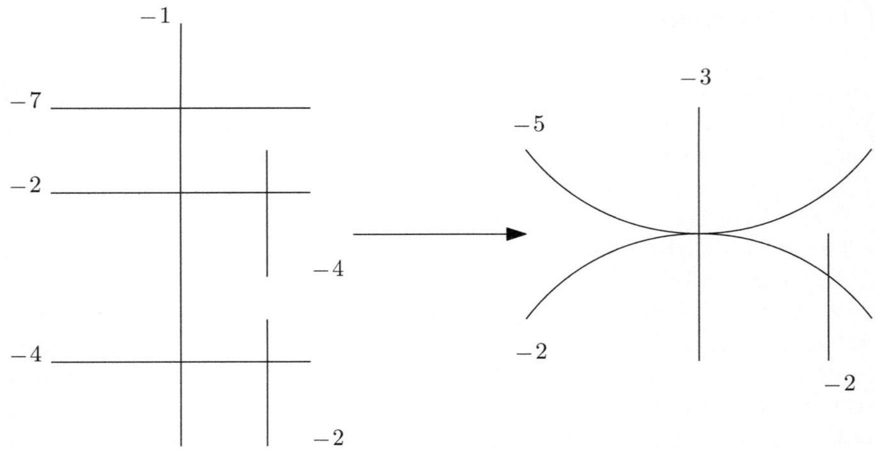

Fig. 3 The rational curves in the example by Polizzi and Mistretta

Contracting the (-1)-curve and the (-2)-curve transversal to it one gets a surface $\bar{S}$ with $K^2_{\bar{S}} = 3$ with exactly 4 rational curves, depicted on the right of Fig. 3. The minimality follows since none has self intersection -1.

Determining the minimal model in the regular case may be much more challenging, since the rational curves are not forced to stay in the fibers of any of the two fibrations $f_i : S \to C_i/G$; indeed we are not able, in the regular case, to compute all rational curves on S. The surface constructed in [9], named "fake Godeaux" there, is a surface with $K^2_S = 1$, $p_g = q = 0$ whose quotient model has exactly the same configuration of singularities, $\frac{1}{7}, \frac{2}{7}, \frac{2}{7}$, of the previous example. Bauer and Pignatelli [9] constructed two rational curves on S with self-intersection -1 intersecting the exceptional divisor of the resolution $S \to X$ as in Fig. 4, on the left. Contracting both we get a surface $\overline{S}$ with $K^2 = 3$, $p_g = q = 0$ with the funny configuration of rational curves on the right of Fig. 4.

Still, we know that ([9, Sect. 4] and [16, Corollary 4.7])

Proposition 5.2 *Let $\mathbb{P}^1 \to X$ be a rational curve on the quotient model of a quasi-étale surface such that C_1 and C_2 have both genus at least 2 (that's a necessary condition for X to be of general type). Then there are at least three distinct points of $\mathbb{P}^1$ mapped to singular points of X.*

This forces every further rational curve in $\overline{S}$ to intersect in at least three distinct points the configuration of curves on the right of Fig. 4. On the other hand ([16, Corollary 4.8])

Proposition 5.3 *Let E be a rational curve with self intersection -1 on a surface of general type, F be a reduced divisor whose support is made by rational curves of self-intersection either -2 or -3. Then $EF \leq 2$.*

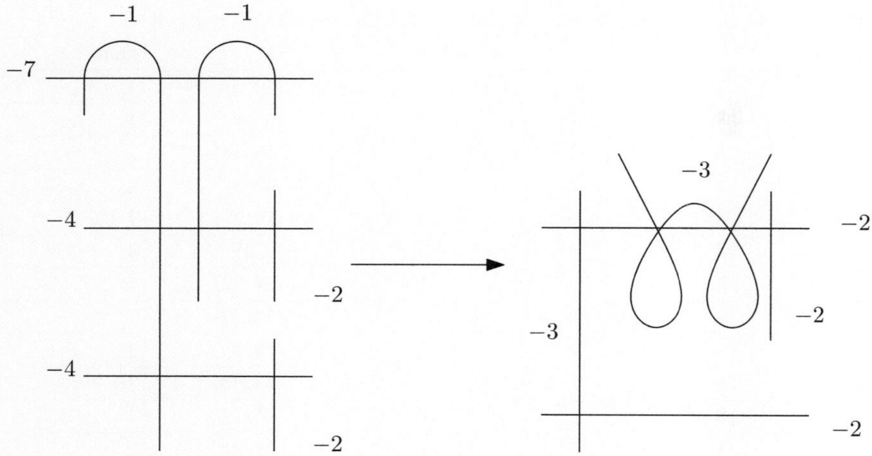

Fig. 4 The rational curves in the fake Godeaux surfaces

We know that $\overline{S}$ is of general type since by the Enriques-Kodaira classification a regular surface with $K^2 > 0$ is either of general type or rational, all rational surfaces are simply connected, and we can compute its fundamental group, which is $\mathbb{Z}_{/6\mathbb{Z}}$. Looking at the configuration of curves on the right of Fig. 4, by Propositions 5.2 and 5.3 we conclude that $\overline{S}$ is minimal.

A similar arguments works for proving the minimality of many of the surfaces constructed with this method, although one may need to substitute Proposition 5.3 with similar statements involving different configurations of rational curves. Some examples are in [9, Proposition 2.7].

We have not mentioned here how the two (-1)-curves on S are constructed; that's a rather complicated construction whose details goes beyond the scopes of this paper; the interested reader may read [9, Sect. 5]. We have recently found ([7]) few more examples of non minimal unmixed quasi-étale surfaces of general type with $p_g = q = 0$, and in all cases (up to now) we were able to determine their minimal model by constructing all the (-1)-curves on them. Still we have to run a different construction in each case: it would be nice to have a general method.

6 Surfaces of General Type with $\chi(\mathcal{O}_S) = 1$

This strategy has been used in the last year by several authors to construct new surfaces of general type, mainly for the "special" region of the geography of the surfaces with $\chi = 1$ (equivalently $p_g = q$). Thank to the contribution of, among others [4, 5, 9, 10, 13, 15, 16, 18–26, 28], we have now the complete list of all

Table 1 Minimal unmixed quasi-étale surfaces of general type with $p_g = q = 0$

Sing X	K_S^2	Sign.		G	#fam
	8	$2^3, 3$	5^3	$\langle 60, 5\rangle$	1
	8	2^5	2^6	$\langle 8, 5\rangle$	1
	8	2^5	2^5	$\langle 16, 14\rangle$	1
	8	2^5	$3^2, 5$	$\langle 60, 5\rangle$	1
	8	$2^2, 4^2$	$2^2, 4^2$	$\langle 16, 3\rangle$	1
	8	3^4	3^4	$\langle 9, 2\rangle$	1
	8	2^6	$3, 4^2$	$\langle 24, 12\rangle$	1
	8	$2^2, 4^2$	$2^3, 4$	$\langle 32, 27\rangle$	1
	8	$2, 5^2$	3^4	$\langle 60, 5\rangle$	1
	8	$2, 4, 6$	2^6	$\langle 48, 48\rangle$	1
	8	$2^3, 4$	2^6	$\langle 16, 11\rangle$	1
	8	5^3	5^3	$\langle 25, 2\rangle$	2
$\frac{1}{2}^2$	6	$2, 4, 6$	$2, 4, 10$	$\langle 240, 189\rangle$	1
$\frac{1}{2}^2$	6	$2^3, 4$	$2^4, 4$	$\langle 16, 11\rangle$	1
$\frac{1}{2}^2$	6	$2, 4, 6$	$2^4, 4$	$\langle 48, 48\rangle$	1
$\frac{1}{2}^2$	6	$2, 3^3$	$2, 5^2$	$\langle 60, 5\rangle$	1
$\frac{1}{2}^2$	6	$2, 7^2$	$3^2, 4$	$\langle 168, 42\rangle$	2
$\frac{1}{2}^2$	6	$2, 5^2$	$3^2, 4$	$\langle 360, 118\rangle$	2
$\frac{1}{2}^4$	4	$2^2, 3^2$	$2^2, 3^2$	$\langle 18, 4\rangle$	1
$\frac{1}{2}^4$	4	$2^3, 4$	$2^3, 4$	$\langle 32, 27\rangle$	1
$\frac{1}{2}^4$	4	$2^2, 4^2$	$2^2, 4^2$	$\langle 8, 2\rangle$	1
$\frac{1}{2}^4$	4	2^5	2^5	$\langle 8, 5\rangle$	1
$\frac{1}{2}^4$	4	$2, 4, 6$	2^5	$\langle 48, 48\rangle$	1
$\frac{1}{2}^4$	4	$2, 5^2$	$2^2, 3^2$	$\langle 60, 5\rangle$	1
$\frac{1}{2}^4$	4	2^5	$3, 4^2$	$\langle 24, 12\rangle$	1
$\frac{1}{2}^4$	4	$2, 4, 6$	$2^2, 4^2$	$\langle 48, 48\rangle$	1
$\frac{1}{2}^4$	4	$2^3, 4$	2^5	$\langle 16, 11\rangle$	1
$\frac{1}{2}^4$	4	$2, 4, 5$	$3, 6^2$	$\langle 120, 34\rangle$	1
$\frac{1}{2}^4$	4	$2^2, 3^2$	$3, 6^2$	$\langle 18, 3\rangle$	1
$\frac{1}{2}^6$	2	$2^3, 4$	$2^3, 4$	$\langle 16, 11\rangle$	1
$\frac{1}{2}^6$	2	$2, 4, 6$	$2^3, 4$	$\langle 48, 48\rangle$	1
$\frac{1}{2}^6$	2	$2, 4, 5$	$2, 6^2$	$\langle 120, 34\rangle$	1
$\frac{1}{2}^6$	2	$2, 3, 7$	4^3	$\langle 168, 42\rangle$	2
$\frac{1}{2}^6$	2	4^3	4^3	$\langle 16, 2\rangle$	1
$\frac{1}{2}^6$	2	$2, 5^2$	$2^3, 3$	$\langle 60, 5\rangle$	1
$\frac{1}{2}^6$	2	$2, 6^2$	$2^3, 3$	$\langle 36, 10\rangle$	1
$\frac{1}{3}, \frac{2}{3}$	5	$2^3, 3$	$3, 4^2$	$\langle 96, 227\rangle$	1

(continued)

Table 1 (continued)

Sing X	K_S^2	Sign.		G	#fam
$\frac{1}{3}, \frac{2}{3}$	5	$2^4, 3$	$3, 4^2$	$\langle 24, 12\rangle$	1
$\frac{1}{3}, \frac{2}{3}$	5	$2^3, 3$	$4^2, 6$	$\langle 48, 48\rangle$	1
$\frac{1}{3}, \frac{2}{3}$	5	$2^3, 3$	$3, 5^2$	$\langle 60, 5\rangle$	2
$\frac{1}{3}, \frac{2}{3}$	5	$2, 5, 6$	$3, 4^2$	$\langle 120, 34\rangle$	1
$\frac{1}{3}, \frac{2}{3}$	5	$2, 4, 6$	$2^4, 3$	$\langle 48, 48\rangle$	1
$\frac{1}{3}, \frac{1}{2}^2, \frac{2}{3}$	3	$2, 4, 6$	$2^2, 3, 4$	$\langle 48, 48\rangle$	1
$\frac{1}{3}^2, \frac{2}{3}^2$	2	$2, 6^2$	$2^2, 3^2$	$\langle 24, 13\rangle$	1
$\frac{1}{3}^2, \frac{2}{3}^2$	2	$3^2, 5$	$3^2, 5$	$\langle 75, 2\rangle$	2
$\frac{1}{3}^2, \frac{2}{3}^2$	2	$2^3, 3$	$3^2, 5$	$\langle 60, 5\rangle$	1
$\frac{1}{3}^2, \frac{2}{3}^2$	2	$2^2, 3^2$	$3, 4^2$	$\langle 24, 12\rangle$	1
$\frac{1}{3}, \frac{1}{2}^4, \frac{2}{3}$	1	$2^3, 3$	$3, 4^2$	$\langle 24, 12\rangle$	1
$\frac{1}{3}, \frac{1}{2}^4, \frac{2}{3}$	1	$2, 4, 6$	$2^3, 3$	$\langle 48, 48\rangle$	1
$\frac{1}{3}, \frac{1}{2}^4, \frac{2}{3}$	1	$2, 3, 7$	$3, 4^2$	$\langle 168, 42\rangle$	1
$\frac{1}{4}, \frac{1}{2}^2, \frac{3}{4}$	2	$2, 4, 7$	$3^2, 4$	$\langle 168, 42\rangle$	2
$\frac{1}{4}, \frac{1}{2}^2, \frac{3}{4}$	2	$2, 4, 5$	$3^2, 4$	$\langle 360, 118\rangle$	2
$\frac{1}{4}, \frac{1}{2}^2, \frac{3}{4}$	2	$2, 4, 5$	$3, 4, 6$	$\langle 120, 34\rangle$	2
$\frac{2}{5}^2$	4	$2^3, 5$	$3^2, 5$	$\langle 60, 5\rangle$	1
$\frac{2}{5}^2$	4	$2, 4, 5$	$3^2, 5$	$\langle 360, 118\rangle$	1
$\frac{2}{5}^2$	4	$2, 4, 5$	$4^2, 5$	$\langle 160, 234\rangle$	3
$\frac{1}{5}, \frac{4}{5}$	3	$2^3, 5$	$3^2, 5$	$\langle 60, 5\rangle$	1
$\frac{1}{5}, \frac{4}{5}$	3	$2, 4, 5$	$3^2, 5$	$\langle 360, 118\rangle$	1
$\frac{1}{5}, \frac{4}{5}$	3	$2, 4, 5$	$4^2, 5$	$\langle 160, 234\rangle$	3

the minimal quasi-étale surfaces of general type with $\chi = 1$ and $K_S^2 > 0$, which includes all the minimal surfaces.

The surfaces of general type with $p_g = q \geq 3$ are classified (see [8] for a more precise account), so we only consider here the case $q \leq 2$. There are only three non minimal surfaces , which we have partially discussed in the previous section: the interested reader will find them in [9, 19].

The minimal quasi-étale surfaces of general type with $p_g = q \leq 2$ form few hundreds of families, whose construction is spread among several papers, sometimes using different notations. We take the opportunity offered by this paper to collect all of them in the same place with a coherent notation, reporting the full list in the Tables 1, 2, 3, 4, 5 and 6.

Table 2 Minimal mixed quasi-étale surfaces of general type with $p_g = q = 0$

Sing X	K_S^2	Sign	$G^{(0)}$	G	#fam
	8	2^5	$\langle 32, 46\rangle$	$\langle 64, 92\rangle$	1
	8	4^3	$\langle 128, 36\rangle$	$\langle 256, 3678\rangle$	3
	8	4^3	$\langle 128, 36\rangle$	$\langle 256, 3679\rangle$	1
$\frac{1}{2}^4$	4	2^5	$\langle 16, 11\rangle$	$\langle 32, 7\rangle$	1
$\frac{1}{2}^4$	4	2^5	$\langle 16, 14\rangle$	$\langle 32, 22\rangle$	1
$\frac{1}{2}^4$	4	4^3	$\langle 64, 23\rangle$	$\langle 128, 836\rangle$	1
$\frac{1}{2}^6$	2	2^5	$\langle 8, 5\rangle$	$\langle 16, 3\rangle$	1
$\frac{1}{2}^6$	2	4^3	$\langle 32, 2\rangle$	$\langle 64, 82\rangle$	1
$\frac{1}{3}^2, \frac{2}{3}^2$	2	$3^2, 4$	$\langle 384, 4\rangle$	$\langle 768, 1083540\rangle$	1
$\frac{1}{3}^2, \frac{2}{3}^2$	2	$3^2, 4$	$\langle 384, 4\rangle$	$\langle 768, 1083541\rangle$	1
$\frac{1}{2}^3, \frac{1}{4}^2$	2	$2^3, 4$	$\langle 64, 73\rangle$	$\langle 128, 1535\rangle$	1
$\frac{3}{8}, \frac{5}{8}$	3	$2^3, 8$	$\langle 32, 39\rangle$	$\langle 64, 42\rangle$	1
$\frac{1}{2}^2, D_{2,1}^2$	1	$2^3, 4$	$\langle 16, 11\rangle$	$\langle 32, 6\rangle$	1
$\frac{1}{2}^2, D_{2,1}^2$	2	$2^3, 4$	$\langle 32, 27\rangle$	$\langle 64, 32\rangle$	1
$\frac{1}{2}^2, D_{2,1}^2$	2	$2^2, 3^2$	$\langle 18, 4\rangle$	$\langle 36, 9\rangle$	1

A short explanation of the notation:

- The column Sing X gives the singularities of the quasi-étale quotient: we use the notation $\frac{q}{n}$ for the cyclic quotient singularities, $D_{n,q}$ for the singularities, in the mixed case, which are branching points of the double cover $Y \to X$. We use exponents for multiplicities: for example $\frac{1}{2}^2, D_{2,1}^2$ means that X has 4 singular points, 2 ordinary nodes and 2 of type $D_{2,1}$ (in particular Y has 6 nodes, and the fixed locus of the involution is given by two of them). In the case of surfaces isogenous of a product, equivalently if X is smooth, we leave this field blank.
- The column K_S^2 is self-explanatory.
- The columns Sign. give the involved signatures, two in the unmixed case, one in the mixed case. We use here, for short, exponents for representing the multiplicity, and omit g' when equal to zero. So $2^3, 3$ is a shortcut for $0; 2, 2, 2, 3$.
- The columns G and $G^{(0)}$ (the latter only in the mixed case) give the corresponding group in the MAGMA/GAP4 notation. So $\langle 60, 5\rangle$, for example, is the 5th group of order 60 in the MAGMA/GAP4 database of finite groups: this is the alternating group in 5 elements $\mathfrak{A}_5$.
- g_{alb} is the genus of the general fibre of the Albanese map, which is very important for the classification of the irregular surfaces. The column is missing for $q = 0$ since there is no Albanese map in that case. When, for $q = 2$, we leave it blank, it means that the Albanese map is not a fibration.
- In few cases there are 2 or 3 different families for which all the previous data coincide: instead of putting more identical rows, we used only one row for all of them, and add a last column, #fam, counting the number of families corresponding

Table 3 Minimal unmixed quasi-étale surfaces of general type with $p_g = q = 1$

Sing X	K_S^2	g_{alb}	Sign.		G	#fam
	8	3	2^6	$1; 2^2$	$\langle 4, 2\rangle$	1
	8	3	2^5	$1; 2^2$	$\langle 8, 5\rangle$	1
	8	3	$2^2, 4^2$	$1; 2^2$	$\langle 8, 2\rangle$	2
	8	3	$2, 8^2$	$1; 2^2$	$\langle 16, 5\rangle$	1
	8	3	$2^2, 4^2$	$1; 2^2$	$\langle 8, 3\rangle$	1
	8	3	$2^3, 6$	$1; 2^2$	$\langle 12, 4\rangle$	1
	8	3	$2^3, 4$	$1; 2^2$	$\langle 16, 11\rangle$	1
	8	3	$2, 4, 12$	$1; 2^2$	$\langle 24, 5\rangle$	1
	8	3	$2, 6^2$	$1; 2^2$	$\langle 24, 13\rangle$	1
	8	3	$3, 4^2$	$1; 2^2$	$\langle 24, 12\rangle$	1
	8	3	$2, 4, 8$	$1; 2^2$	$\langle 32, 9\rangle$	1
	8	3	$2, 4, 6$	$1; 2^2$	$\langle 48, 48\rangle$	1
	8	4	2^6	$1; 3$	$\langle 6, 1\rangle$	1
	8	4	2^5	$1; 3$	$\langle 12, 4\rangle$	1
	8	4	$2^2, 3^2$	$1; 3$	$\langle 18, 3\rangle$	2
	8	4	$3, 6^2$	$1; 3$	$\langle 18, 3\rangle$	1
	8	4	$2^3, 4$	$1; 3$	$\langle 24, 12\rangle$	1
	8	4	$2, 6^2$	$1; 3$	$\langle 36, 10\rangle$	1
	8	4	$2, 6^2$	$1; 3$	$\langle 36, 12\rangle$	1
	8	4	$2, 4^2$	$1; 3$	$\langle 36, 9\rangle$	2
	8	4	$2, 5^2$	$1; 3$	$\langle 60, 5\rangle$	1
	8	4	$2, 3, 12$	$1; 3$	$\langle 72, 42\rangle$	1
	8	4	$2, 4, 5$	$1; 3$	$\langle 120, 34\rangle$	1
	8	5	2^6	$1; 2$	$\langle 8, 3\rangle$	1
	8	5	3^4	$1; 2$	$\langle 12, 3\rangle$	2
	8	5	$2^2, 4^2$	$1; 2$	$\langle 16, 3\rangle$	3
	8	5	$2^2, 3^2$	$1; 2$	$\langle 24, 13\rangle$	2
	8	5	$3, 6^2$	$1; 2$	$\langle 24, 13\rangle$	1
	8	5	$2, 8^2$	$1; 2$	$\langle 32, 5\rangle$	1
	8	5	$2, 8^2$	$1; 2$	$\langle 32, 7\rangle$	1
	8	5	4^3	$1; 2$	$\langle 32, 2\rangle$	1
	8	5	4^3	$1; 2$	$\langle 32, 6\rangle$	1
	8	5	$2, 6^2$	$1; 2$	$\langle 48, 49\rangle$	1
	8	5	$2, 4, 8$	$1; 2$	$\langle 64, 32\rangle$	2
	8	5	$2, 5^2$	$1; 2$	$\langle 80, 49\rangle$	2
$\frac{1}{2}^2$	6	3	$2, 5^2$	$1; 2$	$\langle 24, 3\rangle$	
$\frac{1}{2}^2$	6	3	$2, 5^2$	$1; 2$	$\langle 32, 9\rangle$	
$\frac{1}{2}^2$	6	3	$2, 5^2$	$1; 2$	$\langle 32, 11\rangle$	
$\frac{1}{2}^2$	6	3	$2, 5^2$	$1; 2$	$\langle 48, 33\rangle$	

(continued)

Table 3 (continued)

Sing X	K_S^2	g_{alb}	Sign.		G	#fam
$\frac{1}{2}^2$	6	3	2^6	$1; 2^2$	$\langle 48, 3\rangle$	
$\frac{1}{2}^2$	6	3	2^5	$1; 2^2$	$\langle 168, 42\rangle$	
$\frac{1}{2}^2$	6	4	$2^2, 4^2$	$1; 2^2$	$\langle 8, 3\rangle$	
$\frac{1}{2}^2$	6	4	$2, 8^2$	$1; 2^2$	$\langle 12, 3\rangle$	
$\frac{1}{2}^2$	6	4	$2^2, 4^2$	$1; 2^2$	$\langle 24, 10\rangle$	
$\frac{1}{2}^2$	6	4	$2^3, 6$	$1; 2^2$	$\langle 36, 11\rangle$	
$\frac{1}{2}^2$	6	4	$2^3, 4$	$1; 2^2$	$\langle 72, 40\rangle$	
$\frac{1}{2}^2$	6	4	$2, 4, 12$	$1; 2^2$	$\langle 120, 34\rangle$	
$\frac{1}{3}, \frac{2}{3}$	5	3	$2, 6^2$	$1; 2^2$	$\langle 6, 1\rangle$	
$\frac{1}{3}, \frac{2}{3}$	5	3	$3, 4^2$	$1; 2^2$	$\langle 12, 1\rangle$	
$\frac{1}{3}, \frac{2}{3}$	5	3	$2, 4, 8$	$1; 2^2$	$\langle 12, 4\rangle$	
$\frac{1}{3}, \frac{2}{3}$	5	3	$2, 4, 6$	$1; 2^2$	$\langle 24, 5\rangle$	
$\frac{1}{3}, \frac{2}{3}$	5	3	2^6	$1; 3$	$\langle 24, 12\rangle$	
$\frac{1}{3}, \frac{2}{3}$	5	3	2^5	$1; 3$	$\langle 48, 48\rangle$	
$\frac{1}{3}, \frac{2}{3}$	5	3	$2^2, 3^2$	$1; 3$	$\langle 96, 64\rangle$	
$\frac{1}{3}, \frac{2}{3}$	5	3	$3, 6^2$	$1; 3$	$\langle 168, 42\rangle$	
$\frac{1}{2}^4$	4	2	$2^3, 4$	$1; 3$	$\langle 4, 2\rangle$	
$\frac{1}{2}^4$	4	2	$2, 6^2$	$1; 3$	$\langle 6, 2\rangle$	
$\frac{1}{2}^4$	4	2	$2, 6^2$	$1; 3$	$\langle 6, 1\rangle$	
$\frac{1}{2}^4$	4	2	$2, 4^2$	$1; 3$	$\langle 8, 3\rangle$	
$\frac{1}{2}^4$	4	2	$2, 5^2$	$1; 3$	$\langle 12, 5\rangle$	
$\frac{1}{2}^4$	4	2	$2, 3, 12$	$1; 3$	$\langle 12, 4\rangle$	
$\frac{1}{2}^4$	4	2	$2, 4, 5$	$1; 3$	$\langle 16, 8\rangle$	
$\frac{1}{2}^4$	4	2	2^6	$1; 2$	$\langle 24, 8\rangle$	
$\frac{1}{2}^4$	4	2	3^4	$1; 2$	$\langle 48, 29\rangle$	
$\frac{1}{2}^4$	4	3	$2^2, 4^2$	$1; 2$	$\langle 8, 3\rangle$	
$\frac{1}{2}^4$	4	3	$2^2, 3^2$	$1; 2$	$\langle 12, 3\rangle$	
$\frac{1}{2}^4$	4	3	$3, 6^2$	$1; 2$	$\langle 16, 6\rangle$	
$\frac{1}{2}^4$	4	3	$3, 6^2$	$1; 2$	$\langle 16, 4\rangle$	
$\frac{1}{2}^4$	4	3	$2, 8^2$	$1; 2$	$\langle 24, 13\rangle$	
$\frac{1}{2}^2, \frac{1}{3}, \frac{2}{3}$	3	2	$2, 8^2$	$1; 2$	$\langle 24, 8\rangle$	
$\frac{1}{2}^2, \frac{1}{3}, \frac{2}{3}$	3	2	4^3	$1; 2$	$\langle 48, 29\rangle$	
$\frac{1}{2}^2, \frac{1}{4}, \frac{3}{4}$	2	2	4^3	$1; 2$	$\langle 16, 8\rangle$	
$\frac{1}{2}^2, \frac{1}{4}, \frac{3}{4}$	2	2	4^3	$1; 2$	$\langle 24, 3\rangle$	
$\frac{1}{3}^2, \frac{2}{3}^2$	2	2	$2, 6^2$	$1; 2$	$\langle 6, 1\rangle$	

(continued)

Table 3 (continued)

Sing X	K_S^2	g_{alb}	Sign.		G	#fam
$\frac{1}{3}^2, \frac{2}{3}^2$	2	2	$2, 4, 8$	$1; 2$	$\langle 12, 1\rangle$	
$\frac{1}{3}^2, \frac{2}{3}^2$	2	2	$2, 5^2$	$1; 2$	$\langle 12, 4\rangle$	
$\frac{1}{2}^6$	2	2	$2, 5^2$	$1; 2$	$\langle 8, 4\rangle$	
$\frac{1}{2}^6$	2	2	$2, 5^2$	$1; 2$	$\langle 8, 3\rangle$	

Table 4 Minimal mixed quasi-étale surfaces of general type with $p_g = q = 1$

Sing X	K_S^2	g_{alb}	Sign	$G^{(0)}$	G	#fam
	8	5	$1; 2^2$	$\langle 8, 2\rangle$	$\langle 16, 6\rangle$	1
	8	5	$1; 2^2$	$\langle 8, 3\rangle$	$\langle 16, 8\rangle$	1
	8	5	$1; 2^2$	$\langle 8, 5\rangle$	$\langle 16, 3\rangle$	1
$\frac{1}{2}^2$	6	3	1;2	$\langle 24, 13\rangle$	$\langle 48, 30\rangle$	1
$\frac{1}{2}^2$	6	7	1;2	$\langle 24, 13\rangle$	$\langle 48, 31\rangle$	1
$\frac{1}{2}^4$	4	3	$1; 2^2$	$\langle 4, 1\rangle$	$\langle 8, 1\rangle$	1
$\frac{1}{2}^4$	4	3	$1; 2^2$	$\langle 4, 2\rangle$	$\langle 8, 2\rangle$	1
$\frac{1}{2}^4$	4	2	1;2	$\langle 16, 3\rangle$	$\langle 32, 29\rangle$	1
$\frac{1}{2}^4$	4	3	1;2	$\langle 16, 4\rangle$	$\langle 32, 13\rangle$	1
$\frac{1}{2}^4$	4	3	1;2	$\langle 16, 4\rangle$	$\langle 32, 14\rangle$	1
$\frac{1}{2}^4$	4	2	1;2	$\langle 16, 4\rangle$	$\langle 32, 32\rangle$	1
$\frac{1}{2}^4$	4	2	1;2	$\langle 16, 4\rangle$	$\langle 32, 35\rangle$	1
$\frac{1}{2}^4$	4	3	1;2	$\langle 16, 5\rangle$	$\langle 32, 15\rangle$	1
$\frac{1}{3}, \frac{2}{3}$	5	3	1;3	$\langle 12, 1\rangle$	$\langle 24, 4\rangle$	1
$\frac{1}{3}, \frac{2}{3}$	5	3	1;3	$\langle 12, 4\rangle$	$\langle 24, 5\rangle$	1
$\frac{3}{5}$	6	5	1;5	$\langle 10, 1\rangle$	$\langle 20, 3\rangle$	1
$\frac{1}{2}, D_{2,1}^2$	2	2	$1; 2^2$	$\langle 2, 1\rangle$	$\langle 4, 1\rangle$	1
$\frac{1}{2}, D_{2,1}^2$	2	2	1;2	$\langle 8, 3\rangle$	$\langle 16, 8\rangle$	1
$\frac{1}{2}, D_{2,1}^2$	2	2	1;2	$\langle 8, 4\rangle$	$\langle 16, 9\rangle$	1

to the row. The reader will not find the column #fam in the latter part of Table 3; indeed in that cases the number of families corresponding to each row is not known, at least to the author.

It is worth noticing that all possible values of K_S^2 in the yellow region of Fig. 2 are obtained already by the Table 1, so by surfaces with $p_g = q = 0$. In the irregular case, since an inequality of Debarre shows that for a minimal irregular surface of general type $K_S^2 \geq 2p_g$, also the quasi-étale surfaces with $p_g = q = 1$ and $p_g = q = 2$

Table 5 Minimal unmixed quasi-étale surfaces of general type with $p_g = q = 2$

Sing X	K_S^2	g_{alb}	Sign.		G	#fam
	8	2	2^6	2;	$\langle 2, 1\rangle$	1
	8	2	3^4	2;	$\langle 3, 1\rangle$	1
	8	2	2^5	2;	$\langle 4, 2\rangle$	2
	8	2	$2^2, 4^2$	2;	$\langle 4, 1\rangle$	1
	8	2	5^3	2;	$\langle 5, 1\rangle$	1
	8	2	$2^2, 3^2$	2;	$\langle 6, 2\rangle$	1
	8	2	$3, 6^2$	2;	$\langle 6, 2\rangle$	1
	8	2	$2, 8^2$	2;	$\langle 8, 1\rangle$	1
	8	2	$2, 5, 10$	2;	$\langle 10, 2\rangle$	1
	8	2	$2, 6^2$	2;	$\langle 12, 5\rangle$	2
	8	2	$2^2, 3^2$	2;	$\langle 6, 1\rangle$	1
	8	2	4^3	2;	$\langle 8, 4\rangle$	1
	8	2	$2^3, 4$	2;	$\langle 8, 3\rangle$	2
	8	2	$2^3, 3$	2;	$\langle 12, 4\rangle$	2
	8	2	$3, 4^2$	2;	$\langle 12, 1\rangle$	1
	8	2	$2, 4, 8$	2;	$\langle 16, 8\rangle$	1
	8	2	$2, 4, 6$	2;	$\langle 24, 8\rangle$	2
	8	2	$3^2, 4$	2;	$\langle 24, 3\rangle$	1
	8	2	$2, 3, 8$	2;	$\langle 48, 29\rangle$	1
	8		$1; 2^2$	$1; 2^2$	$\langle 4, 2\rangle$	1
	8		$1; 3$	$1; 2^2$	$\langle 6, 1\rangle$	1
	8		$1; 2$	$1; 2^2$	$\langle 8, 3\rangle$	1
$\frac{1}{2}^2$	6		$1; 2$	$1; 2$	$\langle 12, 3\rangle$	1
$\frac{1}{2}^4$	4		$1; 2^2$	$1; 2^2$	$\langle 2, 1\rangle$	1
$\frac{1}{2}^4$	4		$1; 2$	$1; 2$	$\langle 8, 3\rangle$	1
$\frac{1}{2}^4$	4		$1; 2$	$1; 2$	$\langle 8, 4\rangle$	1
$\frac{1}{3}, \frac{2}{3}$	5		$1; 3$	$1; 3$	$\langle 6, 1\rangle$	1

Table 6 Minimal mixed quasi-étale surfaces of general type with $p_g = q = 2$

Sing X	K_S^2	Sign	$G^{(0)}$	G	#fam
	8	2;	$\langle 2, 1\rangle$	$\langle 4, 1\rangle$	1

reach all values of K_S^2 attained by minimal surfaces of general type: this supports our claim that this method could possibly fill the yellow region.

It would be nice to have a complete classification of all unmixed quasi-étale surfaces of general type with $\chi(\mathcal{O}_S) = 1$; the list is complete only in the mixed irregular case by Theorem 5.1.

The methods developed in the mentioned papers allow in principle to construct the whole list of quasi-étale surfaces with a fixed value of the triple (p_g, q, K_S^2). Since when we blow up a point the value of K_S^2 drops by 1, there is no lower bound for the value of K_S^2 of a surface of general type in terms of birational invariants as p_g and q; in particular these methods can't give a complete classification unless one can prove such a lower bound for quasi-étale surfaces; at the moment we do not have even a reasonable conjecture for that bound.

A different method, which could possibly in the future produce a complete classification of the quasi-étale surfaces with $\chi = 1$, has been recently presented in [7], where K_S^2 is substituted by a different number related to the structure of the Néron-Severi group of a quasi-étale surface.

References

1. M.A. Armstrong, On the fundamental group of an orbit space. Proc. Camb. Philos. Soc. **61**, 639–646 (1965)
2. M.A. Armstrong, The fundamental group of the orbit space of a discontinuous group. Proc. Camb. Philos. Soc. **64**, 299–301 (1968)
3. W. Barth, C. Peters, A. van de Ven, *Compact Complex Surfaces*. Ergebnisse der Mathematik und ihrer Grenzgebiete.3, vol. 4 (Springer, Berlin, 1984), pp. x+304
4. I. Bauer, F. Catanese, Some new surfaces with $p_g = q = 0$. *The Fano Conference* (University of Torino, Turin, 2004), pp. 123–142
5. I. Bauer, F. Catanese, F. Grunewald, The classification of surfaces with $p_g = q = 0$ isogenous to a product of curves. Pure Appl. Math. Q. **4**(2), part 1, 547–586 (2008)
6. I. Bauer, F. Catanese, R. Pignatelli, Surfaces of general type with geometric genus zero: a survey. *Complex and Differential Geometry*. Springer Proceedings in Mathematics, vol. 8 (2011), pp. 1–48
7. I. Bauer, R. Pignatelli, Product-Quotient Surfaces: new invariants and algorithms. To appear on Groups, Geometry and Dynamics. arxiv:math/1308.5508
8. I. Bauer, F. Catanese, R. Pignatelli, Complex surfaces of general type: some recent progress. *Global Aspects of Complex Geometry* (Springer, Berlin, 2006), pp. 1–58
9. I. Bauer, R. Pignatelli, The classification of minimal product-quotient surfaces with $p_g = 0$. Math. Comput. **81**(280), 2389–2418 (2012)
10. I. Bauer, F. Catanese, F. Grunewald, R. Pignatelli, Quotients of a product of curves by a finite group and their fundamental groups. Am. J. Math. **134**(4), 993–1049 (2012)
11. A. Beauville, *Complex Algebraic Surfaces*. Translated from the French by R. Barlow, N.I. Shepherd-Barron, M. Reid. London Mathematical Society Lecture Note Series, vol. 68 (Cambridge University Press, Cambridge, 1983). pp. iv+132
12. W. Bosma, J. Cannon, C. Playoust, The Magma algebra system. I: the user language. J. Symb. Comput. **24**(3–4), 235–265 (1997)
13. G. Carnovale, F. Polizzi, The classification of surfaces with $p_g = q = 1$ isogenous to a product of curves. Adv. Geom. **9**, 233–256 (2009)

14. F. Catanese, Fibred Surfaces, varieties isogeneous to a product and related moduli spaces. Am. J. Math. **122**(1), 1–44 (2000)
15. D. Frapporti, Mixed surfaces, new surfaces of general type with $p_g = 0$ and their fundamental group. To appear on Coll. Math. arXiv:1105.1259
16. D. Frapporti, R. Pignatelli, Mixed quasi-étale quotients with arbitrary singularities. Glasgow Math. J. **57**, 143–165 (2015)
17. A. Garbagnati, M. Penegini, K3 surfaces with a non-symplectic automorphism and product-quotient surfaces with cyclic groups. arXiv:1303.1653
18. M. Mendes Lopes, R. Pardini, The bicanonical map of surfaces with $p_g = 0$ and $K^2 \geq 7$. Bull. Lond. Math. Soc. **33**, 265–274 (2001)
19. E. Mistretta, F. Polizzi, Standard isotrivial fibrations with $p_g = q = 1$ II. J. Pure Appl. Algebra **214**, 344–369 (2010)
20. R. Pardini, The classification of double planes of general type with $K^2 = 8$ and $p_g = 0$. J. Algebra **259**, 95–118 (2003)
21. M. Penegini, The classification of isotrivially fibred surfaces with $p_g = q = 2$, and topics on beauville surfaces. Ph.D Thesis University of Bayreuth (2010)
22. M. Penegini, The classification of isotrivially fibred surfaces with $p_g = q = 2$. With an appendix by S. Rollenske. Collect. Math. **62**, 239–274 (2011)
23. F. Polizzi, Surfaces of general type with $p_g = q = 1$, $K^2 = 8$ and bicanonical map of degree 2. Trans. Am. Math. Soc. **358**, 759–798 (2006)
24. F. Polizzi, On surfaces of general type with $p_g = q = 1$ isogenous to a product of curves. Commun. Algebra **36**, 2023–2053 (2008)
25. F. Polizzi, Standard isotrivial fibrations with $p_g = q = 1$. J. Algebra **321**, 1600–1631 (2009)
26. F. Polizzi, Numerical properties of isotrivial fibrations. Geom. Dedicata **147**, 323–355 (2010)
27. F. Serrano, Isotrivial fibred surfaces. Ann. Mat. Pura Appl. **171**(4), 63–81 (1996)
28. F. Zucconi, Surfaces with $p_g = q = 2$ and an irrational pencil. Can. J. Math. **55**, 649–672 (2003)

Isotrivially Fibred Surfaces and Their Numerical Invariants

Francesco Polizzi

Abstract We give a survey of our previous work on relatively minimal isotrivial fibrations $\alpha\colon X \longrightarrow C$, where X is a smooth, projective surface and C is a curve. In particular, we consider two inequalities involving the numerical invariants K_X^2 and $\chi(\mathcal{O}_X)$ and we illustrate them by means of several examples and counter examples.

2010 Mathematics Subject Classification 14J99, 14J29

1 Introduction

One of the most useful tools in the study of algebraic surfaces is the analysis of *fibrations*, that is morphisms with connected fibres from a smooth surface X to a curve C. A fibration $\alpha\colon X \longrightarrow C$ is called *isotrivial* when all its smooth fibres are isomorphic; a deep investigation of such kind of fibrations can be found in [6].

A special kind of isotrivial fibrations are the standard ones, whose definition is as follows. A smooth surface X is called a *standard isotrivial fibration* if there exists a finite group G, acting faithfully on two smooth curves C_1 and C_2 and diagonally on their product, such that X is isomorphic to the minimal desingularization of $T = (C_1 \times C_2)/G$. Then X has two isotrivial fibrations $\alpha_i : X \longrightarrow C_i/G$, induced by the natural projections $p_i : C_1 \times C_2 \longrightarrow C_i$.

If the action of G on $C_1 \times C_2$ is free, we say that $X = T$ is a *quasi-bundle*. For instance, unmixed Beauville surfaces are defined as rigid quasi-bundles, i.e., quasi-bundles such that $C_i/G \cong \mathbb{P}^1$ and the covers $C_i \longrightarrow \mathbb{P}^1$ are both branched at three points. Therefore the classification of Beauville surfaces can be reduced to combinatorial finite group theory involving triangle groups; we refer the reader to the other papers in this Volume for further details.

Standard isotrivial fibrations were thoroughly investigated by Serrano in [16, 17]; in particular he showed, by a monodromy argument, that every isotrivial fibration

F. Polizzi (✉)
Dipartimento di Matematica e Informatica, Università della Calabria,
Cubo 30B, 87036 Arcavacata Rende, Cosenza, Italy
e-mail: polizzi@mat.unical.it

I. Bauer et al. (eds.), *Beauville Surfaces and Groups*, Springer Proceedings in Mathematics & Statistics 123, DOI 10.1007/978-3-319-13862-6_11

is isomorphic to a standard one. Since then, such fibrations have been widely used in order to produce new examples of minimal surfaces of general type with small birational invariants, in particular with $p_g = q = 0$ [1–5], with $p_g = q = 1$ [7, 10, 12–15] and with $p_g = q = 2$ [11].

In this paper we discuss the following theorem, that was obtained in [15]. Let $f: X \longrightarrow C$ be any relatively minimal, isotrivial fibration with $g(C) \geq 1$. If X is neither ruled nor isomorphic to a quasi-bundle, then $K_X^2 \leq 8\chi(\mathcal{O}_X) - 2$. If, in addition, K_X is ample, then $K_X^2 \leq 8\chi(\mathcal{O}_X) - 5$. This generalizes previous results of Serrano and Tan [17, 18].

This work is organized as follows. In Sect. 2 we set up the notation and the terminology, we state our theorem and we provide a sketch of its proof. In Sect. 3 we exhibit several examples and counterexamples illustrating its meaning. More precisely, Examples 3.1 and 3.2 imply that both the above inequalities involving K_X^2 and $\chi(\mathcal{O}_X)$ are sharp, whereas Examples 3.3 and 3.4 show that, when K_X is not ample, both cases $K_X^2 = 8\chi(\mathcal{O}_X) - 3$ and $K_X^2 = 8\chi(\mathcal{O}_X) - 4$ actually occur. Finally, in Example 3.5 we describe an isotrivially fibred surface X with $K_X^2 = 8\chi(\mathcal{O}_X) - 5$ and K_X *not* ample. In all these examples X is a minimal surface of general type with $p_g = q = 1$, obtained as a standard isotrivial fibration.

Notations and conventions. All varieties in this article are defined over $\mathbb{C}$. If X is a projective, non-singular surface X then K_X denotes the canonical class, $p_g(X) = h^0(X, K_X)$ is the *geometric genus*, $q(S) = h^1(X, K_X)$ is the *irregularity* and $\chi(\mathcal{O}_X) = 1 - q(X) + p_g(X)$ is the *Euler characteristic*.

If G is a finite group and $g \in G$, we denote by $|G|$ and $o(g)$ the orders of G and g, respectively.

2 The Main Result

Definition 2.1 Let X be a smooth, complex projective surface and let $\alpha: X \longrightarrow C$ be a fibration onto a smooth curve C. We say that α is *isotrivial* if all its smooth fibres are isomorphic.

Definition 2.2 A smooth surface S is called a *standard isotrivial fibration* if there exists a finite group G, acting faithfully on two smooth projective curves C_1 and C_2 and diagonally on their product, so that S is isomorphic to the minimal desingularization of $T := (C_1 \times C_2)/G$. We denote such a desingularization by $\lambda: S \longrightarrow T$.

If $\lambda: S \longrightarrow T = (C_1 \times C_2)/G$ is any standard isotrivial fibration, composing the two projections $\pi_1: T \longrightarrow C_1/G$ and $\pi_2: T \longrightarrow C_2/G$ with λ one obtains two morphisms $\alpha_1: S \longrightarrow C_1/G$ and $\alpha_2: S \longrightarrow C_2/G$, whose smooth fibres are isomorphic to C_2 and C_1, respectively. One also has $q(S) = g(C_1/G) + g(C_2/G)$, see [9].

A monodromy argument ([17, Sect. 2]) implies that any isotrivial fibration $\alpha\colon X \longrightarrow C$ is birational to a standard one; in other words, there exists $T = (C_1 \times C_2)/G$ and a birational map $T \dashrightarrow X$ such that the following diagram

$$\begin{array}{ccc} S & & \\ {\scriptstyle\lambda}\downarrow & \searrow^{\psi} & \\ T & \dashrightarrow & X \\ {\scriptstyle\pi_2}\downarrow & & \downarrow{\scriptstyle\alpha} \\ C_2/G & \xrightarrow{\cong} & C \end{array} \tag{1}$$

commutes.

When the action of G is free, then $S = T$ is called a *quasi-bundle*; in this case one has $K_S^2 = 8\chi(\mathcal{O}_S)$. In 1996, F. Serrano and, indipendently, S. L. Tan improved this result, showing that for any isotrivial fibration $\alpha\colon X \longrightarrow C$ one has

$$K_X^2 \leq 8\chi(\mathcal{O}_X) \tag{2}$$

and that the equality holds if and only if X is either ruled or isomorphic to a quasi bundle [17, 18]. Serrano's proof is based on a fine analysis of the projective bundle $\mathbb{P}(\Omega_X^1)$, whereas Tan's proof uses base change techniques.

This paper deals with the following refinement of (2), that we proved in [15].

Theorem 2.3 *Let $\alpha\colon X \longrightarrow C$ be any relatively minimal isotrivial fibration, with $g(C) \geq 1$. If X is neither ruled nor isomorphic to a quasi bundle, then*

$$K_X^2 \leq 8\chi(\mathcal{O}_X) - 2, \tag{3}$$

and if the equality holds then X is a minimal surface of general type whose canonical model has precisely two ordinary double points as singularities. Moreover, under the further assumption that K_X is ample, we have

$$K_X^2 \leq 8\chi(\mathcal{O}_X) - 5. \tag{4}$$

Finally, inequalities 3 and 4 are sharp.

Let us give a sketch of the proof of Theorem 2.3, whose full details can be found in [15]. Let us consider a standard isotrivial fibration S birational to X and the corresponding commutative diagram (1). Since α is relatively minimal and $g(C) \geq 1$, the surface X is a minimal model. Moreover we are assuming that X is not ruled, so K_X is nef and the birational map $\psi\colon S \longrightarrow X$ is actually a morphism [8, Proposition 8], which induces an isomorphism of X with the minimal model S_m of S. Since C_2/G has positive genus, all the (-1)-curves of S are necessarily contained in fibres of $\alpha_2\colon S \longrightarrow C_2/G$, hence our isotrivial fibration $\alpha\colon X \longrightarrow C$ is equivalent to an isotrivial fibration $\alpha_m\colon S_m \longrightarrow C_2/G$.

The surface T contains at most a finite number of isolated singularities that, locally analytically, look like the quotient of $\mathbb{C}^2$ by the action of the cyclic group $\mathbb{Z}/n\mathbb{Z} = \langle \xi \rangle$ defined by $\xi \cdot (x, y) = (\xi x, \xi^q y)$, where $0 < q < n$, $(n, q) = 1$ and ξ is a primitive n-th root of unity. We call this singularity a *cyclic quotient singularity* of type $\frac{1}{n}(1, q)$. The exceptional divisor $\mathcal{E}$ of its minimal resolution is a HJ-string (abbreviation of Hirzebruch-Jung string), that is to say, a connected union $\mathcal{E} = \bigcup_{i=1}^{k} Z_i$ of smooth rational curves $Z_1, \ldots, Z_k$ with self-intersection $-b_i := Z_i^2 \leq -2$, and ordered linearly so that $Z_i Z_{i+1} = 1$ for all i, and $Z_i Z_j = 0$ if $|i - j| \geq 2$. More precisely, given the continued fraction

$$\frac{n}{q} = [b_1, \ldots, b_k] = b_1 - \cfrac{1}{b_2 - \cfrac{1}{\cdots - \cfrac{1}{b_k}}}, \quad b_i \geq 2\,,$$

the dual graph of $\mathcal{E}$ is

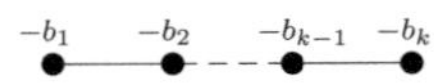

For instance, the cyclic quotient singularities $\frac{1}{n}(1, n-1)$ are precisely the rational double points of type A_{n-1}; in particular, the singularities $\frac{1}{2}(1, 1)$ are the ordinary double points.

Finally, the invariants $K_{S_m}^2$ and $e(S_m)$ can be computed knowing the number and type of the singularities of T. In fact, since $g(C_2/G) = g(C) \geq 1$, all the (-1)-curves of S are components of reducible fibres of $\alpha_2 \colon S \longrightarrow C_2/G$; it follows that it is possible to define, for any such a reducible fibre F, an invariant $\delta(F) \in \mathbb{Q}$ such that

$$K_{S_m}^2 = 8\chi(\mathcal{O}_{S_m}) - \sum_{F \text{ reducible}} \delta(F). \tag{5}$$

The proof of Theorem 2.3 follows from a careful analysis of the possible values of $\delta(F)$, based on some identities on continued fractions which are a consequence of the so-called Riemenschneider's duality between the HJ-expansions of $\frac{n}{q}$ and $\frac{n}{n-q}$.

3 Examples

Let us denote by $\Gamma(0 \mid m_1, \ldots, m_r)$ the abstract group of Fuchsian type with presentation

$$\left\langle g_1, \ldots, g_r \mid g_k^{m_k} = 1,\ \prod_{i=1}^{r} g_i = 1 \right\rangle$$

and by $\Gamma(1 \mid n_1, \ldots, n_s)$ the abstract group of Fuchsian type with presentation

$$\left\langle l_1, \ldots, l_s,\ h_1, h_2 \mid l_k^{n_k} = 1,\ \ [h_1, h_2]\prod_{i=1}^{s} l_i = 1\right\rangle.$$

We call the sets $(m_1, \ldots, m_r)$ and $(n_1, \ldots, n_s)$ the *branching data* of these groups. For convenience we make abbreviations such as $(2^3, 3^2)$ for $(2, 2, 2, 3, 3)$ when we write down the branching data.

Example 3.1 This example appears in [14, Sect. 7].

Take $G = D_8$, the dihedral group of order 8, with presentation

$$G = \langle x, y | x^2 = y^4 = 1, xy = y^3 x\rangle.$$

There are two epimorphisms of groups

$$\varphi\colon \Gamma(0|2^4, 4) \longrightarrow G, \quad \psi\colon \Gamma(1|2) \longrightarrow G$$

defined in the following way:

$$\begin{aligned} &\varphi(g_1) = x,\ \ \varphi(g_2) = xy,\ \ \varphi(g_3) = x,\ \ \psi(g_4) = xy^2,\ \ \psi(g_5) = y,\\ &\psi(\ell_1) = y^2,\ \ \psi(h_1) = y,\ \ \psi(h_2) = x. \end{aligned} \tag{6}$$

By Riemann Existence Theorem [10, Proposition 1.3] they induce two G-coverings

$$f_1\colon C_1 \longrightarrow \mathbb{P}^1 \cong C_1/G, \quad f_2\colon C_2 \longrightarrow E \cong C_2/G,$$

where E is an elliptic curve, $g(C_1) = 4$ and $g(C_2) = 3$. Moreover, f_1 is branched at five points with branching orders 2, 2, 2, 2, 4, whereas f_2 is branched at one point with branching order 2.

By (6) it follows that the unique element of G, different from the identity, which acts with fixed points on both C_1 and C_2 is y^2. Denoting by $|\mathrm{Fix}_{C_i}(g)|$ the number of fixed points of $g \in G$ on the curve C_i, we obtain

$$|\mathrm{Fix}_{C_1}(y^2)| = 2, \quad |\mathrm{Fix}_{C_2}(y^2)| = 4,$$

so we have $4 \cdot 2 = 8$ points in $C_1 \times C_2$ whose stabilizer is non-trivial (it is isomorphic to $\langle y^2\rangle$). The G-orbit of each of them has cardinality $|G|/o(y^2) = 4$, hence we have exactly two singular points in $T = (C_1 \times C_2)/G$. More precisely, since $\langle y^2\rangle$ has order 2, it follows

$$\mathrm{Sing} T = 2 \times \frac{1}{2}(1, 1).$$

Let $\lambda\colon X \longrightarrow T$ be the minimal resolution of singularities of T; then X is a surface of general type, whose numerical invariants can be computed by using [10, Proposition 5.1]; we obtain

$$p_g(X) = q(X) = 1, \quad K_X^2 = 6.$$

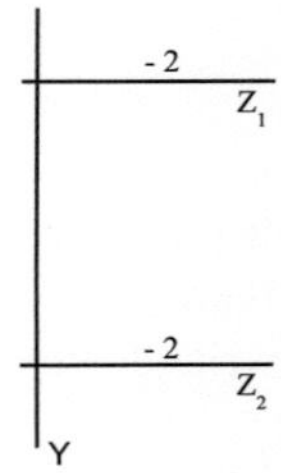

Fig. 1 The unique singular fibre of $\alpha\colon X \longrightarrow E$ in Examples 3.1 and 3.4

The isotrivial fibration $\pi_2\colon T \longrightarrow C_2/G \cong E$ yields, after composition with λ, an isotrivial fibration $\alpha\colon X \longrightarrow E$, which is the Albanese morphism of X. The fibration α has a unique singular fibre $F = 2Y + Z_1 + Z_2$, where the Z_i are two disjoint (-2) curves and $YZ_i = 1$, see Fig. 1.

Since E is an elliptic curve, any (-1)-curve of X must be a component of F. On the other hand, we have $K_X F = 2g(C_1) - 2 = 6$ and $F^2 = 0$, hence $K_X Y = 3$ and $Y^2 = -1$. Thus Y is not a (-1)-curve, so X is a minimal surface of general type satisfying

$$K_X^2 = 8\chi(\mathcal{O}_X) - 2$$

and T is the canonical model of X. This example shows that inequality (3) in Theorem 2.3 is sharp. Notice that K_X is *not* ample, since $K_X Z_i = 0$.

Example 3.2 This example can be found in [10, Sect. 5].

Take as G the semi-direct product $\mathbb{Z}/3\mathbb{Z} \ltimes (\mathbb{Z}/4\mathbb{Z})^2$ whose presentation is

$$G = \langle x, y, z \mid x^3 = y^4 = z^4 = 1, [y, z] = 1, xyx^{-1} = z, xzx^{-1} = (yz)^{-1} \rangle.$$

There are two epimorphisms of groups

$$\varphi\colon \Gamma(0|3^2, 4) \longrightarrow G, \quad \psi\colon \Gamma(1|4) \longrightarrow G$$

defined in the following way:

$$\begin{aligned} &\varphi(g_1) = x, \quad \varphi(g_2) = x^2y^3, \quad \varphi(g_3) = y, \\ &\psi(\ell_1) = y, \quad \psi(h_1) = x, \quad \psi(h_2) = xyxy^2. \end{aligned} \tag{7}$$

By Riemann Existence Theorem they induce two G-coverings

$$f_1\colon C_1 \longrightarrow \mathbb{P}^1 \cong C_1/G, \quad f_2\colon C_2 \longrightarrow E \cong C_2/G,$$

where E is an elliptic curve, $g(C_1) = 3$ and $g(C_2) = 19$. Moreover, f_1 is branched at three points with branching orders 3, 3, 4, whereas f_2 is branched at one point with branching order 4.

By (7) it follows that the nontrivial elements of G having fixed points on $C_1 \times C_2$ are precisely those in the set

$$\Sigma = \bigcup_{\sigma \in G} \langle \sigma y \sigma^{-1} \rangle \setminus \{1\}$$

and the elements of order 4 in Σ are $\{y, z, y^3z^3, y^3, z^3, yz\}$. The product surface $C_1 \times C_2$ contains exactly 48 points with nontrivial stabilizer, and for each of them the G-orbit has cardinality $|G|/o(y) = 12$, thus $T = (C_1 \times C_2)/G$ contains 4 singular points.

Moreover the conjugacy class of y in G is $\{y, z, y^3z^3\}$, whereas the conjugacy class of y^3 is $\{y^3, z^3, yz\}$. Therefore, for any $h \in \Sigma$ with $o(h) = 4$, one has that h is not conjugate to h^{-1} in G. Looking at the local action of G around each of the fixed points, this implies that

$$\operatorname{Sing} T = 4 \times \frac{1}{4}(1, 1).$$

Let $\lambda \colon X \longrightarrow T$ be the minimal resolution of singularities of X; then X is a surface of general type whose invariants are

$$p_g(X) = q(X) = 1, \quad K_X^2 = 2.$$

The isotrivial fibration $\pi_2 \colon T \longrightarrow C_2/G \cong E$ yields, after composition with λ, an isotrivial fibration $\alpha \colon X \longrightarrow E$ which is the Albanese morphism of X. The isotrivial fibration α has a unique singular fibre $F = 4Y + A_1 + A_2 + A_3 + A_4$, where the A_i are disjoint smooth rational curves such that $A_i^2 = -4$ and $YA_i = 1$ (see Fig. 2).

We have $K_X F = 2g(C_1) - 2 = 4$ and $F^2 = 0$, so we deduce $K_X Y = Y^2 = -1$. Thus Y is a (-1)-curve, which is necessarily the unique (-1)-curve in X. Since Y is contained in a fibre of $\alpha \colon X \longrightarrow E$, after blowing down Y we obtain another isotrivial fibration $\alpha_m \colon X_m \longrightarrow E$ such that

$$p_g(X_m) = q(X_m) = 1, \quad K_{X_m}^2 = 3,$$

that is

$$K_{X_m}^2 = 8\chi(\mathcal{O}_{X_m}) - 5. \tag{8}$$

Fig. 2 The unique singular fibre of $\alpha \colon X \longrightarrow E$ in Example 3.2

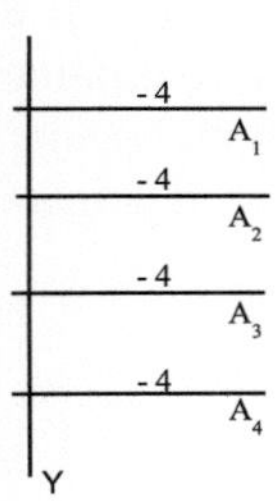

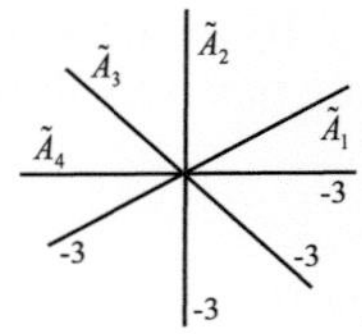

Fig. 3 The unique singular fibre of $\alpha_m : X_m \longrightarrow E$ in Example 3.2

The isotrivial fibration α_m contains a unique singular fibre, namely the image of F in X_m, that is illustrated in Fig. 3.

Here each $\widetilde{A}_i$ is a smooth rational curve with self-intersection -3; in particular X_m contains neither (-1)-curves nor (-2)-curves. This shows that X_m is a minimal model and that K_{X_m} is ample, so (8) implies that inequality (4) in Theorem 2.3 is sharp.

The next two examples show that, if K_X is not ample, then both cases $K_X^2 = 8\chi(\mathcal{O}_X) - 3$ and $K_X^2 = 8\chi(\mathcal{O}_X) - 4$ may actually occur.

Example 3.3 We exhibit an isotrivial fibration with K_X not ample and $K_X^2 = 8\chi(\mathcal{O}_X) - 3$; this example appears in [10, Sect. 5].

Take $G = D_{12}$, the finite dihedral group of order 12, with presentation

$$G = \langle x, y | x^2 = y^6 = 1, xy = y^5 x \rangle.$$

There are two epimorphisms of groups

$$\varphi \colon \Gamma(0|2^3, 6) \longrightarrow G, \quad \psi \colon \Gamma(1|3) \longrightarrow G$$

defined in the following way:

$$\begin{aligned} &\varphi(g_1) = x, \quad \varphi(g_2) = xy^2, \quad \varphi(g_3) = y^3, \quad \psi(g_4) = y, \\ &\psi(\ell_1) = y^2, \quad \psi(h_1) = x, \quad \psi(h_2) = y. \end{aligned} \tag{9}$$

By Riemann Existence Theorem they induce two G-coverings

$$f_1 \colon C_1 \longrightarrow \mathbb{P}^1 \cong C_1/G, \quad f_2 \colon C_2 \longrightarrow E \cong C_2/G,$$

where E is an elliptic curve, $g(C_1) = 3$ and $g(C_2) = 5$. Moreover, f_1 is branched at four points with branching orders 2, 2, 2, 6, whereas f_2 is branched at one point with branching order 3. By (9) it follows that the nontrivial elements of G having fixed points on $C_1 \times C_2$ are precisely those in the set

$$\Sigma = \bigcup_{\sigma \in G} \langle \sigma y^2 \sigma^{-1} \rangle \setminus \{1\} = \{y^2, y^4\}.$$

The product surface $C_1 \times C_2$ contains exactly 8 points with nontrivial stabilizer, and for each of them the G-orbit has cardinality $|G|/o(y^2) = 4$, thus $T = (C_1 \times C_2)/G$ contains two singular points. Since y^2 is conjugate to y^4 in G, looking at the local action of G around each of the fixed points one obtains

$$\text{Sing}\, T = \frac{1}{3}(1, 1) + \frac{1}{3}(1, 2).$$

Let $\lambda \colon X \longrightarrow T$ be the minimal resolution of singularities of X; then X is a surface of general type whose invariants are

$$p_g(X) = q(X) = 1, \quad K_X^2 = 5,$$

that is

$$K_X^2 = 8\chi(\mathcal{O}_X) - 3.$$

The isotrivial fibration $\pi_2 \colon T \longrightarrow C_2/G \cong E$ yields, after composition with λ, an isotrivial fibration $\alpha \colon X \longrightarrow E$ which is the Albanese morphism of X. The isotrivial fibration α has a unique singular fibre $F = 3Y + A + 2B_1 + B_2$, where A is a (-3)-curve and the B_i are (-2)-curves, see Fig. 4.

Using $K_X F = 2g(C_1) - 2 = 4$ and $F^2 = 0$ one obtains $K_X Y = 1$ and $Y^2 = -1$, hence Y is not a (-1)-curve and X is a minimal model. Notice that K_X is not ample, since X contains the (-2)-curves B_i.

Example 3.4 We exhibit an isotrivial fibration with K_X not ample and $K_X^2 = 8\chi(\mathcal{O}_X) - 4$; this example appears in [14, Sect. 6].

Take $G = \mathbb{Z}/2\mathbb{Z} \times \mathbb{Z}/2\mathbb{Z}$, with presentation

$$G = \langle x, y | x^2 = y^2 = 1, [x, y] = 1 \rangle.$$

There are two epimorphisms of groups

$$\varphi \colon \Gamma(0|2^5) \longrightarrow G, \quad \psi \colon \Gamma(1|2^2) \longrightarrow G$$

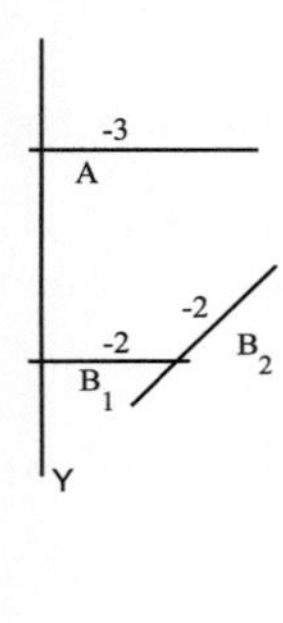

Fig. 4 The unique singular fibre of $\alpha \colon X \longrightarrow E$ in Example 3.3

defined in the following way:

$$\varphi(g_1) = x, \quad \varphi(g_2) = y, \quad \varphi(g_3) = xy, \quad \psi(g_4) = xy, \quad \varphi(g_5) = xy$$
$$\psi(\ell_1) = x, \quad \psi(\ell_2) = x, \quad \psi(h_1) = y, \quad \psi(h_2) = y. \tag{10}$$

By Riemann Existence Theorem they induce two G-coverings

$$f_1 \colon C_1 \longrightarrow \mathbb{P}^1 \cong C_1/G, \quad f_2 \colon C_2 \longrightarrow E \cong C_2/G,$$

where E is an elliptic curve, $g(C_1) = 2$ and $g(C_2) = 3$. Moreover, f_1 is branched at five points, f_2 is branched at two points and all the branching orders are equal to 2. By (10) it follows that the unique nontrivial element of G having fixed points on $C_1 \times C_2$ is x. The product surface $C_1 \times C_2$ contains exactly 8 points with nontrivial stabilizer, and for each of them the G-orbit has cardinality $|G|/o(x) = 2$, thus $T = (C_1 \times C_2)/G$ contains four singular points. More precisely, since $\langle x \rangle$ has order 2, we have

$$\operatorname{Sing} T = 4 \times \frac{1}{2}(1, 1).$$

Let $\lambda \colon X \longrightarrow T$ be the minimal resolution of singularities of X; then X is a surface of general type whose invariants are

$$p_g(X) = q(X) = 1, \quad K_X^2 = 4.$$

The isotrivial fibration $\pi_2 \colon T \longrightarrow C_2/G \cong E$ yields, after composition with λ, an isotrivial fibration $\alpha \colon X \longrightarrow E$ which is the Albanese morphism of X. The isotrivial fibration α has two singular fibres F_1 and F_2, which look like the singular fibre in Fig. 1, i.e. they are of the form $F_i = 2Y_i + Z_{1i} + Z_{2i}$, where the Z_{ji} are disjoint (-2)-curves. Using $2 = K_X F_i = 2K_X Y_i$ and $F_i^2 = 0$ one obtains $K_X Y_i = 1$ and $Y_i^2 = -1$. In particular Y_i is not a (-1)-curve, so X is a minimal surface of general type satisfying

$$K_X^2 = 8\chi(\mathcal{O}_X) - 4.$$

Notice that K_X is not ample, since X contains the four (-2)-curves Z_{ji}.

Looking at Theorem 2.3, one might ask whether any isotrivially fibred surface X with $K_X^2 = 8\chi(\mathcal{O}_X) - 5$ has ample canonical class. The answer is negative, as shown by the following example that can be found in [10, Sect. 6]

Example 3.5 Take $G = D_{3,7,2}$, namely the metacyclic group of order 21 whose presentation is

$$G = \langle x, y \mid x^3 = y^7 = 1, xyx^{-1} = y^2 \rangle.$$

There are two epimorphisms of groups

$$\varphi \colon \Gamma(0|3^2, 7) \longrightarrow G, \quad \psi \colon \Gamma(1|7) \longrightarrow G$$

defined in the following way:

$$\varphi(g_1) = x^2, \quad \varphi(g_2) = xy^6, \quad \varphi(g_3) = y,$$
$$\psi(\ell_1) = y, \quad \psi(h_1) = y, \quad \psi(h_2) = x. \tag{11}$$

By Riemann Existence Theorem they induce two G-coverings

$$f_1 \colon C_1 \longrightarrow \mathbb{P}^1 \cong C_1/G, \quad f_2 \colon C_2 \longrightarrow E \cong C_2/G,$$

where E is an elliptic curve, $g(C_1) = 3$ and $g(C_2) = 10$. Moreover, f_1 is branched at three points with branching orders 3, 3, 7, whereas f_2 is branched at one point with branching order 7.

By (11) it follows that the nontrivial elements of G having fixed points on $C_1 \times C_2$ are precisely those in the set

$$\Sigma = \bigcup_{\sigma \in G} \langle \sigma y \sigma^{-1} \rangle \setminus \{1\} = \{y, y^2, y^3, y^4, y^5, y^6\}.$$

The product surface $C_1 \times C_2$ contains exactly 9 points with non-trivial stabilizer, and for each of them the G-orbit has cardinality $|G|/o(y) = 3$; thus $T = (C_1 \times C_2)/G$ contains 3 singular points.

Moreover, the conjugacy class of y in G is $\{y, y^2, y^4\}$ and the conjugacy class of y^3 is $\{y^3, y^5, y^6\}$. In particular, every element in Σ is conjugate to its inverse. Looking at the local action of G around each of the fixed points, this implies that

$$\operatorname{Sing} T = \frac{1}{7}(1, 1) + \frac{1}{7}(1, 2) + \frac{1}{7}(1, 4).$$

Let $\lambda \colon X \longrightarrow T$ be the minimal resolution of singularities of X; then X is a surface of general type, whose invariants are

$$p_g(X) = q(X) = 1, \quad K_X^2 = 1.$$

The isotrivial fibration $\pi_2 \colon T \longrightarrow C_2/G \cong E$ yields, after composition with λ, an isotrivial fibration $\alpha \colon X \longrightarrow E$, which is the Albanese morphism of X. The fibration α has a unique singular fibre $F = 7Y + 4A_1 + A_2 + 2B_1 + B_2 + C$, which looks as in Fig. 5.

Here A_i, B_i and C are smooth rational curves, and the integer over each of them denotes as usual the corresponding self-intersection.

Using $K_X F = 4$ and $F^2 = 0$ we obtain $K_X Y = Y^2 = -1$, hence Y is the unique (-1)-curve in X. The minimal model X_m of X is obtained by first contracting Y and then the image of A_1. Since Y is contained in a fibre of $\alpha \colon X \longrightarrow E$, we obtain another isotrivial fibration $\alpha_m : X_m \longrightarrow E$ such that

$$p_g(X_m) = q(X_m) = 1, \quad K_{X_m}^2 = 3, \tag{12}$$

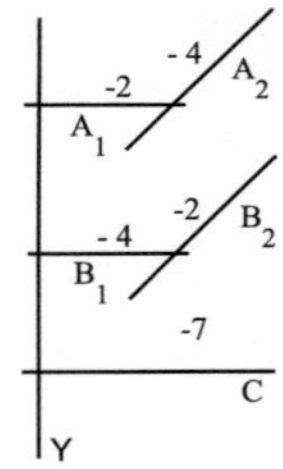

Fig. 5 The unique singular fibre of $\alpha\colon X \longrightarrow E$ in Example 3.5

that is

$$K^2_{X_m} = 8\chi(\mathcal{O}_{X_m}) - 5.$$

However, the canonical class K_{X_m} is *not* ample, as X_m contains two (-2)-curves, namely the images of B_1 and B_2.

Acknowledgments This paper is an expanded version of the talk given by the author at the conference *Beauville surfaces and Groups*, Newcastle University (UK), 7–9th June 2012. The author is grateful to the organizers N. Barker, I. Bauer, S. Garion and A. Vdovina for the invitation and the kind hospitality. He was partially supported by Progetto MIUR di Rilevante Interesse Nazionale *Geometria delle Varietà Algebriche e loro Spazi di Moduli*. He also thanks the referee, whose comments helped to improve the presentation of these results.

References

1. I. Bauer, F. Catanese, Some new surfaces with $p_g = q = 0$, in *Proceedings of the Fano Conference 123142* (University of Turin, Torino, 2004)
2. I. Bauer, F. Catanese, F. Grunewald, The classification of surfaces with $p_g = q = 0$ isogenous to a product of curves. Pure Appl. Math. Q. **4**(2), part 1, 547–5861 (2008)
3. I. Bauer, F. Catanese, R. Pignatelli, Surfaces of general type with geometric genus zero: a survey. *Complex and Differential Geometry*, Springer Proceedings in Mathematics, vol. 8, pp. 1–48 (2011)
4. I. Bauer, F. Catanese, F. Grunewald, R. Pignatelli, Quotient of a product of curves by a finite group and their fundamental groups. Am. J. Math. **134**(4), 993–1049 (2012)
5. I. Bauer, R. Pignatelli, The classification of minimal product-quotient surfaces with $p_g = 0$. Math. Comput. **81**(280), 2389–2418 (2012)
6. F. Catanese, Fibred surfaces, varieties isogenous to a product and related moduli spaces. Am. J. Math. **122**, 1–44 (2000)
7. G. Carnovale, F. Polizzi, The classification of surfaces of general type with $p_g = q = 1$ isogenous to a product. Adv. Geom. **9**, 233–256 (2009)
8. Classification of complex algebraic surfaces from the point of view of Mori theory (2004). http://www-fourier.ujf-grenoble.fr/~peters/surface.f/surf-spec.pdf
9. E. Freitag, Uber die Struktur der Funktionenkörper zu hyperabelschen Gruppen I. J. Reine. Angew. Math. **247**, 97–117 (1971)
10. E. Mistretta, F. Polizzi, Standard isotrivial fibrations with $p_g = q = 1$. *II*. J. Pure Appl. Algebra **214**, 344–369 (2010)
11. M. Penegini, The classification of isotrivially fibred surfaces with $p_g = q = 2$ (with an appendix by Sonke Rollenske). Collect. Math. **62**(3), 239–274 (2011)

12. F. Polizzi, Surfaces of general type with $p_g = q = 1$, $K_S^2 = 8$ and bicanonical map of degree 2. Trans. Am. Math. Soc. **358**(2), 759–798 (2006)
13. F. Polizzi, On surfaces of general type with $p_g = q = 1$ isogenous to a product of curves. Commun. Algebra **36**, 2023–2053 (2008)
14. F. Polizzi, Standard isotrivial fibrations with $p_g = q = 1$. J. Algebra **321**, 1600–1631 (2009)
15. F. Polizzi, Numerical properties of isotrivial fibrations. Geom. Dedicata **147**, 323–355 (2010)
16. F. Serrano, Fibrations on algebraic surfaces, in *Geometry of Complex Projective Varieties (Cetraro 1990)*, ed. by A. Lanteri, M. Palleschi, D.C. Struppa. Mediterranean Press, pp. 291–300 (1993)
17. F. Serrano, Isotrivial fibred surfaces. Annali di Matematica pura e applicata **CLXXI**, 63–81 (1996)
18. S.L. Tan, On the invariant of base changes of pencils of curves, II. Math. Z. **222**, 655–676 (1996)
19. The GAP Group, GAP—Groups, Algorithms, and Programming, Version 4.4 (2006). http://www.gap-system.org

MIX
Papier aus verantwortungsvollen Quellen
Paper from responsible sources
FSC® C105338

If you have any concerns about our products,
you can contact us on
ProductSafety@springernature.com

In case Publisher is established outside the EU,
the EU authorized representative is:
Springer Nature Customer Service Center GmbH
Europaplatz 3, 69115 Heidelberg, Germany

Printed by Libri Plureos GmbH
in Hamburg, Germany